PHYSICS AND APPLICATIONS OF GRAPHENE - THEORY

Edited by **Sergey Mikhailov**

Physics and Applications of Graphene - Theory
Edited by Sergey Mikhailov

Published by InTech
Janeza Trdine 9, 51000 Rijeka, Croatia

Publishing Process Manager Ivana Lorkovic

Technical Editor Teodora Smiljanic

Cover Designer InTech Design Team

Additional hard copies can be obtained from orders@intechopen.com

Physics and Applications of Graphene - Theory, Edited by Sergey Mikhailov
p. cm.
ISBN 978-953-307-152-7

For EU product safety concerns:
IN TECH d.o.o., Prolaz Marije Krucifikse Kozulić 3, 51000 Rijeka, Croatia,
info@intechopen.com or visit our website at intechopen.com.

Meet the editor

Dr. Sergey Mikhailov received his PhD from the Institute of Radioengineering and Electronics of Russian Academy of Sciences and performed his scientific research in different Universities and research institutes in Russia, Germany, Sweden and Japan. He is a coauthor of two patents and has published more than 140 research papers in physical journals and conference proceedings. He was a Fellow of the Alexander von Humboldt Foundation, Germany, and a Senior Fellow of the Japan Society for the Promotion of Science. His research interests are in the theory of electrodynamic and many-body properties of low-dimensional electron systems in semiconductor structures and the theory of semiconductor devices and terahertz-wave generation. At present he is a Project Leader at the Institute of Physics of the University of Augsburg and works on the theory of nonlinear electrodynamic phenomena in graphene.

Contents

Preface

The Stone Age, the Bronze Age, the Iron Age ... Every global epoch in the history of the mankind is characterized by materials used in it. In 2004 a new era in the material science was opened: the era of graphene or, more generally, of two-dimensional materials [K. Novoselov, A. Geim et al. Science 306, 666 (2004)]. Graphene is the one-atom thin layer of sp^2-bonded carbon atoms arranged in a honey-comb lattice. It possesses the unique physical properties: graphene is the strongest and the most stretchable known material, has the record thermal conductivity and the very high intrinsic mobility and is completely impermeable. The charge carriers in graphene are the massless Dirac fermions and its unique electronic structure leads to a number of interesting physical effects, such as the minimal electrical conductivity, anomalous quantum Hall effect, Klein tunneling, the universal optical conductivity and the strong nonlinear electromagnetic response. Graphene offers and promises a lot of different applications, including conductive ink, terahertz transistors, ultrafast photodetectors, bendable touch screens, strain tensors and many other. In 2010 Andre Geim and Konstantin Novoselov were awarded the Nobel Prize in Physics "for groundbreaking experiments regarding the two-dimensional material graphene".

Nowadays, graphene is in the focus of research activity of condensed matter physicists in the whole world. Research articles reporting on different aspects of graphene studies are collected in the present two volumes "Physics and Applications of Graphene". These books cover a broad spectrum of experimental and theoretical studies of graphene and present contributions from research groups and laboratories from the North and South America, Europe, Asia and Australia.

The contributed articles are presented in two volumes. The readers interested in experimental studies of graphene are referred to the first volume. The second volume contains theoretical contributions, divided into five Sections. In Part I *ab initio* studies of the electronic structure of graphene in the presence of defects and impurities are described. In Part II the theory of graphene nano-flakes and nano-ribbons is presented. The magnetic properties of grapheme are discussed in Part III and the transport properties are studied in Part IV. The last Part of the volume is devoted to the linear and nonlinear optical properties of graphene.

Sergey Mikhailov
University of Augsburg
Germany

Part 1

Electronic Structure and Defects

1

Theoretical Studies on Formation, Property Tuning and Adsorption of Graphene Segments

R.Q. Zhang and Abir De Sarkar
City University of Hong Kong,
Department of Physics and Materials Science, Hong Kong SAR,
China

1. Introduction

A single graphene sheet is a planar monolayer of *sp*^{2}-bonded carbon atoms arranged on a two-dimensional honeycomb lattice made of hexagonal rings. Graphene is the basic structural element of some carbon allotropes including graphite, carbon nanotubes and fullerenes. Planar polycyclic aromatic hydrocarbons (PAHs) with only benzenoid hexagonal rings can be viewed as fragments of a graphene sheet with the peripheral atoms saturated with hydrogen, and thus provide molecular models of graphene segments. Graphene segments are of paramount importance both from scientific and technological perspectives. Moreover, the PAHs or graphene segments themselves are of great research interest per se, since they are widely found in the residues of domestic and natural combustion of coal, wood, and other organic materials, and their unique electronic properties provide opportunities for novel functionalized nanomaterials and nanodevices (Wu, Pisula et al. 2007). Understanding the mechanism of formation of graphene segments is necessary to control its formation and in turn, to meet its application requirements. In our work, we have elucidated the role played by H during CVD growth of carbon materials (Zhang, Chu et al. 2000) including graphene and diamond. Graphene materials are endowed with a wealth of properties, including luminescence; which has been frequently reported in various CVD Diamond or a-C:H films (Bergman, McClure et al. 1994; Kania and Oelhafen 1995; Rusli, Amaratunga et al. 1995; Bourée, Godet et al. 1996; Liu, Gangopadhyay et al. 1997). Visible photoluminescence (PL) has been observed in carbon nanoclusters embedded in SiO_2 matrix (Zhang, Bayliss et al. 1996) and from C_{60} thin films (Gapozzi, Casamassima et al. 1996). Unlike a-Si:H, the PL efficiency of a-C:H film is high and it shows luminescence even at room temperatures (Wagner and Lautenschlager 1986; Fang 1991; Schütte, Will et al. 1993; Xu, Hundhausen et al. 1993; Nevin, Yamagishi et al. 1994). A clear understanding of the luminescence phenomenon would help to produce high-quality luminescent films by a proper control of the experimental conditions. Correspondingly, the structure property relationship responsible for broadband luminescence in a:C-H films and carbon nanostructures was clearly pinpointed in our work (Zhang, Bertran et al. 1998; Feng, Lin et al. 2009). To the best of our knowledge, size-dependent PL mechanism was first reported by us (Zhang, Bertran et al. 1998).

The intermolecular, weak π-π interactions are of utmost importance for understanding the various properties of graphene sheets (Novoselov, Geim et al. 2004; Feng, Lin et al. 2010) and other carbon-related nanostructures, including hydrogen-terminated graphene nanoribbons with a finite nanosize width (Barone, Hod et al. 2006). The role of π-π interactions in benzene dimers (Feng, Lin et al. 2010) and in the stacking of graphene sheets (or graphene multilayers) have been thoroughly addressed in our detailed theoretical studies (Feng, Lin et al. 2009). In our investigations (Lin, Zhang et al. 2007; Fan, Zeng et al. 2009), the weak van der Waal's (vdW) interactions were also found to be of pivotal significance in binding bio-molecules to carbon nanotubes (CNTs) from the viewpoints of important biological applications and bio-compatibility. These non-covalent, weak interactions do not affect the chemical and conductive properties of carbon nanotubes unlike covalent bonds and thereby aid an efficient retention of its pristine properties during their actual practical applications.

Water is a universal solvent and plays a crucial role in the mechanism of a variety of chemical and biological processes. The properties of water molecules or clusters in different ambiences can be a lot different from that of its bulk phase. To shed light on that, we have probed theoretically into the interaction of water clusters with graphite (Lin, Zhang et al. 2005). Single-walled carbon nanotubes (SW-CNTs) have novel structural, mechanical, and electronic properties but are hydrophobic. Water encapsulated within hydrophobic SW-CNTs, commonly known as ice nanotubes (INT), provide important clues to the functionality of biological nanopores (Sansom and Biggin 2001). Moreover, INTs have been found to exhibit novel properties such as proton conduction, hydrogen-bond network, phase transitions, etc (Maniwa, Kumazawa et al. 1999; Hummer, Rasaiah et al. 2001; Koga, Gao et al. 2001; Martí and Gordillo 2001; Noon, Ausman et al. 2002; Mann and Halls 2003; Martí and Gordillo 2003; Mashl, Joseph et al. 2003; Wang, Zhu et al. 2004). Through a systematic investigation, we have revealed the geometrical structure adopted by INTs within SW-CNT and the signatures in its vibrational spectra (Feng, Zhang et al. 2007).

Hydrogen is one of the most promising energy fuels for automobiles and can be potentially exploited in smaller portable devices. Due to the large surface area, carbon-based nanostructures, such as CNTs, appear to be ideal storage materials for the hydrogen storage. There is an ongoing debate within the experimental community with regard to the viability of CNTs as hydrogen storage materials. Our effective predictive modelling (Fan, Zhang et al. 2009) provides important and useful pointers to experimentalists on this. The chapter systematically organizes our computational findings pertaining to graphene segments into different sections, which is intended to provide a deep insight into the properties of graphene segments and useful guidance to future research and applications.

2. Graphene synthesis

A scientifically clear understanding of the different methods for graphene synthesis is essential to realize the optimum potentiality of graphene in a large variety of its applications. The size and quality of the graphene produced depends on the techniques used and the next sub-sections are devoted to some of the commonly used methods, their merits and demerits.

2.1 Chemical Vapor Deposition (CVD) approach

Synthesis of graphene by CVD has been introduced recently (Sutter, Flege et al. 2008; Li, Cai et al. 2009; Li, Zhu et al. 2009; Reina, Jia et al. 2009; Reina, Thiele et al. 2009). Graphene acquired from the CVD process has demonstrated large area, high quality, controllable number of layers and low defects. CVD approach has been found to be by far the most effective technique to produce high quality, large scale graphene that can be compatibly integrated into the Si device flows.

The CVD based graphene synthesis process typically involves a thin layer of a transition metal (usually a few hundred nanometers thick) deposited on a substrate e.g. SiO_2. The substrate is then put into a furnace to be heated up to about 1000° C in a hydrocarbon gas (e.g. methane and hydrogen) environment. The transition metallic layer catalyzes the decomposition of hydrocarbon gas and the dissociated carbon atoms gradually absorbs into the metal layer or diffuses/remains on the metal surface depending on the metal. Experimentally, many different transition metal catalysts, (e.g. Ru, Ir, Pd, Ni, Cu) have been used to synthesize graphene and two distinct growth mechanisms have been proposed (Li, Cai et al. 2009). **(I)** Precipitated growth, in which decomposed C atoms dissolve into the catalyst first and then precipitate to the metal surface to form graphene during the subsequent cooling. This is because the solubility of carbon in the metal decreases with temperature and the concentration of carbon decrease exponentially from the surface into the bulk. The follow-up cooling process helps the carbon atoms to segregate to the metal surface to form graphene. **(II).** Diffusive mechanism, in which the decomposed C atoms remain or diffuse on the metal surface and then incorporate into graphene directly. Mechanism **I** corresponds to those metals that interact strongly with C atoms and has the binary phase of metal carbide (e.g., Ni) and growth mechanism **II** corresponds to those which have no metal carbide phase (e.g., Cu). For mechanism **I**, continuous precipitation of C from the interior of catalysts normally leads to the non-uniform, multilayer formation of graphene layer as carbon prefers to segregate at the nickel grain boundaries (Yu, Lian et al. 2008). This problem is alleviated in mechanism **II** (Li, Cai et al. 2009) and it is known to be the best for the synthesis of monolayer graphene. Notably, inch-sized graphene has been demonstrated and synthesized on the Cu foil surface (Li, Cai et al. 2009; Li, Zhu et al. 2009). A rapid cooling rate in mechanism **I** can aid the suppression of preferential segregation of carbon at grain boundaries of the metal (e.g Ni) and thereby control the number of graphene layers (Kim, Zhao et al. 2009), as demonstrated by Kim *et al.* Graphene segregation during cooling is a non-equilibrium process (mechanism **I**). Non-equilibrium segregation in general involves the transport of vacancy-impurity (vacancy-carbon in this case) complexes to sinks, such as grain boundaries and surfaces during cooling, and strongly depends on the cooling rate (Thuvander and Andrén ; Yu, Lian et al. 2008). Different cooling rates lead to different segregation behaviors. Extremely fast cooling rate results in a quench effect in which the solute atoms lose the mobility before they can diffuse. A finite amount of carbon is found to segregate at the surface at medium cooling rates, which is found to be optimal. The extremely slow cooling rate allows carbon with enough time to diffuse into the bulk, so there will not be enough carbon segregated at the surface. Roughness of the metal substrates affects the uniformity of graphene layers synthesized by CVD (Yu, Lian et al. 2008). Thinner and more uniform graphene can be synthesized on smoother Ni substrates.

Metal-catalyzed graphene synthesis has been very well studied. Yet, the role of H_2 in the growth atmosphere, which is also very crucial for graphene growth, needs to be addressed duly.

2.1.1 Role of H_2 during CVD growth

It is well-known that hydrogen gas plays a key role for CVD diamond growth, while it acts as an etchant for amorphous carbon. A hydrogen molecule is very stable at temperatures up to 1000 °C. However, in the presence of a metal catalyst, at about 450 °C, the molecule dissociates and becomes reactive. Dissociation of the H_2 molecules in the presence of Ni and other metals was clearly demonstrated in an experiment done by Haluška et al. (Haluska, Hirscher et al. 2004). Atomic hydrogen then acts as an etching agent reducing preferably amorphous carbon that contains unsaturated dangling bonds. The role of etching by hydrogen during CVD is discussed in the following sub-section.

2.1.1.1 Etching by hydrogen during CVD

Diamond deposition with CVD has been successfully demonstrated using hydrogen as an etchant to remove the non-diamond phase. The selectivity of hydrogen in etching the two carbon phases, graphite and diamond, is considered the key factor for success in the synthesis of high-quality diamond films. The preferential etching of the sp^2 phase over the sp^3 phase by atomic hydrogen has been extensively reported. The role of atomic hydrogen both as sp^2 etchant and sp^3 promoter during the diamond growth is therefore well recognized experimentally and theoretically (Loh, Foord et al. 1996; Mendes, Corat et al. 1997). In addition, the etching selectivity of hydrogen has also been used to control the preferential growth orientation for obtaining oriented diamond crystals (Zhang, Jiang et al. 1997).Our comprehensive theoretical study based on Hartree-Fock (HF) molecular orbital approaches (Zhang, Chu et al. 2000) revealed the role of hydrogen species during the CVD growth and clarified the etching effect of the hydrogen species on sp^2 phase of carbon.

The overlap between the Highest Occupied Molecular Orbital (HOMO) of one molecule and the Lowest Unoccupied MO (LUMO) of another (also known as electron delocalization) determines the nature of chemical reaction between the two molecules (Hoffmann 1988; Fukui and Fujimoto 1997). Because the extent of electron delocalization is inversely proportional to the energy difference between these MOs, a small energy difference between the HOMO of one molecule (electron donor) and the LUMO of the other (electron acceptor) indicates a favorable reaction. This energy difference between HOMO of the electron donor and LUMO of the electron acceptor is hereafter referred to as the HOMO−LUMO difference of the reacting system. The frontier orbitals (HOMO and LUMO) were determined in this work using HF and configuration interaction (CI) instead of density functional theory (DFT) because the latter theory provides too close occupied and unoccupied states to analyze.

Under typical CVD conditions of diamond growth, hydrogen exists as various species of neutrals and ions. The HOMO-LUMO difference (with hydrogen as electron donor) as a function of the cluster size of two types of clusters is shown in Fig. 1. Each curve displays an overall tendency to decrease as the cluster size increases. This result is consistent with our earlier report (Zhang, Bertran et al. 1998) that the HOMO-LUMO gap of the same silicon or carbon cluster decreases as the cluster size increases and levels off when the cluster size reaches about 30 atoms. Thus, the reactivity of hydrogen species is less sensitive to the cluster size for larger clusters. The HOMO-LUMO differences between neutral hydrogen and all the clusters are illustrated in Figure 1(a), in which molecular hydrogen shows larger HOMO-LUMO differences and thus exhibits a lower reactivity than the atomic hydrogen. This relatively low reactivity of molecular hydrogen gives rise to the well-known small etching effect of molecular hydrogen in CVD processes (Harris, Doll et al. 1995). Being an abundant species in some CVD processes, atomic hydrogen is considered to act as the phase etchant. As illustrated in Figure 1(a), the HOMO-LUMO difference of H/sp^2-carbon is significantly different from that of H/sp^3-carbon, indicating that the reactivity of atomic hydrogen with the two carbon phases should be considerably different. The smaller HOMO-

LUMO difference of H/sp^2-carbon implies the preferential etching selectivity for the sp^2 phase, which is in good agreement with experiments (Donnelly, McCullough et al. 1997; Ishikawa, Yoshimi et al. 1997).

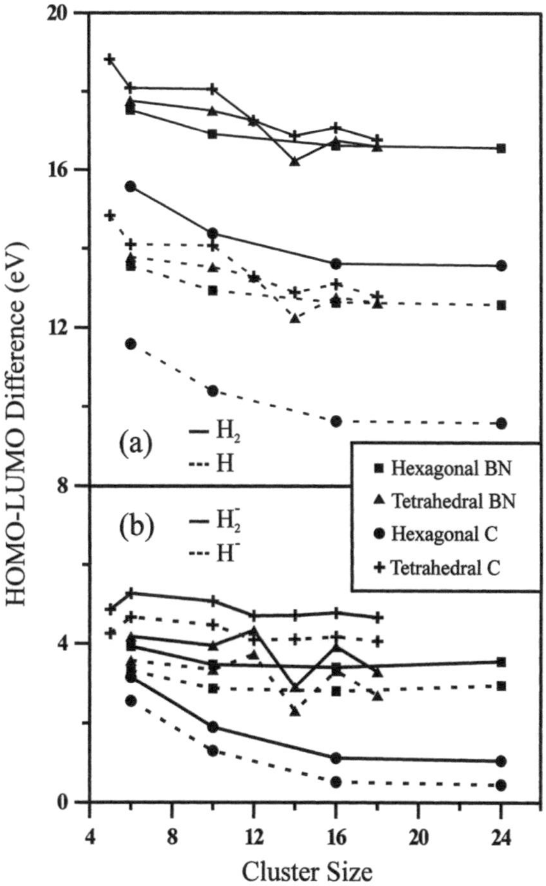

Fig. 1. HOMO-LUMO differences between (a) neutral hydrogen species and (b) negative hydrogen species; and BN, carbon clusters as functions of atomic number (Zhang, Chu et al. 2000).

Fig. 1(b) shows the HOMO-LUMO difference of the H^- ion and the clusters. The result shows that the reactivity of the H^- ion is considerably higher than that of the neutral hydrogen. Similar to the case of neutral hydrogen, the H^- ion shows preferential etching of the sp^2-carbon over the sp^3-carbon phase. In fact, when the charged hydrogen ion approaches the carbon clusters, charge transfer may take place. This may lead to neutralization of the H- ion and charging of the carbon clusters. Finally, the chemical reaction occurs between the charged clusters and neutral hydrogen. Hence, the reactivity of the hydrogen ion can also be explored from the point of view of charge transfer. Although it is impossible to obtain the HOMO-LUMO data for H+, using the concept of charge transfer, the information about H+ may be indirectly studied by considering the reactivity between positively charged clusters and the atomic hydrogen. Table 1 lists the results on small charged clusters. Comparing with the neutral, the negatively charged cluster has its HOMO and LUMO moved up. The energy difference between the HOMO of the negatively charged cluster and the LUMO of hydrogen is relatively small, indicating that the reactivity between the charged cluster and hydrogen is still higher than that between neutrals. For the positively charged cluster, the calculated HOMOs and LUMOs move to lower energies. The reactivity of atomic hydrogen (electron donor now) with the positively charged cluster is still higher than that with neutral cluster. Hydrogen ions also found in experiments have a higher reactivity than their neutral one (Davis, Haasz et al. 1987), which is in good agreement with our results.

	HOMO	HOMO of neutral	LUMO	LUMO of neutral
C_6^+ (sp^2)	−0.55619	−0.33067	−0.23272	0.14906
C_6^+ (sp^3)	−0.59758	−0.42311	−0.36760	0.22666
C_6^- (sp^2)	0.05325	−0.33067	0.36703	0.14906
C_6^- (sp^3)	0.20480	−0.42311	0.37316	0.22666

Table 1. HOMO and LUMO Values of Small Charged Clusters in Comparison with Those of Neutral Ones, Obtained Using HF Method with Basis Set 6-31G** (Unit: au) (Zhang, Chu et al. 2000).

	HOMO			LUMO		
	carried charge (atomic unit)			carried charge (atomic unit)		
	−1	0	+1	−1	0	+1
C_6 (sp^2)	0.05325	−0.33067	−0.55619	0.36703	0.14906	−0.23272
C_{10} (sp^2)	0.02096	−0.28689	−0.48896	0.29973	0.10296	−0.20351
C_6 (sp^3)	0.20480	−0.42311	−0.59758	0.37316	0.22666	−0.36760
C_{10} (sp^3)	0.18870	−0.42209	−0.55917	0.34822	0.21937	−0.33920

Table 2. Trend for HOMO and LUMO with Cluster Size (Unit: au) (Zhang, Chu et al. 2000).

As shown in Table 2, the influence of carried charge on the HOMO and LUMO energies decreases as the cluster size increases. This suggests that charge transfer has only a minor influence on the larger clusters. Accordingly, the charge-transfer effect should have little influence on the conclusions drawn from the calculations for neutral species. We note that while charging might influence the sticking probability of the species with the substrate, the bonding characteristics between the two parties would be mainly determined by the reactivity between their neutrals. The interaction between hydrogen species and the substrate has two meanings: their sticking to the substrate and their chemical reaction with the substrate. The higher reactivity of the hydrogen ion implies a higher sticking probability with the substrate than for atomic hydrogen. In summary, atomic hydrogen and hydrogen ions show a large difference in their reactivities towards the sp^2 and the sp^3 carbon phases. This difference facilitates the diamond growth via CVD methods. Hydrogen ions also show higher reactivity than the neutral.

2.1.1.2 Prevention or minimization of etching

The effects of hydrogen can be turned around to the aid of synthesis and growth of graphene during CVD by a careful control of the growth conditions. Despite the etching action of hydrogen on the sp^2 phase of carbon, it is found that a critical amount of hydrogen is necessary to synthesize a few layers of graphene. This is because hydrogen maintains a balance between the production of reactive hydrocarbonaceous radicals and the etching of the graphene layer during the CVD process. If the ratio of methane to hydrogen is too low, the etching reaction becomes much faster than the formation of graphene layers. This was also experimentally proved in a recent work of Kong's group (Reina, Thiele et al. 2009). As a result, the ratio of gas mixture between methane and hydrogen needs to be optimized in order to obtain a continuous, homogeneous and uniform graphene layer(s) (Park, Meyer et al. 2010).

2.1.1.3 *Beneficial effects of H_2*

If the growth conditions are properly adjusted, H_2 may be utilized to promote the synthesis and growth of graphene during CVD. H_2 in growth atmosphere influences the uniformity of graphene layers synthesized by CVD (Yu, Lian et al. 2008). With a high dosage of H_2 introduced 1 h before introducing the hydrocarbon gases, the uniformity of graphene is significantly enhanced, which suggests an annealing effect of H_2. It is believed that H_2 can eliminate certain impurities (such as *S* and *P*) that may cause local variations in the carbon dissolvability in the metal substrates (Angermann and Hörz 1993). In addition, atomic H can remove defects in carbon (and anneal dangling bonds) at elevated temperatures.

2.1.2 Improvisations on CVD

In general, graphene synthesized by CVD has a high quality and a large area. Yet, they are multi-crystalline in structure (Li, Cai et al. 2009). This is attributable to multiple factors: the epitaxial growth mechanism of graphene on transition metal surfaces, the multi-crystalline nature of the catalyst substrate, and the simultaneous nucleation of C atoms from multiple sites of the substrate surface. Therefore, further scientific research is required for fabricating high quality graphene with large single crystal domains. The recent experimental observation clearly showed the domain formation of CVD synthesized graphene and the defects are lined along the boundaries of the domains (Li, Cai et al. 2009). There are many experimental studies on the growth mechanism of graphene on the catalyst surface (Loginova, Bartelt et al. 2008; Gruneis, Kummer et al. 2009; Loginova, Bartelt et al. 2009; McCarty, Feibelman et al. 2009; Starodub, Maier et al. 2009). McCarty and his co-workers (Loginova, Bartelt et al. 2008; Loginova, Bartelt et al. 2009; McCarty, Feibelman et al. 2009; Starodub, Maier et al. 2009) have highlighted the significance of C dimer super-saturation on the metal surface to initiate the nucleation of graphene. They have also shown that graphene nucleation preferentially occurs at metal steps rather than at terraces. While theoretical studies in this nucleation are still scarce, Chen and his co-workers have recently studied the formation of carbon monomer and dimer on transition metal terraces and steps to probe the epitaxial growth of graphene (Chen, Zhu et al. 2010).
The methods of graphene synthesis which preceded CVD are briefly mentioned below.

2.2 Exfoliation method

One of the earliest and simplest methods consisted in micromechanical exfoliation or cleavage of graphite (Novoselov, Geim et al. 2004; Novoselov, Geim et al. 2005; Novoselov, Jiang et al. 2005). Layer(s) of graphene are peeled off mechanically from highly ordered graphite using a Scotch tape and then deposited on a substrate e.g. SiO_2. This is a simple yet efficient method, as graphene is obtained from highly ordered graphite crystals. Graphene extracted by microexfoliation shows very good electrical and structural quality. However, the shortcoming of this most elementary method is its non-scalability and production of uneven graphene films with small area.

2.3 Epitaxial growth

Graphene is also synthesizable by annealing of SiC crystal (Berger, Song et al. 2004; Berger, Song et al. 2006) at a very elevated temperature (~2000 K) in ultra-high vacuum. Thermal desorption of Si from the top layers of SiC crystalline wafer yields a multilayered graphene structure that behaves like graphene. The number of layers can be controlled by limiting time or temperature of the heating treatment. The quality and the number of layers in the samples depend on the SiC face used for their growth (Castro Neto, Guinea et al. 2009) (and references therein). Although the produced structure has a larger area than that obtainable by the exfoliation technique, still the coverage or area is way below the size required in electronic applications. Moreover, it is difficult to functionalize graphene obtained by this route.

2.4 Wet-chemistry approach

Wet-chemistry based approach is also employed to synthesize graphene by reduction of chemically synthesized graphene oxide (Stankovich, Dikin et al. 2007; Eda, Fanchini et al. 2008; Pichon 2008) (and references therein). Graphite is transformed into acid-intercalated graphite oxide by a severe oxidative treatment in sulphuric and nitric acid (Hummers and Offeman 1958). The intercalant is then rapidly evaporated at elevated temperatures, followed by its exposure to ultrasound or ball milling. Exfoliation of the graphite oxide readily occurs in aqueous medium due to the hydrophilicity of the former. Subsequent reduction of exfoliated graphite oxide sheets by hydrazine results in the precipitation of graphene owing to its hydrophobicity (Stankovich, Dikin et al. 2007). It is more versatile than the methods comprising exfoliation and epitaxial growth on SiC and easier to scale up. Yet, it has a poor control on the number of layers of graphene produced. Graphene synthesized by this method may remain partially oxidized, which potentially changes its electronic, optical, and mechanical properties.

3. Properties of graphene

The distinctive electronic, thermal and mechanical properties of graphene make it a very promising candidate for a wide range of applications in nanoscience and nanotechnology. The versatile properties of graphene are very well documented in the exponentially growing scientific literature. Some of its interesting properties and its technological implications are discussed hereafter.

3.1 Electronic properties

Graphene has immense potential for electronics for its extraordinarily high mobility of its charge carriers at room temperature. When Si-based technology is approaching its fundamental limits, graphene seems to be an ideal candidate to take over from silicon (Geim and Novoselov 2007). Yet, graphene is semi-metallic with no band gap, which severely limits its applications in electronics (Wei, Wang et al. 2010) due to its high leakage current in many applications. The electronic band gap plays a central role in modern device physics and technology and controls the performance of semiconductor devices. Moreover, it is a property inherent to semiconductors and insulators which considerably govern their transport and optical properties (Zhang, Tang et al. 2009). It has been possible to open and tune the band gap of graphene bilayers by applying an electric field (Zhang, Tang et al. 2009) or by doping (Ohta, Bostwick et al. 2006). These results have profound implications for potential utilization of graphene in electronics. The structure of graphene can be tailored to change its electronic properties (spectrum) by several means discussed in the next sub-section. The structural manipulation may induce optical properties (i.e band-gap opening), which in turn gets incorporated into it, resulting in its potential for opto-electronic applications.

3.2 Optical properties

Quantum size effects show up when the π electrons in graphene are confined laterally e.g. in graphene segments (Zhang, Bertran et al. 1998; Feng, Lin et al. 2009) or graphene nanoribbons (GNR) (Han, Ouml et al. 2007). An energy gap opens up when carriers are confined to a quasi-one dimensional system like GNR. GNR shows optical properties which are sensitive to their width, family, crystallographic orientation and edge termination (Nakada, Fujita et al. 1996; Wakabayashi, Fujita et al. 1999); similar to the CNTs. The energy gap of lithographically patterned GNR structures has been found to scale inversely with the ribbon width, which demonstrates the possibility to engineer the band gap of graphene nanostructures by lithographic processes (Han, Ouml et al. 2007). An alternative route to induce the formation of a band gap is through the hydrogenation of graphene (Elias, Nair et al. 2009; Guisinger, Rutter et al. 2009; Samarakoon and Wang 2010). The modification of the carbon bonds associated with the hydrogenation preserves the crystalline order of the lattice but leads to rehybridization of the carbon atoms from a planar sp^2 to a distorted sp^3 state (Sofo, Chaudhari et al. 2007). Recent experimental studies have demonstrated reversible hydrogenation through heating (Elias, Nair et al. 2009). Our theoretical findings relevant to structure and size dependent PL in carbon nanostructures (Zhang, Bertran et al. 1998; Feng, Lin et al. 2009) are discussed in the following sub-section.

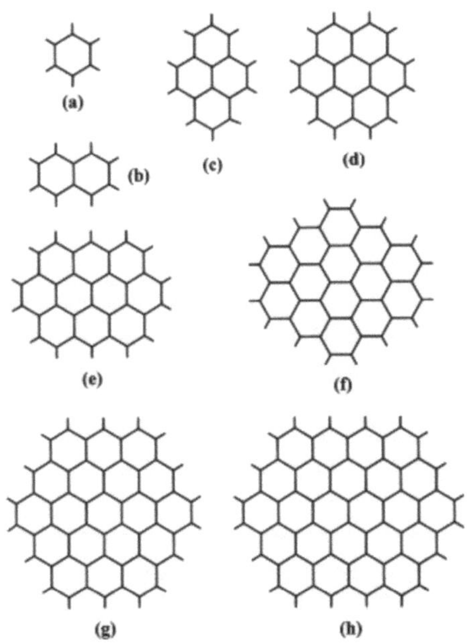

Fig. 2. Models for hexagonal clusters (or graphene segments): (a) C_6H_6, (b) $C_{10}H_8$, (c) $C_{16}H_{10}$, (d) $C_{24}H_{12}$, (e) $C_{32}H_{14}$, (f) $C_{42}H_{16}$, (g) $C_{54}H_{18}$ and (h) $C_{66}H_{18}$. The terminated bonds indicate the sites of boundary hydrogen atoms (Zhang, Bertran et al. 1998).

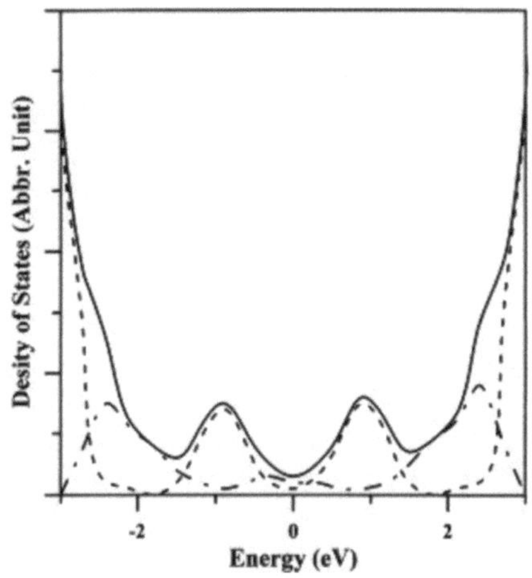

Fig. 3. Schematic diagram illustrates the distribution of band-gap states in a-C:H materials. A solid line indicates the total density of states (DOS); a dashed line represents the unlocalized states; and the dash-dotted line shows the localized states (Zhang, Bertran et al. 1998).

The broadband luminescence between 1.5 and 2.5 eV from carbon-based films has been attributed to the presence of a sp^2 amorphous phase or graphite phase (Badzian, Badzian et al. 1988; Nemanich, Glass et al. 1988). The distribution of states within the energy gap introduced by an sp^2 disordered phase in CVD diamond film has been considered as the origin of the broadband luminescence (Bergman, McClure et al. 1994), based on the theory of amorphous material (Street 1984). Further, the mechanism of luminescence and recombination in a-C:H and its alloys has been proposed (Robertson 1996 (a); Robertson 1996 (b); Silva, Robertson et al. 1996) to be due to a modification of the band edges in hydrogenated amorphous silicon. The band-tail states were assumed to arise from clusters of sp^2 sites. The π-bonding sp^2 phase has frequently been related to the origin of the luminescence (Rusli, Amaratunga et al. 1995; Bourée, Godet et al. 1996; Liu, Gangopadhyay et al. 1997). The broadband luminescence is commonly believed to be related to the gap/tail states produced from small sp^2 clusters with various sizes and/or shapes. A new photoluminescence model taking account of individual cluster has presented a convincing mechanism (Demichelis, Schreiter et al. 1995). Still, clear pictures of which structures or shapes of sp^2 carbon-clusters contribute to the highly efficient luminescence were missing.

For a crystal diamond that is also a pure sp^3-bonded material, the gap is not only indirect but also wide (5.49 eV at 77 K (Collins 1993)). The gap of a nano-structural diamond-like crystallite may be even larger, and thus may not relate to the luminescence that is presently concerned in the range 1.5~2.5 eV generated by means of optical transition between band states. On the other hand, the π-bonding states of graphite lie closest to the Fermi level, so it does not seem realistic to relate the gap formed by these states to the luminescence between 1.5 and 2.5 eV, too. However, a nano-sized π-bonding sp^2 cluster may show a fairly wide energy gap in comparison with that of the graphite material, as shown in our work (Zhang, Bertran et al. 1998), which coincides with the size-dependence rule that normally holds for many other materials. This feature has actually been seen in the calculations with a Hückel approximation for several sp^2 carbon clusters (Robertson and O'Reilly 1987; Robertson 1995). However, the previous study was only qualitative and did not distinguish the

different features of states in the gap-tail caused by the individual sp^2 carbon clusters due to their different sizes and/or shapes. Such a distinction, in our opinion, is very important for understanding the origin of highly efficient luminescence from carbon-based materials.
For the carbon-based materials described here, it is widely believed that both the size and shape determine the energy gap and the broadband luminescence (Demichelis, Schreiter et al. 1995; Robertson 1996 (a); Robertson 1996 (b); Silva, Robertson et al. 1996). However, any structural deviation from the stable configurations (for carbon, the stable structures are tetrahedral or sp^3 and hexagonal or sp^2) may produce localized states so that the energy gap is influenced. Obviously, the localized states do not relate to the efficient, room temperature luminescence. Thus, the small clusters with structures such as a fivefold ring, sevenfold ring and off-plane hexagonal that are not the stable hexagonal structures are doubtful for the main contribution to the efficient broadband luminescence. Also, the main source for such luminescence would be the stable hexagonal carbon clusters. We carried out calculations for a series of tetrahedral and hexagonal atomic clusters, shown in Fig 2 based on semi-empirical molecular orbital and density functional theories to determine the size-dependence effect of the energy gap in such clusters. We have classified the band-tail states into localized and confined, as shown in Fig 3 in order to emphasize the role of hexagonal planar shape of carbon-clusters in the broad band visible PL. The localized states result from the structural deviation from graphite-like configuration, and the associated luminescence may be described by using the conventional theory for amorphous materials. The confined states are generated due to the existence of stable graphite-like local structures with various sizes shown in Fig. 2 and are the main factor for giving efficient, room-temperature luminescence. Our calculations of a series of small hexagonal carbon clusters shown in Fig. 2 demonstrate that the energy-gap distribution, due to the difference in size, is considerably broad, which explains the broadband feature of luminescence. Weak п-п interaction enables multilayer stacking of graphene sheets in different possible ways shown in Fig 4 (Feng, Lin et al. 2009) and thereby helps the formation and stability of large-sized sp^2 C-H films. Our calculations show that the energy gaps of graphene are strongly dependent on their sizes (Feng, Lin et al. 2009), while the stacking order and the number of stacked layers have a minor influence. The energy gap is found to decrease with the increasing size, as shown in Fig 5. It turns out that by controlling the formation of graphene during CVD, it would be possible to control the size of the growing graphene and thereby tune its luminescent properties by utilizing our results on the size dependence of energy gap.

3.3 Non-covalent binding properties

The weak, intermolecular, vdW, п-п interactions (Hunter and Sanders 1990) play a crucial role in the crystal packing of organic molecules containing aromatic rings (Desiraju and Gavezzotti 1989; Hunter, Lawson et al. 2001), the intercalation of certain drug molecules into DNA (Brana, Cacho et al. 2001), the binding affinities of host-guest complexes (Muehldorf, Van Engen et al. 1988; Ferguson, Sanford et al. 1991; Chipot, Jaffe et al. 1996), as well as the three-dimensional structures of biological systems, including proteins and nucleic acids, and their molecular organization and recognition processes (Burley and Petsko 1985; Blundell, Singh et al. 1986; Hobza, Selzle et al. 1994). Graphene sheets can be stacked into bilayers and multilayers by virtue of π-π interactions between the neighboring sheets. The electronic properties of graphene multilayers vary with the stacking order, and rapidly evolve with the number of layers, approaching the 3D limit of graphite (Geim and Novoselov 2007; Castro Neto, Guinea et al. 2009).

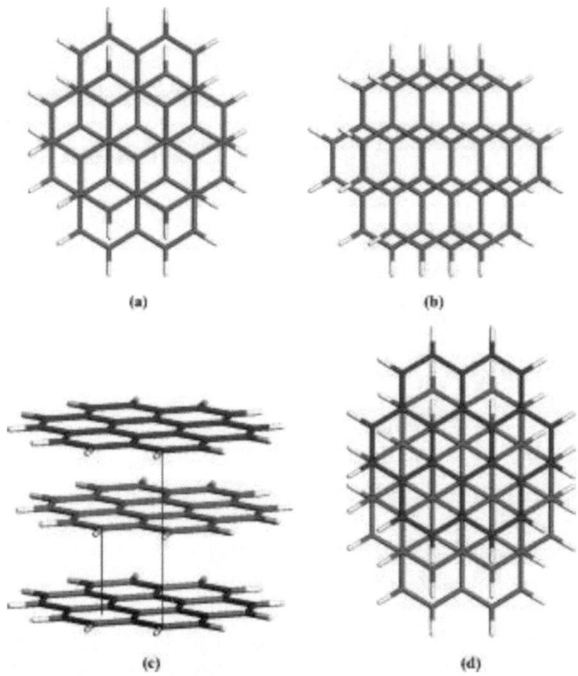

Fig. 4. Top views of the arrangement of the carbon atoms in two adjacent coronene planes in (a) staggered and (b) parallel-displaced stacking, and in coronene trimers with (c) ABA and (d) ABC stacking; gray, red, and blue represent the lowest, middle, and top layers, respectively (Feng, Lin et al. 2009).

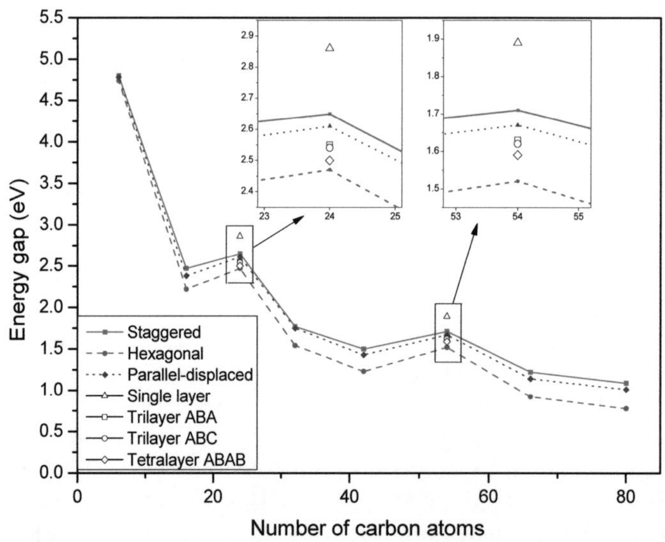

Fig. 5. Variation of the energy gap with the size of graphene sheet model dimers for the staggered, hexagonal, and parallel-displaced stackings, respectively. The open symbols and the insets show the energy gaps of $C_{24}H_{12}$ and $C_{54}H_{18}$ monomers, and their trimers with ABA and ABC stackings, and tetramers with an ABAB sequence (Feng, Lin et al. 2009).

Theoretically, it is challenging to model weak vdW interaction with the conventional DFT unless an appropriate, correction term for vDW interaction is incorporated. With such a treatment, we have shown in our work that the stacked graphene segments can be held together in different orientations by π-π interactions and the binding energy is strongly dependent on the size of the PAH, on stacking order, and on the number of stacked layers (Feng, Lin et al. 2009). A graphene bilayer has very unusual electronic properties, such as the anomalous integer quantum Hall effect that significantly changes with respect to a single layer (McCann and Fal'ko 2006; Novoselov, McCann et al. 2006; Malard, Nilsson et al. 2007), and which can be used to distinguish between a graphene bilayer and a monolayer. Moreover, as already mentioned in a preceding sub-section, the band-gap of graphene bilayer is tunable in different ways (Ohta, Bostwick et al. 2006; Zhang, Tang et al. 2009). The importance of graphene bilayers emphasizes the need to understand the binding between the two sheets in graphene bilayers. We have shown in our work using dimer, bilayer models that the binding energy increases with size until it saturates when it reaches the size of about 80 atoms (Feng, Lin et al. 2009). A clear understanding of the weak vdW interactions discussed in the section is relevant to some potential applications of graphene.

4. Physisorption and related applications

4.1 Water physisorption and novel ice structure formation

A water cluster adsorbed on a graphite surface is a prototypical weakly bound vdW π-system that involves water-graphite and water-water interactions. Our investigations show that the binding energy of water clusters interacting with graphite is dependent on the number of water molecules that form hydrogen bonds, but is independent of the water cluster size. Furthermore, we have found that these physically adsorbed or physisorbed water clusters show little change in their IR peak position and leave an almost perfect planar graphite surface (Lin, Zhang et al. 2005). SW-CNTs provide a well-defined nanoscale cylindrical pore that can serve as a nanometer-sized capillary in the fabrication of quasi-one-dimensional (Q1D) materials by filling SW-CNTs with chosen materials (Saito, Dresselhaus et al. 1998; Ugarte, Stoeckli et al. 1998). Even though water has been extensively studied, some of its properties remain partially unknown. A significant number of them are related to the behavior of water under confinement within nanoscale Q1D channels such as SW-CNTs, and the confined water is expected to exhibit different physical properties from its bulk counterparts. Since many similar scientifically relevant systems can be found in nature, this issue is of great interest to biology, geology, and materials science (Sansom and Biggin 2001).

We have systematically studied using a self-consistent charge density-functional tight-binding method complemented with an empirical vdW force correction to show that water molecules can form cylindrical crystalline structures (see Fig 6), referred to as INTs, by hydrogen bonding under confinement within single-walled carbon nanotubes (Feng, Zhang et al. 2007). Each water molecule in the optimized INTs is hydrogen bonded to its four nearest-neighbor water molecules in a tetra-coordinate configuration, and all water molecules constituted a novel cylindrical ice phase with ordered hydrogen-bond network, which is weaker than that of conventional hexagonal ice.

Our calculations show that for the confining CNTs, e.g., (16,0) SW-CNT, the HOMO-LUMO gap was slightly reduced from 0.567 to 0.554 eV while the Fermi level increased from −4.619 to −4.607 eV after it was filled with 6-gonal INT. Figure 7 also shows that both the HOMO and LUMO of INT−CNT complex are solely composed of the ones of the confining CNT and

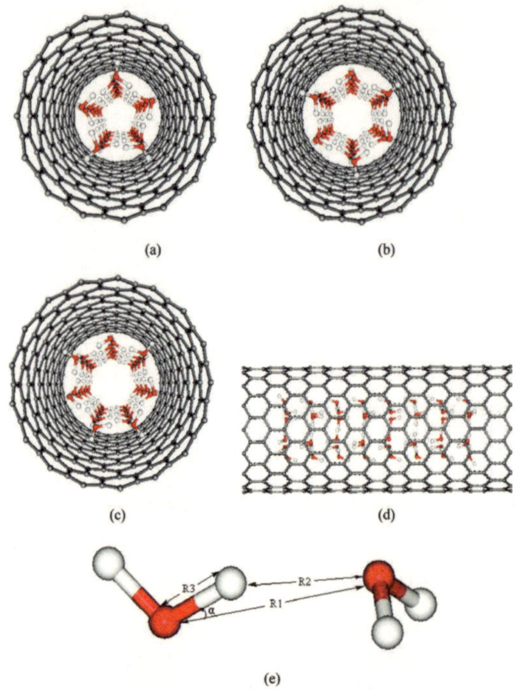

Fig. 6. Optimized (a) pentagonal, (b) hexagonal, and (c) heptagonal INT confined in zigzag (l,0) SW-CNT, l = 15, 16, and 17, respectively, and (d) side view of hexagonal INT model; (e) structural parameters used for quantifying hydrogen bond as discussed in the text. Red, white, and gray spheres represent oxygen, hydrogen, and carbon atoms, respectively (Feng, Zhang et al. 2007).

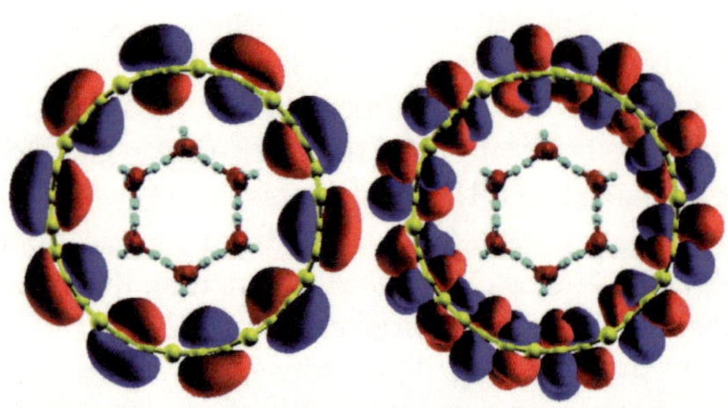

Fig. 7. Two isosurfaces of the wave functions of the HOMO (left) and LUMO (right) derived from bands at the Γ point for the 6-gonal INT confined in the zigzag (16,0) SW-CNT (Feng, Zhang et al. 2007).

that there is no overlap between molecular orbitals from CNT and INT. Consequently, it can be derived that water molecules of INTs can hardly interact with confining CNTs via the ordinary OH/π interactions, which are the main intermolecular forces between the water molecule and aromatic rings, for instance, in water-benzene/graphite complexes (Lin, Zhang et al. 2005). The reason is that each water molecule of INTs is hydrogen bonded to two water molecules with its two O−H bonds as hydrogen donors, and there is no other O−H bond available for a water molecule to interact with the hexagonal carbon rings of confining CNTs via OH/π interaction. We have revealed that the unique crystalline structures of INTs are mainly due to the steric hindrance effect induced by the confining CNT rather than the ordinary OH/π interaction between INT and CNT.

4.2 Hydrogen molecule physisorption and storage

Hydrogen physisorption by carbon-based nanostructures, which has potential for hydrogen storage, still remains challenging. The current preparative methods for CNTs generate heterogeneous tubes varying in length, diameter, and chirality. The presence of multiply (*n*, *m*) chiral SW-CNTs in a given sample poses a major barrier towards realizing many potential applications of CNTs. It may also contribute to the many inconsistencies in the experimental results.

The interaction between H_2 and CNT is a typical π-involved weak interaction. Using an approximate density functional method augmented with a vdW dispersion term, we have systematically investigated the role of CNT's curvature and chirality on the physisorption of H_2. In our work, we have explored the different possible physisorption sites for H_2 molecule on SW-CNT, as shown in Fig 8. Based on our results shown in Fig 9, we propose that CNTs with diameter of 6–7 Å, such as (5, 5), (8, 0), and (6, 3) tubes, are energetically optimal candidates for physisorption of H_2. In this relatively narrow range of diameters, the internal adsorption binding energies are around −0.22 eV, which is three times as large as that of H_2 on graphene surface; for external adsorption, the binding energy of −0.061 eV is just 18% below that of H_2 on graphene surface. It is conceivable that the inconsistencies in the experimental results (in terms of the hydrogen storage capacities of CNTs) were caused in part by the varying diameters of the as-prepared CNT samples used. We have found that nanotube's curvature plays an important role in the physisorption process of hydrogen, while the chirality of the tube has a negligible effect.

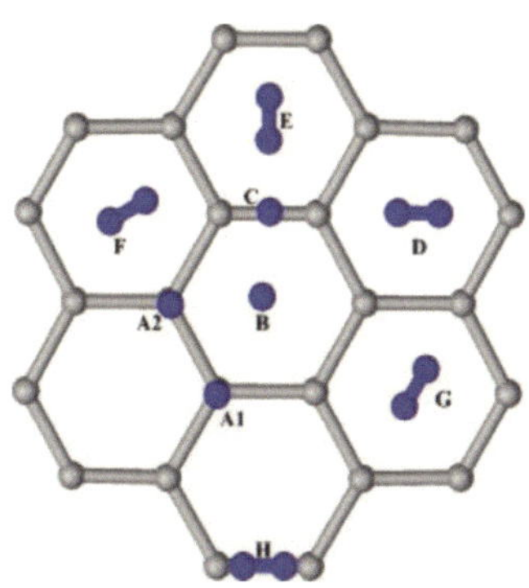

Fig. 8. Different adsorption sites for a hydrogen molecule on a segment of armchair SW-CNTs (Fan, Zhang et al. 2009).

4.3 Interaction with bio molecules, its relevance to biosensing

4.3.1 Flavin Adenine dinucleotide (FAD)

FAD is in the redox active group of flavoenzymes that catalyzes important biological redox reactions and is perhaps the most versatile of all of the redox coenzymes. We simulated the adsorption procedure of the FAD on the semiconducting (10,0) and the metallic (5,5) carbon nanotubes (CNTs) using a density functional tight binding method with the inclusion of an empirical dispersion term in total energy.

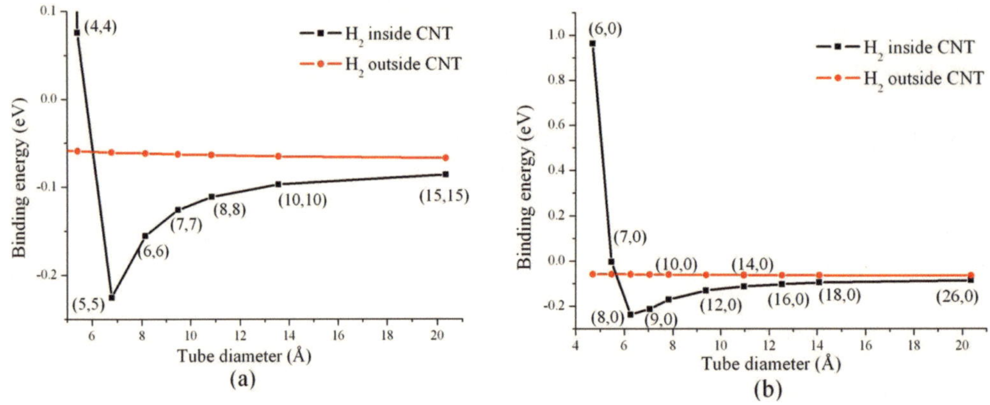

Fig. 9. Optimized binding energies of molecular hydrogen internally and externally adsorbed to CNTs for armchair (a) and zigzag (b) CNTs (Fan, Zhang et al. 2009).

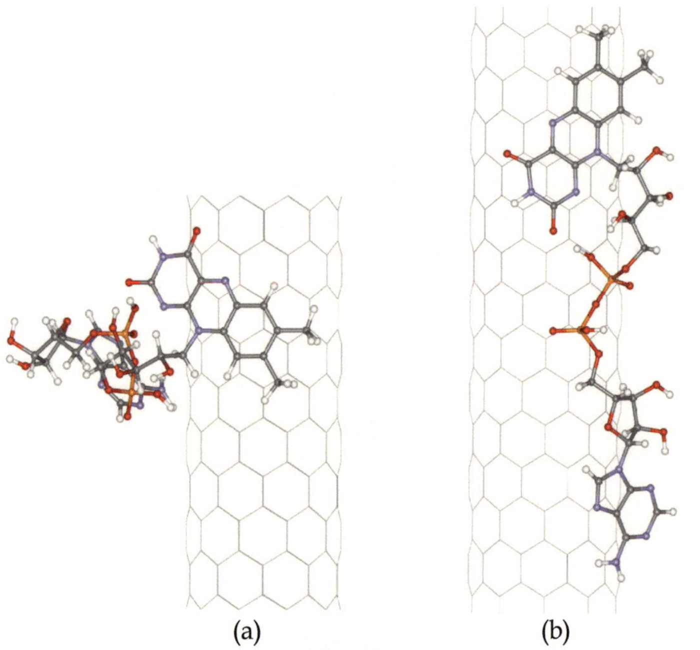

Fig. 10. Optimized structures of the (a) perpendicular and (b) parallel configurations of FAD adsorbed on (10,0) CNT (Lin, Zhang et al. 2007).

The flavin and adenine groups of FAD could be attracted to the CNT surface through π-π stacking but remain at the physisorption distances. The configurations with the FAD long axis perpendicular or parallel to the tube axes of the semiconducting (10,0) and metallic (5,5) CNTs, shown in Fig 10 were almost energetically degenerate. In the FAD/(10,0) system, the FAD flavin group contributed more components in the band structure at the Fermi energy level (see Fig 11), which was responsible for the enhancement of the electronic transferability as observed in a cyclic voltammogram experiment (Guiseppi-Elie, Lei et al. 2002).

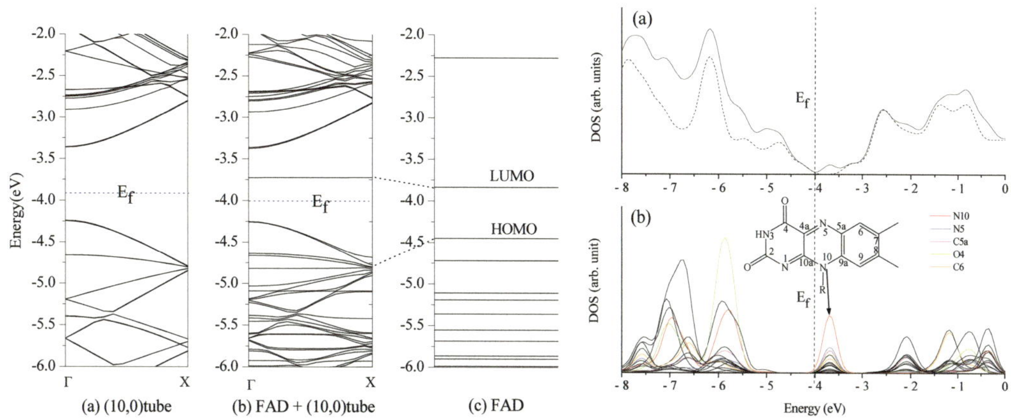

Fig. 11. (Left) Band structure of (a) an isolated (10,0) CNT, (b) FAD/(10,0), and (c) an isolated FAD molecule calculated with the same periodic condition. (Right) Total density of states of (10,0) CNT (dashed line) and FAD/(10,0) (solid line), and (b) projected density of states of O, N, and C atoms of the FAD flavin group in the FAD/(10,0) system. The largest five components were marked with different colors, and those from others atoms are shown as black color (Lin, Zhang et al. 2007).

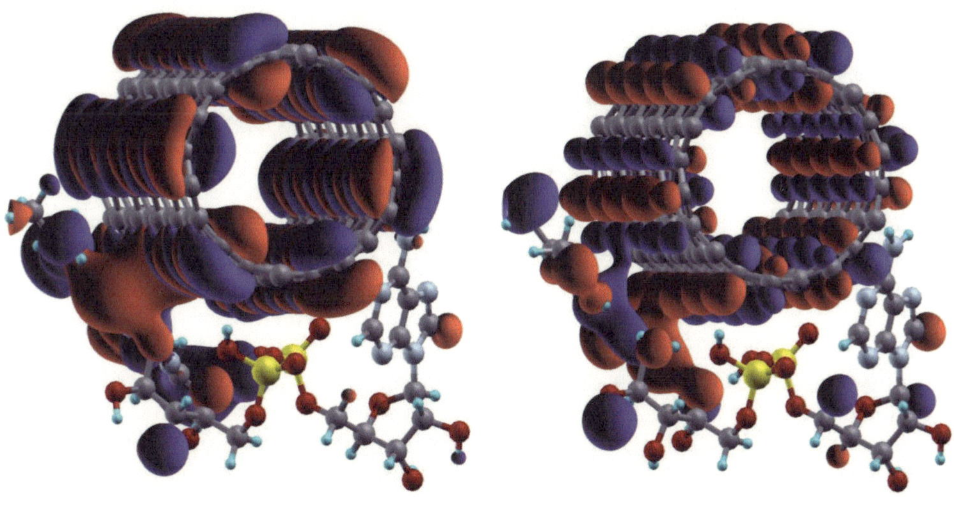

Fig. 12. Isosurfaces of the wave functions of the HOMO (left panel) and LUMO (right panel) derived bands at the Γ point for the FAD adsorbed on the (10,0) CNT surface. The isovalue is 0.02 au (Lin, Zhang et al. 2007).

The total DOS and projected density of states (PDOS) of FAD/(10,0) showed that the flavin group contributed significant components at the Fermi energy while the adenine group had few such components and the phosphate group had none. Hence, the flavin group served as the active unit relating to the electronic mobility of the FAD interacting with the CNT. In the FAD/(10,0) system, both HOMO and LUMO have significant contribution from the FAD flavin group (Fig. 12). The HOMO and LUMO of the FAD/(5,5) system showed their main features contributed by the CNT and the FAD flavin group respectively. This shows that the CNT may have contributed to the electron excitation procedure.

Despite a physisorption, there was a noticeable effect on the CNT electronic structure and mobility. Our results have prompted others (Ju, Doll et al. 2008; Ju and Papadimitrakopoulos 2008; Ju, Kopcha et al. 2009) to experimentally explore the properties of CNTs functionalized by flavin mononucleotide and its analogue.

4.3.2 Peptides

SW-CNTs have potential for biological applications ranging from biomedical sensors to drug delivery (Martin and Kohli 2003; Li, Ng et al. 2005; Contarino, Sergi et al. 2006). Yet, such biological applications have, so far, been limited due to two major obstacles: hydrophobicity and conformational heterogeneity. Although the solubility in water can be improved by chemically modifying the SW-CNTs through covalent bonding of various functional groups to the nanotubes (Hirsch 2002; Huang, Fernando et al. 2003), these modifications can perturb the intrinsic properties of SW-CNTs, such as electrical properties. As a result, alternative approaches using the non-covalent adsorption of surfactants (O'Connell, Bachilo et al. 2002), polymers (Dalton, Blau et al. 2001) and biomolecules (Zheng, Jagota et al. 2003; Zheng, Jagota et al. 2003; Numata, Asai et al. 2005) to solubilize the SW-CNTs have been proposed and tested. To that end, much research attention has been focused on the design and utilization of polypeptide/CNT complexes because of their functionality in biological systems (Dieckmann, Dalton et al. 2003; Zorbas, Ortiz-Acevedo et al. 2004; Ortiz-Acevedo, Xie et al. 2005; Pender, Sowards et al. 2005; Karajanagi, Yang et al. 2006; Su, Leung et al. 2006) recently.

We investigated the binding nature of three peptides (inactive NB1 and active B1 and B3) to single-walled carbon nanotubes (SWCNTs) using a density functional tight-binding (DFTB) method with an empirical vdW force correction (Fan, Zeng et al. 2009). Figure 13 shows the optimized geometries of the three peptides/(5,5) CNT complexes. We have shown that peptides (inactive NB1 and active B1 and B3) could be spontaneously attracted to the sidewall of CNTs through π−π and/or H−π stacking, which is at the physisorption distance. The competition of π−π and/or H−π stacking plays a key role in binding the peptides to the CNTs, thus, determining and stabilizing the binding of the peptide/CNT systems. The preservation of the helical conformation upon peptide B3 adsorption to the side wall of SWCNT is consistent with the experimental observation obtained using CD spectroscopy (Su, Leung et al. 2006). Our results demonstrate that the geometric structure of CNT remains almost unchanged after the adsorption of peptides. Moreover, the isosurfaces of the selected frontier orbitals show that the π-electronic structures of the CNTs are preserved upon the non-covalent adsorption of B3 peptides, which is similar to the features for simple planar organic molecules adsorbed on CNTs (Tournus and Charlier 2005; Tournus, Latil et al. 2005). DOS in Fig 14 furthers our understanding of its electronic properties. Compared with the pristine semiconducting (8,0) CNT, the DOS of the B3/(8,0) CNT shows new states near the Fermi level, which were contributed by the peptide B3. Still, the nature of physisorption is shown since the total DOS preserves the features of the DOS of pristine (8,0) CNT.

We find new DOS, composed mainly of the HOMO of B3 molecular orbitals, between the (8,0) CNT conduction and valence bands. Hence, the band gap of the system has sharply decreased from the pristine (8,0) tube 0.58 to 0.29 eV of the complexes. Similar finding of band gap reduction has been found in the (7,3) polyC−DNA complex compared to the free semiconducing (7,3) tube (Enyashin and et al. 2007). The finding of the new DOS formed between the (8,0) CNT conduction and valence bands is quite similar to our previous study of FAD/(10,0) CNT (Lin, Zhang et al. 2007). This implies that the non-covalent modification

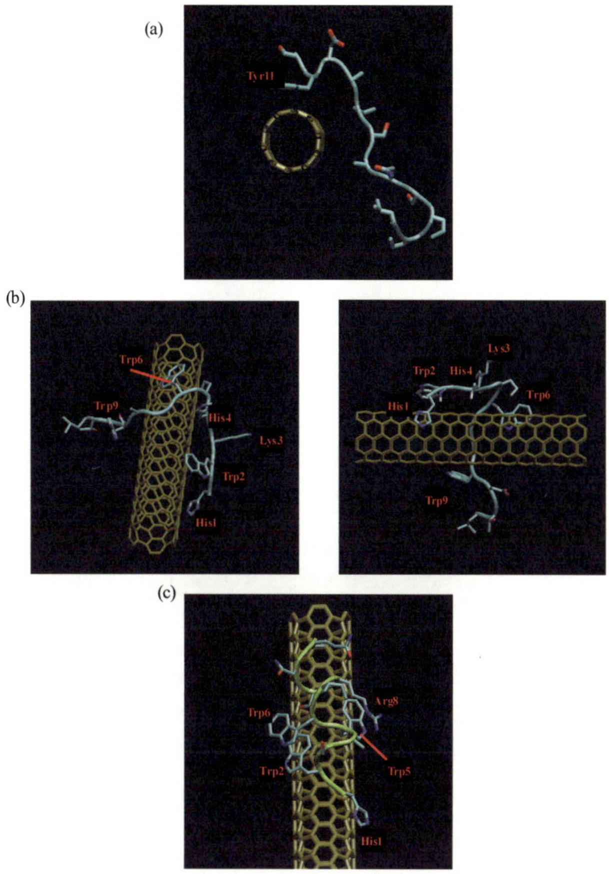

Fig. 13. Illustration of the peptide/(5,5) CNT interactions: (a) inactive peptide NB1, (b) the active peptide B1, and (c) the active peptide B3. The π−π stacking and XH−π (X = C and N) interactions are displayed in the left and right panels for the (b) B1/CNT complex, respectively. Key residues are labelled in red (Fan, Zeng et al. 2009).

of SW-CNTs by the active peptides might increase the former's electron transfer capabilities. Our study confirms the experimental findings (Wang, Humphreys et al. 2003) on the key role of the arene parts, such as His and Trp, and also agree with earlier theoretical reports (Chen, Hong et al. 2006; Tomásio and Walsh 2007).

5. Conclusion

A proper understanding of the growth and properties of graphene is a must for its optimal utilization. We have clarified the functionality of H in etching out the sp^2 phase of carbon nanostructures and thereby in promoting the sp^3 phase during CVD growth of graphene/diamond. The growth atmosphere and conditions needs to be properly adjusted during CVD growth in order to avoid this etching effect of hydrogen and invoke the beneficial effects of H_2. The size of graphene segments needs to be controlled during its growth by CVD in order to tune its luminescent properties as the energy gap scales inversely with the size of graphene segments. The structure of PAHs or graphene segments,

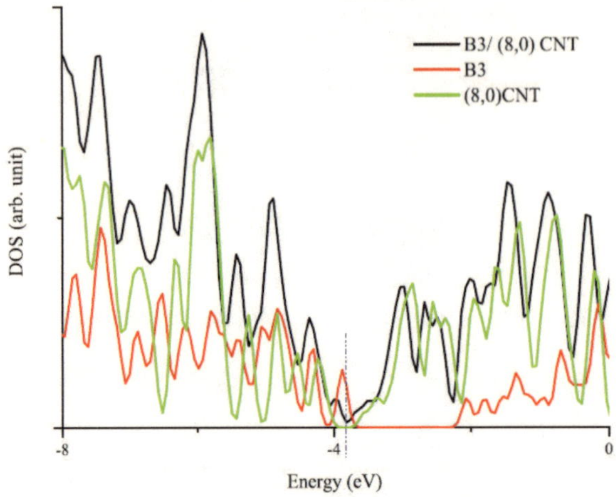

Fig. 14. Total density of states for B3/(8,0) CNT (Fan, Zeng et al. 2009).

both mono and multi layered, is solely attributed to visible and efficient luminescence at room temperature in our comprehensive findings. Weak, intermolecular, vdW interactions and its importance have been thoroughly explored in our work. We have found the π-π weak vdW interaction to bind together the planar graphene segments or PAHs in bilayers and multi-layers over a large size range. Besides, in our theoretical studies, the weak interactions of graphene with some important bio molecules provide crucial clues to its possible bio applications. For instance, the encapsulation of water molecules within hydrophobic CNTs provides pointers to the capillary functionality of biological nanopores. We have practically ruled out the role of chirality of CNTs in binding H_2 molecules, while we have pointed out the role of curvature of CNTs in the same. This has important implications for the hydrogen storage potential of CNTs. Furthermore, we have highlighted the importance of weak interaction of CNTs with a few important biomolecules in preserving the chemical, electrical and other properties of the former. The weak interaction was found to enhance the electron transfer capabilities of some bio molecules in our study. This signifies the bio-comptability of CNTs and its viability in different bio and other practical applications. A good understanding of the binding or interaction between bio-molecules and CNTs obtained from our work provides useful indicators for designing biosensors and drug delivery devices with bio-functionalized CNTs.

6. References

Angermann, H. H. and G. Hörz (1993). Influence of sulfur on surface carbon monolayer formation and graphite growth on nickel. *Applied Surface Science*,70-71,Part 1,(June 02, 1993) (163-168),0169-4332

Badzian, A. R., T. Badzian, et al. (1988). Crystallization of diamond crystals and films by microwave assisted CVD (Part II). *Materials Research Bulletin*,23,4,(April, 1988) (531-548),0025-5408

Barone, V., O. Hod, et al. (2006). Electronic Structure and Stability of Semiconducting Graphene Nanoribbons. *Nano Letters*,6,12,(November 24, 2006) (2748-2754),1530-6984

Berger, C., Z. Song, et al. (2004). Ultrathin Epitaxial Graphite: 2D Electron Gas Properties and a Route toward Graphene-based Nanoelectronics. *Journal of Physical Chemistry B*,108,52,(December 3, 2004) (19912-19916),1520-6106

Berger, C., Z. Song, et al. (2006). Electronic Confinement and Coherence in Patterned Epitaxial Graphene. *Science*,312,5777,(May 26, 2006) (1191-1196), 0036-8075

Bergman, L., M. T. McClure, et al. (1994). The origin of the broadband luminescence and the effect of nitrogen doping on the optical properties of diamond films. *Journal of Applied Physics*,76,5,(May 23, 1994) (3020-3027),0021-8979

Blundell, T., J. Singh, et al. (1986). Aromatic Interactions. *Science*,234,4779,(November 21, 1986) (1005-1005),0036-8075

Bourée, J. E., C. Godet, et al. (1996). Optical and luminescence properties of polymer-like a-C:H films deposited in a dual-mode PECVD reactor. *Journal of Non-Crystalline Solids*,198-200,Part 2,(May 02,1996) (623-627),0022-3093

Brana, M. F., M. Cacho, et al. (2001). Intercalators as anticancer drugs. *Current Pharmaceutical Design*,7,17,(November, 2001) (1745-1780),1381-6128

Burley, S. and G. Petsko (1985). Aromatic-aromatic interaction: a mechanism of protein structure stabilization. *Science*,229,4708,(July 5, 1985) (23-28),0036-8075

Castro Neto, A. H., F. Guinea, et al. (2009). The electronic properties of graphene. *Reviews of Modern Physics*,81,1,(January 14, 2009) (109-162),0034-6861

Chen, H., W. Zhu, et al. (2010). Contrasting behavior of carbon nucleation in the initial stages of graphene epitaxial growth on stepped metal surfaces. *Physical Review Letters*,104,18,(May 7, 2010) (186101-186104),0031-9007

Chen, Q., Y. Hong, et al. (2006). Chaotic behaviors and toroidal/spherical attractors generated by discontinuous dynamics. *Physica A: Statistical and Theoretical Physics*,371,2,(November 15, 2006) (293-302),0378-4371

Chipot, C., R. Jaffe, et al. (1996). Benzene Dimer: A Good Model for π−π Interactions in Proteins? A Comparison between the Benzene and the Toluene Dimers in the Gas Phase and in an Aqueous Solution. *Journal of the American Chemical Society*,118,45,(November 13, 1996) (11217-11224),0002-7863

Collins, A. T. (1993). Intrinsic and extrinsic absorption and luminescence in diamond. *Physica B: Physics of Condensed Matter*,185,1-4,(April, 1993) (284-296)

Contarino, M. R., M. Sergi, et al. (2006). Modular, self-assembling peptide linkers for stable and regenerable carbon nanotube biosensor interfaces. *Journal of Molecular Recognition*,19,4,(July/August, 2006) (363-371),1099-1352

Dalton, A. B., W. J. Blau, et al. (2001). A functional conjugated polymer to process, purify and selectively interact with single wall carbon nanotubes. *Synthetic Metals*,121,1-3,(March 15, 2001) (1217-1218),0379-6779

Davis, J. W., A. A. Haasz, et al. (1987). Flux and energy dependence of methane production from graphite due to H+ impact. *Journal of Nuclear Materials*,145-147,(February 2, 1987) (417-420),0022-3115

Demichelis, F., S. Schreiter, et al. (1995). Photoluminescence in a-C:H films. *Physical Review B*,51,4,(January 15, 1995) (2143-2147),1050-2947

Desiraju, G. R. and A. Gavezzotti (1989). From molecular to crystal structure; polynuclear aromatic hydrocarbons. *Journal of the Chemical Society, Chemical Communications*10, (621-623),0022-4936

Dieckmann, G. R., A. B. Dalton, et al. (2003). Controlled Assembly of Carbon Nanotubes by Designed Amphiphilic Peptide Helices. *Journal of the American Chemical Society*,125,7,(February 19, 2003) (1770-1777),0002-7863

Donnelly, C. M., R. W. McCullough, et al. (1997). Etching of graphite and diamond by thermal energy hydrogen atoms. *Diamond and Related Materials*,6,5-7,(April, 1997) (787-790),0925-9635

Eda, G., G. Fanchini, et al. (2008). Large-area ultrathin films of reduced graphene oxide as a transparent and flexible electronic material. *Nature Nanotechnology* 3,5,(April 6, 2008) (270-274),1748-3387

Elias, D. C., R. R. Nair, et al. (2009). Control of Graphene's Properties by Reversible Hydrogenation: Evidence for Graphane. *Science*,323,5914,(January 30, 2009) (610-613),0036-8075

Enyashin, A. N. and et al. (2007). DNA-wrapped carbon nanotubes. *Nanotechnology*,18,24,(May 25, 2007) (245702-245710),0957-4484

Fan, W. J., J. Zeng, et al. (2009). Quantum Mechanical Quantification of Weakly Interacting Complexes of Peptides with Single-Walled Carbon Nanotubes. *Journal of Chemical Theory and Computation*,5,10,(October, 2009) (2879-2885),1549-9618

Fan, W. J., R. Q. Zhang, et al. (2009). Prediction of energetically optimal single-walled carbon nanotubes for hydrogen physisorption. *Applied Physics Letters*,95,1,(July 10, 2009) (013116-013113),0003-6951

Fang, R. c. (1991). Emission properties of amorphous silicon and carbon films. *Journal of Luminescence*,48-49,Part 2,(January-February, 1991) (IN5-IN6),0022-2313

Feng, C., C. S. Lin, et al. (2009). Stacking of polycyclic aromatic hydrocarbons as prototype for graphene multilayers, studied using density functional theory augmented with a dispersion term. *The Journal of Chemical Physics*,131,19,(November 16, 2009) (194702-194708)

Feng, C., C. S. Lin, et al. (2010). pi-pi INTERACTION IN BENZENE DIMER STUDIED USING DENSITY FUNCTIONAL THEORY AUGMENTED WITH AN EMPIRICAL DISPERSION TERM. *Journal of Theoretical & Computational Chemistry*,9,(June 2, 2009) (109-123),0219-6336

Feng, C., R. Q. Zhang, et al. (2007). Signatures in Vibrational Spectra of Ice Nanotubes Revealed by a Density Functional Tight Binding Method. *The Journal of Physical Chemistry* C,111,38,(July 20, 2007) (14131-14138),1932-7447

Ferguson, S. B., E. M. Sanford, et al. (1991). Cyclophane-arene inclusion complexation in protic solvents: solvent effects versus electron donor-acceptor interactions. *Journal of the American Chemical Society*,113,14,(July, 1991) (5410-5419),0002-7863

Fukui, K. and H. Fujimoto (1997). *Frontier Orbitals and Reaction Paths: Selected Papers of Kenichi Fukui*,World Scientific: Singapore River Edge, NJ

Gapozzi, V., G. Casamassima, et al. (1996). *Solid State Communications*,98, (853)

Geim, A. K. and K. S. Novoselov (2007). The rise of graphene. *Nat Mater*,6,3, (183-191),1476-1122

Gruneis, A., K. Kummer, et al. (2009). *New J. Phys.*,11, (9)

Guiseppi-Elie, A., C. H. Lei, et al. (2002). *Nanotechnology*,13,, (559),

Guisinger, N. P., G. M. Rutter, et al. (2009). Exposure of Epitaxial Graphene on SiC(0001) to Atomic Hydrogen. *Nano Letters*,9,4, (1462-1466),1530-6984

Haluska, M., M. Hirscher, et al. (2004). Interaction of hydrogen isotopes with carbon nanostructures. *Materials Science and Engineering B: Solid-State Materials for Advanced Technology*,108,1-2, (130-133)

Han, M. Y., Ouml, et al. (2007). Energy Band-Gap Engineering of Graphene Nanoribbons. *Physical Review Letters*,98,20, (206805)

Harris, S. J., G. L. Doll, et al. (1995). *Appl. Phys. Lett.*,67,, (2314),

Hirsch, A. (2002). Functionalization of Single-Walled Carbon Nanotubes. *Angewandte Chemie International Edition*,41,11, (1853-1859),1521-3773

Hobza, P., H. L. Selzle, et al. (1994). Structure and Properties of Benzene-Containing Molecular Clusters: Nonempirical ab Initio Calculations and Experiments. *Chemical Reviews*,94,7, (1767-1785),0009-2665

Hoffmann, R. (1988). *Rev. Mod. Phys.*,60,, (601),

Huang, W., S. Fernando, et al. (2003). Preferential Solubilization of Smaller Single-Walled Carbon Nanotubes in Sequential Functionalization Reactions. *Langmuir*,19,17, (7084-7088),0743-7463

Hummer, G., J. C. Rasaiah, et al. (2001). Water conduction through the hydrophobic channel of a carbon nanotube. *Nature*,414,6860, (188-190),0028-0836

Hummers, W. S. and R. E. Offeman (1958). Preparation of Graphitic Oxide. *Journal of the American Chemical Society*,80,6, (1339-1339),0002-7863

Hunter, C. A., K. R. Lawson, et al. (2001). Aromatic interactions. *Journal of the Chemical Society, Perkin Transactions* 25, (651-669),1472-779X

Hunter, C. A. and J. K. M. Sanders (1990). The nature of .pi.-.pi. interactions. *Journal of the American Chemical Society*,112,14, (5525-5534),0002-7863

Ishikawa, Y., H. Yoshimi, et al. (1997). *Jpn. J. Appl. Phys., Part 1*,36,, (1233),

Ju, S.-Y., J. Doll, et al. (2008). Selection of carbon nanotubes with specific chiralities using helical assemblies of flavin mononucleotide. *Nat Nano*,3,6, (356-362),1748-3387

Ju, S. Y., W. P. Kopcha, et al. (2009). Brightly Fluorescent Single-Walled Carbon Nanotubes via an Oxygen-Excluding Surfactant Organization. *Science*,323,5919,(Mar) (1319-1323),0036-8075

Ju, S. Y. and F. Papadimitrakopoulos (2008). Synthesis and redox behavior of flavin mononucleotide-functionalized single-walled carbon nanotubes. *Journal of the American Chemical Society*,130,2,(Jan) (655-664),0002-7863

Kania, P. and P. Oelhafen (1995). Photoluminescence study of <100> textured CVD diamonds. *Diamond and Related Materials*,4,4, (425-428),0925-9635

Karajanagi, S. S., H. Yang, et al. (2006). Protein-Assisted Solubilization of Single-Walled Carbon Nanotubes. *Langmuir*,22,4, (1392-1395),0743-7463

Kim, K. S., Y. Zhao, et al. (2009). Large-scale pattern growth of graphene films for stretchable transparent electrodes. *Nature*,457,7230, (706-710),0028-0836

Koga, K., G. T. Gao, et al. (2001). *Nature*,412,, (802),

Li, J., H. T. Ng, et al. (2005). *Methods Mol. Bio.*,300, (191)

Li, X., W. Cai, et al. (2009). Large-Area Synthesis of High-Quality and Uniform Graphene Films on Copper Foils. *Science*,324,5932,(June 5, 2009) (1312-1314)

Li, X., W. Cai, et al. (2009). *Nano Lett.*,9, (4268-4272)

Li, X., Y. Zhu, et al. (2009). *Nano Lett.*,9, (4359-4363)

Lin, C. S., R. Q. Zhang, et al. (2005). Simulation of Water Cluster Assembly on a Graphite Surface. *The Journal of Physical Chemistry B*,109,29, (14183-14188),1520-6106

Lin, C. S., R. Q. Zhang, et al. (2007). Geometric and Electronic Structures of Carbon Nanotubes Adsorbed with Flavin Adenine Dinucleotide: A Theoretical Study. *The Journal of Physical Chemistry C*,111,11, (4069-4073),1932-7447

Liu, S., S. Gangopadhyay, et al. (1997). Photoluminescence studies of hydrogenated amorphous carbon and its alloys. *Journal of Applied Physics*,82,9, (4508-4514)

Loginova, E., N. C. Bartelt, et al. (2008). *New J. Phys.*,10, (16)

Loginova, E., N. C. Bartelt, et al. (2009). *New J. Phys.*,11, (20)

Loh, K. P., J. S. Foord, et al. (1996). *Diamond Relat. Mater.*,5, (231)

Malard, L. M., J. Nilsson, et al. (2007). Probing the electronic structure of bilayer graphene by Raman scattering. *Physical Review B*,76,20, (201401)

Maniwa, Y., Y. Kumazawa, et al. (1999). *Jpn. J. Appl. Phys.*,38, (668)

Mann, D. J. and M. D. Halls (2003). *Phys. Rev. Lett.*,90, (195503)

Martí, J. and M. C. Gordillo (2001). *Phys. Rev. E*,64, (21504)

Martí, J. and M. C. Gordillo (2003). *J. Chem. Phys.*,119, (12540)

Martin, C. R. and P. Kohli (2003). The emerging field of nanotube biotechnology. *Nat Rev Drug Discov*,2,1, (29-37),1474-1776

Mashl, R. J., S. Joseph, et al. (2003). *Nano Lett.*,3, (589)

McCann, E. and V. I. Fal'ko (2006). Landau-Level Degeneracy and Quantum Hall Effect in a Graphite Bilayer. *Physical Review Letters*,96,8, (086805)

McCarty, K. F., P. J. Feibelman, et al. (2009). *Carbon*,47, (1806)

Mendes, R. C., E. J. Corat, et al. (1997). *Diamond Relat. Mater.*,6,, (490),

Muehldorf, A. V., D. Van Engen, et al. (1988). Aromatic-aromatic interactions in molecular recognition: a family of artificial receptors for thymine that shows both face-to-face and edge-to-face orientations. *Journal of the American Chemical Society*,110,19, (6561-6562),0002-7863

Nakada, K., M. Fujita, et al. (1996). Edge state in graphene ribbons: Nanometer size effect and edge shape dependence. *Physical Review B*,54,24, (17954)

Nemanich, R. J., J. T. Glass, et al. (1988). Raman scattering characterization of carbon bonding in diamond and diamondlike thin films. *Journal of Vacuum Science & Technology A: Vacuum, Surfaces, and Films*,6,3, (1783-1787)

Nevin, W. A., H. Yamagishi, et al. (1994). Emission of blue light from hydrogenated amorphous silicon carbide. *Nature*,368,6471, (529-531)

Noon, W. H., K. D. Ausman, et al. (2002). *Chem. Phys. Lett.*,355, (445)

Novoselov, K. S., A. K. Geim, et al. (2005). *Nature*,438, (197-200)

Novoselov, K. S., A. K. Geim, et al. (2004). Electric Field Effect in Atomically Thin Carbon Films. *Science*,306,5696,(October 22, 2004) (666-669)

Novoselov, K. S., D. Jiang, et al. (2005). *Proc. Natl. Acad. Sci. USA* 102, (10451-10453)

Novoselov, K. S., E. McCann, et al. (2006). Unconventional quantum Hall effect and Berry/'s phase of 2[pi] in bilayer graphene. *Nat Phys*,2,3, (177-180),1745-2473

Numata, M., M. Asai, et al. (2005). Inclusion of Cut and As-Grown Single-Walled Carbon Nanotubes in the Helical Superstructure of Schizophyllan and Curdlan (β-1,3-Glucans). *Journal of the American Chemical Society*,127,16, (5875-5884),0002-7863

O'Connell, M. J., S. M. Bachilo, et al. (2002). Band Gap Fluorescence from Individual Single-Walled Carbon Nanotubes. *Science*,297,5581,(July 26, 2002) (593-596)

Ohta, T., A. Bostwick, et al. (2006). Controlling the Electronic Structure of Bilayer Graphene. *Science*,313,5789,(August 18, 2006) (951-954)

Ortiz-Acevedo, A., H. Xie, et al. (2005). Diameter-Selective Solubilization of Single-Walled Carbon Nanotubes by Reversible Cyclic Peptides. *Journal of the American Chemical Society*,127,26, (9512-9517),0002-7863

Park, H. J., J. Meyer, et al. (2010). Growth and properties of few-layer graphene prepared by chemical vapor deposition. *Carbon*,48,4, (1088-1094),0008-6223

Pender, M. J., L. A. Sowards, et al. (2005). Peptide-Mediated Formation of Single-Wall Carbon Nanotube Composites. *Nano Letters*,6,1, (40-44),1530-6984

Pichon, A. (2008). Graphene synthesis: Chemical peel. *Nat Chem*,1755-4330

Reina, A., X. T. Jia, et al. (2009). *Nano Lett.*,9, (30-35)

Reina, A., S. Thiele, et al. (2009). *Nano Research*,2,6, (509-516)

Robertson, J. (1995). Structural models of a-C and a-C:H. *Diamond and Related Materials*,4,4, (297-301)

Robertson, J. (1996) (a). Photoluminescence mechanism in amorphous hydrogenated carbon. *Diamond and Related Materials*,5,3-5, (457-460)

Robertson, J. (1996) (b). Recombination and photoluminescence mechanism in hydrogenated amorphous carbon. *Physical Review B*,53,24, (16302)

Robertson, J. and E. P. O'Reilly (1987). Electronic and atomic structure of amorphous carbon. *Physical Review B*,35,6, (2946)

Rusli, G. A. J. Amaratunga, et al. (1995). Photoluminescence in amorphous carbon thin films and its relation to the microscopic properties. *Thin Solid Films*,270,1-2, (160-164),0040-6090

Saito, R., G. Dresselhaus, et al. (1998). *Physical Properties of Carbon Nanotube*,Imperial College Press,London

Samarakoon, D. K. and X.-Q. Wang (2010). Tunable Band Gap in Hydrogenated Bilayer Graphene. *ACS Nano*,4,7, (4126-4130),1936-0851

Sansom, M. S. P. and P. C. Biggin (2001). Biophysics: Water at the nanoscale. *Nature*,414,6860, (156-159),0028-0836

Schütte, S., S. Will, et al. (1993). Electronic properties of amorphous carbon (a-C:H). *Diamond and Related Materials*,2,10, (1360-1364),0925-9635

Silva, S. R. P., J. Robertson, et al. (1996). Structure and luminescence properties of an amorphous hydrogenated carbon. *Philosophical Magazine B: Physics of Condensed Matter; Statistical Mechanics, Electronic, Optical and Magnetic Properties*,74,4, (369-386)

Sofo, J. O., A. S. Chaudhari, et al. (2007). Graphane: A two-dimensional hydrocarbon. *Physical Review B*,75,15, (153401)

Stankovich, S., D. A. Dikin, et al. (2007). *Carbon*,45, (1558-1565)

Starodub, E., S. Maier, et al. (2009). *Phys. Rev. B*,80, (8)

Street, R. A. (1984). In: *Semiconductors and Semimetals*. J. I. Pankove,(Ed.). Vol. 84, (197), Academic Press New York.

Su, Z., T. Leung, et al. (2006). Conformational Selectivity of Peptides for Single-Walled Carbon Nanotubes. *The Journal of Physical Chemistry B*,110,47, (23623-23627),1520-6106

Sutter, P. W., J. I. Flege, et al. (2008). *Nat. Mater.*,7, (406-411)

Thuvander, M. and H. O. Andrén APFIM Studies of Grain and Phase Boundaries: A Review. *Materials Characterization*,44,1-2, (87-100),1044-5803

Tomásio, S. D. M. and T. R. Walsh (2007). Atomistic modelling of the interaction between peptides and carbon nanotubes. *Molecular Physics: An International Journal at the Interface Between Chemistry and Physics*,105,2, (221 - 229),0026-8976

Tournus, F. and J. C. Charlier (2005). Ab initio study of benzene adsorption on carbon nanotubes. *Physical Review B*,71,16, (165421)

Tournus, F., S. Latil, et al. (2005). pi -stacking interaction between carbon nanotubes and organic molecules. *Physical Review B*,72,7, (075431)

Ugarte, D., T. Stoeckli, et al. (1998). *Appl. Phys. A.*,7, (101)

Wagner, J. and P. Lautenschlager (1986). Hard amorphous carbon studied by ellipsometry and photoluminescence. *Journal of Applied Physics*,59,6, (2044-2047)

Wakabayashi, K., M. Fujita, et al. (1999). Electronic and magnetic properties of nanographite ribbons. *Physical Review B*,59,12, (8271)

Wang, J., Y. Zhu, et al. (2004). *Phys. Chem. Chem. Phys.*,6, (829)

Wang, S., E. S. Humphreys, et al. (2003). Peptides with selective affinity for carbon nanotubes. *Nat Mater*,2,3, (196-200),1476-1122

Wei, Z., D. Wang, et al. (2010). Nanoscale Tunable Reduction of Graphene Oxide for Graphene Electronics. *Science*,328,5984,(June 11, 2010) (1373-1376)

Wu, J., W. Pisula, et al. (2007). Graphenes as Potential Material for Electronics. *Chemical Reviews*,107,3, (718-747),0009-2665

Xu, S., M. Hundhausen, et al. (1993). Influence of substrate bias on the properties of a-C:H films prepared by plasma CVD. *Journal of Non-Crystalline Solids*,164-166,Part 2, (1127-1130),0022-3093

Yu, Q., J. Lian, et al. (2008). Graphene segregated on Ni surfaces and transferred to insulators. *Applied Physics Letters*,93,11, (113103-113103)

Zhang, Q., S. C. Bayliss, et al. (1996). The stable blue and unstable UV photoluminescence from carbon nanoclusters embedded in SiO2 matrices. *Solid State Communications*,99,12, (883-886),0038-1098

Zhang, R. Q., E. Bertran, et al. (1998). Size dependence of energy gaps in small carbon clusters: the origin of broadband luminescence. *Diamond and Related Materials*,7,11-12, (1663-1668),0925-9635

Zhang, R. Q., T. S. Chu, et al. (2000). A Theoretical Study on the Interactions of Hydrogen Species with Various Carbon and Boron Nitride Phases. *The Journal of Physical Chemistry B*,104,29, (6761-6766),1520-6106

Zhang, W. J., X. Jiang, et al. (1997). *J. Appl. Phys.*,82, (1896)

Zhang, Y., T.-T. Tang, et al. (2009). Direct observation of a widely tunable bandgap in bilayer graphene. *Nature*,459,7248, (820-823),0028-0836

Zheng, M., A. Jagota, et al. (2003). DNA-assisted dispersion and separation of carbon nanotubes. *Nat Mater*,2,5, (338-342),1476-1122

Zheng, M., A. Jagota, et al. (2003). Structure-Based Carbon Nanotube Sorting by Sequence-Dependent DNA Assembly. *Science*,302,5650,(November 28, 2003) (1545-1548)

Zorbas, V., A. Ortiz-Acevedo, et al. (2004). Preparation and Characterization of Individual Peptide-Wrapped Single-Walled Carbon Nanotubes. *Journal of the American Chemical Society*,126,23, (7222-7227),0002-7863

2

Theory of Doping: Monovalent Adsorbates

B. Sachs[1], T. O. Wehling[1], A. I. Lichtenstein[1] and M. I. Katsnelson[2]
[1] *I. Institut für Theoretische Physik, Universität Hamburg, Jungiusstraße 9, D-20355 Hamburg*
[2]*Radboud University of Nijmegen, Institute for Molecules and Materials, Heijendaalseweg 135, 6525 AJ Nijmegen*
[1]*Germany*
[2]*The Netherlands*

1. Introduction

The discovery of graphene (Novoselov et al., 2004) and its remarkable electronic properties (Castro Neto et al., 2009; Geim & Novoselov, 2007; Katsnelson, 2007) initiated great research interest in this material. Particularly prospective for applications is its extraordinarily high charge carrier mobility (Bolotin et al., 2008; Du et al., 2008; Novoselov et al., 2004).

Realistic graphene samples are subject to disorder including ripples, impurities, edges or strains. While these present undesirable obstacles when trying to minimize electron scattering, controlled external perturbations recently evoked broad interest in the context of functionalization of graphene (Elias et al., 2009).

Impurities on a graphene sample are imaginable in various ways. While lattice imperfections like vacancies do not exist in noticeable concentrations unless they are not created artificially (Chen et al., 2009), adatoms or molecules from the experimental environment can be seen as a frequent source of electron scattering. The impact of adsorbate-induced scattering processes on the transport properties has been subject of ongoing discussion since the first fabrication of single graphene sheets in 2004 (see Peres (2010) for a review).

Transport experiments with chemically doped graphene samples yield different results regarding the strength of electron scattering due to the dopants: While room temperature experiments with NO_2, e.g., reported chemical doping without significant loss of carrier mobility (Schedin et al., 2007), the deposition of K at cryogenic temperatures clearly reduced the electron mobility (Chen et al., 2008). Correspondingly, the role of charged impurity scattering as compared to, e.g., scattering by resonant impurities or ripples has been controversially debated: Depending on experimental details both, charged impurities (Adam et al., 2007; Chen et al., 2008; Hwang et al., 2007; Nomura & MacDonald, 2006; Tan et al., 2007), as well as resonant impurities were discussed as dominant scattering sources (Katoch et al., 2010; Katsnelson & Novoselov, 2007; Ni et al., 2010; Ostrovsky et al., 2006; Stauber et al., 2007). Understanding charge redistributions in realistic graphene-adsorbate systems is hence crucial.

The high sensitivity of graphene to adsorbate-induced doping has been proven in numerous experiments (Bostwick et al., 2006; Ohta et al., 2006; Zhou et al., 2008). The two-dimensional nature maximizes surface effects, which even allows the detection of single adsorption events.

Using exfoliated graphene on a SiO_2 substrate with Ti/Au contacts, Schedin et al. (2007) visualized events of single molecule adsorption on graphene. Recently developed gas sensors (Collins et al., 2000; Kong et al., 2000; Robinson, Perkins, Snow, Wei & Sheehan, 2008) raise hope for a future realization of marketable single molecule detectors.

The natural concentration of impurities in graphene devices depends crucially on the sample preparation and the experimental setup. Meyer et al. (2008) reported the detection of single hydrogen adsorbates on graphene on a SiO_2 substrate by transmission electron microscopy (TEM). Under atmospheric conditions at room temperature they estimated the adsorbate concentration to 0.3%, which relates to about one impurity per $10nm^2$.
Especially hydrogenation, as demonstrated by Elias et al. (2009), provides a good prospect for controlled design of graphene's electronic properties. Hence, there is wide interest in fractionally and fully hydrogenated graphene, the *graphane*. Attaching hydrogen on graphene from both sides leads to a change from sp^2 to sp^3 hybridization, which opens a band gap. Through annealing, hydrogenation turns out to be reversible, i.e. the electronic properties of pristine graphene can be restored. Several theoretical works (Lebègue et al., 2009; Liu & Shen, 2009; Sofo et al., 2007) found a band gap of *graphane* between 3.5eV and 5.4eV. While this is slightly too high for electronic applications, partially hydrogenated graphene might be useful (Xiang et al., 2009). Hence, understanding of the adsorption mechanisms of atomic hydrogen is essential in search of new paths towards functionalization.
Equally, fluorination of graphene promises a route towards a graphene-based wide band gap semiconductor (Cheng et al., 2010; Nair et al., 2010; Robinson et al., 2010). At coverages of 70% or more, graphene-fluorine systems with reversible modification of the conductivity by several orders of magnitudes has been achieved.

Motivated by these recent and promising experiments on impurity effects in graphene, a theoretical investigation of doping effects in graphene is given in this chapter; in particular, monovalent adsorbates are considered. Extensive density functional theory (DFT) calculations are presented to derive a theory of doping and charge redistributions in graphene and to identify simple models describing these effects realistically. We concentrate on two issues: charge transfer as relevant for doping, i.e. changes in the number of mobile carriers, as well as charge transfer as relevant for Coulomb scattering. For hydrogen, fluorine, hydroxyl, chlorine and potassium adsorbates we determine the amount of the charge transfer by means of different electrostatic models and compare to band structure based methods (section (3)). Furthermore, by means of a tight-binding model, impurities are illustrated to lead to long-range doping of graphene such that even ultra-low concentrations of contamination do affect the carrier concentration. We investigate the effects of long range Coulomb interaction in this context and show that the Coulomb repulsion plays a minor role in the process of charge redistribution for impurity concentrations higher than 0.007%.

2. Calculation of charge transfer

The investigation of adsorption processes rises the question of the doping and Coulomb scattering due to single adsorbates. For instance, theoretical transport calculations predict a strong dependence of the scattering cross section on the amount of charge transferred between adsorbates and graphene (Robinson, Schomerus, Oroszlány & Fal'ko, 2008). A priori charge transfer is an ambiguous quantity as it comes back to defining the spacial extent of individual atoms within a solid. Therefore, several concepts for the description of charge transfer will be taken into consideration, carefully compared and their implications for experimental observables like doping or scattering properties will be discussed.

2.1 Population analysis and partitioning of the electron density

A widely employed class of approaches to charge transfer analysis like the Mulliken, Bader or Hirshfeld analysis aims at directly partitioning the electronic charge density among the atoms of the system. To this end, a DFT calculation is performed which yields the electronic density and the Kohn-Sham wave functions. Partitioning schemes using projections of the Kohn-Sham wavefunctions onto localized atomic orbitals (Löwdin or Mulliken analysis, see, e.g., Segall et al. (1996)) as well as schemes dealing with the electronic density (Hirshfeld or Bader analysis, see, e.g., Meister & Schwarz (1994)) have been employed in the context of graphene adsorbate systems. While ionically bond systems are likely well suited to be correctly described by this kind of charge transfer analysis, the interpretation of Mulliken, Bader or Hirshfeld charges in physisorbed graphene-impurity systems (Leenaerts et al., 2008) or strongly covalent systems (see section 3.2) can be ambiguous. In the latter case, e.g., charge is smeared out in covalent bonds, and therefore a partition of the interstitial region in solids is hard.

In general, we expect conventional space partition methods to be more precise for ionic than for covalent adsorbates. The Bader analysis (Bader, 1991) of covalently and ionically bond adatoms presented in sections 3.1 and 3.2 will confirm this presumption. On that account, methods to obtain charge transfer based on electrostatic potentials or the band structure will be explained in the following.

2.2 Electrostatic approaches to charge transfer

An alternative way to describe the amount of charge transfer is to utilize electrostatic models and to apply them on output from electronic structure methods like density functional theory. In this sense, the DFT results can be seen as "experimental data" being analysed by theoretical tools and models.

The DFT calculations presented, here, were performed by means of the Vienna Ab-initio Simulation Package (Kresse & Furthmüller, 1996a;b) (VASP) with PW91-GGA functionals; the geometric structure was modelled by a three-dimensional supercell and an interlayer spacing of about 25Å in order to prevent interaction. The Brillouin zones were sampled within the tetrahedron and the Methfessel-Paxton method in combination with carefully chosen k-meshes and cut-off energies. Geometries were relaxed until all forces were smaller than 0.02eV/Å per atom.

Electrostatic potential landscapes, experimentally investigated on graphene in EFM experiments (Moser et al., 2008), can be extracted from DFT simulations (Fig. 1, left) and used to determine impurity induced charge transfer. From the point of view that (doped) graphene is metallic, the sheet can be considered as a grounded metal plate of infinite size, such that — in the simplest model — charged adsorbates can be described by means of an image charge model: the adatoms are assumed as partial point charges which induces image charges in graphene. This model is valid at length scales above the screening length of the doped graphene sheet (Katsnelson (2006)). Then, the electrostatic potential, $V(r)$, in the vacuum region above the graphene sheet and the impurity can be modelled as

$$V(\vec{r}) = V_e(z) + \frac{1}{4\pi\epsilon_0} \sum_{i=1} q_i \left[\frac{1}{| \vec{r} - \vec{r}_i |} - \frac{1}{| \vec{r} - \vec{r}_i^{\prime} |} \right]. \quad (1)$$

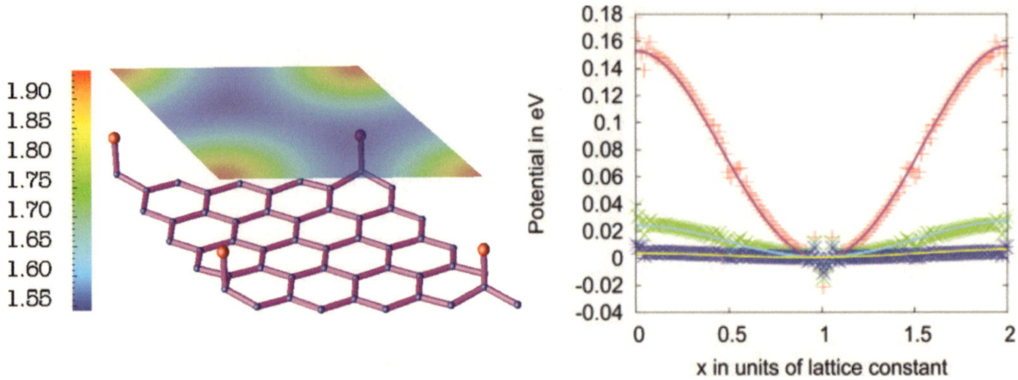

Fig. 1. Left: Periodic 4x4 graphene supercell with fluorine adsorbates (red dots). The contour plot shows the electrostatic potential (in eV) in a height of $z = 2a_{gr}$ (with $a_{gr} \approx 2.46$Å the graphene lattice constant). To calculate the charge transfer, several paths through the cell are considered for different heights z and the image charge potential (Eqn. (1)) is fitted to the data by optimization of the parameters q_i. Right: Potential along paths connecting two adjacent chlorine adatoms on graphene 4x4 for heights $z = 3a_{gr}$ (red), $z = 4a_{gr}$ (green), $z = 5a_{gr}$ (blue) and fitted curves in units of the lattice constant; the curves are vertically shifted to the x-axis. Small noise occurs for certain x where a carbon atom is located.

The fitting parameters of this model are the adsorbate point charges q_i and an offset $V_e(z)$; the charges are centered at the positions of the impurity atoms, $\vec{r}_i = (x_i, y_i, z_i)$, and their mirror images $\vec{r}'_i = (x_i, y_i, -z_i)$. The offset is given by $V_e(z) = V_0 + E_0 z$, where V_0 is a constant and E_0 a constant electric field in z-direction due to the three-dimensional periodicity of the supercell.

The method proves well-suited to fit the charge values for all atomic adsorbates regarded. Merely for adsorbate groups like hydroxyl, additional dipole fields make the fitting procedure error-prone (sec. 3.2). Determining partial charges based on Eq. (1) is similar to analysing dipole moments obtained from the charge density as, e.g., performed for metal adatoms on graphene by Chan et al. (2008).

An alternative approach to charge transfer based on electrostatic potential is to analyse core potential shifts of the carbon atoms. These core level shifts arise from charge rearrangement around the impurity in the graphene sheet and can be calculated within VASP. The analysis of core level shifts allows estimates of the impurity charge and gives qualitative insight into the range of redistributions (Fig. 2). Within VASP, the averaged core potential for an atom sitting at position $\vec{R}_n$ is determined by (Kresse, 2010)

$$\overline{V}_n = \int V_{DFT}(\vec{r})\rho_{test}(|\vec{r} - \vec{R}_n|)d^3r, \quad (2)$$

where V_{DFT} denotes the electrostatic potential from DFT and ρ_{test} a test charge with norm 1 in the core region of each atom. This approach is similar to Adessi et al. (2006), where also atom centered test charges have been employed. The shift of these averaged core potentials, $\Delta\overline{V}$, as function of the distance to the bonding C atom is illustrated in Fig. 2.

Assuming screening within the linear response regime, the analysis of the core potentials allows to obtain the relative strength of charge transfer between the different adsorbates and graphene. If the charge transfer for one reference system is known, also absolute values for the charge transfer of all systems can be obtained. Additionally, the core potential shifts allow to qualitatively characterize the spacial extent of doped regions. A detailed discussion of these issues will be given in sections 3.1, 3.2 and 3.3.

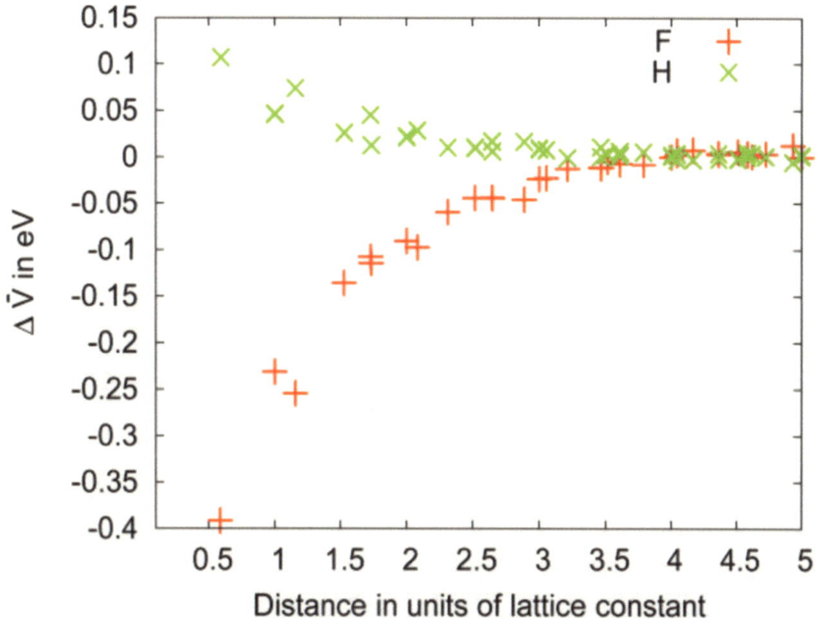

Fig. 2. Average core potential shift with respect to a C atom at large distance from the impurity as function of the in-plane distance for hydrogen and fluorine. The maximum distance considered, here, is about half the way through the supercell (here 4.5 lattice constants). It is visible that the doping of hydrogen is opposite to fluorine.

2.3 Band structure and density of states based determination of charge transfer

Spectroscopy experiments are a common tool to study the electronic structure of solids; in particular, these allow the determination of the density of states of graphene samples. Doping adsorbates donate or accept electrons from the graphene sheet, which leads to a shift of the Fermi level. This shift, the difference between the Fermi level E_F and the Dirac point energy E_D, is denoted by

$$\Delta E_F = E_F - E_D. \tag{3}$$

Note that in the case of pristine graphene, $E_F = E_D$, such that $\Delta E_F = 0$; the sign of ΔE_F denotes p- or n-type doping. Integrating the total density of states per unit cell of pristine graphene from E_D to $E_D + \Delta E_F$

$$\Delta q = e \int_{E_D}^{E_D+\Delta E_F} D(E) dE, \tag{4}$$

hence yields the charge transfer between the adsorbate and the graphene sheet which corresponds to a change in the number of mobile carriers. This method relies in the assumption that the adatoms do not change the density of states in the integration interval. While covalently bond adsorbates induce resonances in the DOS near the Dirac point (sec. 3.2), the method is well legitimate to apply for ionically bond impurities, where resonances only occur far away from the Dirac point ($\Delta E_F \leq 1.5eV$; Fig. 3).

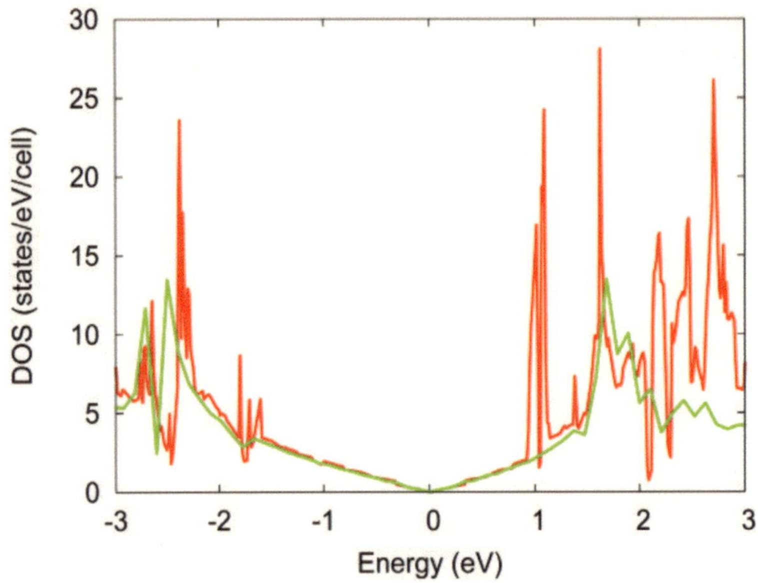

Fig. 3. Total density of states of pristine graphene (green) and of K on a graphene 4x4 supercell (red). The Fermi level is set to zero.

3. Monovalent adsorbates

We now consider monovalent adsorbates and show that these interact strongly with graphene. The charge transfer between the adsorbates and the graphene sheet is calculated by the approaches explained above and its relation to carrier doping as well as electron scattering is pointed out. The electronic structure of a graphene-impurity system features fundamental differences between covalently and ionically bond adsorbates; for instance, ARPES (angle-resolved photoemission spectroscopy) experiments yield that ionic potassium is a strong dopant with rather weak bond (Chen et al., 2008), while weak doping is found for covalent hydrogen in Raman experiments (Ryu et al., 2008). On this account, we will investigate doping processes with regard to the bonding mechanism and briefly point out reasons for the different bonding behaviour. In detail, calculations of monovalent hydrogen, fluorine, chlorine, potassium and hydroxyl adsorbed on graphene within DFT are presented in the following. Next to an extensive analysis of the charge transfer, we will discuss the range of charge redistributions with the help of a tight-binding model.

3.1 Ionically bond impurities

By definition, ionically bond impurites mean high charge transfer and low hybridization with the graphene bands. I.e. monovalent ionic adsorbates are expected to cause charge transfer in the range of $|q| \lesssim e$ [1]. This kind of charge transfer is detectable as a shift of the chemical potential in ARPES experiments (Ohta et al., 2006; Zhou et al., 2008). In the density of states, potassium and chlorine create a sharp resonance, which gives rise to an acceptor level below the Dirac point (Cl) or a donor level above (K; Fig. 3). The weak hybridization of the ionic impurities is also reflected in the migration barrier. These are typically in the range

[1] The reason for having $|q| \lesssim e$ instead of $|q| = e$, in a general case, lies in the fractional covalent character of any ionic bond.

Type	Size (conc.)	Image Charges	Core levels	Bader analysis	Fermi level shifts
K	4x4 (3.1%)	+0.68		+0.83	+0.64
	9x9 (0.6%)	+0.80	+0.80 (ref)	+0.91	+0.97
Cl	4x4 (3.1%)	-0.40		-0.50	-0.38
	9x9 (0.6%)	-0.54	-0.56	-0.57	-0.65

Table 1. Charge transfer between potassium / chlorine and graphene for the image charge model, the averaged core potential method, Bader analysis and the Fermi energy shifts in the DOS (all values in units of *e*) and two different supercell sizes (impurity concentrations). The value named "ref" for potassium is the charge value from the image charge method. It is used as a reference value to be able to extract absolute values for the charges of chlorine, hydrogen, fluorine and hydroxyl from the core levels.

of less than $0.1eV$ and therefore about one order smaller than for neutral covalent impurities (Wehling et al., 2009b).

We calculated charge transfer of potassium and chlorine adatoms by means of Bader analysis, in both electrostatic models and from Fermi level shifts. To this end, we fully relaxed the graphene adsorbate systems and obtained the minimum energy adsorption geometries (c.f. Wehling et al. (2009b)). Chlorine favours a top site (T) bonding (on top of a carbon atom) at 2.7Å above the graphene sheet, whereas potassium prefers the hollow site (H) in the middle of a carbon ring at a height of about 2.6Å.
In table 1 we present the charge transfer for both adsorbates obtained within the different approaches and as function of impurity concentration (supercell size). The signs of the partial charges show that potassium acts as a donor while chlorine behaves as an acceptor. The results for potassium are in rough agreement with Chan et al. (2008); they found a value of 0.76*e* for a 4x4 supercell, as well from Fermi level shifts. Similar findings were published by Lugo-Solis & Vasiliev (2007).
Obviously, the amount of charge transfer is concentration dependent: the strength of the doping decreases with the impurity concentration. This trend is consistently obtained within all methods to calculate charge transfer, here.
For the two ionic impurities all approaches are qualitatively consistent with each other. Depending on the experimental observable to be modelled either the electrostatic potential based methods or the Fermi level shifts should be considered. The values derived from the electrostatic potential should be most useful to discuss contribution to Coulomb scattering while the Fermi level shifts should yield quantitatively the best estimate of doping. The Bader results for ionically bond impurities tend to yield slightly higher charge values than the electrostatic models; covalent adsorbates, presented in the next section, are more problematic within this method.

3.2 Covalently bond impurities

In contrast to ionically bond impurities, covalent adsorbates show strong hybridization with graphene orbitals leading to a formation of stable states with strong bonds. The local density of states (LDOS) of covalently bond impurities is broad and constitutes a midgap state at the Fermi level as well as characteristic resonances at high energies (Wehling et al., 2009b).
All covalent monovalent adsorbates prefer top-site bonding and create impurity states that are localized at the adsorbate and the nearest neighbours of the bonding C atom (Fig. 4).

Type	Size (conc.)	Image Charges	Core level shifts	Bader analysis
H	4x4 (3.1%)	+0.18		+0.04
	5x5 (2.0%)	+0.16		+0.03
	7x7 (1.0%)	+0.14		+0.02
	9x9 (0.6%)	+0.15	+0.12	+0.01
F	4x4 (3.1%)	-0.39		-0.58
	5x5 (2.0%)	-0.39		-0.59
	6x6 (1.4%)	-0.38		-0.58
	9x9 (0.6%)	-0.39	-0.52	-0.57
OH	4x4 (3.1%)			-0.43
	5x5 (2.0%)			-0.45
	9x9 (0.6%)		-0.44	-0.45

Table 2. Charge transfer between hydrogen / fluorine / hydroxyl and graphene for the image charge model, the averaged core potential method and the Fermi energy shifts in the DOS and different different supercell sizes (impurity concentrations). All values in units of *e*.

The bonding partner of the impurity is decoupled from graphene's Dirac bands and scatters electrons similarly to vacancies. The universality of midgap states in graphene has been investigated by several groups (Boukhvalov & Katsnelson, 2009; Casolo et al., 2009; Wehling et al., 2008; 2009a;b).

We now turn to the description of charge redistributions due to covalent adsorbates. Their contribution to Coulomb scattering is directly related to changes in the electrostatic potential and consequently to the partial charges derived from the core level shifts or the image charge model. Our results for charge transfer within different models between graphene and hydrogen, fluorine, and hydroxyl adsorbates are given in table 2.
For hydrogen we obtain modulations in the electrostatic potential and the core levels corresponding to a "charge" as relevant for Coulomb scattering on the order of $0.1 - 0.2e$.

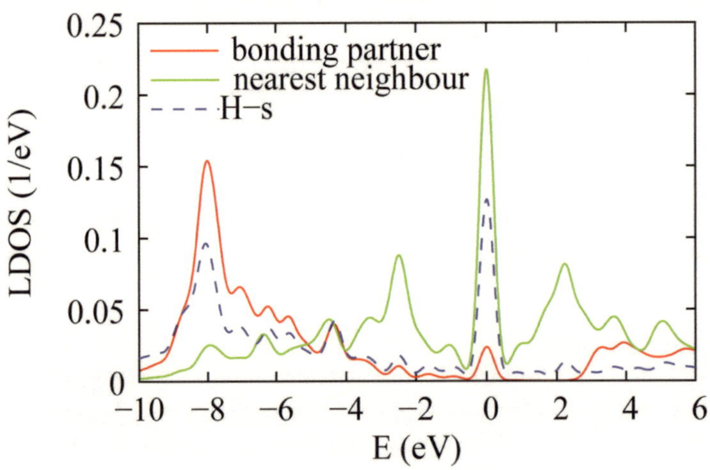

Fig. 4. Local density of states (LDOS) of hydrogen, the bonding partner, and the nearest neighbour on a 4x4 supercell (from Wehling et al. (2009b)).

Bader analysis yields a hydrogen partial charge of $q = 0.01 - 0.04e$. This is qualitatively consistent with the electrostatic models, as $|q| \ll e$ in both cases, but using the Bader charge in the context of Coulomb scattering would lead to an underestimation of the scattering strength. A hydrogen charge $|q| \ll e$ has also been obtained within Mulliken population analysis by Zhu et al. (2005).

Experimentally, the contribution of H adatoms on graphene to Coulomb scattering has not yet been determined. Ryu et al. (2008) investigated hydrogen doping effects by means of Raman spectroscopy. They estimate, under a hydrogen contamination saturating about 13% of carbon bonds (corresponding to approximately one hydrogen atom per 2x2 supercell), a charge donation of $0.003e$ per hydrogen atom. This is qualitatively in line with $|q| \ll e$ as obtained in the context of Coulomb scattering from DFT. However, it also demonstrates that an effective electrostatic charge on the order of $0.1 - 0.2e$ does not necessarily imply doping of the graphene bands by the same amount. As the LDOS in the vicinity of the Fermi level is significantly altered by the covalent adsorbates, Eq. (4) cannot be used to extract the doping from the DFT calculations.

Fluorine adatoms show covalent bonding, in contrast to the other groups VII elements (Wehling et al., 2009b). Due to the strong hybridization, the bonding carbon atom is lifted in z-direction by around 0.5Å. The fluorine adatom sits in a height of $z \approx 1.6$Å. Consistent with the large eletronegativity of F, the partial charge obtained within the electrostatic models (-0.4 — $-0.5e$) as well as by Bader analysis ($\sim -0.6e$) is significantly bigger than for H and correspondingly stronger Coulomb scattering due to F adsorbates is expected. The same holds for hydroxyl adsorbates which have been analysed within Bader and the averaged core potential method. Long range charge redistributions due to hydroxyl and fluorine will be further investigated within a tight-binding model in sec. 3.3.

Our charge analysis shows that the (electrostatic) charge of covalently bond adsorbates is rather constant in the range of impurity concentrations between 0.6% and 3%. This is in contrast to the ionically bond impurities, where charge transfer has been proven to increase with the supercell size.

3.3 Charge redistributions and coulomb interactions in a tight-binding model

To learn more about charge redistributions induced by covalent impurities in graphene we investigate this problem within a tight-binding (TB) model. For pristine graphene a TB model has been first considered by Wallace (1947). Concerning the question of charge rearrangement, the key benefit of this method lies in the possibility to calculate at much lower impurity concentrations than possible in full-potential DFT simulations. Supercells of a size up to around 100x100, thus concentrations of 0.005%, could be taken into account in our TB simulations.

Omitting the spin index and restricting to nearest-neighbour hopping, $t \approx 2.7eV$, the tight-binding Hamiltonian of pristine graphene reads

$$H_0 = -t \sum_{\langle i,j \rangle} \left(a_i^\dagger b_j + h.c. \right). \tag{5}$$

Here, $a_i(a_i^\dagger)$ and $b_i(b_i^\dagger)$ are the annihilation(creation) operators acting on electrons on site $\vec{R}_i$ in sublattice A or B.

In order to take impurity states into account, we extend the Hamiltonian by an orbital with on-site energy ϵ_{imp} which is coupled to a carbon atom orbital via the hopping V:

$$H = H_0 + V\left(a_{i'}^{\dagger}o + h.c.\right) + \epsilon_{imp}o^{\dagger}o. \tag{6}$$

Here, $o^{\dagger}$ (o) denotes the creation (annihilation) operator of an impurity adsorbed at a carbon atom of sublattice A on site $\vec{R}_{i'}$.
In order to simulate realistic impurities, one has to find accurate values for V and ϵ_{imp} first. This can be done by a fit of the tight-binding band structure to the DFT band structure of the considered graphene-impurity system (see e.g. Wehling et al. (2010)). In this section, we consider impurities with the parameters $V = 4.0eV$ and $\epsilon_{imp} = -2.0eV$ which roughly fit, both, the hydroxyl and the fluorine band structures from Wehling et al. (2009b). To investigate different adsorbate concentrations we simulate supercells of size $d \times d$ containing $2d^2$ carbon π orbitals and one impurity.
With the number operator $n_i = a_i^{\dagger}a_i$ and $n_i = b_i^{\dagger}b_i$ for i belonging to sublattice A and B, respectively, we consider the on-site occupancies $\rho_i = 2\langle n_i\rangle$ (the factor 2 is due to spin degeneracy) and its deviation

$$q_i = e(\rho_i - 1) \tag{7}$$

from the pristine graphene value. Analogously, we define for the impurity occupation $\rho_{imp} = 2\langle o^{\dagger}o\rangle$ and charge $q_{imp} = e(\rho_i - 1)$.
Table 3 gives the impurity charges, q_{imp}, obtained for different concentrations; it increases only slightly with the supercell size and saturates around -0.6e.

Partial impurity charge q_{imp} in units of e						
5x5	7x7	11x11	23x23	37x37	61x61	83x83
-0.341	-0.344	-0.431	-0.536	-0.573	-0.588	-0.591

Table 3. Charge transfer to the hydroxyl group within tight-binding.

The impurity charges obtained in the TB model mimicking hydroxyl or fluorine are in qualitative agreement with the strength of the Coulomb potentials extracted from core potential shifts in section (3.2).
The lateral extent of the doped regions can be studied by constructing a circular disk of radius r_d around the impurity and summing up all partial charges of the atoms within the disk, including the impurity. The total disk charge is given by

$$q_{disk} = q_{imp} + \sum_{i\in\{|\vec{R}_0-\vec{R}_i|\leq r_d\}} q_i \tag{8}$$

where $\vec{R}_0$ is the lateral position of the impurity, $\vec{R}_i$ the position of the carbon atom at site i. For different disk radii, the total charge can be obtained from the model and the range of redistributions estimated (Fig. 5, right).

In the region of the impurity, the sign of the on-site partial charge allows clear distinction between the sublattices (Fig. 5, left). With impurity binding to a sublattice A atom, the A sublattice is hole-doped, while the B sublattice is electron-doped near the impurity and slightly hole-doped far away. In close proximity to the impurity, the disk is charged by the adatom and a maximum in the disk charge due to the midgap impurity state occurs within

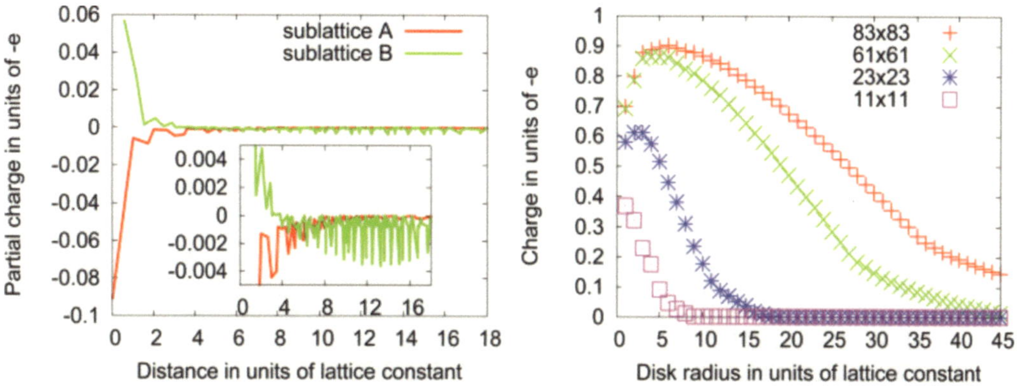

Fig. 5. Hydroxyl in nn TB: Partial charges of graphene atoms on sublattice A (red) and B (green) for a 23x23 supercell as function of the distance (left) together with a zoom (inset). The total disk charge for several supercell sizes is shown right.

some lattice constants off the impurity. For larger supercells, i.e. lower concentrations of impurities, the maximum disk charge increases (Fig. 5, right); the slope of the disk charge is simply given by the single partial charges (Fig. 5, left) which show small Friedel oscillations. Far away from the impurity, the B sublattice is charged negatively such that the total disk charge decreases almost linearly. We note that the range of charge redistributions (region with linear slope in Fig. 5, right) is on the order of the inter impurity distance even for impurity concentrations as low as 0.007% corresponding to the 83×83 supercell.
Concerning long-range doping, the question of a possible oversimplification of the TB model arises. In reality, the Coulomb energy cost might suppress long range charge redistributions. Therefore, we extend the Hamiltonian (6) by an additional term, taking electrostatic interaction between all N electrons into account (Castro Neto et al., 2009):

$$H_C = \frac{1}{2}\sum_{i,j} U_{i,j} n_i n_j. \tag{9}$$

The term $U_{i,j}$ describes the Coulomb repulsion

$$U_{i,j} = \frac{e^2}{4\pi\epsilon_0 |\vec{R}_i - \vec{R}_j|} \tag{10}$$

between two electrons on different sites ($\vec{R}_i \neq \vec{R}_j$). The on-site repulsion was chosen to $U_{ii} = 15eV$. We solve the Hamiltonian $H + H_C$ within the Hartree approximation, wherein the many-body electron-electron interaction (9) is replaced by the electrostatic potential from charge distribution in the system. Hence, we replace (9) by the Hartree Hamiltonian

$$H_H = \sum_i^N V_i n_i \tag{11}$$

with

$$V_i = \sum_j^N U_{i,j} \langle n_j \rangle. \tag{12}$$

The problem is solved self-consistently; after the initial determination of an eigensystem from Hamiltonian (6), the Coulomb part (11) is calculated and the eigensystem updated; this procedure is repeated until a converged solution is obtained.

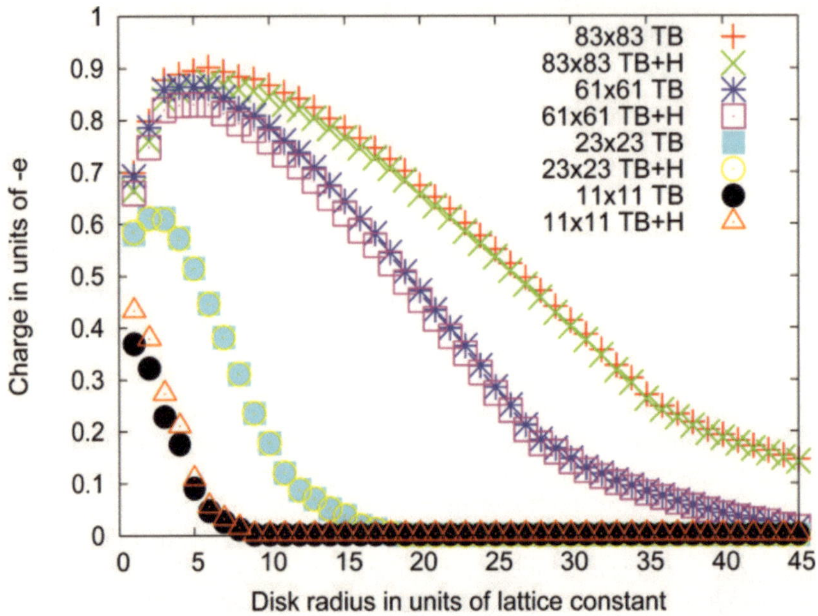

Fig. 6. Charge of the disk around the adsorbate as function of the radius for OH on graphene. The plot shows curves for different supercell sizes A (see legend); "TB" calculations are performed in nn tight-binding without Coulomb repulsion, "TB+H" curves are from nn tight-binding with Coulomb interaction in Hartree approximation.

The results for the impurity with $V = 4.0eV$ and $\epsilon_{imp} = -2.0eV$ show that the long range Coulomb interaction does not significantly affect the charge distribution in the doped region (Fig. 6). For supercells larger than 23x23, the Coulomb repulsion tends to slightly reduce the disk charges, whereas for small supercells, i.e. high concentrations, the charges slightly increase. The transfer between adsorbate and graphene layer only changes insignificantly, such that Coulomb interactions keep charge transfer and redistribution almost unaffected in the window of investigated impurity concentrations from 0.007% to 2%.

4. Conclusions

In this chapter we investigated adsorption processes under the general aspect of charge transfer. Charge transfer can either refer to doping, i.e. from electrons transferred from states localized at the impurity to the host bands, or to the redistribution of charge density associated electrostatic potentials. These two types of charge transfer have to be carefully distinguished. The electrostatic potentials due to charge redistributions are particularly important in the context of Coulomb scattering. We investigated charge transfer of realistic monovalent adsorbates on graphene by electrostatic means, i.e. the image charge method and core level shifts, and derived the effective charge q of the impurities to be used in the context of Coulomb scattering. For ionic impurities we find effective charges on the order of $|q| \sim 0.5 - 1e$

and for covalent impurities $|q|$ in the range of $0.0 - 0.5e$. Hydrogen adatoms which are of particular experimental importance create electrostatic fields corresponding to a partial charge of $q \approx 0.1e$.

For the ionic impurities the graphene bands remain mainly intact and the doping can be estimated from the position of the Fermi level with respect to the Dirac point. Qualitatively, the impurity charges obtained in this method for K and Cl coincide with the respective effective charges in the context of Coulomb scattering. Moreover, the charge transfer of ionic potassium and chlorine proved to be similarly and consistently sensitive to the impurity concentration, both, in the context of doping as well as Coulomb scattering. There are, however, quantitative differences between the charge as relevant for Coulomb scattering and for doping. The latter turned out to be up to 20% bigger than the former.

Further investigated charge redistributions within the graphene sheet by means of a tight-binding describe impurities like hydroxyl or fluorine. The model illustrates that even far away from the impurity and for low impurity concentrations rearrangement of electrons is detectable. These results hold true if Coulomb repulsion is taken into account. The impurity charge in the TB model is in qualitative agreement with the charge transfer obtained from the core potential method applied to the DFT data of hydroxyl or fluorine adsorbates.

In the future, the contribution of different realistic adsorbates to minimum carrier concentrations achievable in graphene would be worth to investigate. Moreover, interfacing the charge transfer with electron transport theory would be desirable and might be a key to a realistic first-principles based theory of electron transport in graphene.

5. Acknowledgement

Support from SFB 668 (Germany), the Cluster of Excellence "Nanospintronics" (LExI Hamburg), the DFG via Priority Programme 1459 "Graphene", FOM (The Netherlands), and computer time at HLRN (Germany) are acknowledged.

6. References

Adam, S., Hwang, E., Galitski, V. & Das Sarma, S. (2007). A self-consistent theory for graphene transport, *Proceedings of the National Academy of Sciences* 104(47): 18392.

Adessi, C., Roche, S. & Blase, X. (2006). Reduced backscattering in potassium-doped nanotubes: Ab initio and semiempirical simulations, *Physical Review B* 73(12): 125414.

Bader, R. (1991). A quantum theory of molecular structure and its applications, *Chemical Reviews* 91(5): 893–928.

Bolotin, K. I., Sikes, K. J., Jiang, Z., Klima, M., Fudenberg, G., Hone, J., Kim, P. & Stormer, H. L. (2008). Ultrahigh electron mobility in suspended graphene, *Solid State Commun.* 146: 351.

Bostwick, A., Ohta, T., Seyller, T., Horn, K. & Rotenberg, E. (2006). Quasiparticle dynamics in graphene, *Nature Physics* 3(1): 36–40.

Boukhvalov, D. & Katsnelson, M. (2009). Enhancement of chemical activity in corrugated graphene, *The Journal of Physical Chemistry C* 113(32): 14176–14178.

Casolo, S., Løvvik, O., Martinazzo, R. & Tantardini, G. (2009). Understanding adsorption of hydrogen atoms on graphene, *The Journal of chemical physics* 130: 054704.

Castro Neto, A., Guinea, F., Novoselov, K. & Geim, A. (2009). The electronic properties of graphene, *Rev. Mod. Phys.* 81(1): 109.

Chan, K., Neaton, J. & Cohen, M. (2008). First-principles study of metal adatom adsorption on graphene, *Phys. Rev. B* 77(23): 235430.

Chen, J.-H., Cullen, W. G., Jang, C., Fuhrer, M. S. & Williams, E. D. (2009). Defect scattering in graphene, *Phys. Rev. Lett.* 102(23): 236805.

Chen, J.-H., Jang, C., Adam, S., Fuhrer, M., Williams, E. D. & Ishigami, M. (2008). Charged-impurity scattering in graphene, *Nature Physics* 4(5): 377–381.

Cheng, S.-H., Zou, K., Okino, F., Gutierrez, H. R., Gupta, A., Shen, N., Eklund, P. C., Sofo, J. O. & Zhu, J. (2010). Reversible fluorination of graphene: Evidence of a two-dimensional wide bandgap semiconductor, *Phys. Rev. B* 81(20): 205435.

Collins, P., Bradley, K., Ishigami, M. & Zettl, A. (2000). Extreme oxygen sensitivity of electronic properties of carbon nanotubes, *Science* 287(5459): 1801.

Du, X., Skachko, I., Barker, A. & Andrei, E. Y. (2008). Approaching ballistic transport in suspended graphene, *Nat. Nano.* 3: 491–495.

Elias, D., Nair, R., Mohiuddin, T., Morozov, S., Blake, P., Halsall, M., Ferrari, A., Boukhvalov, D., Katsnelson, M., Geim, A. et al. (2009). Control of graphene's properties by reversible hydrogenation: evidence for graphane, *Science* 323(5914): 610.

Geim, A. & Novoselov, K. (2007). The rise of graphene, *Nature Materials* 6(3): 183–191.

Hwang, E., Adam, S. & Das Sarma, S. (2007). Carrier transport in two-dimensional graphene layers, *Phys. Rev. Lett.* 98(18): 186806.

Katoch, J., Chen, J., Tsuchikawa, R., Smith, C., Mucciolo, E., Ishigami, M., Chung, S., Qi, X., Maciejko, J., Zhang, S. et al. (2010). Uncovering the dominant scatterer in graphene sheets on SiO2, *Phys. Rev. B* 82(8): 81417.

Katsnelson, M. (2006). Nonlinear screening of charge impurities in graphene, *Physical Review B* 74(20): 201401.

Katsnelson, M. (2007). Graphene: carbon in two dimensions, *Materials Today* 10(1-2): 20 – 27.

Katsnelson, M. & Novoselov, K. (2007). Graphene: new bridge between condensed matter physics and quantum electrodynamics, *Solid State Communications* 143(1-2): 3–13.

Kong, J., Franklin, N., Zhou, C., Chapline, M., Peng, S., Cho, K. & Dai, H. (2000). Nanotube molecular wires as chemical sensors, *Science* 287(5453): 622.

Kresse, G. (2010). VASP the GUIDE. http://cms.mpi.univie.ac.at/vasp/vasp/LVTOT_tag_core_level_shifts.html, accessed 24-Sept-2010.

Kresse, G. & Furthmüller, J. (1996a). Efficiency of ab-initio total energy calculations for metals and semiconductors using a plane-wave basis set, *Computational Materials Science* 6(1): 15–50.

Kresse, G. & Furthmüller, J. (1996b). Efficient iterative schemes for ab initio total-energy calculations using a plane-wave basis set, *Phys. Rev. B* 54(16): 11169–11186.

Lebègue, S., Klintenberg, M., Eriksson, O. & Katsnelson, M. I. (2009). Accurate electronic band gap of pure and functionalized graphane from GW calculations, *Phys. Rev. B* 79(24): 245117.

Leenaerts, O., Partoens, B. & Peeters, F. (2008). Paramagnetic adsorbates on graphene: A charge transfer analysis, *Applied Physics Letters* 92: 243125.

Liu, L. & Shen, Z. (2009). Bandgap engineering of graphene: A density functional theory study, *Applied Physics Letters* 95: 252104.

Lugo-Solis, A. & Vasiliev, I. (2007). Ab initio study of K adsorption on graphene and carbon nanotubes: Role of long-range ionic forces, *Phys. Rev. B* 76(23): 235431.

Meister, J. & Schwarz, W. H. E. (1994). Principal Components of Ionicity, *J. Phys. Chem.* 98: 8245–8252.

Meyer, J., Girit, C., Crommie, M. & Zettl, A. (2008). Imaging and dynamics of light atoms and molecules on graphene, *Nature* 454(7202): 319–322.

Moser, J., Verdaguer, A., Jiménez, D., Barreiro, A. & Bachtold, A. (2008). The environment of graphene probed by electrostatic force microscopy, *Applied Physics Letters* 92: 123507.

Nair, R. R., Ren, W. C., Jalil, R., Riaz, I., Kravets, V. G., Britnell, L., Blake, P., Schedin, F., Mayorov, A. S., Yuan, S., Katsnelson, M. I., Cheng, H. M., Strupinski, W., Bulusheva, L. G., Okotrub, A. V., Grigorieva, I. V., Grigorenko, A. N., Novoselov, K. S. & Geim, A. K. (2010). Fluorographene: Two Dimensional Counterpart of Teflon, *ArXiv preprint arXiv:1006.3016* .

Ni, Z., Ponomarenko, L., Nair, R., Yang, R., Grigorieva, S., Schedin, F., Shen, Z., Hill, E., Novoselov, K. & Geim, A. (2010). On resonant scatterers as a factor limiting carrier mobility in graphene, *Arxiv preprint arXiv:1003.0202* .

Nomura, K. & MacDonald, A. H. (2006). Quantum hall ferromagnetism in graphene, *Phys. Rev. Lett.* 96(25): 256602.

Novoselov, K., Geim, A., Morozov, S., Jiang, D., Zhang, Y., Dubonos, S., Grigorieva, I. & Firsov, A. (2004). Electric Field Effect in Atomically Thin Carbon Films, *Science* 306(5696): 666–669.

Ohta, T., Bostwick, A., Seyller, T., Horn, K. & Rotenberg, E. (2006). Controlling the electronic structure of bilayer graphene, *Science* 313(5789): 951.

Ostrovsky, P. M., Gornyi, I. V. & Mirlin, A. D. (2006). Electron transport in disordered graphene, *Phys. Rev. B* 74(23): 235443.

Peres, N. M. R. (2010). Colloquium: The transport properties of graphene: An introduction, *Rev. Mod. Phys.* 82(3): 2673–2700.

Robinson, J., Perkins, F., Snow, E., Wei, Z. & Sheehan, P. (2008). Reduced graphene oxide molecular sensors, *Nano letters* 8(10): 3137–3140.

Robinson, J., Schomerus, H., Oroszlány, L. & Fal'ko, V. (2008). Adsorbate-Limited Conductivity of Graphene, *Phys. Rev. Lett.* 101(19): 196803.

Robinson, J. T., Burgess, J. S., Junkermeier, C. E., Badescu, S. C., Reinecke, T. L., Perkins, F. K., Zalalutdniov, M. K., Baldwin, J. W., Culbertson, J. C., Sheehan, P. E. & Snow, E. S. (2010). Properties of fluorinated graphene films, *Nano Letters* 10(8): 3001–3005.

Ryu, S., Han, M., Maultzsch, J., Heinz, T., Kim, P., Steigerwald, M. & Brus, L. (2008). Reversible basal plane hydrogenation of graphene, *Nano Lett* 8(12): 4597–4602.

Schedin, F., Geim, A., Morozov, S., Hill, E., Blake, P., Katsnelson, M. & Novoselov, K. (2007). Detection of individual gas molecules adsorbed on graphene, *Nature materials* 6(9): 652–655.

Segall, M. D., Pickard, C. J., Shah, R. & Payne, M. C. (1996). Population analysis in plane wave electronic structure calculations, *Mol. Phys.* 89: 571.

Sofo, J., Chaudhari, A. & Barber, G. (2007). Graphane: A two-dimensional hydrocarbon, *Phys. Rev. B* 75(15): 153401.

Stauber, T., Peres, N. M. R. & Guinea, F. (2007). Electronic transport in graphene: A semiclassical approach including midgap states, *Phys. Rev. B* 76(20): 205423.

Tan, Y., Zhang, Y., Bolotin, K., Zhao, Y., Adam, S., Hwang, E., Das Sarma, S., Stormer, H. & Kim, P. (2007). Measurement of scattering rate and minimum conductivity in graphene, *Phys. Rev. Lett.* 99(24): 246803.

Wallace, P. (1947). The Band Theory of Graphite, *Phys. Rev.* 71(9): 622–634.

Wehling, T., Balatsky, A., Tsvelik, A., Katsnelson, M. & Lichtenstein, A. (2008). Midgap states in corrugated graphene, *EPL (Europhysics Letters)* 84: 17003.

Wehling, T., Katsnelson, M. & Lichtenstein, A. (2009a). Adsorbates on graphene: Impurity states and electron scattering, *Chemical Physics Letters* 476(4-6): 125 – 134.

Wehling, T., Katsnelson, M. & Lichtenstein, A. (2009b). Impurities on graphene: Midgap states and migration barriers, *Phys. Rev. B* 80(8): 085428.

Wehling, T., Yuan, S., Lichtenstein, A., Geim, A. & Katsnelson, M. (2010). Resonant scattering by realistic impurities in graphene, *Phys. Rev. Lett.* 105(5): 056802.

Xiang, H., Kan, E., Wei, S., Whangbo, M. & Yang, J. (2009). "Narrow" Graphene Nanoribbons Made Easier by Partial Hydrogenation, *Nano letters* 9(12): 4025–4030.

Zhou, S., Siegel, D., Fedorov, A. & Lanzara, A. (2008). Metal to Insulator Transition in Epitaxial Graphene Induced by Molecular Doping, *Phys. Rev. Lett.* 101(8): 086402.

Zhu, Z., Lu, G. & Wang, F. (2005). Why H atom prefers the on-top site and alkali metals favor the middle hollow site on the basal plane of graphite, *J. Phys. Chem. B* 109(16): 7923–7927.

3

The Effect of Atomic-Scale Defects and Dopants on Graphene Electronic Structure

Rocco Martinazzo, Simone Casolo and Gian Franco Tantardini
Dipartimento di Chimica Fisica ed Elettrochimica and CIMaINa
Università degli Studi di Milano, Milan
Italy

1. Introduction

Graphene, thanks to its extraordinary electronic and mechanical properties, is a potential candidate for a number of applications. Being one-atom thick, it is extremely sensitive to the presence of adsorbed atoms and molecules (either physisorbed or chemisorbed on the surface) and, more generally, to defects such as vacancies, holes and/or substitutional dopants. This property, apart from being directly usable in molecular sensor devices, can also be employed to tune graphene electronic properties.

In this Chapter we review that basic features of atomic-scale defects that can be useful for material design. After a brief introduction (Section 2) of the main properties determining the peculiar electronic structure of graphene, and the experimental realisation of defective substrates (Section 3), we focus in Section 4 on isolated "p_z defects" such as atom vacancies or adsorbed species which covalently bind carbon atoms. In particular, we discuss in detail the formation of so-called midgap states and the microscopically ordered magnetic structures which give rise to. In Section 5 we analyse the electronic structure of multiple defective graphene substrates and show, in particular, how it is possible to use simple rules to predict the presence of magnetic moments and midgap states by looking at the defect locations on the lattice. Subsequently, we analyse the more complicated situation where the electronic structure, as modified by the presence of some defects, affects chemical reactivity of the substrate towards adsorption (chemisorption) of atomic/molecular species, leading to a preferential sticking on specific lattice positions. In Section 6 we consider the reverse problem, that is how to use defects (vacancies, adsorbed species, substitutional dopants, etc..) to engineer graphene electronic properties. This is possible nowadays since recent advances in lithographic and self-assembling techniques allow one to produce well-ordered structures and thus 'tune' the electronic bands. In this context, we show for instance how it is possible to open a band-gap in graphene and preserve at the same time the pseudo-relativistic behaviour of its charge carriers. We further analyse the case of substitutional dopants (group IIIA/VA elements) which, if periodically arranged, may show a gapped quasi-conical structure corresponding to massive Dirac carriers. All these possible structures might find important technological applications in the development of novel graphene-based logic transistors.

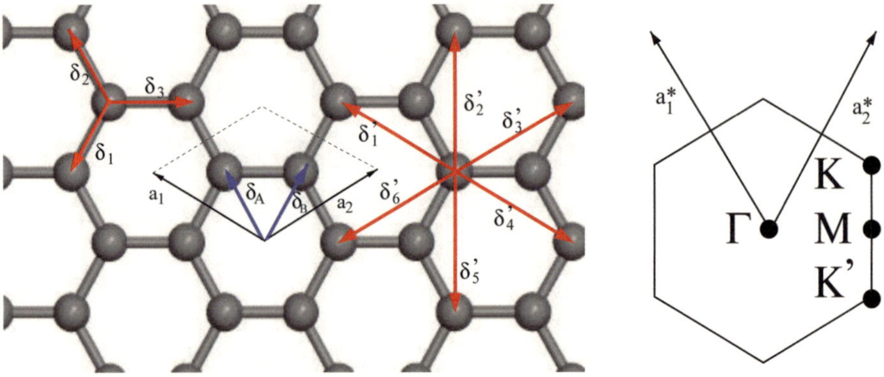

Fig. 1. Left panel: graphene unit cell $(\mathbf{a}_1, \mathbf{a}_2)$, along with the vectors joining nearest- and next-to nearest neighbours, $\boldsymbol{\delta}_i$ and $\boldsymbol{\delta}'_i$ respectively. Also indicated the position vectors $\boldsymbol{\delta}_A$ and $\boldsymbol{\delta}_B$ for A and B sites. The Wigner-Seitz cell is one the hexagons. Right: first Brillouin zone with the highest symmetry point indicated. The arrows are the reciprocal lattice vectors.

2. The π-electron gas

Carbon atoms in graphene are arranged to form a honeycomb lattice tightly held by strong σ bonds between sp^2 orbitals which form occupied σ bands at energies well below the Fermi level. The remaining valence electrons (one for each carbon atom) populate a π band which localises above and below the lattice with a node on the surface plane. An 'antibonding' π^* band is empty when the system is at $T = 0$ K and charge-neutral, but can easily be occupied, *e.g.* by applying a gate potential in a typical field-effect transistor (FET) configuration. Such π/π^* band system governs the low-energy (say up to ~ 2 eV) behaviour of charge carriers in graphene and is responsible for most of the extraordinary properties of this material. This "π cloud" is the focus of this section, where we introduce the main theoretical tools used in this Chapter.

2.1 Tight-binding Hamiltonian

In building up a simple, one-electron model for these π electrons only one writes the wave function as a linear combination of two Wannier basis functions built with p_z orbitals, one for each sublattice (Wallace, 1947),

$$\psi_{\mathbf{k}}(\mathbf{r}) = c_A \psi^A_{\mathbf{k}}(\mathbf{r}) + c_B \psi^B_{\mathbf{k}}(\mathbf{r}) \tag{1}$$

$$\psi^A_{\mathbf{k}}(\mathbf{r}) = \frac{1}{\sqrt{N}} \sum_{j \in S} e^{-i\mathbf{kr}} p_z(\mathbf{r} - \mathbf{R}^A_j) \qquad \psi^B_{\mathbf{k}}(\mathbf{r}) = \frac{1}{\sqrt{N}} \sum_{j \in S} e^{-i\mathbf{kr}} p_z(\mathbf{r} - \mathbf{R}^B_j) \tag{2}$$

where the sums run over lattice vectors $\mathbf{R_j}$ within a large supercell S including N graphene unit cells, $\mathbf{R}^A_j = \mathbf{R}_j + \boldsymbol{\delta}_A$ is the position of A site in the j-th cell, and analogously for $\mathbf{R}^B_j$. Equivalently, in second-quantized form

$$\hat{H}^{TB} = -t_1 \sum_{<i,j>} \sum_{\sigma} \left(\hat{a}^\dagger_{i,\sigma} \hat{b}_{j,\sigma} + h.c. \right) - t_2 \sum_{<<i,j>>} \sum_{\sigma} \hat{a}^\dagger_{i,\sigma} \hat{a}_{j,\sigma} - t_2 \sum_{<<i,j>>} \sum_{\sigma} \hat{b}^\dagger_{i,\sigma} \hat{b}_{j,\sigma} + etc. \tag{3}$$

where $\hat{a}^\dagger_{i,\sigma}$ ($\hat{b}^\dagger_{i,\sigma}$) creates an electron with spin $\sigma = \uparrow, \downarrow$ on the i-th lattice site of the A(B) sublattice, the first two sums run over nearest neighbouring sites (t_1 is the hopping energy) and the second ones over sites which are nearest neighbours in each sublattice (t_2 is the corresponding hopping) [1]. In absence of magnetic fields the hoppings can be chosen real, and the accepted value for t_1 is $\sim$ 2.7 eV while $|t_2| << t_1$ depends on the parametrization used. Neglecting overlap between orbitals on different C atoms, the usual anticommutation rules $[\hat{c}^{\dagger}_{i,\sigma}, \hat{c'}_{j,\sigma'}]_+ = \delta_{c,c'}\delta_{i,j}\delta_{\sigma,\sigma'}$ $(c = a, b)$ hold; hence, introducing the Fourier transformed operators $\hat{a}_{\mathbf{k},\sigma}$ according to

$$\hat{a}_{i,\sigma} = \frac{1}{\sqrt{N}} \sum_{\mathbf{k}\in BZ} e^{-i\mathbf{k}\mathbf{R}_i} \hat{a}_{\mathbf{k},\sigma} \tag{4}$$

where the sum runs over k points in the first Brillouin zone (BZ) (analogously for $\hat{b}_{\mathbf{k},\sigma}$) the above Hamiltonian can be rewritten as

$$\hat{H}^{TB} = -t_1 \sum_{\mathbf{k},\sigma} f(\mathbf{k}) \hat{a}^{\dagger}_{\mathbf{k},\sigma} \hat{b}_{\mathbf{k},\sigma} + h.c. - t_2 \sum_{\mathbf{k},\sigma} g(\mathbf{k}) \hat{a}^{\dagger}_{\mathbf{k},\sigma} \hat{a}_{\mathbf{k},\sigma} - t_2 \sum_{\mathbf{k},\sigma} g(\mathbf{k}) \hat{b}^{\dagger}_{\mathbf{k},\sigma} \hat{b}_{\mathbf{k},\sigma} \tag{5}$$

or, in matrix notation,

$$\hat{H}^{TB} = -\sum_{\mathbf{k},\sigma} \left[\hat{a}^{\dagger}_{\mathbf{k},\sigma}, \hat{b}^{\dagger}_{\mathbf{k},\sigma}\right] \begin{bmatrix} t_2 g(\mathbf{k}) & t_1 f(\mathbf{k}) \\ t_1 f^*(\mathbf{k}) & t_2 g(\mathbf{k}) \end{bmatrix} \begin{bmatrix} \hat{a}_{\mathbf{k},\sigma} \\ \hat{b}_{\mathbf{k},\sigma} \end{bmatrix}$$

Here $f(\mathbf{k})$ and $g(\mathbf{k})$ are 'structure factors' for the nearest- and next-nearest neighbours,

$$f(\mathbf{k}) = \sum_{i=1,3} e^{-i\mathbf{k}\boldsymbol{\delta}_i}$$

$$g(\mathbf{k}) = \sum_{i=1,6} e^{-i\mathbf{k}\boldsymbol{\delta}'_i}$$

Diagonalization is trivial and gives the energy bands,

$$\epsilon(\mathbf{k})_{\pm} = -t_2 g(\mathbf{k}) \pm t_1 |f(\mathbf{k})| = -t_2 g(\mathbf{k}) \pm t_1 \sqrt{3 + g(\mathbf{k})} \tag{6}$$

where $|f(\mathbf{k})|^2 = 3 + g(\mathbf{k})$ has been used and the minus (plus) sign solution correspond to the π (π^*) band (see *e.g.* Bena & Montambaux (2009); Castro Neto et al. (2009); Wallace (1947)). Close to the $K(K')$ point $|f(\mathbf{K}+\mathbf{q})|^2 \sim v_F^2 q^2$ and the dispersion is conical, giving rise to the so-called Dirac cones. Here $v_F = \frac{\sqrt{3}}{2}a = \frac{3}{2}d$, where d is the carbon-carbon distance, $\sim$ 1.42 Å, and a the lattice constant. Consequently, the density-of-states (DOS) is linearly vanishing at zero energy, $\rho(\epsilon) \sim 2|\epsilon|/\pi\sqrt{3}t^2$, one of the fingerprints of massless Dirac electrons. Its vanishing value challenges one's intuition since experiments find a finite, non-zero minimum conductivity at this energy (Peres, 2010).

Albeit simple, this tight-binding model is accurate enough to correctly represent graphene π bands, at least close to the high symmetry points K and K'. The latter control the low-energy physics of charge carriers, and are the source of the exceptional interest in graphene. If only nearest-neighbours interaction is allowed the two sublattices form two disjoint sets where A-type sites connect to B-type sites only and *vice versa*. The Hamiltonians is said *bipartitic* and displays an interesting symmetry: for each non-zero energy level ϵ and eigenfunction

[1] Notice that the on-site energies (the energy of carbon p_z orbitals) have been set equal to zero, but additional terms of the form $\sum_i \epsilon_i \hat{c}^{\dagger}_i \hat{c}_i$ would appear if graphene were subjected to an inhomogeneous external potential.

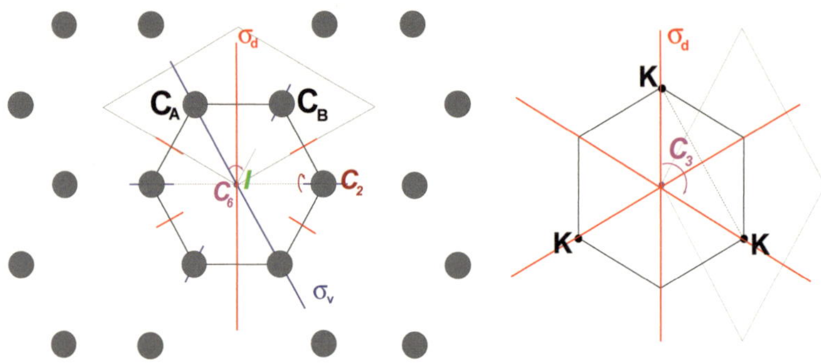

Fig. 2. Left panel: Point symmetry elements in graphene lattice. Reflection planes orthogonal to the page, σ_v and σ_d, are replicated by the six-fold rotation axis C_6, along with the two-fold rotation axis on the page plane, C_2. I is the inversion center, at the center of the Wigner-Seitz cell (solid line). The page plane is, of course, a reflection plane (σ_h). Right panel: symmetry elements of the k-group at the K point (D_{3h}). Black dots mark the K point and its images.

$|\psi_+\rangle = c_A |A\rangle + c_B |B\rangle$ (where $|A\rangle/|B\rangle$ is non-zero on A/B lattice sites only), there exists a 'conjugate' level with energy $-\epsilon$ and wavefunction $|\psi_-\rangle = c_A |A\rangle - c_B |B\rangle$. This is called *electron-hole* ($e-h$) symmetry since at *half-filling* (as it is case of graphene with one electron per site), the Fermi level lies at zero energy, and the above symmetry relates electron and holes. For a proof, just apply a phase-change to one of the two sets of sublattices states [2], *e.g.* $\hat{b}_{i,\sigma} \rightarrow -\hat{b}_{i,\sigma}$, as this converts H into $-H$.

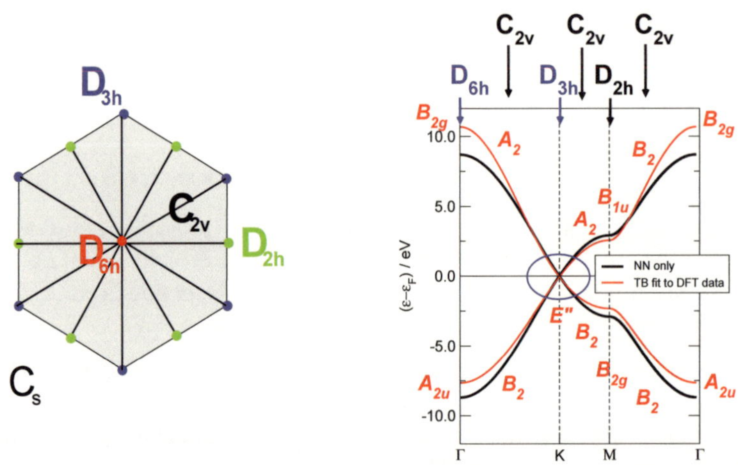

Fig. 3. Left panel: point-symmetry classification of Bloch states in graphene. k-groups are color coded as indicated. Right panel: symmetry labels for graphene electronic states on the highest-symmetry lines of the BZ.

Electron-hole symmetry, as we shall see in the following, plays an important role in graphene, even if it holds only approximately (*i.e.* with nearest-neighbour interaction only and neglecting orbital overlap). Here we just notice that, because of such symmetry, the

[2] Interestingly, this operation corresponds to an operator $\hat{\pi}$, $\hat{\pi}\hat{c}_{i,\sigma} = (-)^{\tau}\hat{c}_{i,\sigma}\hat{\pi}$ ($\tau = 1,2$ for $c = a, b$), which reduces to the z component of the pseudospin in spinor notation.

band-structure is expected to have a gap at the Fermi level unless there are specific reasons for having energy levels exactly at *zero*. As we now show, the specific reason is provided by the *spatial* symmetry of the substrate.

Graphene lattice is highly symmetric. Its Wigner-Seitz unit cell has the same point symmetry of benzene, namely it belongs to the D_{6h} point group, see fig.2, which is the point group for symmetry operations in real-space. For Bloch electronic states with k-vector $\mathbf{k}$, symmetry is reduced to that subgroup of D_{6h} which either leaves $\mathbf{k}$ invariant or transform it into one its images, *i.e.* $\mathbf{k} \rightarrow \mathbf{k} + \mathbf{G}$ with $\mathbf{G}$ a reciprocal lattice vector (Mirman, 1999). Such subgroup is known as k-group at $\mathbf{k}$, $G(\mathbf{k})$, and determines the possible symmetry of the electronic states. For instance, at the K point the k-group is D_{3h} since only three-fold rotation axes and σ_d planes transform the K images into themselves [3]. A full analysis of the symmetry properties of Bloch electrons is given in fig.3, left panel, where the k-groups are color-coded, grey for C_s, black for C_{2v}, green for D_{2h}, blue for D_{3h} and red for D_{6h}. The main point here is that graphene is sufficiently symmetric that allows for k-groups supporting *two*-dimensional irreducible representations (E irreps), namely D_{6h} at Γ and D_{3h} at K, K'. As spatial symmetry is (almost) compatible with $e-h$ symmetry, a zero energy state results whenever the electronic wavefunctions span a two-dimensional irreducible representation (odd in number, in general), *i.e.* they give rise to a doubly-degenerate level. This is exactly the case of the K (K') point, where Wannier functions built with p_z orbitals of the A and B sublattice span the E'' irrep of the above D_{3h} k-group. Notice also that this symmetry argument is enough to explain the conical dispersion of the energy at the K (K') point which makes graphene so attractive: without an inversion center, degeneracy is lifted already at first order in $\mathbf{k} \cdot \mathbf{p}$ perturbation theory when moving away from the BZ corners.

While spatial symmetry is exact, $e-h$ symmetry holds in the nearest-neighbour approximation only. Nevertheless, since inclusion of higher order hopping terms does not modify the level ordering (*i.e.* the minimum of the π^* band lies always above the top of the π band) the Fermi level at charge neutrality matches exactly the energy where the E irrep is found (called Dirac point as it is the cone apex).

2.2 Hubbard Hamiltonian

The tight-binding (TB) Hamiltonian is a model in which each electron moves independently from the others. Despite it represents a good approximation for graphene energy spectrum, such a simple picture will necessarily fail in computing spin properties in all but the simplest situations. A simple way to include electron-electron interactions is given by the Hubbard model

$$\hat{H} = \hat{H}^{TB} + U \sum_i \hat{n}_{i,\uparrow} \hat{n}_{i,\downarrow} \tag{7}$$

where $\hat{H}^{TB}$ is the tight-binding Hamiltonian of eq.3, the sum runs over all carbon sites and $\hat{n}_{i,\sigma} = \hat{c}^\dagger_{i,\sigma} \hat{c}_{i,\sigma}$ are the corresponding number operators. This Hamiltonian combines the tendency of electrons to delocalize onto the lattice due to their kinetic (hopping) energy together with an "on-site" Coulomb repulsion that tends (for $U > 0$) to localize them to minimize double orbital occupation.

The Hubbard model is a very useful tool for the study of magnetism in complex materials.

[3] The remaining symmetry elements determine the so-called *star* of the given k point, which is the set of points generated by these elements once applied to $\mathbf{k}$. Such set of physically distinct points in k space are degenerate in energy; this is the case of K' and K, for instance, since they belong to the star of each other.

It has long been used in the chemical community[4] -and proved to be rather accurate for such systems- to investigate excitation spectra in polycyclic aromatic hydrocarbons (today, graphene dots). Though simple, the model requires quite a large effort for its solution. Therefore, one often resorts to its mean-field approximation,

$$\hat{H}^{mf} = \hat{H}^{TB} + U\sum_i \hat{n}_{i,\uparrow} < \hat{n}_{i,\downarrow} > + U\sum_i < \hat{n}_{i,\uparrow} > \hat{n}_{i,\downarrow} - U\sum_i < \hat{n}_{i,\uparrow} >< \hat{n}_{i,\downarrow} > \quad (8)$$

where the average occupation number of one spin-species at given site tunes an effective on-site energy for the other spin-species, *e.g.* $\epsilon_{i,\uparrow}^{eff} = +U < \hat{n}_{i,\downarrow} >$. This is essentially equivalent to an (unrestricted) Hartree-Fock approach to the π electrons and is useful, as compared with density functional theory (DFT) methods applied to the exact Hamiltonian, to study very large systems, of dimension comparable to those experimentally realized. Though we will not solve the Hubbard model in the following, there are some exact, analytic results that can be obtained from it and that turn out to be an important tool in discussing defects in graphene.

2.3 Valence bond picture

An alternative, easy-to-use way of looking at graphene electronic structure is provided by the 'chemical picture'. With this we mean the traditional picture of chemical bonds as given by the Lewis structures and modified to account for the 'chemical resonance'. In this picture, electrons are mostly localized on atomic orbitals (usually hybridzed) of the atoms forming the molecule, and couple in singlet pairs to form bonds and lone-pairs. For carbon atoms in graphene the three sp^2 orbitals (with one electron each) are singlet-coupled with electrons in sp^2 orbitals of neighbouring sites. The remaining electron (the one described by the TB Hamiltonian above) can couple with its counterpart of one of the three neighbours. The state of the system is a superposition of these different ways of binding, and the system gains energy from such a *resonance* phenomenon.

This naïve picture finds its root in the *Valence Bond* (VB) theory of chemical bond, which developed from the Heitler-London study of the H_2 molecule, soon after the foundation of quantum mechanics. The theory, as intensively pushed forward by Slater and Pauling, is a practical way of looking at the chemical bond and at the bond-breaking, bond-forming processes which are essential for chemical reactivity. It can also be turned into a variational method for the many-electron problem which uses a correlated wavefunction *ansatz* and captures the important part of the electron correlation [5](Cooper, 2002; Cooper et al., 1987; Raimondi et al., 1985; Shaik & Hiberty, 2007). In many respects, it has to be considered complementary to the *Molecular Orbital* (MO) approach, though the latter proved to be more efficient.

Valence Bond theory focuses on spin and builds the singlet wavefunction of an even-numbered ground-state molecule as a 'product' of singlet pairs, one for each bond (pairs of orbitals), thereby identifying a chemical formula. For less standard species such as graphene, different products are equally likely and the correct wavefunction is the linear combinations of all the possible structures. For instance, let us look at the benzene molecule

---

[4] In the chemical community is dubbed Parisier-Parr-Pople approximation, after Pariser-Parr and Pople who first introduced it in the early fifties.

[5] This is so because even the simplest VB wavefunctions can be re-written as linear combinations of Slater determinants, and include the so-called 'static' correlation. The latter is essential for describing bond formation and near-degeneracies; in extended systems is responsible for Mott transitions.

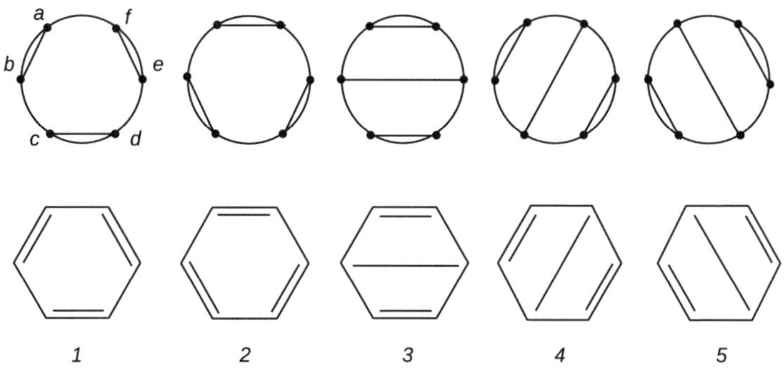

Fig. 4. The five possible perfect pairing Rumer diagrams for the benzene molecule with their correspondence with Kekulé (1-2) and Dewar (3-5) resonance structures.

as prototypical case of aromatic compounds. Considering only the six π electrons localized in their respective p_z orbitals, the possible linearly independent ("perfect pairing") functions can be schematically depicted as in figure 4; for six electron and an overall singlet state there are five couplings[6]. In the graphical representation of Fig.4, known as Rumer diagrams, one represents each atomic center with a dot and uses a line for singlet coupling between them. The chemical picture (bottom row of the same figure) uses only the 2 Kekulé structures on the left, since the 3 Dewar structures gives a negligible contribution to the energy, as can be guessed from the bond pattern.

Notice that VB theory is closely related to the Hubbard model discussed in the previous section: the atomic orbitals housing the electrons need not be those of the free atoms. If they are 'polarized' by the environment (*e.g.* they are linear combinations of free-atomic orbitals) the VB *ansatz* accounts both for localization and band-like behaviour, as in the Hubbard model. It is not hard to show, indeed, that the Hubbard model for the H_2 molecule can be obtained from a simple[7] VB *ansatz* to the two-electron wavefunction. In the following we will make a qualitative use of this chemical picture, as it provides insights into defect formation and properties; see also Wassmann et al. (2010) for its role in interpreting details of STM pictures.

3. Defect formation

In this Section we describe defect formation[8]. As we shall see in the following, even though vacancies and adatoms turn out to have very similar effects on the π electron system, we distinguish them here according to their preparation methods, *i.e.* high (e^-, ions, etc.) *vs.* low (neutrals) energy beams. The reason is that only in the first case defects can be considered randomly arranged. Adatoms at all but very low concentration tend to cluster on the surface, and understanding this phenomenon requires knowledge of how the electronic structure of the substrate is modified upon formation of the very first defects.

Notice, however, that defects are also naturally present in graphene as in any common

[6] The number of linearly independent spin-function for N electrons in the S spin state, usually denoted as f_S^N, can be easily obtained by angular momentum coupling rules. The properties of the corresponding spin spaces stem from their deep connection to the group of permutations of N objects.

[7] It is enough to consider the so-called Coulson-Fisher wavefunction.

[8] With this we mean point defects, *i.e.* adatoms, substituted or displaced atoms. It should be noticed, however, that other defects may also be important for the transport properties.

material. Adatoms as hydrogen or hydrocarbons, for instance, can be introduced by the preparation method, while other point defects affecting transport properties, *e.g.* charge-impurities, local potentials, etc., may result from imperfections on the substrates where graphene is accommodated.

3.1 Electron and ion bombardment

The irradiation by high energy particles is the main tool for creating defects in graphene and in other carbon nanostructures. When the projectile particle impinges on the structure it transfers energy to the lattice. In bulk materials (*e.g.* in graphite) energy dissipation is rather effective, up to eventually stop the projectile, and it occurs through *nuclear* and *electronic* stopping mechanisms. Nuclear stopping is due to the collisions between the projectile and the carbon nuclei, an essentially classical process governed by momentum transfer and Coulomb interaction. On the other hand, (inelastic) electronic stopping occurs by the many possible electron transitions in the material, hence promotion into conduction band (hot electrons), ionizations, but also through plasmon excitations, photoemissions, etc. The relative importance of the two mechanisms depends on the beam energy, on the projectile mass and on the electronic structure of the target material. Nevertheless a microscopic theory of energy dissipation in nanostructures is still under study since the models developed for bulk materials cannot be easily applied in a reduced dimensionality material such as graphene (Krasheninnikov & Nordlund, 2010).

The mechanism for the defect formation has been studied intensively in the last decade. In brief, when the energy transferred to an atom is larger than the so-called displacement threshold ($\sim$20 eV in case of graphite) this can leave its equilibrium position and move trough the bulk to form, for instance, a Frenkel pair or, for single layer graphene, a vacancy. Large ions can produce multiple vacancies up to small holes in the lattice depending on their size. Electron beams produced in transmission electron microscopes (TEM) can instead be focused down to scales comparable to the carbon-carbon distance, giving a precise control of the induced damage up to form single vacancies. Moreover TEMs allow a real-time imaging of the damage process and of the chemical reaction that follows the vacancies formation (Meyer et al., 2008; 2010).

The formation a single vacancy in graphene leaves three σ dangling bonds and it removes a π electron. The first span a low-energy, one-dimensional irreducible representation (A irrep in the following) of the (local) D_{3h} point group and an E irrep. Therefore, the ground-state is degenerate and undergoes a Jahn-Teller distortion: the closure of two dangling bonds to form a pentagon, with an energy gain of about 0.2 eV. The strain induced by the other hexagons in the lattice prevents further distortions of the third unsaturated atom out of plane (El-Barbary et al., 2003) and the final magnetic moment for such a structure has been reported to be between 1.0 and 1.5 μ_B (Lehtinen et al., 2004; Yazyev & Helm, 2007), localized on the unpaired site. When exposed to a hydrogen flux, the vacancy rapidly saturates its dangling bonds, with H atoms pointing slightly out of the graphene plane (Lehtinen et al., 2004).

In the case of neutral-atom bombardment, the projectile can also react to form a covalent bond with a carbon atom. This is what happens by irradiating samples with low energies hydrogen atoms. It has been shown that at very low densities the chemisorbed H atom defect produces STM images very similar to the single vacancy case. As already mentioned, at higher densities H atoms tend instead to cluster in dimers or larger structures due to electronic effects that will be discussed in the following. Nevertheless, when considering π electrons only, vacancy and singly-bond chemisorbed species are equivalent, since a single electron is removed from the

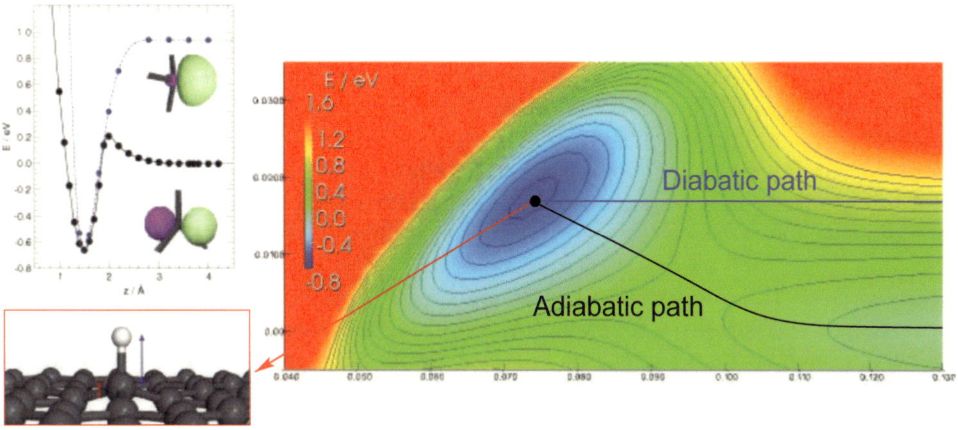

Fig. 5. Left Panel: chemisorption curves for a H atom on a graphene *top* site as obtained from periodic DFT calculations using a 2x2 supercell. In the adiabatic path (black) the carbon atom is allowed to relax at each C-H distance, while in the diabatic path (blue) the top site is kept fixed in the puckered geometry. z is the adsorbate height on the surface. The inset on the bottom shows the equilibrium geometry. Right panel: map of potential energy surface as a function of the heights of H and C atoms above the surface, for a collinear geometry. Also indicated the diabatic and adiabatic paths. Units are eV and Å.

aromatic network of graphene. As an example, chemisorption of a single H atom is detailed in the following section.

3.2 Sticking of atomic and molecular species

A hydrogen atom impinging on graphene with a low collision energy can either physisorb or chemisorb. The physisorption regime has long been probed with the help of selective-adsorption resonances in H atom scattering off graphite (Ghio et al., 1980). The extrapolated value for the physisorption binding energy ($\sim$40 meV) to a single layer is in very good agreement with recent theoretical studies (Bonfanti et al., 2007). Physisorbed species are highly mobile and easily desorb from the surface since they couple only weakly with the substrate. For this reason, chemisorption turns to be more interesting for graphene electronic structure engineering.

Chemisorption of single H atoms on *graphite* has been studied since the first theoretical works of Jeloaica & Sidis (1999) and Sha & Jackson (2002), who first showed that it indeed occurs if the substrate is allowed to relax. Among the four possible adsorption sites the *hollow* and *bridge* were found not binding while the two kinds of *atop* sites (with or without an carbon atom on the layer underneath for graphite) give essentially the same behaviour, since graphene layers in graphite lay $\sim$ 3.4 Å apart. This implies that the (surface) chemistry of graphene is very similar to that of graphite.

Adsorption on the top site induces a surface reconstruction ('puckering'). Such a reconstruction consists in the outward motion of the carbon atom beneath the adsorbed hydrogen, and occurs as a consequence of $sp^2 - sp^3$ re-hybridization of the carbon valence orbitals needed to form the CH bond. The re-hybridization induces a change in geometry of the substrate site, from a planar (sp^2) to a tetrahedral (sp^3) form, thereby leading to the surface puckering. The energy required for such a process, defined as the energy difference

between the relaxed and the puckered configuration, is substantial ($\sim$ 0.8 eV) and this explains why binding energies to graphene are typically smaller than for other carbon species. If the graphene layer is kept flat the carbon - hydrogen bond is metastable only (Casolo et al., 2010; Sha & Jackson, 2002), while allowing surface relaxation chemisorption becomes an activated process with stable products ($\sim$ 0.80 eV).
When the hydrogen atom collides on a carbon site already puckered, *i.e.* already in the sp^3 form, chemisorption is a barrierless process. Otherwise, following an adiabatic path (hence allowing the carbon atom relaxation to its equilibrium position at every point along the reaction coordinate) an energy barrier $\sim$0.2 eV high is found, as a consequence of the re-hybridization. These adsorption curves are shown in fig.5. The barrier (which is also present when the substrate is kept planar) has an important, purely electronic origin. Indeed, it has been shown (Bonfanti et al., 2008) that it results from an avoided crossing between a repulsive interaction with the Kekulé-like ground-state and an attractive interaction with the low-lying, Dewar-like excited state (see Fig.4 in Casolo et al. (2009a)). This can be nicely understood in terms of the chemical picture above since the Kekulé-like structures do not have unpaired electrons which can readily couple with that of the incoming H atom.
The overall binding picture of H atoms has found substantial experimental proof for graphite surfaces. Hot hydrogen atoms produced by dissociating H_2 molecules at $\sim$ 2000 K are required to overcome the barrier and observe sticking. Thus, chemisorption is under *kinetic control*[9], in marked contrast with vacancy formation through e^-/ion bombardment discussed above. Indeed, as we show below, H atoms do *not* adsorb completely random on the surface. A number of TPD, AES, EELS and HREELS spectroscopy data (Andree et al., 2006; Güttler et al., 2004a;b; Zecho et al., 2004; 2002) is available, along with detailed kinetic Monte Carlo simulations (Cuppen & Hornekær, 2008; Gavardi et al., 2009) of TP desorption curves and accurate studies of vibrational relaxation dynamics (Sakong & Kratzer, 2010) and reaction dynamics to form H_2 (Casolo et al., 2009b; Jackson & Lemoine, 2001; Martinazzo & Tantardini, 2005; 2006a;b; Morisset et al., 2004; 2005). Notice that even though we focused here on adsorption of H atoms the same holds for other simple, monovalent chemical species.

4. Low density: π-defect structure

4.1 The appearance of midgap state

The effect of atomic scale defects in graphite, and later on in graphene, has been experimentally studied since the late eighties, when scanning tunneling microscopy (STM) allowed to capture images on solid surfaces at atomic scale resolution. It appeared immediately that when a vacancy was created by irradiating the sample, a bright $\sqrt{3}$x$\sqrt{3}R30°$ charge density reconstruction appears (Mizes & Foster, 1989; Ruffieux et al., 2000; Ugeda et al., 2010).
A carbon vacancy or a defect in the π-network due to a monovalent chemisorbed species creates in graphene an *imbalance* between the number of sites in each sublattices. This lowers the overall lattice symmetry, up to eventually remove the Dirac cones and open a band gap. Looking at the tight-binding Hamiltoninan in equation 3 the introduction of a π-defect in the graphene lattice reads as the removal of the basis function corresponding to the defect site, and the system eigenstates become necessarily odd-numbered. Therefore, in the nearest-neighbour approximation, because of the electron-hole symmetry, one of the eigenvalues in the energy spectrum necessarily lies at the Fermi level. This zero-energy state

[9] H diffusion is largely impeded by electronic/geometrical effects, see below.

is a singly occupied molecular orbital called *midgap* state, even when a gap is not really present. When relaxing the nearest-neighbour approximation such states move from the Fermi level, but remain close to it. Their presence is important for the transport properties being responsible for resonant scattering mechanisms.

The appearance of midgap states in bipartitic systems has been intensively studied in solid state physics because of the implications they have for the appearance of magnetism. Inui *et al.* (Inui et al., 1994) formulated a useful theorem for bipartitic tight-binding models with a sublattice imbalance. According to their result, in any bipartite lattice in which the numbers of sublattices sites N_A and N_B are not equal, there are at least $\eta = |n_A - n_B|$ linearly independent eigenfunctions of the Hamiltonian at zero energy, all with null amplitudes on the minority sublattice sites. The proof is simple: for let $N_A > N_B$ and $|\psi\rangle = \sum_i \alpha_i |a_i\rangle$ be a trial solution at zero energy. The coefficients α_i need to satisfy $\sum_i \langle b_j|H|a_i\rangle \alpha_i = 0$ for $j = 1, ..N_B$ which is a set of N_B equations for the $N_A > N_B$ coefficients, with η linearly independent solutions. This also shows that ψ's localize on the A lattice sites.

Analogous results have been already known in hydrocarbon chemistry for some time. The tight-binding approach described above has been used for decades in quantum chemistry to study aromatic hydrocarbons, under the name of Hückel method. The mathematical properties of the Hückel Hamiltonian have been formalized in a series of theorems and corollaries in a famous book of Dewar (Dewar, 1969). Bipartite lattices were listed there as "alternant" hydrocarbons, and the emergence of midgap states formally predicted in case of odd-numbered alternant hydrocarbons.

Calculations, both at tight-binding and at higher levels of theory (DFT), confirm these expectations: in graphene the zero-energy states originated in this way correspond to semilocalized modes around the defect which decay slowly with the distance, *i.e.* with a r^{-1} power law (Pereira et al., 2006; 2008), a result which has been recently confirmed by experiments (Ugeda et al., 2010). Pereira et al. (2008) performed a comprehensive analysis of the effect low-density defects have on the graphene DOS, by using numerical tight-binding calculations for $\sim 4\text{x}10^6$ lattice sites and analytic results. Analogous results have been found in DFT studies[10] of isolated vacancies (Yazyev & Helm, 2007) and adatoms (Boukhvalov et al., 2008; Casolo et al., 2009a).

4.2 Chemical resonance formula

In the case of a single π-defect, or a random distribution of them, the appearance of midgap states may be easily understood by applying the resonance-based VB picture described in Section 2. Considering benzene as the simplest building block of graphene, it is easy to realize how adsorption of a H atom breaks the aromatic network and leaves one unpaired electron free to move on the lattice by *bond switching*: spin-recoupling with a neighbouring double bond creates an unpaired electron in one every two lattice sites. *Ab-initio* VB calculations (Bonfanti et al., 2008) show that this indeed the case: the 5 π electrons have 5 different ways of couplings (Fig.6) but only those with the unpaired electron in the so-called *ortho* and *para* positions are relevant; an electron in *meta* position would involve a Dewar-like structure, which has a high energy bond-pattern (see Fig.6). The bond switching mechanism is very useful and well known in basic organic chemistry, where it easily allows predictions for orientation effects, *e.g.* in electrophilic aromatic substitutions. In contrast to the full analysis

[10] The approach used is intrinsically periodic. Therefore, the results are best viewed as referring to defects which are periodically arranged on superlattices with large unit cells.

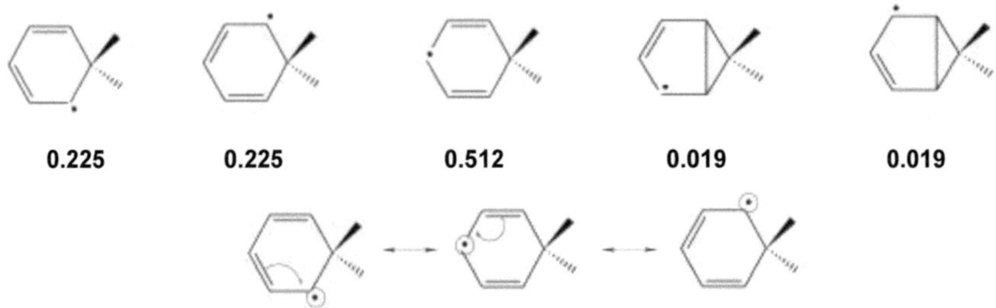

Fig. 6. Valence Bond model for the binding of a radical species (a H atom) on benzene. The numbers give the weights of the corresponding VB structures (Bonfanti et al., 2008). The resulting unpaired electron localizes mostly in *ortho* (two leftmost structures on the top row) and *para* (mid panel) position, as emphasized by the bond-switching mechanism reported in the chemical formula of the bottom row.

of possible spin-couplings, exporting this model to graphene is rather straightforward[11]. A picture of the mechanism is shown in fig.7 for a coronene model, that is meant to represent the whole graphene lattice. The itinerant electron hops between sites of one type only, thereby occupying a delocalized state which is the midgap state described previously in the tight-binding (MO) picture.
Whatever picture we use the result is a *spin density* (magnetization) localized close to the defect, on the sites of the hexagonal sublattice not housing it [12]. At low density, where hybridization does not occur, such spin-density thus determines the appearance of (microscopic) magnetically ordered domains. It further influences reactivity of the substrate with foreign species, which can readily 'saturate' (singlet-couple) this electron if they land on the correct sites, as will be shown in the next section.
Before concluding this Section, we can now understand why simple adatoms do not move on the surface. Indeed, for the H atom to hop on the neighbouring site the unpaired spin has to move from one sublattice to the other and this requires breaking completely the existing CH bond and forming a new one: the barrier to diffusion, then, matches the desorption

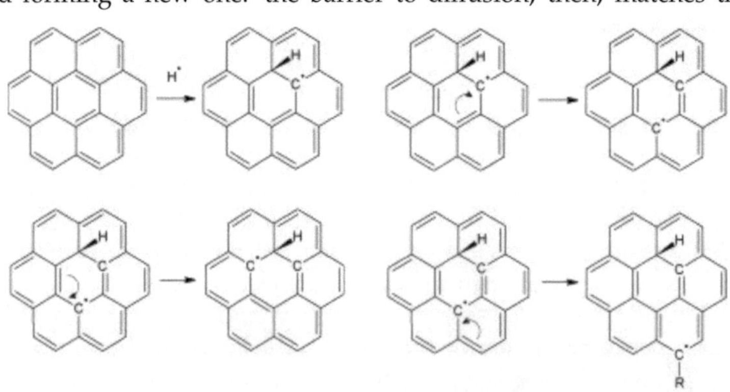

Fig. 7. Itinerant electron model for the p_z-vacancy-induced midgap state

[11] This amounts to consider Kekulé structures only, which are much fewer than the whole set of f_S^N couplings for all but small N values.
[12] It can also be turned into a *charge-density* by addition/removal of one electron.

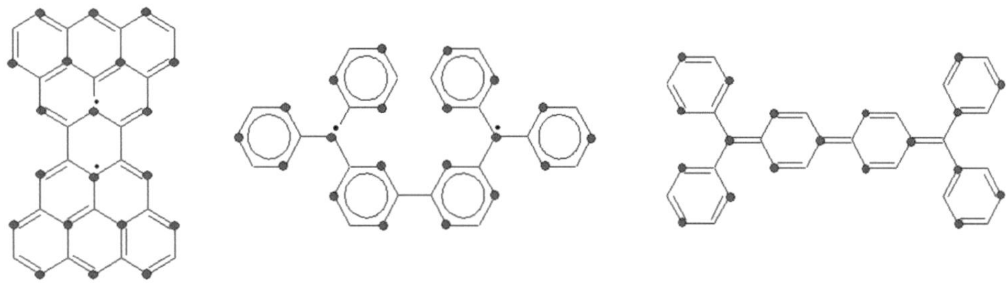

Fig. 8. Molecules with no sublattice imbalance. The first two molecules on the left are di-radical species, *i.e.* they have two midgap states. The Schlenk-Brauns hydrocarbon shown in the middle panel differs from the Chichibabin hydrocarbon shown in the right panel by the connectivity only. Nevertheless, the latter does not present unpaired electrons. In the first two molecules, the largest set of non-adjacent sites is marked by red dots.

energy. This explains the experimental observation that H atoms are immobile on the surface (Hornekær et al., 2006b). For more complex species, *e.g.* O atoms, spin-recoupling *on* the adatom may help the diffusion (isomerization) process *via* formation of a 'bridge' between the two sites. This would explains why DFT computed barriers for diffusion of OH species are definitely smaller than the desorption energy (Ghaderi, 2010).

5. High density: spin-ordering, clustering and related issues

5.1 Predicting midgap states and magnetism

The very simple counting rule for midgap states introduced above usually works fine for graphene, but fails to be predictive for some class of finite size graphenes or analogous (complementary) holes on the graphene sheet. For instance, the first two molecules in Fig.8 have no sublattice imbalance (being symmetric) but are radical species, *i.e.* they necessarily have midgap states (Pogodin & Agranat, 2003). To a closer inspection, sublattice imbalance is indeed only a sufficient condition for midgap states to appear.

To rationalize the situation, it is necessary to introduce the concept of *non-adjacent* sites in a N-site bipartitic system. We say that two sites are non-adjacent if they are not bound (connected) to each other; for instance, two sites on the same sublattice are non-adjacent. Clearly, there exists a maximal set of non-adjacent sites and we call α the sites in this set, and β the remaining ones ($N_\alpha, N_\beta = N - N_\alpha$ in number, respectively). Each site α binds at least to one site β, otherwise it would represent a completely isolated site. Arranging one electron per site α, however, we can form at most N_β bonds at a time, and therefore we are left with $\eta = N_\alpha - N_\beta = 2N_\alpha - N$ unpaired electrons, *i.e.* midgap states. The case of a sublattice imbalance discussed above is a special result of this rule: when $N_A > N_B$, since the A lattice sites are always non-adjacent and $N_\alpha \geq N_A$, we have $\eta \geq N_A - N_B$. Fig.8 shows molecules having $\eta > 0$ and no sublattice imbalance, with the indicated α sets. As it is evident from its derivation, this result can be equivalently re-phrased by defining η to be the number of unpaired electrons in the Lewis structure(s) with the maximum number of π (*i.e.* double) bonds.

Notice that, since the spectrum of the Hamiltonian is determined by the system topology the whole set of counting rules for midgap states can be derived entirely from graph theory. In particular, midgap states appears as zeros of the characteristic polynomial of the adjacency matrix **A**, that defines the connectivity of the graph (Bonchev & Rouvray, 1991; Randić, 2003)).

In this context, the above result is known as *graph nullity* theorem.
Having derived the exact conditions determining the appearance of midgap states, the question arises of how spins couple when a number η of unpaired electrons are present. The determination of the spin state cannot come, of course, from the simple tight-binding Hamiltonian, since in these open-shell configurations energy ordering is mainly determined by electron correlation. At first glance, it can be guessed that electrons occupying quasi-degenerate midgap states tend to keep their spins parallel, in a sort of molecular Hund's rule, as this reduces Coulomb repulsion, *i.e.* system's total spin should always be $\eta/2$ (Longuet-Higgins, 1950). This is actually the case only when midgap states originate from a sublattice imbalance, since in such instance they are forced to stay on the same sublattice. When midgap states (unpaired electrons) lie on different sublattice they best couple at *low* spin. This result can be shown to be exact for the realistic model provided by the (repulsive) Hubbard Hamiltonian: Lieb (1989) showed that for any bipartitic system at half-filling the ground-state spin S is given by the sublattice imbalance $S = \frac{1}{2}|N_A - N_B|$. This is a subtle effect of electron correlation, which would lead to an energetically unfavourable spin polarization of the remaining occupied orbitals if the above Hund rule were followed in absence of sublattice imbalance [13]. From a different perspective, it has been associated with the most "spin-alternant" structure (Ovchinnikov's rule, Ovchinnikov (1978)).
According to the rules above it is now possible to predict the number of midgap states and the spin state of a number of complex graphene structures without relevant exceptions. We only note that the theorems stated above for bipartite lattices do not apply for topological defects that destroy bipartitism. Nevertheless, it has been noticed that the Ovchinnikov's rule can be usually extended to non-bipartite systems (Shaik & Hiberty, 2007), although some care has to be paid (López-Sancho et al., 2009). For instance, the ground-state multiplicity of Stone-Wales defects is correctly predicted to be zero by this rule.

5.2 Preferential sticking

When adsorbing hydrogen atoms on graphite or graphene under kinetic control STM images clearly show the formation of dimers and clusters (Hornekær et al., 2006b). Since H atoms are immobile on the surface this must be due to a *preferential sticking* mechanism. This mechanism was first suggested by (Hornekær et al., 2006b) who looked at the STM images formed by exposing Highly Oriented Pyrrolitic Graphite (HOPG) samples to a H atom beam, and observed formation of stable pairs, also confirmed by *first-principles* calculations (Hornekær et al., 2006b; Rogeau et al., 2006). Later Casolo et al. (2009a) showed that the preference for certain lattice sites comes from the spin density localized on one of the two sublattices (the midgap state), as generated by the first adsorbate.
The overall picture (Casolo et al., 2009a) is consistent with the VB chemical model: when a first H atom is on an A-type site, the unpaired electron localizes on the B sublattice and bond formation easily occurs on its sites. An "AB dimer" (which has no sublattice imbalance) is formed and a singlet ground-state is obtained where aromaticity is partially restored. Conversely, if adsorbtion occurs on the same sublattice, *i.e.* to form "A_2" dimers, the incoming H atom does *not* make use of the available spin-density, and adsorption energies are comparable to that of the first H atom. Furthermore, as another electron is set free on the *same* B sublattice occupied by the unpaired electron, the ground state is a triplet ($\eta = N_B - N_A = 2$).

[13] Notice that for two electrons in different zero energy state, first-order perturbation theory always gives a triplet ground-state. The nature of the midgap states and the ensuing interactions with the doubly occupied orbitals play a decisive role.

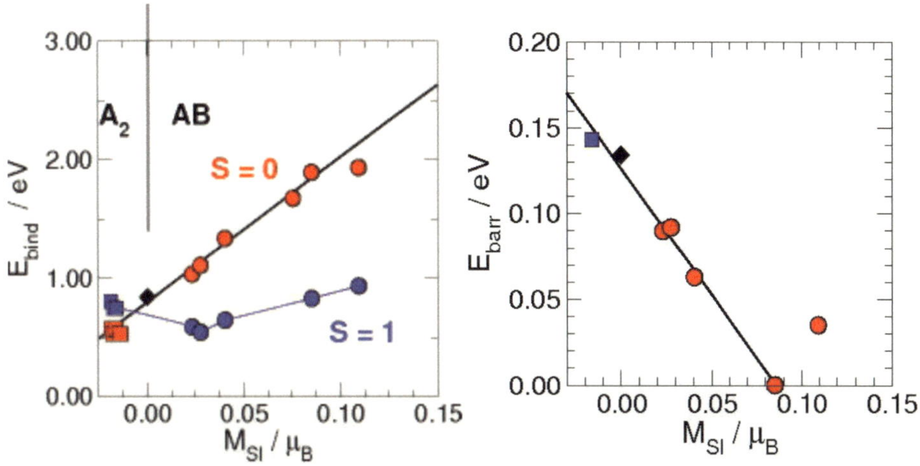

Fig. 9. Left panel: Binding energies for secondary H adsorption as a function of the site-integrated magnetization (M_{SI}), for AB (circles) and A_2 (squares) dimers. Both singlet (red) and triplet (blue) solutions are shown in red and blue respectively. Also shown the data point for single H adsorption (black diamond) and a linear fit to the data set (solid line). Right panel: corresponding barrier energies for secondary atom adsorption (ground-state only). Data point at M_{SI}=0 is for single H adsorption.

The results of DFT calculations (Casolo et al., 2009a) on a number of dimers are shown in Fig.9 as function of the *site-integrated* magnetization, *i.e.* the average number of unpaired electrons in each site as results from the first adsorption event. It should be noted that substrate relaxation effects, though substantial ($\sim$ 0.8 eV), are site-independent for all but the *ortho* dimer[14]; thus the curves in the graphs of Fig. 9 reflects purely electronic effects. Binding and barrier energies both depend linearly on the local magnetization, thereby implying a linear relationship between them; this is a common tendency in activated chemical reactions known as Brønsted-Evans-Polayni rule. An exception is provided by the *ortho* dimer (rightmost data point in the graphs of Fig. 9), whose formation requires further rearrangement in the first C-H neighbourhood. This is shown in figure 10 where the equilibrium geometry of the dimer is reported in the left panel. It is clear from the figure that the two H atoms point in opposite directions (as in a H-C-C-H eclipsed conformation of an alkane), which suggests that, despite their proximity, they would not easily desorb to form H_2 upon heating the substrate. This is indeed what has been found by a combined theoretical and experimental study by Hornekær et al. (2006a): upon heating, the *ortho* dimer prefers to isomerize to the *para* dimer, which dehydrogenates easier (*i.e.* at a lower temperature). The highest temperature peak in the TPD spectra corresponds then to this isomerization process. The *para* dimer itself, whose equilibrium geometry is shown in the right panel of Fig.10, forms abundantly when exposing graphene to a H atom beam, since its formation is *barrierless* (see fig.9). This forms the basis for the preferential sticking mechanism first suggested by Hornekær et al. (2006b), which is here summarized with the results of fig.9, namely formation of AB dimers is both thermodynamically *and* kinetically favoured over formation of A_2 dimers and single atom adsorption.

[14] Surface puckering upon adsorption, to a good approximation, involves nearest neighbouring C atoms only.

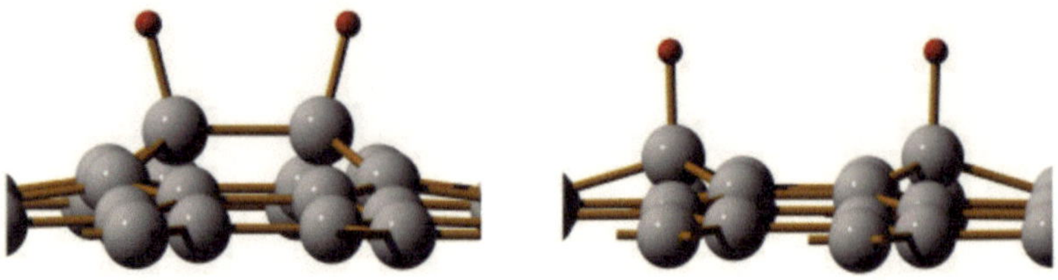

Fig. 10. Equilibrium geometry for ortho (left) and para (right) dimer structures.

Unfortunately[15], the preferential sticking mechanism above works only for dimers. Once an AB dimer is formed (A_2 is unfavoured) there are no further unpaired electrons available, and no bias on the adsorption of additional H atoms. This is confirmed by DFT calculations (Casolo et al., 2009a) on a number of larger A_2B_2, A_2B, A_3B_1 and A_3 clusters. As expected from the VB model above, adsorption of a third hydrogen atom to a stable AB dimer parallels that of the first H, with essentially no preference towards any specific sublattice position, and always produce doublet structures ($M = 1\,\mu_B$) (see fig.11). Similar conclusions hold when adding a third H atom to the (magnetic) *meta* dimer A_2: adsorption on B lattice sites is strongly favoured ($E_{bind} = 1.2 - 1.9$ eV) and produces doublet structures ($M = 1\,\mu_B$), whereas H atoms bind to A lattice sites with an energy $\sim 0.7 - 0.8$ eV and produce highly magnetic structures ($M = 3\,\mu_B$) (see fig.11). Energy barriers to adsorption follow the same trend: calculations show that, with few exceptions, barriers to sticking a third H atom compare rather well with that for single H atom adsorption for the processes AB$\rightarrow A_2B$ and $A_2 \rightarrow A_3$, and may be considerably smaller for $A_2 \rightarrow A_2B$ ones. These three-atom clusters, similarly to the single H atom, necessarily bias the adsorption of a fourth atom. The computed binding and barrier energies for this process have been found to compare rather well with the dimer values, actually they nicely fit to the *same* linear trends shown in Fig.9. Finally, all the considered A_2B_2, A_2B, A_3B_1 and A_3 clusters have been found to have 0, 1, 2 and 3 unpaired electrons in their ground-state, respectively, in agreement with expectations (*i.e.* either the VB model or the Lieb theorem).

Few exceptions to this picture are for compact clusters where substrate relaxation does play some role, see *e.g.* the structures on the right of fig. 11. Compared to other trimers, these structures are favoured because of the substrate *softening* occurring after formation of the *para* dimer, which is in a typical *boat* configuration (fig.10). Such softening is expected to reduce the relaxation energy needed for the binding of the additional atom, with a gain of some tenths of eV on the overall energetic balance[16]. This would explain why experiments at intense H atom flux do not find a random distribution of *dimers*, as would be expected on basis of electronic effects only, rather clusters made up of a number of atoms (Ferro et al., 2009; Hornekær et al., 2007). Though a detailed analysis would require the knowledge of the adsorption *barrier* for a rather large number of clusters, the linear relationship shown above may help in making educated guesses on the basis of the binding energies only.

[15] There is a long open search for efficient pathways leading to H_2 formation on graphitic surfaces, because of its importance in explaining the observed abundance of molecular hydrogen in the interstellar medium.

[16] Remember that the relaxation energy for the single H atom ($\sim$ 0.8 eV) has the same magnitude as the overall binding energy, *i.e.* about half of the bond formation energy is spent for relaxing the substrate.

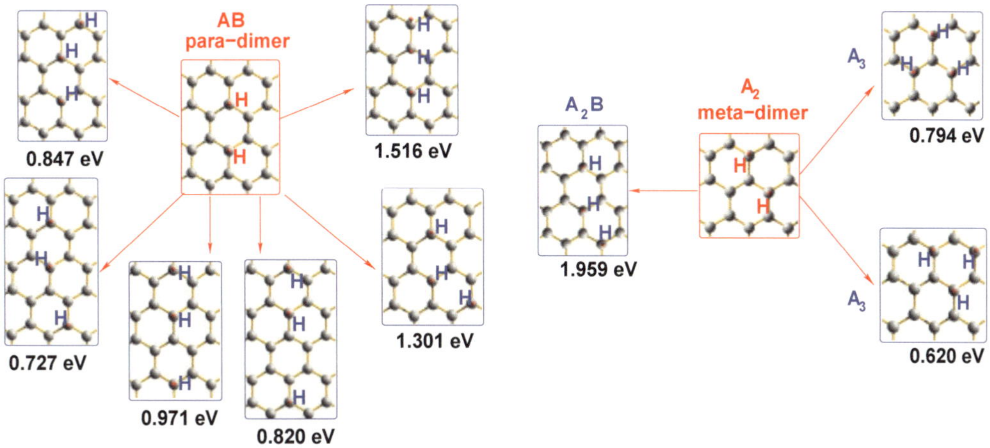

Fig. 11. Some of the possible hydrogen trimers structures obtainable from the AB *para* dimer (left diagram) and from the A_2 *meta* dimer (right). Binding energies are also shown.

5.3 A route to graphane?

Graphane is a novel two-dimensional material, namely the fully hydrogenated graphene sheet. The structure is still bipartitic and each sublattice bears all hydrogens on the *same* side of the lattice plane (in *meta* to each other), in such a way to have a chair-like configuration as in cyclohexane. It is an insulating material with no π electrons (Lebègue et al., 2009; Sofo et al., 2007), that might have been recently obtained by simply exposing graphene to cold hydrogen plasma (Elias et al., 2009). This result is very interesting in light of what has been shown in the previous sections. In order to produce graphane by simple hydrogen exposure it is necessary that either H atoms adsorb selectively on one sublattice *only* for a given graphene face or hydrogen diffuses (isomerize) to occupy the sites on the right *face*.

According to the discussion of previous section, formation of *meta* dimers is unlikely, *ortho*, *para* positions being highly favoured for adsorption. According to DFT calculations on free-standing graphene (Casolo, 2009), this is true both for the *syn*- (on the same face) and the *anti*- (on opposite faces) dimers. However, even if hydrogen atoms were likely to form *anti-para* dimers also in supported graphene (and they are required for graphane production) an efficient *syn*- to *anti*- conversion mechanism would be needed to convert those dimers already formed on the same graphene face. Unfortunately, as we have seen in the previous section, even in this case, no true preferential sticking can occur after dimer formation and it will be very unlikely that all the other hydrogens will chemisorbed in the correct sites and face. Indeed, recent molecular dynamics simulations showed that disordered, frustrated hydrogenated domains would rather form (Flores et al., 2009).

If graphane has been really formed by hydrogen exposure some other effect has to play a role. Curved graphene areas might help this process. Graphene is a very elastic membrane that naturally exhibit ripples that tend to lay down along steps and kinks of the supporting substrate on which it was grown. Indeed recent experimental findings suggested that hydrogen chemisorbs more efficiently on the ridges of the silicon carbide substrate surface onto which graphene usually lays (Balog et al., 2009). Moreover, it has been also shown that hydrogenation of single-layer graphene is easier than for many-layer graphene, likely as a consequence of the higher corrugation displayed by the graphene surface (Jung et al., 2009; Luo et al., 2009). This is reasonable, as for nanotubes the curvature reduces the p_z-p_z

overlap, *i.e.* aromaticity, thereby lowering the barrier energies for H chemisorption (Ruffieux et al., 2002a;b). Still, there is no clear evidence that local curvature plays a role in graphane formation, and more investigations in this direction are needed.

6. Defect-based material design

When it comes to device fabrication only few of the many extraordinary properties of graphene are relevant, at least for the chip-makers (Schwierz, 2010). Among them, its thickness allows the thinnest possible gate-controlled regions in transistors and, according to scaling theories, should reduce electrostatic problems if short channels have to be built. Mobility is an important factor as it allows for instance high-performance interconnects and fast response to external (gate) potentials. It becomes of secondary importance in short channels, where high fields builds up and carrier velocity saturates, but also in this respect graphene proved to have superior properties than conventional materials. Indeed, high-performance transistors for frequency applications have been realized (Lin et al., 2010), and record cut-off frequencies are being continuously scored. However, for its usage in logic applications the absence of a band-gap is a major problem (Avouris et al., 2007; Schwierz, 2010): even when the Fermi level crosses the charge neutrality point a non-zero residual conductivity avoids the complete current pinch-off [17]. The absence of a band-gap, indeed, prevents the achievement of the high current on-off ratios required for logic operations.

Graphene can be turned into a true semiconductor by properly engineering it. Electron confinement, though in general not trivial for massless, pseudorelativistic carriers, can be obtained by cutting large-area graphene to form narrow nanoribbons. Apart from related fabrication issues, one main drawback of such an approach is the removal of the Dirac cones and the resulting band-bending. This is expected to increase the effective mass of the carriers, thereby reducing their mobility. Indeed, it has been generally found that mobility is a decreasing function of the gap (Schwierz, 2010), and this is an undesirable by-side effect worth considering with such a traditional approach. Alternatively, *symmetry breaking* is known to turn the massless Dirac carriers into massive (yet pseudorelativistic) carriers. This can be realized by depositing or growing graphene on a substrate that renders inequivalent the two sublattice positions. For instance, boron nitride has the same honeycomb lattice as graphene and a similar cell parameter, but presents two inequivalent sublattices. When graphene is in contact with such a surface B and N interact differently with the carbon atoms of the graphene sheet, breaking its sublattice equivalence and lifting the degeneracy of the two bands. A similar situation is achieved for graphene grown on silicon carbide surfaces, where a gap has been observed by angle-resolved photoemission spectroscopy though subtle electron correlation effects may play a role in such case (Bostwick et al., 2007; Zhou et al., 2007).

In the following sections we describe alternative possibilities for opening a gap in graphene band structure, namely those offered by superlattices of defects and dopants. One interesting finding in this context is the proof that a band-gap can be opened in graphene *without* breaking its symmetry, with the advantage the new Dirac cones (massless carriers) appear right close to the gapped region (Martinazzo et al., 2010). For this reason we start introducing some symmetry considerations, extending the arguments given in section 2.

[17] The defects discussed in this chapter, along with charged scatterer, are ascending as the most likely origin of the conductivity minimum (Peres, 2010). The counter-intuitive role of defects in *increasing* the conductivity finds its origin in the modification of the graphene DOS close to the Dirac point.

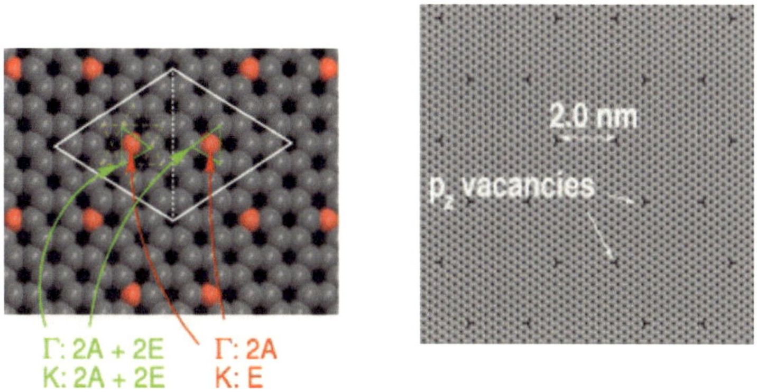

Fig. 12. Left: counting the number of irreps generated by the atomic basis in a 4x4 supercell: indicated are the irreps generated by the atoms at the center of the half-cells (red balls) and by green triangles. Right: a simple "supergraphene", the simplest defective 14x14 honeycomb.

6.1 Symmetry considerations

As we have seen in section 2 graphene's unconventional electronic properties are strictly related to its D_{6h} point symmetry. The k-group at the K-K′ high-symmetry points (D_{3h}) allows for doubly degenerate irreducible representations, and Bloch functions built with p_z orbitals of A and B sublattices span just one of its two-dimensional irreps. As $e-h$ symmetry does not mix one- (A) and two- (E) dimensional representation this level has to lie at zero energy, where the Fermi level (ϵ_F) is located.

Were not there such degenerate level, graphene would be, as any other bipartitic system at half-filling, semiconducting. Graphene can be forced to be so by either lowering the symmetry (*i.e.* changing the *k group* at K(K′) to a simpler one), or changing the *number* of E irreps at the special points while keeping the overall symmetry. In the latter, more intriguing case, since the overall point symmetry is preserved, degeneracies may still occur at energies different from ϵ_F, and new Dirac cones are to be expected. The "recipe" for doing that is very simple (Martinazzo et al., 2010): nxn graphene superlattices have the same symmetry properties and $2n^2$ atoms per cell; by symmetrical removing a number of C atoms [18] is possible to change the number of irreps and turn, in particular, the E ones to be *even* at every, highly symmetric special point (Γ, K, K'). With few exceptions of residual accidental degeneracies, this opens a gap in the band structure.

The approach is made effective by counting the number of irreps generated by $2n^2$ atoms in a nxn unit supercell. The results of this calculation (Martinazzo et al., 2010) can be grouped into three different sequences, $n = 3m, 3m+1, 3m+2$ (m integer), according to the BZ folding properties. In two thirds of the cases, *i.e.* when $n = 3m+1, 3m+2$, removal of the atoms at the center of the two-half cells (red balls in the left panel of fig.12) is sufficient for opening a gap. Figure 12 (right panel) shows one of the simplest resulting semiconducting superlattice, namely with simple, atomic-scale defects arranged in a *honeycomb* lattice. Because of their nature, they are best considered as *supergraphenes*.

6.2 Superlattices of vacancies or holes

Tight-binding and DFT calculations on the simple structures identified in the previous section show indeed a sizable band-gap. The gap size approximately scales as v_F/l_n where v_F is

[18] As shown in the previous sections one can equivalently introduce either a vacancy or an adatom.

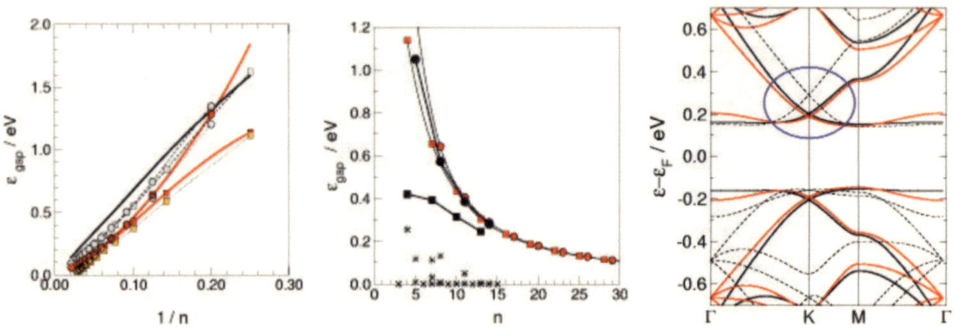

Fig. 13. Energy gaps in simple supergraphenes made with p_z vacancies. Left: results of TB calculations as functions of $1/n$. The symbols are for different parameters of the TB hamiltonian, and the solid line is the result of a perturbative calculation at the K point. See Martinazzo et al. (2010) for details. Middle: tight-binding (red) *vs.* DFT (black) results. In the latter case, defects have been modelled as H atoms. Stars represent the results for asymmetric dimers placed in the same nxn supercells. Right: energy bands for the $n = 13, 14$ supergraphenes.

the Fermi velocity in pristine graphene and l_n is the distance between defects ($l_n = na/\sqrt{3}$), as can be guessed from a dimensional analysis or obtained from a perturbative calculation within the tight-binding approach (Martinazzo et al., 2010). Both the size and the scaling compare favourably with the gap in armchair nanoribbons (Son et al., 2006). However, one distinctive feature of such structures is the additional presence of new Dirac cones right close to the gapped region (blue circle in fig.13). This might be important in charge transport, since they can sustain massless carriers when the Fermi level, as tuned by a gate potential, is swept across the gap.

In practice, it is still experimentally challenging to realize the atomic-scale patterned structures introduced above. It is however sufficient to consider similar superlattices of *holes* analogously to the graphene *antidots* superlattices investigated by Fürst et al. (2009); Liu et al. (2009); Pedersen et al. (2008). The resulting structures are *honeycombs antidots* as the one shown in fig.14 They are experimentally feasible, as Bai et al. (2010); Eroms & Weiss (2009); Fischbein & Drndic (2008); Shen et al. (2008) have shown that is possible to create circular holes with diameters as small as $2-3$ nm and periodicity ~ 5 nm. Analogous patterns of H adatoms have also been realized thanks to the interaction between graphene and an underlying metal surface that creates Moire patterns activating chemisorption in specific areas (Balog et al., 2010).

Tight-binding calculations on the honeycomb antidots show that the band-gap is quite large for reasonable values of the superlattice constant and of the hole diameter and, as before, new Dirac cones appear at low energies, close to the gapped region (Martinazzo et al., 2010). The gap size is even larger than for the simplest structures considered above, though the latter remain *optimal* in this context: when the gap size is renormalized to the number of defects per unit cell, honeycombs such as that reported in fig. 12 prove to be "magic".

Notice that previously suggested antidot superlattices (Fürst et al., 2009; Pedersen et al., 2008) show comparable gaps, and are therefore equally valid candidates for turning graphene into a true semiconductor. The only difference is a subtle symmetry-related issue. These hexagonal superlattices are all of $\sqrt{3}n$x$\sqrt{3}n$ type, hence with the same D_{6h} symmetry of the honeycomb lattices considered here, and this would suggest that exactly the same results hold for them. A

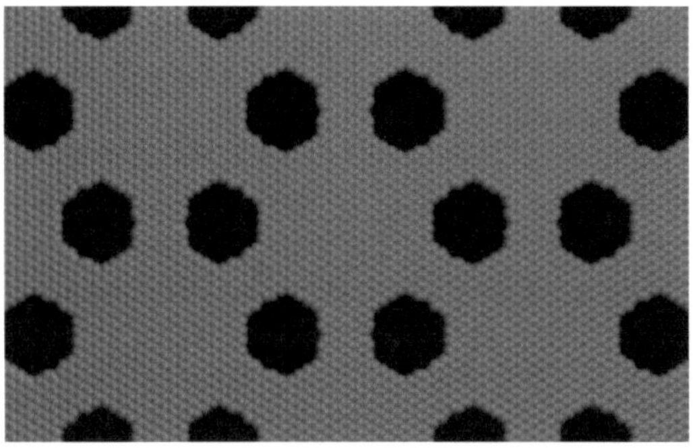

Fig. 14. A honeycomb antidot superlattice presenting a gap by the symmetry preserving approach discussed in the text.

closer inspection, however, reveals that in the $\sqrt{3}n$x$\sqrt{3}n$ case, the K, K′ points of pristine graphene always fold to Γ. This is advantageous for the band-gap opening, since these structures are generally semiconducting if sufficiently defective[19]. However, as in Γ the k group has the inversion symmetry, residual degeneracies at $\epsilon \neq \epsilon_F$, is *not* lifted at first order, and therefore no linear dispersion is present.

6.3 Superlattices of substitutional atoms

Symmetry arguments similar to the one given above apply as well to the case where C atoms are replaced (rather than removed) by other species, in such a way to form superlattices substitutional of dopants. The only difference is that now foreign species are present and point symmetry can be altered. Here we focus on group IIIA and VA elements, mainly because of the fast progresses in methods for the controlled synthesis of N- and B- doped graphenes. For instance, Panchakarla et al. (2009) have recently shown how it is possible to insert B or N dopants in graphene by adding the correct precursors in the arc discharge chamber, while Ci et al. (2010) have reported the synthesis of large islands of boron nitride embedded in graphene by atomic layer deposition techniques. Methods to selectively replace C atoms from graphene lattice have also been proposed by Pontes et al. (2009).

Substitutional defects behave similarly to p_z vacancies (to which they reduce when the hoppings become zero) but introduce impurity bands which partially hybridize with those of the substrate. In addition, the diagonal disorder they introduce breaks $e-h$ symmetry giving rise to a Fermi level shift, *i.e.* to $p-$ and $n-$ doping for group IIIA and VA elements, respectively. If superstructures are only weakly defective, however, the Fermi level shift scales as $1/n$, since the linear-energy dispersion implies $E_F = v_F\sqrt{\pi n_e}$ (here n_e is the electron (hole) excess density, $n_e \propto 1/n^2$). Thus, analogously to the superlattices of the previous section, the defect-induced perturbation affects the electronic structure *close* to the Fermi level, and symmetry arguments may be used to establish whether degeneracy occurs at the special points in the important low-energy region.

It has been shown that, depending on the overall symmetry, superlattices of N and B defects (and mixture thereof) can either preserve the Dirac cones (D_{6h} superlattices) or open a band gap (D_{3h}) (Casolo et al., 2011). For instance, honeycomb superlattices of B (or N) dopants only (D_{6h} symmetry) are found to preserve the Dirac cones, and to be only $p-$ ($n-$) doped (the cone-apex shift being $\propto 1/n$). Indeed, the low-energy band structure in these superlattice is

[19] The number of E irreps is always even because of the 'coalescence' of the two valleys. Thus, one only needs to remove the accidental degeneracy created by such folding.

very similar to that of graphene, with a group velocity at the shifted Dirac cones depending linearly on $1/n$ too. As the Fermi level can be tuned by a gate potential, these systems offer the possibility of investigating the role that the effective speed of light (the above velocity at the Dirac cones) has on the transport and optical properties of graphene. Conversely, BN-honeycomb superlattices (D_{3h} symmetry), thanks to the balanced doping, are found to develop a gap right at the Fermi level. The resulting dispersion relation is found to be quasi-conical, corresponding to *massive* Dirac fermions. The resulting gaps are found sizable and, furthermore, the effective rest masses are rather small, $m^* \sim 0.01 m_e$ or smaller. This suggests that these structures might have good field-switching and transport properties.

7. References

Andree, A., Le Lay, M., Zecho, T. & Küppers, J. (2006). Pair formation and clustering of d on the basal plane of graphite, *Chem. Phys. Lett.* 425: 99.

Avouris, P., Chen, Z. & Perebeinos, V. (2007). Carbon based electronics, *Nat. Nanotech.* 2: 605.

Bai, J. W., Zhong, X., Jiang, S., Y, Y. H. & Duan, X. F. (2010). Graphene nanomesh, *Nature Nanotech.* 5: 190.

Balog, R., Jorgensen, B., Nilsson, L., Andersen, M., Rienks, E., Bianchi, M., Fanetti, M., Laegsgaard, E., Baraldi, A., Lizzit, S., Sljivancanin, Z., Besenbacher, F., B, B. H., Pedersen, T. G., Hofmann, P. & Hornekær, L. (2010). Bandgap opening in graphene induced by patterned hydrogen adsorption, *Nat. Mater.* 4: 315.

Balog, R., Jørgensen, B., Wells, J., Lægsgaard, E., Hofmann, P., Besenbacher, F. & Hornekær, L. (2009). Atomic hydrogen adsorbate structures on graphene, *J. Am. Chem. Soc.* 25: 131.

Bena, C. & Montambaux, G. (2009). Remarks on the tight binding model for graphene, *New J. Phys.* 11: 095003.

Bonchev, D. & Rouvray, D. H. (eds) (1991). *Chemical Graph Theory*, Gordon and Breach, chapter 6.

Bonfanti, M., Casolo, S. & Martinazzo, R. (2008). Unpublished.

Bonfanti, M., Martinazzo, R., Tantardini, G. F. & Ponti, A. (2007). Physisorption and diffusion of hydrogen atoms on graphite from correlated calculations on the h-coronene model system, *J. Phys. Chem. C* 111: 5825.

Bostwick, A., Ohta, T., Seyller, T., Horn, K. & Rotenberg, E. (2007). Quasi-particle dynamics in graphene, *Nat. Phys.* 3: 36.

Boukhvalov, D. W., Katsnelson, M. I. & Lichtenstein, A. I. (2008). Hydrogen on graphene: Electronic structure, total energy, structural distortions and magnetism from first-principles calculations, *Phys. Rev. B* 77: 035427.

Casolo, S. (2009). *Adsorption, Clustering and reaction of H atoms in graphene*, PhD thesis, University of Milan.

Casolo, S., Flage-Larsen, E., Løvvik, O. M., Darling, G. R. & Tantardini, G. F. (2010). The role of self-interaction error in computing radical adsorption on graphene, *Phys. Rev. B* 81: 205412.

Casolo, S., Løvvik, O. M., Martinazzo, R. & Tantardini, G. F. (2009)a. Understanding adsorption of hydrogen atoms on graphene, *J. Chem. Phys.* 130: 054704.

Casolo, S., Martinazzo, R., Bonfanti, M. & Tantardini, G. F. (2009)b. Quantum dynamics of the eley-rideal hydrogen formation reaction on graphite at typical interstellar cloud conditions, *The Journal of Physical Chemistry A* 113(52): 14545–14553.

Casolo, S., Martinazzo, R. & Tantardini, G. F. (2011). Band Gap engineering in graphene with substitutional defects superlattices. In press .

Castro Neto, A. H., Guinea, F., Peres, N. M. R., Novoselov, K. S. & Geim, A. K. (2009). The electronic properties of graphene, *Rev. Mod. Phys.* 81: 109.

Ci, L., Song, L., Jin, C., Jariwala, D., Wu, D., Li, Y., Srivastava, A., Wang, Z. F., Storr, K., Balicas, L., Liu, F. & Ajayan, P. M. (2010). Atomic layers of hybridized boron nitride and graphene domains, *Nature Mater.* 9: 430.

Cooper, D. L. (2002). *Valence Bond Theory - Theoretical and computational chemistry 10*, Elsevier.

Cooper, D. L., Gerratt, J. & Raimondi, M. (1987). Modern valence bond theory, *in* K. P. Lawley (ed.), *Ab initio methods in quantum chemistry II*, John Wiley & Sons Ltd.

Cuppen, H. M. & Hornekær, L. (2008). Kinetic monte carlo studies of hydrogen abstraction reaction from graphite, *J. Chem. Phys.* 128: 174707.

Dewar, M. J. S. (1969). *The Molecular Orbital theory of organic chemistry*, McGraw-Hill.

El-Barbary, A. A., Telling, R. H., Ewels, C. P. & nd P R Briddon, M. I. H. (2003). Structure and energetics of the vacancy in graphite, *Phys. Rev. B* 68: 144107.

Elias, D. C., Nair, R. R., Mohiuddin, T. M. G., Morozov, S. V., Blake, P., Halsall, M. P., Ferrari, A. C., Boukhvalov, D. W., Katsnelson, M. I., Geim, A. K. & Novoselov, K. S. (2009). Control of graphene's properties by reversible hydrogenation: Evidence for graphane, *Science* p. 5914.

Eroms, J. & Weiss, D. (2009). Weak localization and transport gap in graphene antidot lattices, *New J. Phys.* 11: 095021.

Ferro, Y., Morisset, S. & Allouche, A. (2009). Evidence of hydrogenated hexamers on graphene, *Chem. Phys. Lett.* 478: 42.

Fischbein, M. D. & Drndic, M. (2008). Electron beam nanosculpting of suspended graphene sheets, *Appl. Phys. Lett.* 93: 113107.

Flores, M. Z. S., Autreto, P. A. S., Legoas, S. B. & Galvao, D. S. (2009). Graphene to graphane: a theoretical study, *Nanotechnology* 20: 465704.

Fürst, J. A., Pedersen, T. G., Brandbyge, M. & Jauho, A.-P. (2009). Density functional study of graphene antidot lattices: Roles of geometrical relaxation and spin, *Phys. Rev. B* 80(11): 115117.

Gavardi, E., Cuppen, H. M. & Hornekær, L. (2009). A kinetic monte carlo study of desorption of h2 from graphite (0001), *Chem. Phys. Lett.* 477: 285.

Ghaderi, N. (2010). Private communication.

Ghio, E., Mattera, L., Salvo, C., Tommasini, F. & Valbusa, U. (1980). Vibrational spectrum of h and d on the (0001) graphite surface from scattering experiment, *J. Chem. Phys.* 73: 556.

Güttler, A., Zecho, T. & Küppers, J. (2004a). Adsorption of d(h) atoms on ar ion bombarded (0001) graphite surfaces, *Surf. Sci.* 570: 218.

Güttler, A., Zecho, T. & Küppers, J. (2004b). A leed and stm study of h(d) adsorption on c(0001) surfaces, *Chem. Phys. Lett.* 395: 171.

Hornekær, L., Sljivančanin, Ž., Xu, W., Otero, R., Rauls, E., Stensgaard, I., Lægsgaard, E., Hammer, B. & Besenbacher, F. (2006)a. Metastable structure and recombination pathways for atomic hydrogen on the graphite (0001) surface, *Phys. Rev. Lett.* 96: 156104.

Hornekær, L., Xu, W., Otero, R., Lægsgaard, E. & Besenbacher, F. (2007). Long range orientation of metastable atomic hydrogen adsorbate clusters on the graphite (0001) surface, *Chem. Phys. Lett.* 446: 237.

Hornekær, L., Rauls, E., Xu, W., Šljivancanin, Ž., Otero, R., Stensgaard, I., Laegsgaard, E., Hammer, B. & Besenbacher, F. (2006)b. Clustering of chemisorbed h(d) atoms on the graphite (0001) surface due to preferential sticking, *Phys. Rev. Lett.* 97: 186102.

Inui, M., Trugman, S. A. & Abrahams, E. (1994). unusual properties on midband states with off diagonal disorder, *Phys. Rev. B* 49: 3190.

Jackson, B. & Lemoine, D. (2001). Eley-rideal reactions between h atoms on metal and graphite surfaces: The variation of reactivity with substrate, *J. Chem. Phys.* 144: 474.

Jeloaica, L. & Sidis, V. (1999). Dft investigation of the adsorption of atomic hydrogen on a cluster-model graphite surface, *Chem. Phys. Lett.* 300: 157.

Jung, N., Kim, N., Jockusch, S., Turro, N. J., Kim, P. & Brus, L. (2009). Reversible basal plane hydrogenation of graphene, *Nano Lett.* 8: 4597.

Krasheninnikov, A. V. & Nordlund, K. (2010). Ion and electron irradiation-induced effects in nanostructured materials, *J. Appl. Phys.* 107: 071301.

Lebègue, S., Klintenberg, M., Eriksson, O. & Katsnelson, M. I. (2009). Accurate electronic band gap of pure and functionalized graphane from gw calculations, *Phys. Rev. B* 79: 245117.

Lehtinen, P. O., Foster, A. S., Ma, Y., Krasheninnikov, A. V. & Nieminen, R. M. (2004). Irradiation induced magnetism in graphene: a density functional study, *Phys. Rev. Lett.* 93: 187202.

Lieb, E. H. (1989). Two theorems on the hubbard model, *Phys. Rev. Lett.* 62: 1201.

Lin, Y.-M., Dimitrakopoulos, C., Jenkins, K. A., Farmer, D. B., Chiu, H.-Y., Grill, A. & Avouris, P. (2010). 100-GHz Transistors from Wafer-Scale Epitaxial Graphene, *Science* 327(5966): 662–.

Liu, W., Wang, Z. F., Shi, Q. W., Yang, J. & Liu, F. (2009). Band-gap scaling of graphene nanohole superlattices, *Phys. Rev. B* 80: 233405.

Longuet-Higgins, H. C. (1950). Some studies in molecular orbital theory i. resonance structures and molecular orbitals in unsaturated hydrocarbons, *J. Chem. Phys.* 18: 265.

López-Sancho, M. P., de Juan, F. & Vozmediano, M. A. H. (2009). Magnetic moments in the presence of topological defects in graphene, *Phys. Rev. B* 79: 075413.

Luo, Z., Yu, T., Kim, K., You, Y., Lim, S., Shen, Z., Wang, S. & Lin, J. (2009). Thickness dependent reversible hydrogenation graphene layers, *ACS Nano* 3: 1781.

Martinazzo, R., Casolo, S. & Tantardini, G. F. (2010). Symmetry-induced band-gap opening in graphene superlattices, *Phys. Rev. B* 81: 245420.

Martinazzo, R. & Tantardini, G. F. (2005). Quantum effects in an exoergic barrierless reaction at high collision energies, *J. Phys. Chem. A* 109: 9379.

Martinazzo, R. & Tantardini, G. F. (2006a). Quantum study of eley-rideal reaction and cid of hydrogen on a graphite surface. i chemisorbed case, *J. Chem. Phys.* 124: 124702.

Martinazzo, R. & Tantardini, G. F. (2006b). Quantum study of eley-rideal reaction and cid of hydrogen on a graphite surface. ii physisorbed case, *J. Chem. Phys.* 124: 124703.

Meyer, J. C., Girit, C. O., Crommie, M. F. & Zettl, A. (2008). Hydrocarbon lithography on graphene membranes, *Appl. Phys. Lett.* 92: 123110.

Meyer, J. C., Kisielowski, C., Erni, R., Rossell, M. D., Crommie, M. F., & Zettl, A. (2010). Direct imaging of lattice atoms and topological defects in graphene membranes, *Nano Lett.* 8: 3582.

Mirman, R. (1999). *Point groups, space groups, crystals, molecules*, World Scientific, Ltd, Singapore.

Mizes, H. A. & Foster, J. S. (1989). Long-range electronic perturbations caused by defects using stm, *Science* 244: 559.

Morisset, S., Aguillon, F., Sizun, M. & Sidis, V. (2004). Quantum dynamics of h2 formation on a graphite surface through the langmuir hinshelwood mechanism, *J. Chem. Phys.* 121: 6493.

Morisset, S., Aguillon, F., Sizun, M. & Sidis, V. (2005). Wave-packet study of h2 formation on a graphite surface through the langmuir-hinshelwood mechanism, *J. Chem. Phys.* 122: 194702.

Ovchinnikov, A. A. (1978). Multeplicity of the ground state of large alternant organic moleculaes with conjugated bonds, *Theoret. Chim. Acta* 47: 297.

Panchakarla, L. S., Subrahmanyam, K. S., Saha, S., Govindaraj, A., Krishnamurthy, H. R., Wagmre, U. V. & Rao, C. N. R. (2009). Synthesis, structure and propertie of boron and nitrogen doped graphene, *Adv. Mater.* 21: 4726.

Pedersen, T. G., Flindt, C., Pedersen, J., Mortensen, N. A., Jauho, A. & Pedersen, K. (2008). electronic properties graphene antidot lattice, *Phys. Rev. Lett.* 100: 136804.

Pereira, V. M., Guinea, F., Lopes dos Santos, J. M. B., Peres, N. M. R. & Castro Neto, A. H. (2006). Disorder localized states in graphene, *Phys. Rev. Lett.* 96: 036801.

Pereira, V. M., Lopes dos Santos, J. M. B. & Castro Neto, A. H. (2008). Modeling disorder in graphene, *Phys. Rev. B* 77: 115109.

Peres, N. M. R. (2010). The transport properties of graphene: an introduction, *Rev. Mod. Phys.* 82, 2673.

Pogodin, S. & Agranat, I. (2003). Clar goblet and related non-kekulé benzenoid lpahs. a theoretical study, *J. Org. Chem.* 68: 2720.

Pontes, R. B., Fazzio, A. & Dalpian, G. M. (2009). Barrier-free substitutional doping of graphene sheets with boron atoms: ab initio calculations, *Phys. Rev. B* 79: 033412.

Raimondi, M., Simonetta, M. & Tantardini, G. F. (1985). Ab initio valence bond theory, *Comp. Phys. Rep.* 2: 171.

Randić, M. (2003). Aromaticity of polycyclic conjugated hydrocarbons, *Chem. Rev.* 103: 3449.

Rogeau, N., Teillet-Billy, D. & Sidis, V. (2006). Double h atom adsorpition on a cluster model of a graphite surface, *Chem. Phys. Lett.* 431: 135.

Ruffieux, P., Gröning, O., Bielmann, M., Mauron, P., Schlapbach, L. & Gröning, P. (2002a). Hydrogen adsorption on sp2-bonded carbon: Influence of the local curvature, *Phys. Rev. B* 66: 245416.

Ruffieux, P., Gröning, O., Bielmann, M., Mauron, P., Schlapbach, L. & Gröning, P. (2002b). hydrogen adsorption on sp2 bonded carbon: the effect of local curvature, *Phys. Rev. Lett.* 66: 245416.

Ruffieux, R., Gröning, O., Schwaller, P. & Schlapbach, L. (2000). Hydrogen atoms cause long range electronic effects on graphite, *Phys. Rev. Lett.* 84: 4910.

Sakong, S. & Kratzer, P. (2010). Hydrogen vibrational modes on graphene and relaxation of the c-h stretch excitation from first-principles calculations, *The Journal of Chemical Physics* 133(5): 054505.

Schwierz, F. (2010). Graphene transistors, *Nature Nanothech.* 5: 487.

Sha, X. & Jackson, B. (2002). First-principles study of the structural and energetic properties of h atoms on a graphite (0001) surface, *Surf. Sci.* 496: 318.

Shaik, S. & Hiberty, P. C. (2007). *A chemist's guide to valence bond theory*, Wiley.

Shen, T., Wu, Y. Q., Capano, M. A., Rokhinson, L. P., Engel, L. W. & Ye, P. D. (2008). Magnetoconductance oscillations in graphene antidot arrays, *Appl. Phys. Lett.* 93: 122102.

Sofo, J. O., Chaudhari, A. S. & Barber, G. D. (2007). Graphane: A two-dimensional hydrocarbon, *Phys. Rev. B* 75: 153401.

Son, Y.-W., Cohen, M. L. & Louie, S. G. (2006). Energy gaps in graphene nanoribbons, *Phys. Rev. Lett.* 97(21): 216803.

Ugeda, M. M., Brihuega, I., Guinea, F. & Gómez-Rodríguez, J. M. (2010). Missing atom as a source of carbon magnetism, *Phys. Rev. Lett.* 104: 096804.

Wallace, P. R. (1947). The band theory of graphite, *Phys. Rev.* 71: 622.

Wassmann, T., Seitsonen, A. P., Saitta, A. M., Lazzeri, M. & Mauri, F. (2010). Clar's Theory, pi-Electron Distribution, and Geometry of Graphene Nanoribbons, *J. Am. Chem. Soc.* 132(10): 3440–3451.

Yazyev, O. V. & Helm, L. (2007). Defect induced magnetism in graphene, *Phys. Rev. B* 75: 125408.

Zecho, T., Güttler, A. & Küppers, J. (2004). A tds study of d adsorption on terraces and edges of graphite surface, *Carbon* 42: 609.

Zecho, T., Güttler, A., Sha, X., Jackson, B. & Küppers, J. (2002). Adsorption of hydrogen and deuterium atoms on (0001) graphite surface, *J. Chem. Phys.* 117: 8486.

Zhou, S. Y., Gweon, G.-H., Fedorov, A. V., First, P. N., de Heer, W. A., Lee, D.-H., Guinea, F., Castro Neto, A. H. & Lanzara, A. (2007). Substrate-induced bandgap opening in epitaxial graphene, *Nat. Mater.* 6: 770.

4

The Non-Neighbor Effect in Graphene Ribbons

Tong Guo-Ping
Zhejiang Normal University
People's Republic of China

1. Introduction

Graphene is the name given to a flat monolayer of carbon atoms tightly packed into a two-dimensional honeycomb lattice (Novoselov et al., 2004), and is a rapidly rising star on the horizon of materials science and condensed matter physics. This two-dimensional material exhibits exceptionally high crystal and electronic quality and has already revealed a cornucopia of new physics and potential applications. Charge transport properties in graphene are greatly different from that of conventional two-dimensional electronic systems as a consequence of the linear energy dispersion relation near the charge neutrality point (Dirac point) in the electronic band structure (Geim & Novoselov, 2007; Novoselov et al., 2005; Zhang et al., 2005).
Theoretically, the energy band structure of a graphite monolayer had been investigated using the tight-binding approximation (Wallace, 1947). In the work of Wallace, the nearest- and next-nearest-neighbor interaction for the $2p_z$ orbitals in graphene were considered, but the wave function overlap between carbon atoms was neglected. Since his aim is to show how the π-electron distribution is related to the electrical conductivity of graphite, he did not attempt to draw the band distribution. In 1952, Coulson & Taylor considered the overlap integrals between atomic orbitals in studying the band structure of the graphite monolayer. Their work suggested that the overlap was important for the electronic density of states and referred mainly to the π states, leading to a description of the conduction band (Coulson & Taylor, 1952). To study the valence bands in graphene, Lomer used the group-threoretical method to deal with the electronic energy bands based on the three atomic orbitals $2s$, $2p_x$, and $2p_y$ (Lomer, 1955). Because there are two atoms per unit cell, there are six basis functions to be considered, and in general the tight binding model must lead to a 6×6 determinantal secular equation for the energy. The method used group theory is able to solve it easily. Slonczewski and Weiss found that the Lomer's work can be simplified greatly by a different choice of the location of the origin (Slonczewski & Weiss, 1958). A better tight-binding description of graphene was given by Saito et al. (Saito et al., 1998), which considers the nonfinite overlap between nearest neighbors, but includes only interactions between nearest neighbors. To understand the different levels of approximation, Reich et al. started from the most general form of the secular equation, the tight binding Hamiltonian, and the overlap matrix to calculate the band structure (Reich et al., 2002). But their work did not involve the effect of the non-nearest-neighbor interaction on the band structure. This work will be discussed in details in Section 2.
Because there is no energy gap, perfect graphene sheets are metallic. How open the gap of graphene? According to the quantum size effect, graphene nanoribbons maybe achieve this

goal. Early in 1996, Fujita et al. started to study the electronic structure of graphene ribbons (Fijita et al.,1996; Nakada et al., 1996) by the numerical method. For perfect graphene ribbons, the armchair shaped edge ribbons can be either semiconducting (n=3m and n=3m+1, where m is an integer) or metallic (n=3m+2) depending on their widths. First-principles calculations show that the origin of the gaps for the armchair edge nanoribbons arises from both quantum confinement and the deformation caused by edge dangling bonds (Son et al., 2006; Rozhkov et al., 2009). In low-energy approximation, the analytical electronic states of the nanoribbons are studied based on the Dirac equation (Brey & Fertig, 2006). In 2007, Zheng et al. got an analytical expression of the electronic structure, including the boundary relaxation, for the armchair nanoribbon by the tight-binding approximation and hard-wall boundary condition, which only involves the nearest-neighbor hopping integrals (Zheng et al., 2007). In this chapter, we focus on the effects of the non-nearest-neighbor hopping integrals and atomic wave function overlap on the electronic structure, and on the competition between the non-neighbor interaction and edge deformation. The tight-binding energy dispersion relations of graphene nanoribbons, including up to third neighbors, are introduced in section 3. In Section 4, the competition of both is discussed. The stretching deformation of graphene ribbons based on the elastic theory is presented in last section.

2. The non-nearest-neighbor effect in graphene sheets

In this section the tight-binding method is used to study the band structure of the π electrons in graphene. Although this method is simple, it provides a lot of important information for understanding the π electronic band structure. The first tight-binding description for graphene was given by Wallace (Wallace, 1947). He considered nearest- and next-nearest-neighbor interaction for the graphene $2p_z$ orbitals, but neglected the overlap between wave functions centred at different atoms. To compensate for the lack of this work, the non-finite overlap between the basis functions was considered (Saito et al., 1998), but the interaction between nearest neighbors was taken only into account. A better tight-binding description including up to third-nearest neighbors for graphene was given by Reich et al. (Reich et al., 2002).

Let us now consider the band structure from the viewpoint of the tight-binding approximation. The structure of graphene is composed of two types of sublattices A and B as shown in Fig.1. If $\varphi(\boldsymbol{r})$ is the normalized orbital $2p_z$ wave function for an isolated carbon atom, then the wave function of graphene has the form

$$|\psi\rangle = C_A|\psi_A\rangle + C_B|\psi_B\rangle, \tag{1}$$

where

$$|\psi_A\rangle = \frac{1}{\sqrt{N}}\sum_A e^{i\boldsymbol{k}\cdot\boldsymbol{R}_A}\varphi(\boldsymbol{r}-\boldsymbol{R}_A),$$

and

$$|\psi_B\rangle = \frac{1}{\sqrt{N}}\sum_B e^{i\boldsymbol{k}\cdot\boldsymbol{R}_B}\varphi(\boldsymbol{r}-\boldsymbol{R}_B). \tag{2}$$

The first sum is taken over A and all the lattice points generated from it by primitive lattice translation; the second sum is similarly over the points generated from B. Here C_A and C_B are coefficients to be determined, $\boldsymbol{R}_A$ and $\boldsymbol{R}_B$ are the positions of atoms A and B, respectively, and N is the number of the unit cell in a graphene sheet.

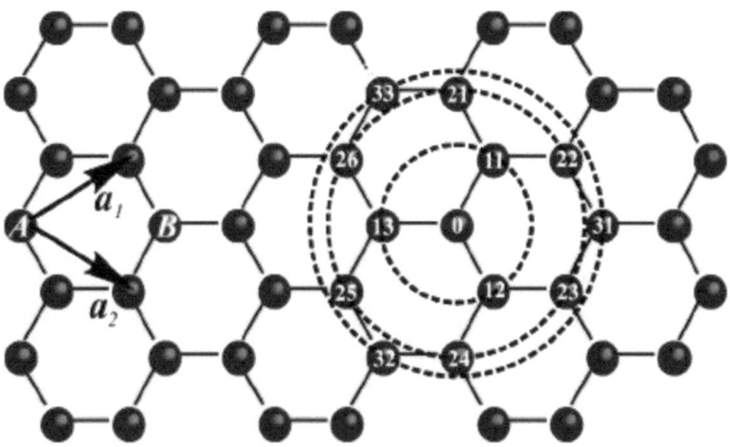

Fig. 1. Structure of a graphene sheet, consisting of sublattices A and B. $\boldsymbol{a}_1$ and $\boldsymbol{a}_2$ are the unit-cell vectors of graphene with a lattice constant a = 0.246 nm. A carbon atom A_0 has three nearest neighbors B_{1i}, six next-nearest neighbors A_{2i}, and three second-nearest neighbors B_{3i}.(Reich et al., 2002)

Under the tight-binding approximation, the Hamiltonian of the perfect system is

$$H = \sum_i \varepsilon |i\rangle\langle i| - \sum_{(i,j)} \gamma_0 |i\rangle\langle j| - \sum_{((i,j))} \gamma_1 |i\rangle\langle j| - \sum_{(((i,j)))} \gamma_2 |i\rangle\langle j| , \tag{3}$$

where (i,j), $((i,j))$, and $(((i,j)))$ denote the nearest, next-nearest, and third neighbors, respectively, ε is the self-energy of the site atom, and γ_0, γ_1, and γ_2 are the nearest-, next-nearest-, and third-neighbor hopping energies. Substituting Eq. (1) in

$$H|\psi\rangle = E|\psi\rangle , \tag{4}$$

and carrying out a simple derivation and finishing, we obtain the secular equation

$$\begin{vmatrix} H_{AA}(\boldsymbol{k}) - E(\boldsymbol{k})S_{AA}(\boldsymbol{k}) & H_{AB}(\boldsymbol{k}) - E(\boldsymbol{k})S_{AB}(\boldsymbol{k}) \\ H^*_{AB}(\boldsymbol{k}) - E(\boldsymbol{k})S^*_{AB}(\boldsymbol{k}) & H_{AA}(\boldsymbol{k}) - E(\boldsymbol{k})S_{AA}(\boldsymbol{k}) \end{vmatrix} = 0 , \tag{5}$$

where $E(\boldsymbol{k})$ are the electronic energy eigenvalues and $\boldsymbol{S}$ is the overlap matrix. In Eq.(5), we have made use of the equivalence of the A and B carbon atoms in the graphene sheet. The solution to Eq.(5) is

$$E_{\pm}(\boldsymbol{k}) = \frac{-(-2E_0 + E_1) \pm \sqrt{(-2E_0 + E_1)^2 - 4E_2E_3}}{2E_3} , \tag{6}$$

with

$$E_0 = H_{AA}S_{AA},\quad E_1 = S_{AB}H^*_{AB} + H_{AB}S^*_{AB},$$

$$E_2 = H^2_{AA} - H_{AB}H^*_{AB},\quad E_3 = S^2_{AA} - S_{AB}S^*_{AB}. \tag{7}$$

The plus sign denotes the conduction band and the minus sign is the valence band. To calculate the Hamiltonian and overlap matrix elements, we derive the third-neighbor tight-binding description. We see from Fig.1 that a carbon atom A has three nearest neighbors B_{1i}, six next-nearest neighbors A_{2i}, and three second-nearest neighbors B_{3i}, all of which belong to the other sublattice.

For third-neighbor interaction the Hamiltonian matrix element H_{AA} can be written as

$$\begin{aligned} H_{AA} &= \langle \psi_A | H | \psi_A \rangle \\ &= \frac{1}{N}\sum_{R_A}\sum_{R'_A} e^{i\mathbf{k}\cdot(\mathbf{R}'_A - \mathbf{R}_A)} \langle \varphi(\mathbf{r}-\mathbf{R}_A) | H | \varphi(\mathbf{r}-\mathbf{R}'_A)\rangle \\ &= \frac{1}{N}\sum_{\mathbf{R}_A} \langle \varphi(\mathbf{r}-\mathbf{R}_A) | H | \varphi(\mathbf{r}-\mathbf{R}_A)\rangle \\ &+ \frac{1}{N}\sum_{\mathbf{R}_A,\mathbf{R}_{Ai}} e^{i\mathbf{k}\cdot\mathbf{R}_{Ai}} \langle \varphi(\mathbf{r}-\mathbf{R}_A) | H | \varphi(\mathbf{r}-\mathbf{R}_A-\mathbf{R}_{Ai})\rangle \\ &= \varepsilon + \frac{1}{N}\sum_{\mathbf{R}_A}\gamma_1\left\{e^{i\mathbf{k}\cdot\mathbf{R}_{A1}} + e^{i\mathbf{k}\cdot\mathbf{R}_{A2}} + e^{i\mathbf{k}\cdot\mathbf{R}_{A3}} + e^{i\mathbf{k}\cdot\mathbf{R}_{A4}} + e^{i\mathbf{k}\cdot\mathbf{R}_{A5}} + e^{i\mathbf{k}\cdot\mathbf{R}_{A6}}\right\} \end{aligned} \tag{8}$$

where $\mathbf{R}_{Ai}$ is the position vectors of six next-nearest neighbor atoms A_i with respect to atom A. Here ε is called the $2p$ orbital energy, or self-energy, of the site atom, which is given by

$$\varepsilon = \langle \varphi(\mathbf{r}-\mathbf{R}_A) | H | \varphi(\mathbf{r}-\mathbf{R}_A)\rangle, \tag{9}$$

and γ_1 is the next-nearest-neighbor hopping integral of π electrons and defined as follows

$$\gamma_1 = \langle \varphi(\mathbf{r}-\mathbf{R}_A) | H | \varphi(\mathbf{r}-\mathbf{R}_A-\mathbf{R}_{Ai})\rangle,\ (i = 1,2,3,4,5,6). \tag{10}$$

In Eq.(8) the maximum contribution to the matrix element H_{AA} is the first term, which comes from the orbital energy of $\mathbf{R}'_A = \mathbf{R}_A$. The next order contribution to H_{AA} is the second term coming from terms of $\mathbf{R}'_A = \mathbf{R}_A + \mathbf{R}_{Ai}$. The other order contribution to this matrix element is very small compared to the first term, which can be neglected. If we define the function

$$\begin{aligned} u(\mathbf{k}) &= 2\cos\mathbf{k}\cdot\mathbf{a}_1 + 2\cos\mathbf{k}\cdot\mathbf{a}_2 + 2\cos\mathbf{k}\cdot(\mathbf{a}_1-\mathbf{a}_2) \\ &= 2\cos 2\pi k_1 a + 2\cos 2\pi k_2 a + 2\cos 2\pi a(k_1 - k_2), \end{aligned} \tag{11}$$

where $k_i = \mathbf{k}\cdot\mathbf{a}_i/2\pi$ are the components of a wave vector $\mathbf{k}$ in units of the reciprocal lattice vectors $\mathbf{k}_1$ and $\mathbf{k}_2$, and $a = |\mathbf{a}_1| = |\mathbf{a}_2| = 0.246$ nm is the lattice constant of graphene, then Eq.(8) can be rewritten as

$$H_{AA} = \varepsilon + \gamma_1 u(\mathbf{k}). \tag{12}$$

Let us next calculate H_{AB}. For this, we shall consider interactions between nearest, and third neighbors in the lattice, the nearest and third neighbors of atoms of type A being always atoms of type B. Therefore, the matrix element H_{AB} is

$$\begin{aligned} H_{AB} &= \langle \psi_A | H | \psi_B \rangle \\ &= \frac{1}{N} \sum_{\boldsymbol{R}_A} \sum_{\boldsymbol{R}_B} e^{i\boldsymbol{k}\cdot(\boldsymbol{R}_B - \boldsymbol{R}_A)} \langle \varphi(\boldsymbol{r} - \boldsymbol{R}_A) | H | \varphi(\boldsymbol{r} - \boldsymbol{R}_B) \rangle \\ &= \gamma_0 (e^{i\boldsymbol{k}\cdot\boldsymbol{R}_{B11}} + e^{i\boldsymbol{k}\cdot\boldsymbol{R}_{B12}} + e^{i\boldsymbol{k}\cdot\boldsymbol{R}_{B13}}) + \gamma_2 (e^{i\boldsymbol{k}\cdot\boldsymbol{R}_{B31}} + e^{i\boldsymbol{k}\cdot\boldsymbol{R}_{B32}} + e^{i\boldsymbol{k}\cdot\boldsymbol{R}_{B33}}), \end{aligned} \tag{13}$$

where the hopping energies are given by

$$\gamma_0 = \langle \varphi(\boldsymbol{r} - \boldsymbol{R}_A) | H | \varphi(\boldsymbol{r} - \boldsymbol{R}_A - \boldsymbol{R}_{B1i}) \rangle \quad (i = 1,2,3), \tag{14}$$

$$\gamma_2 = \langle \varphi(\boldsymbol{r} - \boldsymbol{R}_A) | H | \varphi(\boldsymbol{r} - \boldsymbol{R}_A - \boldsymbol{R}_{B3i}) \rangle \quad (i = 1,2,3). \tag{15}$$

Here $\boldsymbol{R}_{B1i}$ and $\boldsymbol{R}_{B3i}$ are the vectors pointing from atom A to atoms B_{1i} and B_{3i}, respectively. Using the same treatment we can obtain the overlap matrix element

$$\begin{aligned} S_{AA} &= \langle \psi_A | \psi_A \rangle \\ &= \frac{1}{N} \sum_{\boldsymbol{R}_A} \sum_{\boldsymbol{R}'_A} e^{i\boldsymbol{k}\cdot(\boldsymbol{R}'_A - \boldsymbol{R}_A)} \langle \varphi(\boldsymbol{r} - \boldsymbol{R}_A) | \varphi(\boldsymbol{r} - \boldsymbol{R}'_A) \rangle \\ &= 1 + \frac{1}{N} \sum_{\boldsymbol{R}_A} \sum_{i} e^{i\boldsymbol{k}\cdot\boldsymbol{R}_{A2i}} \langle \varphi(\boldsymbol{r} - \boldsymbol{R}_A) | \varphi(\boldsymbol{r} - \boldsymbol{R}_A - \boldsymbol{R}_{A2i}) \rangle \\ &= 1 + s_1 u(\boldsymbol{k}) \end{aligned} \tag{16}$$

with

$$s_1 = \langle \varphi(\boldsymbol{r} - \boldsymbol{R}_A) | \varphi(\boldsymbol{r} - \boldsymbol{R}_A - \boldsymbol{R}_{A2i}) \rangle \quad (i = 1,2,3,4,5,6), \tag{17}$$

where $\boldsymbol{R}_{A2i}$ are the vectors pointing from atom A to atoms A_{2i}, and s_1 is the overlap of atomic wave functions between next-nearest neighbors. Similarly, we have

$$\begin{aligned} S_{AB} &= \langle \psi_A | \psi_B \rangle \\ &= \frac{1}{N} \sum_{\boldsymbol{R}_A} \sum_{\boldsymbol{R}_B} e^{i\boldsymbol{k}\cdot(\boldsymbol{R}_B - \boldsymbol{R}_A)} \langle \varphi(\boldsymbol{r} - \boldsymbol{R}_A) | \varphi(\boldsymbol{r} - \boldsymbol{R}_B) \rangle \\ &= s_0 (e^{i\boldsymbol{k}\cdot\boldsymbol{R}_{B11}} + e^{i\boldsymbol{k}\cdot\boldsymbol{R}_{B12}} + e^{i\boldsymbol{k}\cdot\boldsymbol{R}_{B13}}) + s_2 (e^{i\boldsymbol{k}\cdot\boldsymbol{R}_{B31}} + e^{i\boldsymbol{k}\cdot\boldsymbol{R}_{B32}} + e^{i\boldsymbol{k}\cdot\boldsymbol{R}_{B33}}), \end{aligned} \tag{18}$$

where s_0 and s_2 are the overlap integrals between nearest, and third neighbors, which are given by

$$s_0 = \langle \varphi(\boldsymbol{r} - \boldsymbol{R}_A) | \varphi(\boldsymbol{r} - \boldsymbol{R}_A - \boldsymbol{R}_{B1i}) \rangle \quad (i = 1,2,3), \tag{19}$$

and

$$s_2 = \langle \varphi(\boldsymbol{r} - \boldsymbol{R}_A) | \varphi(\boldsymbol{r} - \boldsymbol{R}_A - \boldsymbol{R}_{B3i}) \rangle \quad (i = 1,2,3). \tag{20}$$

Substituting Eqs.(12), (13), (16), and (18) into Eq.(7) yields

$$E_0 = [\varepsilon + \gamma_1 u(\boldsymbol{k})][1 + s_1 u(\boldsymbol{k})]\,, \tag{21}$$

$$E_1 = 2s_0\gamma_0[3 + u(\boldsymbol{k})] + (s_0\gamma_2 + s_2\gamma_0)g(\boldsymbol{k}) + 2s_2\gamma_2[3 + u(2\boldsymbol{k})]\,, \tag{22}$$

$$E_2 = [\varepsilon + \gamma_1 u(\boldsymbol{k})]^2 - \gamma_0^2[3 + u(\boldsymbol{k})] - \gamma_0\gamma_2 g(\boldsymbol{k}) - \gamma_2^2[3 + u(2\boldsymbol{k})]\,, \tag{23}$$

$$E_3 = [1 + s_1 u(\boldsymbol{k})]^2 - s_0^2[3 + u(\boldsymbol{k})] - s_0 s_2 g(\boldsymbol{k}) - s_2^2[3 + u(2\boldsymbol{k})]\,, \tag{24}$$

where

$$g(\boldsymbol{k}) = 2u(\boldsymbol{k}) + u(2k_1 - k_2, k_1 - 2k_2)\,.$$

Inserting E_0 to E_3 into Eq.(6) we can obtain the tight-binding energy dispersion relation in the third-neighbor approximation.

To give the numerical results of energy dispersion, we must know the values of the hopping energies and overlap integrals. We take the parameters $\varepsilon = -0.28$ eV, $\gamma_0 = -2.97$ eV, $\gamma_1 = -0.073$ eV, $\gamma_2 = -0.33$ eV, $s_0 = 0.073$, $s_1 = 0.018$, and s_2=0.026 (Reich et al., 2002). The computed results for some high-symmetry points (*KΓM*) are shown in Fig. 2, where the solid line denotes the nearest-neighbor result, the dashed line represents the next-nearest-neighbor, and the dotted line is the third-neighbor. It is clear that the next-nearest-neighbor hopping integrals and overlap between atomic wave functions will play an important role on the band width at *Γ* point, which can largely reduce the bandwidth, and the third-neighbor interaction can slightly enhance the bandwidth. But the role of both is just opposite for *M* point. It is worth pointing out that when we take only into account the nearest neighbor hopping integral and let both the overlap s_0 and the site energy ε be zero, the energy bands are symmetric with respect to the Fermi level. The nearest neighbor result in Fig. 2 is to include the overlap s_0, so the energy bands become asymmetric, leading to the

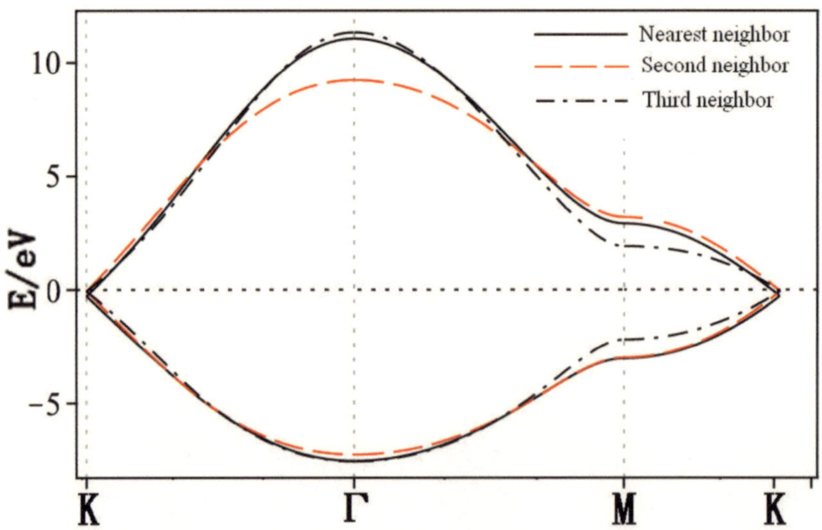

Fig. 2. Tight-binding energy bands of graphene for high-symmetry points. The solid line denotes the nearest neighbor, the dashed line represents the next-nearest neighbor, and the dotted line is the third neighbor.

Fermi level down slightly. This means that the overlap integral is important for the band structure. Hence, the non-nearest neighbor hopping and overlap integral need to be considered in calculations of the energy band. Since there is no energy gap at K point (Dirac point), the graphene sheet is metallic.

3. The non-nearest-neighbor effect in graphene nanoribbons

3.1 Armchair nanoribbons

As mentioned in Section 2, for a graphene sheet there are no energy gaps at the Dirac points. How to open the energy gaps of graphene? One method is to deduce the size of graphene and let it become a narrow ribbon, compared to the length of the ribbon. From this, the band gaps will change with their widths (Son et al., 2006), and the more narrow the width is, the larger the gap. Another effective approach is to change the bond length of graphene by exerting a strain force. Besides, we can also open the gap by using absorption atoms on graphene or doping impurity in. For an armchair ribbon, the analytical solution of electronic dispersion has been given based on the tight-binding approach, but the dispersion obtained is in the framework of the nearest-neighbor interaction (Zheng et al, 2007). In this section we only discuss the electronic dispersion of perfect graphene nanoribbons without any edge deformation within the tight-binding approximation and the third-neighbor interaction is taken into account.

We choose the ribbon to be macroscopically large along the x direction but finite in the y direction, which leads to a graphene nanoribbon with armchair edges. Since the ribbon is very long compared to its width and has the translational symmetry along the x direction, we can choose the plane-wave basis along the x direction and take the stationary wave in the y direction because the electronic bahavior is limited to the space between two edges. The structure of armchair nanoribbons consists of two types of sublattices A and B, and the unit cell contains n A-type carbon atoms and n B-type atoms as illustrated in Fig.3.

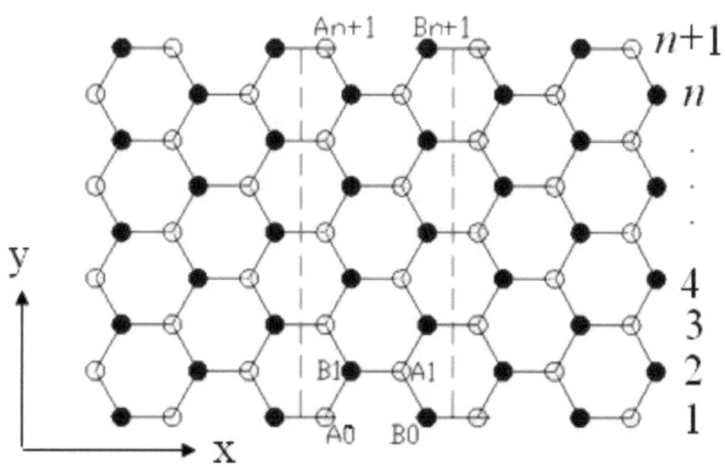

Fig. 3. Structure of an armchair graphene ribbon with sublattices A (empty) and B (solid). The ribbon width is denoted by numbers n. Every unit cell has n numbers of A and B sublattices. Assume that the edges of the ribbon are a hard wall.

Within the tight-binding approximation, the wave functions of A and B sublattices can be written as

$$|\psi_A\rangle = \frac{1}{N_A}\sum_{j=1}^{n}\sum_{x_{A_j}} e^{ik_x x_{A_j}}\phi_A(j)|\varphi(\boldsymbol{r}-\boldsymbol{R}_{A_j})\rangle, \tag{25}$$

$$|\psi_B\rangle = \frac{1}{N_B}\sum_{j=1}^{n}\sum_{x_{B_j}} e^{ik_x x_{B_j}}\phi_B(j)|\varphi(\boldsymbol{r}-\boldsymbol{R}_{B_j})\rangle, \tag{26}$$

where N_A and N_B are the normalized coefficients, $\phi_A(j)$ and $\phi_B(j)$ are the components for A and B sublattices in the y direction and satisfy the following hard-wall boundary conditions

$$\phi_A(0) = \phi_B(0) = 0\,,$$

$$\phi_A(n+1) = \phi_B(n+1) = 0\,. \tag{27}$$

Assume that the stationary wave has the form

$$\phi_A(j) = \phi_B(j) = \sin\left(\frac{1}{2}\sqrt{3}ak_y j\right),\ (j = 1,2,\cdots,n) \tag{28}$$

leading to a discretized wave vector in the y direction

$$k_y(q) = \frac{2\pi q}{\sqrt{3}a(n+1)},\ (q = 1,2,\cdots,n)\,. \tag{29}$$

To find out the normalized coefficients in Eqs.(25) and (26), we introduce the normalization condition

$$\langle\psi_A|\psi_A\rangle = \langle\psi_B|\psi_B\rangle = 1\,, \tag{30}$$

from which, we get

$$N_A = N_B = \sqrt{N_x(n+1)/2}\,,$$

where N_x is the number of unit cells along the x direction. Therefore, the total wave function of the graphene ribbon can be written as

$$\begin{aligned}|\psi\rangle &= C_A|\psi_A\rangle + C_B|\psi_B\rangle \\ &= C_A\sqrt{\frac{2}{N_x(n+1)}}\sum_{j=1}^{n}\sum_{x_{A_j}} e^{ik_x x_{A_j}}\sin\left(j\sqrt{3}k_y a/2\right)|\varphi(\boldsymbol{r}-\boldsymbol{R}_{A_j})\rangle \\ &+ C_B\sqrt{\frac{2}{N_x(n+1)}}\sum_{j=1}^{n}\sum_{x_{B_j}} e^{ik_x x_{B_j}}\sin\left(j\sqrt{3}k_y a/2\right)|\varphi(\boldsymbol{r}-\boldsymbol{R}_{B_j})\rangle\,.\end{aligned} \tag{31}$$

Substituting Eqs.(3) and (31) into the Schrodinger equation leads to an energy dispersion relation of the form

$$E_{\pm}(k_x,q)=\frac{-(-2E_0+E_1)\pm\sqrt{(-2E_0+E_1)^2-4E_2E_3}}{2E_3}, \tag{32}$$

where

$$E_0(k_x,q)=\{\varepsilon+\gamma_1[f(k_x,q)-3]\}\{[1+s_1[f(k_x,q)-3]\}, \tag{33}$$

$$E_1(k_x,q)=2s_0\gamma_0 f(k_x,q)+(s_0\gamma_2+s_2\gamma_0)g(k_x,q)+2s_2\gamma_2 h(k_x,q), \tag{34}$$

$$E_2(k_x,q)=\{\varepsilon+\gamma_1[f(k_x,q)-3]\}^2-\gamma_0^2 f(k_x,q)-\gamma_0\gamma_2 g(k_x,q)-\gamma_2^2 h(k_x,q), \tag{35}$$

$$E_3(k_x,q)=\{1+s_1[f(k_x,q)-3]\}^2-s_0^2 f(k_x,q)-s_0 s_2 g(k_x,q)-s_2^2 h(k_x,q), \tag{36}$$

and

$$f(k_x,q)=1+4\cos\left(\frac{q\pi}{n+1}\right)\cos\left(\frac{3k_x a}{2}\right)+4\cos^2\left(\frac{q\pi}{n+1}\right),$$

$$h(k_x,q)=1+4\cos\left(\frac{2q\pi}{n+1}\right)\cos(3k_x a)+4\cos^2\left(\frac{2q\pi}{n+1}\right),$$

$$g(k_x,q)=2f(k_x,q)+2\cos\left(\frac{3k_x a}{2}+\frac{3q\pi}{n+1}\right)+2\cos\left(\frac{3k_x a}{2}-\frac{3q\pi}{n+1}\right)+2\cos(3k_x a)-6.$$

The electronic dispersion given by Eq. (32), in form, is exactly the same as that found for a graphene sheet, but both have the difference in nature (Jin et al., 2009). The region $-\pi/3\le k_x a\le\pi/3$ is within the first Brillouin zone. These results are valid for various energy ranges.

Since the electronic structure of perfect armchair graphene nanoribbons depends strongly on the width of the ribbon, the system, for instance, is metallic when n=3m+2 (m is an integer) and is insulating otherwise. To give a graph of energy bands, we still use the same parameter values as taken in a graphene sheet. The electronic energy bands of the armchair nanoribbons with three different widths are plotted in Fig.4, where (a) is n=3m=6, (b) n=3m+1=7, and (c) n=3m+2=8. Labels (1), (2), and (3) denote the nearest, next-nearest, and third neighbors, respectively. As $n=6$ and $n=7$, the armchair ribbons appear insulating. Fig.4 shows that the next-nearest-neighbor hopping and overlap integral would give rise to change of the energy band width. This is because the energy levels of the conduction band top are squeezed, which correspond to the stationary waves with small q values. The third neighbors not only affect the energy gaps, such as n=6 and n=7, but also the band widths. The influence on the band width mainly is because the bands related to the standing waves with small q values in conduction and valence band produce a larger bend and this effect was particularly evident when n=7,8. However, the effect on the energy gaps is because the bands corresponding to larger q go down slightly. It is worth noting that when $n=7$, there is a flat conduction or valence band, taking not into account the third neighbors, which corresponds to the quantum number q=(n+1)/2. Such a flat band is independent of wave vector k_x and in general exists only when n is equal to odd. As for $n=8$, the lowest conduction band and the upmost valence band touch at Dirac point, which leads to the metallic behavior of the armchair ribbon.

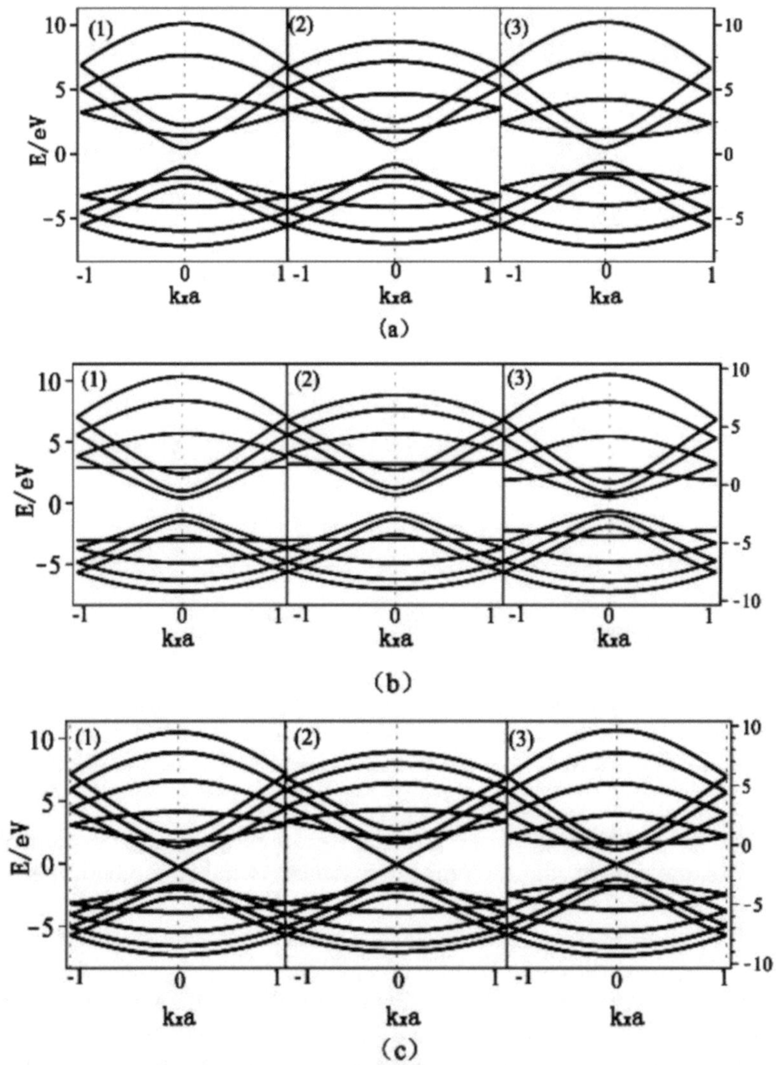

Fig. 4. Energy bands of perfect armchair graphene nanoribbons with widths (a) n=6, (b) n=7, and (c) n=8. Labels (1), (2), and (3) represent the nearest, next-nearest, and third neighbors, respectively.

3.2 Zigzag nanoribbons

The spectrum of graphene nanoribbons depends on the nature of their edges: zigzag or armchair. In Fig. 5, we show a honeycomb lattice having zigzag edges along the x direction and armchair edges along the y direction, where the solid cycles denote the sublattice A and the empty is B. If we choose the ribbon to be infinite in the x direction, we produce a graphene nanoribbon with zigzag edges. It is interesting to note that the atoms at each edge are of the same sublattice (B on the top edge of Fig. 5 and A on the bottom).

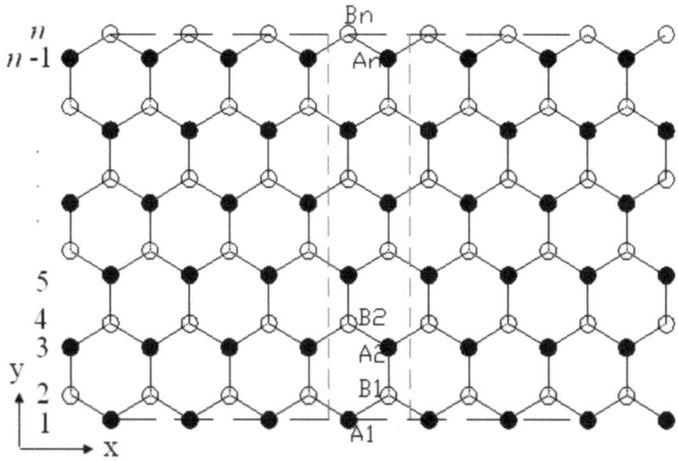

Fig. 5. The structure of a zigzag graphene nanoribbon, having the translational symmetry along the *x* direction. Atoms enclosed in the vertical rectangle represent the unit cell. *n* is the row number of atoms. The solid cycles denote the sublattice *A* and the empty is *B*.

We now calculate the electronic energy bands of the zigzag ribbon as shown in Fig.5 by using the tight-binding approach including up to third-nearest neighbors. To do this, let us label the sublattices *A* and *B* with number, respectively, and let A_1, A_2,..., A_n correspond to odd labels 1,3,5,...,2*n*-1 and B_1, B_2,..., B_n to even labels 2,4,...,2*n*. From this, the Hamiltonian becomes a 2*n*×2*n* matrix, which is given by

$$\mathbf{H}=\begin{bmatrix} f_2 & f_0 & f_1 & \gamma_2 & 0 & 0 & \cdots & \cdots \\ f_0 & f_2 & f_3 & f_1 & 0 & 0 & \cdots & \cdots \\ f_1 & f_3 & f_2 & f_0 & f_1 & \gamma_2 & \cdots & \cdots \\ \gamma_2 & f_1 & f_0 & f_2 & f_3 & f_1 & \cdots & \cdots \\ 0 & 0 & f_1 & f_3 & f_2 & f_0 & f_1 & \gamma_2 \\ 0 & 0 & \gamma_2 & f_1 & f_0 & f_2 & f_3 & f_1 \\ \cdots & \cdots & \cdots & \cdots & \cdots & \cdots & \cdots & \cdots \end{bmatrix}, \tag{37}$$

where f_0, f_1, f_2,and f_3 are the Hamiltonian matrix elements, which are given by

$$f_0=\langle\psi_j|H|\psi_i\rangle=2\gamma_0\cos\left(\frac{\sqrt{3}k_xa}{2}\right)\qquad(j=i-1), \tag{38}$$

$$f_1=\langle\psi_j|H|\psi_i\rangle=2\gamma_1\cos\left(\frac{\sqrt{3}k_xa}{2}\right)\qquad(j=i\pm2), \tag{39}$$

$$f_2=\langle\psi_j|H|\psi_i\rangle=\varepsilon+2\gamma_1\cos\left(\sqrt{3}k_xa\right)\qquad(j=i), \tag{40}$$

$$f_3=\langle\psi_j|H|\psi_i\rangle=\gamma_0+2\gamma_2\cos\left(\sqrt{3}k_xa\right)\qquad(j=i+1). \tag{41}$$

γ_0, γ_1, and γ_2 are the nearest-, next-nearest-, and third-neighbor electronic hopping amplitudes, respectively. Similarly, the overlap matrix *S* can be written as

$$S=\begin{bmatrix} g_2 & g_0 & g_1 & s_2 & 0 & 0 & \cdots & \cdots \\ g_0 & g_2 & g_3 & g_1 & 0 & 0 & \cdots & \cdots \\ g_1 & g_3 & g_2 & g_0 & g_1 & s_2 & \cdots & \cdots \\ s_2 & g_1 & g_0 & g_2 & g_3 & g_1 & \cdots & \cdots \\ 0 & 0 & g_1 & g_3 & g_2 & g_0 & g_1 & s_2 \\ 0 & 0 & s_2 & g_1 & g_0 & g_2 & g_3 & g_1 \\ \cdots & \cdots & \cdots & \cdots & \cdots & \cdots & \cdots & \cdots \end{bmatrix}, \quad (42)$$

where

$$g_0=\langle\psi_j|\psi_i\rangle=2s_0\cos\left(\frac{\sqrt{3}k_xa}{2}\right)\quad (j=i-1), \quad (43)$$

$$g_1=\langle\psi_j|\psi_i\rangle=2s_1\cos\left(\frac{\sqrt{3}k_xa}{2}\right)\quad (j=i\pm 2), \quad (44)$$

$$g_2=\langle\psi_j|\psi_i\rangle=1+2s_1\cos\left(\sqrt{3}k_xa\right)\quad (j=i), \quad (45)$$

$$g_3=\langle\psi_j|\psi_i\rangle=s_0+2s_2\cos\left(\sqrt{3}k_xa\right)\quad (j=i+1). \quad (46)$$

Here s_0, s_1, and s_2 are the nearest-, next-nearest-, and third-neighbor overlap integrals between the $2p_z$ orbitals, respectively. Substituting Eqs.(36) and (37) into the following secular equation

$$\det\left[\mathbf{H}-E\mathbf{S}\right]=0, \quad (47)$$

we can obtain all *n* eigenvalues of E_i (k_x) (i=1,…,2n) for a given wave vector k_x. The electronic dispersion relations (or energy bands) of zigzag nanoribbons are shown in Fig. 6.

In order to conveniently compare with the third-neighbor result, we also give the nearest- and next-nearest-neighbor electronic energy bands together with it. In Fig. 6(a) and (b), the left is the nearest-neighbor result, the middle is the next-nearest-neighbor, and the right is the third-neighbor for the ribbon widths *n*=4 and *n*=10. We see from Fig.6 that the zigzag graphene nanoribbons are metallic and the energy bands are wide (more than 10eV), and the spacing between the energy bands is decreased as increasing of the width *n*. When the nearest neighbor interaction is taken only into account, the energy band structure is symmetrical (see Fig.6 (1)). But the next-nearest-neighbor hopping and overlap can make the energy bands become nonsymmetrical, i.e. the conduction band becomes narrowed and the valence band is widened. It is obvious that the top of the conduction band is pressed downward and the bottom of the valence band is pulled downward. However, the effect of the third neighbors on the band structure is the same as that of the next-nearest neighbors, but the latter is stronger than the former.

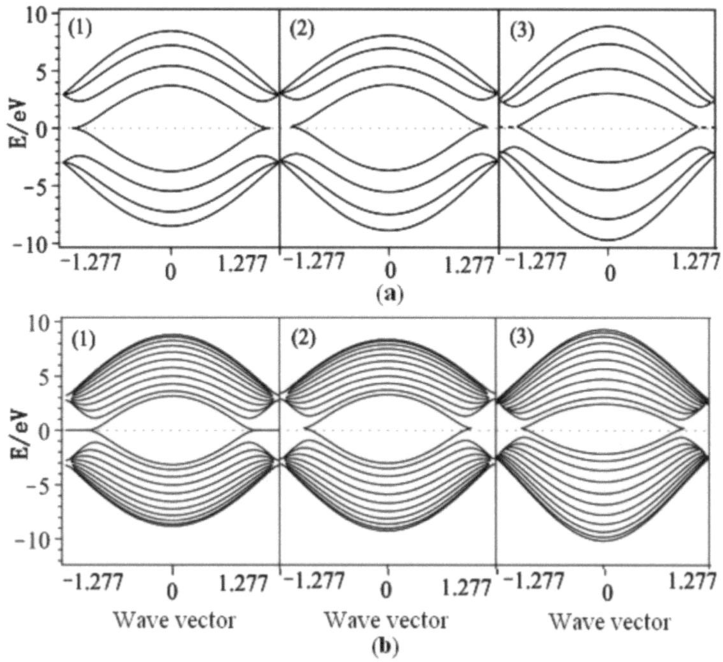

Fig. 6. Energy bands of zigzag nanoribbons with widths (a) n=4 and (b) n=10. Labels (1), (2), and (3) refer to the nearest, next-nearest, and third neighbors, respectively.

On the other hand, we see that the highest valence band state and the lowest conduction band state for the zigzag ribbons are always degenerate at k_x = 1.277. A pair of almost flat bands appears within the region of $0.851 \leq |k_x| \leq 1.277$ where the bands sit in the very vicinity of the Fermi level. This phenomenon arises only in the nearest neighbor result and does not occur in the non-nearest neighbor case. The degeneracy of the center bands at k_x = 1.277 does not originate from the intrinsic band structure, and the corresponding wave functions are completely localized on the edge sites (Nakada et al., 1996).

Based on the above discussion, we conclude that the effect of the third-neighbor terms on the energy band of the zigzag ribbon is large compared to that of the next-nearest-neighbor terms. Therefore, it is important to include the third neighbors when we calculate the bands. This is because the distance between the next-neighbor carbon atoms is very close to that between the third-neighbor atoms.

4. Competition between the non-neighbor interaction and edge deformation

4.1 Energy gaps in armchair nanoribbons

We now discuss the change of the energy gaps in armchair graphene nanoribbons. The results of first-principles calculations (Son et al., 2006) show that the differences among three widths (n=3m, n=3m+1, and n=3m+2) are quite apparent.

Our aim is to take into account the non-neighbor hopping integral and overlap and to understand their contribution to the band gap. To do this, we need to derive the formulas of the band gaps from Eq. (31). After a simple derivation, we easily obtain the following band gap formulas (Jin et al., 2009)

$$\Delta^0 = -\frac{2\sqrt{(-\gamma_0 + \varepsilon s_0)^2 f}}{s_0^2 f - 1}, \tag{48}$$

$$\Delta^1 = -\frac{2\sqrt{f(\gamma_0 f s_1 + 3\gamma_1 s_0 - \varepsilon s_0 - f\gamma_1 s_0 - 3\gamma_0 s_1 + \gamma_0)^2 f}}{s_1^2 f^2 - 6 f s_1^2 + 2 f s_1 + 9 s_1^2 - 6 s_1 + 1 - s_0^2 f}, \tag{49}$$

$$\Delta^2 = \frac{\sqrt{4u_1^2u_2^2 - 4u_1u_2u_5 + u_5 - 4(u_1^2 - f\gamma_0^2 - \gamma_0\gamma_2 g - \gamma_2^2 h)(u_2^2 - u_3)}}{(u_2^2 - u_3)}, \tag{50}$$

where

$$u_1 = \varepsilon + \gamma_1(f - 3),$$

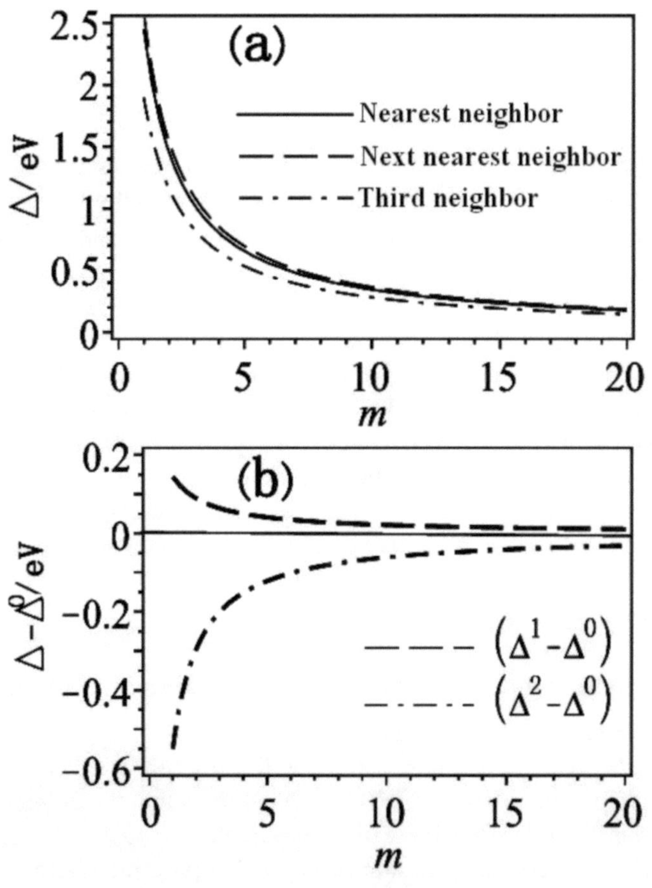

Fig. 7. The variation of band gaps of armchair graphene nanoribbons for n=3m as a function of width m. (a) Band gap curves for the nearest, next nearest, and third neighbors. (b) Band gap differences between the nearest and next nearest, and third neighbors.

$$u_2 = 1 + s_1(f - 3),$$

$$u_3 = s_0^2 f + s_0 s_2 g + s_2^2 h,$$

$$u_4 = (s_0 \gamma_2 + s_2 \gamma_0) g,$$

$$u_5 = 2 s_0 \gamma_0 f + u_4 + 2 s_2 \gamma_2 h.$$

Here Δ^0, Δ^1, and Δ^2 represent the band gaps of the nearest-, next-nearest-, and third-neighbor interaction, respectively. Quantities *f*, *h*, and *g* have been defined in Section 3. Since *f*, *h*, and *g* depend on the ribbon width *n*, the different width has the different band gap.
In order to illustrate the problem, we take *n*=3*m* as an example to discuss the change of the gap with width. The results are summarized as function of width *m*, as shown in Fig.7, where the energy gap differences between the nearest and next-nearest neighbors and third neighbors are given together with. If we only consider the second-nearest-neighbor hopping and overlap, the difference between the nearest and next nearest neighbors is small for large *m*, but larger for small *m*. However, when we account to third neighbors, their contribution to the energy gap is large compared to that of the second neighbor, especially for small *m*.

4.2 Effect of the edge deformation on the energy gap

Because every carbon atom on the edge has one dangling bond unsaturated, the edge atoms of armchair nanoribbons are passivated by hydrogen atoms in general so that the σ bonds between hydrogen and carbon and the on-site energies of the carbons at the edges would be different from those in the middle of the ribbon. The bonding distances and angles between carbon atoms at the edges are also expected to change dramatically, which leads to considerable variations of electronic structure, especially within the low-energy range (Son et al., 2006). The bond lengths between carbon atoms at the edges are predicted to vary about 3-4% when hydrogenerated. Correspondingly, the hopping integral increases about 12% extracted from the analytical tight-binding expression (Son et al., 2006; Porezag et al., 1995).
To see the consequence of such effects more clearly, we introduce a simpler edge-deformed model, in which the Hamiltonian of the ribbon with deformation on the edge can be written as

$$H = \sum_i \varepsilon_i |i\rangle\langle i| - \sum_{(i,j)} (\gamma_0 + \delta\gamma_{0ij})|i\rangle\langle j| - \sum_{((i,j))} (\gamma_1 + \delta\gamma_{1ij})|i\rangle\langle j| - \sum_{(((i,j)))} (\gamma_2 + \delta\gamma_{2ij})|i\rangle\langle j|. \quad (51)$$

As mentioned above, the variation of the next-nearest and third neighbor hopping integrals can be neglected for smaller deformation, i.e., $\delta\gamma_{1ij} = \delta\gamma_{2ij} = 0$. Let the variation of the hopping integral and the on-site energy of the *i*th carbon atom be $\delta\gamma_{0ij}$ and ε_i, respectively. Therefore, Eq.(51) cab be rewritten as

$$H = H_{0i} + H_i, \quad (52)$$

where

$$H_{0i} = \sum_i \varepsilon_i |i\rangle\langle i| - \sum_{(i,j)} \gamma_0 |i\rangle\langle j| - \sum_{((i,j))} \gamma_1 |i\rangle\langle j| - \sum_{(((i,j)))} \gamma_2 |i\rangle\langle j|, \quad (53)$$

$$H_i = -\sum_{(i,j)} \delta\gamma_{0ij} \left|i\right\rangle\left\langle j\right| . \tag{54}$$

The energy dispersion relation corresponding to the Hamiltonian H_{0i} still is the same as Eq.(32) in form, where ε is replaced by ε_i . For convenience, we rewrite Eq.(32) as follows

$$E_{\pm}^{0}(k_x, q) = \alpha \pm \left|\lambda\right| , \tag{55}$$

where α and λ are dependent of the parameters $\varepsilon_i, \gamma_0, s_0, \gamma_1, s_1$, and so on. Since H_i is small compared to H_{0i}, we can solve Eq.(52) by using the perturbation approach. Thus, a new dispersion relation is

$$E_{\pm}(k_x, q) = \beta \pm \left|\lambda + \delta\lambda\right| , \tag{56}$$

where β is the energy shift originating from the variation of the on-site energy and $\delta\lambda$ is the shift originating in the hopping integral variation. If the nearest neighbor interaction is involved only (Zheng et al., 2007), then β and $\delta\lambda$ are given by

$$\beta = \frac{2}{n+1}\sum_{i=1}^{n} \varepsilon_i \sin^2\left(\frac{q\pi}{n+1} i\right) , \tag{57}$$

$$\begin{aligned} \delta\lambda = -\frac{2}{n+1}\sum_{i=1}^{n} \Bigg[& \delta\gamma_{0A(i)B(i)} \sin^2\left(\frac{q\pi}{n+1} i\right) e^{-ik_x a} \\ & + \delta\gamma_{0A(i)B(i-1)} \sin\left(\frac{q\pi}{n+1} i\right) \sin\left(\frac{q\pi}{n+1}(i-1)\right) e^{ik_x a/2} \\ & + \delta\gamma_{0A(i)B(i+1)} \sin\left(\frac{q\pi}{n+1} i\right) \sin\left(\frac{q\pi}{n+1}(i+1)\right) e^{ik_x a/2} \Bigg] . \end{aligned} \tag{58}$$

These results are valid for small edge deformations, atoms or molecules attached to edge carbon atoms. As long as given the deformation distribution function, we can obtain the energy dispersion relation of the edge deformation.

Assume that the deformation is very small and localized along two edges (Son et al., 2006), from Eqs.(55) and (56), we can obtain the differences between the energy gaps to the first order in $\delta\gamma_0$ and ε for different width ribbons as follows(Son et al., 2006; Zheng et al., 2007)

$$\Delta_{e3m} = \Delta_{3m} - \Delta_{3m}^{0} = -\frac{8\delta\gamma_0}{3m+1}\sin^2\frac{m\pi}{3m+1} , \tag{59}$$

$$\Delta_{e3m+1} = \Delta_{3m+1} - \Delta_{3m+1}^{0} = \frac{8\delta\gamma_0}{3m+2}\sin^2\frac{(m+1)\pi}{3m+2} , \tag{60}$$

$$\Delta_{e3m+2} = \Delta_{3m+2} - \Delta_{3m+2}^{0} = \frac{2\delta\gamma_0}{m+1} . \tag{61}$$

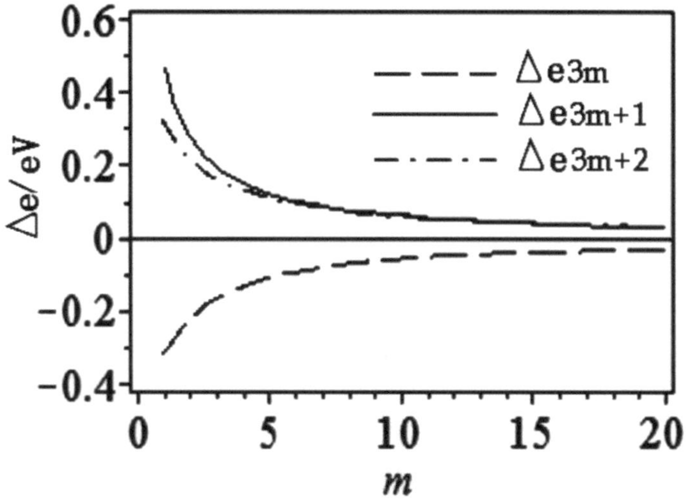

Fig. 8. The energy gap differences before and after deformation for armchair graphene ribbons. Solid, dashed, and dotted lines are the results of n=3m+1, n=3m, and n=3m+2, respectively.

Here the energy gap refers to the difference between the lowest conduction band and the highest valence band. Δ_{3m}^{0}, Δ_{3m+1}^{0}, and Δ_{3m+2}^{0} are the energy gaps of non-deformed ribbons. This result shows that all armchair graphene ribbons with edge deformation have nonzero energy gaps.

For smaller deformation, we set the hopping integral change $\delta\gamma_0/\gamma_0 = 0.12$ (Son et al., 2006). A graph of the gap difference vs. width m is shown in Fig.8. This implies that the 12% increase of the hopping integrals between carbon atoms at the edges opens the gaps of the (3m+2) armchair ribbons and decreases (increases) the gaps of 3m-armchair ribbons ((3m+1)-armchair ribbons). In order to facilitate comparison, we take n=3m as an example. By comparing Fig.8 with Fig.7 (b), we see that the next-nearest neighbor effect is able to make the gap increase slightly with respect to the nearest neighbor case and the third-neighbor interaction would lead to decrease of the gap, and the smaller edge deformation would reduce the gap. Therefore, the competition results of both are that the effect of the boundary relaxation opposes the change of the next-nearest-neighbor interaction and strengthens the change of the third-neighbor interaction. The n=3m+1 situation is just opposite to the n=3m. For n=3m+2, the non-neighbor interaction does not change the gap and keeps this zero gap unchanged. Hence, there is no competition between the both. In fact, the edge deformation would have a penetration depth (Zheng et al., 2007). Since the depth is very small, our conclusions obtained above still are valid for this case.

5. Stretching deformation of graphene ribbons

In this section, we discuss the deformation of graphene due to an external force and effect of the deformation on the band gap. Assume that the length L of a graphene sheet is long compared to its width W, i.e. $L > W$, a wider ribbon satisfying translational symmetry in the length and width directions, and the force between carbon atoms satisfies Hook's law. We exert a tension force on the two edges of the graphene, as illustrated in Fig.9.

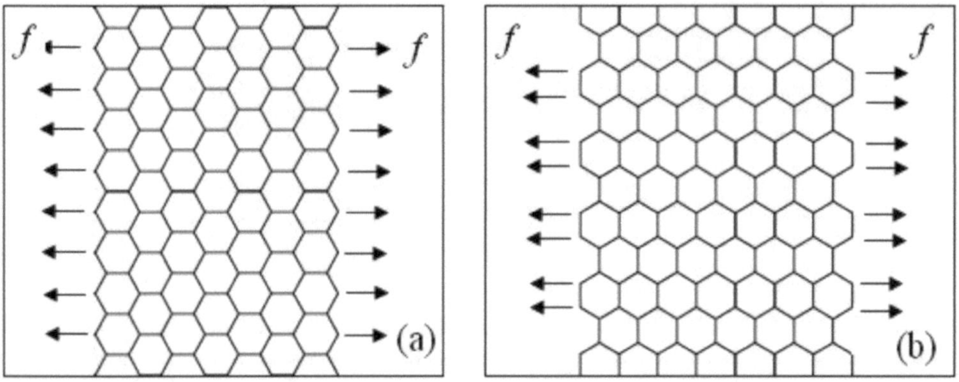

Fig. 9. Force on wider graphene ribbons with (a) zigzag and (b) armchair edges.

Let the force on each atom and lattice spring constant be f and k, respectively. The bond lengths and angles will change with the force and thereby lead to the change of hopping integrals, which causes the variation or opening of the energy gap. For zigzag edge ribbons, when an external force is much less than the stationary spring force between two neighbor atoms, the relation between the force f on each atom and bond half-angle α can be obtained

$$f_N = \frac{f}{kd_0} = \sqrt{3}\cot\alpha - 2\cos\alpha \tag{62}$$

Here d_0 is the original bond length, f_N is the dimensionless force on each atom. For small deformation, the bond half-angle is given by

$$\alpha = \frac{\pi}{3} + \frac{\sqrt{3}}{6}\left[(1+f_N) - \sqrt{1+14f_N+f_N^2}\right]. \tag{63}$$

Based on elastic mechanics, the deformed bond lengths are written as

$$d_1 = d_0(1+f_N),$$

$$d_2 = d_3 = d_0\frac{\sqrt{3}}{2}\Big/\sin\left(\frac{\pi}{3} + \frac{\sqrt{3}}{6}\left[(1+f_N) - \sqrt{1+14f_N+f_N^2}\right]\right). \tag{64}$$

Here d_1 is the bond length parallel to the direction of force f. Similarly, for armchair edge ribbons, we have

$$d_1' = d_0(1-\frac{1}{6}f_N),$$

$$d_2' = d_3' = \frac{d_0}{2}\Big/\cos\left(\frac{\pi}{3} - \frac{1}{2}\left[(\sqrt{3}+2f_N) - \sqrt{(\sqrt{3}+2f_N)^2 + 8f_N/\sqrt{3}}\right]\right). \tag{65}$$

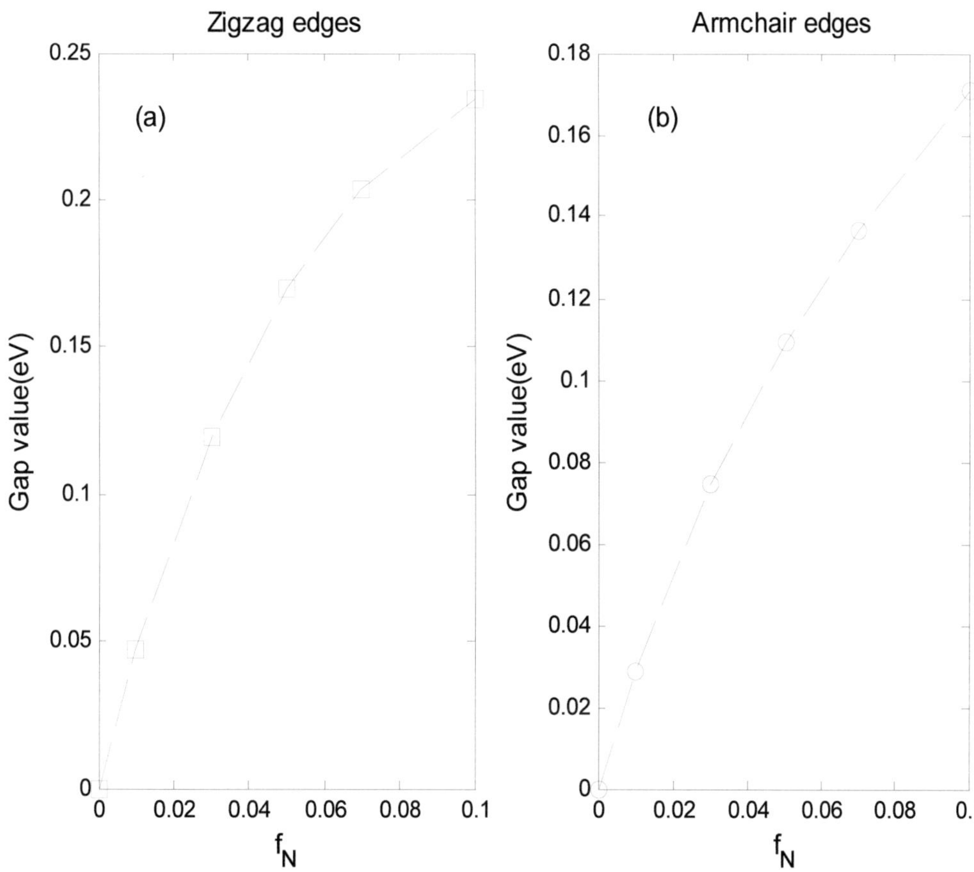

Fig. 10. Energy gaps of wider graphene ribbons with (a) zigzag and (b) armchair edges at K points as function of the dimensionless tensile force f_N.

Here d_1' is the bond length perpendicular to the direction of force f. To calculate hopping integrals, we choose the hydrogen-like atom wave functions as $2p_z$ orbitals, which is given by

$$\varphi(\boldsymbol{r}) = \sqrt{\frac{\lambda^5}{\pi}}\, r\cos\theta\, e^{-\lambda r}. \tag{66}$$

Here λ is the Slater orbital index, taken to be 2.18 in calculations. Substituting Eq.(66) into Eqs. (10), (14), (15), (17), (18) and (19) and noting that the bond lengths between carbon atoms are different from the undeformed graphene, we can obtain the analytical expressions for the hopping and overlap integrals (Wei & Tong, 2009). Because the bond lengths are dependent of force *f*, the hopping and overlap integrals depend on the force. Fig.10 shows the change of the energy gap with tensile force f_N at Dirac points *K*, where the third neighbor is included. The Dirac points will vary with the force, for zigzag ribbons given by

$$K = \left(\frac{\pi}{d_2 \cos\alpha + d_1}, \frac{\pi}{2d_2 \sin\alpha} - \frac{\pi d_2 \sin\alpha}{2(d_2 \cos\alpha + d_1)^2} \right),$$

$$K' = \left(\frac{\pi}{d_2 \cos\alpha + d_1}, \frac{\pi d_2 \sin\alpha}{2(d_2 \cos\alpha + d_1)^2} - \frac{\pi}{2d_2 \sin\alpha} \right). \quad (67)$$

It is clear that a pulling force may make Dirac points opening an energy gap, and which varies nonlinearly as the force. When the force is small, the change of the gap nearly is linear. But as the force becomes large, this change appears nonlinear. By comparison, we see that the gap of zigzag edges is more than that of armchair edges under the same force. This means that the gap of wider graphene ribbons with zigzag shaped edges is easily opened by an external force with respect to the armchair edges.

6. Conclusion

In this chapter, we study in details the electronic energy dispersion relations of graphene and its nanoribbons within the tight binding model, including up to the third-neighbor interaction. For a graphene sheet, there are no energy gaps at high-symmetry points *K*. The next-nearest-neighbor hopping integrals and wave function overlap between carbon atoms impact strongly on the bandwidth, i.e., their effects make the bandwidth become narrow with respect to the nearest neighbor result. The third neighbors can increase the bandwidth slightly and decrease the energy difference between the lowest conduction and highest valence bands greatly. The electronic dispersion of armchair edge graphene nanoribbons is given analytically based on the tight binding approach and hard-wall boundary condition. For the armchair nanoribbon, different widths have different dispersion relations. When $n=3m$ and $n=3m+1$, the second neighbor terms are able to reduce the bandwidth and slightly increase the band gap at Γ point. In general, smaller quantum number q impacts on the bandwidth and larger q affects the band gap. The effect of the third neighbor interaction is opposite to that of the second neighbor, but a flat band disappears when we involve the third neighbors. As for $n=3m+2$, the non-neighbor interaction can not open the gap at Dirac point. We also evaluate the influence of the edge deformation on this ribbon and compare the competition between both the non-neighbor interaction and edge deformation in energy gaps. For zigzag nanoribbons, there is no energy gap and the non-neighbor interaction impacts only on the bandwidth. In addition, the energy gaps of graphene ribbons with armchair or zigzag edges can be opened by an external force. Opening the gap of the zigzag edge ribbon is easier with respect to the armchair ribbon.

The problem we discussed above is the ideal graphene nanoribbons. If we consider the warping of the edges and the non-flat ribbon, the energy dispersion would how to change? These issues are worthy of further study.

7. References

Brey, L. & Fertig, H. A. (2006). Electronic states of graphene nanoribbons studied with the Dirac equation. *Phys. Rev.* B73, 23, 235411(1-5), 1550-235x

Castro Neto, A. H.; Guinea, F.; Peres, N. M. R.; Novoselov, K. S. & Geim, A. K. (2009). The electronic properties of graphene. *Reviews of Modern Physics,* 81, 1, 109-162, 1539-0756

Fujita, M.; Wakabayashi, K.; Nakada, K. & Kusakabe, K. (1996). Peculiar localized state at zigzag graphite edge. *J. Phys. Soc. Jpn.*, 65, 7, 1920-1923

Geim, A. K. & Novoselov, K. S. (2007). The rise of graphene. *Nature Materials* , 6, 183-191, 1476-4660

Jin, Z. F.; Tong, G. P. & Jiang, Y. J. (2009). Effect of the non-nearest-neighbor hopping on the electronic structure of armchair graphene nanoribbons. *Acta Physica Sinica*, 58, 12, 8537-8543, 1000-3290

Lomer, W. M. (1955). The Valence Bands in Two-Dimensional Graphite. *Proc. Roy. Soc.* (London), A227, 1170, 330-349, 1471-2946

McClure, J. W. (1956). Diamagnetism of graphite. *Phys. Rev.* 104, 3, 666-671,

Nakada, K.; Fujita, M.; Dresselhaus, G. & Dresselhaus, M. S. (1996). Edge state in graphene ribbons: Nanometer size effect and edge shape dependence. *Phys. Rev.* B54, 24, 17954-17961, 1550-235x

Novoselov, K. S.; Geim, A. K.; Morozov, S. V.; Jiang, D.; Zhang, Y.; Dubonos, S. V.; Grigorieva, I. V. & Firsov , A. A. (2004) . *Science*, 306, 5696, 666-669, 1095-9203

Novoselov, K. S.; Geim, A. K.; Morozov, S. V.; Jiang, D.; Katsnelson, M. I.; Grigorieva, I. V.; Dubonos, S. V. & Firsov, A. A. (2005). Two-dimensional gas of massless Dirac fermions in graphene. *Nature*, 438, 197-200, 0028-0836

Porezag, D.; Frauenheim, T.; Kohler, T.; Seifert, G. & Kaschner, R. (1995). Construction of tight-binding-like potentials on the basis of density-functional theory: Application to carbon. *Phys. Rev.* B51, 19, 12947-12957, 1550-235x

Reich, S; Maultzsch, J. & Thomsen, C. (2002). Tight-binding description of graphene. *Phys. Rev.* B66, 3, 035412(5), 1550-235x

Rozhkov, A. V.; Savel'ev, S. & Nori, F. (2009). Electronic properties of armchair graphene nanoribbons. *Phys. Rev.*, B79, 12, 125420(10), 1550-235x

Saito, R.; Dresselhaus, D. & Dresselhaus, M. S. (1998). *Physical Properties of Carbon Nanotubes*, Imperial College Press, 1-86094-093-5, London

Slonczewski, J.C. & Weiss, P.R. (1958). Band structure of graphite. *Phys. Rev.* 109, 2, 272-279,

Son, Y. W.; Cohen, M. L. & Louie, S. G. (2006). Energy gaps in graphene nanoribbons. *Phys. Rev. Lett.*, 97, 21, 216803(1-4), 1079-7114

Wallace, P. R. (1947). The band theory of graphite. *Phys. Rev.* 71, 9, 622-634

Wei, Y. & Tong, G. P. (2009). Effect of the tensile force on the electronic energy gap of graphene sheets. *Acta Physica Sinica*, 58, 3, 1931-1934, 1000-3290

Zheng, H. X.; Wang, Z. F.; Luo, T.; Shi, Q. W. & Chen, J. (2007). Analytical study of electronic structure in armchair graphene nanoribbons. *Phys. Rev.* B75, 16, 165411(1-6), 1550-235x

5

Quasi-Particle Spectra on Substrate and Embedded Graphene Monolayers

Alexander Feher[1], Eugen Syrkin[2], Sergey Feodosyev[2], Igor Gospodarev[2] and Kirill Kravchenko[2]

[1]*Centre of Low Temperature Physics Faculty of Science P. J. Šafárik University & Institute of Experimental Physics SAS,*

[2]*B.I.Verkin Institute for Low Temperature Physics and Engineering NASU,*

[1]*Slovakia*

[2]*Ukraine*

1. Introduction

Graphite, graphene, and compounds based on them are of great interest both as objects of fundamental research and as some of the most promising materials for modern technologies. The two-dimensional form of graphite - graphene - was prepared only very recently, immediately attracting a great deal of attention. Graphene can be deposited on solid substrates and has been shown to exhibit remarkable properties including large thermal conductivity, mechanical robustness and two-dimensional electronic properties. Note that electrons in graphene obey linear dispersion relation resulting in the observation of a number of very peculiar electronic properties. These properties are essentially changed when different defects are introduced into material. Special interest is devoted to graphite intercalated by metals, since in such graphitic systems the temperature of superconducting transition essentially depends on the type of intercalating metal. Besides, the discovery of superconductors as MgB_2 and iron pnictides intensified the search for high-temperature superconductivity in materials other than copper oxides. It is known that in the formation of the superconducting state the electron-phonon interaction plays a crucial role (according to the Bardeen-Cooper-Schrieffer theory). Therefore it is necessary to analyze in detail the phonon spectra of pure graphite and to find out how these spectra are influenced by different defects and by intercalation.

This chapter consists of three sections. The first section is devoted to the calculation of the local electronic density of graphene containing a substitutional impurity, vacancy defects due to the substrate surface roughness and adsorbed atoms. The local densities of states for atoms of the sublattice which not contains the vacancy show sharp peaks at energy $\varepsilon = \varepsilon_F$ (ε_F is the energy of the Dirac singularity for ideal graphene). Local spectral densities of atoms of the sublattice which contains the vacancy conserve the same Dirac singularity as is observed in an ideal graphene.

The second section will present our model, which allows to quantitatively describe the phonon spectrum of graphite and to determine the relaxation of force constants for the formation of the surface of the sample and the formation of thin films (bigraphene, trigraphene, etc.). Bending stiffness is calculated at the microscopic level for graphene monolayers, the characteristic features of the phonon density of states are identified and atomic displacements along the graphene monolayers and perpendicularly to them are determined. The mean square amplitude of atomic vibrations will be calculated, allowing us to make conclusions on the stability of the bigraphene lattice up to room temperatures. Note that the phonon spectra of graphene monolayers deposited on the substrate are determined by the substrate and they bear very little information on the vibrational characteristics of the carbon atoms.

Finally, the third section will deal with the changes introduced into phonon spectrum of graphite intercalated by different metals. In graphite the spectral density of phonons polarized along the **c**-axes shows a *V*-type singularity analogous to the so-called Dirac singularity typical for the graphene electron density states. We study the appearance of the quasilocalized states which increase the number of phonons near this singularity. Our investigations make it possible to predict the general properties of phonon and electron spectra for graphite intercalated with different metals. In addition, analyzing the changes in phonon spectra of graphite intercalated with various metals, using the BCS theory, we have proposed, taking into account dynamic properties of these layered systems, a possible way leading to the increase of T_c.

2. Electronic spectra of graphene with defects

It is well known (Kossevich, 1999) that in flat monoatomic layers the mean-square amplitudes of the atoms in the normal direction to the layer plane diverge even at $T=0$. Therefore, graphene monolayers can not exist as a flat formation in the free state. From the substrate only small scales can break off, which immediately become crimped (Meyer et al., 2007). So we can study and practically apply only such a graphene, which is deposited on a certain substrate, which provides for the stability of the plane graphene monolayers (Novoselov et al., 2005; Castro et al., 2009; Peres et al., 2007). In the study of the electronic properties of graphene a dielectric (often silicon) substrate is used, since it does not change its electronic spectrum, but greatly increases the possibility of introducing different kind of defects into it. For example, in graphene spray-deposited on silicon, the vacancies can appear (Shangduan et al., 2008), whereas in graphite (a set of weakly interacting graphene monolayers) and carbon nanotubes vacancies are "healed", forming a stacking fault with local fivefold symmetry axis (Chen et al., 2007).

This section presents the results of the calculations of local electron densities of atoms of graphene containing a substitutional impurity, vacancy defects due to the irregularities of the substrate and adsorbed atoms.

According to its electronic properties graphene is a two-dimensional semiconductor with zero gap. The fact that the charge carriers in graphene are formally described by the Dirac equation rather than the Schrödinger equation is caused by the symmetry of the crystal lattice of graphene, which consists of two equivalent carbon sublattices *A* and *B* (left part of Fig. 1). Electronic subbands formed by the combination of symmetric and antisymmetric wave functions for the two sublattices intersect at the edge of the Brillouin zone, which leads to a cone-shaped energy spectrum near the Dirac points *K* and *K'* (right part of Fig. 1), so that the dynamics of electrons is described by the linear dependence of energy on the quasi-momentum (in ordinary metals and semiconductors the dispersion dependence is parabolic).

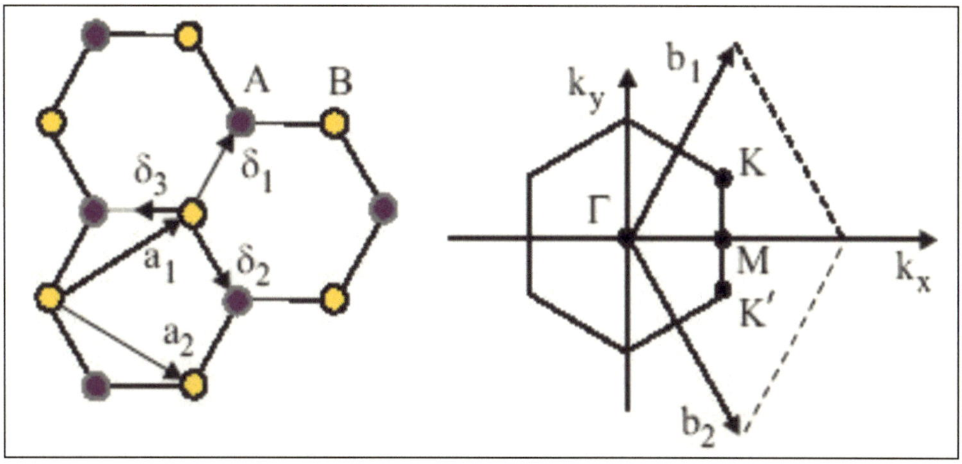

Fig. 1. Structure of the crystal lattice of graphene (left) and its first Brillouin zone (right)

The electronic spectrum of graphene can be described in the strong coupling approximation, taking into account only the interaction of nearest neighbours (Castro et al., 2009; Peres et al., 2007; Skrypnyk & Loktev, 2006; Skrypnyk & Loktev, 2008; Bena & Kivelson, 2005). The corresponding Hamiltonian has the form

$$H = \sum_i \varepsilon_i |i\rangle\langle i| - \sum_{i,j} J_{ij} |i\rangle\langle j| , \qquad (1)$$

where the indices i and j denote the sites of a two-dimensional lattice, ε_i is the energy of a particle at the i site, and J_{ij} is the so-called overlap integral. Fig. 2a presents the density of electronic states of graphene (curve 1, red line) and the real part of the Green's function (curve 2, blue line). These calculations were made using the method of Jacobi matrices (Peresada et al., 1975). Note that in a perfect graphene, due to the physical equivalence of the atoms of both sublattices, the local Green's function $G(\varepsilon,i) = \langle i|\left(\varepsilon\hat{I} - \hat{H}\right)^{-1}|i\rangle$ coincides with the full function $G(\varepsilon) = \lim\limits_{N\to\infty} \frac{1}{N}\sum\limits_{i=1}^{N}\langle i|\left(\varepsilon\hat{I} - \hat{H}\right)^{-1}|i\rangle$.

A feature on the density of states for $\varepsilon = \varepsilon(K)$ (namely, that the value $\varepsilon(K)$ corresponds to ε_F which is the Fermi energy in graphene) determines the behavior of the real part of the Green's function in the vicinity of ε_F. For a wide class of perturbations caused by defects we can find quasilocalized states, using the Lifshits equation in the interval $\left[-\varepsilon(M),\varepsilon(M)\right]$ (in this model $\varepsilon(M) = J$) (Lifshits, 1945). This equation, which determines their energy, can be written as (Kossevich, 1999; Peresada et al., 1975)

$$\mathrm{Re}G(\varepsilon) = S(\varepsilon,\Lambda_{ik}) , \qquad (2)$$

where the $S(\varepsilon,\Lambda_{ik})$ function is determined by the operator of the perturbation $\hat{\Lambda}$ (Λ_{ik} are matrix elements of this operator on defined basis).

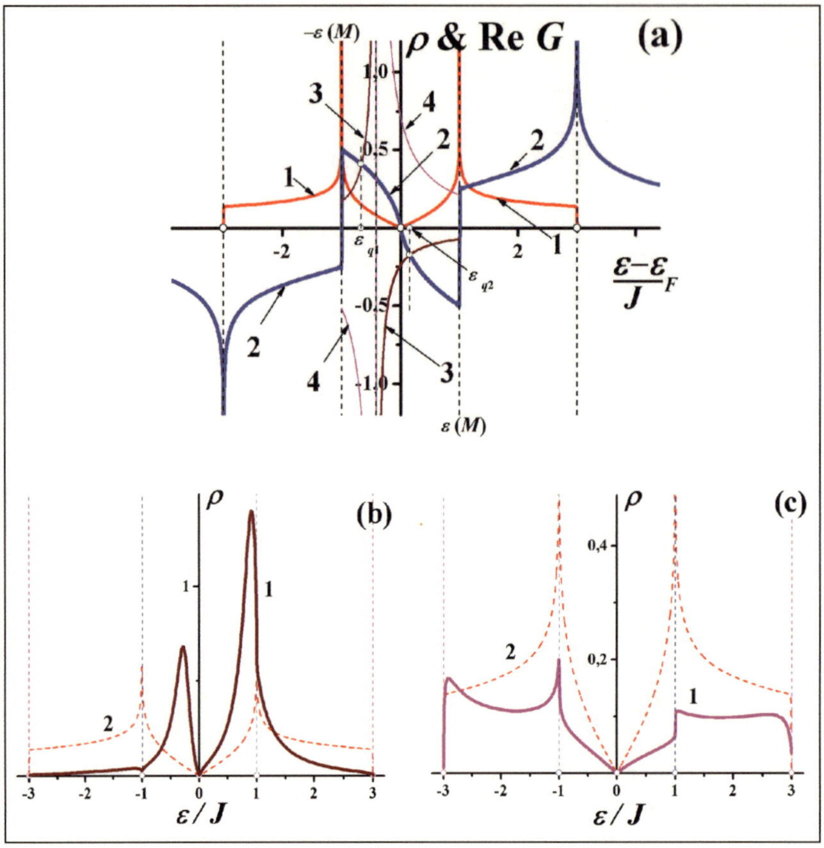

Fig. 2. **(a)** Density of states (curve 1) and real part of the Green's function (curve 2) in ideal graphene, curves 3 and 4 represent functions $S(\varepsilon)$ for substitutional impurities of nitrogen and boron; **(b)** and **(c)** are spectral densities of nitrogen and boron impurities, respectively (curves 1), curves 2 in these figures are the densities of states for ideal graphene

Fig. 2a illustrates the graphic solutions of the Lifshits equation for graphene containing isolated substitutional impurities, namely the atoms of nitrogen and boron. Local spectral densities $\rho(\varepsilon,i) \equiv \frac{1}{\pi} \lim_{\gamma \downarrow 0} \operatorname{Im} G(\varepsilon + i\gamma, i)$ of impurity atoms are calculated in (Peres et al., 2007). For an isolated substitutional impurity, different from the host lattice atom, the energy of the impurity site $i=0$ ($\varepsilon_0 = \tilde{\varepsilon}$) and the overlap integral $J_{i0} = (1+\eta)J$, the function $S(\varepsilon,\tilde{\varepsilon},\eta)$ has the form

$$S(\varepsilon,\tilde{\varepsilon},\eta) = \frac{(1+\eta)^2}{\tilde{\varepsilon} + \varepsilon\eta(2+\eta)}. \tag{3}$$

For nitrogen impurity (according to (Peres et al., 2007) in such a case $\tilde{\varepsilon} - \varepsilon(K) \approx -0.525J$, and $\eta \approx -0.5$ (the dependence $S(\varepsilon)$ is shown in Fig. 2a, curve 3). For equation (2), as seen from the figure, the solution for interval $[-\varepsilon(M), \varepsilon(K)]$ is point ε_{q1} and for interval

$[\varepsilon(K),\varepsilon(M)]$ it is the point ε_{q2}. Local densities of states $\rho(\varepsilon,i)$ of nitrogen impurities calculated in (Peres et al., 2007) have quasi-local maxima in both of these intervals (Fig. 2b). Although, because of differences in these intervals, the imaginary part of the Green's function is different from zero, the location of quasi-local maxima are different from ε_{q1} and ε_{q2}, and the presence or absence of solutions of the Lifshits equation in the interval under the given parameters of the defect determines the presence or absence of quasilocalized states in this interval.

So, as discussed in (Peres et al., 2007), for the impurity boron ($\tilde{\varepsilon}-\varepsilon(K)\approx 0.525J$; $\eta \approx 0.5$) the quasilocalized states are absent in the interval $[-\varepsilon(M),\varepsilon(M)]$, as seen in Fig. 2c. In this case equation (2) also has no solutions in the interval $[-\varepsilon(M),\varepsilon(M)]$ (corresponding dependence $S(\varepsilon)$ is shown in Fig. 2a as curve 4). Local Green's function of the boron impurity has a peculiarity outside the band of quasi-continuous spectrum, corresponding to a so-called local level in the energy spectrum. Therefore, the area under the curve 2 is smaller than the area under the curve 1, the difference being the residue at local level, what is clearly seen in Fig. 2c.

The function $\mathrm{Re}G(\varepsilon)$ allows us to conclude that the solution of equation (2) in the interval $[-\varepsilon(M),\varepsilon(M)]$ exists for a wide class of perturbations. Fig. 3 shows the local density of states of an atom adsorbed on a graphene monolayer for different values $\tilde{\varepsilon}$ and η, characterizing the interaction of this atom with carbon atoms. It is clear that the weakening of the adsorbed atom interaction with its nearest neighbors causes that on the local density of states sharp resonance peaks are formed near the $\varepsilon = \varepsilon_F$.

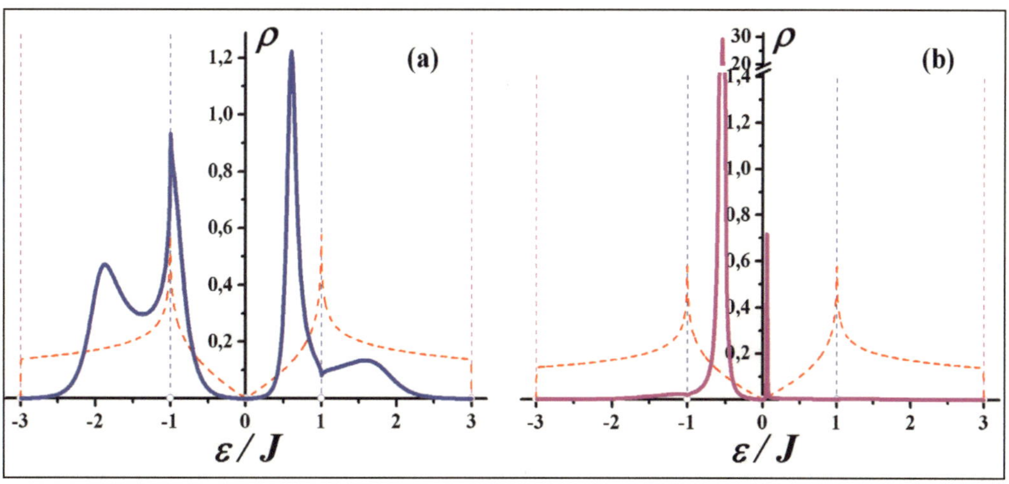

Fig. 3. Local density of states of atoms adsorbed on graphene film: **(a)** $\tilde{\varepsilon}=-0.5J$, $\eta=-0.5$; **(b)** $\tilde{\varepsilon}=-0.5J$, $\eta=-0.9$; dashed lines represent the density of states of ideal graphene

Thus, the behaviour of the real part of the Green's function in graphene for $-\varepsilon(M)\le\varepsilon\le\varepsilon(M)$ shows high sensitivity of the density of states at given energies to various perturbations caused by defects and other changes in the crystal structure, in particular, the possible formation of localized excitations with energies close to the Fermi energy. Generally speaking, it is not important whether this perturbation is degenerate

(function $S(\varepsilon,\Lambda_{ik})$ can be written explicitly) or nondegenerate: then quasilocalized states will arise enriching the electronic spectrum near $\varepsilon = \varepsilon_F$. As an example of the influence of such nondegenerate perturbation we consider the local spectral density of atoms of different coordination spheres at an isolated vacancy in the graphene monolayer.
The calculation results are partially presented in Fig. 4, showing the local density $\rho(\varepsilon,i)$ of states for the nearest, second, seventh and tenth neighbours of vacancies (Feher et al., 2009).

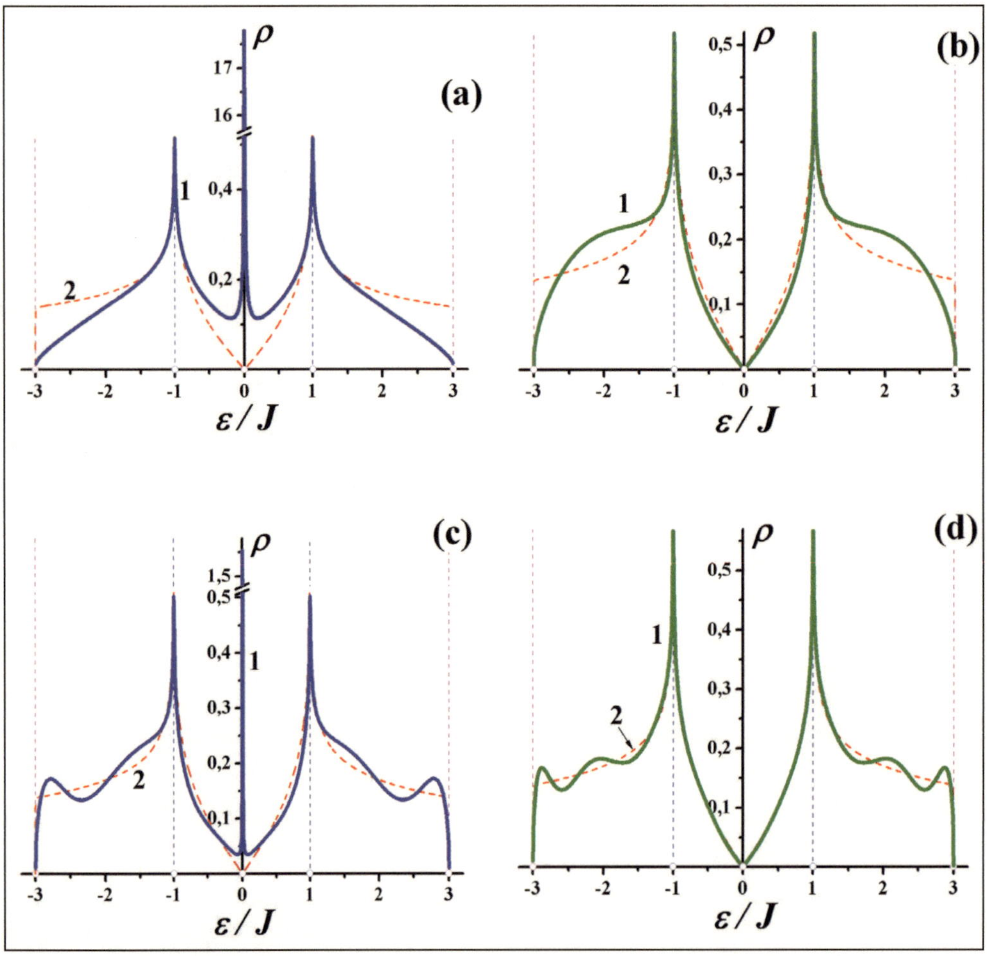

Fig. 4. Local density of electronic states of neighboring vacancies (curves 1); **(a)** the nearest, **(b)** the second, **(c)** the seventh, **(d)** the tenth; curve 2 (dashed lines) represent the density of electronic states of perfect graphene

Odd neighbours of a vacancy belong to the sublattice which does not contain vacancies (let this be the sublattice B in Fig.1). In the local densities of atoms of this sublattice $\rho(\varepsilon,B)$ a sharp peak occurs at $\varepsilon = \varepsilon(K) = \varepsilon_F$. Local density of states in the sublattice A retains, for $\varepsilon = \varepsilon(K) = \varepsilon_F$, the same Dirac singularity as in the perfect graphene. This was proved in (Feher et al., 2009) by using the formula obtained in (Kotlyar & Feodosyev, 2006), which relates an arbitrary matrix element of the Green's function $G(\varepsilon,i)$ in the formalism of the method of Jacobi matrices to the first diagonal element of this function, i.e. to the local Green's function $\rho(\varepsilon, A)$. Interaction with second neighbours would change the local density of states $\rho(\varepsilon, A)$ near $\varepsilon = \varepsilon_F$, but the magnitude of this change is in the order of the interaction, which is small.

The surface roughness of the substrate may cause the anisotropy of the interatomic interaction, since the overlap integrals J_{ik} will be different for the interaction of the same atom with its various neighbours. Note that in models of "anisotropic graphene" (when the anisotropy of overlap integrals is retained over long-range order) a gap between the bands of the electronic spectrum as well as the formation of additional logarithmic singularities due to the displacement of the intersection of electronic branches inside the first Brillouin zone may appear. Here we consider the case when the atom with the anisotropy of the overlap integrals is an isolated defect.

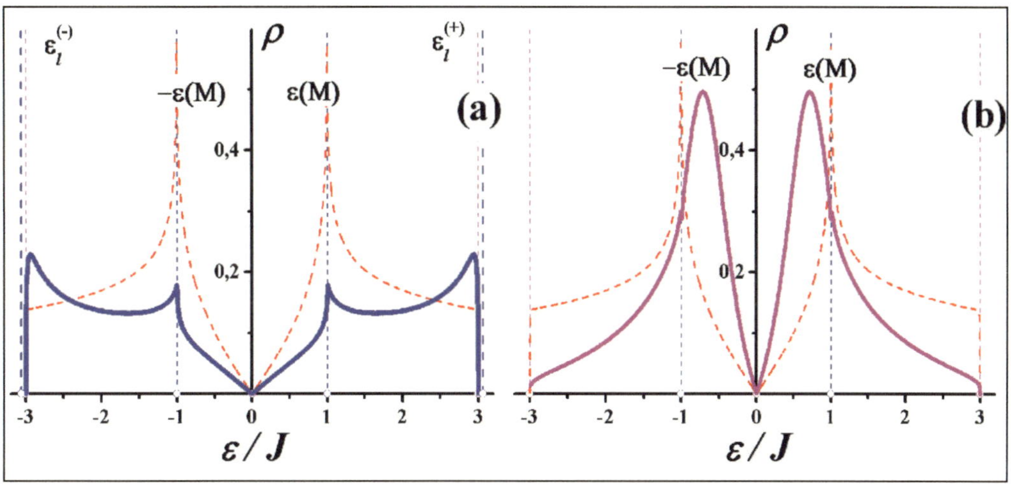

Fig. 5. Local density of states of an isolated defect in lattice of carbon atoms with an anisotropic interaction between nearest neighbours: **(a)** $\eta = 0.5$; **(b)** $\eta = -0.5$; dashed lines are density of states of a perfect isotropic graphene

Let the interaction of the atom with one of the nearest neighborus be described by the overlap integrals J and $\tilde{J} = (1+\eta)J$. Fig. 5 shows the corresponding local densities of states of the atom. Fig. 5a shows the density of states for $\eta = 0.5$, corresponding to the enhanced that is, the interaction increased. In this case the electronic spectrum contains, along with the quasi-continuous part, also two symmetric discrete levels $\varepsilon_l^{(\pm)} \approx \pm 3.0698J$ (quasi-continuous band of spectrum in this case is $-3J \le \varepsilon \le 3J$). The values of $\varepsilon_l^{(\pm)}$ denote the local levels of spectra. Residues at these points (so-called intensities of discrete levels) are $\mu_l^{(\pm)} \approx 0.139$. The area under the curve $\rho(\varepsilon)$ is in this case less than unity, the rest being in the sum of the intensities of discrete levels. For $\eta = -0.5$ the electronic states of an anisotropic defect lie in the band of a quasi-continuous spectrum and the interval $-\varepsilon(M) \le \varepsilon \le \varepsilon(M)$ contains essentially more electronic states as compared to the perfect isotropic graphene. Thus, controlling the roughness of the substrate could be a promising method for tuning the electronic spectrum of graphene.

3. Phonon spectra and vibrational characteristics of carbon nanolayers

The phonon spectra of graphene monolayers deposited on the substrate are determined by the substrate and have very little value for the determination of the vibrational characteristics of carbon atoms. At the same time, carbon nanofilms, consisting of several graphene monolayers (starting with two, i.e. bigraphene) are stable up to room temperature. This section describes the phonon spectra and vibrational characteristics of bulk samples of graphite and their transformation for decreasing number of layers, down to carbon nanofilms. It outlines our model which allows to quantitatively describe the phonon spectrum of graphite and to determine the relaxation of the force constants in the formation of the sample surface and the formation of ultrathin films (bigraphene, trigraphene, etc.). Flexural stiffness was calculated at the microscopic level for graphene monolayers, as well as the characteristic features of the phonon density of states and the contributions to the atomic displacements along the graphene monolayers and in the perpendicular direction to them were identified. We also calculated the mean square amplitudes of atomic vibrations, explaining the reason of the stability of the lattice bigraphene up to the room temperature.

It is known that the crystal of graphite is a strongly anisotropic layered crystal. Strong anisotropy is, on the one hand, due to a significant difference in the interatomic distances between nearest neighbours in the layer plane and in adjacent layers, and, on the other hand, due to different types of force bonds (covalent, metallic and van der Waals bonds).

Graphite consists of graphene monolayers, the atoms of which form a regular hexagon. The atoms of one of the sublattices are located in the centers of the triangles of the other sublattice. Bravais vectors, lying in the basal plane, can be selected as follows: $R_1 = a_0\left(\frac{\sqrt{3}}{2};\frac{1}{2};0\right)$ and $R_2 = a_0\left(\frac{\sqrt{3}}{2};-\frac{1}{2};0\right)$, where $a_0 \approx 2.45$ Å. The period of the graphite lattice along the axis *c*, in the direction perpendicular to the graphene layers, is equal to twice the interlayer distance, i.e. $R_3 = c_0(0;0;1)$, the parameter $c_0 \approx 6.7$ Å (Fig. 6). The atoms of different sublattices of the graphite basal plane are differently situated with respect to the atoms of neighboring planes, leading to the differences in the interaction between atoms belonging to different sublattices and to their physical inequivalence. Thus, there will be different local Green's functions corresponding to these atoms and determining their vibrational characteristics, such as the mean-square displacements of atoms along different crystallographic directions.

Strong anisotropy of interatomic interaction causes a number of typical distinguishing features in the behaviour of the phonon spectra and vibrational characteristics (Kosevich et al., 1994). For example, the elastic moduli C_{33} and C_{44} associated with displacements along the ***c*** axis and determining the speed of sound propagating or polarized along this direction, are from about 30 for about 300 times smaller than the elastic moduli C_{11} and C_{66}, determining the speed of sound propagating and polarized in the basal planes (Nicklow et al., 1972; Belen'kii, et al., 1988). Therefore, if the propagation of the vibrations polarized along ***c*** in the basal ***ab*** plane would have the nature of a sound wave and not a quasi-flexural wave, the mean-square displacements of atoms in a given direction would attain, even at low temperatures, the values corresponding to the melting of the crystal. That is, the mere existence of solid graphite at room temperature indicates that the fluctuations are essentially determined by the restoring forces acting on the atom from other atoms, lying in the same layer. These restoring forces are due to the noncentral interatomic interaction and indicate the presence of elastic stresses in the graphene layers that form the crystal lattice of graphite.

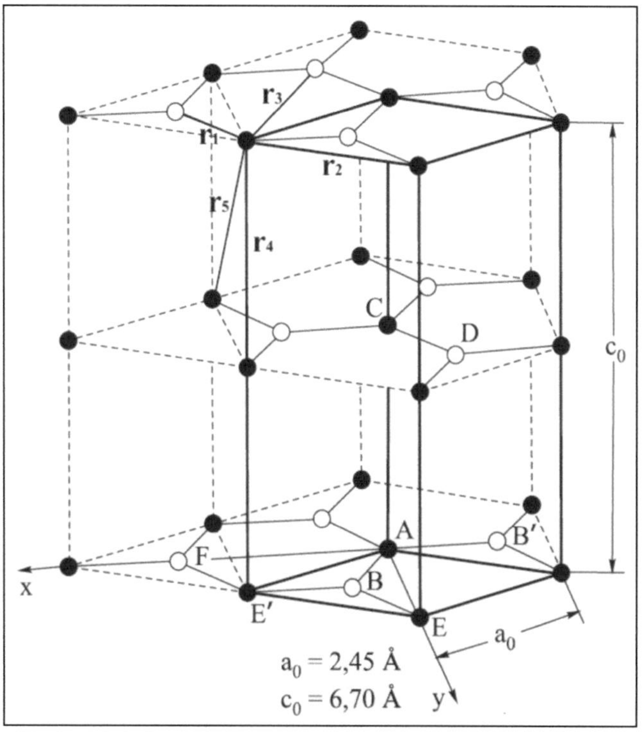

Fig. 6. The structure of the crystal lattice of graphite

In the long-wave region the dispersion law of the transverse phonon mode (TA) of a strongly anisotropic layered crystal, whose frequency vanishes at $k=0$ and which corresponds to the waves propagating in the ***ab*** plane and polarized along ***c*** (i.e. quasi-flexural mode) has the form

$$\omega_{T_3}^2(\vec{k}) = \frac{C_{44}}{\rho}k^2 + \frac{\kappa}{m}k^4 . \tag{4}$$

Here ρ is the density of graphite, m is the mass of the carbon atom and the coefficient κ characterizes the bending stiffness of layers. The deviation described by the second term of (4), characterising quasi-flexural wavelength of the (TA) dispersion curve is clearly seen on dispersion curves obtained from inelastic neutron scattering data (Kosevich et al., 1994).
The fact that we can examine each layer of graphite as plane means that in the crystal the condition

$$\sigma_{iz}n_{iz} = 0 \tag{5}$$

is fulfilled, where the coordinate z is chosen along the ***c*** axis and the coordinates x and y in the basal plane.
Bending stiffness of flat graphene monolayers, which provides a characteristic distortion of the dispersion curve (Kosevich et al., 1994), does not depend on the interlayer interaction and can be determined from the consideration of only one isolated graphene monolayer. When the condition of equilibrium of a flat layer, which has for an isolated graphene

monolayer the form $\beta_{1z} + 6\beta_2 + 4\beta_3 = 0$, is fulfilled, the bending stiffness κ can be written as (Syrkin et al., 2009)

$$\kappa = \left(\frac{a_0}{8}\right)^2 \sqrt{\frac{6\beta_3}{m} \frac{\beta_{1z} + 6\beta_3}{\beta_{1z} + \beta_3}}. \tag{6}$$

In (Syrkin et al., 2009) the expressions for the flexural rigidity in some other layered structures are also given. For the layered hexagonal close-packed crystal

$$\kappa = a^2 \sqrt{\frac{3}{2}(\beta_{1z} + 2\beta_{2z})}. \tag{7}$$

In the layered tetragonal lattice the flexural rigidity is anisotropic (depending on the direction of propagation of flexural waves in the layer plane)

$$\kappa = \left(\frac{a}{2}\right)^2 \sqrt{-(\beta_{1z} + 2\beta_{2z}) - 2(\beta_{1z} + 4\beta_{2z})\sin^2 2\phi}, \tag{8}$$

where ϕ is the polar angle in the basal plane, measured from the x-axis in the counterclockwise direction.

Therefore, when describing the intralayer interaction in the graphite, the central and noncentral interactions between atom and its first, second and third neighbours must be considered. Accounting for more distant neighbours in the basal plane has no meaning as to the accuracy of the adiabatic approximation. In describing the weak interlayer interaction it is natural to limit of only to the interaction between atoms from neighbouring layers, being at distances r_4 or r_5, where the value r_5 is only slightly larger than r_4 (see Fig. 6). For the nearest neighbours in the basal plane ($\Delta = r_1$), whose interaction is determined by the superposition of covalent and metallic bonds, the force matrix is characterized by all three parameters of interaction. The bond between more distant neighbours ($\Delta = r_2, r_3, r_4, r_5$) can be regarded as a van der Waals one and described by an isotropic pair potential.

The proposed model involves eleven force constants. These constants can be found using follows data: four moduli of elasticity, the condition of symmetry of these moduli under the permutation of pairs of indices, the neutron diffraction data (Kosevich et al., 1994), the inelastic X-ray scattering data (Maultzsch et al., 2004) and the Raman scattering data (Dresselhaus et al., 2002). The values of force constants (see Table 1) that characterize the interaction between the atoms of graphite can be thus unambiguously obtained and checked.

	Δ				
	r_1	r_2	r_3	r_4	r_5
$\alpha.10^3$ N/m	337.882	50.476	19.647	2,581	0.371
$\beta.10^3$ N/m	$\beta_x = 170.864$ $\beta_y = 96.375$	10.149	8.661	0.0654	0.0353

Table 1. The force constants of graphite

Interlayer van der Waals interaction can be described by the Lennard-Jones potential, which allows, together with the help of expressions for C_{33} and C_{44}, to find constants responsible for the interaction between the nearest atoms of neighbouring layers. Moreover, from the values of force constants the parameters of this potential can be determined, which is important for the calculations of the surface relaxation in graphite ($\sigma \approx 3.092$ Å, $\varepsilon \approx 152.3$ K).
It is obvious that the flat form of a free graphene monolayer is not stable. Therefore, we analysed the phonon spectrum and the rms (root mean squares) amplitudes of displacements in the films consisting of two and three graphene monolayers. Results of the experimental studies of graphene bilayer films are given in (Morozov & Firsov, 2009).
It was shown that in the graphite the interlayer interaction contains both central and noncentral forces, therefore the formation of the surface can not be described by the Lifshitz-Rosenzweig model (as a solution of the stochastic boundary problem) and will be characterized by the reconstruction and relaxation processes. At the same time, it is natural to assume that the breakage of weak interlayer van der Waals bonds will change neither the distance between atoms in graphene layers nor the force constants that characterize the intralayer interaction.
In fact, the surface reconstruction and relaxation will lead to changes in the interlayer distances and force constants α_4, β_4, α_5 and β_5 characterizing the interlayer interaction.
Condition (5) leads to the same ratio between the force constants and lattice parameters as the condition $C_{13} = C_{31}$. For thin films with N monolayers this condition takes the form (Gospodarev et al., 2010)

$$\beta_1 + 6\beta_2 + 4\beta_3 = \frac{N}{N-1}\frac{2}{3a^2}\left[r_4^2\beta_4 - 9\left(r_1^2 - 2r_4^2\right)\beta_5\right]. \tag{9}$$

From the parameters of the Lennard-Jones potential for the considered graphite thin films the interlayer distance and the force constants describing the interlayer interaction can be easily found:
for two-layer film (bigraphene):

$$r_4 \approx 3.636 \text{ Å};\ \alpha_4 \approx 0.373\ N/m;\ \beta_4 \approx 0.0035\ N/m;$$

$$r_5 \approx 3.902 \text{ Å};\ \alpha_4 \approx -0.009\ N/m;\ \beta_4 \approx 0.004\ N/m;$$

for three-layer film (trigraphene):

$$r_4 \approx 3.453 \text{ Å};\ \alpha_4 \approx 1.585\ N/m;\ \beta_4 \approx -0.0015\ N/m;$$

$$r_5 \approx 3.713 \text{ Å};\ \alpha_4 \approx 0.016\ N/m;\ \beta_4 \approx 0.004\ N/m.$$

Fig. 7 presents the phonon densities of states of bigraphene (Fig. 7a) as well as the contributions to them from the atomic displacements along the layers (Fig. 7b) and perpendicular to them (Fig. 7c).
In each figure the dashed line shows the corresponding characteristics of an infinite graphite sample (Gospodarev et al., 2009). Densities of states of the film and bulk samples are practically the same, significant differences were observed only in the frequency range in which the phonon spectrum of graphite resembles that of a three-dimensional system and the interaction between the vibrational modes polarized in the plane of the layers and perpendicularly to the layers is sufficiently large.

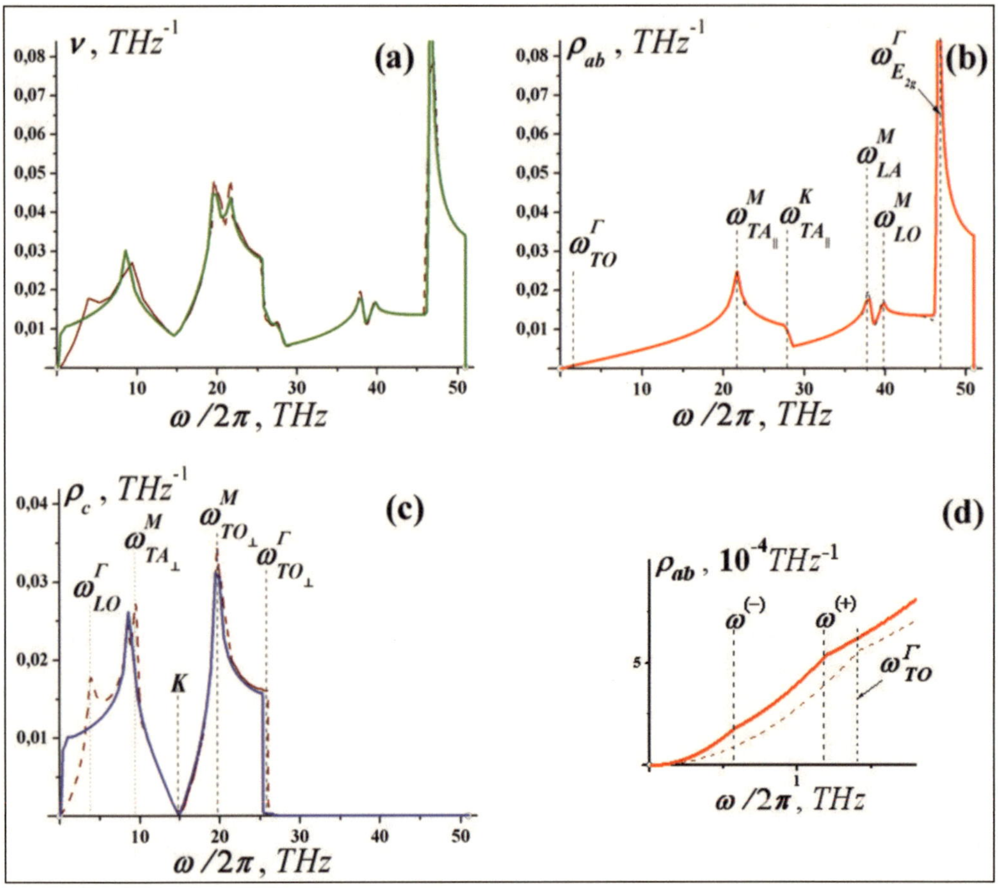

Fig. 7. The phonon density of states of a thin film of graphite, consisting of two graphene monolayers (a) ; the partial contributions to it from the atomic displacement along the basal plane and perpendicular to its direction (b, c, respectively); (d) is low frequency part of (b).

In bigraphene, the transverse phonon modes TA and TO, typical for graphite, degenerate into two discrete levels corresponding to the symmetrical and antisymmetrical displacements of layers. The frequencies of these levels are marked in Fig. 7d as $\omega^{(-)}$ and $\omega^{(+)}$. The frequency $\omega^{(+)}$ in bigraphene corresponds to the same atomic displacements as the frequency $\omega_{TO}(\Gamma)$ on bulk sample, a lower value of $\omega^{(+)}$ as compared with $\omega_{TO}(\Gamma)$ is caused by the surface relaxation. The bigraphene spectral density $\rho_{ab}(\omega)$ (Fig. 7d) shows kinks at $\omega = \omega^{(-)}$ and $\omega = \omega^{(+)}$. For $\omega > \omega^{(+)}$ the spectral density acquires the characteristic two-dimensional appearance $\rho_{ab}(\omega) \sim \omega$.

Starting from very low frequencies, the spectral density $\rho_c(\omega)$ acquires the form characteristic of a two-dimensional scalar model, leading to very high values of the mean-square displacements of atoms in the direction perpendicular to layers.

Fig. 8 shows the temperature dependence of the rms amplitudes of atomic displacements of bigraphene (2c), trigraphene (3c for surface layers, 3c' for central layer) and bulk sample (∞c) in the perpendicular direction to the layers, and it also shows that the amplitude of atomic displacements along the graphene layers is almost independent of sample thickness (*ab*).

The rms amplitude of atomic displacements along the ***c*** axis strongly increases with the decreasing film thickness. At room temperature the amplitude of transverse vibrations of an atom of the trigraphene central layer (curve 3c') is about twice the value for the bulk sample (curve ∞ c).

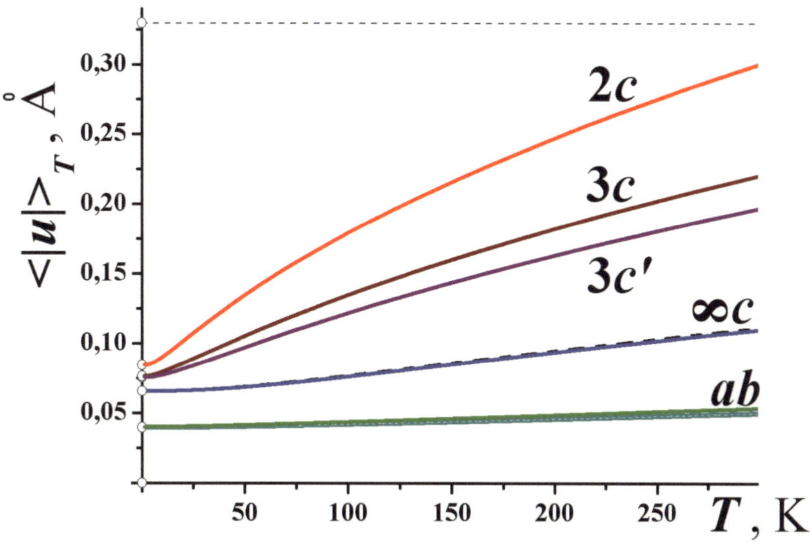

Fig. 8. Temperature dependence of the mean amplitudes of atoms along different crystallographic directions in the films of graphite, consisting of two and three graphene monolayers

The horizontal dashed line in Fig. 8 indicates the mean value of the amplitude of atomic vibrations along the ***c*** axis for bulk graphite at $T \approx 3000$ *K*. This temperature is about 1000 *K* below the melting temperature of graphite ($T \approx 3800 \pm 50$ *K*), therefore at $T \approx 3000$ *K* the crystal lattice of graphite still has sufficient margin of stability. Bigraphene and trigraphene also have a sufficient margin of stability at room temperature, since the mean square amplitudes of their atomic vibrations are markedly below the dashed line.

4. Phonon and electron spectra of metal intercalated graphite

Special interest has been devoted to graphite intercalated by metals, since in such graphitic systems the superconducting transition temperature T_c essentially depends on the type of intercalated metal. For example T_c for C_6Yb is 6.5 *K*, while for C_6Ca it is 11.5 *K* (Weller et al., 2005; Emery et al., 2005). It is known that for the formation of the superconducting state the electron-phonon interaction plays crucial role. Since the electronic spectra of these compounds probably do not depend on the type of intercalated metal, T_c variations in such compounds are dominantly determined by the peculiarities of their phonon spectra. Therefore it is necessary to analyse in detail the phonon spectra of pure graphite (see section 2) and to find out how these spectra vary due to the intercalation with various metals and to determine how these variations depend on the dynamic parameters of both carbon and intercalating metal.

Both structures mentioned above consist of graphite monolayers between which two dimensional triangle lattices of metals with period $a\sqrt{3}$ are placed. The lattice constant in the direction perpendicular to layers is, in both compounds, $c' = 4.5$ Å (Weller et al., 2005; Emery et al., 2005). The absence of data about acoustical, optical and other properties, which would enable to determine the parameters of interatomic interactions, prompted us to make some assumptions about the values of these interactions. We neglect the carbon-carbon and metal-metal interactions across layers. The interaction of metal atoms within one layer will be considered as a central force, i.e. matrices of force constants have form given in (Feher et al., 2009).

The distance between nearest atoms of metal and carbon is equal to

$$r_{\text{C-Me}} \equiv \sqrt{\left(\frac{c'}{2}\right)^2 + \frac{a^2}{3}} \approx 2.66\,\mathring{A}$$

i.e. it is larger than the distance between second nearest carbon neighbours in graphene monolayer ($a \approx 2.45$ Å), but shorter than between third neighbours ($2a/\sqrt{3} \approx 2.83\,\mathring{A}$). It can be therefore assumed that the potential describing this interaction may be considered as pair and isotropic, i.e. the force constants of interatomic interaction fulfill condition $\beta_z(r_{\text{C-Me}}) = \beta_x(r_{\text{C-Me}}) = \beta(r_{\text{C-Me}})$. Since interatomic distances in graphene monolayer do not change due to intercalation, force constants also do not change. Therefore the value of $\beta(r_{\text{C-Me}})$ may be found from the condition of the symmetry of the elastic modulus tensor with respect to a transposition of index pair. This condition has form

$$\beta\left(\frac{a}{\sqrt{3}}\right) + 6\beta(a) + 4\beta\left(\frac{2a}{\sqrt{3}}\right) = 2\left[\left(\frac{c'}{2a}\right)^2 - \frac{1}{3}\right]\beta(r_{\text{C-Me}}) \cdot$$

From this condition we get $\beta(r_{\text{C-Me}}) \approx 0.31$ N/m, both for carbon-ytterbium and carbon-calcium interactions. Unfortunately, we do not have any data from which force constant $\alpha(r_{\text{C-Me}})$, characterizing central interaction between metal and carbon atoms, may be directly determined. However, on the basis of the values of distances between carbon and metal atoms we estimated that the force constant value lies somewhere in the range from $\alpha(r_{\text{C-Me}}) \approx 20$ N/m to $\alpha(r_{\text{C-Me}}) \approx 50$ N/m (Syrkin et al., 2009). We picked up four $\alpha(r_{\text{C-Me}})$ values: 20, 30, 40, and 50 N/m.

Fig. 9 shows the frequency dependences of partial contributions to the density of phonon states from displacements of metal and carbon atoms in the direction perpendicular to layers. The areas below dependences corresponding to intercalating metal are hatched. In Fig. 9, the left set show dependences for C_6Ca, the right one for C_6Yb, the force constant $\alpha(r_{\text{C-Me}})$ increases from top to bottom. We see that for C_6Ca sharp resonance peaks appear on partial contributions from both intercalating metal and carbon. These peaks are shifted, with the increase of $\alpha(r_{\text{C-Me}})$, towards the centre of the frequency range, leading to the increase of density of phonon states near the Brillouin zone's *K*-point, through which the Fermi level of electrons in graphene passes. For the C_6Yb compound (*Yb* has more than four times larger atomic mass than *Ca*) the resonance peaks appear at lower frequencies and an apparent

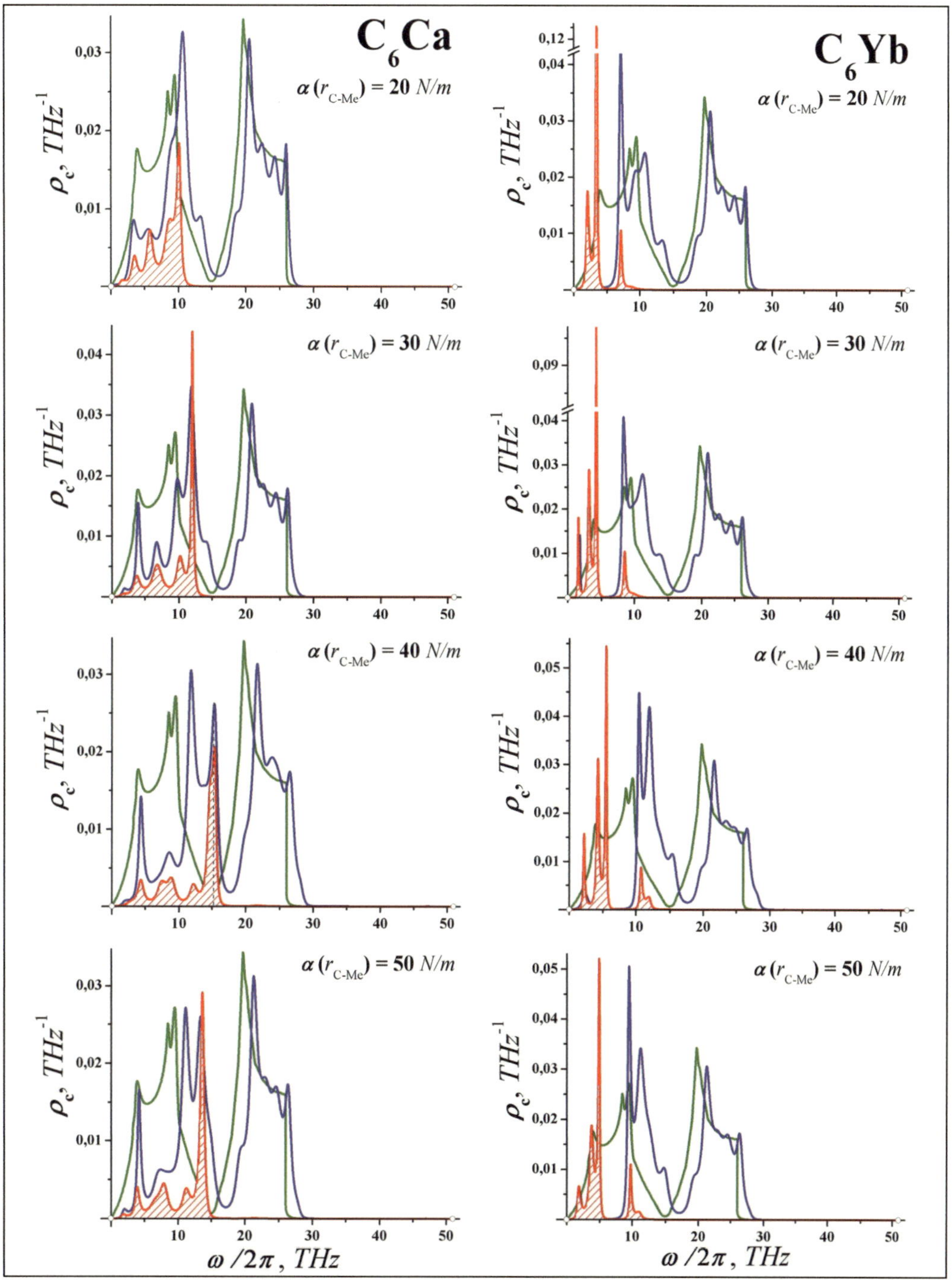

Fig. 9. The partial contributions to the phonon density of states of intercalated graphite (blue lines) from the displacements along the c axis of the carbon atoms and the metal (red lines with hatched area under the curves). Top to bottom $\alpha(r_{C\text{-}Me}) =$ 20, 30, 40, 50 N/m. Green lines in all the fragments correspond to pure graphite.

increase of the phonon states density at frequencies near the Brillouin zone's *K*-point is observed only for anomalously large values of α ($r_{C\text{-}Me}$)≈ 40 ÷ 50 *N/m* (note that T_c for C_6Ca is almost *1.8* times higher than that for C_6Yb). Note that the sharp resonance peaks in the density of states, which are characteristic for states localized near the isolated impurity atom, appear for metal atoms which form periodic structure. Such a behavior of the spectral densities is due to the weakness of the interaction between remotely spaced atoms. The interatomic distances within the metal layers differ slightly from distances between the atoms of neighbouring layers.
Fig. 10 presents the total phonon density of states for both C_6Ca (left) and C_6Yb (right) and partial contributions from intercalating metal (hatched area).

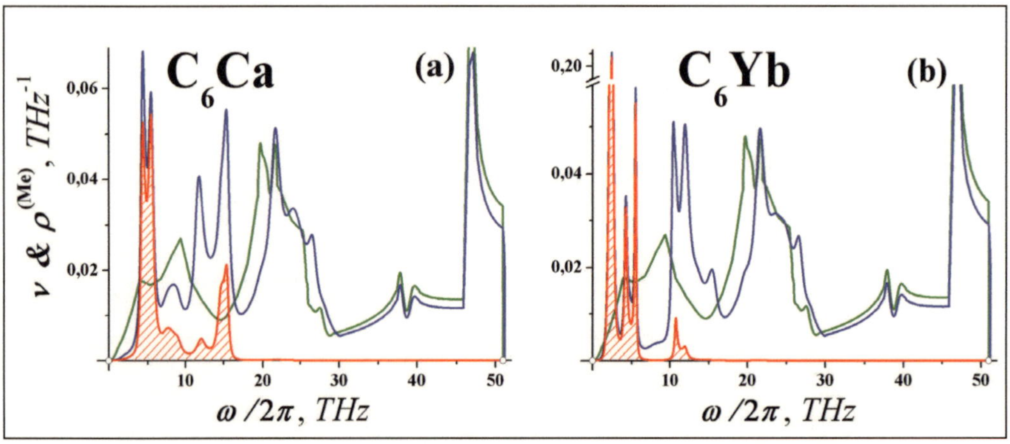

Fig. 10. The phonon density of states of intercalated graphite (blue lines) and the contribution to them of metal atoms (red lines with hatched area under the curves) for α = 50 *N/m*. Green lines are the phonon density of states of pure graphite.

Unconventional behavior of phonon subsystem in graphite containing metal intercalated layers may substantially influence its electronic properties and the electron-phonon interaction (EPI), including the superconducting transition temperature. At present time there is much convincing evidence that EPI is principal mechanism of the transition into the superconducting state and this interaction should be taken into account for developing a consistent theory for the description of superconductivity in different types of compounds, including high-T_c superconductors (Maksimov, 2008). It follows from the BCS theory (Bardeen et al., 1957) (see also (Maksimov & Dolgov, 2007)) that the value of T_c depends on the basic characteristics of electron and phonon spectra in the following way

$$T_c \propto \langle \omega_{ph} \rangle \exp\left(-\frac{1}{\lambda}\right), \qquad (10)$$

where $\langle \omega_{ph} \rangle$ is the mean phonon frequency and λ is the EPI constant, which is proportional to the density of electron states on the Fermi surface.
Microscopic analysis of EPI was not performed within this work, but we note that a consistent many-particle theory of EPI already exists, describing both normal and superconducting states of metals (Rainer, 1986; Allen & Mitrovic, 1982; Maksimov &

Dolgov, 2007). However, we can state that to increase T_c high phonon frequencies $\langle \omega_{ph} \rangle$, large values of the constant of the interaction of electrons with these phonons λ and high density of electron states on the Fermi surface are needed. Such properties are manifested by, for example, compounds of metals with light elements, such as hydrides (Ashcroft, 2004; Tse et al., 2007), borides (Nagamutsi et al., 2001) carbides and nitrides (Maksimov et al., 2007), since their phonon spectra show high-frequency modes corresponding to the vibrations of light atoms (H, B, C, N). For graphene doped by hydrogen T_c above the boiling point of liquid nitrogen was recently predicted using the BCS theory (Savini et al., 2010).
Moreover, a detailed analysis has been carried out (see review Maksimov, 2008), showing that electrons in MgB_2 exhibit very strong interaction with quasi-flexural modes. This is very interesting for describing the evolution of temperature of superconducting transition in the intercalating graphite.
Indeed, the frequency $\langle \omega_{ph} \rangle$ decreases at intercalating the graphite by metals which have atomic masses essentially larger than carbon. So, the temperature of superconducting transition for graphite intercalated by *Li* is 1.9 *K*, which is lower than for graphite intercalated by *Ca* or *Yb*. At the same time the intercalation by *Yb* and *Ca* increases the mean vibration frequency of quasi-flexural branch. Besides, it is especially interesting that the intercalation by *Ca* and *Yb* leads to an essential increase of the quasi-flexural phonons number. The quasi-momentum of these phonons corresponds to the *K* point of the first Brillouin zone, in the vicinity of which the Fermi level is lying in the electron spectra.
The role of defects is very important for the electron-phonon interaction formation, and consequently, for superconducting properties in three dimensional layered crystals. So, in a new type of high temperature layered superconductors *ROFeAs* (where *R* is rare-earth element) both the superconductive state and T_c are determined by defects (Hosono, 2008). These materials are characterized by layered structure consisting of alternating molecular layers *R-O* and *Fe-As*. The superconductivity in such compounds originates from electron doping by F or hole doping by introducing oxygen vacancies (Sadovskii, 2008; Izyumov &Kurmaev, 2008; Ivanovskii, 2008).

5. Conclusion

The results of this chapter allow us to draw following conclusions concerning the electronic spectrum of graphene with defects, the phonon quasi-particle spectra of graphene, bigraphene and metal-intercalated graphite:

i. It was shown that a vacancy in graphene conserves the Dirac singularity of the local density of electronic states in one of the sublattices. Moreover, a quasi-Dirac singularity was also observed in phonon spectra of graphene for atom displacements in the direction perpendicular to layers.
ii. Calculations and description of the phonon spectrum and vibrational characteristics of the bulk graphite and carbon nanofilms (bigraphene and trigraphene) lead to the explanation of the plane shape stability of bigraphene and trigraphene nanofilms at room temperatures.
iii. Analyzing the changes in phonon spectra of graphite intercalated with various metals, using the BCS theory, we have proposed, taking into account dynamic properties of these layered systems, a possible way leading to the increase of T_c.

Graphite and metallic compounds on its basis seem to be very attractive for complex investigations of electron and phonon spectra, the electron-phonon interaction, and for the study of superconductive transition conditions. The reasons for this attractiveness are, among other:

i. well pronounced flexural modes of graphite, practically coincident (in the same value of quasi-momentum) with the Dirac peculiarities in phonon and electron spectra;
ii. an increased quantity of phonon and electron states near the Fermi level caused by intercalating metals, and
iii. the existence of highly frequency phonon modes, pointing out to their attractiveness.

It should be noted, that in the review (Maksimov, 2008) author argues that in the MgB_2 compound electrons most strongly interact with flexural modes. This seems to be very interesting for the description of the evolution of the superconducting transition temperature in intercalated graphite, since flexural modes play an essential role in graphite phonon spectra.

Finally, for further analysis of the mechanism of superconducting transition in graphite intercalated with metals more complex investigations of the phonon and electron spectra and of the dependence of the electron-phonon interaction on intercalating metal are needed.

6. Acknowledgments

This work was supported by the grants of the Ukrainian Academy of Sciences under the contracts No. 23/07-H, No. 4/10-H and No. 4119 UNTS and by the grants of the Slovak Research and Development Agency under No. APVV-VVCE-0058-07, No. APVV-0006-07 and VEGA No. 1/0159/09.

7. References

Allen, P.B. & Mitrovic, B. (1982). Theory of Superconducting T_c. In: *Solid State Physics*, Seitz, F.; Turnbull, D. & Ehrenreich, H., (Ed.), 1 - 92 Academic, New York

Ashcroff, N.W. (2004). Hydrogen Dominant Metallic Alloys: High Temperature Superconductors? *Phys. Rev. Lett.*, Vol. 92, (2004) 187002 - 6

Bardeen, J.; Cooper, J. L.N. & Schriffer, J.R. (1957). Theory of Superconductivity. *Phys. Rev.*, Vol. 108, (1957) 1175 - 1204

Belen'kii, G.L.; Belen'kii, E.L.; Salaev, E.Yu. & Suleimanov, R.A. (1988). Deformation effects in layer crystals. *Physics-Uspekhi (Advances in Physical Science)* Vol. 31, (1988) 434 - 456

Bena, C. & Kivelson, S.A. (2005). Quasiparticle scattering and local density of states in graphite. *Phys. Rev. B.*,Vol. 72, (2005) 125432 - 7

Castro Neto, A.H.; Guinea, F.; Peres, N.M.; Novoselov, K.S. & Geim, A.K. (2009). The electronic properties of graphene. *Rev. Mod. Phys.*,Vol 81, (2009) 110 -155

Chen, Li; Zhang, Y. & Shen, Yui (2007). Self-healing in defective carbon nanotubes: a molecular dynamic study. *J. Phys.: Cond. Matter,* Vol. 19, (2007) 386212-38616

Dresselhaus, M. S. & Dresselhaus, G. (2002). Intercalation Compounds of Graphite. *Adv. in Phys.*,Vol. 51, (2002) 1 - 186

Emery, N.; Herold, C.; d'Astuto, M.; Garcia, V.; Bellina, Ch.; Mareche, J.F.; Lagrange, P. & Loupias, G. (2005). Superconductivity of Bulk CaC_6 . *Phys. Rev. Lett.*, Vol. 95, (2005) 087003 - 7

Ivanovskii, A.L. (2008). New high-temperature superconductors based on rare-earth and transition metal oxyarsenides and related phases. *Physics-Uspekhi (Advances in Physical Science),* Vol. 51, (2008) 1229 - 1260

Izyumov, Yu.A. & Kurmaev, E.Z. (2008). FeAs systems: a new class of high-temperature superconductors . *Physics-Uspekhi (Advances in Physical Science),* Vol. 51, (2008) 1261 - 1286

Feher, A.; Gospodarev, I. A.; Grishaev, V. I.; Kravchenko, K. V.; Manzheli , E. V. & Syrkin, E. S. (2009). Effect of defects on the quasiparticle spectra of graphite and graphene. *Low Temp. Phys.,* Vol. 35, (2009) 679 – 686

Gospodarev, I. A.; Eremenko V. V.; Kravchenko, K. V.; Sirenko V. A.; Syrkin, E. S. & Feodosyev, S. B. (2010). Vibrational characteristics of niobium diselenide and graphite nanofilms. *Low Temp. Phys.,* Vol. 36, (2010) 344 - 350

Gospodarev, I. A.; Kravchenko, K. V.; Syrkin, E. S. & Feodosyev, S. B. (2009). Quasi-two-dimensional features in the phonon spectrum of graphite. *Low Temp. Phys.,* Vol. 35, (2009) 589 - 595

Hosono, H. (2008). Layered Iron Pnictide Superconductors: Discovery and Current Status. *J. Phys. Soc. Japan, Suppl. C,* Vol. 77, (2008) 1 - 8

Kosevich, A.M.; Syrkin, E.S. & Feodosyev, S.B. (1994). Peculiar Features of Phonon Spectra of Low-Dimensional Crystals. *Phys. Low-Dim. Struct.,* Vol. 3, (1994) 47 - 51

Kossevich, A.M. (1999). The Crystal Lattice (Phonons, Solitons, Dislocations), WILEY-VCH Verlag Berlin GmBH, Berlin

Kotlyar, A.V. & Feodosyev, S.B. (2006). Local oscillations in crystal lattices with a single connected region of the quasi-continuous phonon spectrum. *Low Temp. Phys.,*Vol. 32, (2006) 256 - 262

Lifshits, I.M. (1945). On the theory of regular pertubations. *Reports of AS SSSR,* Vol. 48, (1945) 83 - 86

Maksimov, E.G. (2008). Room-temperature superconductivity: myth or reality? *Physics-Uspekhi (Advances in Physical Science),* Vol. 51, (2008) 167 - 170

Maksimov, E.G. & Dolgov, O.V. (2007). *A* note on the possible mechanisms of high-temperature superconductivity. *Physics-Uspekhi (Advances in Physical Science),* Vol. 50, (2007) 933 - 937

Maksimov, E.G.; Ebert S.V.; Magnitskaya M.V. & Karakozov, A. E. (2007). Ab-initio calculations of the physical properties of transition metal carbides and nitrides and possible routes to high-T_c superconductivity. *J. Exp. Theor. Phys.,* Vol. 105, (2007) 642 – 651

Maultzsch, J.; Reich, S.; Thomsen, C.; Requardt, H. & Ordejön, P. (2004). Phonon Dispersion in Graphite. *Phys. Rev. Lett.,* Vol. 92, (2004) 075501 - 4

Meyer, J. C.; Geim, A. K.; Katsnelson, M. I.; Novoselov, K. S.; Booth, T. J. & Roth, S. (2007). The structure of suspended graphene sheets. *Nature,* Vol. 446, (2007) 60 - 63

Morozov, S.V. & Firsov, A.A. (2009). Transport properties of bigraphene, *Abstract of Reports on the 35th Conference on Low-Temperature Physics,* 198. Chernogolovka, Russia, September 29 - October 2, 2009

Nagamutsu, J.; Nagakawa, N.; Muranaka, T.; Zenitani, Y. & Akimutsu J. (2001). Superconductivity at 39 K in magnesium diboride. *Nature,* Vol. 410, (2001) 63 - 65

Nicklow, R.; Wakabayashi, N. & Smith, H.G. (1972). Lattice Dynamics of Pirolitic Graphite. *Phys. Rev. B,* Vol. 5, (1972) 4951 – 4962

Novoselov, K. S.; Geim, A. K.; Morozov S. V.; Diang, D.; Katsnelson, M. I.; Grigorieva, I. V.; Dubonos, S. V. & Firsov, A. A. (2005). Two-dimensional gas of massless Dirac fermions in graphine. *Nature*, Vol. 438, (2005) 197 - 201

Peresada, V.I.; Afanasyev, V.N. & Borovikov, V.S. (1975). On calculation of density of states of single-magnon perturbations in ferromagnetics. *Sov. Low Temp. Phys.*, Vol. 1, (1975) 227 - 232 (and see references here)

Peres, N.M.R; Klironomos, F.D.; Tsai, S.-W.; Lopes dos Santos, J.M.D. & Castro-Neto, A.H. (2007). Electrom waves in chemistry substituted graphene. *Europhys. Lett.*, Vol. 80, (2007) 67007 - 67012

Rainer, D. (1986). Principles of ab-initio calculations of superconducting transition temperatutes, In: *Progress in Low Temperature Physics Vol. X*, Brewer, D. F. (Ed.), 371 – 421 Elsevier Science Publisher B.V., Amsterdam

Sadovskii, M.V. (2008). High-temperature superconductivity in iron-based layered compounds. *Physics-Uspekhi (Advances in Physical Science)*, Vol. 51, (2008) 1201 – 1227

Savini, G.; Ferrari, A.C. & Giustino, F. (2010). First-Principles Prediction of Doped Graphane as a High-Temperature Electron-Phonon Superconductor. *Phys. Rev. Lett.*, Vol. 105, 037002 - 4

Skrypnyk, Yu.V. & Loktev, V.M. (2006). Impurity effects in a two-dimensional systems with the Dirac spectrum. *Phys. Rev. B*, Vol. 73, (2006) 241402 - 6

Skrypnyk, Yu.V. & Loktev, V.M. (2008). Spectral function of graphene with short-range impurity centers. *Low Temp. Phys.*, Vol. 34, (2008) 818 - 825

Shangduan Wu; Lei Jing; Qunxiang Li; Shi Q. W.; Chen, Jie; Xiaoping Wang & Jinlong Ya (2008). Average Density of States in Disordered Graphene Systems. *Phys. Rev. B*, Vol. 77, (2008) 195411 - 7

Syrkin, E. S.; Feodosyev, S. B.; Kravchenko, K. V.; Eremenko, A. V.; Kantor, B. Ya. & Kosevich, Yu. A. (2009). Flexural rigidity of layers and its manifestation in the vibrational characteristics of strongly anisotropic layered crystals. *Low Temp. Physics*, Vol. 35, (2009) 158 - 162

Tse, J.S.; Yao, Y. & Tanaka, K. (2007). Novel Superconductivity in Metallic SnH_4. *Phys. Rev. Lett.*, Vol. 98, (2007) 117004 - 8

Weller, T.E.; Ellerby, M.; Saxena, S.S.; Smith, R.P. & Skipper, N.T. (2005). Superconductivity in the intercalated graphite compounds C_6Yb and C_6Ca. *Nat. Phys.*, Vol. 1, (2005) 39 - 41

6

Structural and Electronic Properties of Hydrogenated Graphene

Duminda K. Samarakoon and Xiao-Qian Wang
Department of Physics and Center for Functional Nanoscale Materials, Clark Atlanta University, Atlanta, Georgia 30314
USA

1. Introduction

1.1 Overview

Graphane is a two-dimensional system consisting of a single planar layer of fully saturated carbon atoms, which has recently been realized experimentally through hydrogenation of graphene membranes. In this chapter, we categorize theoretical approaches using the first-principles calculations, and we discuss in some detail our applications of calculation approaches to graphane systems. Specifically, we have studied the stability of chair, boat, and twist-boat graphane structures. Our results indicate that locally stable twist-boat membranes significantly contribute to the experimentally observed lattice contraction. The band gaps of graphane nanoribbons decrease monotonically with the increase of the ribbon width and are insensitive to the edge structure. We have also studied the electronic structural characteristics in a hydrogenated bilayer graphene under a perpendicular electric bias. The bias voltage applied between the two hydrogenated graphene layers allows continuously tuning the band gap and leads a transition from semiconducting to metallic state. Desorption of hydrogen from one layer in the chair conformation yields a ferromagnetic semiconductor with tunable band gap. Finally, we offer some views on the strength and weakness of the approaches that are discussed, and touch upon some of the challenging problems that need to be addressed in the future.

1.2 Graphene and graphane

Graphene, a single layer of all-carbon hexagonal network, is an emerging material for applications in electronics and photonics Berger et al. (2006); Geim & Novoselov (2007); Geim et al. (2007); Gilje et al. (2007); Novoselov et al. (2005); Wang et al. (2008); Zhang et al. (2006). As a truly two-dimensional system and a zero-gap semiconductor where the carriers behave as massless fermions, graphene possesses a number of outstanding electronic properties such as tunable carrier type and density Berger et al. (2006), exceptionally high carrier mobility Zhang et al. (2006), quantization of the conductivity Novoselov et al. (2005), and fractional quantum Hall effect (QHE) even at room temperature Zhang et al. (2005). These phenomena, particulary the QHE Zhang et al. (2005), have elucidated many important aspects of quantum many-body systems Zhang et al. (2005). The corresponding electronic states in graphene promote theoretical advances Ciftja & Fantoni (1996); Halperin et al. (1993); Jain (1989); Laughlin (1983) in studying strongly correlated Dirac fermions. Functionalizing graphene by reversible hydrogenation can change the electronic properties from metallic to semiconducting owing to the induced changes of functionalized carbon from sp^2 to sp^3 hybridization Geim et al. (2007); Gilje et al. (2007); Wang et al. (2008). The resultant hydrocarbon compound, graphane, can be modified into new materials, fine tuning its electronic properties, and has opened up increasingly fertile possibilities in hydrogen storage and two dimensional electronics Berger et al. (2006); Geim & Novoselov (2007); Geim et al. (2007); Gilje et al. (2007); Novoselov et al. (2005); Wang et al. (2008); Zhang et al. (2006).

1.3 Energy gaps

To develop increasingly small and fast transistors, it is desirable to have an energy gap Berger et al. (2006); Zhang et al. (2006). In contrast to complicated graphene-based structures like quantum point contacts and quantum dots, chemical derivatives of graphene provide a unique tool for controlling electronic properties Gilje et al. (2007). In order to utilize their remarkable electronic characteristics, it would be highly desirable to understand the associated electronic structures. Based on first-principles density-functional calculations, the stability and semiconducting behavior of graphane, an extended two dimensional fully saturated hydrocarbon derived from a single graphene sheet Boukhvalov et al. (2008); Sofo et al. (2007), was predicted. Recent experiments Elias et al. (2009) demonstrated the graphane formation by exposing pristine graphene to atomic hydrogen. There exist drastic changes in the crystal structure of graphane such that the lattice spacing shrinks by as much as 5% whereas the hexagonal symmetry remains intact.

The experimentally observed lattice spacing Elias et al. (2009) has a significantly broader variation than theoretically studied conformations of graphane Boukhvalov et al. (2008); Sofo et al. (2007). Theoretical work has considered two conformations: a chair like conformer in which hydrogen atoms are alternating on both sides of the plane, and a boat like conformer in which hydrogen atoms are alternating in pairs Boukhvalov et al. (2008); Sofo et al. (2007). In the ground state chair conformation of graphane, hydrogen attaches to graphene sublattices from two opposite sides and carbon atoms in the sublattices move out of the plane that yields the shrinkage of the in-plane periodicity. However, the change in hybridization from sp^2 to sp^3 leads to longer C-C bonds, which surpasses the lattice shrinkage by chair membrane buckling. The experimental observation of more compressed areas implies the existence of alternative membranes in the crystal structures of graphane, which results in stronger membrane buckling and shorter in-plane lattice spacing Elias et al. (2009).

We have performed a comprehensive investigation of structural and electronic properties of the graphane and graphane nanoribbons. We employ a combination of classical molecular dynamics Tersoff (1988) and first-principles density-functional approach. Kresse & Furthmuller (1996) Classical molecular dynamics was used to pre-screen molecular geometries, and first-principles calculations were employed to determine the electronic structure. Our results indicate that the locally stable twist-boat membranes lead to pronounced lattice shrinkage, and thus contribute to the broader distribution of lattice spacing observed experimentally Elias et al. (2009). The incorporation of twist-boat membranes into the crystal structure of graphane is shown to preserve the semiconducting feature of graphane. Furthermore, a systematic study of the graphane nanoribbons shows the band gaps of graphane nanoribbons decrease monotonically with the increase of the ribbon width and are insensitive to the edge structure.

1.4 Bilayer graphane

Graphene is a one-layer sheet of carbon with a structure that resembles chicken wire. Graphene has proven to possess unique electronic and physical properties, such as the unconventional quantum Hall effect Zhang et al. (2006), and high carrier mobility at room temperature Berger et al. (2006); Neto et al. (2009); Novoselov et al. (2004), thereby holding potential for a wide range of applications including graphene transistors, integrated circuits, and biosensors Berger et al. (2006); Neto et al. (2009); Novoselov et al. (2004); Zhang et al. (2006). The quantum-Hall effect Tsui et al. (1982) and the associated strongly correlated electron systems have generated a tremendous impetus on the development of novel ideas in many-body physics like the existence of fractionally charged quasiparticles Laughlin (1983), topological quantum numbers Thouless (1998), chiral Luttinger liquids Wen (1990; 1991), composite fermion particles Ciftja (2000); Jain (1989), and Chern-Simmons effective-field theories Ciftja & Wexler (2001); Halperin et al. (1993). Bringing graphene up to the level of technologically relevant material, however, depends on improved understanding and control of the structural and electronic properties. Specifically, an energy gap can be engineered by introducing lateral confinement such as in graphene nanoribbons Li et al. (2008); Nduwimana & Wang (2009), hydrogenated graphene Elias et al. (2009); Guisinger et al. (2009); Sofo et al. (2007), or in biased bilayer graphene Castro et al. (2007; 2008); Mak et al. (2009); McCann (2006); Min et al. (2007); Nilsson et al. (2008); Ohta et al. (2006); Oostinga et al. (2008); Zhang et al. (2009). The engineering of band gaps generates a pathway for possible graphene-based nanoelectronic and nanophotonic devices.

The extremely high carrier mobility makes graphene an ideal material for nanoelectronic applications, especially in field effect transistors Berger et al. (2006); Neto et al. (2009); Novoselov et al. (2004). Although graphene nanoribbon field effect transistors have been shown to exhibit excellent properties Li et al. (2008), mass production of graphene nanoribbon-based devices is beyond the capability of current lithography technology Nilsson et al. (2008). An alternative route to induce the formation of a band gap is through the hydrogenation of graphene Elias et al. (2009); Guisinger et al. (2009). The modification of the carbon bonds associated with the hydrogenation preserves the crystalline order of the lattice, but leads to re-hybridization of the carbon atoms from a planar sp^2 to a distorted sp^3 state Sofo et al. (2007). Recent experimental studies have demonstrated reversible hydrogenation through heating and proceeding with de-hydrogenation of the graphane to graphene Elias et al. (2009). On the other hand, bilayer graphene has attracted a great deal of attention recently. In bilayer graphene, the low energy excitations are one of the characteristics of massive chiral fermions, unlike Dirac fermions in graphene Novoselov et al. (2005). Most importantly, bilayer graphene can have a tunable gap *via* chemical doping or by applying an external gate voltage. In lieu of the increasing amount of experimental and theoretical studies of the bilayer graphene transistors Xia et al. (2010), the exploration of various modified bilayer systems could play a crucial role in future nanoelectronics applications.

Experimental advances have motivated our study of what could emerge if bilayer graphene were subjected to hydrogenation and electric bias. We have investigated the corresponding bilayer systems based on first-principles density-functional calculations. Fully-hydrogenated bilayer graphene is similar to the one-layer graphane in that the electronic properties change from metallic to the semiconductive due to the induced changes of functionalized carbon from sp^2 to sp^3 hybridization, and the interlayer chemical bonding that stabilizes the hydrogenated structure Leenaerts et al. (2009). We show that with applied electric bias, the resultant energy gap can be tuned. Of particular interest are the effects associated with symmetry breaking due to the presence of an external electric field perpendicular to the hydrogenated bilayer graphene. Our theoretical study suggests a unique opportunity to tune the band gap of a ferromagnetic semiconductor with desorption of hydrogen from one layer in the chair conformation.

The chapter is organized as follows. In Section 2 we discuss the first-principles calculation method employed. In Section 3, we describe in detail the study of lattice contraction observed experimentally, through careful examinations of chair, boat, and twist-boat membranes. We then investigate the effect of electric bias on the bilayer graphane. Finally, in Section 4 we summarize our theoretical simulation work.

2. Method

2.1 First-principles calculations

Although first-principle methods are generally more reliable than the empirical methods, they are currently limited to small systems (a few hundred atoms). Recently, several groups have developed methods for performing first-principles electronic structure calculations that scale linearly with system size (the $O(N)$ methods). These methods are now applicable to systems that could only be studied by means of empirical and semi-empirical methods a decade ago. Moreover, the relative reduction in computational cost enables the molecular-dynamics simulations and therefore the investigation of complicated physical and chemical systems.

Recent advances in *ab initio* methods have experienced a considerable amount of success in predicting ground-state structural and cohesive properties of condensed-matter systems. The pioneering work of Car and Parrinello based on dynamical simulated annealing promoted a new type of approach applicable to density-functional theory within the local-density approximation (LDA). Density-functional molecular dynamics (Car-Parrinello) and other iterative methods based on plane-wave basis have made such calculations possible for systems consisting of several hundred atoms.

Most of contemporary LDA calculations are based on the Kohn-Sham formulation. LDA provides structural and elastic data in good agreement with experiment; lattice constants, bulk modules, elastic constants and phonon frequencies are usually predicted within 5% of experimental values. For binding energies, LDA consistently overestimates experimental values by approximately 10-20%. This error is attributed to the incomplete cancelation of errors within the LDA method. While these methods have been very successful, several difficulties arise when they are extended to systems with large length scales or those containing transition-metal atoms.

The first-principles methods based on plane-wave basis sets require many components in the expansion to keep track of the locality of the electronic wave functions. In the plane-wave basis, the kinetic energy operator is diagonal while the potential energy matrix is not sparse. In contrast, both the kinetic energy and potential energy matrix are approximately band-diagonal in the wavelet basis. Moreover, the wavelet transform, along with the associated multiresolution analysis, does not involve long-range operations and is thus particularly suitable for parallelization and wave-function-based $O(N)$ algorithms, since every operation can be partitioned into hierarchical real-space domains. In the following, we briefly discuss the density functional theory, the first-principles molecular dynamics, code developments, and the wavelet bases for electronic structure calculations.

2.2 Density functional theory

To properly handle a many-electron system so that one can derive its various properties from fundamental quantum mechanics is a constant challenge in theoretical physics and chemistry.

Although the interaction between electrons is well known, the facts that electrons, with a spin quantum number of 1/2, have to obey specific statistical rules and that one normally has to deal with quite a few of them at the same time make this problem immensely formidable. One approach that has become the standard one for large-scale electronic simulations is the density functional theory in the so-called Kohn-Sham framework. It is based on a theorem stating that the ground-state energy of a many-electron system can be represented as a functional of the electron density only. As a result, one can obtain the electronic energy without dealing with the many-body wave function which is highly multidimensional with the notorious property of being antisymmetric with respect to particle exchange. Being a scalar in the real space, the electron density is a much simpler quantity to manage, making it possible to investigate more complex systems. By minimizing the energy functional with respect to possible density distributions one can then determine the ground-state electronic energy for a given atomic arrangement.

The energy minimization procedure is most conveniently carried out by a mapping of the truly interacting system to an auxiliary system of noninteracting particles with the same density distribution. The resulting total-energy functional

$$E[n] = T_0[n] + \int d^3r\, v_{ext}(\mathbf{r})\, n(\mathbf{r}) + E_h[n] + E_{xc}[n] \tag{1}$$

includes the kinetic energy functional of the noninteracting system $T_0[n]$, external potential energy, Hartree energy $E_h[n]$, and the so-called exchange-correlation energy functional $E_{xc}[n]$. $E_{xc}[n]$ includes all the many-body effect as well as the difference in the kinetic energies of the interacting and noninteracting systems.

The direct variation of energy with respect to the density is replaced by finding the noninteracting orbitals ψ self-consistently in the local Kohn-Sham equations

$$(-\frac{\hbar^2}{2m}\nabla^2 + V_{ext} + V_{eff})\psi_i = \varepsilon_i \psi_i\,, \tag{2}$$

where the effective one-particle potential V_{eff} includes the Hartree potential and the exchange-correlation potential derived from a functional derivative $V_{xc} = \delta E_{xc}/\delta n$. The density is calculated from all occupied one-particle orbitals. The fact that the effective potential is a simple local function makes a tremendous difference in practical calculations. Other quantum-chemistry schemes such as the Hartree-Fock method commonly involves nonlocal operators which require much more computational resources. It is ok that the density-functional theory has become the prevailing approach in modern electronic-structure calculations with wide applications in quantum chemistry and materials physics.

Inarguably one could not have solved the exact many-body problem by regrouping energy terms. As a matter of fact, although the existence of the exchange-correlation energy functional E_{xc} is fully established, its exact form remains unknown and contains integrals of nonlocal quantities. It is therefore a challenging many-body problem to investigate this important quantity in real materials. In practical calculations, approximations to the energy functional E_{xc} are required. Commonly used ones include the local-density approximation (LDA), in which the density is assumed to be locally uniform and the result for a homogeneous electron gas is used point by point based on the local density, and the generalized gradient approximation (GGA), in which the gradient correction to the LDA is added.

In order to study systems of hundreds of atoms, one focuses on the properties of the valence electrons and employ norm-conserving pseudopotentials to model the effects of core electrons. The one-particle orbitals will be expanded in terms of plane waves to eliminate any bias in the basis functions. The self-consistent solution of the corresponding Kohn-Sham orbitals will be carried out by evaluating relevant quantities in either the real space or momentum space.

We have performed first-principles calculations for various graphane structures. The structure and electronic properties of all conformations were investigated using first-principles density-functional calculations. Perdew-Burke-Ernzerhof parametrization Perdew et al. (1996) of the generalized gradient approximation were used in all the calculations. A kinetic energy cutoff of 280 eV in the plane-wave basis and appropriate Monchorst-Pack *k*-points ($6 \times 6 \times 1$ for graphane and $10 \times 1 \times 1$ for graphane nanoribbons) were sufficient to converge the grid integration of the charge density. Although the first-principles approach systematically underestimates the band gaps Yang et al. (2008), we are interested primarily in the general feature of the conformations or membranes. The initial search for stable structures was carried out through classical molecular dynamics by means of Tersoff potential Tersoff (1988). The obtained local energy-minimum structures were further optimized through first-principles calculations with forces less than 0.01 eV/Å.

The structural and electronic properties were investigated using first-principles density-functional calculations $DMol_3$ (2010). Our first-principles calculations are based on spin-polarized density functional theory with local density approximation (LDA) for exchange-correlation potential Vosko et al. (1980). A supercell with a vacuum space of 16 Å normal to graphene plane was used. A kinetic energy change of 3×10^{-4} eV in the orbital basis and appropriate Monchorst-Pack *k*-point grids of $6 \times 6 \times 1$ were sufficient to converge the integration of the charge density. The optimization of atomic positions proceeds until the change in energy is less than 1×10^{-6} eV per cell. Although the LDA approach systematically underestimates the band gaps, we are primarily interested in the relative stability of the conformations and the electric field effects. While calculations based on hybrid functionals or many-body GW approaches can rectify the gaps (the rectified gap is 5.2−5.4 eV *vs.* the LDA result of 3.6 eV for graphane) Samarakoon & Wang (2009); Zhang et al. (2009), the implementation of the corresponding electric-field effect is cumbersome. The LDA approach is expected to provide qualitatively correct pictures and remains the popular choice for investigations of electric-field effects Zhang et al. (2009). Another reason for choosing LDA is attributed to the fact that generalized-gradient-approximation (GGA) leads to weak bonding between graphene layers and yields excessively large values of bilayer distance. By contrast, LDA calculation gives rise to a bilayer distance of $\sim$3.3 Å in good conformity with the results of graphite Partoens & Peeters (2006).

3. Results and discussion

3.1 Cyclohexane membranes

In order to effectively search for stable crystal structures of graphane, it is instructive to make reference to distinctive configurations of cyclohexane (see top panel of Figure 1) referred to as chair, boat, twist-boat and chair-twist-boat, respectively. Due to the inherent tendency of the sp^3 hybridization on tetravalent carbons, cyclohexane does not form a planar hexagonal arrangement. The chair isomer is the ground state configuration, while twist-boat is the second lowest-energy isomer. The chair conformation changes in the ring-flipping process, leading to the axial hydrogens becoming equatorial. Between the two stable chair states (with D_{3d} symmetry), the twist-boat (with D_2 symmetry), boat (with C_{2v} symmetry), and chair-twist-boat isomers can be constructed. The boat and chair-twist-boat forms are metastable states of the twist and chair forms, respectively. The twist-boat form

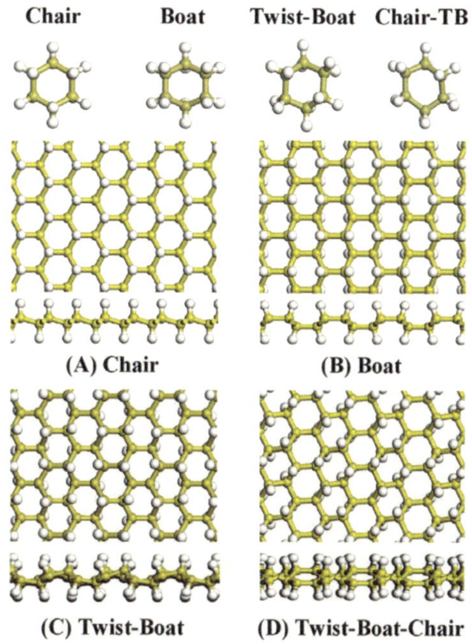

Fig. 1. Top panel: top view of chair, boat, twist-boat, and chair-twist-boat conformations of cyclohexane. Carbon atoms are shown as gold and hydrogen atoms are white. Middle and bottom panels: top and side views of graphane conformations for (A) chair, (B) boat, (C) twist-boat, and (D) twist-boat-chair, respectively. Reprinted with permission from Ref. Samarakoon & Wang (2009), Copyright © 2009 American Chemistry Society.

may be isolable because—like the chair form—it stands for an energy minimum. The boat conformation is free from angle strain, but has a higher energy than the chair form due to steric strain in connection to the flagpole interaction. The torsional strain in the boat conformation has a maximum value since two of the carbon bonds are fully eclipsed. This is to be contrasted to the chair conformer in which all bonds are staggered and complete absence of torsional strain, while the twist-boat has four partially eclipsed bonds.

3.2 Chair, boat, and twist-boat conformations in graphane

The counterparts of chair, boat, and twist-boat conformers of cyclohexane in two-dimensional structures of graphane can be constructed accordingly. We illustrate in Figure 1 the structures of chair, boat, and twist-boat conformations of graphane, along with a twist-boat-chair structure. The chair and boat structures coincide with configurations previously obtained using density functional calculations Boukhvalov et al. (2008); Sofo et al. (2007). The unit cell of chair and boat conformation has $P\bar{3}m1$ and $Pmmn$ symmetry, respectively. Consistent with the energy order of cyclohexane, the chair configuration is lower in energy and has less membrane buckling than those of boat conformer.

The twist-boat configuration of graphane has more in-plane shrinkage than either chair or boat conformation. However, the twist-boat structure becomes not stable against the boat conformation in geometry optimization using first-principles calculations. Closer scrutiny of the geometry optimization process from a twist-boat structure of graphane to the boat configuration reveals that the additional energy cost is attributed to the fact that all carbon

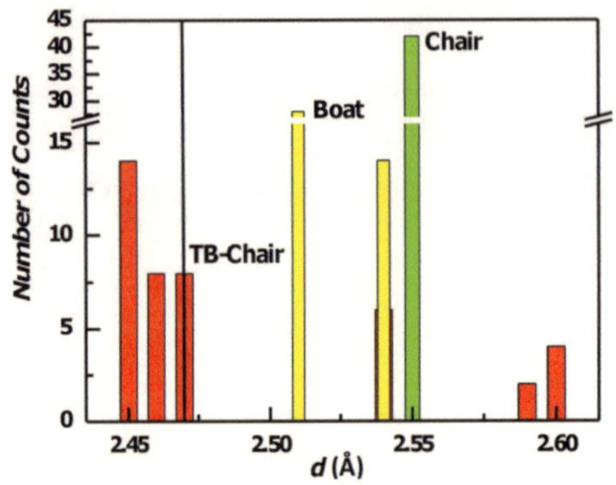

Fig. 2. The number of counts (collected from a sample with 42 lattice spacings) of in-plane lattice spacing for chair (green), boat (yellow), and twist-boat-chair (red) conformations of graphane, respectively. The vertical solid line corresponds to theoretical in-plane lattice spacing of graphene. Reprinted with permission from Ref. Samarakoon & Wang (2009), Copyright © 2009 American Chemistry Society.

Structure	E_B (eV)	E_g (eV)	a (Å)	a_0 (Å)	d (Å)
Graphene	-8.57	0	1.42	1.42	2.47
Chair	-12.23	3.5	1.54	1.47	2.55
Boat	-12.06	3.5	1.54, 1.58	1.40, 1.58	2.51, 2.54
TB-Chair	-11.97	3.8	1.53-1.57	0.89-1.51	2.45-2.60

Table 1. Calculated binding energy per carbon atom E_B, band gap E_g, the bond length a and the associated planar projection a_0, the in plane lattice spacing d for chair, boat, twist-boat(TB)-chair conformations of graphane, respectively. Reprinted with permission from Ref. Samarakoon & Wang (2009), Copyright © 2009 American Chemistry Society.

atoms in the unit cell participate in the bond twisting process. By contrast, in cyclohexane only four out of six carbon atoms mimicking a pair of twist bonds are involved in the optimization between boat and twist-boat forms.

3.3 Lattice contraction in graphane

It becomes clear that the experimentally observed graphane is unlikely to be in the single crystal form of chair, boat, or twist-boat, since each of those structures has only one or two distinctive in-plane lattice spacing, in contrast to a wide range of distribution observed experimentally Elias et al. (2009). Moreover, the twist-boat configuration is no longer stable against the boat structure. However, the instability of the twist-boat crystal structure does not preclude the existence of locally stable twist-boat membranes. To pursue this scenario, we show in Figure 1 a twist-boat-chair configuration, which consists of adjacent twist-boat and chair membranes. Our first-principles calculation shows that the twist-boat-chair configuration is a stable structure of graphane, although the energy is slightly higher than the boat and chair conformations. Summarized in Table I are the structural and electronic properties of the conformation as compared to those for graphene as well as chair and boat conformations of graphane. It is worth noting that in the twist-boat-chair configuration, locally stable twist-boat membranes are favored over boat ones since two carbon atoms in the membrane serve as linkage atoms for the neighbor chair membranes. The unit cell of twist-boat-chair structure has a P_2/c symmetry with monoclinic angle $\gamma = 138.5°$.

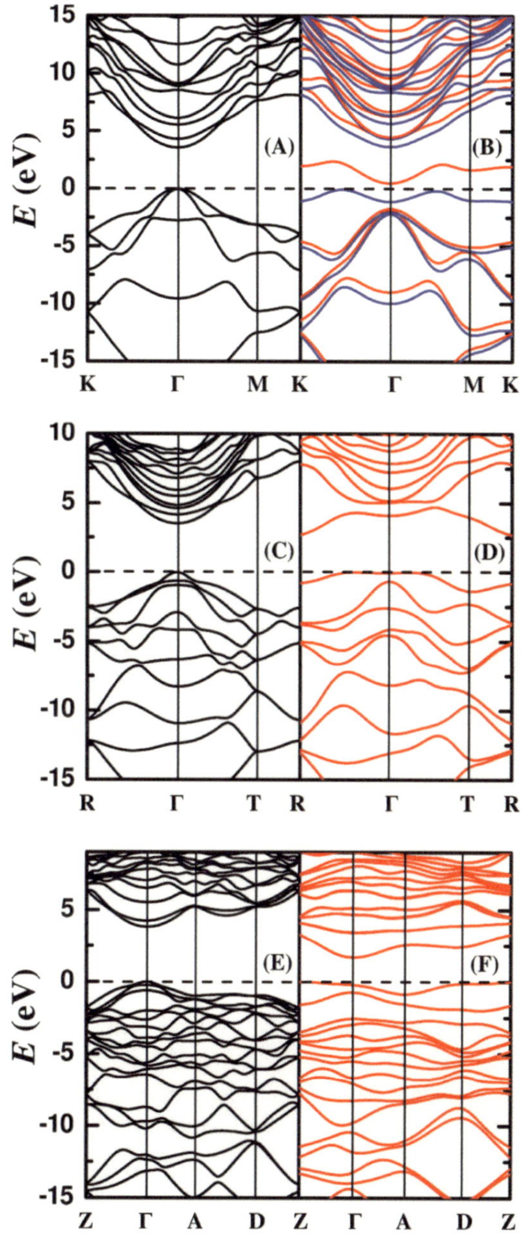

Fig. 3. Calculated band structure for (A) chair, (C) boat, and (E) twist-boat-chair conformations of graphane, and their counterparts (B), (D), and (F) for one-side half-hydrogenated graphene (graphone), respectively. The red and blue curves represent spin-up and down components, respectively. For chair conformation, **K**=$(\pi/3a, 2\pi/3a)$, **M**=$(0, \pi/2a)$, where $a = 2.55$ Å. For boat conformation, **R**=$(\pi/2b_1, \pi/2b_2)$, **T**=$(\pi/2b_1, 0)$, where $b_1 = 2.55$ Å and $b_2 = 4.33$ Å. For twist-boat-chair conformation, **Z**=$(0, \pi/2c_2)$, **A**=$(\pi/2c_1, 0)$, **D**=$(\pi/2c_1, \pi/2c_2)$, where $c_1 = 6.63$ Å and $c_2 = 4.92$ Å. The valence band maximum is set to 0 eV. Reprinted with permission from Ref. Samarakoon & Wang (2009), Copyright © 2009 American Chemistry Society.

An important ramification of our investigation of various graphane crystal structures is that the low energy conformations have no more than one parallel aligned nearest neighbor hydrogens, as the hydrogen in chair and boat has zero and one parallel aligned neighbor, respectively. On the other hand, the chair membranes in the twist-boat-chair structure may have two neighboring hydrogens. This trend is also observable for various other stable structures of graphane. We depict in Figure 2 the distribution of lattice spacing for various crystal structures of graphane. For the chair conformation, there exists only one distinctive lattice spacing. For the boat conformer, there are two distinctive in-plane lattice spacings. The twist-boat-chair conformation has a broad range of in-plane lattice spacings from 2.45−2.60 Å. The lowest lattice spacing of 2.45 Å is about 1% of lattice contraction as compared with the value obtained for graphene (2.47 Å), which is about 5% contraction of the value for chair (2.55 Å). This contraction can be correlated to short in-plane bond lengths which can be as small as 0.89 Å (see Table 1).

The transformation among various graphane structures amounts to flipping hydrogens from one side of the plane to another, along with the associated strain relaxation for the carbon atoms attached. The transition states are characterized with distortions of the hexagonal network with elongated carbon bonds in order to accommodate hydrogens that are in the network plane during the flipping process. This implies that once the hydrogens are absorbed onto graphene with fully saturation, it becomes difficult for the system to adopt the ground-state chair conformation. As a result, the sequence of alternating hydrogens on both sides of the plane is broken, introducing other type of membranes into the crystal structure of graphane. Consequently, this leads to out of plane distortions that induce in-plane shrinkage and results in a decrease of the in-plane lattice spacing in relation to that of the chair conformation.

We believe that the experimentally observed broad distribution of lattice spacings can be attributed, to a large extent, to the existence of membranes other than the chair form. In this regard, locally stable twist-boat membrane, as exemplified in twist-boat-chair, is the prototype of low energy configurations with paralleled aligned nearest-neighbor hydrogens. Furthermore, as exemplified in the twist-boat-chair structure, the twist-boat conformer serves as effective link to neighboring chair conformers.

3.4 Electronic band structures

The band gap of the twist-boat-chair structure, together with that of the chair and boat conformations can be extracted from the corresponding band structures illustrated in Figure 3. The band structure for twist-boat-chair resembles an interpolation of those of chair and boat ones, and the corresponding band gap of 3.8 eV is very close to those obtained for chair and boat structures Boukhvalov et al. (2008); Sofo et al. (2007). In all the cases, the graphane structures have direct gaps at the band center (Γ point).

3.5 Graphone

Recently, structures with one-sided hydrogenation that are reminiscent of hydrogenation on epitaxial graphene Guisinger et al. (2009) or graphene on a substrate have attracted a great deal of attention Elias et al. (2009). Of particular interest is the recent theoretical prediction that semi-hydrogenated graphene (graphone) becomes a ferromagnetic semiconductor with a small indirect gap. Zhou et al. (2009) The half-hydrogenation in the chair conformation breaks the delocalized π bonding network of graphene, leaving the electrons in the unhydrogenated carbon atoms localized and unpaired. While the idea of a ferromagnetic semiconducting graphone is extremely provocative, a careful examination of various graphone configurations

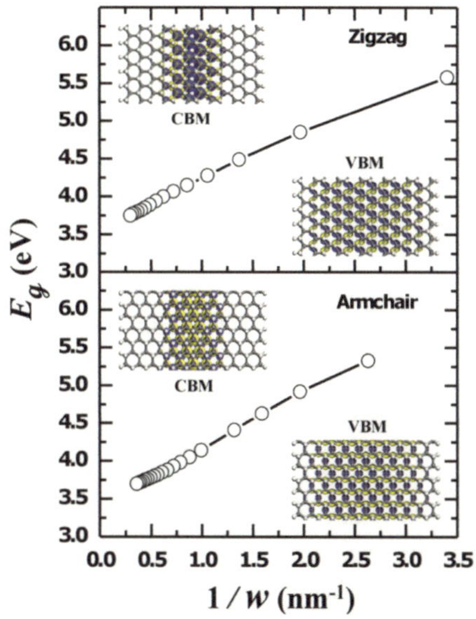

Fig. 4. Calculated band gaps of nanoribbons of graphane with zigzag (top panel) and armchair (bottom panel) edges, respectively. Insets: extracted charge density distribution at the band center (Γ point) of the corresponding CBM and VBM states, respectively. Reprinted with permission from Ref. Samarakoon & Wang (2009), Copyright © 2009 American Chemistry Society.

is necessary. Our calculation of the chair, boat, and twist-boat-chair conformations of graphone reveals that the ferromagnetic state with an indirect gap of 0.51 eV (see Figure 3B) is very fragile. An implication is that spin-polarized valence band maximum states show long-range correlations and thus depend on the size of the unit cell studied. On the other hand, the ground state of boat and twist-boat-chair conformations is nonmagnetic (Figures 3D and 3F). Moreover, in contrast to graphane in that the chair configuration is the lowest energy configuration, the energy for boat-graphone of -9.88 eV is lower than that of -9.42 eV for chair-graphone.

3.6 Graphane nanoribbons

The reason for a band gap opening up in hydrogenated graphene can be attributed to the changes from sp^2 bonded C atoms to sp^3 bonded ones Sofo et al. (2007). We have found that all the fully saturated graphane structures have a wide gap, including graphane nanoribbons. Shown in Figure 4 is the dependence of band gaps on the width of the graphane nanoribbon with armchair and zigzag edges. The naming of the armchair and zigzag nanoribbons follows the edge structure nomenclature, such that an armchair (zigzag) tube unfolds into a zigzag (armchair) ribbon Baron et al. (2006); Son et al. (2006). The ribbons involved in the present study were constructed based on chair conformations and were neutral bond saturated with hydrogen passivation at edges.

As is readily observable in Figure 4, the gaps of the ribbon decrease with increasing width w in an approximate $1/w$ fashion. The extracted charge density distribution of the conduction band minimum (CBM) and valence band maximum (VBM), as seen from insets of Figure 4, indicates predominantly confined electrons and holes in the proximity of ribbon center. The

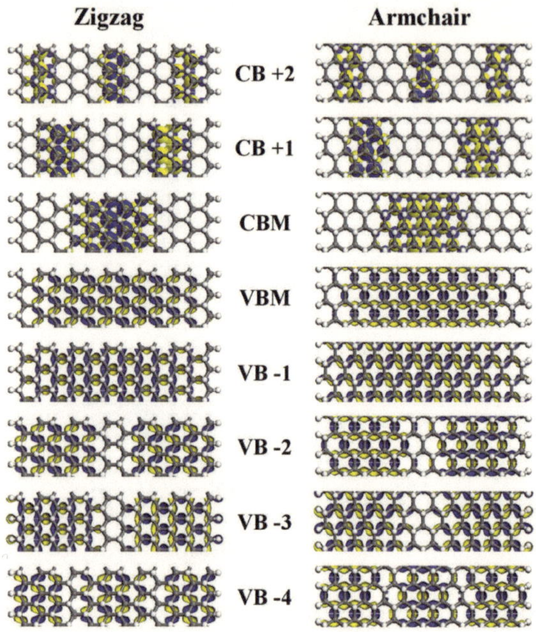

Fig. 5. Charge density distributions of near gap conduction and valence states for armchair (left panels) and zigzag (right panels) edged graphane nanoribbons, respectively. The sign of the wave function is indicated by light blue and yellow regions, respectively. The isovalue is 0.025 au. Reprinted with permission from Ref. Samarakoon & Wang (2009), Copyright © 2009 American Chemistry Society.

calculated gaps are insensitive to the details of the ribbon edge geometry and termination, in sharp contrast to sp^2 bonded graphene nanoribbons (GNRs). The band gap of ultra-thin GNR with armchair edges generally opens up due to the quantum confinement and the edge bond relaxation. The oscillatory band gap for GNR with armchair edges can be explained by the Fermi wavelength in the direction normal to the ribbon direction Baron et al. (2006); Son et al. (2006). Similar to zigzag SWNTs, the band gaps of AGNRs are divided into three groups, with the $3p+2$, $3p+1$, and $3p$ group (p is a positive integer) having a small, medium, and large gap, respectively Baron et al. (2006); Son et al. (2006). It is worth noting that the oscillatory behavior of sp^2 bonded GNRs completely disappears for sp^3 bonded graphane nanoribbons Li et al. (2009). Furthermore, our calculations based on spin-polarized calculations confirmed that the ground state of graphane nanoribbons with zigzag edges is not magnetic, in contrast to the staggered antiferromagnetic state for zigzag GNRs Son et al. (2006).
The extracted density distribution of holes and electrons for graphane nanoribbons with zigzag and armchair edges is illustrated in Figure 5. For conduction bands, the near gap states exhibit s, p, d, ... characters, in conformity with predictions from one-dimensional particle-in-a-box model Nduwimana et al. (2008). The s, p, d, ... features are also observable for valence states, but for a pair of nearly degenerate (at the band center) valence bands. The close resemblance of the charge distributions for zigzag and armchair edged graphane nanoribbons indicates that edge effect becomes dormant for sp^3 bonded ribbons. These results suggest that by tailoring the effective ribbon width it is feasible to design semiconductor graphane nanoribbons with a tunable band gap, which is advantageous over GNRs in that one can

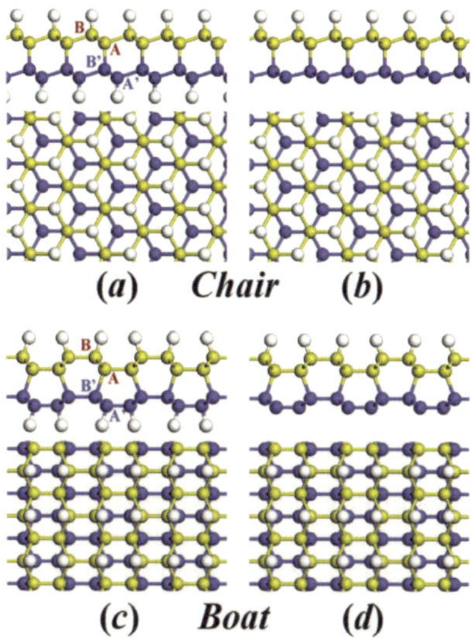

Fig. 6. Top panel: top and side views of (a) fully hydrogenated and (b) semi-hydrogenated chair conformations of bilayer graphene. Bottom panels: top and side views of (c) fully hydrogenated and (d) semi-hydrogenated boat structures of bilayer graphene. Carbon atoms on top and on bottom layers are colored with gold and blue, respectively. Hydrogen atoms are colored with white. Reprinted with permission from Ref. Samarakoon & Wang (2009), Copyright © 2009 American Chemistry Society.

avoid the daunting task of identifying the edge structures. Accordingly, the control of proper nanostructures may offer avenues for the design of highly effective nanodevices.

3.7 Patterned hydrogenation

The approach described in the present work can be employed to investigate hydrogenated graphene systems, such as patterned graphene nanoroads Singh & Yakobson (2009) that are composed of GNRs with fully saturated hydrogenation at the edges and well-defined sharp interfaces between sp^2 and sp^3 bonded membranes. Of particular interest is the interplay between sp^2 and sp^3 hybridizations that can be systematically investigated via band alignment analysis. The band alignment for patterned graphene nanoroads is based on the lineup of charge neutrality levels Nduwimana & Wang (2009) for sp^2 and sp^3 bonded components. Since sp^3 hybridization leads to wide gap semiconducting behavior, the electronic properties of patterned nanoroads are primarily determined by the sp^2 components. For instance, for graphene nanoroads with armchair edges, the band gaps are in accordance with results of GNRs with the effective sp^2 width and show oscillatory behavior Singh & Yakobson (2009).

3.8 Bilayer graphane

While the opening and external tuning of energy gap between valence and conduction bands in Bernal stacking bilayer graphene McCann (2006); Ohta et al. (2006) hold great potential for logic applications, switching off the conduction to a desirable level remains challenging in epitaxial graphene Neto et al. (2009). In this regard, it is of interest to investigate bilayer hydrogenated graphene that is semiconducting from the onset. Figure 6 depicts the fully and half hydrogenated chair and boat conformations. As can be seen from Figure 6, for the

Structure	E_B (eV)	E_g (eV)	l (Å)
Fully-hydrogenated Chair	-12.00	3.24	1.54
Fully-hydrogenated Boat	-11.93	2.92	1.54
Semi-hydrogenated Chair	-10.55	0.54	1.65
Semi-hydrogenated Boat I	-10.79	2.35	1.63
Semi-hydrogenated Boat II	-10.85	0.50	3.26

Table 2. Calculated binding energy per carbon atom E_B, band gap E_g, and the inter-layer bond length l for chair and boat conformations of fully hydrogenated and semi-hydrogenated graphene, respectively. Label I and II refers to boat conformations with and without interlayer bonding, respectively. Reprinted with permission from Ref. Samarakoon & Wang (2010), Copyright © 2009 American Chemistry Society.

fully-hydrogenated structures a chemical bonding between the $A - B'$ sites stabilizes both chair and boat conformations. In the latter case, the chemical bonding induces a structural transformation that deviates from the pattern of Bernal stacking.

3.9 Structural properties of bilayer graphane

The geometry details are listed in Table 2 along with the calculated binding energy and band gap. Analogous to graphane from the one-layer fully-hydrogenated graphene, the chair conformation Flores et al. (2009); Samarakoon & Wang (2009) is the lowest energy conformation for fully hydrogenated bilayer, in agreement with previous first-principles density-functional predictions Sofo et al. (2007). Furthermore, the corresponding chemical bonding between the bilayer remains stable with the desorption of hydrogen in one layer, resulting in a slight increase of the inter-layer bonding distance from 1.54 Å to 1.65 Å (Table 2). It is important to remark that while the inter-layer chemical bonding remains intact after desorption of hydrogens in one layer, the lowest energy configuration for one-sided hydrogenation is a boat conformation without the chemical bonding (Table 2). The crucial difference between the hydrogen desorption in one layer and the one-sided hydrogenation should be of particular interest in the forthcoming discussions.

3.10 Electronic properties of bilayer graphane under electric bias

There have been a number of theoretical studies on opening up a band gap in the gapless bilayer graphene if an electric field is applied perpendicularly Avetisyan et al. (2009); Grüneis et al. (2008); Liu & Shen (2009). The effect of the electric field can be studied by adding a potential *via* the nuclear charges. Our calculations show that the bilayer graphene opens a gap of $\sim$ 0.23 eV by an electric bias of $\sim$0.51 V/Å. This is in agreement with other theoretical predictions and experimental observations Avetisyan et al. (2009); Zhang et al. (2009). However, when the electric bias is further increased, the gap in the bilayer system collapses, and the system turns back to metallic with induced interlayer bonding $A - B'$ reminiscent of the hydrogenated bilayer graphene. We show in Figure 7 the calculated band structures for bilayer graphane for chair conformation. As is readily observable from Figure 7, the band gap decreases monotonically from about 3.24 to 0 eV with increase of electric bias. The critical bias for the semiconducting to metallic transition is estimated to be 1.05 V/Å.
Shown in Figure 8 are the corresponding charge densities. In the absence of bias, the charge density distributions are symmetrical both in conduction and valence bands. With the application of an electric bias, charges transfer in the conduction and valence bands acts in

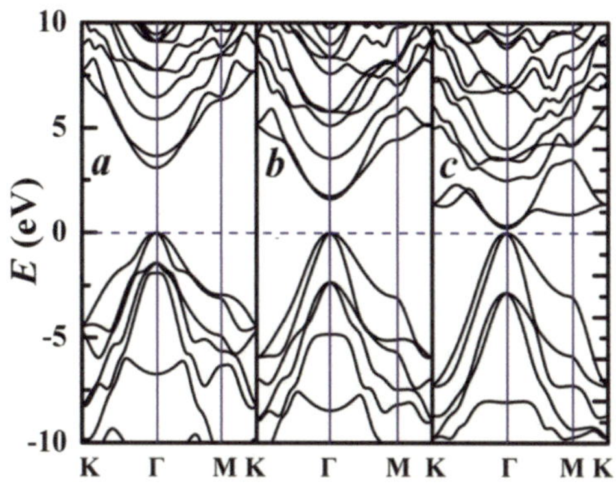

Fig. 7. Calculated band structure of fully-hydrogenated graphene in chair conformation with (a) no electric bias, (b) 0.39 V/Å electric bias, and (c) 1.03 V/Å electric bias, respectively. K $=(\pi/3a, 2\pi/3a)$, M $=(0, \pi/2a)$, where $a = 2.50$ Å. The valence band maximum is set to 0 eV. Reprinted with permission from Ref. Samarakoon & Wang (2010), Copyright © 2009 American Chemistry Society.

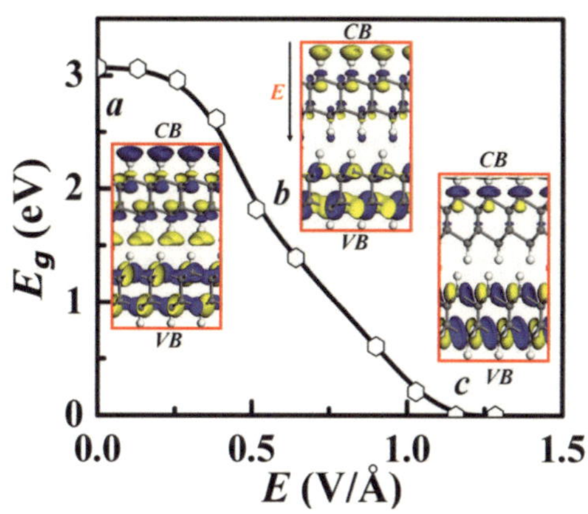

Fig. 8. Calculated dependence of band gap on perpendicular applied electric bias for the bilayer graphane in chair conformation: (a) no bias, (b) 0.39 V/Å electric bias, and (c) 1.03 V/Å electric bias, respectively. Insets: extracted charge density distribution at the band center (Γ point) of the corresponding conduction and valence band states. The isovalue is 0.025 au. Reprinted with permission from Ref. Samarakoon & Wang (2010), Copyright © 2009 American Chemistry Society.

a concerted fashion, resulting in charge accumulation and depletion in the conduction and valence bands, respectively.

3.11 Bilayer graphone

It is worth noting that there is no explicit magnetic states in fully-hydrogenated bilayer graphene. This indicates that the chemical bonds and the electric-field induced dipole-dipole interaction do not lead to unpaired spins. The unpaired spins can be generated through desorption of the hydrogen in one layer or through one-sided absorption. The latter scenario is particulary interesting in that one can take advantage of the electric field that generates chemical bonding prior to the hydrogenation. However, the chemical bonding is

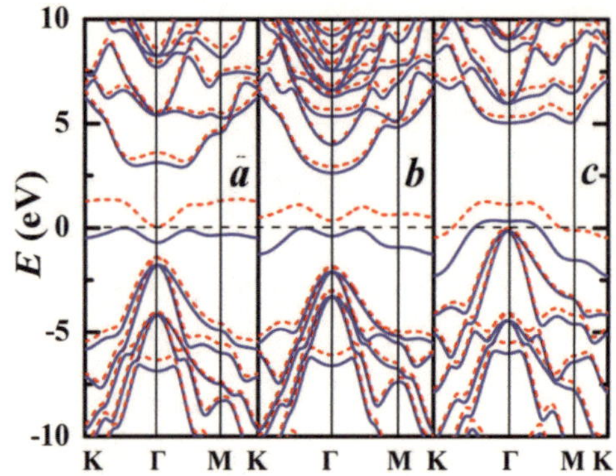

Fig. 9. Calculated band structure for semi-hydrogenated bilayer graphene in chair conformation with (a) -0.26 V/Å electric bias, (b) no electric bias, and (c) 0.39 V/Å electric bias, respectively. The red and blue curves represent spin-up and down components, respectively. Reprinted with permission from Ref. Samarakoon & Wang (2010), Copyright © 2009 American Chemistry Society.

simultaneously breaking when the electric bias is switched off. We have carefully studied both scenarios. The desorption of the fully-hydrogenated chair conformation can be readily confirmed. However, the one-sided hydrogenation is much more involved due to the crucial dependence of the hydrogenation patterns, which favors a boat conformation at large hydrogen coverage that is non-magnetic.

Hydrogenation of graphene is reversible, providing the flexibility to manipulate its coverage Elias et al. (2009). The desorption of the hydrogen atoms from one side of graphane will result in a semi-hydrogenated bilayer graphene which is the counterpart of the monolayer "graphone" Zhou et al. (2009). Graphone is a ferromagnetic semiconductor with a small indirect gap attributed to the breaking of the delocalized π-bonding network of graphene delocalization, which is associated with localized and unpaired electrons Boukhvalov et al. (2008); Zhou et al. (2009). Shown in Figure 9 are the calculated band structures for one side hydrogenated bilayer graphene under electric bias. For semi-hydrogenated bilayer graphene there is an indirect band gap about 0.54 eV (Figure 9b). This changes to metallic for biased voltages bellow -0.26 V/Å (Figure 9a) or above 0.39 V/Å (Figure 9c).

Our results show that the bilayer counterpart of graphone is ferromagnetic. Partial saturation of carbon atoms in hydrogenated graphene breaks its π-bonding network resulting in localized and unpaired electrons Zhou et al. (2009). The magnetic moments couple ferromagnetically with the semi-hydrogenated chair conformation. Electronic structure changes by partial hydrogenation as well. The semi-hydrogenated graphene of chair conformation is an indirect band gap semiconductor with a small band gap, very different from the original graphene and graphane. We illustrate in Figure 10 the dependence of the spin-polarized bands of semi-hydrogenated bilayer graphene with the positive and negative bias. The energy gap decreases monotonically with the electric field by characterizing the properties from magnetic semiconductor with a small gap, to a metal with a zero gap. In contrast to fully-hydrogenated bilayer graphene, the changes in the gap are no longer symmetrical with the negative and positive bias. Apart from the partial shifts of the spin density to the bottom layer, closer scrutiny reveals a paucity of modifications of the spin-density distribution, when the applied electric field goes from negative to positive bias. In connection to the spin density shift, the band dispersion changes from nearly flat to

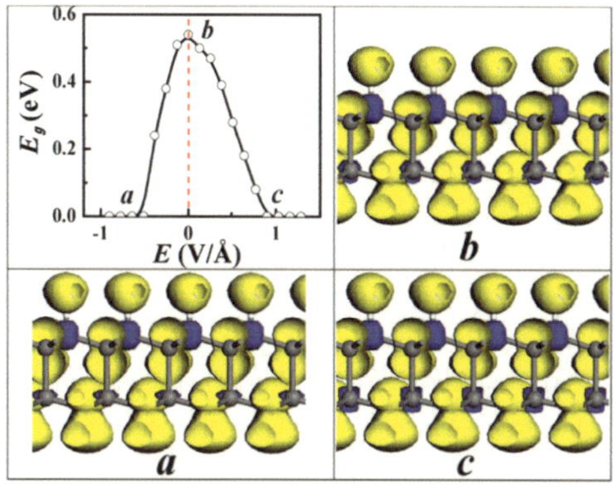

Fig. 10. Calculated band gaps of semi-hydrogenated bilayer graphene in chair structure with (a) -0.26 V/Å electric bias, (b) no bias, and (c) 0.39 V/Å electric bias, respectively. Insets: extracted spin density distribution at the band center (Γ point) of the corresponding conduction and valence states. The isovalue is 0.025 au. Reprinted with permission from Ref. Samarakoon & Wang (2010), Copyright © 2009 American Chemistry Society.

pronounced dispersion near the band edge at *K*, which leads to the change of the indirect gaps.

4. Conclusion

4.1 Chair, boat, and twist-boat membranes in graphane

We have studied various stable crystal structures of graphane and demonstrated that locally stable twist-boat membranes significantly contribute to the experimentally observed lattice contraction. The first-principles results shed considerable light on the electronic characteristics associated with the sp^3 hybridization. Moreover, the first-principles approach can be employed to structural and electronic properties of hydrogenated graphene derivatives. The understanding of structural and electronic stability thus provides a useful means for future development of graphane-based nanodevices.

4.2 Bilayer graphane under electric bias

We have studied the electronic characteristics of biased bilayer graphane. The resultant hydrocarbon compound, bilayer graphane, can be modified into new materials, fine-tuning its electronic properties. These studies have revealed increasingly fertile possibilities in hydrogen storage and two-dimensional electronics. These novel semiconducting behaviors result from a peculiar, effective transformation of sp^2 to sp^3 carbon and allow a continuously tunable band gap in biased bilayer graphane. A bilayer version can deliver yet another interesting feature of tunable band gap. This discovery paves the way for new electronic devices, from lasers that change color to electronic circuits that can rearrange themselves. The tunable band gap, which generally determines transport and optical properties, will enable flexibility and optimization of graphene-based nanodevices. Moreover, our proposed desorption of hydrogen from one layer, coupled with controlled hydrogen-vacancy distribution and patterned hydrogenation, could provide a promising route to realize a ferromagnetic semiconductor in view of the crucial structural difference between monolayer graphone and the bilayer semi-hydrogenated graphene.

5. References

a. Avetisyan, A. A.; Partoens, B.; Peeters, F. M. (2009). Electric Field Tuning of the Band Gap in Graphene Multilayers, *Phys. Rev. B* 79 (3): 035421 – 035427.

b. Avetisyan, A. A.; Partoens, B.; Peeters, F. M. (2009). Electric-Field Control of the Band Gap and Fermi Energy in Graphene Multilayers by Top and Back Gates, *Phys. Rev. B* 80 (19): 195401 – 195421.

Barone, V.; Hod, O.; Scuseria, G. E. (2006). Electronic Structure and Stability of Semiconducting Graphene Nanoribbons, *Nano. Lett.* 6 (12): 2748 – 2754.

Berger, C.; Song, Z.; Li, X.; Wu, X.; Brown, N.; Naud, C.; Mayou, D.; Li, T.; Hass, J.; Marchenkov, A. N.; Conrad, E. H; First, P. N.; de Heer, W. A. (2006). Electronic Confinement and Coherence in Patterned Epitaxial Graphene, *Science* 312 (5777): 1191 – 1196.

Boukhvalov, D. W.; Katsnelson, M. I.; Lichtenstein, A. I. (2008). Hydrogen on graphene: Electronic Structure, Total Energy, Structural Distortions and Magnetism from First-Principles Calculations, *Phys. Rev. B* 77 (3): 035427 – 035433.

Castro, E. V.; Novoselov, K. S.; Morozov, S. V.; Peres, N. M. R.; dos Santos, J. M. B. L.; Nilsson, J.; Guinea, F.; Geim, A. K.; Neto, A. H. C. (2007). Biased Bilayer Graphene: Semiconductor with a Gap Tunable by the Electric Field Effect, *Phys. Rev. Lett.* 99 (21): 216802-216805f.

Castro, E. V.; Peres, N. M. R.; dos Santos, J. M. B. L.; Guinea, F.; Neto, A. H. C. (2008). Bilayer Graphene: Gap Tunability and Edge Properties, *J. Phys.: Conf. Ser.* 129 (1): 012002.

Ciftja, O. & Fantoni, S. (1996). A New Hypernetted-Chain Treatment for Laughlin Quantum Hall States, *Europhys. Lett.* 36 (9): 663 – 667.

Ciftja, O. (2000). The Fermi-Sea-Like Limit of the Composite Fermion Wave Function, *Eur. Phys. J. B* 13 (4): 671 – 677.

Ciftja, O. & Wexler, C. (2001). Energy Gaps for Fractional Quantum Hall States Described by a Chern-Simons Composite Fermion Wavefunction, *Eur. Phys. J. B* 23 (4): 437 – 440.

$DMol_3$ (2010). Accelrys Software Inc., San Diego, CA.

Elias, D. C.; Nair, R. R.; Mohiuddin, T. M. G.; Morozov, S. V.; Blake, P.; Halsall, M. P.; Ferrari, A. C.; Boukhvalov, D. W.; Katsnelson, M. I.; Geim, A. K.; Novoselov, K. S. (2009). Control of GrapheneŠs Properties by Reversible Hydrogenation: Evidence for Graphane, *Science* 323 (5914): 610 – 613.

Flores, M. Z. S.; Autreto, P. A. S.; Legoas, S. B.; Galvao, D. S. (2009). Graphene to Graphane: a Theoretical Study, *Nanotechnology* 20 (46): 465704 – 465709.

a. Geim, A. K. & Novoselov, K. S. (2007). The Rise of Graphene, *Nature Mater.* 6: 183 – 191.

b. Geim, A. K.; Morozov, S.V.; Hill, E. W.; Blake, P.; Katsnelson, M. I.; Novoselov, K. S. (2007). Detection of Individual Gas Molecules Adsorbed on Graphene, *Nat. Mater.* 6: 652 – 655 .

Gilje, S.; Han, S.; Wang, M., Wang, K. L.; Kaner, R. B. (2007). A Chemical Route to Graphene for Device Applications, *Nano Lett.* 7 (11): 3394 – 3398.

Giovannetti, G.; Khomyakov, P. A.; Brocks, G.; Kelly, P. J.; van den Brink, J. (2007). Substrate-Induced Band Gap in Graphene on Hexagonal Boron Nitride: *Ab Initio* Density Functional Calculations, *Phys. Rev. B* 76 (7): 073103 – 073106.

Guisinger, N. P.; Rutter, G. M.; Crain, J. N.; First, P. N.; Stroscio, J. A. (2009). Exposure of Epitaxial Graphene on SiC(0001) to Atomic Hydrogen, *Nano Lett.* 9 (4): 1462 – 1466.

Grüneis, A.; Attaccalite, C.; Wirtz, L.; Shiozawa, H.; Saito, R.; Pichler, T.; Rubio, A. (2008). Tight-Binding Description of the Quasiparticle Dispersion of Graphite and Few-Layer Graphene, *Phys. Rev. B* 78 (20): 205425 – 205440.

Halperin, B.; Lee, P. A.; Read, N. (1993). Theory of the Half-Filled Landau Level, *Phys. Rev. B* 47 (12): 7312 – 7343.
Jain, J. K. (1989). Composite-Fermion Approach for the Fractional Quantum Hall Effect, *Phys. Rev. Lett.* 63 (2): 199-202.
Kresse, G. & Furthmuller, J. (1996). Efficient Iterative Schemes for *Ab Initio* Total-energy Calculations Using a Plane-wave Basis Set, *Phys. Rev. B* 54 (16): 11169 – 11186.
Laughlin, R. B. (1983). Anomalous Quantum Hall Effect: An Incompressible Quantum Fluid with Fractionally Charged Excitations, *Phys. Rev. Lett.* 50 (18): 1395 – 1398.
Leenaerts, O.; Partoens, B.; Peeters, F. M. (2009). Hydrogenation of Bilayer Graphene and the Formation of Bilayer Graphane from First Principles, *Phys. Rev. B* 80 (24): 245422 – 245427 .
Li, X.; Wang, X.; Zhang, L.; Lee, S.; Dai, H. (2008). Chemically Derived, Ultrasmooth Graphene Nanoribbon Semiconductors, *Science* 319 (5867): 1229 – 1232.
Li, Y.; Zhou, Z.; Shen, P.; Chen, Z. (2009). Structural and Electronic Properties of Graphane Nanoribbons, *J. Phys. Chem. C* 113 (33): 15043 – 15045.
Liu, L. & Shen, Z. (2009). Bandgap Engineering of Graphene: A Density Functional Theory Study, *Appl. Phys. Lett.* 95 (25): 252104 – 252107.
Mak, K. F.; Lui, C. H.; Shan, J.; Heinz, T. F. (2009). Observation of an Electric-Field-Induced Band Gap in Bilayer Graphene by Infrared Spectroscopy, *Phys. Rev. Lett.* 102 (25): 256405 – 256408.
McCann, E. (2006). Asymmetry Gap in the Electronic Band Structure of Bilayer Graphene, *Phys. Rev. B* 74 (16): 161403 – 161406.
Min, H.; Sahu, B.; Banerjee, S. K.; MacDonald, A. H. (2007). *Ab Initio* Theory of Gate Induced Gaps in Graphene Bilayers, *Phys. Rev. B* 75 (15): 155115 – 155121.
Nduwimana, A.; Musin, R. N.; Smith, A. H.; Wang, X.-Q. (2008). Spatial Carrier Confinement in Core-Shell and Multishell Nanowire Heterostructures, *Nano Lett.* 8 (10): 3341 – 3344.
Nduwimana, A. & Wang, X.-Q. (2009). Charge Carrier Separation in Modulation Doped Coaxial Semiconductor Nanowires, *Nano Lett.* 9 (1): 283 – 286.
Nduwimana, A. & Wang, X.-Q. (2009). Energy Gaps in Supramolecular Functionalized Graphene Nanoribbons, *ACS Nano* 3 (7): 1995 – 1999.
Neto, A. H. C.; Guinea, F.; Peres, N. M. R.; Novoselov, K. S.; Geim, A. K. (2009). The Electronic Properties of Graphene, *Rev. Mod. Phys.* 81: 109 – 162.
Nilsson, J.; Neto, A. H. C.; Guinea, F.; Peres, N. M. R. (2008). Electronic Properties of Bilayer and Multilayer Graphene, *Phys. Rev. B* 78(4): 045405 – 045438.
Novoselov, K. S.; Geim, A. K.; Morozov, S. V.; Jiang, D.; Zhang, Y.; Dubonos, S. V.; Grigorieva, I. V.; Firsov, A. A. (2004). Electric Field Effect in Atomically Thin Carbon Films, *Science* 306 (5696): 666 – 669.
Novoselov, K. S.; Geim A. K.; Morozov, S. V.; Jiang, D.; Katsnelson, M. I.; Grigorieva, I. V.; Dubonos, S. V.; Firsov, A. A. (2005). Two-dimensional Gas of Massless Dirac Fermions in Graphene, *Nature* 438: 197-200.
Ohta, T.; Bostwick, A; Seyller, T.; Horn, K.; Rotenberg, E. (2006). Controlling the Electronic Structure of Bilayer Graphene, *Science* 313 (5789): 951 – 954.
Oostinga, J. B.; Heersche, H. B.; Liu, X.; Morpurgo, A. F.; Vandersypen, L. M. (2008). Gate-Induced Insulating State in Bilayer Graphene Devices, *Nat. Mater.* 7 : 151- 157.
Partoens, B. & Peeters, F. M. (2006). From Graphene to Graphite: Electronic Structure Around the K Point, *Phys. Rev. B* 74 (7): 075404 – 075414.

Perdew, J. P.; Burke, K.; Ernzerhof, M. (1996). Generalized Gradient Approximation Made Simple, *Phys. Rev. Lett.* 77 (18): 3865 – 3868.

Samarakoon, D. K. & Wang, X.-Q. (2009). Chair and Twisted-Boat Membranes in Hydrogenated Graphene, *ACS Nano* 3 (12): 4017 – 4022.

Samarakoon, D. K. & Wang, X.-Q. (2010). Tunable Band Gap in Hydrogenated Bilayer Graphene, *ACS Nano* 4 (7): 4126 – 4130.

Singh, A. K. & Yakobson, B. I. (2009). Electronics and Magnetism of Patterned Graphene Nanoroads, *Nano Lett.* 9 (4): 1540 – 1543.

Sofo, J. O.; Chaudhari, A. S.; Barber, G. D. (2007). Graphane: A Two-Dimensional Hydrocarbon, *Phys. Rev. B* 75 (15): 153401 – 153404.

Son, Y.-W.; Cohen M. L.; Louie, S. G. (2006). Energy Gaps in Graphene Nanoribbons, *Phys. Rev. Lett.* 97 (21): 216803 – 216806.

Tersoff, J. (1988). New Empirical Approach for the Structure and Energy of Covalent Systems, *Phys. Rev. B* 37(12): 6991 – 7000.

Thouless, D. J. (1998). *Topological Quantum Numbers in Nonrelativistic Physics*, World Scientic, Singapore.

Tsui, D. C.; Stormer, H. L.; Gossard, A. C. (1982). Two-Dimensional Magnetotransport in the Extreme Quantum Limit, *Phys. Rev. Lett.* 48 (22): 1559 – 1562.

Vosko, S. H.; Wilk, L.; Nusair, M. (1980). Accurate Spin-Dependent Electron Liquid Correlation Energies for Local Spin Density Calculations: a Critical Analysis, *Can. J. Phys.* 58 (8): 1200 – 1211.

Wang, X.; Zhi, L.; Mullen, K. (2008). Transparent, Conductive Graphene Electrodes for Dye-Sensitized Solar Cells, *Nano Lett.* 8 (1): 323 – 327.

Wen, X. G. (1990). Electrodynamical Properties of Gapless Edge Excitations in the Fractional Quantum Hall States, *Phys. Rev. Lett.* 64 (18): 2206 – 2209.

Wen, X. G. (1991). Gapless Boundary Excitations in the Quantum Hall States and in the Chiral Spin States, *Phys. Rev. B* 43 (13): 11025 – 11036.

Xia, F.; Farmer, D. B.; Lin, Y.-M.; Avouris, P. (2010). Graphene Field-Effect Transistors with High On/Off Current Ratio and Large Transport Band Gap at Room Temperature, *Nano Lett.* 10 (2): 715 – 718.

Yang, L.; Cohen, L. M.; Louie, G. S. (2008). Magnetic Edge-State Excitons in Zigzag Graphene Nanoribbons, *Phys. Rev. Lett.* 101 (18): 186401 – 186404.

Zhang, Y.; Tan, Y. W.; Stormer, H. L.; Kim, P. (2005). Experimental Observation of the Quantum Hall Effect and Berry's Phase in Graphene, *Nature* 438: 201 – 204.

Zhang, Y.; Jiang, Z.; Small, J. P.; Purewal, M. S.; Tan, Y.-W.; Fazlollahi, M.; Chudow, J. D.; Jaszczak, J. A.; Stormer, H. L.; Kim, P. (2006). Landau-Level Splitting in Graphene in High Magnetic Fields, *Phys. Rev. Lett.* 96 (13): 136806 – 136809.

Zhang, Y.; Tang, T.-T.; Girit, C.; Hao, Z.; Martin, M. C.; Zettl, A.; Crommie, M. F.; Shen, Y. R.; Wang, F. (2009). Direct Observation of a Widely Tunable Bandgap in Bilayer Graphene, *Nature* 459: 820-823.

Zhou, J.; Wang, Q.; Sun, Q.; Chen, X. S.; Kawazoe Y.; Jena P. (2009). Ferromagnetism in Semihydrogenated Graphene Sheet, *Nano Lett.* 9 (11): 3867 – 3870.

7

Applications of Al Modified Graphene on Gas Sensors and Hydrogen Storage

Zhimin Ao, Jack Yang and Sean Li
School of Materials Science and Engineering, The University of New South Wales, NSW 2052 Australia

1. Introduction

Graphene is a single layer of graphite with a hexagonal structure, or an individual sheet of sp^2-hybridized carbon bound in two dimensions. It was experimentally fabricated for the first time through mechanical exfoliation of small mesas from highly oriented pyrolytic graphite (HOPG) in 2004 and subsequently was found with excellent electrical properties (Novoselov et al., 2004). This discovery has triggered enormous amount of interest on graphene both in fundamental and applied research. Graphene has shown promising applications as ultra-sensitive gas sensors, transparent electrodes in liquid crystal display devices, and large capacity electrodes in Li batteries (Rao et al., 2009).

Various methods have been reported for the synthesis of single-layer graphene: (1) Mechanical exfoliation from HOPG (Novoselov et al., 2004; Zhang et al., 2005), which is deposited on to a silicon substrate. This method is low cost, but the graphene produced is of poor quality with the limited area. It is particularly difficult and time-consuming to synthesize single-layer graphene in large scale with this method (Yuan et al., 2009). (2) Chemical exfoliation from bulk graphite (Li et al., 2008a;b). In this case, oxidized graphite (by using strong acids) was cleaved via rapid thermal expansion or ultrasonic dispersion, and subsequently the graphene oxide sheets were reduced to graphene. A serious drawback of this method is that the oxidation process induces a variety of defects which would degrade the electronic properties of graphene. (3) Epitaxial growth on an insulator surface (such as SiC) (Zhou et al., 2007). The graphene obtained showed poor uniformity and contained a multitude of domains. (4) Chemical vapour deposition (CVD) on the surfaces of metals (e.g., Ni) (Reina et al., 2009). Recently, N substitutionally doped graphene was first synthesized by a CVD method with the presence of CH_4 and NH_3 (Wei et al., 2009). As doping accompanies with the recombination of carbon atoms into graphene in the CVD process, dopant atoms can be substitutionally doped into the graphene lattice, which is hard to realize by other synthetic methods.

Several unique electronic properties associated with these 2D crystals have been discovered (Beenakker, 2008). In addition, it is known that carbon nanotubes have good sensor properties (Geim & Novoselov, 2007). Recently, graphenes as highly sensitive gas sensors were also reported (Ao et al., 2008; Bunch et al., 2005; Yang et al., 2010). It was shown that the increase in graphene charge carrier concentration induced by adsorbed gas molecules could be utilized to make highly sensitive sensors, even with the possibility of detecting individual molecules. The sensing property is based on the changes in the resistivity due to molecules adsorbed on

graphene sheet that act as donors or acceptors. The sensitivity of NH_3, CO, and H_2O up to 1 ppb (parts per 10^9) was demonstrated, and even the ultimate sensitivity of an individual molecule was suggested for NO_2. Furthermore, the preliminary works (Ao et al., 2009a) indicated that graphene have promising physisorption properties for hydrogen.
With state-of-art computer simulations, it is believed that the properties of condensed matters can be understood at the atomic level. In the simulation, the simulator builds a model of a real system and explores its properties. In many ways, simulation studies share the same mentality as experimental ones. However, simulations allow absolute control over the experimental parameters and access to the outcomes in details. These strengths have been exploited for the last fifty years since the introduction of computation algorithms that allows one to calculate the properties of materials based on the first-principles in light of fundamental physics outlined in Schrödinger equation without free parameters. In this chapter, the applications of Al modified graphene as gas sensor and hydrogen storage materials are developed and optimized with density functional theories (DFT).

2. Application for ultra-sensitive carbon oxide detection

2.1 Emerging ultra-high sensitive CO detection

It was reported that the detectable range and sensitivity of the single wall carbon nanotubes (SWCNTs) can be widened and enhanced substantially through either doping technology or surface engineering (Kong et al., 2001; Peng & Cho, 2003; Wei et al., 2004). For example, SWCNT coated with Pb nanoparticles has high sensitivity to H_2 (Kong et al., 2001), SnO_2/SWCNTs hybrid material shows an enhanced sensitivity to NO_2 (Wei et al., 2004). The high sensitivity of boron doped SWCNT to CO and H_2O absorptions has been also demonstrated (Peng & Cho, 2003). Most recently, Al-cluster and Al doped SWCNT assembly were suggested to be promising systems for novel molecular sensors to NH_3 (Zhao et al., 2005) and CO (Wang et al., 2007), and the B doped SWCNTs are highly sensitive to the gaseous cyanide and formaldehyde molecules (Zhang et al., 2006). However, the devices with higher sensitivity to these toxic gases are apparently required for environmental safety issues both in workplaces and residential areas, especially in some industrial and military fields.
Graphene based device may be a solution for ultra-high sensitivity gas sensor (Leenaerts et al., 2008; Schedin et al., 2007; Wehling et al., 2008). Similar to CNT, the working principle of graphene devices as gas sensors is based on the changes of their electrical conductivity induced by surface adsorbates, which act as either donors or acceptors associated with their chemical natures and preferential adsorption sites (Collins et al., 2000; Kong et al., 2000; Moseley, 1997). Graphene is considered to be an excellent sensor material and the sensitivity of Al doped graphene system to CO gas detection is investigated using DFT calculations.

2.2 Rationale for choosing Al as candidate dopant: a quantum field theoretical perspective

Since graphene is a stable 2D structure, gas molecules tend to be adsorbed onto pristine graphene weakly through physisorption. This has brought a large disadvantage for using pristine graphene to adsorb toxic gases such as CO. By introducing substituent impurities into graphene through chemical doping, the local electronic structures around the dopants could be modified. Typically, as confirmed by the first principle studies in the next section, incorporating Al into graphene will cause a distortion to the electron density distribution around the dopant. In this case, C-atoms surrounding the Al dopant will attract electrons due to their high electron affinity, whereas on the Al dopant, a decrease in electron density can be observed. The charge redistribution makes the Al to be an active site for CO adsorption. This

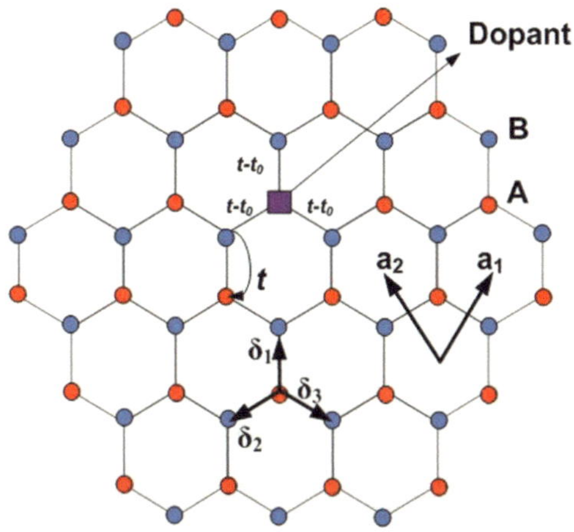

Fig. 1. The honecomb graphene lattice with a substituting atom replacing a carbon atom. The substituent changes the local hopping parameter from t to $t - t_0$. The honeycomb lattice is can be considered as made up with two sublattices with atomic types A and B. In this case, an A-type atom has been substituted with an impurity atom.

charge re-distribution effect can be confirmed with advanced field theoretical methods (Peres et al., 2006; 2007; 2009).

Here, the electronic properties caused by a single dopant in graphene layer can also be studied with the field theory.

Under the tight-binding model, the graphene honeycomb lattice can be considered as being constructed from unit cells consisting two types of atoms A and B (Fig. 1). The unit cell vectors in Cartesian coordinates are $\vec{a}_1 = \frac{a_0}{2}(3, \sqrt{3})$ and $\vec{a}_2 = \frac{a_0}{2}(3, -\sqrt{3})$, with a_0 to be the unit cell parameter. $\vec{\delta}_n$ $(n = 1, 2, 3)$ are vectors connecting a given atom to its three direct neighbours. Introducing a phase factor $\varphi(k) = \sum_{i=1}^{3} e^{ik\delta_i}$, the tight-binding Hamiltonian for graphene, with a single impurity substituting an A-type atom, can be expressed as

$$\hat{\mathcal{H}} = -t \sum_{k,\sigma} \left(\varphi(k) \hat{a}^{\dagger}_{k,\sigma} \hat{b}_{k,\sigma} + h.c. \right) + \frac{t_0}{N_c} \sum_{k,k',\sigma} \left(\varphi(k') \hat{a}^{\dagger}_{k,\sigma} \hat{b}_{k',\sigma} + h.c. \right) + \sum_{k} \frac{\varepsilon_0}{N_c} \hat{a}^{\dagger}_{k,\sigma} \hat{a}_{k,\sigma}, \quad (1)$$

in the reciprocal space. In the above Hamiltonian,$\hat{a}^{\dagger}_{k,\sigma}(\hat{b}^{\dagger}_{k,\sigma})$ and $\hat{a}_{k,\sigma}(\hat{b}_{k,\sigma})$ creates and destroys an electron with wavevector k and spin on A(B) lattice site, respectively. N_c is the number of atoms in a unit cell. The spin degree of freedom is summed over spin index σ. The hopping parameter between two neighboring C atoms is t, where upon dopant substitution, the hopping parameters between the dopant and its neighboring C atoms are changed to $t - t_0$. The dopant atom also introduces a local potential ε_0. By changing the impurity hopping parameters and local potential, different impurity types can be modeled, which will become apparent later.

Our aim is to calculate the electronic density of states (DOS) on the impurity and its nearest neighboring sites. This requires the knowledge of the system's Green's function. Under the Matsubara formalism, it reads

$$G_{\sigma}(\omega_n, k, p) = \begin{pmatrix} G_{aa}(\omega_n, k, p) & G_{ab}(\omega_n, k, p) \\ G_{ba}(\omega_n, k, p) & G_{bb}(\omega_n, k, p) \end{pmatrix}, \quad (2)$$

in which the matrix elements $G_{\alpha\beta}(\omega_n,k,p)$ are defined as the Fourier transforms of $G_{\alpha\beta}(k,p,\tau) = -\left\langle T_\tau\left[\hat{\alpha}_k^\dagger(\tau)\hat{\beta}_p(0)\right]\right\rangle$ with $a,\beta = a,b$. k and p denotes the electronic wavevectors τ is the complex time variable and ω_n are the fermionic Matsubara frequencies.
The system Green's function (Eq. 2) can be solved from the equations-of-motion derived based on Hamiltonian (Eq. 1). The rationale behind the solution procedure is to seek for the relationship behind $G_\sigma(\omega_n,k,p)$ and $G_\sigma^0(\omega_n,k)$. the Matsubara Green's function for pristine graphene, where the later can be expressed analytically as (Peres et al., 2006):

$$G_\sigma^0(\omega_n,k) = \begin{pmatrix} G_{aa}^0(\omega_n,k) & G_{ab}^0(\omega_n,k) \\ G_{ba}^0(\omega_n,k) & G_{bb}^0(\omega_n,k) \end{pmatrix} = \sum_{j=\pm1} \frac{1/2}{i\omega_n - j|\varphi(k)|} \begin{pmatrix} 1 & je^{i\delta(k)} \\ je^{-i\delta(k)} & 1 \end{pmatrix}, \tag{3}$$

with $e^{i\delta(k)} = \varphi(k)/|\varphi(k)|$.
Since we are interested in the electronic DOS on the impurity (A) site and its nearest-neighboring (B) sites, therefore, it would be sufficient to solve for $G_{aa}(\omega_n,k,p)$ and $G_{bb}(\omega_n,k,p)$ in Eq. 2. The electronic DOS can then be found from the imaginary parts of the retarded Green's functions $G_{aa}^r(\omega,k,p)$ and $G_{bb}^r(\omega,k,p)$ through analytical continuation of the Matsubara Green's functions. The presence of both diagonal and nondiagonal disorders means that the solutions will be of a more complex form than the usual T-matrix for a single Anderson impurity scattering problem, and the results are

$$G_{aa}(\omega_n,k,p) = \delta_{k,p}G_{aa}^0(\omega_n,k) + g(\omega_n) + h(\omega_n)\left[G_{aa}^0(\omega_n,k) + G_{aa}^0(\omega,p)\right] + G_{aa}(\omega_n,k)T(\omega_n)G_{aa}^0(\omega_n,p), \tag{4}$$

$$G_{bb}(\omega_n,k,p) = \delta_{k,p}G_{bb}^0(\omega_n,k) + \frac{t^2\varphi^*(k)\varphi(p)}{(i\omega)^2}G_{bb}^0(\omega_n,k)T(\omega_n)G_{bb}^0(\omega_n,p), \tag{5}$$

where

$$g(\omega_n) = t_0^2\bar{G}_{aa}^0(\omega_n)/[N_cD(\omega_n)], \tag{6}$$

$$h(\omega_n) = t_0(t-t_0)/[N_cD(\omega_0)], \tag{7}$$

and

$$T(\omega_n) = -[i\omega_n t_0(2t-t_0) - \varepsilon_0 t^2]/[N_cD(\omega_n)], \tag{8}$$

with

$$D(\omega_n) = (t-t_0)^2 + \left[i\omega_n t_0(t-t_0) - \varepsilon_0 t^2\right]\bar{G}_{aa}^0(\omega_n), \tag{9}$$

and

$$\bar{G}_{aa}^0(\omega_n) = \frac{1}{N_c}\sum_k G_{aa}^0(\omega_n,k). \tag{10}$$

The important term is the $g(\omega_n$ whose double Fourier transform gives $G_{aa}(\omega_n,0,0)$ which is the return (back-scattering) amplitude of the electron wave to the impurity site. Its magnitude, which depends on $D(\omega_n)$, depicts the electronic DOS on the impurity sites.
In the case of Al doping, where the dopant has a larger atomic radius than carbon, we can let $t_0 = -t$, with $\varepsilon_0 < 0$, as a limiting case. This gives,

$$g_1(\omega_n) = \frac{\bar{G}^0(\omega_n)}{(4 - 4i\omega_n\bar{G}^0(\omega_n)) + (i\omega_n - \varepsilon_0\bar{G}^0(\omega_n))}. \tag{11}$$

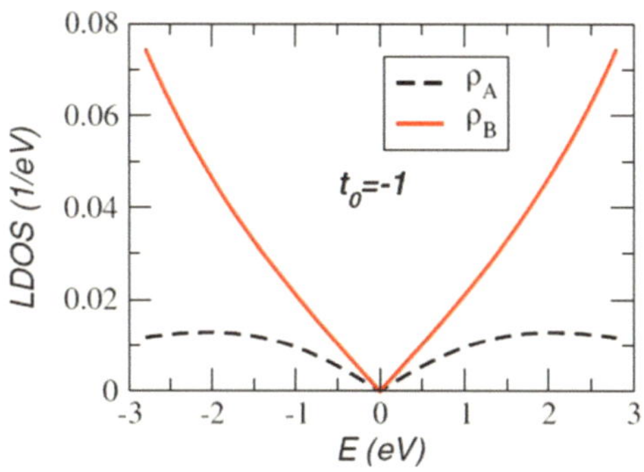

Fig. 2. Local density of states (LDOS) on the impurity atom at A site and surrounding carbon atoms at B sites for the case where $t_0 = -1, t = 1$, mimicking the situation of Al doping in graphene. (Reproduced with permission from Ref. (Peres et al., 2009). Copyright 2009, APS)

On the other hand, for a dopant with smaller ionic radius, we would have $\varepsilon_0 > 0$ and $t_0 = |\gamma|t$. The most interesting case is for a physisorbed species on graphene layer, in which $t_0 = t$, indicating that the molecule would have no interaction with the graphene layer. Under such a condition, the Matsubrara Green's function for a free particle can be reclaimed as,

$$g_2(\omega_n) = \frac{1}{i\omega_n - \varepsilon_0}. \tag{12}$$

Fig. 2 shows the resulting electronic DOS on the impurity site and its neighboring carbon atoms with $t_0 = -t$. In this case, the ε_0 is chosen to be zero, whereas a nonzero ε_0 will simply modify the DOS diagram by destroying the particle-hole symmetry. It is hence evident from Fig. 2 that effect of an Al–like dopant is to deplete electrons from the impurity and causing electron accumulation in the surrounding atoms ($\rho_A < \rho_B$).

2.3 CO adsorption effects on atomic configuration, electronic energy and bond lengthes of Al doped graphene

Above analysis established a solid theoretical framework for further investigation of the Al doped graphene for application of CO sensor with DFT calculations. In this work, all DFT calculations were performed in Dmol^3 code (Delley, 1990). It is widely known that calculations limited at the local density approximation (LDA) overestimate bond energy E_b and underestimate equilibrium distances (Jeloaica & Sidis, 1999; Lugo-Solis & Vasiliev, 2007). Thus, a GGA with the RPBE method is used as the exchange correlation function (Hammer et al., 1999). The DFT semicore pseudopotential (DSPP) core treatment (Delley, 2002) was implemented for relativistic effects, which replaces core electrons by a single effective potential. To ensure that the results of the calculations were comparable, identical conditions had been employed for the isolated CO molecule, the original Al doped graphene and also the adsorbed graphene system. The *k*-point was set to $6 \times 6 \times 2$ for all slabs, which brought out the convergence tolerance of energy of 1.0×10^{-5} hartree (1 hartree = 27.21 eV), and that of maximum force of 0.002 hartree.
In the simulation, three-dimensional periodic boundary condition had been adopted and C-O bond length was set to $l_{C-O} = 1.13$Å, which is consistent with experimental results (Lide, 2000). For the graphene, a single layer 2×2 supercell with a vacuum width of 12 Å above had been constructed, which ensured that the interaction between repeated slabs in a direction

normal to the surface was small enough. All atoms were allowed to relax for all energy calculations. The adsorption energy E_b between the CO gas molecule and graphene is defined as,

$$E_b = E_{\text{CO+graphene}} - (E_{\text{graphene}} + E_{\text{CO}}), \tag{13}$$

where the subscripts CO+graphene, graphene, and CO denote the adsorbed system, isolated graphene and CO molecules, respectively.

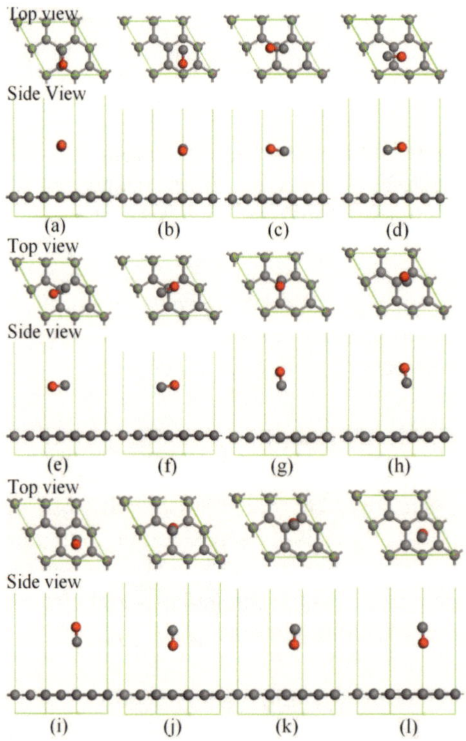

Fig. 3. Twelve available binding sites for CO adsorbed on intrinsic graphene (top and below images show the top and side view, respectively). (a) T-B-T, (b) T-H-T, (c) H-T-H, (d) H-B-H, (e) B(C atom)-T-H, (f) B(O atom)-T-H, (g) T-(O atom upward), (h) B (O atom upward), (i) H (O atom upward), (j) T (C atom upward), (k) B (C atom upward), (l) H (C atom upward). T, B and H denote top site of C atoms, bridge site of C-C bond and hollow site of carbon hexagon, respectively. Gray, pink and red spheres are denoted as C, Al and O atoms, respectively. (Reproduced with permission from Ref. (Ao et al., 2008). Copyright 2008, Elsevier)

To search for the most stable structure between a CO molecule and the intrinsic graphene, E_b described in Eq. 13 and the binding distance, d, for all possible configurations were calculated. Twelve possible binding sites for the CO adsorbed on graphene layer were considered as initial structures as shown in Fig. 3(a)-(l). After full structural relaxation, no distinctive structural change has been found. All of the results are displayed in Table 1. It is found that adsorption configuration shown in Fig. 3(f) has the smallest d value and the largest E_b value among all the possible configurations. This indicates that the configuration shown in Fig. 3(f) is the most stable atomic arrangement with the strongest interaction between CO and graphene with $E_b = 0.016$ eV and $d = 3.768$ Å, which are consistent with the other simulation results of $E_b = 0.014$ eV and $d = 3.740$ Å (Leenaerts et al., 2008). However, in this particular adsorption configuration, the E_b value is still considered to be too small and d too large, even though they are the most favorable one for adsorption, reflecting that CO undergoes weak physisorption on the intrinsic graphene. This indicates that the intrinsic graphene is insensitive to CO molecules.

		Intrinsic graphene		Al doped graphene	
Initial binding configurations		E_b(eV)	d(Å)	E_b(eV)	l(Å)
CO$\parallel$graphene	T-B-T	-0.011	3.839	-4.978	1.964
	T-H-T	-0.012	3.805	-4.973	1.968
	H-T-H	-0.014	3.826	-4.613	3.755
	H-B-H	-0.009	3.857	-4.599	3.814
	B(C atom)-T-H	-0.011	3.855	-4.609	3.800
	B(O atom)-T-H	-0.016	3.768	-4.616	3.821
CO$\perp$graphene	T(O upwards)	-0.007	3.938	-4.979	1.961
	B(O upwards)	-0.007	3.935	-4.978	1.964
	H(O upwards)	-0.003	3.982	-4.975	1.965
	T(C upwards)	-0.004	3.952	-4.629	3.781
	B(C upwards)	-0.003	3.981	-4.607	3.783
	H(C upwards)	-0.005	3.942	-4.609	3.457

Table 1. Summary of results for CO adsorption on intrinsic graphene and Al doped graphene on different adsorption sites. The meaning of T, B and H are given in the caption of Fig. 3. In the table headings, d represents the distance between CO gas molecule and graphene layer, l represents the bond length of Al and C atom in CO gas molecule.

Upon substituting one carbon atoms by Al, the geometric structure of the doped graphene changes dramatically, as shown in Fig. 4. Figs. 4(a) and 4(b) represent the geometries of intrinsic and Al doped graphene after relaxation. As shown in Table 2 and Fig. 4(b), the Al doping results in l elongation from $l_{C-C} = 1.420$ Å to l_{Al-C}=1.632 Å. This is associated with the distortion of hexagonal structures adjacent to the larger Al atom, similar to the restructuring in Al doped SWCNTs (Wang et al., 2007).

When a CO molecule is adsorbed on the Al-substituted graphene, there also exists twelve possible adsorption sites similar to the CO absorption on intrinsic graphene shown in Fig. 3. These are taken as initial configurations. After relaxation, the configuration in Fig. 3(d) has the most stable relaxed structure. The adsorption of CO causes a structure change in the Al doped graphene dramatically, resulting in an expansion of l_{Al1-C2} from 1.632 to 1.870 Å while l_{Al1-C4} elongates from 1.632 to 1.915 Å. The corresponding distance between the CO molecule and Al atom in the Al doped graphene is 1.964 Å, being much shorter than 3.767 Å in the intrinsic graphene system. Moreover, the E_b of CO in the Al doped graphene system is 4.979 eV, which is over 60 times larger than that of CO in the intrinsic graphene system. Comparing with the E_b in other systems, such as E_b = 1.280 eV for CO adsorbed in the Al doped SWCNT systems (Wang et al., 2007), E_b = 0.986 eV in the B doped SWCNT systems (Wang et al., 2007) and E_b = 0.201 eV for CO adsorbed in B doped graphene etc, the Al doped graphene is energetically more favorable for CO adsorption. In other words, the Al doped graphene is much more sensitive to the CO adsorption among the aforementioned systems.

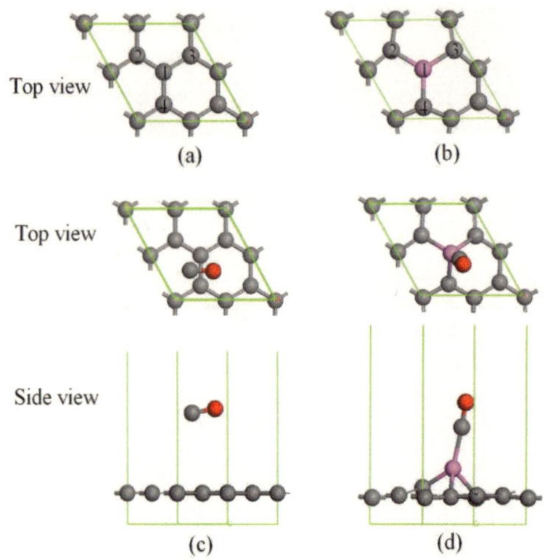

Fig. 4. Atomic configurations of intrinsic graphene and Al doped graphene before and after adsorption of CO gas molecule where one Al atom dopes in site 1, and sites 2, 3 and 4 are C atoms near the doped Al atom. (a) and (b) are the relaxed configurations of intrinsic graphene and Al doped graphene without adsorption. (c) and (d) are the preferred configurations after CO adsorption for intrinsic graphene and Al doped graphene, respectively. (Reproduced with permission from Ref. (Ao et al., 2008). Copyright 2008, Elsevier)

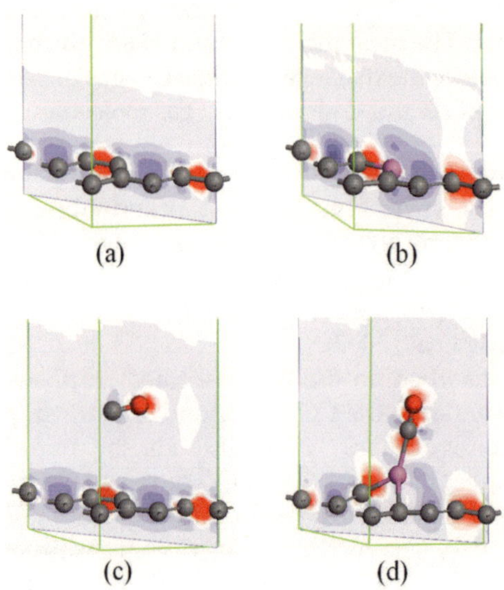

Fig. 5. Images of the electronic density difference for intrinsic graphene (a), Al doped graphene (b), CO-graphene system with preferred configuration (c) and CO-Al doped graphene system with preferred configuration (d). The red region shows the electron accumulation, while the blue region shows the electron loss. (Reproduced with permission from Ref. (Ao et al., 2008). Copyright 2008, Elsevier)

Furthermore, in order to investigate the changes of electronic structures in graphenes caused by the physi- or chemisorption of CO molecule, the net electron transfer (Q) from either the

System	Configuration	Bond	Bond length l (Å)	Q (e)
Intrinsic graphene	Fig. 4(a)	C1-C2	1.420	
		C1-C3	1.420	
		C1-C4	1.420	
	Fig. 4(c)	C1-C2	1.420	0.003
		C1-C3	1.421	
		C1-C4	1.421	
Al doped graphene	Fig. 4(b)	Al1-C2	1.632	
		Al1-C3	1.632	
		Al1-C4	1.632	
	Fig. 4(d)	Al1-C2	1.870	0.027
		Al1-C3	1.910	
		Al1-C4	1.915	

Table 2. Some structure parameters of intrinsic graphene and Al doped graphene before and after adsorption of CO molecule. Q denotes electrons transferred from the graphene layer to CO molecule, measured in the electronic charge e.

intrinsic or the Al doped graphene to the polar CO molecules had been calculated by Mulliken analysis, where Q is defined as the charge variation caused by the CO absorption. As listed in Table 2, $Q = 0.027\ e$ in the Al doped graphene is almost an order of magnitude larger than 0.003 e in the intrinsic graphene. This supports the notion that the Al doping influences the electronic properties of graphene substantially. This can also be verified by the difference of electronic densities between the intrinsic and Al doped graphenes with and without the CO adsorption as shown in Fig. 5. In the figure, the red and blue regions represent the areas of electron accumulation and the electron loss, respectively. Fig. 5(a) indicates the bond in the intrinsic graphene is of covalent nature because the preferential electron accumulation sites are mainly located within the bond rather than heavily centered on a particular atom. However, the electron density distribution along the covalent Al-C bonds has been significantly altered due to the difference in electron affinity of Al and C atom [Fig. 5(b)]. Physisorption of CO on the intrinsic graphene does not alter the electron distribution for both CO molecule and graphene, implying the weak bonding characteristics. It is discernable that electronic polarization is induced by the preferential accumulation of electrons on O in CO molecules [Fig. 5(c)]. As distinct from the CO absorption on the intrinsic graphene, the chemisorption of CO on Al doped graphene leads to significant electron transfer from the graphene to CO molecule [Fig. 5(d)]. In this case, the electrons not only accumulate on the O atom but also on the C atom of the molecule bond with the doped Al atom. The final position of Al atom in the chemisorbed CO-Al-graphene complex is thus a direct consequence of the maximized degree of sp^3 orbital hybridization with neighboring C atoms from both the graphene layer and CO molecule. This is evidential because the red lobes around C atoms in Fig. 5(d) are both pointing towards Al atom.

To further determine the effects of CO absorption on electrical conductivity, DOS for the both systems with and without the absorption were calculated. As shown in Figs. 6(a) and (b), the Al doping in graphene enhances its electrical conductivity by shifting the highest DOS peak to just below the Fermi level E_f, which also leads to the reduction of band gap E_g. This indicates that the doped Al atom induces shallow acceptor states in graphene like B atom in SWCNs, thus enhancing its extrinsic conductivity (Peng & Cho, 2003). When the CO molecule adsorbed on the intrinsic and doped graphene surfaces, the total DOSs are shown in Figs. 6(c) and 6(d). In the intrinsic graphene, the DOS of CO-graphene system near E_f have no distinct change, and the conductivity change is barely observable. It implies that the intrinsic graphene would not be an ideal CO gas sensor. However, for the Al doped graphene with the most stable chemisorbed CO configuration [Fig. 6(d)], not only the highest DOS peak shifts over the E_f, but also the DOS value increases dramatically. This results in an E_g closure [Fig. 6(d)] where E_g of the Al doped graphene is 0.18 eV without adsorption and the E_g becomes zero with adsorption. It suggests that extra number of shallow acceptor states have been introduced when the Al doped graphene interacts with the highly polar CO molecule. As a result, the chemisorbed CO on the Al doped graphene gives rise to a large increase in the electrical conductivity of the doped graphene layer. By detecting the conductivity change of the Al doped graphene systems before and after the adsorption of CO, the presence of this toxic molecule can be detected sensitively. Therefore, the Al doped graphene is a promising sensor material for detecting CO molecules. However, desorption of CO molecule from the Al doped graphene is difficult due to the strong bonding of Al-CO (Peng et al., 2004). This can be solved by applying an electric field F to reactivate the sensor materials (Hyman & Medlin, 2005).

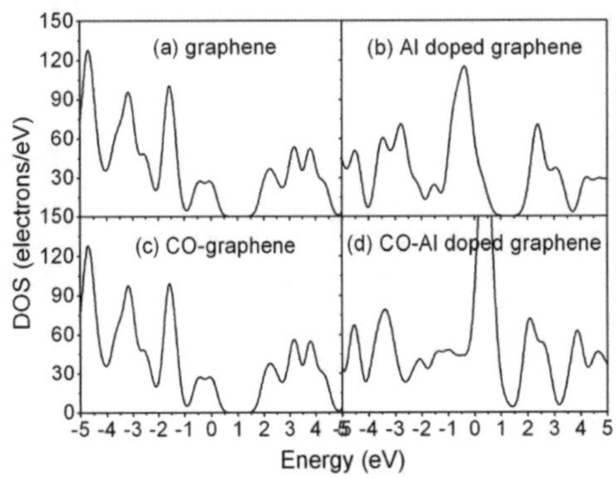

Fig. 6. Electronic density of state (DOS) of intrinsic graphene (a), Al doped graphene (b), CO-graphene system with preferred configuration (c) and CO-Al doped graphene system with preferred configuration (d).(Reproduced with permission from Ref. (Ao et al., 2008). Copyright 2008, Elsevier)

2.4 The effect of electric field on the adsorption/desorption behaviours of CO molecules

The first theoretical work with quantum mechanical calculations on electric field F inducing adsorption/desorption was studied for N_2 molecule on Fe(111) surface (Tomanek et al., 1985). Recent simulation works on the effects of F on: (1) the adsorption and dissociation of oxygen on Pt(111) (Hyman & Medlin, 2005), (2) electronic structure of Au-XO(0,-1,+1) (X = C, N and O) (Tielens et al., 2007), and (3) vibrational frequencies of CO on Pt(111) (Lozovoi & Alavi, 2007) showed that F could induce some new physical phenomena by changing their electronic properties (McEwen et al., 2008).

Therefore, it is of interest to investigate how F influences the adsorption/desorption behaviours of CO on Al-doped graphene. Here, the favorable adsorption configurations of CO on Al-doped graphene under different F had been determined by DFT calculation, and the effects of F on the corresponding interaction between CO and Al-doped graphene will be further discussed. All DFT calculations were performed using $Dmol^3$ code with the same settings as above in the section 2.3 (Delley, 1990; 2000).

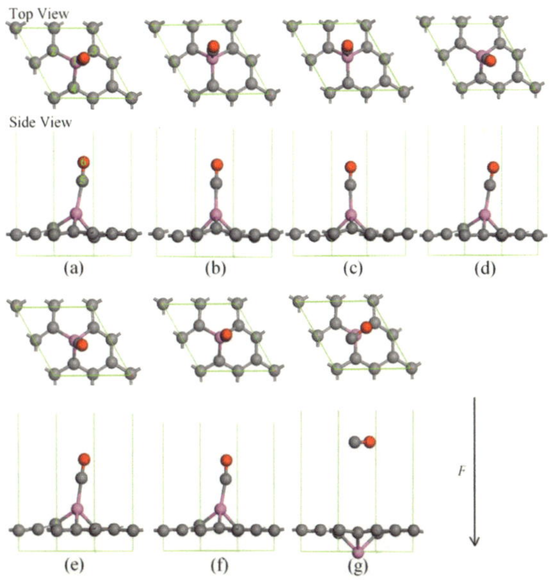

Fig. 7. The favorite adsorption configurations under different F. Atomic structures when $F = -0.03$ a.u. (a), $F = -0.02$ a.u. (b),$F = -0.01$ a.u. (c),$F = 0$ (d), $F = 0.01$ a.u. (e), $F = 0.02$ a.u. (f), (g) stable structure cannot be gotten when $F = 0.03$ a.u. and this structure is the configuration after 200 geometry optimization steps. The direction of the positive F is pointed out by the arrow. In the figure, gray, pink and red spheres are C, Al and O atoms, respectively. One Al atom dopes in site 1 and sites 2, 3 and 4 are C atoms near the doped Al atom, sites 5 and 6 are C and O atoms in the CO molecule. (Reproduced with permission from Ref. (Ao et al., 2010a). Copyright 2010, Elsevier)

In the calculations, all atoms were allowed to relax. Al-doped graphene structures were obtained through substituting one C atom in the graphene supercell by an Al atom as shown in Fig. 7. In this case, the concentration of the doped Al in graphene is 12.5% atomic ratio. For CO adsorption on Al-doped graphene, there are two highly symmetric adsorption configurations: (1) CO molecule resides parallel to the graphene surface, and (2) CO molecule resides perpendicular to the graphene surface. The detailed structures are similar as in the literature [Fig. 1 in Ref. (Ao et al., 2008)]. The $E_b(F)$ of CO molecule on Al-doped graphene under F can be determined by (Acharya & Turner, 2007),

$$E_b(F) = E_{CO+graphene}(F) - [E_{prot}(F) + E_{CO}(F)]. \tag{14}$$

where the subscripts CO+graphene, prot, and CO denote the adsorbed system, the initial isolated graphene with Al atom protruding from the graphene surface and the CO molecule, respectively. In the simulation, F had been chosen in the range of -0.03~0.03 a.u. (1 a.u. = 51 V/Å) and its positive direction is pointed out by the arrow in Fig. 7. Note that the length of the vacuum layer along the direction of normal to the graphene layer in the simulation system is about 15 Å. Thus, the maximum voltage required to induce the electric field with intensity of 0.03 a.u. is about 23 V, which can be easily realized in actual applications.

E_b of the CO/graphene systems with all possible adsorption configurations in the presence of F is listed in Table 3. Based on the calculated E_b values, the corresponding favourite adsorption configurations under different F are present in Fig. 7 where the CO molecule always takes

Initial configurations		F (a.u.)					
		-0.03	-0.02	-0.01	0	0.01	0.02
CO‖graphene	T-B-T	-0.966	-0.740	-0.631	-0.654		
	T-H-T	-0.906	-0.773	-0.664	-0.559	-0.370	
	H-T-H	-0.977	-0.746	-0.566	-0.199	-0.354	
	H-B-H	-0.952	-0.740	-0.618	-0.185	-0.356	
	B(C atom)-T-H	-0.865	-0.719	-0.650	-0.195		
	B(O atom)-T-H	-0.982	-0.692	-0.580	-0.202	-0.354	
CO⊥graphene	T(O upwards)	-0.895	-0.692	-0.593	-0.565	-0.379	-0.291
	B(O upwards)	-0.958	-0.776	-0.664	-0.564	-0.419	-0.280
	H(O upwards)	-0.968	-0.763	-0.648	-0.561	-0.372	
	T(C upwards)	-0.171	0.066	0.212	-0.215	-0.389	
	B(C upwards)	-0.242	0.068	0.253	-0.193		
	H(C upwards)	-0.245	0.111	-0.208	-0.195	-0.416	

Table 3. Summary of adsorption energy E_b of CO adsorption in eV on Al-doped graphene with different adsorption sites under different F where an Al atom replaces one C atom of the unit cell. There are twelve possible configurations for CO adsorption on Al-doped graphene, similar as in Ref. (Ao et al., 2008). T, B and H denote top site of C atoms, bridge site of C-C bond and hollow site of carbon hexagon, respectively. The blank space in the table denotes that there is no equilibrium structure in this situation.

the top site of the doped Al atom. From Table 3, the most stable structures were obtained from the initial arrangements of T-B-T, H-T-H, H-B-H, B(O atom)-T-H, B(O upwards) and H(O upwards) when F = -0.03 a.u., T-B-T, T-H-T, H-T-H, H-B-H, B(C atom)-T-H, B(O upwards) and H(O upwards) when F = -0.02 a.u., T-B-T, T-H-T, H-B-H, B(C atom)-T-H, B(O upwards) and H(O upwards) when F = -0.01 a.u., T-B-T, T-H-T, T(O upwards), B(O upwards) and H(O upwards) when F = 0, H-T-H, H-B-H, B(O atom)-T-H and B(O upwards) when F = 0.01 a.u., T(O upwards) and B(O upwards) when F = 0.02 a.u. where the letters T, B and H denote the sites of atom and CO molecule center on the graphene ring, respectively. Note that in Table 3, the error range of E_b for the favourite adsorption configurations from different initial arrangements above under a given F is within 5%.

	$F=-0.03$	$F=-0.02$	$F=-0.01$	$F=0$	$F=0.01$	$F=0.02$
l_{Al1-C2}	1.863	1.911	1.913	1.872	1.883	1.898
l_{Al1-C3}	1.930	1.907	1.907	1.910	1.915	1.930
l_{Al1-C4}	1.883	1.861	1.865	1.916	1.922	1.921
l_{Al1-C5}	2.046	2.009	1.985	1.964	1.950	1.950
l_{O6-C5}	1.136	1.145	1.153	1.164	1.174	1.188

Table 4. Structure parameters of the favorite adsorption configuration under different F shown in Fig. 7. The unit of F and bond length l are respectively a.u. and Å, which is not shown in the table for clarity.

The corresponding parameters of the stable atomic structures after adsorption in Fig. 7 are listed in Table 4. The results show that the field induces slight atom structure deformation and the top site of the doped Al atom is always the favourable adsorption site. However, the desorption of CO from the Al-doped graphene occurs when $F = 0.03$ a.u. due to the large upward force on CO molecule induced by electrical field as shown in Fig. 7(g), while the Al atom goes below the graphene layer due to the downward force on the positive charged Al atom. The configuration in Fig. 7(g) was obtained after 200 geometry optimization steps to demonstrate the desorption configuration.

Fig. 7 presents that the Al atom protrudes from the graphene surface after CO adsorption. However, the Al-doped graphene was found as a planer configuration before the adsorption. After the final adsorption configuration for CO on the Al-doped graphene was obtained, we removed the adsorbed CO and optimized the geometry of the Al-doped graphene. The result showed that the protruded Al-doped graphene layer could not return to the initial planar state and has a lower energy. This means that the planar Al-doped graphene is metastable, and there should be an energy barrier that prevents the Al-doped graphene transiting from the planer state to the protruding configuration. A force in the vertical direction would induce the transition. In this way, the adsorption energies of CO on Al-doped graphene layer in the presence of different electrical field intensities are defined as Eq. 14 and are given in Fig. 8.

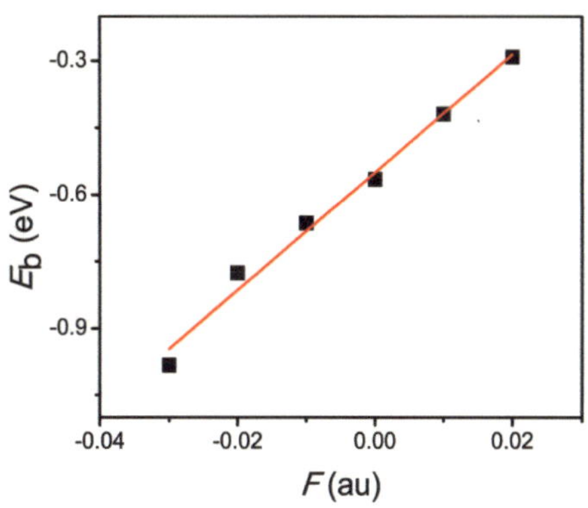

Fig. 8. $E_b(F)$ function of CO molecule adsorbed on Al-doped graphene. The squares are DFT results and the line is the fitted line to guide the eyes. (Reproduced with permission from Ref. (Ao et al., 2010a). Copyright 2010, Elsevier)

In Fig. 8, $E_b(F)$ increases nearly linearly as F increases. Therefore, the adsorption can be significantly strengthened by the negative F, while be weakened by the positive F. As F further increases, such as when $F \geq 0.03$ a.u., desorption occurs due to the electrostatic interaction as shown in Fig. 7(g). The approximate linear relationship between E_b and F can be explained by the first-order Stark effect (Hyman & Medlin, 2005). Thus, the highly positive F can be used to reactivate the sensor material for repetitious application.

On the other hand, in Table 4, as F increases, bond length l_{O6-C5} increases while l_{Al1-C5} reduces where the atom index is shown in Fig. 7. Such variations can be explained by Blyholder model (Koper & van Santen, 1999). The isolated CO molecule is bonded through sp_z hybrid orbital of a C atom and p_z orbital of an O atom. Therefore, a σ bond is formed while the interaction between p_x and p_y orbitals of the C and O atoms produces two π bonds (Blyholder, 1964). However, a lone pair of electrons on the O 2s orbital and a lone pair in the C sp_z hybrid orbital are left, forming a coordinate bond in a complex with a suitable acceptor orbital such as a p orbital on Al with a σ bond. It is believed that the back donation from a metal p orbital to the antibonding π^* molecular orbital of the CO ligand stabilizes the bond by removing the excess electrons. Such a σ orbital can be considered to be a bonding orbital for

C-O, but as an antibonding orbital for the Al-CO bond. Therefore, depopulating this orbital (donation) elongates l_{O6-C5} but shortens l_{Al1-C5}. In general, Fermi level E_f of a neutral Al lies between σ and π^* levels of CO (Wang et al., 2007). Applying a positive F lowers the energy levels of CO comparing to those of the Al-doped graphene (Lozovoi & Alavi, 2007), resulting in the π^* energy level moving towards E_f, whereas the σ level moves away. Thus, a positive F suppresses donation but enhances back donation, leading to a smaller l_{Al1-C5} and a larger l_{O6-C5}.

	$F = -0.03$	$F = -0.02$	$F = -0.01$	$F = 0$	$F = 0.01$	$F = 0.02$
Al1	0.968	0.957	0.927	0.896	0.866	0.829
C2	-0.341	-0.313	-0.286	-0.363	-0.358	-0.332
C3	-0.362	-0.317	-0.294	-0.275	-0.260	-0.229
C4	-0.323	-0.370	-0.370	-0.267	-0.249	-0.267
C5	-0.040	-0.056	-0.076	-0.101	-0.124	-0.147
O6	0.078	0.008	-0.058	-0.128	-0.197	-0.273
Q	0.038	-0.048	-0.144	-0.229	-0.312	-0.410

Table 5. Charges of atoms near the doped Al atom in CO/graphene system as well as charge transfer Q between the graphene and the CO molecules under different F, obtained by Mulliken analysis. The unit of the atom charge is one electron charge $|e|$ and that of F is a.u.

Table 5 lists: (1) the charges of C and Al atoms as well as the CO molecule, and (2) the charge transfer Q between the graphene and the CO molecule under different F obtained by Mulliken analysis. As F increases, the electron numbers of Al1, C5, and O6 increase while those of atoms C2, C3, and C4 decrease. This is because electrons flow from the C atoms in the graphene layer to atoms of Al1, C5 and O6 above the graphene layer with increasing F. This agrees with the reported phenomena where the negative F enhances adsorption while the positive F has a counter effect (Hyman & Medlin, 2005). As F increases, the effects result in a shorter l_{Al1-C5}, and a longer l_{O6-C5} due to the reduction of the attraction between C5 and O6. The trends of l_{Al1-C5} and l_{O6-C5} variations are consistent to the results shown in Table 4.
The electronic distribution under different F is displayed in Fig. 9. The bond between the CO and the graphene system in the figure is covalent because the preferential electron accumulation sites mainly localize in the bond rather than centralize on a particular atom. However, due to the different electronegativity of C, Al and O atoms, electrons lean to the C atom for Al1-C5 bond and O atom for C5-O6 bond. Furthermore, with increasing F, more electrons transfer from the doped graphene to the CO where the lost electrons are all from Al (Table 5). The images in Fig. 9, where the red regions around CO become larger and thicker, also confirm that Q increases with increasing F.

2.5 The effect of temperature on the adsorption/desorption behaviours of CO molecules

The favourable CO adsorption site on the Al doped graphene was identified through DFT calculations (Ao et al., 2008). But it was still far to be optimized for actual applications. In particular, the effect of temperature T on the adsorption/desorption behaviours on the CO/graphene system is still unclear. Based on the DFT results and thermodynamic analysis at 0 K, the adsorption phase diagrams can be established. The Gibbs free energy of adsorption, ΔG_{ads} can be expressed as:

$$\Delta G_{ads}(T) = G_{ads}(T) - G_g(T) - G_{CO}(T), \tag{15}$$

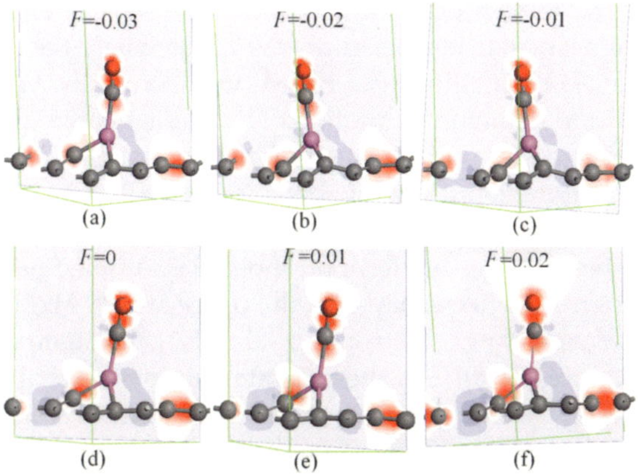

Fig. 9. Images of electron density difference of CO/graphene system under different F, the unit of F is a.u. The red region means electronic accumulation, while the blue region means electronic loss. (Reproduced with permission from Ref. (Ao et al., 2010a). Copyright 2010, Elsevier)

where $G_{ads}(T)$, $G_g(T)$ and $G_{CO}(T)$ are the corresponding Gibbs free energies of the adsorbed system, the Al doped graphene and the CO gas molecule at a particular T, respectively. Since the Gibbs free energy of the system before and after adsorption can be obtained with the DFT results, the adsorption-desorption transition could be determined with Eq. 15 theoretically.

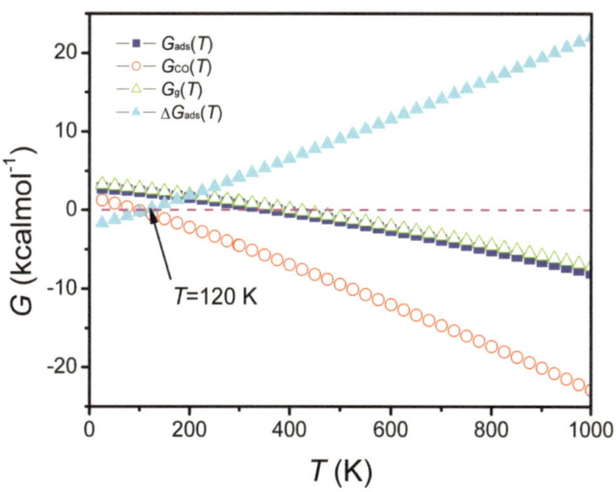

Fig. 10. The temperature dependent Gibbs free energy $G(T)$ functions where subscripts ads, g, and CO denote the adsorbed system, the isolated graphene and the CO molecules, respectively. And $G_{ads}(T)$ denotes Gibbs free adsorption energy, which is obtained in terms of Eq. 14. The symbols are all DFT results. (Reproduced with permission from Ref. (Ao et al., 2009b). Copyright 2009, RSC)

Subsequently, *ab initio* molecular dynamics (MD) calculations were performed under constant volume and constant temperature conditions (NVT) adopting GGA with the revised PBE method. The temperature effects on the atomic and electronic structures were calculated with

a time step of 1 fs at the temperatures from 300 to 450 K with an interval of 50 K. The simulation time t at the particular temperature was 2.5 ps where the total energy fluctuated in the range of 0.01%. MD calculation was based on the velocity Verlet algorithm (Verlet, 1967) for integration of the equation of motion. The implemented algorithm performs the Yoshida-Suzuki multiple-step numerical integration of varying quantity, depending on the choice of interpolation parameters (Suzuki, 1991; Yoshida, 1990). A key parameter in the integration algorithms is the integration time step. A common rule-of-thumb used to set the time step is that the highest frequency vibration should be sampled between 10 and 20 times in one cycle. In this system, the frequency is in the order of 10^{13} Hz, the time step was thus set as 1 fs within a reasonable range (Seitsonen et al., 2001). The temperature was controlled by algorithm of Nose (Nose, 1984). The thermostate employs a feedback loop between the instantaneous kinetic energy and the set temperatures. The rate of feedback is determined by the mass parameter, Q ($Q = 2$) (Loffreda, 2006; Spencer & Yarovsky, 2007; Todorova et al., 2007).
With the thermal desorption method, T dependent desorption time $\tau(T)$ function can be expressed as (Peng et al., 2004; Raaen & Ramstad, 2005)

$$\tau(T) = \nu_0^{-1} \exp[-E_b(T)/k_B T] \tag{16}$$

where k_B is the Boltzmann's constant (8.62×10^{-5} eV/K), and ν_0 is the attempt frequency of 10^{13} Hz for CO (Seitsonen et al., 2001). This thermal desorption method is close to the experimental conditions and it can be used to determine the thermodynamical properties of the adsorption systems (Raaen & Ramstad, 2005).
With the adsorption structures determined by the DFT calculations at an ideal condition, the phase diagram of adsorption/desorption for the CO adsorbed on the Al doped graphene as a function of temperature can be established with the atomistic thermodynamics described in Eq. 15. Such a simple approach allows the exploration of $\Delta G_{ads}(T)$ in an actual condition with respect to experiments. $\Delta G_{ads}(T)$, $G_{ads}(T)$, $G_g(T)$ and $G_{CO}(T)$ functions are plotted in Fig. 10. The results show that $\Delta G_{ads}(T)$ increases as T increases, and eventually becomes positive at $T_d = 120$ K where T_d is defined as the desorption temperature. In another word, the desorption of CO from the Al doped graphene occurs when $T_d > 120$ K at the ideal state with $\tau \to \infty$.

	$T = 0$	$T = 300$ K	$T = 400$ K	$T = 450$ K
l_{Al1-C2}	1.872	1.880	1.946	1.973
l_{Al1-C3}	1.910	1.961	1.972	1.993
l_{Al1-C4}	1.916	1.923	1.929	1.989
l_{Al1-C5}	1.964	1.982	2.097	4.590
l_{C5-O}	1.164	1.161	1.159	1.157

Table 6. Some structure parameters of CO molecule adsorbed on Al doped graphene at different temperature, where l is bond length in Å.

However, with *ab initio* MD calculation at $T = 300$, 350, 400 and 450 K for 2.5 ps to reach the equilibrium at each temperature, it was found that the desorption occurred at 450 K. The atomic configurations at the different temperatures are shown in Fig. 11 and their corresponding atomic structural parameters are listed in Table 6. The results show that T_d is between 400 and 450 K. Since both the data for MD simulation and atomistic thermodynamics come from the simulation, the difference of T_d caused by the simulation methodologies is

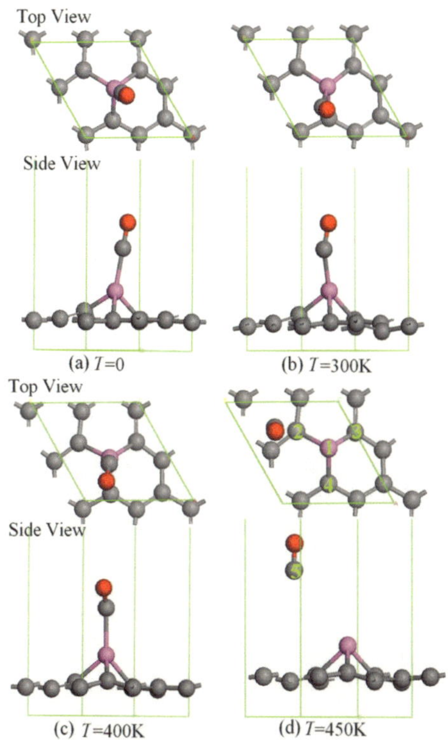

Fig. 11. Atomic structure of CO molecule adsorption in Al doped graphene at different temperature. For $T \neq 0$, the images are the configurations at simulation time $t = 2.5$ ps. In the figure, the uppermost atom is O atom, the third uppermost atom is Al atom, and the others are all C atoms. One Al atom dopes in site 1 and sites 2, 3 and 4 are C atoms near the doped Al atom, and site 5 is C atom in CO gas molecule.(Reproduced with permission from Ref. (Ao et al., 2009b). Copyright 2009, RSC)

limited and it could be ignored. It is believed that the difference is mainly induced by the short equilibrium time of $t = 2.5$ ps used in the MD simulation, which is much shorter than the actual situation.

$E_b(T)$ and $\tau(T)$ determined by Eqs. 13 and 16 are plotted in Figs. 12 and 13, respectively. Fig. 12 shows that $E_b(T)$ decreases linearly with T increasing. This is consistent with the classic Readhead formula (Readhead, 1962). In Fig. 13, $\tau(T)$ decreases exponentially with T increasing, showing $T_d \approx 420$ K at $\tau = 2.5$ ps. This is in agreement with the results obtained from MD simulation, in which the desorption occurs at $400 < T < 450$ K. In the experimental environment, the optimal τ is in an order of microsecond (μs) (Peng et al., 2004), and τ(400K)$\approx$1 μs from Fig. 13, indicating that the gas sensor can be reactivated for repetitious applications by heating the materials up to 400 K. Note that the adsorption-desorption process is dynamic. Once the CO molecule is adsorbed, the adsorption state would be remained for a time of τ until the desorption occurs. During the adsorption period, electrical conductivity changes of the Al doped graphene can be detected.

Except for T_d, the temperature dependence of atomic structure and electrical properties are also critical information for gas detection. Table 6 lists the structural parameters calculated by *ab inito* MD at 300, 400 and 450 K, respectively. As T increases, Al-CO bond length l_{Al1-C5}

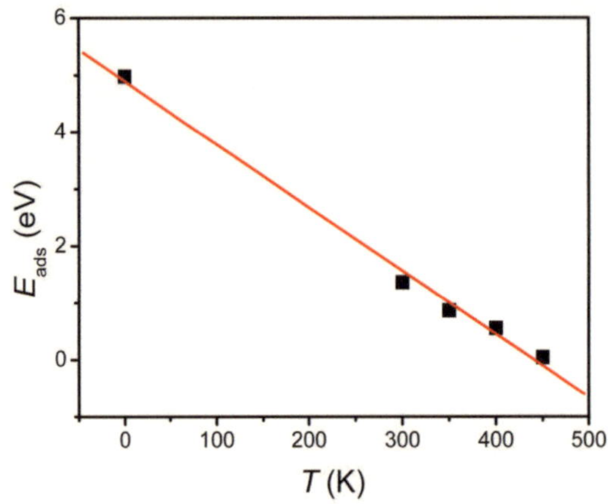

Fig. 12. Temperature dependent adsorption energy of CO molecule in Al doped graphene $E_b(T)$ function. The symbol ■ is the MD simulation result at T =0, 300, 350, 400, 450 K. The solid line is the fitted linear function with the calculated data. (Reproduced with permission from Ref. (Ao et al., 2009b). Copyright 2009, RSC)

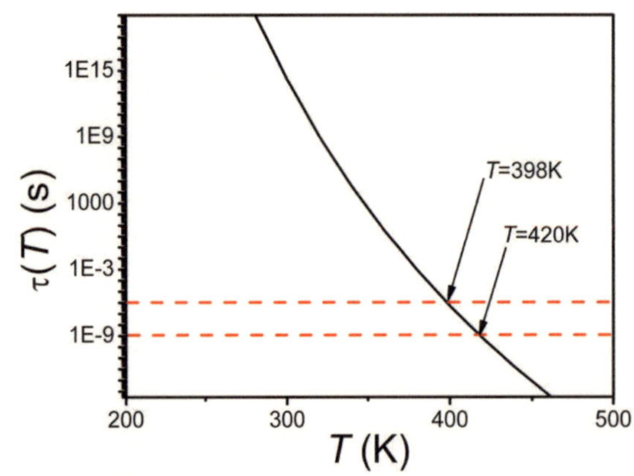

Fig. 13. Temperature dependent desorption time function $\tau(T)$ in terms of Eq. 16 where $E_b(T)$ function needed is from Fig. 12. The two temperatures 398 and 420 K are corresponding desorption temperature in MD simulation and actual situation.(Reproduced with permission from Ref. (Ao et al., 2009b). Copyright 2009, RSC)

increases, or the corresponding bond strength decreases. This is also evidenced by the $E_b(T)$ declination as shown in Fig. 12. When T further increases, the desorption of CO from the Al doped graphene occurs [Fig. 11(d)] where bond length of l_{Al1-C5} changes sharply from 2.097 Å at 400 K to 4.590 Å at 450 K.

In order to better understand the results, Table 7 lists the charges of C atoms surrounding the doped Al atom, the doped Al atom and the CO molecule as well as charge transformation Q between the doped graphene and the CO molecule, which were obtained by Mulliken analysis. It exhibits that Q decreases as T increases and the Al atom loses electrons. The negative charges of the C atoms surrounding the doped Al also decrease. It results in the charge difference between the C and Al atoms decreases and the Al-C bond length in the

	$T=0$	$T=300$ K	$T=400$ K	$T=450$ K
Al1	0.896	0.833	0.765	0.613
C2	-0.363	-0.303	-0.324	-0.274
C3	-0.275	-0.287	-0.245	-0.276
C4	-0.267	-0.239	-0.255	-0.216
C5	-0.101	-0.089	-0.029	0.111
O	-0.128	-0.132	-0.127	-0.123
Q	0.229	0.221	0.156	0.012

Table 7. Charges of atoms surrounding doped Al atom and doped Al atom, and changes transferred Q form graphene to CO gas molecule at different temperature.

graphene layer elongates as T increases. This is consistent with the structure parameters change listed in Table 6. On the other hand, the electrons in C5 and O6 of the CO molecule also decrease with T increasing, and even C5 is positive at $T=450$ K. Due to the static interaction, l_{C5-O6} decreases as T increases as shown in Table 6.

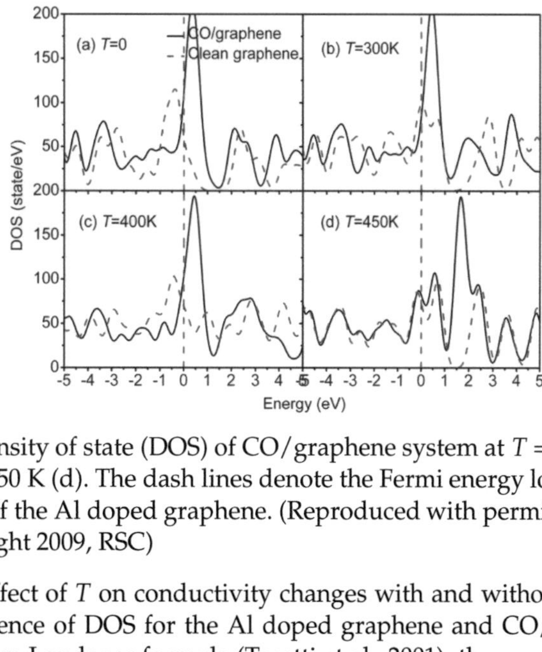

Fig. 14. Electronic density of state (DOS) of CO/graphene system at T = 0 (a), T = 300 K (b), T = 400 K (c) and T = 450 K (d). The dash lines denote the Fermi energy location and the dash curves are the DOS of the Al doped graphene. (Reproduced with permission from Ref. (Ao et al., 2009b). Copyright 2009, RSC)

To understand the effect of T on conductivity changes with and without the adsorption, the temperature dependence of DOS for the Al doped graphene and CO/graphene system are shown in Fig. 14. From Laudauer formula (Tosatti et al., 2001), the number of bands crossing E_f determines the number of conduction channels or the conductivity of CO/graphene system (He et al., 2008; Mares & Van Ruitenbeek, 2005). Therefore, the largest conductivity change induced by the adsorption is found at T =400 K. The performance of CO detection with this material is the best at T =400 K based on the $\tau(T)$ function and conductivity change.

3. Application for efficient hydrogen storage

3.1 Hydrogen storage in Al subtitutionally doped graphene

In recent years, hydrogen-based fuel systems have been considered to be a highly important topic of research for future energy schemes as hydrogen is a more efficient fuel in comparison to the existing carbonaceous fossil fuels (Coontz & Hanson, 2004; Schlapbach & Züttel, 2001). Despite many recent technological developments in the hydrogen-based fuel systems, it is still an enormous challenge to have a safe and efficient reversible hydrogen storage system at ambient conditions (Schlapbach & Züttel, 2001). One possible way for hydrogen storage is an efficient and controllable adsorption/desorption system. Carbon based materials appear promising for such a purpose. Although several mechanisms of hydrogen storage through both physisorption and chemisorption have been proposed (Chandrakumar & Ghosh, 2008; Deng et al., 2004; Klontzas et al., 2008; Mpourmpakis et al., 2007; Nikitin et al., 2008), most of these efforts are far to reach the target of 6 wt% and binding strength of $-0.2 \sim -0.4$ eV/H_2 at ambient temperature and modest pressure for commercial applications specified by U.S. Department of Energy (DOE).

With DFT simulations, it was predicted that a single ethylene molecule can form a stable complex with two transition metals, thus adsorbing ten H_2 molecules and lead to a high storage capacity of $\sim$14 wt% ((Durgun et al., 2006). In addition, the highest H_2 storage capacity of 13 wt% in a fullerene cage with twelve Li atoms capped onto the pentagonal faces was calculated (Sun et al., 2006). This system has average adsorption energy $E_b = -0.075$ eV/H_2. However, all the DFT results are in the ideal condition at the temperature of $T = 0$ K, their performances at the DOE specified operation conditions are unclear.

Since carbon nanostructures have high surface areas and thermal stability along with unique mechanical properties, improvement of their adsorption capacity by suitable modification would be of immense interest (Chandrakumar & Ghosh, 2008; Deng et al., 2004; Durgun et al., 2006; Klontzas et al., 2008; Mpourmpakis et al., 2007; Nikitin et al., 2008; Sun et al., 2006). Thus, hydrogen storage using carbon nanostructures is still an important research topic and deserves more attention. In this section, the potential of graphene as hydrogen storage materials through doping is investigated. The advantages of graphene are: (1) a large surface for hydrogen adsorption, (2) economical and scalable production (Li et al., 2008a), and (3) the strongest material ever measured (Lee et al., 2008).

AlH_3 and related aluminum hydrides as hydrogen storage materials have recently become the focus of renewed interest (Graetz et al., 2006; Li et al., 2007) due to their potentially large hydrogen capacity of $\sim$10 wt%. These materials are thermodynamically unstable in ambient, but it is kinetically stable without much lost of hydrogen for years. Despite these excellent properties, extremely high pressure (exceeding 2.5 GPa) is required for hydrogen adsorption. While these hydrides possess a small negative enthalpy of formation (Graetz et al., 2006), for practical applications the large hydrogen desorption energy proves impractical. The origin of this energy barrier lies in the rather strong mixed ionic and covalent bonds (Graetz et al., 2006) formed between Al and H. Thus it is essential to significantly reduce the desorption energy.

There appears another way for Al atoms to store hydrogen i.e. to further decrease the interaction between Al and H. In this way, the weak chemisorption can be changed into strong physisorption. For hydrogen storage through physisorption, strong interaction between the H_2 molecule and the surfaces along with a large surface area for adsorption are required. The unique characteristics of graphene and Al for hydrogen storage lead to an investigation of the properties of Al doped graphene as a possible hydrogen storage candidate. It would be intriguing to understand the interaction between graphene, Al and H. In this work, the adsorption behaviour of H_2 in Al doped graphene was studied by DFT calculation. In addition, we processed the *ab initio* MD calculation to investigate the effects of temperature and pressure on the corresponding adsorption and desorption behaviours of this system.

All DFT calculations were performed with Dmol[3] code (Delley, 1990). Previous studies (Cabria et al., 2008; Okamoto & Miyamoto, 2001) had shown that the LDA prediction of the physisorption energies of H_2 on the surface of graphite and carbon nanotubes were in good agreement with experiments. The reliability of LDA can be ascribed to the following facts (Cabria et al., 2008): (1) When the electron densities of H_2 and graphene overlap weakly, the nonlinearity of the exchange-correlation energy density functional produces an attractive interaction even in the absence of electron density redistribution; (2) The overestimated binding energy by LDA (Leenaerts et al., 2008; Lugo-Solis & Vasiliev, 2007) may compensate for the insufficient account of van der Waals interactions (Cabria et al., 2008). In contrast, DFT calculation using GGA produced a purely repulsive interaction. Using a GGA-PW91 functional, a repulsive interaction between H_2 and a graphene layer and also between H_2 and a $(6,6)$ carbon nanotube was obtained (Tada et al., 2001). This contradicts the experimental findings (Sahaym & Norton, 2008). It was noted that LDA calculations well reproduce the empirical interaction potentials between graphitic layers and also in the other graphitic systems for distances near to the equilibrium separation although the LDA is not able to reproduce the long-range dispersion interaction (Girifacol & Hodak, 2002). Therefore, LDA had been selected in this work. To ensure that the calculated results were comparable, identical conditions were employed for the isolated H_2 molecules and the graphene, and also the adsorbed graphene system. The k-point was set to $6 \times 6 \times 2$ for all slabs, which brought out the convergence tolerance of energy of 1.0×10^{-5} hartree (1 hartree = 27.21 eV), and that of maximum force is 0.002 hartree/Å.

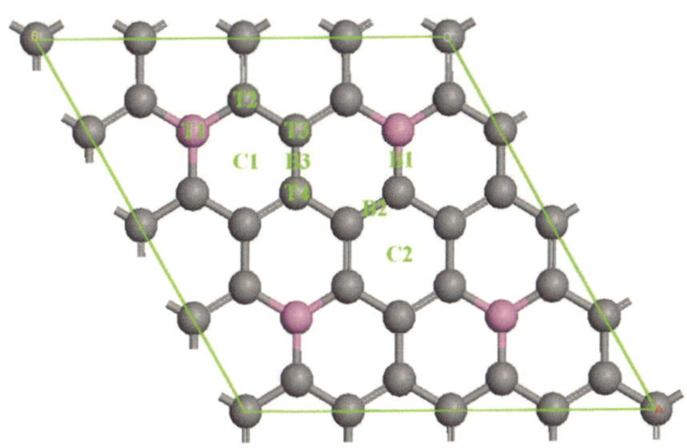

Fig. 15. Eight different adsorption sites on Al doped graphene. The gray and pink balls are respectively C and Al atoms. (Reproduced with permission from Ref. (Ao et al., 2009a). Copyright 2009, AIP)

In the simulation, three-dimensional periodic boundary condition was taken and H-H bond length was set to $l_{H-H} = 0.74$ Å, which is consistent with the experimental results (Lide, 2000). The graphene used in our simulation consist of a single layer of 2×2 supercell with a vacuum width of 12 Å to minimize the interlayer interaction. All atoms were allowed to relax in all energy calculations. The adsorption energy E_b between the H_2 gas molecule and graphene is defined as,

$$E_b = E_{H2+graphene} - (E_{graphene} + E_{H2}) \tag{17}$$

where the subscripts H2+graphene, graphene, and H2 denote the adsorbed system, isolated graphene and H2 molecules, respectively.

For the Al doped graphene, the concentration of Al is 12.5 at% with the additional constrain that there is only one Al atom per graphene hexagonal ring (Fig. 15) to avoid Al atoms clustering on graphene (Krasnov et al., 2007). For H_2 adsorption on the Al doped graphene, there are 4 top sites of T1, T2, T3 and T4, and 3 bridge sites of B1, B2 and B3, and 2 center sites of C1 and C2, as shown in Fig. 15. (In this figure, a larger simulation cell is given in order to better display the different adsorption sites on the Al doped graphene. Fig. 16 reflects the actual simulation cell size.) At each adsorption site, there are two highly symmetrical adsorption configurations, namely H_2 molecule resides parallel or perpendicular to the graphene surface. Therefore, a total of 18 adsorption configurations for H_2 on the Al doped graphene are present.

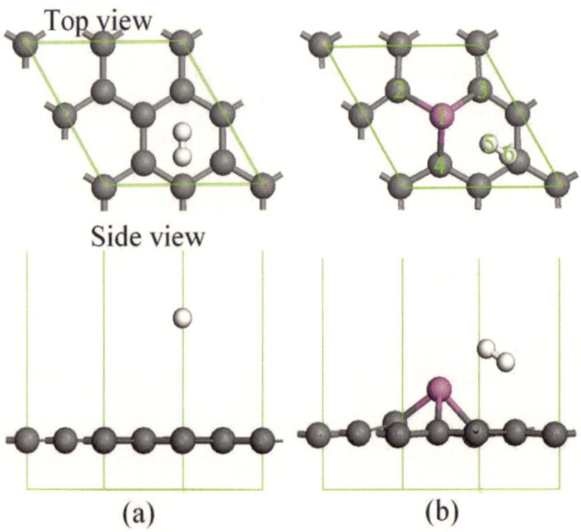

Fig. 16. The favorite adsorption configurations with 1 H_2 molecule adsorbed in intrinsic graphene (a), and in Al doped graphene (b). The white balls are H atoms. (Reproduced with permission from Ref. (Ao et al., 2009a). Copyright 2009, AIP)

Due to the periodicity of H_2 adsorbed in intrinsic graphene or Al doped graphene systems, we had selected the unit cell with the following conditions: eight C atoms and one H_2, or seven C atoms, one Al atom and one H_2 (see Fig. 16). If we placed a H_2 at any location of the cell, the distance from this H_2 to other H_2 molecules in the nearest cells is 4.920 Å. This large separation, compared to the bond length of H_2 (0.740 Å), would ensure that there is no interaction between H_2 molecules in the different cells (Arellano et al., 2000).

To calculate the H_2 adsorption capability of Al doped graphene at room temperature and modest pressure, we performed *ab initio* MD calculation with CASTEP (Cambridge Sequential Total Energy Package) code based on the structure obtained by DFT above, which utilizes plane-wave pseudopotential to perform the first principle quantum mechanics calculations (Degall et al., 2002). LDA with the Ceperley-Alder-Perdew-Zunger (CAPZ) function (Ceperley & Alder, 1980; Perdew & Zunger, 1981) was employed as exchange-correlation functions, cutoff energy $E_c = 280$ eV and *k*-points is $6 \times 6 \times 2$. In this work, the *k*-points of $6 \times 6 \times 2$ for all slabs have the energy convergence tolerance of 1.0×10^{-6} eV/atom. Such energy tolerance is small enough to ensure establishment of the actual equilibrium structure.

Each MD simulation was performed in NPT statistical ensemble, i.e. constant numbers of atoms N, pressure P and T, with $T = 300$ K and $P = 0.0001 \sim 1$ GPa. Time step of 1 fs was selected and simulation time t at a particular T was 2.5 ps where the total energy fluctuation was in the range of 0.01%. The same t was selected for H_2S dissociation on the Fe(110) surface (Spencer & Yarovsky, 2007). A Verlet algorithm (Verlet, 1967) was used to integrate the equations of motion, with T controlled by algorithm of Nose (Nose, 1991), and P was controlled according to the Parrinello-Rahaman algorithm (Parrinello & Rahaman, 1981).

Initial configurations		Intrinsic graphene E_b (eV)	d (Å)	Al doped graphene E_b (eV)	l (Å)	d (Å)
	T1	-0.136	2.845	-0.209	2.762	
	T2			-0.34	2.526	2.682
	T3			-0.407	2.588	2.486
	T4			-0.361	2.942	2.537
H_2 ‖graphene	B1	-0.139	2.817	-0.21	2.757	
	B2			-0.411	2.527	2.575
	B3			-0.411	2.506	2.563
	C1	-0.159	2.635	-0.427	2.083	2.073
	C2			-0.188		2.657
	T1	-0.141	2.615	-0.153	2.622	
	T2			-0.284	2.427	2.749
	T3			-0.406	2.367	2.524
	T4			-0.33	2.976	2.179
H_2 ⊥graphene	B1	-0.142	2.620	-0.206	2.271	3.732
	B2			-0.412	2.468	2.595
	B3			-0.426	3.196	2.074
	C1	-0.148	2.425	-0.426	2.092	2.104
	C2			-0.24	3.117	2.468

Table 8. Summary of results for H_2 adsorption on intrinsic graphene and Al doped graphene on different adsorption sites. For H_2 adsorption on intrinsic graphene, there are 6 different adsorption sites as listed in the table. For H_2 adsorption on Al doped graphene, there are 18 different adsorption configurations as shown in Fig. 15. l represents the distance between Al and H_2. d represents the distance between H_2 molecule and graphene or Al-doped graphene.

After geometry relaxation, E_b values and the corresponding structural parameters of the 18 adsorption configurations for H_2 adsorbed in the intrinsic graphene are listed in Table 8. It was found that the most favorable configuration is H_2 adsorbed on the center site of the carbon ring with E_b = -0.159 eV as shown in Fig. 16(a) and the distance between H_2 and the graphene $d = 2.635$ Å. The results are consistent with other reported results of $E_b = -0.133$ eV and $d \approx 2.8$ Å (Okamoto & Miyamoto, 2001). The small magnitude of E_b (<0.1 eV) shows that the system is in the weak physisorption regime. It indicates that the intrinsic graphene is not suitable for hydrogen storage.

For the adsorption of H_2 on the Al doped graphene, the corresponding results are also listed in Table 8. In light of Table 8, the most favourable position with $E_b = -0.427$ eV for the H_2 molecule is shown in Fig. 16(b). The distance between H_2 and the doped Al, $l = 2.083$Å while that between H_2 and carbon layer, $d = 2.073$Å. As seen from Table 8, the interaction reaches the strongest when both l and d are minimized. The adsorption of H_2 in the Al doped graphene is much larger than that in other systems, such as E_b = -0.41 eV/H_2 in Ti-C_2H_4-graphene system (Durgun et al., 2006), and $E_b = -0.08$ eV/H_2 in

12-Li-doped fullerene (Sun et al., 2006). However, it still falls into the physisorption regime as the long distance between the doped graphene and the adsorbed H_2. Therefore, this strong physisorption interaction would be ideal for hydrogen storage, which adsorbs more H_2 molecules.

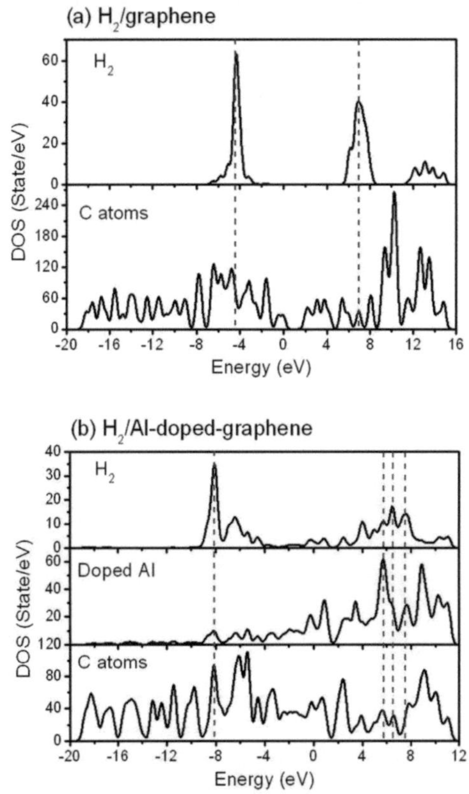

Fig. 17. Electronic density of states (DOSs) of adsorbed H2, doped Al and graphene for both the H_2/graphene and H_2/Al-doped-graphene systems as shown in panel (a) and panel (b), respectively.(Reproduced with permission from Ref. (Ao et al., 2009a). Copyright 2009, AIP)

To understand the enhancement effect of the doped Al on the H_2 adsorption, the DOSs of the adsorbed H_2, the doped Al and the C atoms in both H_2/graphene and H_2/Al-doped-graphene systems were plotted and shown in Fig. 17. Fig. 17 (a) shows the DOSs of H_2+graphene system. The main peaks of H_2 are located at -4.37 eV and 6.92 eV. However, the main peaks of intrinsic graphene are located between 9 and 13 eV. Therefore, the interaction between H_2 molecule and the intrinsic graphene is very weak because of non-overlapping of electrons in these substances, where E_b is small. On the other hand, for the H_2/Al-doped-graphene system shown in Fig. 17(b), the main peaks of H_2 are located at -8.15 eV, 5.74 eV, 6.52 eV, and 7.51 eV, respectively. The bands of H_2 interact with both the doped Al and the C atoms synchronously at the positions indicated by the dash lines, showing a strong interaction between H_2 and the Al doped graphene where E_b is the largest. In addition, the doped Al changes the electronic structures of both H_2 and the graphene, and both their DOSs shift towards the lower energy. It exhibits that the H_2/Al-doped-graphene configuration is a much more stable system.

Atom	Intrinsic graphene	Al doped graphene
Al1(C1)	0.001	0.292
C2	-0.002	-0.228
C3	0	-0.193
C4	0	-0.193
H5	-0.001	-0.001
H6	-0.001	0.021
Q	-0.002	0.019

Table 9. Charges of atoms in H_2 adsorbed in graphene system as well as charge transfer Q between graphene and H_2 molecule, obtained by Mulliken analyse. The unit of the atom charge is one electron charge e, which is elided here for clarity.

Table 9 shows the charge distribution in both the H_2/graphene and H_2/Al-doped-graphene systems using Mulliken analysis. Before and after H_2 adsorption, the charge variation for the former is little while it is significant for the latter. In addition, H6 has much more positive charge than H5. Thus, the interaction between H_2 and the Al doped graphene is mainly achieved through H6. The interaction between the band at the location of the highest peak of DOS plot of H_2 and that of C atoms implies a strong interaction between the H_2 and C atoms, as shown in Fig. 17(b).

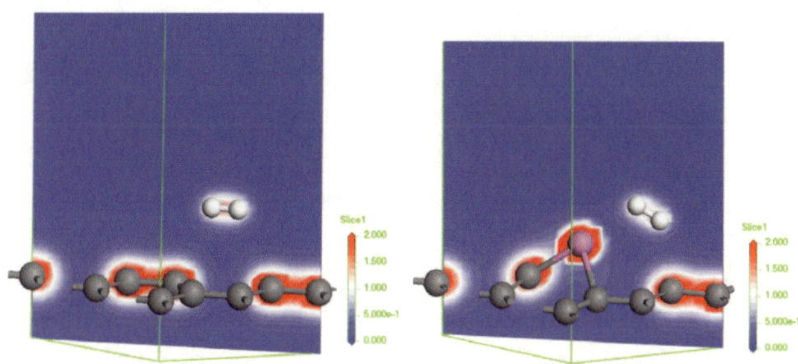

(a) (b)

Fig. 18. Electron density distributions in the H_2/graphene [panel (a)] and H_2/Al-doped-graphene [panel (b)] systems. (Reproduced with permission from Ref. (Ao et al., 2009a). Copyright 2009, AIP)

The illustrations of electron density distribution for the H_2/graphene and H_2/Al-doped-graphene systems are shown in Fig. 18. In the system of H_2/graphene [Fig. 18(a)], no electron exists in the region between H_2 and C layer while some electrons appear in the region among H_2, Al atom and C layer in the system of H_2/Al-doped-graphene [Fig. 18(b)]. This supports the notion that the H_2/Al-doped-graphene possesses a much stronger H_2 adsorption ability.

After understanding the mechanism of the enhancement for H_2 adsorption in the Al doped graphene, it is important to determine how much H_2 molecules can be adsorbed on the 2×2 layer surface. We constructed an adsorption configuration with 3 H_2 molecules adsorbed in the three favourable C1 adsorption positions on the topside of the doped system. After geometry relaxation, the atomic structure is shown in Fig. 19(a). It has $E_b = -0.303$ eV/H_2, which satisfies the requirement of $E_b = -0.20 \sim -0.40$ eV/H_2 at room temperature (Chandrakumar & Ghosh, 2008; Deng et al., 2004; Klontzas et al., 2008; Mpourmpakis et al.,

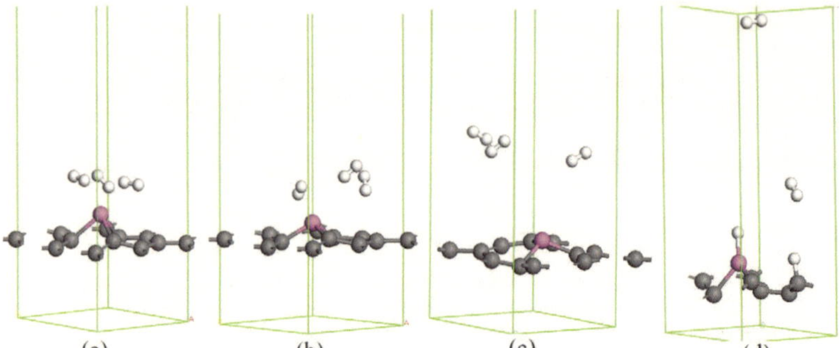

Fig. 19. Atomic configurations H_2/Al-doped-graphene system at different temperature and pressure. (a) In the ideal condition with $T = 0$ K, (b) in the condition with $T = 300$ K and $P = 0.1$ GPa, (c) in the condition with $T = 300$ K and $P = 0.0001$ GPa, and (d) in the condition with $T = 300$ K and $P = 1$ GPa. (Reproduced with permission from Ref. (Ao et al., 2009a). Copyright 2009, AIP)

2007) set by DOE although the value of 5.1 wt% of H_2 adsorbed is slightly below the DOE's 6 wt% target.

In order to understand the effect of the adsorbed H_2 molecule number on the E_b, the configuration with 6 H_2 molecules adsorbed in the Al doped graphene in the favorable C1 adsorption positions on both sides was calculated. It is found that $E_b = -0.164$ eV/H_2, which is almost half of the E_b for above the case where the Al doped graphene adsorbed 3 H_2 on one side of graphene. In addition, the adsorption with 8 H_2 molecules in the Al doped graphene was also calculated, and it is found 2 H_2 molecules were released. In the other words, the interaction between H_2 molecules would weaken the adsorption on the doped graphene and the saturated number of H_2 molecules adsorption is 6. Note that E_b for the cases of 3 H_2 and 6 H_2 are respectively -0.303 eV/H_2 and -0.164 eV/H_2, which is about twice for the case of 3 H_2 comparing with the case of 6 H_2. This is because H_2 molecules were very weakly adsorbed below the graphene layer where the doped Al atom locates above the graphene layer.

It is well known that T and P have essential effects on hydrogen storage, where increasing P and decreasing T enhance the capacity of hydrogen storage. Thus, most studied systems are either under high P or at very low T (Sahaym & Norton, 2008), which may not be viable for mobile applications. For example, a storage capacity of 8 wt% for purified single wall carbon nanotubes (SWNTs) at 80 K with a hydrogen pressure of 13 Mpa (Ye et al., 1999) and a lower hydrogen storage capacity of 2.3 wt% at 77 K were reported (Panella et al., 2005). The hydrogen storage capacities in other carbon related materials, such as activated carbon (AC), single walled carbon nanohorn, SWNTs, and graphite nanofibers (GNFs) were also investigated (Xu et al., 2007). Although the AC had a capacity of 5.7 wt% at 77 K with $P = 3$ MPa, its capacity is $< 1\%$ at 300 K (Xu et al., 2007). Recent experimental results demonstrated that the intrinsic graphene has hydrogen storage capacity of 1.7 wt% under 1 atm at 77 K, and 3 wt% under 100 atm at 298 K (Ghosh et al., 2008). Thus, to meet the DOE target, it is necessary to study the adsorption and desorption behaviours of H_2 in the Al doped graphene at $T = 300$ K with different P. Therefore, the adsorption behaviours of $3H_2$/Al-doped-graphene and $6H_2$/Al-doped- graphene systems were calculated under 0.0001, 0.01, 0.1 and 1 GPa using *ab initio* MD simulation. For both the $3H_2$/Al-doped-graphene and $6H_2$/Al-doped- graphene systems, we found that all H_2 molecules were released at 0.0001 GPa [Fig. 19(c)]. However, there was only one H_2 molecule adsorbed in both the systems at 0.01 GPa, while the structure of the doped graphene was completely destroyed with H and Al forming covalent bond at 1 GPa [Fig. 19(d)]. When P = 0.1 GPa, there are three H_2 left on the top side of the two Al doped systems [Fig. 19(b)]. Therefore, the Al doped graphene for hydrogen storage capacity at room temperature and 0.1 GPa is 5.13 wt% with $E_b = -0.260$ eV/H_2, satisfying the requirements of actual application. In addition, all the adsorbed H_2 molecules can be released when $P = 0.0001$ GPa.

3.2 Hydrogen storage in graphene with Al atom adsorption

Very recently, based on DFT calculations, Ca atoms adsorbed on graphene layers and fullerenes were found to result in high-capacity hydrogen storage mediums, which could be recycled at room temperature (Ataca et al., 2009; Yoon et al., 2008). In these systems, the adsorbed Ca atoms become positively charged and the semimetallic graphene changes into a metallic state, while the hydrogen storage capacity (HSC) can be up to 8.4 wt %. However, a recent report claimed that DFT calculations overestimated significantly the binding energy between the H_2 molecules and the Ca^{+1} cation centers (Cha et al., 2009). On the other hand, Al-doped graphene where one Al atom replaces one C atom of a graphene layer was reported as a promising hydrogen storage material at room temperature with HSC of 5.13 wt % (Ao et al., 2009a).

In this work, DFT was applied for studying the hydrogen adsorption on graphene with Al atom adsorption. The favourite adsorption configuration of Al atoms on single side and on both sides of a graphene layer have been determined. The obtained materials were studied for adsorption of H_2 molecules and we discuss its hydrogen storage properties.

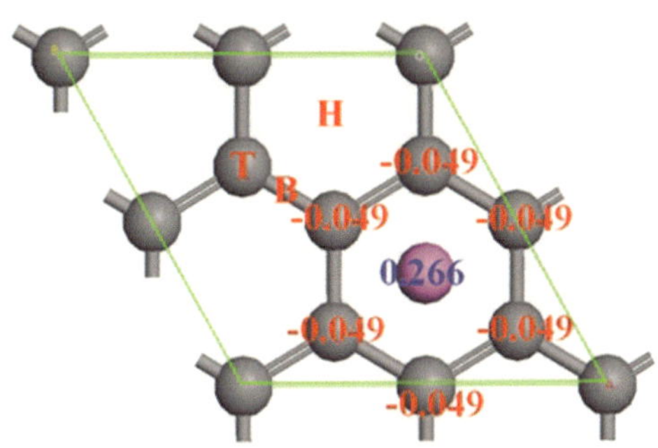

Fig. 20. Three different sites for an Al atom adsorbed on graphene. H, B and T denote the hollow of hexagon, bridge of C-C bond and top site of C atom, respectively. In addition, the charges of atoms near the adsorbed Al atom are also given, where the unit of charge is one electron charge e which is not given in the figure for clarity. The gray and pink balls in this figure and figures below are C and Al atoms, respectively. (Reproduced with permission from Ref. (Ao & Peeters, 2010b). Copyright 2010, APS)

LDA was used for all the calculations in this section. All DFT calculations were performed using the $Dmol^3$ code (Delley, 1990). Double Numerical Plus polarization (DNP) was taken as the basis set. In this case, three-dimensional periodic boundary conditions were applied and the H-H bond length was set to $l_{H-H} = 0.74$ Å identical to the experimental value (Lide, 2000). The computational unit cell consists of a 2×2 graphene supercell with a vacuum width of 18 Å to minimize the interlayer interaction. As shown in Fig. 20, the supercell contains 8 C atoms. All atoms were allowed to relax in all calculations.

The binding energy of Al atoms onto graphene E_{b-Al} is defined as,

$$E_{b-Al} = [E_{nAl-graphene} - (E_{graphene} + nE_{Al})]/n \quad (18)$$

where $E_{nAl-graphene}$, $E_{graphene}$ and E_{Al} are the energy of the system with n Al atoms adsorbed on the graphene layer, the energy of the pristine graphene layer and the energy of one Al atom in the same slab, respectively. The binding energy of H_2 molecules onto Al-adsorbed graphene layer E_{b-H2} is defined as,

$$E_{b-H2} = [E_{iH2+Al-graphene} - (E_{Al-graphene} + iE_{H2})]/i \quad (19)$$

where the subscripts iH2+Al-graphene, Al-graphene, and H2 denote the system with i H_2 molecules adsorbed, isolated Al-adsorbed graphene and a H_2 molecule, respectively.

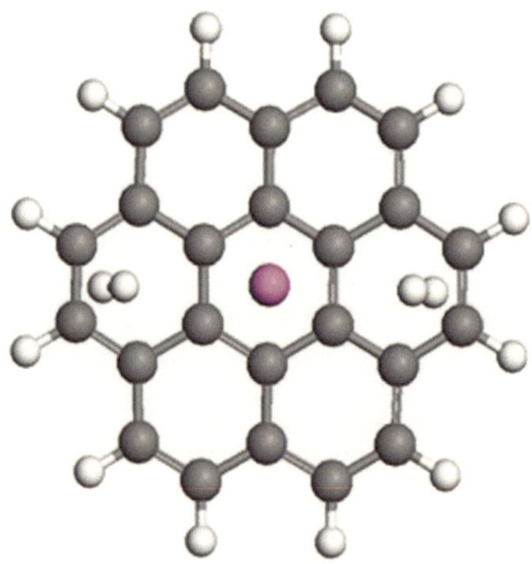

Fig. 21. A cluster model for 2 H_2 molecules adsorbed on graphene with an Al atom adsorbed on its one side. The white balls are hydrogen atoms in this figure and figures below. (Reproduced with permission from Ref. (Ao & Peeters, 2010b). Copyright 2010, APS)

To investigate the potential effects of different methodologies on our results, a calculation using the cluster model was carried out with both LDA and wave function approaches with the Møller-Plesset second order perturbation (MP2) within the Gaussian modules where the $6-331++G^*$ basis set was taken and maximum step size was set to 0.15 Å. Note that the cluster configuration shown in Fig. 21 was used because of the requirement of Gaussian modules, and the system was recalculated by LDA for purposes of comparison. In this calculation, a cluster with 24 carbon atoms and with 1 Al atom and 2 H_2 molecules adsorbed over the carbon surface was simulated where the dangling bonds of the C atoms at the boundary are terminated with H atoms.
On the basis of the published results, one may assume that the uptake capacity of hydrogen would increase if more metal atoms were adsorbed on the surface of a graphene nanostructure (Ataca et al., 2009; Liu et al., 2009). Furthermore, the binding between metal atoms and a surface would be strengthened if more charge is transferred between the metal atoms and the graphene nanostructure. Obviously, the binding can also be enhanced by adding more metal atoms with concomitant additional charges available for electronic transfer. However, metal atoms intend to aggregate into clusters when their concentration is large due to their high cohesive energies compared with those of metal atoms adsorbed on graphene, which

may significantly reduce the hydrogen uptake (Krasnov et al., 2007). For the Al, the cohesive energy is −3.39 eV (Gaudoin et al., 2002). To examine the validity of this assumption, a unit cell with eight C atoms and one Al atom was used in the present study, which is shown in Fig. 20. The ratio Al:C = 1:8 is quite moderate and moreover strictly obeys the doping rules for high coverage metals (Froudakis, 2001; Gao et al., 1998), which makes it possible for us to achieve a relatively high storage capacity. This rule ensures that the Al-Al distance is sufficiently large avoiding clustering of Al on graphene.

The favourite adsorption position of this Al atom on graphene is then determined. There are three different adsorption sites as shown in Fig. 20, which are the hollow of the carbon hexagon (H), the bridge of C-C bond (B), and the top site of the C atom (T), respectively. The Al-Al interaction is indeed negligible owing to the large distance of about 4.92 Å. It is found that the Al adsorbed at the H site has the lowest energy and is therefore the favourite adsorption configuration with a binding energy of -0.824 eV and the distance between Al and the graphene layer d_1 is about 2.079 Å. In Fig. 20, the charges of atoms near the adsorbed Al atom are given, which were obtained by Mulliken analysis. The adsorbed Al atom has a positive 0.266 *e* charge, while each C atom nearby has a negative charge −0.049 *e*. Note that the other two C atoms in the simulation cell contribute the rest of the electron charge to the negative C atoms. Therefore, the long distance of Al-Al, the relative strong bonding between the Al atom and the graphene layer, and the Coulomb repulsion between the Al atoms prevent metal aggregation on graphene.

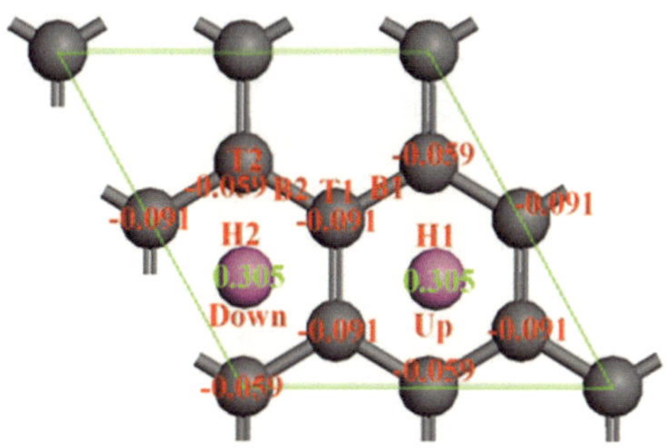

Fig. 22. Six different adsorption sites for the second Al atom on the other side of the graphene layer. The charges of atoms near the adsorbed Al atoms are also given, where the unit of charge is one electron charge *e*. (Reproduced with permission from Ref. (Ao & Peeters, 2010b). Copyright 2010, APS)

Due to the positive charge on the Al atoms and the negative charge on the carbon atoms, an electric field is induced between the Al atoms and the graphene layer, which in turn leads to a back transfer of charge from the graphene layer to the Al atom. Hence, by increasing Al coverage, adsorbed Al atoms would become less positively charged, which would decrease the Coulomb repulsion between the Al atoms, and eventually this may lead to metal aggregation. This also agrees with the doping rules (Froudakis, 2001; Gao et al., 1998).

To further confirm the stability of Al atoms on graphene, the diffusion behavior of an Al atom on graphene was studied by the transition search (TS) method in order to obtain the diffusion barrier. It has been shown above that the most stable configuration of an Al atom on graphene corresponds to adsorption on the H site of graphene. Consequently, the diffusion scenario of an Al atom on graphene between two H sites is considered in order to study surface diffusion. Based on the TS calculation, it is found that the classical barrier for surface diffusion is 0.104 eV. Notice that the calculated diffusion barrier corresponds only to a classical hopping model of diffusion. In practical cases, quantum tunneling effects should also be considered (Wu et al., 2009). In addition, because only a single Al atom is involved in the simulation cell, the Al-Al distance is kept unchanged. While in actual diffusion, the Al-Al distance would be shortened and repulsive Coulomb interaction among positively charged Al atoms would increase, leading to a significant increase of the diffusion barrier, which will prevent aggregation of adsorbed Al atoms on graphene.

Next the adsorption of Al atoms on both sides of the graphene layer is considered in order to increase the available surface area for hydrogen storage, since the charged metal atoms are the nucleation centres for hydrogen adsorption (Ataca et al., 2009; Liu et al., 2009; Yoon et al., 2008). As shown in Fig. 22, there are six different sites for the second Al atom to be positioned on the other side of the graphene layer. After geometry optimization of the six configurations, we found that the lowest energy configuration is realized for the second Al atom adsorbed on the H2 site with energy $E_{b-Al} = -1.096$ eV and the average E_{b-Al} for the two Al atoms is -0.960 eV. As shown in Fig. 22, the two Al atoms are positioned on two shoulder-by-shoulder carbon hexagons but on opposite sides of the graphene layer. The repulsive Coulomb interaction between the positively charged Al atoms on the upper and lower parts of the graphene plane is screened by the negative charge on the C-atoms. The graphene layer is now more negatively charged as compared to the previous single Al atom case, while the adsorbed Al atoms are more positively charged (the charges of the atoms on the Al and C atoms are given in Fig. 22). It leads to a stronger binding energy for the Al atoms on the graphene. In addition, $d_1 \approx 2.138$ Å which is slightly larger as compared to the case of single side adsorption which is counter intuitive. The reason is that the small increase of d_1 is a result of the Coulomb repulsion between the two positively charged Al atoms located above and below the graphene layer, which is screened by the charged graphene layer.

For the case of one H_2 molecule adsorbed on graphene with Al atoms adsorbed on a single side of graphene, the configuration after relaxation is shown in Fig. 23(a) where a 4×4 supercell is taken in order to better display the atomic structure, especially the adsorption site of the H_2 molecule. It indicates that the H_2 molecule would take the center site of equilateral triangles formed by adsorbed Al atoms. The vertical distance between the H_2 molecules and the graphene layer is $d_2 = 2.830$Å, while d_1 decreases slightly to 2.060 Å, and the adsorption energy for the first H_2 molecule is $E_{b-H2} = -0.182$ eV/H_2. In the figure, it shows that a parallelogram formed by the adsorbed Al atoms has two centre sites of equilateral triangles. However, due to the limitation of interaction among adsorbed H_2 molecules, H_2 would take just one of the two centre sites. When more H_2 molecules are adsorbed, the two centre sites would be both occupied as shown in Fig. 23(b) where two H_2 molecules are adsorbed. The adsorption energy for the second H_2 molecule is $E_{b-H2} = -0.273$ eV/H_2, which gives an average adsorption energy for the two H_2 molecules of -0.227 eV/H_2. Fig. 23(c) gives the atomic structure of three adsorbed H_2 molecules. Two H_2 molecules take the two centre sites as in Fig. 23(b), the other H_2 molecule would take the top site of the Al atom. The distance of the three H_2 to the Al atom are respectively 2.786, 2.879 and 2.903 Å with average binding energy of -0.176 eV/H_2. If we further increase the number of H_2 molecules, after relaxation, the result shows that the fourth H_2 molecule cannot be adsorbed. Therefore, it is concluded that the maximum number of H_2 molecules adsorbed on a single side of a 2×2 graphene unit cell is three.

For the cases of one and two adsorbed H_2 molecules, it was found that the H_2 molecules are parallel to the graphene layer and all H_2 molecules are equidistant from the Al atoms. Once

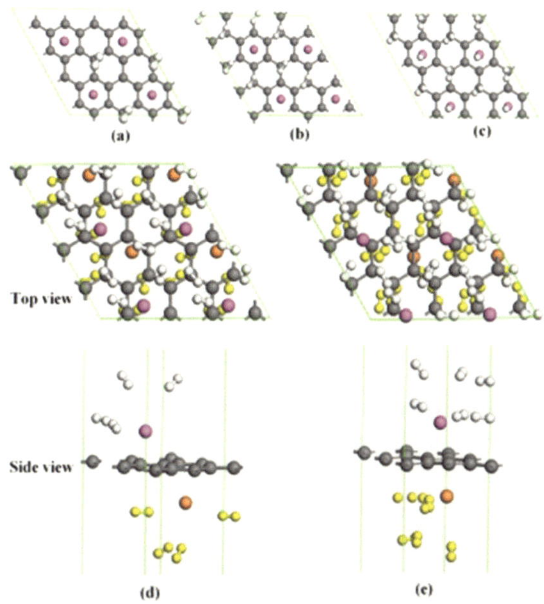

Fig. 23. Atomic structures of H_2 molecules adsorbed on Al-adsorbed graphene. (a) One H_2 molecule adsorbed on graphene with Al adsorbed on the single side, (b) two H_2 molecules adsorbed on graphene with Al adsorbed on a single side of graphene, (c) three H_2 molecules adsorbed on graphene with Al adsorbed on one side of graphene, (d) four H_2 molecules adsorbed on each side of graphene with Al adsorbed on its both sides, (e) six H_2 molecules adsorbed on each side of graphene with Al adsorbed on its both sides. In this figure, 4×4 supercells are plotted to better display the adsorption sites of the H_2 molecules. In (d) and (e), due to the Al atoms and H_2 molecules adsorbed on both sides of graphene, Al atoms and H_2 molecules below the graphene layer are shown as orange and yellow, respectively. Meanwhile, in order to show the two-layer adsorption arrangement of H_2 molecules, initial simulation cells of side view are also given in the nether part of (d) and (e). (Reproduced with permission from Ref. (Ao & Peeters, 2010b). Copyright 2010, APS)

the number of H_2 absorbed on each Al atom exceeded two, the absorbed H_2 molecules tend to tilt towards the Al atoms because of the increased positive charge of the Al atoms and the symmetry of the bonding configuration of the H_2 molecules. This phenomenon is similar to the case of adsorption of H_2 molecules on Ca-adsorbed graphene (Ataca et al., 2009).
In addition, it was noted that the E_{b-H2} of the second H_2 molecule is much larger than that of the first one, i.e. it is about 50% larger. In order to understand this enhancement, PDOS of Al, C atoms and H_2 molecules are plotted and shown in Fig. 24. It was reported that the band broadening of the molecular level of H_2 below the Fermi energy indicates a significant H_2-H_2 interaction that in turn increases its binding energy to the substrate (Ataca et al., 2009). In this work, the same mechanism is found where the band broadening of about -6 eV appears in Fig. 24. In fact, the binding energy of the first H_2 molecule to the Al atom which prefers to be parallel to the graphene layer is generally small (Ataca et al., 2009).
Figure 25 displays the electron density of the system with one and two adsorbed H_2 molecules. Notice that there is non-zero electron density in the region between the graphene layer and the adsorbed Al atom. This is the reason why Al atoms are strongly adsorbed on the graphene

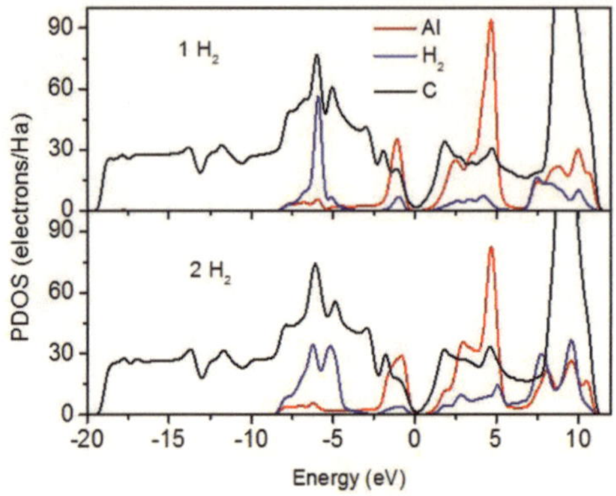

Fig. 24. PDOS of Al, H_2 and C in the systems of one and two H_2 molecules adsorbed on graphene with Al adsorbed on the single side. The Fermi level is at 0. (Reproduced with permission from Ref. (Ao & Peeters, 2010b). Copyright 2010, APS)

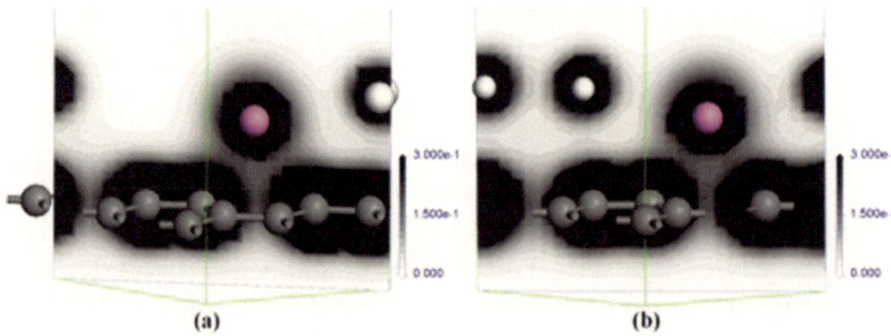

Fig. 25. Electron density distribution in the systems of one and two H_2 molecules adsorbed on graphene with Al adsorbed on a single side of graphene. (Reproduced with permission from Ref. (Ao & Peeters, 2010b). Copyright 2010, APS)

layer. In addition, some electronic distribution also appears among the H_2 molecules, the Al atom and the graphene layer. For H_2 molecules adsorbed on pristine graphene, no electron density was found between the H_2 molecules and the graphene layer (Ao et al., 2009a). Therefore, H_2 adsorption is enhanced in the Al-adsorbed graphene system due to the adsorbed Al atoms that act as bridges to link the electron clouds of the H_2 molecules and the graphene layer. Furthermore, Fig. 25(b) also shows that there is some electron distribution between the two adsorbed H_2 molecules. This means that the interaction between the H_2 molecules will change the electron distribution and may induce an enhancement of the adsorption energy as found in Fig. 24.

Very recently, the mechanism of H_2 adsorption onto Ca cation centers was investigated using both DFT and wave function approaches (Cha et al., 2009). It was found that DFT calculations overestimated the binding energy between the H_2 molecules and the Ca^{1+} cation centers significantly. Similarly, a calculation was carried out on 2 H_2 molecules adsorbed on an Al coated graphene using the cluster model with both LDA and wave function approaches with MP2 within the Gaussian modules. The average binding energies for H_2 in this cluster system were found to be -0.196 and -0.185 eV/H_2 with LDA and MP2, respectively. Thus the two values differ by less than 6% giving some credibility to our numerical obtained binding energy.

In other words, the significant overestimation by DFT as found earlier for the binding of H_2 molecules onto Ca^{1+} system does not occur for our system.
In addition, in order to investigate the effect of the simulation cell size on the results calculations were also performed using a 4×4 supercell with H_2 molecules adsorbed as shown in Fig. 23(a). We found almost the same results as obtained with the 2×2 supercell. In the 4×4 system, the H_2 molecules are adsorbed on the center sites of the equilateral triangles of Al atoms, as shown in Fig. 23(a). The distance between the H_2 molecules and the graphene surface are, respectively, 2.884 and 2.825 Å in 2×2 and 4×4 systems, while E_{b-H2} in 4×4 system is -0.190 eV/H_2 and $E_{b-H2} = -0.182$ eV/H_2 in 2×2 system.
For the case of hydrogen adsorption on Al that is adsorbed on both sides of graphene, the situations of one, two and three H_2 molecules adsorbed on each side of graphene are rather similar to the above case of adsorption on a single side of graphene. In other words, two H_2 molecules will take the center sites of equilateral triangles formed by the adsorbed Al atoms as shown in Figs. 23(a) and 23(b), and the third H_2 will take the top site of the Al atom as in Fig. 23(c). Previously, it is shown that a maximum of three H_2 molecules per 2×2 unit cell can be adsorbed on one side of graphene. However, for the case of adsorption on both sides of the graphene layer, each side can absorb more than 3 H_2 molecules. In Fig. 23(d) with 4 H_2 molecules adsorbed on each side, we show a 4×4 supercell. Two of them take the centre sites of equilateral triangles, and the other two are located on the bridge sites of two Al atoms. However, the four H_2 molecules are in two different planes with distances to the graphene layer being 2.672 and 4.675 Å. The distances of the four H_2 molecules to the Al atom are respectively 2.444, 2.531, 2.918, and 2.947 Å. The average E_{b-H2} is -0.209 eV/H_2. If further increasing the number of H_2 molecules, the two H_2 molecules in the center sites of the equilateral triangles will hop to the bridge sites of the two Al atoms while keeping the two-layer structure. Therefore, each Al atom can absorb a maximum of six H_2 molecules, due to the two-layer adsorption structure and each Al atom has six nearest Al atoms with each adsorbed H_2 molecule shared by two Al atoms.
Fig. 23(e) gives the corresponding atomic structure with H_2 molecules fully adsorbed. It shows that all the H_2 molecules are located at the bridge sites of Al-Al and are arranged into two layers on each side of graphene. Note that the adsorption of H_2 on both sides of graphene will automatically change the sites of adsorbed Al atoms from the centre site of the carbon hexagon to nearly the bridge site of the C-C bond as shown in Figs. 23(d) and 23(e). The different location of the Al atoms in the presence of adsorbed H_2 for single side and both sides of graphene is a consequence of: (1) the different charges of Al atoms adsorbed on one side of graphene and on both sides of graphene, and (2) the different number of adsorbed H_2 molecules. Therefore, HSC is up to 13.79 wt% with an average $E_{b-H2} = -0.193$ eV/H_2. Note that the obtained HSC is in excess of 6 wt%, surpassing DOE's target, and the obtained E_{b-H2} is almost within the required range of -0.2 to -0.4 eV/H_2 (Li et al., 2003).
For the practical purpose, E_{b-H2} is required to be a weak function of the adsorption coverage X of H_2 molecules on graphene, so that the adsorbed H_2 molecules can be desorbed to almost zero X. In this work, E_{b-H2} is about -0.2 eV/H_2 and it is found that the amount of coverage has only a weak effect on E_{b-H2}. The coverage dependence of $E_{b-H2}(X)$ is shown in Fig. 26 with $E_{b-H2}(X)$ varying within 15%. Note that E_{b-H2} is the lowest when 4 H_2 molecules were adsorbed. This is because adsorption is strongest when H_2 molecules are located on the centre sites of equilateral triangles formed by the adsorbed Al atoms. This was confirmed above in Fig. 23(a) where one H_2 molecule was first adsorbed at the centre sites of the equilateral triangles. Due to the interaction between the H_2 molecules as shown in the Figs. 24, 25 and

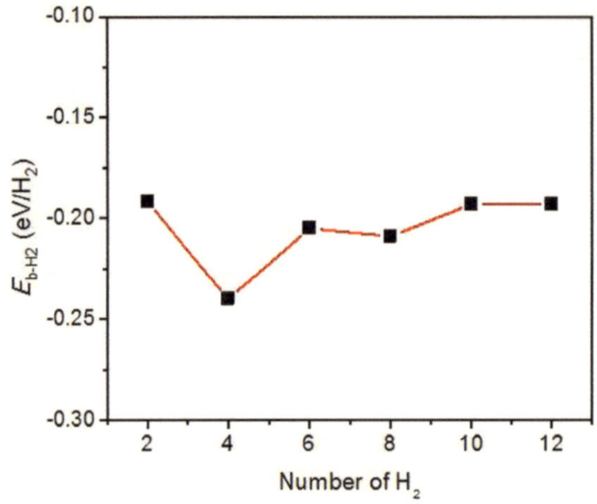

Fig. 26. X dependent average adsorption energy $E_{b-H2}(X)$ of H_2 on graphene with Al adsorbed on both sides of graphene. (Reproduced with permission from Ref. (Ao & Peeters, 2010b). Copyright 2010, APS)

discussed above, adsorption with 2 H_2 molecules on each side on the centre sites of equilateral triangles is strongest.

When 12 H_2 molecules are adsorbed on both sides of a 2×2 supercell of graphene, the H_2 molecules on each side of graphene will be arranged into two layers as shown in Fig. 23(e), the distances of each layer to the graphene surface are respectively about 2.5 and 5.0 Å, while d_1 is about 2.2 Å. As discussed above, the adsorption energy E_{b-Al} for adsorption on both sides of graphene is larger than that for single side adsorption. At the same time, the Al atoms are more positively charged, and the C atoms are more negatively charged when the Al atoms are adsorbed on both sides of graphene. As found previously hydrogen adsorption is mainly induced by charged metal atoms, and the strength of the adsorption depends on the amount of the transferred charge (Liu et al., 2009; Sun et al., 2006). Thus, the graphene layer when Al is adsorbed on both sides of graphene has a larger capacity for H_2 storage. However, due to the limited space between the Al atoms and the repulsive interaction between the adsorbed H_2 molecules, some adsorbed H_2 molecules move upwards, over the Al atoms. This is also the reason why H_2 molecules can form a two-layer arrangement in the case of Al adsorbed on both sides of graphene, only a single H_2 layer is found for the corresponding single side system.

To test the stability of the hydrogen storage system, *ab initio* MD simulations was performed on a $12H_2$-Al-graphene system which is shown in Fig. 23(e). The MD simulation in the NVT ensemble, i.e., constant number of atoms N, volume V, and temperature T, was performed over a time of 1 ps with a massive GGM thermostat at 300 K and without external pressure. We found that only the outer 2 H_2 molecules are escaping from each side of the graphene layer, because they are more weakly bound than the other H_2 molecules. For example, the first H_2 molecule that is released has a binding energy of -0.129 eV. In this case, the HSC becomes 9.64 wt%, which is still much higher than DOE's target. While *ab initio* MD simulation is quite computationally time consumption, 1 ps is not enough to get statistically meaningful values for the desorption temperature. However, it does suggest that the system keeps a rather high hydrogen storage capacity at room temperature. This is even the case in the absence of external pressure, and it is thus possible to release H_2 molecules without removing the Al atoms. Note that the system stability of hydrogen storage in Ti-decorated carbon nanotubes was tested in similar conditions, where the MD calculations lasted 1.5 ps (Yildirim & Ciraci, 2005). In addition, the release of H_2 molecules can be further prevented by decreasing the temperature or increasing the pressure of storage to increase its HSC.

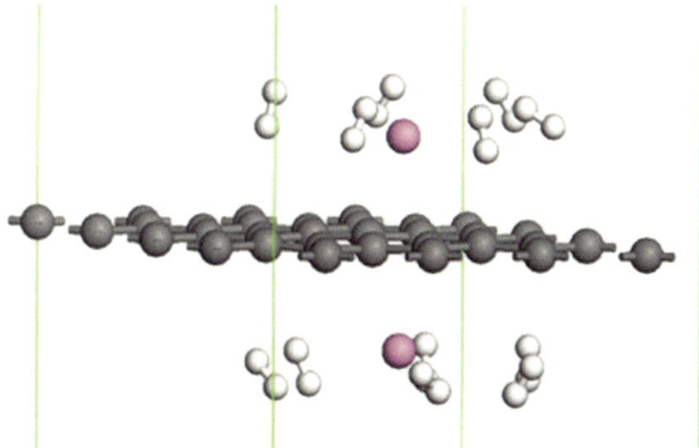

Fig. 27. The configuration of H_2 molecules adsorbed in a 4 supercell system. (Reproduced with permission from Ref. (Ao & Peeters, 2010b). Copyright 2010, APS)

To investigate the effect of the concentration of adsorbed Al atoms on its hydrogen storage capacity, a 4×4 graphene supercell was considered with one Al atom on the centre site of the carbon hexagon above and below the graphene layer. It is found that each Al atom can maximally adsorb 6 H_2 molecules with average $E_{b-H2} = -0.172$ eV/H_2 resulting in a HSC of 5.19 wt%. The adsorption configuration is shown in Fig. 27. Note that the HSC is much lower than 13.79 wt% found for the 2×2 supercell system above. In the case of H_2 adsorbed in the 2×2 system, the H_2 molecules are adsorbed on the bridge sites of Al-Al and are arranged into a two layer configuration. Thus, each adsorbed H_2 molecule interacts with the nearest two Al atoms. In the 4×4 system, which corresponds to a lower density of adsorbed Al, the distance between two Al atoms is very long, up to 9.84 Å. Thus, each H_2 interacts with one Al atom and the graphene layer, and there is more space available for the adsorbed H_2 molecules which are located in a single layer. The corresponding adsorption energies E_{b-H2} also decrease slightly as the Al-Al distance increases. For single H_2 molecule and two H_2 molecules adsorbed on a 4×4 supercell, E_{b-H2} are -0.169 and -0.178 eV/H_2. In case of a 2×2 supercell we found that E_{b-H2} are -0.182 and -0.227 eV/H_2, respectively.

For the practical applications, it is desirable to know the exact charge status of the hydrogen storage material. From it the information whether the hydrogen storage material is fully charged or the adsorbed H_2 molecules are completely released can be obtained. The charge exchanged with the graphene layer can be determined by the conductivity of the graphene layer, which is strongly determined by the DOS at the Fermi level (He et al., 2008; Schedin et al., 2007). The X-dependence of the latter quantity is given in Fig. 28. The result shows that the DOS at the Fermi level decreases as X increases and this dependence becomes weaker at high X.

4. Conclusion

A principle of CO adsorption enhancement was developed theoretically by using density functional theory through doping Al into graphene. The results show that the Al doped graphene has strong chemisorption of CO molecule by forming Al-CO bond, where CO onto intrinsic graphene remains weak physisorption. Furthermore, the enhancement of

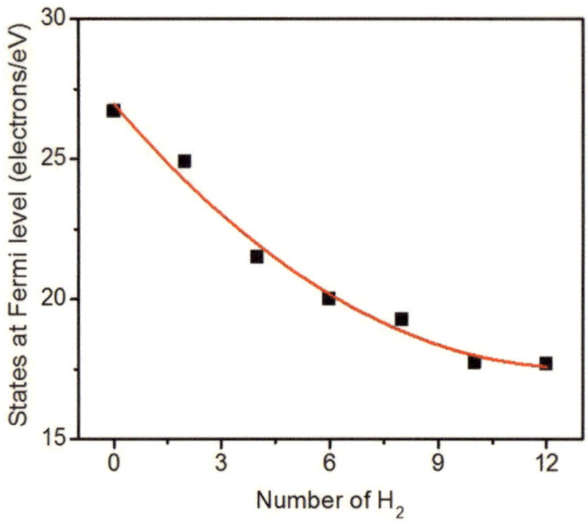

Fig. 28. X dependent number of band states at the Fermi level. (Reproduced with permission from Ref. (Ao & Peeters, 2010b). Copyright 2010, APS)

CO sensitivity in the Al doped graphene is determined by a large electrical conductivity change after adsorption, where CO absorption leads to increase of electrical conductivity via introducing large amount of shallow acceptor states. Therefore, this newly developed Al doped graphene would be an excellent candidate for sensing CO gas. After that, the correlation of the applied electric field F and adsorption/desorption behaviors of CO molecule in the Al doped graphene was studied. The results indicate that the positive F reduces the adsorption energies E_b of the CO adsorbed onto the doped graphene, while E_b increases under the negative F. Furthermore, desorption commences when a large positive F ($F \geq 0.03$ au) is applied. Finally, the thermal stability of interaction between the CO molecules and the Al doped graphene is studied with *ab initio* molecular dynamics calculation to reveal the adsorption/desorption behaviours of the system. Based on the results of the calculations, the adsorption/desorption phase diagram was established by the atomic thermodynamics and the temperature dependent desorption time $\tau(T)$ was determined with thermal desorption method. The results show that the optimal desorption temperature is 400 K. Meanwhile, the effect of T on atomic structure parameters and electrical properties were analyzed, and the results show that the greatest conductivity change before and after adsorption is at $T = 400$ K. Therefore, this sensor material has the best sensing performance with appropriate τ and the biggest conductivity change at 400 K.

Furthermore, the promising hydrogen storage mediums, Al-modified graphenes, are proposed through density functional theory calculations. Hydrogen molecule is predicted to be strongly physically adsorbed on Al substitutionally doped graphene with adsorption energy -0.427 eV/H_2 that is in a reasonable range, so that hydrogen storage can be recycled at near ambient conditions, while a graphene layer with Al adsorbed on both sides can store hydrogen up to 13.79 wt% with a two-layer arrangement of H_2 molecules formed on each side. Its hydrogen storage capacity is in much excess of 6 wt% of DOE's target. In the Al substitutionally doped graphene system, it is believed that the doped Al alters the electronic structures of both C and H_2, and the bands of H_2 overlapping with those of Al and C simultaneously are the underlying mechanism of the hydrogen adsorption enhancement from -0.159 eV/H_2 in pristine graphene. In the Al adsorbed on both sides of graphene system, this high-capacity hydrogen storage is due to the adsorbed Al atoms that act as bridges to link the electron clouds of the H_2 molecules and the graphene layer. In addition, we find that the H_2 concentration in the hydrogen storage medium can be measured by the change in the conductivity of the graphene layer.

There has been an explosion of ideas that suggest graphene for potential applications. This is often led by analogies with carbon nanotubes that continue to serve a guide in searching for new applications. Except for its excellent electronic properties, graphene also displays several unusual attributions. Graphene is a giant aromatic macromolecule that conducts both electricity and heat well in two dimensions. Their mechanical strength of graphene is comparable to that of CNTs, while CNTs can be considered as a rolled up graphene. The shape, size, and chemical structure of graphene sheets can be further modified by engineering. Nonetheless, research toward the application of graphene-based materials has just begun. Many challenges and opportunities remain. For examples, applications for batteries and supercapacitors, in separation technologies, as supports for catalysts, and filler for composition materials, and so on, are widely expected in the recent future.

5. Acknowledgement

This work was financially supported by the Vice-Chancellor's postdoctoral research fellowship program of the University of New South Wales (SIR50/PS19184) and the Australian Research Council Discovery Programs (DP1096769 and DP0988687).

6. References

Acharya, C. & Turner, C. (2007). Effect of an electric field on the adsorption of metal clusters on boron-doped carbon surfaces, *J. Phys. Chem. C* 111: 14804–14812.

Ao, Z., Jiang, Q., Zhang, R., Tan, T. & Li, S. (2009a). Al doped graphene: A promising material for hydrotgen storage at room temperature, *J. Appl. Phys.* 105: 074307–074312.

Ao, Z., Li, S. & Jiang, Q. (2009b). Thermal stability of interaction between the co molecules and the al doped graphene, *Phys. Chem. Chem. Phys.* 11: 1683–1687.

Ao, Z., Li, S. & Jiang, Q. (2010a). Correlation of the applied electrical field and CO adsorption/desorption behavior on al-doped graphene, *Solid State Commun.* 150: 680–683.

Ao, Z. & Peeters, F. (2010b). High-capacity hydrogen storage in Al-adsorbed graphene, *Phys. Rev. B* 81: 205406–205412.

Ao, Z., Yang, J., Li, S. & Jiang, Q. (2008). Enhancement of CO detection in Al doped graphene, *Chem. Phys. Lett.* 461: 276–279.

Arellano, J., Molina, L., Rubio, A. & Alonso, J. (2000). Density functional study of molecular hydrogen on graphene layers, *J. Chem. Phys.* 112: 8114–8119.

Ataca, C., Aktürk, E. & Ciraci, S. (2009). Hydrogen storage of calcium atoms adsorbed on graphene: First-principle plane wave calculations, *Phys. Rev. B* 79: 041406–041409.

Beenakker, C. (2008). Colloquium: Andreev reflection and Klein tunneling in graphene, *Rev. Mod. Phys.* 80: 1337–1354.

Blyholder, G. (1964). Molecular orbital view of chemisorbed carbon monoxide, *J. Phys. Chem.* 68: 2772–2777.

Bunch, J., Yaish, Y., Brink, M., Bolotin, K. & McEuen, P. (2005). Coulomb oscillations and hall effect in quasi-2D graphite quantum dots, *Nano Lett.* 5: 287–290.

Cabria, I., LóPez, M. & Alonso, J. (2008). Hydrogen storage in pure and Li-doped carbon nanopores: Combined effects of concavity and doping, *J. Chem. Phys.* 128: 144704–144711.

Ceperley, D. & Alder, B. (1980). Ground state of the electron gas by a stochastic method, *Phys. Rev. Lett.* 45: 566–569.

Cha, J., Lim, S., C.H., C., Cha, M.-H. & Park, N. (2009). Inaccuracy of density functional theory calculations for dihydrogen binding energetic onto Ca cation centers, *Phys. Rev. Lett.* 103: 216102–216105.

Chandrakumar, K. & Ghosh, S. (2008). Alkali-metal-induced enhancement of hydrogen adsorption in C_{60} fullerence: An ab initio study, *Nano Lett.* 8: 13–19.

Collins, P., Bradley, K., Ishigami, M. & Zettl, A. (2000). Extreme oxygen sensitivity of electronic properties of carbon nanotubes, *Science* 287: 1801–1804.

Coontz, R. & Hanson, B. (2004). Not so simple, *Science* 305: 957.

Degall, M., Lindan, P., Probert, M., Pickard, C., Hasnip, P., Clark, S. & Payne, M. (2002). First-principles simulation: ideals, illustrations and the CASTEP code, *J. Phys.: Condens. Matter* 14: 2717–2744.

Delley, B. (1990). An all-electron numerical method for solving the local density functional for polyatomic molecules, *J. Chem. Phys.* 92: 508–517.

Delley, B. (2000). From molecules to solids with the $DMol^3$ approach, *J. Chem. Phys.* 113: 7756–7764.

Delley, B. (2002). Hardness conserving semilocal pseudopotentials, *Phys. Rev. B* 66: 155125.

Deng, W., Xu, X. & Goddard, W. (2004). New alkali doped pillared carbon materials designed to achieve practical reversible hydrogen storage for transportation, *Phys. Rev. Lett.* 92: 166103–166106.

Durgun, E., Ciraci, S., Zhou, W. & Yildirim, T. (2006). Transition-metal-ethylene complexes as high-capacity hydrogen-storage media, *Phys. Rev. Lett.* 97: 226102–226105.

Froudakis, G. (2001). Why alkali-metal-doped carbon nanotubes possess high hydrogen uptake, *Nano Lett.* 1: 531–533.

Gao, G., Gagin, T. & Goddard, I. W. (1998). Inaccuracy of density functional theory calculations for dihydrogen binding energetic onto Ca cation centers, *Phys. Rev. Lett.* 80: 5556–5559.

Gaudoin, R., Foulkes, W. & Rajagopal, G. (2002). Ab initio calculations of the cohesive energy and the bulk modulus of aluminium, *J. Phys.: Condens. Matter* 14: 8787–8793.

Geim, A. & Novoselov, K. (2007). The rise of graphene, *Nat. Mater.* 6: 183–191.

Ghosh, A., Subrahmanyam, K., Krishna, K., Datta, S., Govindaraj, A., Pati, S. & Rao, C. (2008). Uptake of H_2 and CO_2 by graphene, *J. Phys.Chem. C* 112: 15704–15707.

Girifacol, L. & Hodak, M. (2002). Van der waales binding energies in graphitic structures, *Phys. Rev. B* 65: 125404–125408.

Graetz, J., Chaudhuri, S., Lee, Y., Vogt, T., Muckerman, J. & Reilly, J. (2006). Pressure-induced structural and electronic changes in α-AlH_3, *Phys. Rev. B* 74: 214114–214120.

Hammer, B., Hansen, L. & Nørskov, J. (1999). Improved adsorption energetics within density-functional theory using revised perdew-burke-ernzerhof functionals, *Phys. Rev. B* 59: 7413–7421.

He, C., Zhang, P., Zhu, Y. & Jiang, Q. (2008). Structures and quantum conduction of copper nanowires under electric field using first principles, *J. Phys. Chem. C* 112: 9045–9049.

Hyman, M. & Medlin, J. (2005). Theoretical study of the adsorption and dissociation of oxygen on Pt(111) in the presence of homogeneous electric fields, *J. Phys. Chem. B* 109: 6304–6310.

Jeloaica, L. & Sidis, V. (1999). DFT investigation of the adsorption of atomic hydrogen on a cluster-model graphite surface, *Chem. Phys. Lett.* 300: 157–162.

Klontzas, E., Mavrandonakis, A., Tylianakis, E. & Froudakis, G. (2008). Improving hudrogen storage capacity of MOF by functionalization of the organic linker with lithium atoms, *Nano Lett.* 8: 1572–1576.

Kong, J., Chapline, M. & Dai, H. (2001). Functionalized carbon nanotubes for molecular hydrogen sensors, *Adv. Mater.* 13: 1384–1386.

Kong, J., Franklin, N., Zhou, C., Chapline, M., Peng, S., Cho, K. & Dai, H. (2000). Nanotube molecular wires as chemical sensors, *Science* 287: 622–625.

Koper, M. & van Santen, R. (1999). Electric filed effects on CO and NO adsorption at the Pt(111) surface, *J. Electroanal. Chem.* 476: 64–70.

Krasnov, P., Ding, F., Singh, A. & Yakobson, B. (2007). Clustering of Sc on SWNT and reduction of hydrogen uptake: Ab-initio all electron calculations, *J. Phys. Chem. C* 111: 17977–17980.

Lee, C., Wei, X., Kysar, J. & Hone, J. (2008). Measurement of the elastic properties and intrinsic strength of monolayer graphene, *Science* 321: 385–388.

Leenaerts, O., Partoens, B. & Peeters, F. (2008). Adsorption of H_2O, NH_3, CO, NO_2, and NO on graphene: A first-principles study, *Phys. Rev. B* 77: 125416–125422.

Li, D., Müller, M., Gilje, S. & Kaner, R. (2008a). Processable aqueous dispersions of graphene nanosheets, *Nat. Nanotechnol.* 3: 101–105.

Li, J., Furuta, T., Goto, H., Ohashi, T., Fujiwara, Y. & Yip, S. (2003). Theoretical evaluation of hydrogen storage capacity in pure carbon nanostructures, *J. Chem. Phys.* 119: 2376–2385.

Li, X., Grubisic, A., Stokes, S., Cordes, J., Ganteför, G., Bowen, K., Kiran, B., Willis, M., Burgert, P. & Schnöckel, H. (2007). Unexpected stability of Al_4H_6: A borane analog?, *Science* 315: 356–358.

Li, X., Wang, X., Zhang, L., Lee, S. & Dai, H. (2008b). Chemically derived, ultrasmooth graphene nanoribbon semiconductors, *Science* 319: 1229.

Lide, D. (2000). *CRC Handbook of Chemistry and Physics*, CRC Press, Boca Raton, FL.

Liu, W., Zhao, Y., Li, Y., Jiang, Q. & Lavernia, E. (2009). Enhanced hydrogen storage on Li-dispersed carbon nanotubes, *J. Phys. Chem. C* 113: 2028–2033.

Loffreda, D. (2006). Theoretical insight of adsorption thermodynamics of multifunctional molecules on metal surfaces, *Surf. Sci.* 600: 2103–2112.

Lozovoi, A. & Alavi, A. (2007). Vibrational frequencies of CO on Pt(111) in electric field: A periodic DFT study, *J. Electroanal. Chem.* 607: 140–146.

Lugo-Solis, A. & Vasiliev, I. (2007). Ab initio study of K adsorption on graphene and carbon nanotubes: Role of long-range ionic forces., *Phys. Rev. B* 76: 235431–235438.

Mares, A. & Van Ruitenbeek, J. (2005). Observation of shell effects in nanowires for the noble metals Cu, Ag, and Au, *Phys. Rev. B* 72: 205402–205408.

McEwen, J.-S., Gaspard, P., Mittendorfer, F., Visart de Bocarmé, T. & Kruse, N. (2008). Field-assisted oxidation of rhodium, *Chem. Phys. Lett.* 452: 133–138.

Moseley, P. (1997). Solid state gas sensors, *Meas. Sci. Technol.* 8: 223–227.

Mpourmpakis, G., Tylianakis, E. & Froudakis, G. (2007). Carbon nanoscrolls: A promising material for hydrogen storage, *Nano Lett.* 7: 1893–1897.

Nikitin, A., Li, X., Zhang, Z., Ogasawara, H., Dai, H. & Nilsson, A. (2008). Hydrogen storage in carbon nanotubes through the formation of stable C-H bonds, *Nano Lett.* 8: 162–167.

Nose, S. (1984). A molecular dynamics method for simulations in the canonical ensemble, *Mol. Phys.* 52: 255–268.

Nose, S. (1991). Constant temperature molecular dynamics methods, *Prog. Theor. Phys. Supplement* 103: 1–46.

Novoselov, K., A.K., G., Morozov, S., Jiang, D., Zhang, Y., Dubonos, S., Grigorieva, I. & Firsov, A. (2004). Electric field effect in atomically thin carbon films, *Science* 306: 666–669.

Okamoto, Y. & Miyamoto, Y. (2001). Ab initio investigation of physisorption of molecular hydrogen on planar and curved graphenes, *J. Phys. Chem. B* 105: 3470–3474.

Panella, B., Hirscher, M. & Roth, S. (2005). Hydrogen adsorption in different carbon nanostrcutures, *Carbon* 43: 2209–2214.

Parrinello, M. & Rahaman, A. (1981). Polymorphic transitions in single crystals: A new molecular dynamics method., *J. Appl. Phys.* 52: 7182–7190.

Peng, S. & Cho, K. (2003). Ab initio study of doped carbon nanotubes sensors, *Nano Lett.* 3: 513–517.

Peng, S., Cho, K., Qi, P. & Dai, H. (2004). Ab initio study of CNT NO_2 gas sensor, *Chem. Phys. Lett.* 387: 271–276.

Perdew, J. & Zunger, A. (1981). Self-interaction correction to density-functional approximations for many-electrons systems, *Phys. Rev. B* 23: 5048–5079.

Peres, N., Guinea, F. & Castro Neto, A. (2006). Electronic properties of disordered two-dimensional carbon, *Phys. Rev. B* 73: 125411.

Peres, N., Klironomos, F., Tsai, S.-W., Santos, J., Lopes dos Santos, J. & Castro Neto, A. (2007). Electron waves in chemically substituted graphene, *Euro. Phys. Lett.* 87: 67007.

Peres, N., Tsai, S.-W., Santos, J. & Ribeiro, R. (2009). Scanning tunneling microscopy currents on locally disordered graphene, *Phys. Rev. B.* 7879: 155442.

Raaen, S. & Ramstad, A. (2005). Monte-Carlo simulations of thermal desorption of adsorbed molecules from metal surface, *Energy* 30: 821–830.

Rao, C., Sood, A., Subrahmanyam, K. & Govindaraj, A. (2009). Graphene: The new two-dimensional nanomaterial, *Angew. Chem. Int. Ed.* 48: 7752–7778.

Readhead, P. (1962). Thermal desorption of gases, *Vacuum* 12: 203–211.

Reina, A., Jia, X., Ho, J., Nezich, D., Son, H., Bulovic, V., Dresselhaus, M. & Kong, J. (2009). Large area, few-layer graphene films on arbitrary substrates by chemical vapor deposition, *Nano Lett.* 9: 30–35.

Sahaym, U. & Norton, M. (2008). Advances in the application of nanotechnology in enabling a Śhydrogen economyŠ, *J. Mater. Sci.* 43: 5395–5429.

Schedin, F., G. A., Morozov, S., Hill, E., Blake, P., Katasnelson, M. & Novoselov, K. (2007). Detection of individual gas molecules adsorbed on graphene, *Nat. Mater.* 6: 652–655.

Schlapbach, L. & Züttel, A. (2001). Hydrogen-storage materials for mobile applications, *Nature* 414: 353–358.

Seitsonen, A., Kim, Y., Knapp, M., Wendt, S. & Over, H. (2001). CO adsorption on the reduced $RuO_2(110)$ surface: Energetics and structure, *Phys. Rev. B* 65: 035413–035421.

Spencer, M. & Yarovsky, I. (2007). Ab initio molecular dynamics study of H_2S dissociation on the Fe(110) surface, *J. Phys. Chem. C* 111: 16372–16378.

Sun, Q., Jena, P., Wang, Q. & Marquez, M. (2006). First-principles study of hydrogen storage on $Li_{12}C_{60}$, *J. Am. Chem. Soc.* 128: 9741–9745.

Suzuki, M. (1991). General theory of fractal path integrals with applications to many-body theories and statistical physics, *J. Math. Physics* 32: 400–407.

Tada, K., Furuya, S. & Watanabe, K. (2001). Ab initio study of hydrogen adsorption to single-walled carbon nanotubes, *Phys. Rev. B* 63: 155405–155408.

Tielens, F., Gracia, L., Polo, V. & Andrés, J. (2007). A theoretical study on the electronic structure of Au-$XO^{(0,-1,+1)}$ (X=C, N, and O) complexes: effect of an external electric field, *J. Phys. Chem. A* 111: 13255–13263.

Todorova, N., Spencer, M. & Yarovsky, I. (2007). Ab initio study of s dynamics on iron surface, *Surf. Sci.* 601: 665–671.

Tomanek, D., Kreuzer, H. & Block, J. (1985). Tight-binding approach to field desorption: N_2 on Fe(111), *Surf. Sci.* 157: L315–322.

Tosatti, E., Prestipino, S., Kostlmeier, S., Dal Corso, A. & Di Tolla, F. (2001). String tension and stability of magic tip-suspended nanowires, *Science* 291: 288–290.

Verlet, L. (1967). Computer experiments on classical fluids. I. thermodynamical properties of Lennard-Jones molecules, *Phys. Rev.* 159: 98–103.

Wang, R., Zhang, D., Sun, W., Han, Z. & Liu, C. (2007). A novel aluminum-doped carbon nanotubes sensor for carbon monoxide, *J. Molecular Struc.: Theochem.* 806: 93–97.

Wehling, T.O., N. K., Morozov, S., Vdovin, E., Katsnelson, M., Geim, A. & Lichtenstein, A. (2008). Molecular doping of graphene, *Nano Lett.* 8: 173–175.

Wei, B., Hsu, M., Su, P., Lin, H., Wu, R. & Lai, H. (2004). A novel SnO_2 gas sensor doped with carbon nanotubes operating at room temperature., *Sensor Actuat.* 101: 81–89.

Wei, D., Liu, Y., Wang, Y., Zhang, H., Huang, L. & Yu, G. (2009). Synthesis of N-doped graphene by chemical vapor deposition and its electrical properties, *Nano Lett.* 9: 1752–1758.

Wu, G., Zhang, J., Wu, Y., Li, Q., Chou, K. & Bao, X. (2009). Adsorption and dissociation of hydrogen on MgO surface: A first-principles study, *J. Alloy Compd.* 480: 788–793.

Xu, W., Takahashi, K., Matsuo, Y., Hattori, Y., Kumagai, M., Ishiyama, S., Kaneko, K. & Iijima, S. (2007). Inverstigation of hydrogen storage capacity of various carbon materials, *Int. J. Hydrogen Energy* 32: 2504–2512.

Yang, W., Rantinac, K., Ringer, S., Thordarson, P., Gooding, J. & Braet, F. (2010). Carbon nanomaterials in biosensors: Should you use nanotubes or graphene?, *Angew. Chem. Int. Ed.* 49: 2114–2138.

Ye, Y., Ahn, C., Witham, C., Fultz, B., Liu, J., Rinzler, A., Colbert, D., Smith, K. & Smalley, R. (1999). Hydrogen adsorption and cohensive energy of single-walled carbon nanotubes, *Appl. Phys. Lett.* 74: 2307–2309.

Yildirim, T. & Ciraci, S. (2005). Titanium-decorated carbon nanotubes as a potential high Űcapacity hydrogen storage medium, *Phys. Rev. Lett.* 94: 177501–177504.

Yoon, M., Yang, S., Hicke, C., Wang, E. & Zhang, Z. (2008). Calcium as the superior coating metal in functionalization of carbon fullerences for high-capacity hydrogen storage., *Phys. Rev. Lett.* 100: 206806–206809.

Yoshida, H. (1990). Construction of higher order symplectic integrators, *Phys. Lett. A* 150: 262–268.

Yuan, G., Zhang, W., Yang, Y., Tang, Y., Li, Y., Wang, J., Meng, X., He, Z., Wu, C., Bello, I., Lee, C. & Lee, S. (2009). Graphene sheets via microwave chemical vapor deposition, *Chem. Phys. Lett.* 467: 361–364.

Zhang, Y., Tan, Y. & Stormer, H. (2005). Experimental observation of the quantum hall effect and berryŠs phase in graphene, *Nature* 438: 201–204.

Zhang, Y., Zhang, D. & Liu, C. (2006). Novel chemical sensor for cyanides: Boron-doped carbon nanotubes, *J. Phys. Chem. B* 110: 4671–4674.

Zhao, Q., Nardelli, M., Lu, W. & Bernholc, J. (2005). Carbon nanotube-metal cluster composites: A new road to chemical sensors?, *Nano Lett.* 5: 847–851.

Zhou, S., Gweon, G.-H., Fedorov, A., First, P., de Heer, W., Lee, D.-H., Guinea, F., Castro, A. & Lanzara, A. (2007). Substrate-induced bandgap opening in epitaxial graphene, *Nat. Mater.* 6: 770–775.

8

Graphene-Based Devices Based on Topological Zero Modes and Pseudospin Degree of Freedom

J. C. Martinez[1], M. B. A. Jalil[1,3] and S. G. Tan[2,3]
[1]*Information Storage Materials Laboratory,*
Electrical and Computer Engineering Department, National University of Singapore,
[2]*Data Storage Institute, A*STAR (Agency for Science, Technology and Research),*
[3]*Computational Nanoelectronics and Nano-device Laboratory,*
Electrical and Computer Engineering Department, National University of Singapore,
Singapore

1. Introduction

Lately we have witnessed a resurgence of interest in some exotic but hitherto unobserved predictions of relativistic quantum mechanics, particularly in the phenomena of zitterbewegung (Schliemann et al., 2006) and the Klein paradox (Katsnelson et al., 2006). In the former case this interest has been fueled by the realization that bound electrons in suitable semiconductor nanostructures are expected to display zitterbewegung at much lower frequencies and with much larger amplitudes than free electrons. In the latter case, the recently established conducting properties of graphene, an atomic layer of graphite, point to charge carriers moving at speeds close to the Fermi speed ($10^6 \mathrm{m\,s}^{-1}$) which plays the role of light speed for this system (Kane, 2005). In pursuit of these tantalizing developments one might also consider the possibility of harnessing the novel features of graphene for device applications. In this article, we will introduce two concepts which are unique to graphene systems which we believe could be potentially useful for nanodevices.

a. Topological Zero Modes
b. Pseudospin Orbital Coupling

1.1 Topological zero modes

We conjecture that with present day technologies, one might be able to observe the interactions between particle and antiparticle in the low-energy context of bilayer graphene. To introduce the phenomenon we wish to describe let us think for a moment of the game of bowling. One can enjoy it in two ways: a normal person can hurl the ball and be thrilled by strikes and spares, while a child can gently slide the ball on the sidetrack and watch it roll back from the return rack. But another possibility occurs if the ball neither goes for the pins nor comes back but stays put somewhere. Paradoxically this scenario can be more intriguing as one tries to analyze the resting position of the ball. An analogous situation occurs in quantum mechanical tunneling, where electrons are either transmitted or reflected if their kinetic energies are greater or less than the strength of the step-potential barrier they are incident upon [Landau & Lifshitz, 1968]. However in relativistic quantum mechanics, a third scenario can theoretically arise, namely, spatially-bound particles due to the interaction between particle and antiparticle. Such interplay is unknown in semi-classical or non-relativistic quantum mechanics, and is rarely observed even in relativistic quantum mechanics since the requisite electric fields far exceed available technologies. In this article we will show in Section 3 that this effect can be reproduced in bilayer graphene in the presence of an antisymmetric potential kink.

1.2 Pseusospin orbital coupling

It is well known that electron spin is coupled to its momentum due to the spin orbit coupling effect which can be understood via classical electrodynamics or relativistic quantum mechanics. Dirac's equation predicts that in the presence of electromagnetic fields, a single spin particle experiences the Zeeman and the spin orbit coupling effects. In the latter, one can visualize that an electron traveling with a non-vanishing speed in the electric field, will in its rest frame "see" an effective magnetic field. The magnetic field strength depends on the angle between the momentum and the electric field in the plane which contains both and the field direction is perpendicular to this plane. It is only natural to envisage that the electron spin will precess about this effective magnetic field and that the spin precession would be tightly coupled to electron scattering, due to the dependence of the field strength on the electron motion. What follows is the realization that this phenomenon has useful device properties; indeed, in the last twenty years, numerous transistor designs based upon the Rashba and Dresselhaus spin orbit coupling in semiconductors (Supriyo Datta & B. Das, 1988) have emerged. Bilayer graphene has a Hamiltonian that resembles the massive Dirac system. This led to the idea that an analogy of the above might lead to the design of devices similar to spintronics but in the context of graphene. However, in graphene our focus lies in the pseudospin rather than the real electron spin. A scientific imperative here is to theoretically ascertain the possible existence of such pseudospin orbit coupling and details of this would be presented in Section 4 of this article.

2. Brief introduction to bilayer dynamics

Much has been said about the novelty of graphene which promises new electronics with applications wild and aplenty. The reduced Hamiltonian of bilayer graphene has been the center stage for trapping the "bowling ball" and generating a slew of topological dynamics, as described in Section 1. We will begin with a brief introduction to the full bilayer graphene Hamiltonian leading to its reduced form. As is well known, the bilayer graphene comprises two monolayer graphene stacked vertically and has a more complicated energy structure. The bilayer graphene Hamiltonian (McCann & Fal'ko, 2006) has been expressed as:

$$H_1\psi_1 = \begin{pmatrix} u/2 & v_3\pi & 0 & c\pi^+ \\ v_3\pi^+ & -u/2 & c\pi & 0 \\ 0 & c\pi^+ & -u/2 & t \\ c\pi & 0 & t & u/2 \end{pmatrix} \begin{pmatrix} \varphi_{A1} \\ \varphi_{B2} \\ \varphi_{A2} \\ \varphi_{B1} \end{pmatrix} \tag{1}$$

where $\pi = p_x + ip_y$ and $t = \varepsilon\gamma$. In graphene it is known that interlayer coupling between A_1 and B_2 is strong. Subscripts A, B refer to the sublattice index, while 1, 2 refer to the layer index. Alternatively there are different versions of the Hamiltonian found in the literature. The differences arise mainly from symmetry as well as variations due to approximations.

Bilayer Graphene Hamiltonian

1.

$$H_1\psi_1 = \begin{pmatrix} u/2 & v_3\pi & 0 & c\pi^+ \\ v_3\pi^+ & -u/2 & c\pi & 0 \\ 0 & c\pi^+ & -u/2 & t \\ c\pi & 0 & t & u/2 \end{pmatrix}\begin{pmatrix} \varphi_{A1} \\ \varphi_{B2} \\ \varphi_{A2} \\ \varphi_{B1} \end{pmatrix}$$

(McCann, E et al, 2006)

2.

$$H_2\psi_2 = \begin{pmatrix} -u/2 & c\pi^+ & 0 & 0 \\ c\pi & -u/2 & t & 0 \\ 0 & t & u/2 & c\pi^+ \\ 0 & 0 & c\pi & u/2 \end{pmatrix}\begin{pmatrix} \varphi_{A1} \\ \varphi_{B1} \\ \varphi_{A2} \\ \varphi_{B2} \end{pmatrix}$$

(Martin, Ivar et al 2008); (Manes, J.L. et al, 2007)

3.

$$H_3\psi_3 = \begin{pmatrix} -u/2 & c\pi^+ & 0 & v_3\pi \\ c\pi & -u/2 & t & 0 \\ 0 & t & u/2 & c\pi^+ \\ v_3\pi^+ & 0 & c\pi & u/2 \end{pmatrix}\begin{pmatrix} \varphi_{A1} \\ \varphi_{B1} \\ \varphi_{A2} \\ \varphi_{B2} \end{pmatrix}$$

Table 1. Hamiltonian of the bilayer graphene resembles the massive Dirac Hamiltonian

The spinor wavefunction of (1) in Table I above is different from those in (2) and (3). A transformation can be performed as follows

$$\psi = \begin{pmatrix} \psi_{A1} \\ \psi_{B2} \\ \psi_{A2} \\ \psi_{B1} \end{pmatrix} \to \psi' = U\psi = \begin{pmatrix} \varphi_{A1} \\ \varphi_{B1} \\ \varphi_{A2} \\ \varphi_{B2} \end{pmatrix}, \quad \text{where } U = \begin{pmatrix} 1 & 0 & 0 & 0 \\ 0 & 0 & 0 & 1 \\ 0 & 0 & 1 & 0 \\ 0 & 1 & 0 & 0 \end{pmatrix} = U^+ \tag{2}$$

It is not hard to see that U defined above has the property of: Hermicity, i.e. $U = U^+$ and Unitarity, i.e. $UU^+ = 1$. The Hamiltonian can then be transformed using $H^t = UHU^+$. In summary we have

$$H_1\psi_1 = \begin{pmatrix} u/2 & v_3\pi & 0 & c\pi^+ \\ v_3\pi^+ & -u/2 & c\pi & 0 \\ 0 & c\pi^+ & -u/2 & t \\ c\pi & 0 & t & u/2 \end{pmatrix}\begin{pmatrix} \phi_{A1} \\ \phi_{B2} \\ \phi_{A2} \\ \phi_{B1} \end{pmatrix} \to H_1^t\psi_1^t = \begin{pmatrix} u/2 & c\pi^+ & 0 & v_3\pi \\ c\pi & u/2 & t & 0 \\ 0 & t & -u/2 & c\pi^+ \\ v_3\pi & 0 & c\pi & -u/2 \end{pmatrix}\begin{pmatrix} \phi_{A1} \\ \phi_{B1} \\ \phi_{A2} \\ \phi_{B2} \end{pmatrix} \tag{3}$$

The Hamiltonian H_1^t is identical to H_3 and is very close to that of H_2 (see Table 1) except that $v_3\pi$ and $v_3\pi^+$ have been dropped in H_2. It will be interesting to inspect if the above Hamiltonian fit the Dirac's equation exactly, which in its standard form is given by

$$\begin{pmatrix} E-mc & -\boldsymbol{\sigma}\cdot\boldsymbol{p} \\ \boldsymbol{\sigma}\cdot\boldsymbol{p} & -E+mc \end{pmatrix}\psi = 0 \tag{4}$$

To achieve greater transparency, we manipulate the Dirac's equation slightly by simply multiplying both sides of the Hamiltonian $\alpha_0 \frac{i\hbar\partial_0}{c} = \gamma\cdot p + mc$ with γ_0, which then leads to $\frac{i\hbar\partial_0}{c} = \gamma_0\vec{\gamma}\cdot\vec{p} + \gamma_0 mc = \hbar\gamma_5\vec{\sigma}\cdot\vec{\partial} + \gamma_0 mc$. Recalling that $\alpha_0 = \gamma_0 = \begin{pmatrix} I & 0 \\ 0 & -I \end{pmatrix}, \gamma_0\gamma_0 = 1;\ i\gamma_5 = \begin{pmatrix} 0 & 1 \\ 1 & 0 \end{pmatrix}$, one could obtain straightforwardly that $i\gamma_5\boldsymbol{\sigma}\cdot\boldsymbol{p} + \gamma_0 mc = \begin{pmatrix} mc & \boldsymbol{\sigma}\cdot\boldsymbol{p} \\ \boldsymbol{\sigma}\cdot\boldsymbol{p} & -mc \end{pmatrix}$. The energy equation can thus be written as follows

$$\begin{pmatrix} E-mc & -\boldsymbol{\sigma}\cdot\boldsymbol{p} \\ -\boldsymbol{\sigma}\cdot\boldsymbol{p} & E+mc \end{pmatrix}\psi = 0 \tag{5}$$

which seems to be a closer fit with H_1. The above shows that bilayer graphene Hamiltonian mimics that of the massive Dirac systems to some extent. Future nanofabrication technologies, which afford us greater control over the various interlayer and sublattice coupling strengths, might allow us to produce a graphene-like system with closer correspondence to the massive Dirac system.

Reduction of the Bilayer Graphene Hamiltonian

In bilayer graphene, the 4x4 Hamiltonian is cumbersome, and one often uses a 2x2 simplified form. We will in this section simplify the larger bilayer graphene Hamiltonian. We will use H_3 as an example

$$H_3\psi_3 = \begin{pmatrix} -u/2 & c\pi^+ & 0 & v_3\pi \\ c\pi & -u/2 & t & 0 \\ 0 & t & u/2 & c\pi^+ \\ v_3\pi^+ & 0 & c\pi & u/2 \end{pmatrix}\begin{pmatrix} \varphi_{A1} \\ \varphi_{B1} \\ \varphi_{A2} \\ \varphi_{B2} \end{pmatrix} = \begin{pmatrix} -\frac{u}{2}\varphi_{A1} + c\pi^+\varphi_{B1} + v_3\pi\varphi_{B2} \\ c\pi\varphi_{A1} - \frac{u}{2}\varphi_{B1} + t\varphi_{A2} \\ t\varphi_{B1} + \frac{u}{2}\phi_{A2} + c\pi^+\varphi_{B2} \\ v_3\pi^+\varphi_{A1} + c\pi\varphi_{A2} + \frac{u}{2}\varphi_{B2} \end{pmatrix} \tag{6}$$

where u is the applied electrical voltage between the two layers; t is the interlayer coupling. One assumes here that in the limit of strong interlayer coupling, i.e. $t >> u$, occupancy at sites A2 and B1 become zero. It thus follows from Eq. (6) that

$$\begin{pmatrix} -\frac{u}{2}\varphi_{A1} + c\pi^+\varphi_{B1} + v_3\pi\varphi_{B2} \\ c\pi\varphi_{A1} - \frac{u}{2}\varphi_{B1} + t\varphi_{A2} \\ t\varphi_{B1} + \frac{u}{2}\varphi_{A2} + c\pi^+\varphi_{B2} \\ v_3\pi^+\varphi_{A1} + c\pi\varphi_{A2} + \frac{u}{2}\varphi_{B2} \end{pmatrix} = E\begin{pmatrix} \varphi_{A1} \\ \varphi_{B1} \\ \varphi_{A2} \\ \varphi_{B2} \end{pmatrix} = E\begin{pmatrix} \varphi_{A1} \\ 0 \\ 0 \\ \varphi_{B2} \end{pmatrix} \tag{7}$$

resulting in the following set of equations: $c\pi\varphi_{A1} - \frac{u}{2}\varphi_{B1} + t\varphi_{A2} = 0$ and $t\varphi_{B1} + \frac{u}{2}\varphi_{A2} + c\pi^{+}\varphi_{B2} = 0$. Solving these in the limit of $t >> u$, one obtains:

$$\phi_{A2} = -\frac{2c\pi t\varphi_{A1} + c\pi^{+}u\varphi_{B2}}{2t^2}; \quad \phi_{B1} = \frac{c\pi u\varphi_{A1} - 2c\pi^{+}t\varphi_{B2}}{2t^2} \tag{8}$$

and with proper substitution, one obtains:

$$\begin{pmatrix} -\frac{u}{2}(1-\frac{c^2p^2}{t^2})\varphi_{A1} - \frac{c^2\pi^{+2}}{\mathrm{t}}\varphi_{B2} + v_3\pi\varphi_{B2} \\ 0 \\ 0 \\ -\frac{c^2\pi^2}{t}\varphi_{A1} + v_3\pi^{+}\varphi_{A1} + \frac{u}{2}(1-\frac{c^2p^2}{t^2})\varphi_{B2} \end{pmatrix} = E\begin{pmatrix} \varphi_{A1} \\ \varphi_{B1} \\ \varphi_{A2} \\ \varphi_{B2} \end{pmatrix} \tag{9}$$

which leads to the simplified Hamiltonian for $H_3\psi_3$ of

$$H_3^S\psi_3^S = \begin{pmatrix} -\frac{u}{2}(1-\frac{c^2p^2}{t^2}) & 0 & 0 & (-\frac{c^2\pi^{+2}}{t} + v_3\pi) \\ 0 & 0 & 0 & 0 \\ 0 & 0 & 0 & 0 \\ (-\frac{c^2\pi^2}{t} + v_3\pi^{+}) & 0 & 0 & \frac{u}{2}(1-\frac{c^2p^2}{t^2}) \end{pmatrix}\begin{pmatrix} \varphi_{A1} \\ 0 \\ 0 \\ \varphi_{B2} \end{pmatrix} \tag{10}$$

from which one could write in 2x2 form

$$H\begin{pmatrix} \varphi_{A1} \\ \varphi_{B2} \end{pmatrix} = \begin{pmatrix} -\frac{u}{2}(1-\frac{c^2p^2}{t^2}) & (-\frac{c^2\pi^{+2}}{t} + v_3\pi) \\ (-\frac{c^2\pi^2}{t} + v_3\pi^{+}) & \frac{u}{2}(1-\frac{c^2p^2}{t^2}) \end{pmatrix}\begin{pmatrix} \varphi_{A1} \\ \varphi_{B2} \end{pmatrix} \tag{11}$$

The energy equation of the reduced bilayer graphene can now be written as

$$E = -\frac{c^2}{t}\begin{pmatrix} 0 & \pi^{+2} \\ \pi^2 & 0 \end{pmatrix} - \frac{u}{2}(1-\frac{c^2p^2}{t^2})\begin{pmatrix} 1 & 0 \\ 0 & -1 \end{pmatrix} + v_3\begin{pmatrix} 0 & \pi \\ \pi^{+} & 0 \end{pmatrix} \tag{12}$$

The approximation which has been adopted above is related to present-day fabrication technologies. Future advances in technologies might allow us to engineer the various coupling parameters in graphene. It is conceivable that we may be able to pre-design a useful Hamiltonian and employ the nanofabrication technologies to realize that desired form of Hamiltonian. The Hamiltonian above has been used to understand the presence of topological objects like monopoles in graphene and their associated Berry's phase which may have great implications to electron, or perhaps we may just call them (Dirac) particle dynamics or conductance (Novoselov, et al.,2006; Tan, et al, 2010).

3. Topological zero mode device

The phenomenon we briefly described in Section 1 requires a particle-antiparticle pair to be held apart by an external electrostatic field and yet be strongly correlated. For Dirac electrons the typical energy is $O(mc^2)$ and the corresponding correlation length is the Compton wavelength $\hbar/mc$ (Itzykson & Zuber, 1980), implying a restraining electric field of the order of $10^{17}\ \mathrm{V\,m^{-1}}$, a field far beyond present-day capabilities. But we may use bilayer graphene whose top speed is 300 times smaller than c and the excitation mass one-twentieth of the electron's (Castro Neto et al. 2009). (As we will explain below, here chiral pairs replace particle-antiparticle pairs.) Then the electric field needed is 10 orders of magnitude smaller, or $10^{7}\ \mathrm{V\,m^{-1}}$, which is accessible with present technology. Monolayer graphene under these circumstances would generate a different mechanism, more akin to the Zener breakdown, and quite unrelated to our purpose (Martinez et al. 2010).

We therefore study a gated bilayer graphene configuration with an impressed voltage kink V to provide a restraining potential for a particle and its chiral partner. Such kinks can be produced for instance in a graphene p-n junction (Abanin & Levitov, 2007). Recently it has been shown that such configurations can support zero modes and chiral states in the vicinity of the domain wall separating the insulating regions (Martin et al., 2008). If the bias $V(x)$ in the form of a kink is applied between the layers then the motion in the y-direction is that of a free particle and the dynamics in the x-direction will be the one of interest. We will show that a charged particle with energy less than the bias will not undergo total reflection as expected quantum mechanically but will remain in the vicinity of the kink and can manifest itself as charge bound to the kink. The system turns out to be able to support zero modes, occurring *always* at the same value of the particle's y-momentum, and the Hall conductivity plateaus in its vicinity correspond to those of the graphene monolayer. Moreover we find other bound states as well as scattering states as the energy of the particle is raised above the kink strength. All these imply that the kink introduces new and unexpected features into the bilayer dynamics which are externally adjustable. Due to the relativistic and topological nature of our results, we expect them to be of general interest in other non-graphene areas of investigation, e.g., particle physics (Horava, 2005) and superconductivity (Lu & Yip, 2008).

As we saw in Section 2 the low-energy Hamiltonian for the graphene bilayer is a 4×4 matrix in the space spanned by the four-component wave function. It can be reduced further into a reasonably accurate effective 2×2 matrix if we are interested only in the lowest-energy bands, i.e., when the interlayer hopping between nearest neighbors is much larger than the electron energy (measured from zero momentum). If we model the kink potential as an anti-symmetric tanh profile imposed by the electrostatic bias and introduce appropriate scales, the effective 2×2 wave equation is (Martin et al., 2008; cf. Eq. (12))

$$\begin{pmatrix} -(e + r\ \tanh x) & \left(\frac{d}{dx} + p_y\right)^2 \\ \left(\frac{d}{dx} - p_y\right)^2 & -(e - r\ \tanh x) \end{pmatrix} \begin{pmatrix} u(x) \\ v(x) \end{pmatrix} = 0\,, \tag{13}$$

where r ($0 \le r \le 1$) denotes the bias and p_y the particle momentum in the y-direction, e a scaled energy (Fig. 1). In this section only e denotes energy. This effective Hamiltonian involves the atomic sites (A_1, B_2). Thus the 'spinor' structure of Eq. (13) has nothing to do with spin; rather it refers to the electron occupancy at these atomic sites. The qualification 'chiral' is thus applied to this case. We will refer to the equations obtained from Eq. (13) as the first and second component equations, respectively. Formally Eq. (13) can be reduced into a single equation because it is clear that one component converts to the other through the replacement $x \to -x$. Thus, a possible set of solutions to Eq. (13) can be obtained, in which their two components are related by $u(x) = \pm v(-x)$. Equation (13) has been treated as two separate problems, one for $\Psi = (u(x),\, u(-x))^T$ and another for $\Phi = (u(x),\, -u(-x))^T$. However, we will not adopt this approach here, since the resulting solutions would appear to imply a *nonlocal* relation between $u(x)$ and $u(-x)$ for the *entire* range of x: in other words, we would obtain differential equations involving $u(x)$ and $u(-x)$ (and similarly for $v(x)$) simultaneously and for every x. In this system, as defined by Eq. (13), there is no underlying symmetry to support such a relation. In the absence of bias ($r = 0$), the above system can easily be solved using local relations. In the presence of bias, which is local as well, there is no reason to introduce a nonlocality whereby the dynamics at $+x$ (for given x) is directly related to that at $-x$, and for all x. Thus, we follow an alternative method below, in which *local* relations (i.e. occurring at the *same* spatial point) between $u(x)$ and $v(x)$ will suffice to obtain an exact solution: hence *all* dynamical relations will connect phenomena at the *same* spatial point. This point will be reinforced at several places in this section. Hence, our result differs from those of Martin et al. (2006).

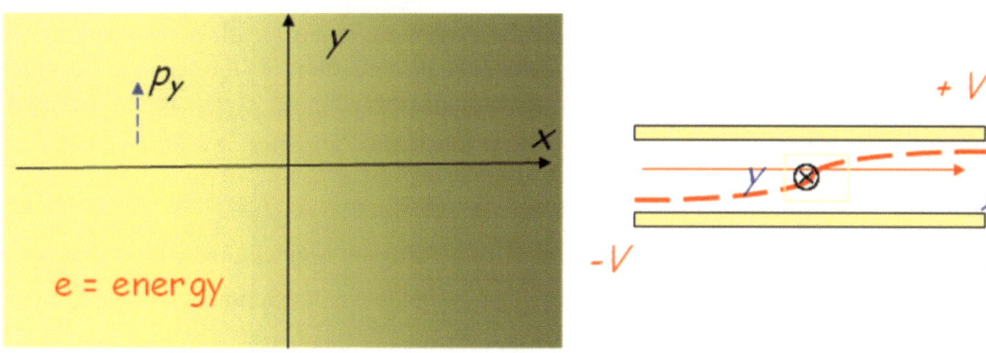

Fig. 1. Views of the bilayer system top (left) and side (right) with kink potential $V(x) = r \tanh x$.

To solve Eq. (13) exactly we introduce the ansatz for $x \geq 0$:

$$\begin{pmatrix} u \\ v \end{pmatrix} = e^{-\xi\, x} \begin{pmatrix} U(x) \\ V(x) \end{pmatrix}, \tag{14}$$

where ξ is a complex number and we replace x with the auxiliary variable $z = -e^{-2x}$. Our focus is principally, though not exclusively, on the intragap solutions, $|e| \leq r$, and we give explicit expressions for $r = 1$. We obtain for the first component of Eq. (13):

$$(1-z)\,z\,V'' + (1-z)(1-p_y+\xi)\,V' - \tfrac{1}{2}U = 0\,, \tag{15}$$

where $'$ denotes d/dz. In arriving at Eq. (15) we had imposed the eigenvalue relation $(p_y-\xi)^2 V - (e+1)U = 0$, whose consistency will be verified later. Plugging this relation for U into Eq. (15) we obtain a hypergeometric equation for $V(x)$ alone with the solution (Abramowitz & Stegun, 1964)

$$V(x) = {}_2F_1(a,\ b,\ c,\ z), \tag{16}$$

$a = \frac{1}{2}(p_y - \xi)(-1 + \sqrt{\frac{e-1}{e+1}}), b = \frac{1}{2}(p_y - \xi)(-1 - \sqrt{\frac{e-1}{e+1}}), c = 1 - p_y + \xi$, and ${}_2F_1(a, b, c, z)$ is the Gauss hypergeometric series and the parameters a, b, c depend only on the energy and p_y. (Useful properties of the hypergeometric equation are listed in the Appendix.) The second equation can be solved similarly:

$$V(x) = \frac{(p_y + \xi)^2}{e-1}\, {}_2F_1(A,\ B,\ C,\ z), \tag{17}$$

$A = \frac{1}{2}(p_y + \xi)(-1 + \sqrt{\frac{e+1}{e-1}}), B = \frac{1}{2}(p_y + \xi)(-1 - \sqrt{\frac{e+1}{e-1}}), C = 1 + p_y + \xi$, with corresponding eigenvalue relation $(p_y + \xi)^2 U - (e-1)V = 0$. (A technical point is in order: when solving the second equation one finds the roles of U and V reversed. This is a veiled signal that locality is crucial because any assumed nonlocal relation, except possibly for overall sign, invoked for the first equation would not generally be consistent for the second.) Together the two relations yield a consistency condition implying an independent equation relating ξ, e and p_y: writing $\xi = a + i\beta$, we solve this equation to obtain $a(\beta)$ = $\frac{1}{\sqrt{2}}[(p_y^4 + 1 - e^2)^{1/2} + (-)p_y^2]^{1/2}$. This ensures the consistency of the ansatz (14) and the eigenvalue relations: any dependence on x, U, or V has now been eliminated in favor of an algebraic one involving e and p_y alone. Implicit in the above development is the locality of all the intervening relations, that is, all dynamical relations between u and v occur at the *same* spatial point x. The complex conjugates of (16) and (17) are easily seen to be solutions of Eq. (15) also so we can form a linear combination of these to arrive at the complete ($x \geq 0$) solution. Next the above procedure can be repeated for $x < 0$. Replacing ξ with -ξ in the ansatz (14) and employing a new auxiliary variable, $y = -e^{2x}$, we discover that the solutions are exactly the same ones (15) and (17) above but with the order reversed and z replaced by y. The numbers a and β are given by the same relation above. (Thus there are four solutions of which, for $x \geq 0$ ($x \leq 0$), we choose a positive (negative). There is no such restriction on β). We summarize these results: corresponding to Eq. (16)

$$V(x) = {}_2F_1(a',\ b',\ c',\ y), \quad x < 0 \tag{16a}$$

$a' = \frac{1}{2}(p_y + \xi)(1 + \sqrt{\frac{e+1}{e-1}}), b' = \frac{1}{2}(p_y + \xi)(1 - \sqrt{\frac{e+1}{e-1}}), c' = 1 + p_y + \xi$, and to Eq. (17),

$$V(x) = \frac{(p_y - \xi)^2}{e+1} \; {}_2F_1(A', B', C', y), \quad x < 0\,, \tag{17a}$$

$A' = \frac{1}{2}(p_y - \xi)(-1 + \sqrt{\frac{e-1}{e+1}}),\ B' = \frac{1}{2}(p_y - \xi)(-1 - \sqrt{\frac{e-1}{e+1}}),\ C' = 1 - p_y + \xi$. Once again the eigenvalue relations enforce a local connection between $V(x)$ and $U(x)$. With these, we have evaluated the complete intragap solutions for $e < 1$.

A glance at Eq. (17) suggests that it may not be valid for $e = 1$. However, on writing the eigenvalue relation in the form $p_y + \xi = \sqrt{e^2 - 1}/(p_y - \xi)$, we note that as $e \to 1^-$, both sides approach zero. Employing this result to carry out a limiting procedure on Eq. (17), we obtain the $e = 1$ solution: $\frac{1}{2p_y^2} e^{p_y x} {}_2F_1(1/2p_y, -1/2p_y, 1, z)$. For this solution to remain bound for $x \geq 0$ we must have $p_y < 0$, which can be verified later once the dispersion graphs have been obtained (see Fig. 1).

There remains the problem of enforcing continuity between the $x \geq 0$ and $x \leq 0$ solutions. Restricting to the first equation, we form the linear combination $g_>(x) = e^{-ax} {}_2F_1(a, b, c, z)\,(c_1 e^{i\beta x} + c_2 e^{-i\beta x})$, for $x \geq 0$ and c_1 and c_2 complex constants; a similar expression, $g_<(x)$, holds for $x \leq 0$ with the signs of x reversed, a replaced by $-a$, and with coefficients c_3 and c_4 instead. As intragap solutions, these functions are normalizable, and the coefficients c_i are determined by demanding that $g_>(x)$, $g_<(x)$ and their first three derivatives be continuous at $x = 0$. These yield four homogeneous simultaneous equations and we use the requirement that the determinant of the system must vanish to derive the dispersion relation, $e = e(p_y)$. Clearly, the only independent dynamical parameter is the energy, e. The same procedure is repeated for the second equation (parameters a, b, c are exchanged with A, B, C and vice versa), in which case we use the label $h(x)$ instead of $g(x)$, and we obtain a pair of complete solutions for the two components of Eq. (13), respectively which we write in schematic form (for $x \geq 0$)

$$\begin{pmatrix} u(x) \\ v(x) \end{pmatrix} = g_>(x) \begin{pmatrix} (p-\xi)^2/(e+1) \\ 1 \end{pmatrix}, h_>(x) \begin{pmatrix} 1 \\ (p+\xi)^2/(e-1) \end{pmatrix}. \tag{18}$$

The procedure outlined above produced *two* pairs (18) but there is really only *one* pair of solutions. Examining these solutions, we notice that one set can be obtained from the other by the formal substitution $p_y \to -p_y$ and $e \to -e$ and multiplying the spinors by $-i\sigma_y$. (We had also made use of the symmetry of the hypergeometric solutions given in the Appendix.) Hence the solutions (18) are really chiral conjugates of the each other. This can be shown formally by transforming Eq. (13) into its chiral conjugate form, wherein the sign of the energy *and* p_y are reversed. First we write Eq. (13) as

$$\begin{pmatrix} -\varphi(x) & \pi^{+2} \\ \pi^{-2} & \varphi(x) \end{pmatrix} \begin{pmatrix} u(x) \\ v(x) \end{pmatrix} = e \begin{pmatrix} u(x) \\ v(x) \end{pmatrix} \tag{19}$$

where $\varphi(x)$ denotes the kink potential and $\pi^{\pm} = d/dx \pm p_y$. Now we multiply both sides by σ_y and cast the result in the form

$$\begin{pmatrix} -\varphi(x) & \pi^{-2} \\ \pi^{+2} & \varphi(x) \end{pmatrix} \sigma_y \begin{pmatrix} u(x) \\ v(x) \end{pmatrix} = (-e) \begin{pmatrix} u(x) \\ v(x) \end{pmatrix} \tag{20}$$

This tells that $(u,v)^T$ is conjugate to $\sigma_y\ (u,v)^T$, in the sense that one is obtained from the other by the substitution $p_y \to -p_y$ and $e \to -e$, consistent with the results above. By virtue of the electron-hole symmetry of the graphene Hamiltonian, this results in the chiral symmetry between the solutions.

Recall that the original graphene Hamiltonian was derived within the tight-binding model (Gonzalez et al., 1993). One can view our system in terms of a set of coupled oscillators, so the states $(1,\,1)^T$ and $-i\sigma_y(1,\,1)^T = (-1,\,1)^T$ represent the two distinct and independent normal modes, one being symmetric, the other antisymmetric. These symmetries are global and affect the entire system. The procedure outlined above to solve Eq. (13) assumes no global symmetry. The conclusion from this comparison is not a relation between $u(x)$ and $u(-x)$ (implying nonlocality) but the chiral nature of these solutions (which is local). This harks back at the local 'spinor' structure of Eq. (13).

3.1 Discussion

We have plotted the dispersion relations $e = e(p_y)$ in Fig. 2. Notice the $e \to -e,\ p_y \to -p_y$ symmetry of the curves. In Fig. 2 (a) the two initial ($r = 0$) parabolic bands of the bilayer have the form $e = \pm p_y^2$ and these are separated as the biasing potential grows (Fig. 2 (b) and (c)) thus creating a gap between the bands. The parabolic bands then take the form $e = \pm\sqrt{p_y^4 + r^2}$. As we will see below these separated bands correspond to over-barrier scattering states which do not decay on account of the potential. For the intragap states, two zero modes appear symmetrically about the origin. A more detailed examination of the zero modes in Fig. 2(b) and (c) suggests that they occur at the same p_y value for *any* non-zero bias $(r \neq 0)$. This is indeed correct and can be shown explicitly by making use of the method presented in Section 3. We can give for instance the zero-mode wave functions V obtained from the first equation for any value of r

$$\begin{aligned} V_{>}^{e=0}(x) &= c_1\, e^{-\xi x}\, {}_2F_1\left(\frac{p_y-\xi}{\sqrt{2}} e^{3i\pi/4},\ \frac{p_y-\xi}{\sqrt{2}} e^{-3i\pi/4},\ 1-p_y+\xi,\ z\right) + cc \\ V_{<}^{e=0}(x) &= c_3\, e^{-\xi x}\, {}_2F_1\left(\frac{p_y+\xi}{\sqrt{2}} e^{i\pi/2},\ \frac{p_y+\xi}{\sqrt{2}} e^{-i\pi/2},\ 1+p_y+\xi,\ 1/z\right) + cc \end{aligned} \tag{21}$$

where $p_y = \pm 0.1255$, cc denotes the complex conjugate of the function to the left of it and ξ is defined in Section 2. Although there is no r dependence in Eq. (21), we find that the expressions for U, namely, $U_{>(<)}(x) = \pm\frac{1}{r}(p_y \mp \xi)^2 V_{>(<)}(x)$, contain an r-dependence.

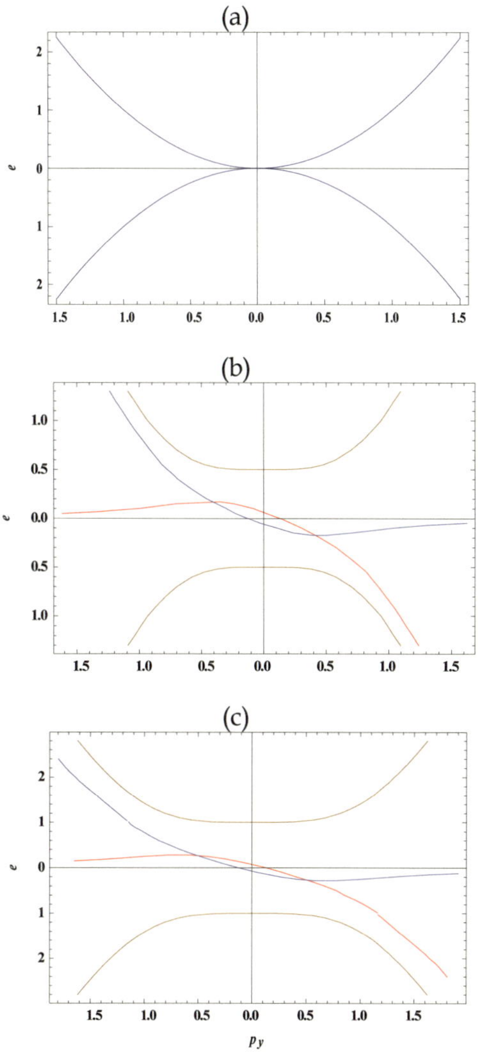

Fig. 2. Dispersion relations for (a) $r = 0$, (b) $r = ½$ and (c) $r = 1$. Zero modes occur at $e = 0$, $p_y = \pm 0.1255$. Note that the $e = 1$ solution has $p_y < 0$. The brown curves correspond to over-barrier scattering states.

The zero modes of our system are not Dirac fermions but chiral modes specific to the bilayer. The fact that these modes occur provided a bias of any strength is present is an indication of the topological character of the kink. That a kink is dynamically necessary is clear because the decay of the wave functions as $x \to \pm\infty$ requires the chiral pair to be close to each other near the origin. Each particle of the pair in turn is held in place by this electrostatic bias (with opposite signs on both sides of $x = 0$) along with the interaction with its conjugate. We can check the consistency of the above from a computation of the topological charge of the Fermi point. Writing the Hamiltonian (13) (without e) as $\vec{\varphi}(\vec{p}, x) \cdot \vec{\sigma}$, this topological charge is given by (Volovik, 2003)

$$N_3 = \frac{1}{8\pi}\varepsilon^{abc}\varepsilon_{ijk}\int_\Sigma dS^k \hat{\varphi}^a \partial_{p_i}\hat{\varphi}^b \partial_{p_j}\hat{\varphi}^c \quad (22)$$

where Σ is a surface enclosing the origin of the *p*-plane and may be taken to be two infinite planes parallel to the $p_x - p_y$ plane, one to the right and the other to the left of the origin (the separation between the planes being infinitesimal). See Fig. 3. This charge gives the difference between the number of right-moving and left-moving zero modes. Then, $N_3 = 2$, since we must sum the contributions from the right side and left sides of Σ. This is clearly consistent with our results. The existence of the zero modes can be exploited in some applications, as we will point out below.

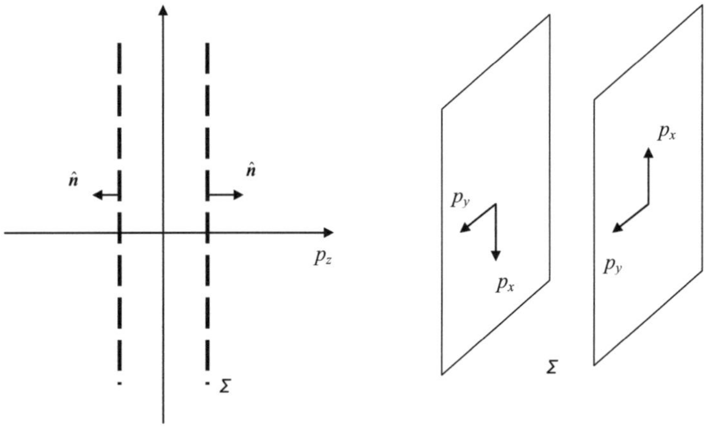

Fig. 3. Two views of the surface Σ with outward normals $\hat{\boldsymbol{n}}$ shown. Note that the p_z-separation between planes is non-zero provided $r \neq 0$. The contribution to the integrand (22) vanishes as $p_x, p_y \to \pm\infty$.

We can show that the Hamiltonian (13) reduces to the monolayer case in the vicinity of the zero mode, i.e., $|e| << 1$. Take for definiteness $U = V$ and assume $U = f\,g$, f being the zero-mode solution. g is a slowly varying function of e which has the value of 1 at $e = 0$, and varies over a length scale which is much larger than the kink width. If we substitute U into the Eq. (13) we obtain $(ip_x + p_y)^2 fg = g(ip_x + p_y)^2 f + 2(ip_x + p_y)f(ip_x + p_y)g + f(ip_x + p_y)^2 g$. The first term, when combined with the kink potential $-V\,f\,g$, vanishes for the zero mode *f*. The second contains the factor $(ip_x + p_y)f = (p_y - (a \pm i\beta))f + e^{-(\alpha \pm i\beta)x}\frac{d}{dx}{}_2F_1(a,b,c,e^{-2x})$. We can neglect the derivative because it will yield an additional factor e^{-2x} (which is small over the length scale of *g*), while the eigenvalue condition allows us to replace $p_y - (a \pm i\beta))$ with unity for a zero mode. Thus the Hamiltonian for g is $\begin{pmatrix} 0 & ip_x + p_y \\ -ip_x + p_y & 0 \end{pmatrix}$ correct to $O(e^2)$, which is of the form $\vec{\sigma} \cdot \vec{p}_\perp$, where $\vec{p}_\perp$ is perpendicular to the zero-mode momentum. Thus, at the vicinity of the zero mode (around $e = 0$), the Hamiltonian mimics that of monolayer graphene, with its energy spectrum having the characteristic $\sqrt{N}$ signature (Li & Andrei, 2007; Gusynin & Sharapov, 2005).

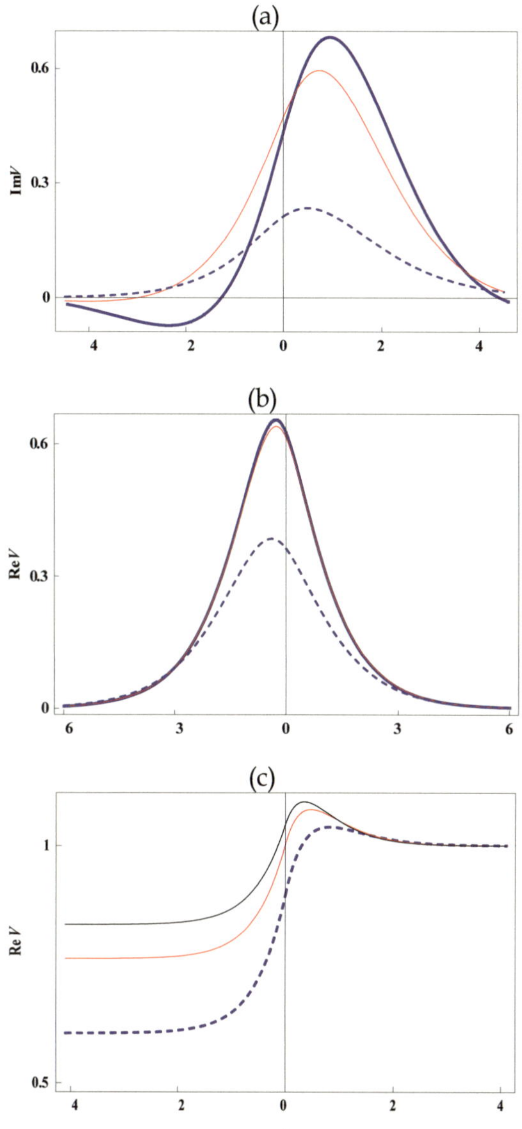

Fig. 4. Wave functions. (a) Imaginary part of the intragap wave function (unnormalized) for $r = 1$: thick blue (e = 0, $p_y = 0.1255$, i.e. the zero mode); dashed, blue ($e = -0.8$, $p_y = 1.03$); red ($e = 0.28$, $p_y = -0.58$). The real parts are much smaller (about a tenth) than the imaginary parts. (b) Real part of wave functions for the $r = 1$, $e > 1$, bound case: dashed blue ($e = 1.1$, $p_y = -1.2$); red ($e = 2.1$, $p_y = -1.69$); thick blue ($e = 2.2$, $p_y = 1.73$). (c) Over-barrier scattering wave functions (unnormalized): black ($e = 2.5$, $p_y = 1.514$); red ($e = 2$, $p_y = 1.316$), blue, dashed ($e = 1.5$, $p_y = 1.057$). The imaginary parts are very small.

Graphs of the dominant (real or imaginary) part of the wave functions are shown in Fig. 4. These figures imply that both the real and imaginary parts of the wave function are important. (By contrast, Martin *et al.* (2008) only find real wave functions.) As displayed in

Fig. 4, the intragrap states appear to always penetrate into the potential so their maxima are found inside it. Although all the intragap states are always bound, they are not the only bound states. In general, whenever the kinematic inequality $p^4 \neq e^2 - 1$ holds, we have bound states because in this case what we really have is $p^4 > e^2 - 1$: this is always the case for the intragap states, but it is also true of the blue and red $|e| > 1$ spiral arms in Figs. 2(b) and (c), a consequence of the fact that not enough of the total energy has been allotted to the motion in the *x*-direction. For these bound $|e| > 1$ states we see in Fig. 4 that their maxima lie to the left of the origin: these states do not penetrate into the kink as deeply as the intragap states do.

When $|e| > 1$ *and* $p^4 = e^2 - 1$, the particles have sufficient energy to overcome the barrier and we have the overbarrier scattering case (the particles go $x \rightarrow \pm\infty$). The same procedure for intragap states can be applied, with the factor $e^{-\xi x}$ in Eq. (14) excluded since we are looking for undamped solutions. Some wave functions of this type are given in Fig. 4(c). One can calculate the transmission and reflection amplitudes for them. These suggest greater transmission than reflection.

Because any negative energy eigenstate is related to a positive energy eigenstate by a unitary transformation, the local density of states $\rho(r) = \sum_e \psi_e^+(r)\,\psi_{e'}(r)\,\delta(e-e')$ is symmetric about $e = 0$ and the negative and positive energy eigenstates contribute equally. Including the zero modes, the conservation of the total number of states implies that the difference in densities with and without the kink $\delta\rho$ is

$$0 = \int d^2r \left(2\int_{-\infty}^{0^-} \delta\rho(r,e)de + |\psi_0(r)|^2 \right) \tag{23}$$

The two zero modes are normalized, so the integral of $\delta\rho$ over energy and space is - 1. Taking electron spin into account this means a charge of - 2*e* and total spin zero for the valence band. If the zero modes are unoccupied, the charge and spin for the kink are, respectively, Q = -2*e* and S = 0; if the zero modes are singly (doubly) occupied, then Q = -*e* (0), S = ½ (0). These serve as the signature of the presence of charge and the confinement of the chiral pair in the vicinity of the kink. Since the zero modes occur in pairs we do not see charge fractionalization here (Hou et al., 2007). Moreover, we need not be concerned here with a violation of Kramer's theorem (Su et al., 1980).

4. Pseudospin orbit coupling (pseudo-SOC) device

By the Dirac equation, spin orbit coupling can be derived in the vacuum with applied electric fields; one thus visualizes that similar effects should arise for the pseudospin of graphene particles governed by the Dirac equations. Thus pseudo-spin orbit coupling (Tan et al, 2010) can be derived by applying external electric field to a system governed by the graphene Hamiltonian. We will start with H_3 which is not a comparable form to standard Dirac Hamiltoninan. But by transformation e. g. using a unitary matrix U, one can obtain a more compatible form (McCann, E et al, 2006) of H_1

$$H_3\psi_3 = \begin{pmatrix} 0 & c\pi^+ & 0 & v_3\pi \\ c\pi & 0 & t & 0 \\ 0 & t & 0 & c\pi^+ \\ v_3\pi^+ & 0 & c\pi & 0 \end{pmatrix}\begin{pmatrix} \varphi_{A1} \\ \varphi_{B1} \\ \varphi_{A2} \\ \varphi_{B2} \end{pmatrix} \rightarrow H_1\psi_1 = \begin{pmatrix} 0 & v_3\pi & 0 & c\pi^+ \\ v_3\pi^+ & 0 & c\pi & 0 \\ 0 & c\pi^+ & 0 & t \\ c\pi & 0 & t & 0 \end{pmatrix}\begin{pmatrix} \varphi_{A1} \\ \varphi_{B2} \\ \varphi_{A2} \\ \varphi_{B1} \end{pmatrix} \quad (24)$$

The transformation which has been performed by U determines the type of pseudo-spin to be investigated, e.g. by the above method which results in $\chi = (A_1\ B_2)^T$ pseudo-spin is defined by the linear superposition of wavefunction amplitude between site A_1 and B_2. In application to particles which mimic Dirac fermions due to material bandstructure, it would be instructive to replace the coupling mass term of mc^2 for particles in vacuum with a coupling term Δ which arises due to material bandstructure but plays the same role as the mass term, as far as the Dirac matrix is concerned. The coupling term Δ gives rise to the energy dispersion where the effective mass of particles in the materials can be derived; in other words, particle effective mass is a function of Δ but not vice versa. For monolayer graphene, Δ vanishes and it can be derived from the energy dispersion relation that particles behave like massless Dirac fermions. In graphene-like materials, χ can be described as a pseudo-spin which consists of a linear combination of waves due to different sub-lattice sites. The above is, however, merely one example of graphene-like materials which can also be written as

$$\begin{pmatrix} E - \Delta_a & -\boldsymbol{\sigma}\cdot \boldsymbol{p}c \\ -\boldsymbol{\sigma}\cdot \boldsymbol{p}c & E - \Delta_b \end{pmatrix} \cdot \begin{pmatrix} \chi \\ \theta \end{pmatrix} = 0 \quad (25)$$

where in this specific case, $\Delta_a = \begin{pmatrix} 0 & v_3\pi \\ v_3\pi^+ & 0 \end{pmatrix}$ and $\Delta_b = \begin{pmatrix} 0 & t \\ t & 0 \end{pmatrix}$ Multiplying by γ^0 to the left, one obtains the graphene Hamiltonian comparable in form to the general Dirac equation

$$\begin{pmatrix} (E - eV) - \Delta_a & -\boldsymbol{\sigma}\cdot \boldsymbol{p}c \\ \boldsymbol{\sigma}\cdot \boldsymbol{p}c & -(E - eV) + \Delta_b \end{pmatrix} \cdot \begin{pmatrix} \chi \\ \theta \end{pmatrix} = 0 \quad (26)$$

where V is the external potential.
The above can be written in terms of Dirac matrices as follows

$$\gamma^\mu D_\mu - R = 0 \quad (27)$$

where $\gamma^0 = \begin{pmatrix} 1 & 0 \\ 0 & -1 \end{pmatrix}$, $\gamma^j = \begin{pmatrix} 0 & \sigma^j \\ -\sigma^j & 0 \end{pmatrix}$, $R = \begin{pmatrix} \Delta_a & 0 \\ 0 & -\Delta_b \end{pmatrix}$, and $D_\mu = \partial_\mu - \frac{ie}{\hbar}A_\mu$. Multiplying the above to the left with $\gamma^\mu D_\mu + R$ yields

$$\begin{pmatrix} (E - eV)^2 - p^2c^2 - \Delta_a^2 & -ie\hbar c\boldsymbol{\sigma}\cdot \boldsymbol{\mathcal{E}} \\ -ie\hbar c\boldsymbol{\sigma}\cdot \boldsymbol{\mathcal{E}} & (E - eV)^2 - p^2c^2 - \Delta_b^2 \end{pmatrix} \cdot \begin{pmatrix} \chi \\ \theta \end{pmatrix} = 0 \quad (28)$$

where $\mathcal{E}^j/c = -\partial_0 A_j$ is the electric field. We provide the above to merely illustrate the close connection between the bilayer graphene and the vacuum Dirac Hamiltonian, such that useful analogies of pseudo-spin orbit coupling to the vacuum spin orbit coupling can be drawn. Thus, to simplify matter, we temporarily disregard the fact that $\Delta_a = \begin{pmatrix} 0 & v_3\pi \\ v_3\pi^+ & 0 \end{pmatrix}$ and $\Delta_b = \begin{pmatrix} 0 & t \\ t & 0 \end{pmatrix}$. Rather we replace $\Delta_a^2 = \begin{pmatrix} (v_3 p)^2 & 0 \\ 0 & (v_3 p)^2 \end{pmatrix}$ with $\Delta_a^2 = \begin{pmatrix} mc^2 & 0 \\ 0 & mc^2 \end{pmatrix}$ so that one can write $\Delta_a^2 = \Delta I$ where $\Delta = mc^2$. We make an even more drastic assumption that one can write $\Delta_b = \begin{pmatrix} mc^2 & 0 \\ 0 & mc^2 \end{pmatrix} = \Delta I$. With this, Eq.(26) can be reduced to

$$\theta = \frac{\boldsymbol{\sigma} \cdot \boldsymbol{p} c}{E - eV - [\Delta_b]} \chi \tag{29}$$

where $[\Delta_b] = \sqrt{[\Delta_a^2]} = \Delta$. The relativistic energy equation which could be used to describe the analogous effect of pseudo spin orbit coupling, i.e. the coupling of pseudo spin to particle momentum in the presence of electric fields, for Dirac fermions in graphene-like material systems is thus

$$\left[(E - eV)^2 - \Delta_a^2 - p^2 c^2 - \frac{ie\hbar c^2}{E - eV - [\Delta_b]} \boldsymbol{\mathcal{E}} \cdot \boldsymbol{p} - \frac{e\hbar c^2}{E - eV - [\Delta_b]} \boldsymbol{\sigma} \cdot (\boldsymbol{\mathcal{E}} \times \boldsymbol{p}) \right] \chi = 0 \tag{30}$$

To avoid excessive details into the material science and band structure of graphene, we will take the liberty of assuming that the relation $\Delta_a^2 = \Delta^2 I; \Delta_b = \Delta I$ is satisfied in bilayer graphene or, at least, can be realized by material engineering.

We will now investigate the effects of pseudo SO coupling on the pseudo spin χ. As is well-known, a particle in a SO coupling system experiences an effective magnetic field $\boldsymbol{B}_E = \boldsymbol{p} \times \boldsymbol{\mathcal{E}}$ which couples directly to its momentum vector, thus preserving time-reversal symmetry. In the technology-relevant field of spintronics, such $\boldsymbol{B}_E$ can be used to control the precession of spin when coupled with appropriate momentum constraints (e.g. single mode one-dimensional ballistic transport), similar to gate bias-controlled spin precession via Rashba or Dresselhaus SO coupling in the so-called Datta-Das spin transistor (Supriyo, Datta et al, 1989). On the other hand, spin relaxation is related to electron precession about $\boldsymbol{B}_E$ (D'yakonov, M.I. et al, 1971), which suggests that pseudospin relaxation can be analyzed in analogy to spin relaxation under spin orbit coupling, but in the relativistic limit. In typical graphene-like materials, $\Delta = \sqrt{[\Delta_a^2]} = [\Delta_b]$ is small (10-300 meV for massive fermion, 0 meV for massless fermion). Since kinetic energy $KE << \Delta$ (5 order of magnitude smaller) for the non-relativistic approximation to apply, the corresponding number of non-relativistic particles is very small. For particles confined to energy range 1-2 order of magnitude smaller than Δ, we consider these particles as relativistic; this prompts the need to analyze the pseudo SO effect in the relativistic limit. One could visualize the pseudo spin precessing about an effective magnetic field which could only be "seen" by the pseudo spin, at a precessional frequency which could be deduced from $\hbar\omega = \sqrt{\hbar e p|\mathfrak{E}|c^2/(E-[\Delta_b])}$. With the average velocity given by $\langle\chi|v_\mu|\chi\rangle = \frac{c^2 p}{\sqrt{p^2c^2+[\Delta_a^2]}}$, the precession angle over the Bloch sphere of the pseudospin, for a unit of particle travel length in the relativistic regime is given by:

$$\frac{\Omega}{l} = \frac{\omega}{\langle\chi|v_\mu|\chi\rangle} = \sqrt{\frac{e|\mathfrak{E}|(p^2c^2+[\Delta_a^2])}{\hbar(E-[\Delta_b])pc^2}} \tag{31}$$

A series of pseudospin relaxation has been predicted and analyzed for different energy regimes. In summary, it has been studied that Dirac particles in the energy range of $pc \approx 1\,\text{meV}$ (which although is relatively small compared to the energy gap of $\Delta \approx 200$ – 300 meV, it is large enough to be within the relativistic regime), Eq. (31) reduces to $\frac{\Omega}{l} = \frac{\Delta}{p}\sqrt{\frac{e|\mathfrak{E}|}{\hbar c^3}}$. Increasing particle's momentum reduces the precessional angle for a fixed travel length. By contrast, in the ultra-relativistic limit (i.e. massless Dirac particle), Eq.(31) reduces to $\Omega/l = \sqrt{e|\mathfrak{E}|/\hbar c}$, which predicts that massless Dirac particle has a constant Ω/l. This can be understood as typically, massless particle travels at the effective speed of light in the medium. In this limit, pseudo spin relaxation becomes independent of particle momentum. In the non-relativistic limit, where $\left(\frac{p^2}{2m} + eV + \frac{\hbar ec^2}{4\Delta^2}\boldsymbol{\sigma}\cdot(\boldsymbol{\mathfrak{E}}\times\boldsymbol{p})\right)\boldsymbol{\chi} = \boldsymbol{\varepsilon\chi}$, it is found that $\langle\chi|H_{SC}^{nr}|\chi\rangle = \frac{ehc^2}{4\Delta^2}|\mathfrak{E}|p$. The average particle velocity can be approximated as $\langle\chi|v_\mu|\chi\rangle = p/m$, and in a similar manner, the precession angle Ω per unit travelling distance is given by $\Omega/l = e|\mathfrak{E}|/4\Delta$, which is independent of particle momentum. Therefore in both non-relativistic and ultra-relativistic limits, Ω/l is independent of the particle momentum. But in the former, Ω/l depends inversely on Δ; such a dependence obviously cannot exist in the ultra-relativistic limit where the coupling mass term vanishes.

Based on the above understanding, we briefly propose that a nanoscale device which consists of a graphene ring and a charged nano-sized dot at the centre would be a suitable platform to utilize the pseudospin orbit coupling of the graphene Dirac particles. The pseudospin orbit strength can be calculated in the relativistic and low energy limits in analogy to spin orbit coupling in semiconductors. Pseudospin orbit coupling strength can be enhanced by accelerating the Dirac particles around the ring, due to the small energy gap in graphene and the large radial electric field due to the charged quantum dot.

5. Conclusions

The zero modes (21) may find possible application in two ways: (a) one can take advantage of the relation (18) applied to the zero modes as a switching indicator and (b) in so far as zero modes are of two types associated with the chiral functions $(1, 1)^T$ and $(-1, 1)^T$, they might store information much as binary bits do. The topological properties of the charge and spin of these zero modes confer a certain degree of robustness to these binary states. Even the presence of some disorder would not alter this conclusion provided the kink retains its topological character. An indication of this is that the zero modes appear even for small r value. Further application of the zero modes can be derived from utilizing the valley degree of freedom (or "valleytronics" (Rycerz et al., 2007), which can be modified along the kink direction. Our results would also be of interest in brane theory (Horava, 2005) and superconductivity (Lu & Yip, 2008). We also describe another relativistic effect in graphene, namely, pseudospin orbit coupling (pseudo-SOC) effect. Potentially the pseudo-SOC effect can be used for pseudospin field effect transistor (FET) in much the same way that the physical spin orbit coupling is used for semiconductor spin-FET. The pseudospin orbit coupling strength has to be further enhanced for it to be comparable to the conventional semiconductor-based Rashba effect. Future work which focuses on modifying the graphene structure can potentially enhance this useful pseudo-SOC effect within experimentally accessible parameters.

The support of NRF/NUS under Grants Nos. R-143-000-357-281 and R-263-000-482-112 are gratefully acknowledged.

Appendix

The hypergeometric differential equation is (in general z is complex)

$$z(1-z)\frac{d^2w}{dz^2}+[c-(a+b+1)z]\frac{dw}{dz}-ab\,w=0\,,$$

whose solution is the Gauss series

$$_2F_1(a,b,c;z)={}_2F_1(b,a,c;z)=\frac{\Gamma(c)}{\Gamma(a)\Gamma(b)}\sum_{n=0}^{\infty}\frac{\Gamma(a+n)\Gamma)\Gamma+n)}{\Gamma(c+n)}\frac{z^n}{n!}$$

with circle of convergence on the unit circle $|z|=1$. The series is not defined when $c = 0$ or a negative integer. We also have applied the useful differential relation

$$\frac{d}{dz}\,{}_2F_1(a,b,c,z)=\frac{ab}{c}\,{}_2F_1(a+1,b+1,c+1,z)$$

When $c - a - b$ is an integer (as in overbarrier scattering), extra care is required to solve the hypergeometric equation. In this case the following is true: two linearly independent solutions of

$$z(1-z)\frac{d^2w}{dz^2}+(1-z)(1-p)\frac{dw}{dz}-\frac{1}{2}q\,w=0\,,$$

are the pair

$$ {}_2F_1(-\tfrac{1}{2}(p+\sqrt{p^2-2q}),\ -\tfrac{1}{2}(p-\sqrt{p^2-2q}),\ 1-p,\ z), $$

$$ (-z)^p\,{}_2F_1(\tfrac{1}{2}(p-\sqrt{p^2-2q}),\ \tfrac{1}{2}(p+\sqrt{p^2-2q}),\ 1+p,\ z). $$

6. References

Abanin, D. A. & Levitov, L. S. (2007) Quantized transport in graphene p-n junctions in a magnetic field *Science*, 317, 5838 (August, 2007) 641 - 643, ISSN 0036-8075

Abramowitz, M. & Stegun, I. (1964) *Handbook of Mathematical Functions*, Dover, New York.

Castro Neto, A. H., Guinea, F., Peres, N. M. R., Novoselov, K. S. & Geim, A. K. (2009) The electronic properties of graphene *Rev. Mod. Phys.* 81, 1 (January, 2009) 109-162, ISSN 0034-6861

D'yakonov M.I., Perel V.I., 1971, *Zh. Eksp. Teor. Fiz.* 60, 1954 (Sov. Phys. JETP 38, 1053 (1971) ISSN 0038-5648).

Gonzalez, J., Guinea, F. & Vozmediano, M. A. H. (1993) The electronic spectrum of fullerenes from the Dirac equation, *Nucl. Phys.* B 406, 3 (October, 1993) 771 - 794, ISSN 0550-3213

Gusynin, V. P. & Sharapov, S. G. (2005) Unconventional integer quantum Hall effect in graphene, *Phys. Rev. Lett.* (for McCann Edward) 95, 14 (September, 2005) 146801, ISSN 0031-9007

Horava, P. (2005) Stability of Fermi surfaces and K theory, *Phys. Rev. Lett.* (for McCann Edward) 95, 1 (July, 2005) 016495, ISSN 0031-9007

Hou, C.-Y. , Chamon, C. & Mudry, C. (2007) Electron fractionalization in two-dimensional graphenelike structures, *Phys. Rev. Lett.* (for McCann Edward) 98, 18 (May, 2007) 186809, ISSN 0031-9007

Itzykson, C. & Zuber, J.-B. (1980) *Quantum Field Theory*, Dover, 10: 0486445682 New York.

Kane, C. L. (2005) Erasing electron mass, *Nature* 438, 7065 (November, 2005) 168 - 170 ISSN 0028-0836

Katsnelson, M. I., Novoselov, K. S. & Geim, A. K. (2006) Chiral tunneling and the Klein paradox in graphene, *Nature Phys.* (for Noveselov) 2, 9 (September, 2006) 620 - 625, ISSN 1745-2473

Landau, L. D. & Lifshitz, E. M. (1968). *Quantum Mechanics*, Elsevier-Science, 0 7506 3539 8, London.

Li, G. & Andrei, E. (2007) Observation of Landau levels of Dirac fermions in graphite, *Nature Phys.* 3, 9 (September 2007) 623 - 627, ISSN 1745-2473

Lu, C.-K. & Yip, S. (2008) Zero-energy vortex bound state in noncentrosymmetric superconductors, *Phys. Rev.* (for Manes) B78, 13 (October, 2008) 132502, ISSN 1098-0121

Manes J.L. et al., (2007) Existence and topological stability of Fermi points in multi-layer graphene, *Phys. Rev.* (for Manes) B 75 (April, 2007) 155424, ISSN 1098-0121

Martin, I, Blanter, Ya. M. & Morpurgo, A. F. (2008) Topological confinement in bilayer graphene *Phys. Rev. Lett.* (for McCann Edward) 100, 3 (January, 2008) 036804, ISSN 0031-9007

Martinez, J. C., Jalil, M. B. A. & Tan, S. G. (2010) Klein tunneling and zitterbewegung and the formation of a polarized *p-n* junction, *Appl. Phys. Lett.* 97, 6 (August, 2010) 062111, ISSN 0003-6951

McCann Edward, Fal'ko Vladimir I., Landau-level degeneracy and quantum Hall effect in a graphite bilayer, Phys. Rev. Lett. 96 (March 2005), 086805, ISSN 0031-9007.

Novoselov, K. S., McCann, E., Morozov, S. V., Fal'ko, V. I., Katsnelson, M. I., Zeitler, U., Jiang, D., Schedin, F. & Geim, A. K. (2006) Unconventional quantum Hall effect and Berry's phase of 2π in bilayer graphene, *Nature Phys.* (for Noveselov) 2, 3 (March, 2006) 177- 180, ISSN 1745-2473.

Rycerz, A., Tworzydlo, J. & Beenaker, C. W. J. (2007) Valley filter and valley valve in graphene, *Nature Phys.* (for Noveselov) 3, 3 (March, 2007) 172 - 175, ISSN 1745-2473

Schliemann, J., Loss, D. & Westervelt, R. M. (2006) Zitterbewegung of electrons and holes in III-V semiconductors quantum wells, *Phys. Rev.* (for Manes) B 73, 8 (February, 2006) 085323, ISSN 1098-0121

Su, W. P., Schrieffer, J. R. & Heeger, A. J. (1980) Soliton excitations in polyacetylene, *Phys. Rev.* B 22, 4 (August, 1980), 2099 - 2111, ISSN 1098-0121.

Supriyo Datta & B. Das, Electronic analog of the electro-optic modulator (1989), Appl. Phys. Lett. 56 (December, 1989) 665 - 667, ISSN 0003-6951.

Tan S. G. et al. (2010) Pseudospin-orbital coupling for pseudospntronic device in graphene, *Magnetism and Mag. Materials* (August, 2010) 322, 2390 - 2394, ISSN 0304-8853.

Tan. S. G., Jalil, M. B. A., & Fujita, T. (2010) Monopole and topological electron dynamics in adiabatic spintronic and graphene systems, *Ann. Phys.* (N.Y.) 325, 8 (August, 2010) 1537-1549, ISSN 0003-4916

Volovik, G. E. (2003). *The Universe in a Helium Droplet,* Oxford, ISBN 13: 978-0-19-956484-2, Clarendon

9

Simulated Nanoscale Peeling Process of Monolayer Graphene Sheet - Effect of Edge Structure and Lifting Position

Naruo Sasaki[1], Hideaki Okamoto, Shingen Masuda, Kouji Miura* and Noriaki Itamura
Seikei University,
**Aichi University of Education,*
Japan

1. Introduction

Adhesion and peeling phenomena play important roles for connecting two objects regardless of whether they are inorganic, organic or biological materials, which contributes to buliding up microscopic devices. The carbon nanostructures such as carbon nanotube (CNT) and graphene have recently attracted great interests as the components of the electronic, magnetic and optical devices. We have so far studied the peeling mechanics of the carbon nanotube (CNT) adsorbed onto the graphite surface both theoretically (Sasaki et al., 2006, 2008, 2009b) and experimentally (Ishikawa et al., 2008, 2009). It is clarified that the transition from the line- to the point-contact between the CNT and the graphite surface occurs during the peeling process. The CNT on the sub-microscale has the same size as the spatulae of the microscopic hairs aligned on the gecko foot (Autumn et al., 2000, 2002). Therefore the study of the peeling process of the nano-scale objects such as CNT is useful for not only developing the gecko-foot-mimic adhesives (Qu et al., 2008) but also understanding the elementary process of adhesion.

On the other hand, since the success of its experimental isolation (Novoselov et al., 2004), the potential of various application of the graphene such as the components of the electronic devices (Novoselov et al., 2005, Geim et al., 2007), has been discussed by many researchers. There is also a possibility that its adhesion with the substrate is applied to the adhesive tape at nanoscale. Therefore the peeling mechanics of the graphene sheet is very important, which can be regarded as the elementary process of the macroscopic sticky tape such as the gecko-foot-mimic adhesives (Autumn et al., 2000, 2002, Qu et al., 2008), or that of the microscopic extension of the crack in the fracture process. In our preliminary experiments, we have already succeeded in peeling the multilayered graphene plate with a thickness of several μm by using atomic-force microscopy tip. Here the two-component epoxy resin adhesive is used to bond the graphene plate to the AFM tip. Here the standard Si_3N_4 tip for the contact AFM experiment is used. The junction formed between the AFM tip and the graphene should be mechanically rigid enough to measure the elasticity of the graphene

---

[1]Corresponding author: naru@st.seikei.ac.jp

sheet during the peeling process. The two-component epoxy resin adhesive satisfies the above condition. If the thickness of the peeled graphene plate is reduced, the comparison between the present simulation and the experiment will become possible.
Therefore, in this chapter, ahead of experiment, we have theoretically reported the nanoscale peeling behaviors of the monolayer graphene sheet based on the molecular mechanics simulation (Sasaki et al., 2009a, 2010). The peeling force curve exhibits the nanoscale change of the graphene shape from the surface to the line contact. The center position and the left edge are chosen as the lifting position. In Section 3, the peeling of the monolayer graphene sheet with the armchair edge for lifting the center position is discussed. In Secs. 4 and 5, the peeling of the monolayer graphene sheet with the armchair- and zigzag-edge for lifting the edge position is discussed, respectively.

2. Model and method of simulation

In the simulation, a rectangular-shaped monolayer graphene sheet with each side of 38 Å × 20 – 21 Å, comprised of 310 carbon atoms, is peeled from the rigid rectangular graphene sheet (which is called, the 'graphite surface,' hereafter) with each side of 164-165 Å × 58 Å, comprised of 3536 carbon atoms [Fig. 1(a)]. First, both the above graphene sheets are separately optimized by minimizing the covalent bonding energy described by the Tersoff potential energy (Tersoff, 1988), V_{cov}, using the Polak-Rebiere-type conjugate gradient (CG) method (Press et al., 1999). Here the convergence criterion is set so that the maximum of absolute value of all the forces acting on the movable atoms, becomes lower than 10^{-5} eV/Å. Next, the graphene sheet is put and adsorbed onto the graphite surface, so that the AB stacking registry between the graphene sheet and the graphite surface is satisfied as shown in Figs. 1(b) and 1(c). Here the green-colored six-membered ring at the center position or the outermost left edge of the graphene sheet is assumed to be attached to the AFM tip apex [Fig. 1(a)], and then it is gradually moved upward along the z direction, parallel to the [0001] axis, by 0.1 Å. For each lifting position of the graphene sheet, z, the total energy $V_{total} = V_{cov} + V_{vdW}$, is minimized using the CG method, where V_{vdW} is the nonbonding vdW interaction described by the modified Lennard-Jones (LJ) potential energy (Lu et al., 1988, Stoddard and Ford, 1973), acting between the graphene sheet and the graphite surface. Thus the optimized positions of the movable carbon atoms of the graphene sheet, (x, y, z), the vertical peeling force F_z, and the lateral sliding forces F_x and F_y, acting on the lifting center, are calculated during the peeling process. In this paper, the graphene sheets with armchair- [Fig. 1(b)] and zigzag-edges [Fig. 1(c)] are discussed.

3. Center-lifting case of armchair-edge graphene

When the six-membered ring located at the center position of the monolayer graphene sheet is lifted, the graphene sheet exhibits the characteristic transition of its shape during the peeling process within the x-z plane as illustrated in Figs. 2A-2J, corresponding to Figs. 3A-3J, the vertical force acting on the lifting center position F_z plotted as a function of the displacement from the initial position along z-direction, z. At first the monolayer graphene sheet takes an initial planar structure parallel to the rigid graphite surface [Fig. 2A: z = 0 Å]. Here the surface contact is formed between the graphene sheet and the graphite surface. The vertical force F_z is zero [Fig. 3A]. Just after the beginning of the peeling [Fig. 2B: z = 2.0 Å], the attractive interaction force takes the minimum value, –3.1eV/Å [Fig. 3B].

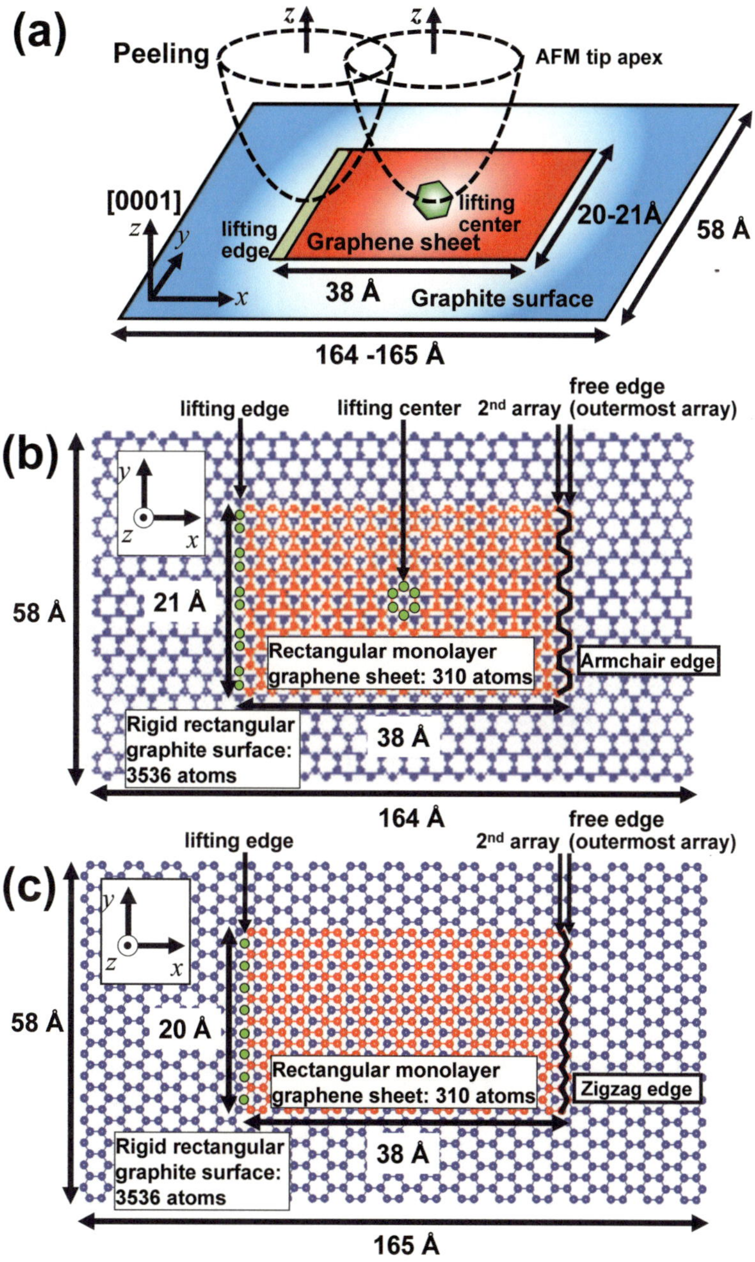

Fig. 1. (a) The schematic illustration of the model of the monolayer graphene sheet physically adsorbed onto the rigid graphite surface used in the simulation. The green-colored six-membered ring at the center position or left edge of the graphene sheet is assumed to be adsorbed onto the atomic force miroscopy tip apex indicated by broken lines, and it is moved upward along the z (or [0001]) direction, by z = 0.1 Å. Initial AB stacking registry of the red-colored graphene sheet with (b) armchair and (c) zigzag edge adsorbed onto the blue-colored graphite surface within the x - y plane.

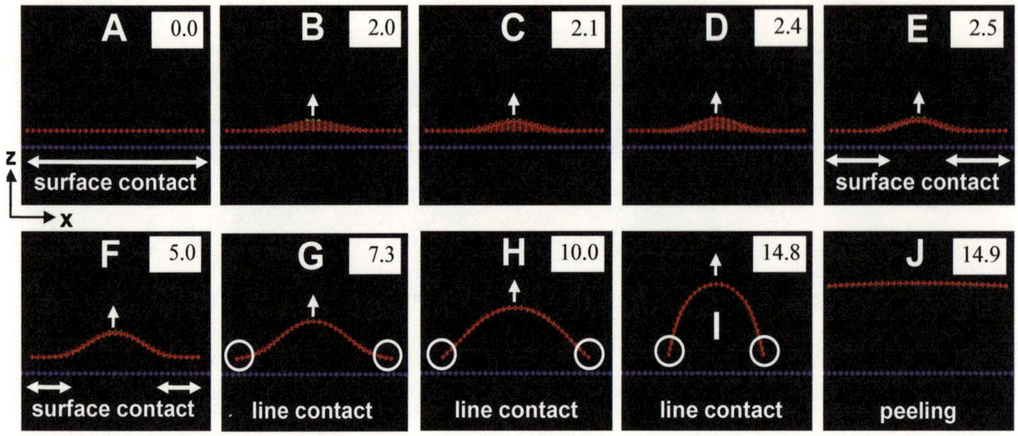

Fig. 2. The transition of the shape of the monolayer graphene sheet during the peeling process from A to J within the x - z plane. The red-colored graphene sheet and blue-colored graphite surface are shown. The displacement of the lifting center position from the initial position, z [Å], is indicated on the upper-right positions of each picture.

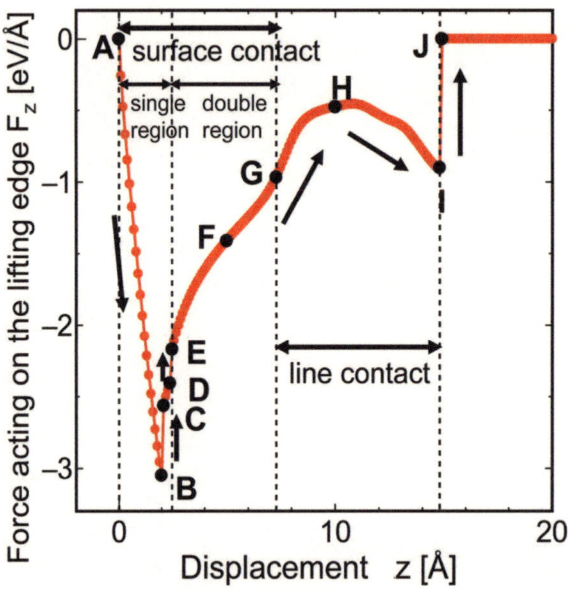

Fig. 3. The vertical force, F_z, acting on the center six-membered ring, plotted as a function of the lifting displacement z. The positions A-J correspond to those of Fig.2.

Between z = 2.0 Å and 2.1 Å, the first discrete partial peeling of the graphene occurs [Figs. 2B→2C], which produces the 1st discontinuous jump in the force curve [Figs. 3B→3C]. The partial peeled area around the lifting center of the graphene is shown in Figs. 4B→4C. Then, between z = 2.4 Å and 2.5 Å, the second discrete partial peeling of the graphene occurs [Figs. 2D→2E], which produces the 2nd discontinuous jump in the force curve [Figs. 3D→3E]. The partial peeled area of the graphene is shown in Figs. 4D→4E. Which of these two areas is peeled first is expected to be actually the stochastic process under the room temperature condition. Now the surface contact region is split into the left and right sections [Fig. 2E]. After the two discrete jumps, F_z increases as the peeling proceeds, since the attractive surface contact region gradually decreases [Fig. 2F: z = 5.0 Å]. Then the surface contact continuously turns into the line contact at z = 7.3 Å [Fig. 2G]. Here the 'line contact' is defined by the following two criteria: 1) The carbon atoms on the left and right outermost arrays of the graphene sheet [Fig. 1(b)] receive the repulsive interaction force from the graphite surface. 2) The carbon atoms on the second arrays [Fig. 1(b)] next to the outermost arrays receive the attractive interaction force. As illustrated in Fig. 5, the average forces acting on one carbon atom on the outermost and the second arrays satisfy the above criteria at z = 7.3 Å, which corresponds to Fig. 2G.

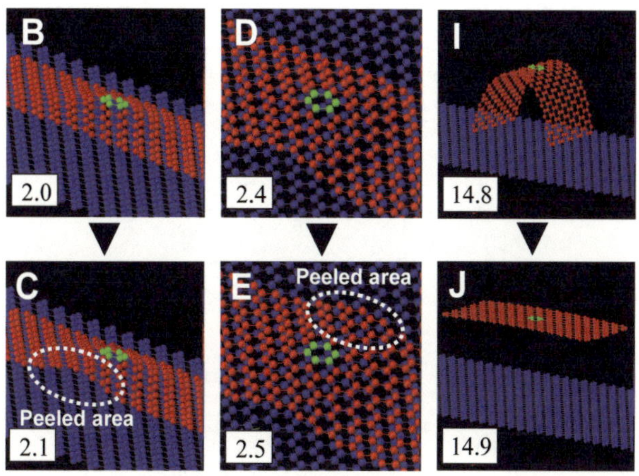

Fig. 4. The atomic structures of the graphene sheet just before and after the discrete change, B→C, D→E, and I→J. The regions surrounded by dotted ellipses show the partial peeled areas.

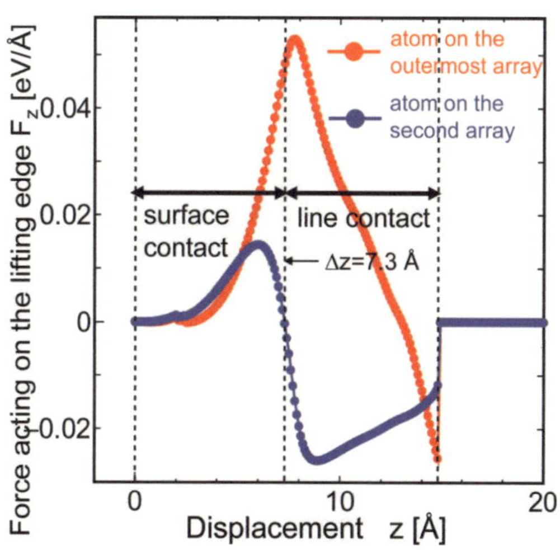

Fig. 5. The averaged forces acting on one atom on the left and right outermost arrays (red-colored) and those on the left and right second arrays (blue-colored), as a function of the displacement of the lifting center position from the initial position, z [Å].

Once the line contacts are formed between the free edges (outermost arrays) of the peeled graphene sheet and the graphite surface, they clearly slide on the graphite surface as indicated by a circle in Figs. 2G→2H→2I with a rapid increase of the bending of the graphene sheet. Within *x-y* plane, the right outermost array of the graphene sheet slides nearly straightforward along *-x* direction, not so sensitive to the lattice structure of the surface as illustrated in Figs. 6(a) and 6(b), which show the trajectories of the two carbon atoms on the right outermost array illustrated in Fig. 1(b). The sliding of the outermost arrays during G and I appears much more clearly than that during A and G, During H and I, the decrease of F_z [Figs. 3H and 3I] can be explained by the decrease of the repulsive force acting on the carbon atoms on the left and right edges of the graphene sheet as shown in Fig. 5, that's to say, the relative increase of the effect of the attractive interaction force.

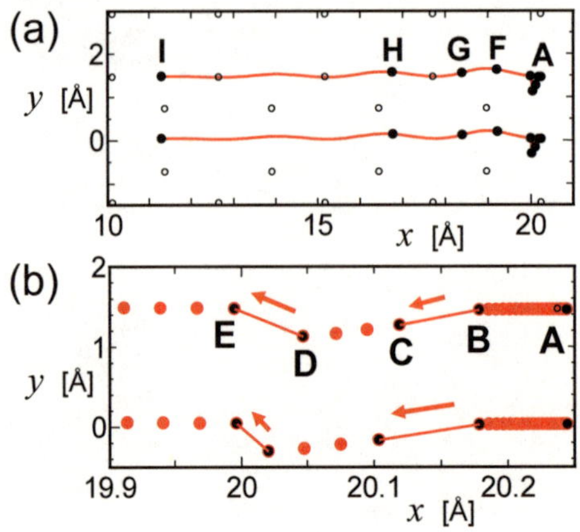

Fig. 6. The trajectory of the two carbon atoms on the right free edge (outermost array) indicated by white circles in Fig. 1(b). (a) The whole trajectory A→I, and (b) the part of the trajectory A→E, including the discrete jumps, B→C and D→E, are indicated. White circles mean carbon atoms of the graphite surface. The indices A – I correspond to those in Fig. 2.

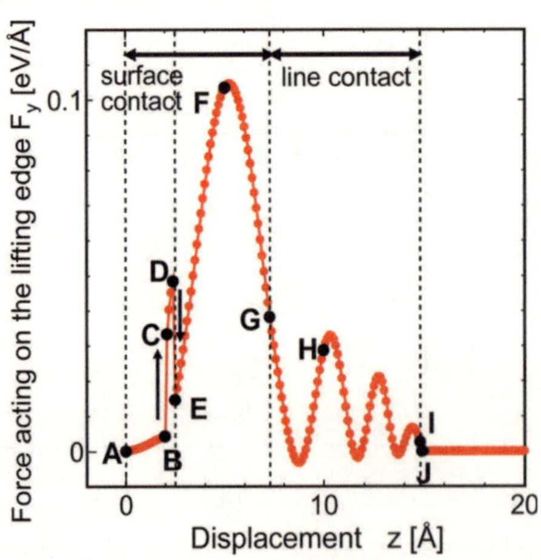

Fig. 7. The lateral force, F_y, acting on the center six-membered ring, plotted as a function of the displacement *z*. The positions A-J correspond to those of Fig. 2.

When the bending of the graphene sheet becomes larger than a certain range, both the left and right line contacts break and the graphene sheet is completely peeled from the surface [Figs. 2I→2J: z = 14.8 Å→14.9 Å], which produces the 3rd discontinuous jump in the force curve [Figs. 3I→3J]. As illustrated in Figs. 2I→2J, and Figs. 4I→4J, the graphene sheet exhibits the transition from the arched shape to the planer shape.
Thus the vertical peeling force F_z exhibits the characteristic shape as shown in Fig. 3, which reflects the transition from the surface to the line contact between the graphene sheet and the graphite surface. On the other hand, the lateral sliding force F_x is zero due to the structural symmetry of the system. However, the lateral sliding force F_y shows a finite value with an oscillation whose period and amplitude decreases as z increases. This oscillation of Fig. 7 reflects the trajectory of the graphene edges illustrated in Fig. 6 at the graphene - substrate interface during the peeling process. The maximum lateral force $F_y \approx 0.1$ eV/Å which is only about 3 % of the absolute value of the maximum adhesion force $|F_z|$ = 3.1 eV/Å.

4. Edge-lifting case of armchair-edge graphene

4.1 Nano-scale peeling along vertical direction

When the left edge of the monolayer graphene sheet with armchair edge [Fig. 1(b)] is lifted, the shape of the graphene sheet markedly changes during the peeling process within the x-z plane as illustrated in Figs. 8A-8J, corresponding to Figs. 9(a)A-9(a)J, the vertical force acting on the lifting edge F_z plotted as a function of the edge height z.

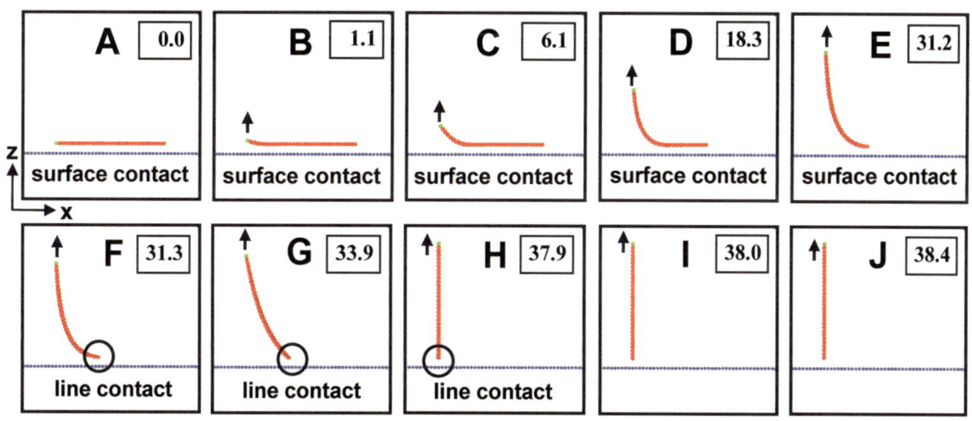

Fig. 8. The transition of the shape of the monolayer graphene sheet with free edge of armchair-type during the peeling process from A to J within the x - z plane. The red-colored graphene sheet and blue-colored graphite surface are shown. The height of the lifting left edge z [Å] are indicated on the upper-right positions of each picture.

Surface-contact region At first the monolayer graphene sheet takes an initial planar structure parallel to the rigid graphite surface [Fig. 8A: z = 0 Å], and the vertical force F_z is zero [Fig. 9(a)A], which means the graphene sheet completely takes the surface contact with the graphite surface. Just after the start of the peeling [Fig. 8B: z = 1.1 Å], the attractive interaction force $|F_z|$ becomes the maximum, 0.74 eV/Å [Fig. 9(a)B]. After that the surface contact area gradually decreases as the peeling proceeds [Figs. 8C-8E], where Fig. 9(a) exhibits the atomic-scale zigzag structures [Figs. 9(a)C-E], which will be explained in Section 4.2.

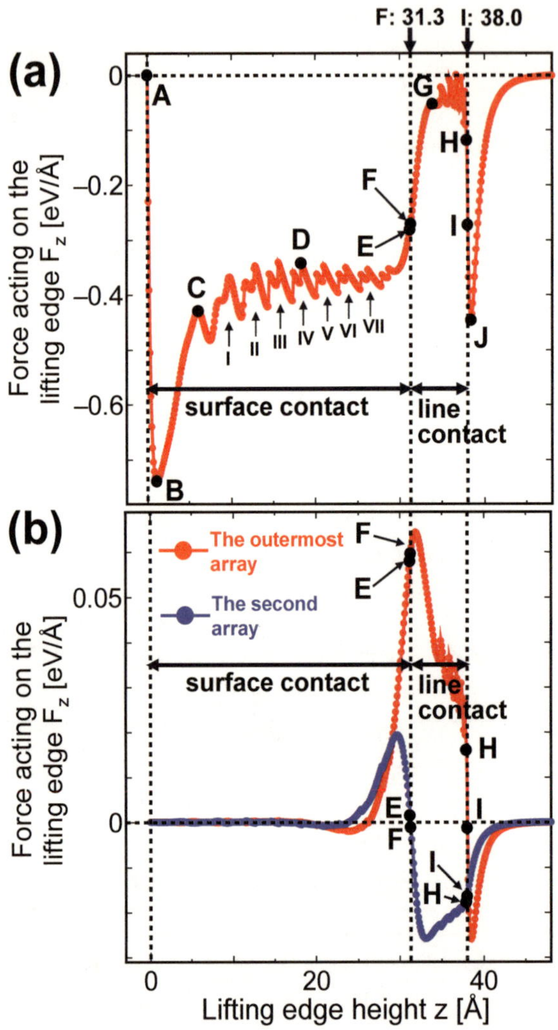

Fig. 9. (a) The vertical force F_z acting on the lifting edge, plotted as a function of the lifting edge height z for the graphene with armchair-type free edge. The indices A-J correspond to those of Fig. 8. (b) The red-colored averaged force per one atom acting on the outermost array, and the blue-colored one acting on the second array, as a function of the lifting edge height z [Å]. The indices E, F, H and I correspond to those of Figs. 8 and 9(a).

Line-contact region After the surface contact vanishes, the line contact appears [Fig. 8F]. Here the 'line contact' is defined by the following two criteria similar to the case of the center-lifting peeling: 1) The outermost array of the free edge of the graphene sheet [Fig. 1(b)] receives the averaged repulsive interaction force per one carbon atom from the graphite surface. 2) The second array next to the outermost array [Fig. 1(b)] receives the averaged attractive interaction force per one carbon atom. As illustrated in Fig. 9(b), the edge height z = 31.3 Å satisfies the above criteria [Fig. 8F]. Here the free edge adsorbed onto the graphite surface is nearly fixed even if the left edge is lifted [Fig. 8F $\rightarrow$ Fig. 8G: z = 33.9 Å]. As a result the in-plane bending of the graphene sheet markedly decreases, and $|F_z|$ decreases to zero

[Figs. 9(a)F-G]. After that the free edge slides on the surface until the graphene sheet takes planar structure within the *y-z* plane [Fig. 8H: z = 37.9 Å], where $|F_z|$ increases again [Fig. 9(a)H].

Toward complete peeling Then the graphene sheet is continuously moved upward [Fig. 8I: z = 38.0 Å $\rightarrow$ Fig. 8J: z = 38.4 Å], and is completely peeled from the surface. It is noted the line contact clearly vanishes at z = 38.0 Å [Fig. 9(b)I]. The attractive interaction force $|F_z|$ increases to take the maximum value [Figs. 9(a)I$\rightarrow$J], and then it gradually decreases to zero toward the complete peeling.

4.2 Atomic-scale sliding within lateral plane

Fig. 9(a) shows the atomic-scale zigzag structures within the surface- and line-contact regions, which can be explained by the following atomic-scale sliding motions of the graphene sheet within the *x-y* plane.

Surface-contact region During the surface contact region between C and E in Fig. 9(a), z-F_z curve takes the atomic-scale zigzag structures from I to VII. The zigzag behaviors exhibit the transition from the continuous [Figs. 10(a)I - II] to the sawtooth shapes [Figs. 10(a)III - VII]. First Figs. 10(a)1$\rightarrow$2$\rightarrow$3$\rightarrow$4$\rightarrow$5 correspond to Figs. 10(b) 1$\rightarrow$2$\rightarrow$3$\rightarrow$4$\rightarrow$5, which show that the graphene sheet continuously slides passing over the nearest neighboring AB stacking sites with the graphite surface. The trajectories of the graphene sheet exhibit the continuous zigzag paths as shown in Fig. 10(b)5. Next Figs. 10(a) 6$\rightarrow$7$\rightarrow$8$\rightarrow$9$\rightarrow$10 correspond to Figs. 10(c) 6$\rightarrow$7$\rightarrow$8$\rightarrow$9$\rightarrow$10 which show that the graphene sheet takes the zigzag stick-slip motions between the nearest neighboring AB stacking sites. Just before the slip, the graphene sheet deviates quite a little from the AB-stacking site [Fig. 10(c)6: z = 17.3 Å]. Then it discretely jumps or slips to the neighboring AB-stacking site [Fig. 10(c)7: z = 17.4 Å]. As the peeling proceeds, the graphene sheet continuously slides quite a little [Fig. 10(c)8: z = 18.2 Å], then it discretely slips again to the neighboring AB stacking site [Fig. 10(c)9: z = 18.3 Å]. After that the graphene sheet continuously slides quite a little again [Fig. 10(c)10: z = 20.1 Å] until the next slip toward the neighboring AB stacking site occurs. As a result the trajectories of the graphene sheet exhibit the discrete zigzag paths connecting the nearest neighboring AB-stacking sites as shown in Fig. 10(c)10. The period of the zigzag behavior of the F_z curve decreases from 3.7Å to 2.5 Å as shown in Fig. 10(a) as the peeling proceeds. The lattice spacing of the graphite surface, 2.5 Å, appears in the peeling force curve particularly for the stick-slip region.

Line-contact region During the line contact region between G and H in Fig. 9(a), z-F_z curve takes another atomic-scale zigzag structures as shown in Fig. 11(a). One of the zigzag behaviors in the force curve [Figs. 11(a) 1$\rightarrow$2$\rightarrow$3] corresponds to the stick-slip sliding motions of the graphene sheet [Figs. 11(b) 1$\rightarrow$2$\rightarrow$3]. Here the free edge of the graphene sheet slides with nearly the straight stick-slip motions. One of the carbon atoms on the free edge passes over the carbon-carbon bonds as shown in Figs. 11(b)1 and 3.

5. Edge-lifting case of zigzag-edge graphene

Recently it has been reported that the edge structure of the graphene sheet plays quite an important role in electronic, magnetic and optical properties of graphene, which can be also expected to give influences on the mechanical properties such as the peeling process. Therefore, in this section, the peeling process of the graphene sheet with zigzag edge is

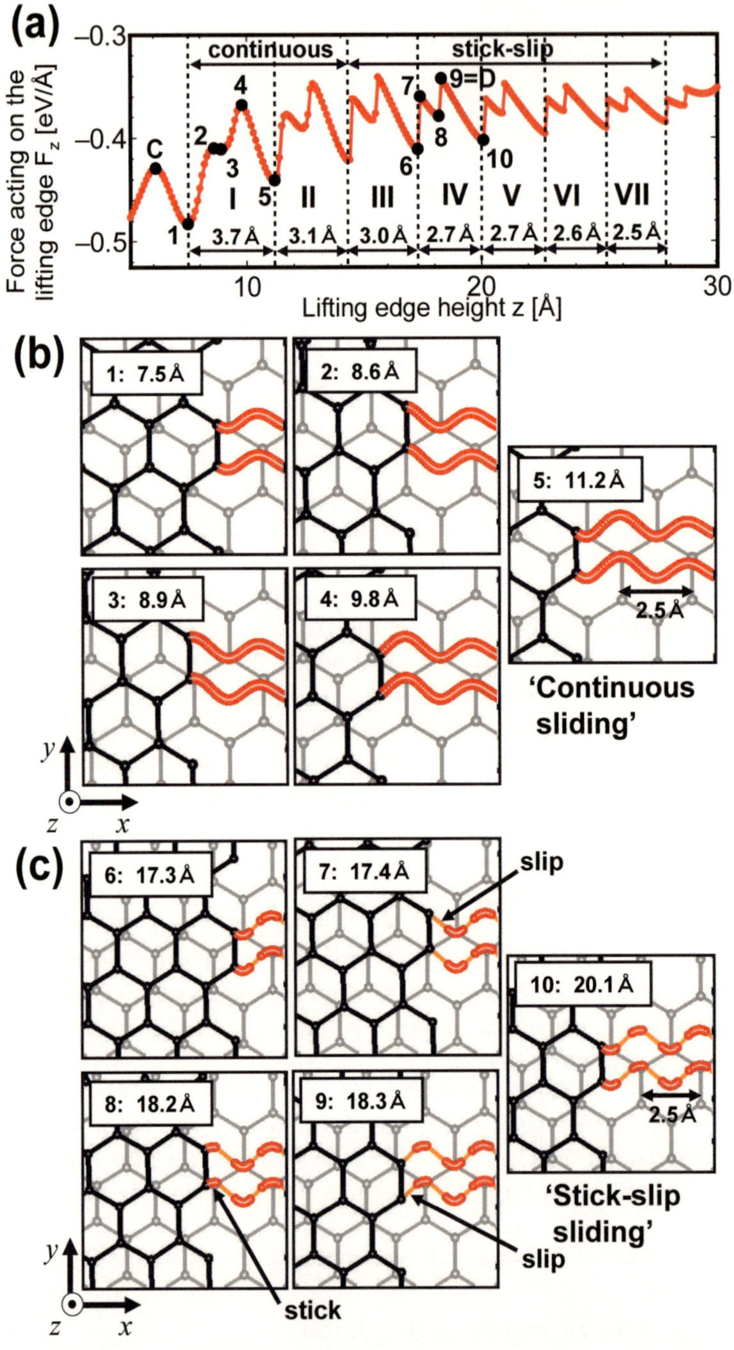

Fig. 10. (a) Enlargement of part of the z - F_z curve (Fig. 9(a)) corresponding to the continuous and stick-slip process during the surface contact. (b) The trajectories of the two carbon atoms on the free edge from 1 to 5 indicated in (a). (c) The trajectories of the two carbon atoms on the free edge from 6 to 10 indicated in (a).

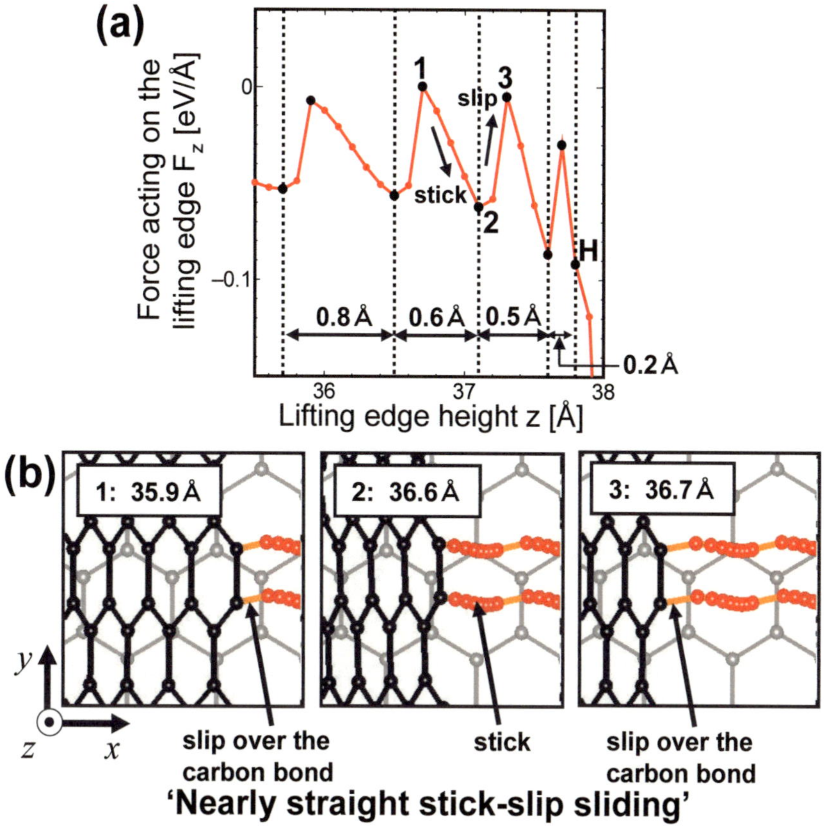

Fig. 11. (a) Enlargement of part of the z - F_z curve (Fig. 9(a)) corresponding to the nearly straight stick-slip region during the line contact. (b) The trajectories of the two carbon atoms on the free edge from 1 to 3 indicated in (a).

discussed. In the simulation, the model obtained by rotating Fig. 1(b) by 30° is used [Fig. 1(c)], and the left zigzag edge is lifted to simulate the peeling process, while the right free edge is zigzag type. As a result, the nanoscale peeling process within the x-z plane and the global shape of the force curve [Fig. 12] is similar to Figs. 8 and 9, respectively. The qualitative tendency of the decrease of the period and amplitude of the force curve [Fig. 12I-IV] is similar to that for Fig. 9(a)I-VII. However the details of the atomic-scale mechanics of the zigzag edge are clearly different from those of the armchair edge as follows:

During the surface contact, the graphene sheet first takes zigzag [Figs. 13(b)1-6] and then straight stick-slip motions [Figs. 13(c)7-11], passing over the nearest neighboring AB-stacking site along $[10\bar{1}0]$ direction. It is noted, to avoid AA-stacking registry, the graphene sheet takes zigzag slip toward the nearest neighboring AB-stacking site as shown in Fig. 13(b)1$\rightarrow$2, although it then takes straight slip as shown in Fig. 13(c)7$\rightarrow$8. The minimum period of the force curve of 4.4 Å [Fig. 13(a)IV] reflects the lattice period of the graphite surface along the $[10\bar{1}0]$ direction, while 2.5 Å for the armchair-type edge [Fig. 10(a)VII] reflects that along the $[12\bar{3}0]$ direction. Thus the edge structure gives the marked effects on the atomic-scale dynamics depending on the lattice orientation of the surface.

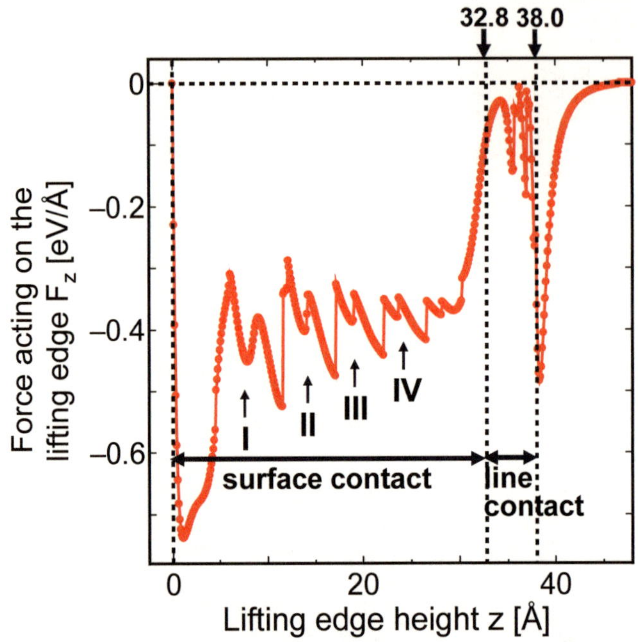

Fig. 12. The vertical force F_z acting on the lifting edge, plotted as a function of the lifting edge height z for the graphene with zigzag-type free edge

During the line contact, the difference between the armchair- and zigzag-type edge is enhanced. Fig. 14(a) reflects the zigzag stick-slip motion of the graphene sheet [Figs. 14(b)1-9] unlike nearly the straight stick-slip motion [Figs. 10(b)1-3]. Important point of the line-contact sliding is that each carbon atom on the free edge takes stick-slip motion between the nearest neighboring six-membered rings. When each atom is located on the hollow site of the six-membered ring, the graphene sheet does not deform along the y direction [Figs. 14(b)1, 4-5, and 8-9]. However, when each atom is located a little far from the hollow site or near the carbon bond, the graphene sheet bends toward the y direction to decrease the total interaction energy [Figs. 14(b)2-3 and 6-7]. Thus, in the case of the zigzag-type edge, collective motion of the single carbon 'atom' on the free edge nearly dominates the graphene mechanics together with its deformation. On the other hand, for the armchair-type edge, collective motion of the single carbon 'bond' is dominant.

6. Discussions and conclusions

In this chapter molecular mechanics study of the nanoscale peeling of the monolayer graphene sheet has been performed. The peeling force curve clearly exhibits the change of the graphene shape from the surface- to the line-contact.

In Section 3, the peeling of the monolayer graphene sheet with the armchair edge for lifting the center position is discussed. It is noted that the maximum lateral sliding force $F_x \approx$ 0.1 eV/Å is only about 3 % of the absolute value of the maximum adhesion force $|F_z| \approx$ 3.1 eV/Å. This small sliding force F_x is derived from the superlubricity at the interface between the graphene sheet and the graphite surface (Sasaki et al., 2007) and atomic-scale wear (Sasaki et al., 2009c). There is possibility that such anisotropy between the vertical

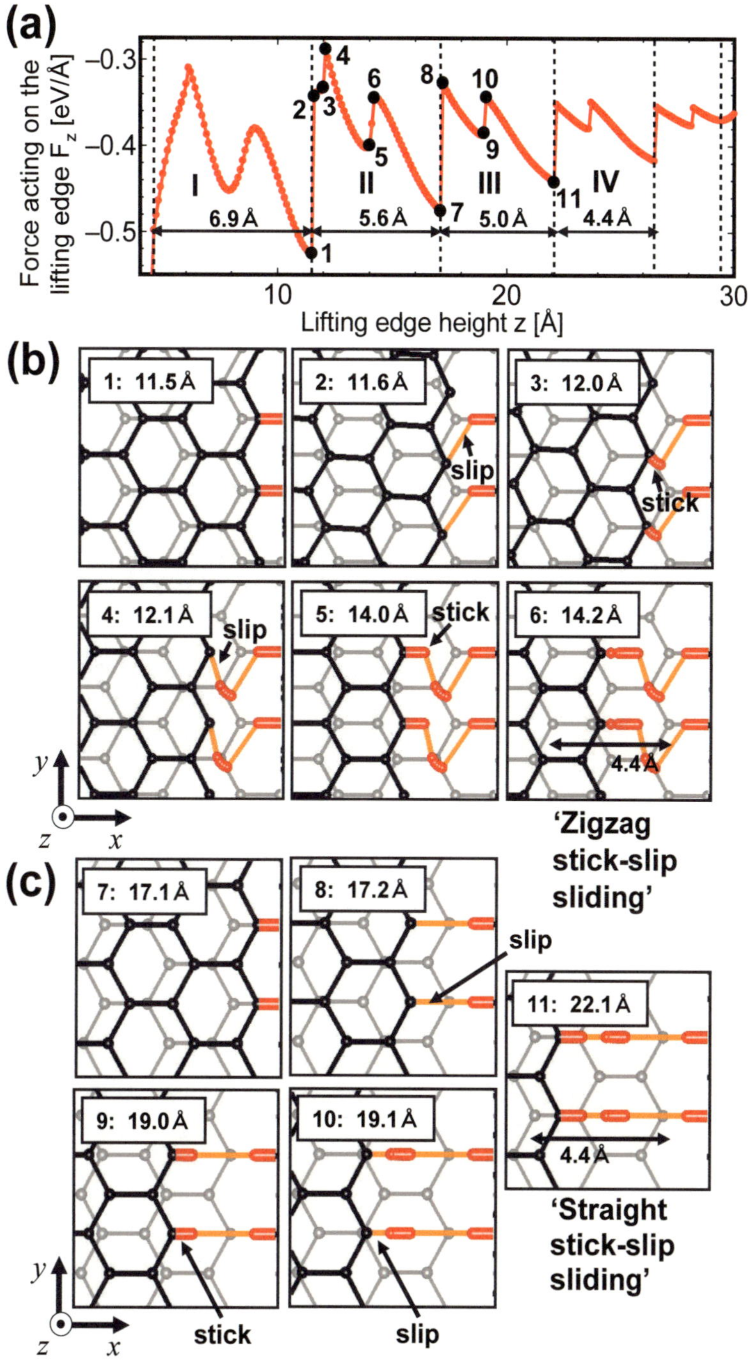

Fig. 13. (a) Enlargement of part of the z - F_z curve (Fig. 12(b)) corresponding to the zigzag and straight stick-slip process during the surface contact. (b) The trajectories of the two carbon atoms on the free edge from 1 to 6 indicated in (a). (c) The trajectories of the two carbon atoms on the free edge from 7 to 11 indicated in (a).

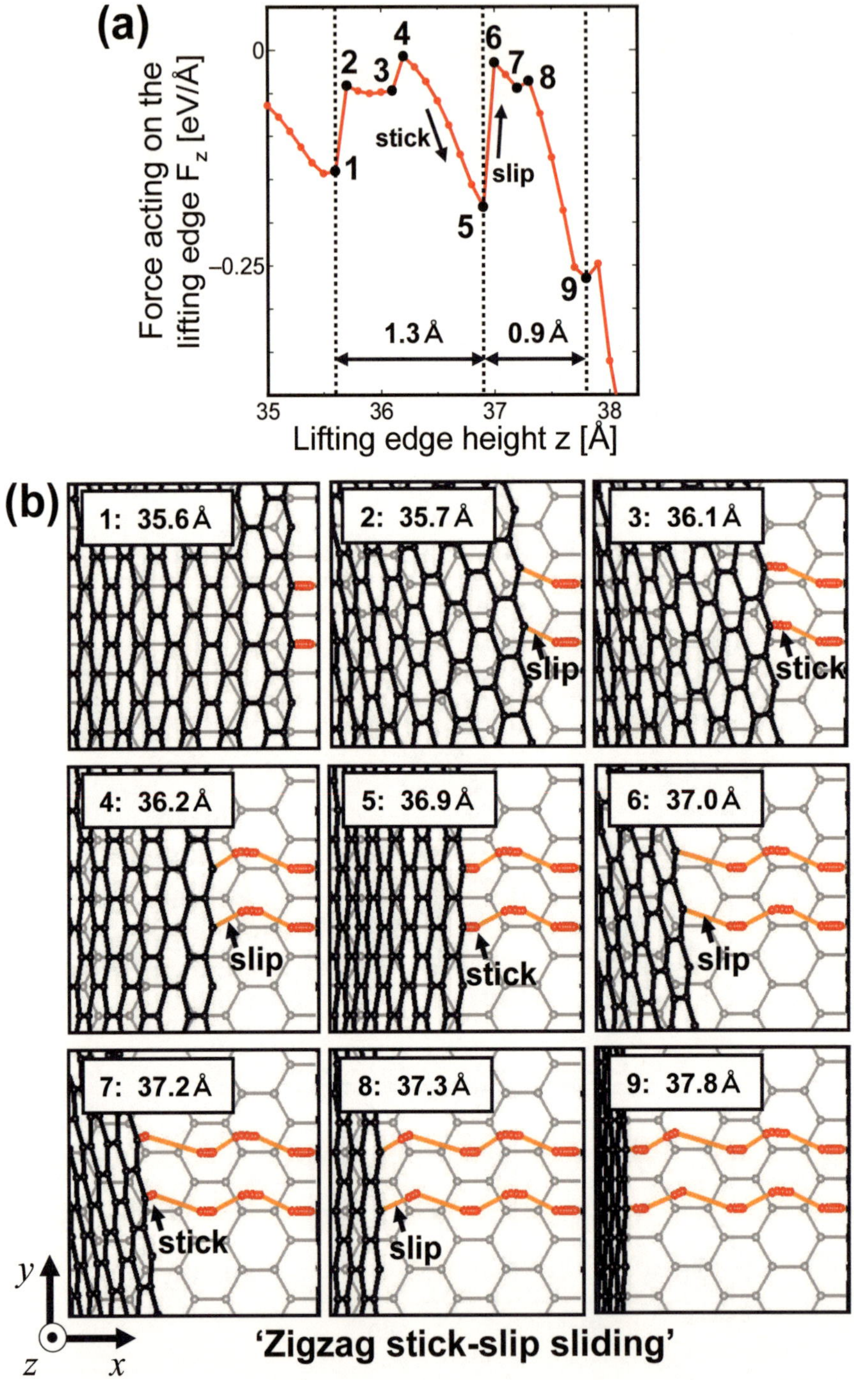

Fig. 14. (a) Enlargement of part of the z - F_z curve (Fig. 12(b)) corresponding to the zigzag stick-slip process during the line contact. (b) The trajectories of the two carbon atoms on the free edge from 1 to 9 indicated in (a).

force F_z and the lateral sliding force F_x can be applied to the adhesives, which can be strongly adhered to the substrate but can easily slide on it. Our AFM measurement exhibits that the maximum pull-off force is about several hundreds of nN, which is clearly much larger than the binding force, 3 eV/Å≈ 4.8 nN, assumed in our simulation.

In Secs. 4 and 5, the peeling of the monolayer graphene sheet with the armchair- and zigzag-edge for lifting the edge position is discussed, respectively. The atomic-scale sliding motion of the monolayer graphene sheet during the peeling process is found. For the graphene sheet with armchair edge, the transition from the continuous to the stick-slip motion of the graphene sheet is found, which can be explained as follows: The peeling process induces the increase of the peeled area of the graphene sheet, and the decrease of the surface contact area. Considering the peeled area of the graphene sheet acts as an effective spring as shown in Fig. 15, the increase of the peeled area makes the effective spring softer, and the decrease of the surface contact area decreases the energy barrier to slide the graphene sheet. Finally the peeling process induces the transition from the continuous to the stick-slip sliding motion of the graphene sheet, together with the decrease of the period and amplitude of the z-F_z curve. Important point is that the period of the peeling force curve for the armchair-edge graphene for the surface contact region corresponds to the lattice spacing of the graphite surface along $[12\bar{3}0]$ direction, 2.5 Å. On the other hand, for the zigzag-edge graphene, the period becomes the lattice spacing along $[10\bar{1}0]$ direction, 4.4 Å. This means the sliding length of the graphene sheet along x direction becomes nearly equal to the peeled length along z direction. The zigzag structures of the peeling force curve with the same period of about several Å have been also observed by our preliminary experiments using the multilayered graphene, which will be reported elsewhere. Of course, if the number of the peeled graphene sheets is reduced, the direct comparison between the present simulation and the experiment will become possible.

As a result, the center-lifting case requires the largest attractive peeling force, -3.1 eV/ Å, in order to peel the graphene sheet as shown in Fig. 16(a). On the other hand, the edge-lifting case requires only -0.74 eV/Å, about 20 % of that for the center-lifting case as shown in Figs. 16(b) and 16(c). The edge structures give little influences on the basic features of the force curve. However, the sliding direction and the edge structure clearly gives marked influences on the surface- and line-contact regions, respectively.

Another important point is that the behavior of the lateral force curve $F_x(z)$ is qualitatively the same as that of the vertical force curve $F_z(z)$ during the surface contact as shown in Fig. 9(b). Therefore it can be said that the peeling force curve, $F_z(z)$, directly reflects the atomic-scale friction force, $F_x(z)$, which decreases to 0.019 eV/Å≈ 30 pN for z = 27.8 Å [Fig. 9(b)]. This ultralow friction force F_x is derived from the superlubricity at the interface between the graphene sheet and the graphite surface (Sasaki et al., 2007, 2009c, Itamura 2009a, 2009b) . Furthermore effect of the edge structure on the peeling process is clarified by comparison of the free edge between the armchair- and zigzag-types. As mentioned above, the atomic-scale structure of the force curve during the surface contact reflects the lattice spacing of the graphite surface. So the minimum period of the atomic-scale structure of the force curve can tell us the atomic-scale lattice orientation and structure of the free edge of graphene. Such information can be used for the control of the electronic properties of the graphene sheet adsorbed onto the substrate. Therefore this paper indicates the possibility of the identification of the lattice orientation and the edge structure of the graphene sheet.

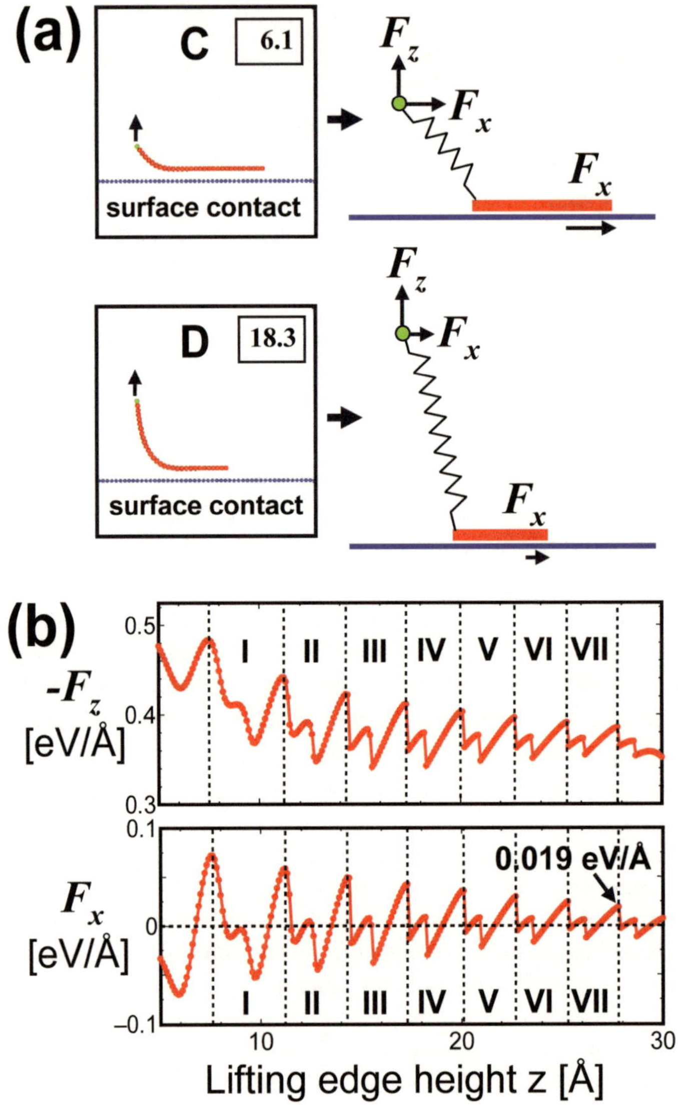

Fig. 15. (a) Schematic illustration of the increase of the peeled area and the decrease of the surface contact area from C (z = 6.1 Å) to D (z = 18.3 Å) for the graphene sheet with armchair-type free edge. (b) – F_z and F_x plotted as a function of the edge height z, show qualitatively the same behavior to each other for the graphene sheet with armchair-type free edge.

In this chapter, we discussed the importance of the dynamics of the free edge during the peeling process. On the other hand, we also found the importance of the shape of the graphene sheet. Additional simulated model and results are shown in Fig. 17. As shown in Fig. 17(a), rectangular graphene sheet whose aspect ratio is different from that of the graphene

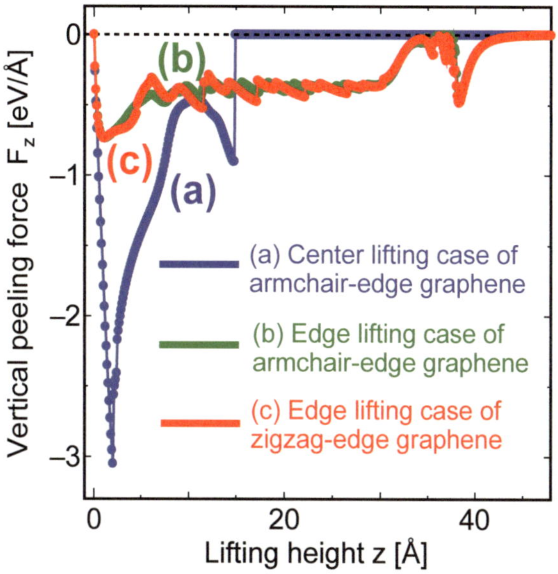

Fig. 16. Comparison of the vertical forces, F_z, among (a) center-lifting case of the armchair-edge graphene sheet, (b) edge-lifting case of the armchair-edge graphene sheet, and (c) edge-lifting case of the zigzag-edge graphene sheet.

sheet of Fig. 1(b) is used. The basic shape of the vertical force curve, Fig. 17(b), is similar to that of Fig. 3. However the armchair type edge is peeled first for Fig. 17(a), although the zigzag type edge is peeled first for Fig. 1(b) as discussed in Section 3. This means that the shape of the graphene sheet plays an important role for deciding which edge is peeled first. Effect of the graphene shape on the peeling process will be discussed in detail somewhere in the near future.

Lastly it should be noted that the peeling process discussed in this chapter is closely related to the atomic-scale wear of the graphite and the graphene tip formation in the friction force microscopy (Sasaki et al., 2009c). When the tip is pushed onto the surface for less than the critical tip height, the outermost graphene layer is attached to the FFM tip, which results in the formation of the graphene tip. In that case, the graphene sheet takes the surface contact with the second layer graphene, and it takes the two-dimensional stick-slip motion. However it is difficult to observe directly the stick-slip motion during the scan process, due to the very small gap between the FFM tip and the graphite surface. On the other hand, if the peeling process is used, it can be expected that the contact at the AFM tip/graphite interface has a wider space to be observed directly by ex. Transmission Electron Microscopy (TEM). This paper indicates the possibility of a direct observation of the stick-slip motion of the graphene sheet, that's to say, the elementary process of the atomic-scale friction or superlubricity which occurs at the tip/graphite surface interface.

This research was supported by a Grant-in-Aid for Scientific Research (B) (No. 20360022) and for Specially Promoted Research (No. 21000008) from the Japan Society for the Promotion of Science. This work was also supported by the Ministry of Education, Culture, Sports, Science and Technology through a Grant-in-Aid for Building Strategic Research Infrastructures.

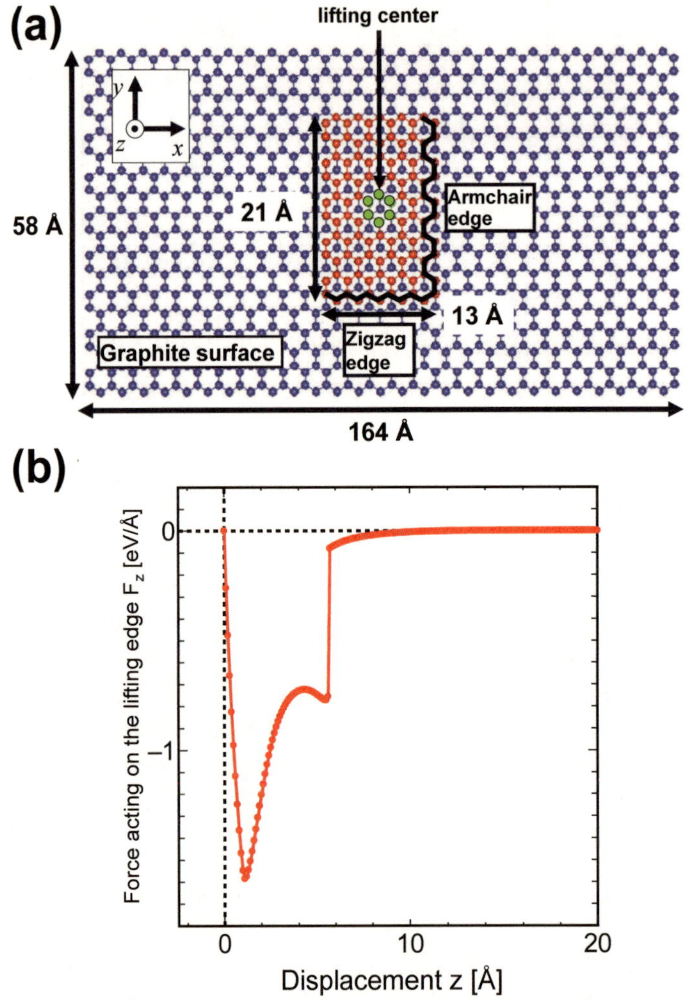

Fig. 17. (a) The model of the red-colored monolayer graphene sheet physically adsorbed onto the blue-colored rigid graphite surface within the $x - y$ plane. The green-colored six-membered ring at the center position is moved upward along the z (or [0001]) direction, by z = 0.1 Å. Initial AB stacking registry of the red-colored graphene sheet with the blue-colored graphite surface is assumed. (b) The vertical force F_z acting on the center six-membered ring, plotted as a function of the lifting displacement z.

7. References

Autumn, K., Liang, Y. A., Hsieh, S. T., Zesch, W., Chan, W. P., Kenny, T. W., Fearing, R., & Full, R. J. (2000). Adhesive force of a single gecko foot-seta. *Nature*, Vol. 405, No. 6 (June, 2000) 681-685, ISSN 0028-0836 (print), 1476-4687 (online)

Autumn, K., Sitti, M., Liang, Y. A., Peattie, A. M., Hansen, W. R., Sponberg, S., Kenny, T. W., Fearing, R., Israelachvili, J. N. & Full, R. J. (2002). Evidence for van der Waals adhesion in gecko setae. *Proc. Natl. Acad. Sci. U.S.A.*, Vol. 99, No. 19 (September, 2002) 12252-12256

Geim, A. K. and Novoselov, K. S. (2007). The rise of graphene. *Nature Mat.*, Vol. 6, No. 3 (March, 2007) 183-191, ISSN 1476-1122 (print), 1476-4660 (online)

Ishikawa, M., Kato, M., Harada, R., Sasaki, N. & Miura, K. (2008). Visualization of nanoscale peeling of carbon nanotube on graphite. *Appl. Phys. Lett.*, Vol. 93, No. 8 (August, 2008) 0831221-0831223, ISSN 0003-6951 (print), 1077-3118 (online)

Ishikawa, M., Harada, R., Sasaki, N. & Miura, K. (2009). Adhesion and peeling forces of carbon nanotubes on a substrate. *Phys. Rev. B.*, Vol. 80, No. 19 (November, 2009) 1934061-1934064, ISSN 1098-0121 (print), 1550-235x (online)

Itamura, N., Miura, K., & Sasaki, N., (2009a). Analysis of Mechanism of Low Lateral Stiffness of Superlubric C_{60} Bearing System. *Jpn. J. of Appl. Phys.*, Vol. 48, (March, 2009) 0302141-0302143, ISSN 0021-4922 (print), 1347-4065 (online)

Itamura, N., Miura, K., & Sasaki, N., (2009b). Simulation of Scan-Directional Dependence of Superlubricity of C_{60} Molecular Bearings and Graphite. *Jpn. J. of Appl. Phys.*, Vol. 48, (June, 2009) 0602071-0602073, ISSN 0021-4922 (print), 1347-4065 (online)

Lu, J. P., Li, X. -P. & Martin, R. M. (1988). Ground state and phase transitions in solid C_{60}. *Phys. Rev. Lett.*, Vol. 68, No. 10 (March, 1992) 1551-1554, ISSN 0031-9007 (print), 1079-7114 (online)

Novoselov, K. S., Geim, A. K., Morozov, S. V., Jiang, D., Zhang, Y., Dubonos, S. V., Grigorieva, I. V. & Firsov, A. A. (2004). Electric Field Effect in Atomically Thin Carbon Films. *Science*, Vol. 306, No. 5696 (October, 2004) 666-669, ISSN 0036-8075 (print), 1095-9203 (online)

Novoselov, K. S., Geim, A. K., Morozov, S. V., Jiang, D., Katsnelson, M. I., Grigorieva, I. V., Dubonos, S. V. & Firsov A. A. (2005). Two-dimensional gas of massless Dirac fermions in graphene. *Nature*, Vol. 438, No. 11 (November, 2005) 197-200, ISSN 0028-0836 (print), 1476-4687 (online)

Press, W. H., Teukolsky, S. A., Vetterling, W. T. & Flannery, B. P. (1999). Sec. 10.6 Minimization or Maximization of Functions, In: *Numerical Recipes in FORTRAN : The Art of Scientific Computing, 2nd ed.*, 413-418, Cambridge Univ. Press, ISBN 0-521-43064-X, New York

Qu, L., Dai, L., Stone, M., Xia, Z. & Wang, Z. L. (2008). Carbon Nanotube Arrays with Strong Shear Binding-On and Easy Normal Lifting-Off. *Science*, Vol. 322, No. 5899 (October, 2008) 238-242, ISSN 0036-8075 (print), 1095-9203 (online)

Sasaki, N., Toyoda, A., Saitoh, H., Itamura, N., Ohyama, M. & Miura, K., (2006). Theoretical Simulation of Atomic-Scale Peeling of Single-Walled Carbon Nanotube from Graphite Surface. *e-J. Surf. Sci. Nanotech.*, Vol. 4, (January, 2006) 133-137, ISSN 1348-0391

Sasaki, N., Itamura, N. & Miura, K. (2007). Simulation of Atomic-Scale Ultralow Friction of Graphite/C_{60}/Graphite Interface along [1010] Direction. *Jpn. J. of Appl. Phys.*, Vol. 46, No. , (December, 2007) L1237-L1239, ISSN 0021-4922 (print), 1347-4065 (online)

Sasaki, N., Toyoda, A., Itamura, N. & Miura, K., (2008). Simulation of Nanoscale Peeling and Adhesion of Single-Walled Carbon Nanotube on Graphite Surface. *e-J. Surf. Sci. Nanotech.*, Vol. 6, (February, 2008) 72-78, ISSN 1348-0391

Sasaki, N., Okamoto, H., Itamura, N. & Miura, K., (2009a). Peeling of Graphene Sheet - Simulation Study. *e-J. Surf. Sci. Nanotech.*, Vol. 7, (July, 2009) 783-786, ISSN 1348-0391

Sasaki, N., Saitoh, H., Itamura, N. & Miura, K., (2009b). Analysis of Lateral Orientation of Single-Walled Carbon Nanotube on Graphite. *e-J. Surf. Sci. Nanotech.*, Vol. 7, (January, 2009) 48-52, ISSN 1348-0391

Sasaki, N., Saitoh, H., Terada, K., Itamura, N. & Miura, K., (2009c). Simulation of Atomic-Scale Wear of Graphite - Nanotip Induced Graphene Formation. *e-J. Surf. Sci. Nanotech.*, Vol. 7, (March, 2009) 173-180, ISSN 1348-0391

Sasaki, N., Okamoto, H., Itamura, N. & Miura, K., (2010). Atomic-Scale Friction of Monolayer Graphenes with Armchair- and Zigzag-Type Edges During Peeling Process. *e-J. Surf. Sci. Nanotech.*, Vol. 8, (March, 2010) 105-111, ISSN 1348-0391

Stoddard, S. D. and Ford, J. (1973). Numerical Experiments on the Stochastic Behavior of a Lennard-Jones Gas System. *Phys. Rev. A*, Vol. 8, No. 3 (September, 1973) 1504-1512, ISSN 1050-2947 (print), 1094-1622 (online)

Tersoff, J., (1988). Empirical Interatomic Potential for Carbon, with Applications to Amorphous Carbon. *Phys. Rev. Lett.*, Vol. 61, No. 25 (December, 1988) 2879-2882, ISSN 0031-9007 (print), 1079-7114 (online)

10

Magnetoexciton Binding Energy in Graphene Bilayers

Zlatko Koinov
The University of Texas at San Antonio
San Antonio, Texas, USA

1. Introduction

A lot of experimental and theoretical studies in recent years are focusing on the unusual relativistically, kinematic properties of the electronic states in graphene predicted theoretically decades ago (Haldane, 1988; Semenoff, 1984). Later, it was confirmed that the graphitic monolayer have anomalous relativistically properties (Novoselov et al., 2005; 2004). Because electrons and holes in a graphene behave like massless Dirac particles, there is a number of unusual properties, such as high charge carrier mobility (Novoselov et al., 2005), the graphene's conductivity never falls below a minimum value (Nomura & MacDonald, 2007; Ziegler, 2006), and an anomalous quantum Hall effect.(Zhang et al., 2005)
Bilayer graphene systems, where carriers in one layer are electrons and carriers in the other are holes, have been considered as ideal candidates for observing superfluid properties at room temperatures (Lozovik & Sokolik, 2008; Min et al., 2008; Zhang & Jorlecar, 2008). It is expected that the excitons will behave as neutral bosons at low densities, and therefore, they can undergo Bose-Einstein condensation (BEC) when the interlayer distance is comparable to the distance between the particles within each layer. However, when we separate electrons and holes by introducing a dielectric between them we reduce the exciton binding energy, and so the critical temperature for condensation decreases. A possible way to increase the binding energy is to apply magnetic field perpendicular to the layers. As we shall see, the calculations predict the existence of a condensate of magnetoexcitons with superfluid properties under the Kosterlitz-Thouless critical temperature T_{KT} which in graphene bilayers decreases in the limit of large interlayer separation as $T_{KT} \sim B^{-2}$ (as $T_{KT} \sim B^{-1/2}$ in the limit of small interlayer separation).
Magnetoexcitons are bound states between two charged fermions (an electron from the conductive band and a hole from the valence band) in the presence of a magnetic field. The calculations are much more complicated compare to the corresponding calculations in the absence of a magnetic field because even a small transverse exciton velocity (or small transverse wave vector $\mathbf{Q}$) will induce an electric field in the rest frame of the exciton. This electric field will push the electron and the hole apart, so the magnetoexciton binding energy must decrease as the transverse velocity increases. In other words, the magnetic field induces a coupling between the center-of-mass and the relative internal motions, and therefore, the correct description of excitons in a strong magnetic field should take into account this coupling effect.

Turning our attention to magnetoexciton dispersion in non-relativistic systems, such as coupled quantum wells (CQW's) with parabolic dispersions ($E_{c,v} = \hbar^2 k^2/2m_{c,v}$) we find that the following Hamiltonian (Shevchenko, 1997)

$$\widehat{H} = -\frac{\hbar^2}{2\mu}\nabla_{\mathbf{r}}^2 + \frac{\imath e\gamma\hbar}{2\mu c}(\mathbf{B}\times\mathbf{r}).\nabla_{\mathbf{r}} + \frac{e^2B^2}{8\mu c^2}\mathbf{r}^2 - V(\mathbf{r}+\mathbf{R}_0)$$

is used to obtain the magnetoexciton dispersion. Here μ is the exciton reduced mass, $\gamma = (m_v - m_c)/(m_c + m_v)$, $\mathbf{R}_0 = R^2\mathbf{Q}_0$, where $\mathbf{Q}_0 = (-Q_y, Q_x, 0)$, and $R = (\hbar c/eB)^{1/2}$ is the magnetic length. $V(\mathbf{r}) = e^2/(\varepsilon_0\sqrt{|\mathbf{r}|^2 + d^2})$ represents the electron-hole Coulomb attraction screened by the dielectric constant ϵ_0. Since the Coulomb term in the Hamiltonian is the only term which depends on the exciton momentum $\mathbf{Q} = (Q_x, Q_y, 0)$, the magnetoexciton dispersion does not depend on the electron and hole masses and the magnetoexciton mass is determined only by Coulomb interaction. In strong magnetic fields one can apply the lowest Landau level (LLL) approximation. In the LLL approximation the magnetoexciton mass M_{CQW} and the binding energy E_{CQW} are as follows:

$$\frac{M_{CQW}}{M_{2D}} = \left[(1+\frac{d^2}{R^2})e^{\left(\frac{d^2}{2R^2}\right)}Erfc\left(\frac{d}{\sqrt{2}R}\right) - \sqrt{\frac{2}{\pi}}\frac{d}{R}\right]^{-1}$$

$$E_{CQW} = E_b \exp\left(\frac{d^2}{2R^2}\right) Erfc\left(\frac{d}{\sqrt{2}R}\right).$$

Here $Erfc(x)$ is the complementary error function, $M_{2D} = 2^{3/2}\varepsilon_0\hbar^2/(\sqrt{\pi}e^2R)$ and $E_b = \sqrt{\pi}e^2/(\sqrt{2}\varepsilon_0 R)$ are the magnetoexciton mass and the two-dimensional ($d = 2$) magnetoexciton binding energy, respectively.

Strictly speaking, the excitons are bound states between two charged fermions, and therefore, the appropriate framework for the description of the bound states is the Bethe-Salpeter (BS) formalism (Salpeter & Bethe, 1951). In the case of parabolic band quantum-well structures we find that beyond the LLL approximation, the BS equation contains an extra term (BS term) (Koinov, 2008). This term takes into account the transitions to the Landau levels with indexes $n \geq 1$. The contributions to the magnetoexciton binding energy and mass can be obtained by applying a variational procedure. In the non-relativistic case the results are as follows: in a strong magnetic field, the ground-state energy is very close to that obtained by means of the Schrödinger equation, but the magnetoexciton dispersion is determined by the BS term rather than the electron-hole Coulomb term in the Schrödinger equation.

Since the unique electronic behaviors of graphene is a result of the unusual quantum-relativistic characteristics of the so-called Dirac fermions, we shall study magnetoexciton binding energy in graphene bilayers embedded in a dielectric by applying the relativistic BS equation in the LLL approximation (Koinov, 2009). Several non-trivial effects produced by magnetic fields have been recently predicted in quantum field theories. For example, in the massless QED, the analysis based on the BS equation has predicted that the external constant magnetic field generates an energy gap (dynamical mass) in the spectrum of massless fermions for any arbitrary weak attractive interaction between fermions (Gusynin et al., 1994; 1995). It is expected that the effect is model independent (universal), because the physical reason of this effect lies in the dimensional reduction in the dynamics of fermion pairing in the presence of a the constant magnetic field.

In what follows we examine how both the magnetoexciton binding energy and magnetoexciton mass in graphene bilayer systems vary with the magnetic field and the separation d between the layers in the LLL approximation. The solution of this problem is related to the continued activities to observe superfluid properties of excitons in a bilayer graphene. It is expected that the excitons will behave as neutral bosons at low densities, and therefore, they can undergo Bose-Einstein condensation when the interlayer distance is comparable to the distance between the particles within each layer. The condensate of neutral bosons (excitons) should have superfluid properties under the Kosterlitz-Thouless critical temperature T_{KT}. It is worth mentioning that the calculations done by treating the Coulomb interaction as a perturbation (Berman et al., 2008) provide in the LLL approximation *a number of extra terms which do not exist in the case of CQW's.* From a general point of view, we have to expect that the binding energy is *exactly* four times higher than E_{CQW}, while the magnetoexciton mass is *exactly* four times lower than M_{CQW}. The physical reason for the above statement lies in the fact that in the LLL approximation we have a dimensional reduction in the dynamics of the electron-hole pairing from two space variables plus a time variable to zero space variable and a time variable. Because of this $2+1 \to 0+1$ reduction the results should be insensitive to the type of the band dispersion. The factor four is due to the four-component-spinor description used in the relativistic case.

2. Bethe-Salpeter equation

The system under consideration is made from two graphene sheets embedded in a dielectric and separated by distance d. Each of the two graphene layers has two Dirac-like linear dispersion $\hbar v_F k$ bands centered at two non-equivalent points $\mathbf{K}$ and $\mathbf{K}\prime$, where v_F is the Fermi velocity of electrons in graphene. Since the layers are embedded in a dielectric, there is no hopping of π-electrons between the layers. There is a potential difference $\pm V_g/2$ (gate voltage) applied to each of the two layers which allows us to adjust the charge density in the layers. We assume that the potential difference is chosen in a manner that the electrons are in the top layer (pseudospin index $\tau = 1$) and the same number of holes in the bottom layer ($\tau = 2$).

The unit cell of graphene has two atoms, A and B, each belonging to the different sublattice. The operator $\psi^{(\tau)\dagger}_{\sigma,A,\alpha}(\mathbf{r})$ ($\psi^{(\tau)\dagger}_{\sigma,B,\alpha}(\mathbf{r})$) creates an electron of spin $\sigma = \uparrow, \downarrow$ on the atom A (atom B) of the unit cell in layer τ defined by the position vector $\mathbf{r}$. We introduce four component spinors:

$$\Psi^{(\tau)}_{\sigma}(\mathbf{r}) = \begin{pmatrix} \psi^{(\tau)}_{\sigma,A,\mathbf{K}}(\mathbf{r}) \\ \psi^{(\tau)}_{\sigma,B,\mathbf{K}}(\mathbf{r}) \\ \psi^{(\tau)}_{\sigma,B,\mathbf{K}'}(\mathbf{r}) \\ \psi^{(\tau)}_{\sigma,A,\mathbf{K}'}(\mathbf{r}) \end{pmatrix}, \qquad (1)$$

$$\overline{\Psi}^{(\tau)}_{\sigma}(\mathbf{r}) = \Psi^{(\tau)\dagger}_{\sigma}(\mathbf{r})\gamma^0,$$

where the following representation of the Dirac matrices is chosen:

$$\gamma^0 = \begin{pmatrix} 1 & 0 & 0 & 0 \\ 0 & -1 & 0 & 0 \\ 0 & 0 & -1 & 0 \\ 0 & 0 & 0 & 1 \end{pmatrix}, \quad \gamma^1 = \begin{pmatrix} 0 & 1 & 0 & 0 \\ -1 & 0 & 0 & 0 \\ 0 & 0 & 0 & -1 \\ 0 & 0 & 1 & 0 \end{pmatrix},$$
$$\gamma^2 = \begin{pmatrix} 0 & -\imath & 0 & 0 \\ -\imath & 0 & 0 & 0 \\ 0 & 0 & 0 & \imath \\ 0 & 0 & \imath & 0 \end{pmatrix}. \tag{2}$$

In continuum approximation the non-interacting quasiparticles in the layers are described by the Hamiltonian:

$$H_0 = \sum_{\sigma,\tau} \int d^2\mathbf{r}\overline{\Psi}_\sigma^{(\tau)}(\mathbf{r})\widehat{H}^{(\tau)}\Psi_\sigma^{(\tau)}(\mathbf{r}), \tag{3}$$

where

$$\begin{aligned} \widehat{H}^{(\tau)} &= v_F\left(\gamma^1\widehat{p}_x + \gamma^2\widehat{p}_y\right), \\ \widehat{p}_x &= -\imath\hbar\frac{\partial}{\partial x}, \\ \widehat{p}_y &= -\imath\hbar\frac{\partial}{\partial y}. \end{aligned}$$

The action that describes the non-interacting quasiparticles in a layer τ is:

$$S_0^{(\tau)} = \int d^2\mathbf{r}dt\overline{\Psi}_\sigma^{(\tau)}(\mathbf{r},t)\left[\gamma^0\imath\hbar\frac{\partial}{\partial t} - v_F\left(\gamma^1\widehat{p}_x + \gamma^2\widehat{p}_y\right)\right]\Psi_\sigma^{(\tau)}(\mathbf{r},t) \tag{4}$$

In the presence of a perpendicular magnetic field $\mathbf{B} = (0,0,B)$ and a potential difference $\pm V_g/2$ (gate voltage) applied to each of the two layers, the action (4) assumes the form:

$$S_0^{(\tau)} = \int d^2\mathbf{r}dt\overline{\Psi}_\sigma^{(\tau)}(\mathbf{r},t)\left[\gamma^0\left(\imath\hbar\frac{\partial}{\partial t} - V_g^{(\tau)}\right) - v_F\left(\gamma^1\widehat{\pi}_x + \gamma^2\widehat{\pi}_y\right)\right]\Psi_\sigma^{(\tau)}(\mathbf{r},t), \tag{5}$$

where $\widehat{\pi}_{x(y)} = \widehat{p}_{x(y)} \mp (e/c)\mathbf{A}_{x(y)}(\mathbf{r})$, and $\mathbf{A}(\mathbf{r}) = (1/2)\mathbf{B}\times\mathbf{r}$ is the vector potential in a symmetric gauge.

In what follows we assume that the interaction between an electron with a position vector $\mathbf{r}_1$ from the top layer ($\tau = 1$) and a hole with a position vector $\mathbf{r}_2$ from the bottom layer ($\tau = 2$) is described by the Coulomb potential $V(\mathbf{r}_1 - \mathbf{r}_2) = e^2/\varepsilon_0\sqrt{|\mathbf{r}_1 - \mathbf{r}_2|^2 + d^2}$.

Instead of two position vectors $\mathbf{r}_1$ and $\mathbf{r}_2$, we introduce the center-of-mass $\mathbf{R} = \alpha(\mathbf{r}_1 + \mathbf{r}_2)$ and the relative $\mathbf{r} = \mathbf{r}_1 - \mathbf{r}_2$ coordinates ($\alpha = 1/2$).

The basic assumption in our BS formalism is that the electron-hole bound states are described by the BS wave function (BS amplitude). This function determines the probability amplitude to find the electron at the point $\mathbf{r}_1$ at the moment t_1 and the hole at the point $\mathbf{r}_2$ at the moment t_2. The BS amplitude depends on the relative internal time $t - t'$ and on the "center-of-mass" time:

$$\Phi^{\mathbf{Q}}(\mathbf{r},\mathbf{R};t,t') = \exp\left(-\frac{\imath E(\mathbf{Q})\alpha}{\hbar}(t+t')\right)\phi^{\mathbf{Q}}(\mathbf{r},\mathbf{R};t-t'), \tag{6}$$

where $E(\mathbf{Q})$ is the exciton dispersion. The BS equation for the equal-time BS amplitude in the center-of-mass and reduced coordinates is (Kouzakov & Studenikin, 2005):

$$\begin{aligned}&\Phi^{\mathbf{Q}}(\mathbf{r},\mathbf{R};t,t)=\int d^2\mathbf{r}'d^2\mathbf{R}'dt'G^{(1)}(\mathbf{R}+\alpha\mathbf{r},\mathbf{R}'+\alpha\mathbf{r}';t-t')\gamma^0\times\\&G^{(2)}(\mathbf{R}'-\alpha\mathbf{r}',\mathbf{R}-\alpha\mathbf{r};t'-t)\gamma^0V(\mathbf{r}')\Phi^{\mathbf{Q}}(\mathbf{r}',\mathbf{R}';t',t').\end{aligned} \tag{7}$$

The Fourier transforms of the electron and hole propagators $G^{(\tau)}(\mathbf{r},\mathbf{r}';t)$ are define in terms of the Dirac four component spinors $\psi^{\kappa}(\mathbf{r})$ and the corresponding eigenvalues $E_n=\hbar v_F\sqrt{2n}/R$ (Kouzakov & Studenikin, 2005):

$$G^{(\tau)}(\mathbf{r},\mathbf{r}';\omega)=\sum_{\kappa}\frac{\psi^{\kappa}(\mathbf{r})\overline{\psi}^{\kappa}(\mathbf{r}')}{\hbar\omega-E_n\pm\imath0^+} \tag{8}$$

Here we keep only the positive energy pole contributions, $n=0,1,2,...$, and $\kappa=(n,j_z,\sigma)$, where j_z is the z component of the total angular momentum.
When the translation symmetry is broken by the magnetic field, the Green's functions can be written as a product of phase factors and translation invariant parts. The phase factor depends on the gauge. In the symmetric gauge the Green's functions are:

$$G^{(\tau)}(\mathbf{r},\mathbf{r}';\omega)=\exp\left[\imath\frac{e}{\hbar c}\mathbf{r}.\mathbf{A}(\mathbf{r}')\right]\widetilde{G}^{(\tau)}(\mathbf{r}-\mathbf{r}';\omega). \tag{9}$$

The broken translation symmetry requires a phase factor for the BS amplitude:

$$\phi^{\mathbf{Q}}(\mathbf{r},\mathbf{R};\Omega)=\exp\left[\imath\frac{e}{\hbar c}\mathbf{r}.\mathbf{A}(\mathbf{R})\right]\chi^{\mathbf{Q}}(\mathbf{r},\mathbf{R};\Omega). \tag{10}$$

The BS equation (7) admits translation invariant solution of the form:

$$\chi^{\mathbf{Q}}(\mathbf{r},\mathbf{R};\omega)=\exp\left[-\imath\left(\mathbf{Q}.\mathbf{R}\right)\right]\widetilde{\chi}^{\mathbf{Q}}(\mathbf{r};\omega). \tag{11}$$

The Fourier transform of the function $\widetilde{\chi}^{\mathbf{Q}}(\mathbf{r};\omega)$ satisfies the following BS equation:

$$\begin{aligned}&\widetilde{\chi}^{\mathbf{Q}}(\mathbf{k};\omega)=\int\frac{d^2\mathbf{q}}{(2\pi)^2}\frac{d^2\mathbf{p}}{(2\pi)^2}d^2\mathbf{R}\int_{-\infty}^{\infty}\frac{d\Omega}{2\pi}e^{-\imath(\mathbf{q}+\mathbf{Q}).\mathbf{R}}\\&\widetilde{G}^{(1)}\left(\frac{1}{2}\mathbf{q}+\mathbf{k}-\frac{e}{\hbar c}\mathbf{A}(\mathbf{R});\hbar\omega+\alpha(E-V_g)\right)\gamma^0\times\\&\widetilde{G}^{(2)}\left(-\frac{1}{2}\mathbf{q}+\mathbf{k}-\frac{e}{\hbar c}\mathbf{A}(\mathbf{R});\hbar\omega-\alpha(E-V_g)\right)\gamma^0V\left(\mathbf{p}-\left[\mathbf{k}-\frac{2e}{\hbar c}\mathbf{A}(\mathbf{R})\right]\right)\widetilde{\chi}_{\mathbf{Q}}(\mathbf{p};\Omega),\end{aligned} \tag{12}$$

where $\widetilde{G}^{(\tau)}(\mathbf{k};\hbar\omega)$ are the Fourier transforms of $\widetilde{G}^{(\tau)}(\mathbf{r};\hbar\omega)$.
In the effective-mass approximation the exact fermion Green's functions $G^{(\tau)}$ are replaced by the corresponding propagator of the free fermions. The translation invariant parts of the free fermion propagators can be decomposed over the Landau level poles (Gorbar et al., 2002):

$$\widetilde{G}^{(\tau)}(\mathbf{k};\hbar\omega) = 2\imath \sum_{n=0}^{\infty}(-1)^n e^{-R^2\mathbf{k}^2} \frac{\hbar\omega\gamma^0 f_1(k) + f_2(\mathbf{k})}{\hbar^2\omega^2 - 2n\hbar v_F^2 eB/c},$$
$$f_1(k) = \frac{1}{2}(1-\imath\gamma^1\gamma^2)L_n(2R^2k^2) - \frac{1}{2}(1+\imath\gamma^1\gamma^2)L_{n-1}(2R^2k^2),$$
$$f_2(\mathbf{k}) = 2v_F\hbar(k_x\gamma^1 + k_y\gamma^2)L^1_{n-1}(2R^2k^2).$$

Here $L_n^1(x)$ are the generalized Laguerre polynomials, $L^1_{-1}(x) = L_{-1}(x) = 0$ and $L_n(x)$ are the Laguerre polynomials. In strong magnetic fields the probability for transitions to the excited Landau levels due to the Coulomb interaction is small. Thus, the contributions to the Green's functions from the excited Landau levels is negligible, and therefore, one can apply the LLL approximation, where we keep only $n = 0$ term:

$$\begin{aligned}\widetilde{G}^{(1)}(\mathbf{k};\hbar\omega) &\approx \imath\exp\left(-R^2\mathbf{k}^2\right)\frac{\gamma^0(1-\imath\gamma^1\gamma^2)}{\hbar\omega+\imath 0^+},\\ \widetilde{G}^{(2)}(\mathbf{k};\hbar\omega) &\approx \imath\exp\left(-R^2\mathbf{k}^2\right)\frac{\gamma^0(1-\imath\gamma^1\gamma^2)}{\hbar\omega-\imath 0^+}.\end{aligned} \tag{13}$$

The infinitesimal imaginary parts in our case reflect the fact that there are holes in layer number 2 (in electron-hole representation poles of the holes are above the real axis) and electrons in layer number 1.

The solution of the BS equation (12) in the LLL approximation can be written in the following form:

$$\widetilde{\chi}^{\mathbf{Q}}(\mathbf{k};\omega) = \exp\left[-R^2\mathbf{k}^2 - \imath\mathbf{R}_0.\mathbf{k}\right]\Phi_E(\omega). \tag{14}$$

Here $\Phi_E(\omega)$ is a 4×4 matrix. Thus, the LLL approximation reduces the problem from $2+1$-dimensions to $0+1$-dimension problem. The matrix $\Phi_E(\omega)$ and the magnetoexciton dispersion $E(\mathbf{Q})$ are determined by the solutions of the following equation:

$$\Phi_E(\omega) = -I(|\mathbf{Q}|)\int_{-\infty}^{\infty}\frac{d\Omega}{2\pi}\frac{\gamma^0(1-\imath\gamma^1\gamma^2)\gamma^0\Phi_E(\Omega)\gamma^0(1-\imath\gamma^1\gamma^2)\gamma^0}{\left(\hbar\omega+\alpha(E-V_g)+\imath 0^+\right)\left(\hbar\omega-\alpha(E-V_g)-\imath 0^+\right)}. \tag{15}$$

The solution of (15) is given by:

$$\begin{aligned}\Phi_E(\omega) &= \begin{pmatrix} 0&0&0&0\\ 0&1&0&1\\ 0&0&0&0\\ 0&1&0&1\end{pmatrix}\frac{1}{\left(\hbar\omega+\alpha(E-V_g)+\imath 0^+\right)\left(\hbar\omega-\alpha(E-V_g)-\imath 0^+\right)},\\ E(\mathbf{Q}) &= V_g - 4I(\mathbf{Q}).\end{aligned} \tag{16}$$

Thus, in the LLL approximation, the magnetoexciton dispersion is determined by the Coulomb interaction term $I(\mathbf{Q}) = \int d^2\mathbf{r}\varphi_{00}^2(r)V(\mathbf{r}+\mathbf{R}_0)$, where $\varphi_{00}(r) = (\sqrt{2\pi}R)^{-1}\exp\left(-r^2/4R^2\right)$ is the ground-state wave function of an electron in a magnetic field. For small wave vectors we calculate:

$$E(\mathbf{Q}) \approx V_g - 4E_{CQW} + \frac{\hbar^2Q^2}{2M(B)},\quad \frac{M(B)}{M_{CQW}} = \frac{1}{4}. \tag{17}$$

In graphene bilayer structures the magnetoexciton mass (binding energy) is four times lower (higher) than the corresponding magnetoexciton mass (binding energy) in coupled quantum wells with parabolic dispersion and the same d, ε_0 and B. In the limit of very small interlayer separation $d << R$ the asymptotical values of the binding energy and the effective magnetic mass of magnetoexciton in bilayer graphene are $4E_b$ and $M_{small}(B) = M_{2D}/4 \propto B^{1/2}$, respectively. In the limit of large interlayer separation $d >> R$ the asymptotical values of the magnetic mass is $M_{large}(B) = M_{2D}\pi^{1/2}d^3/(2^{7/2}R^3) \propto B^2$. As we mentioned above, under the certain critical temperature T_c a Bose-Einstein condensation might be expected to occur. The condensate of magnetoexcitons should have superfluid properties under the Kosterlitz-Thouless critical temperature T_{KT} (Kosterlitz & Thouless, 1973). A possible verification of our predictions could be based on the fact that at a fixed superfluid density the phase stiffness and the KT critical temperature, both are inversely proportional to the magnetoexciton mass, i.e. $T_{KT} \propto M^{-1}(B)$. Since the effective mass increases as a function of the magnetic field, one should expect that at a fixed superfluid density the KT critical temperature decreases in the limit of large interlayer separation as $T_{KT} \propto B^{-2}$ (as $T_{KT} \propto B^{-1/2}$ in the limit of small interlayer separation).

3. Conclusion

We have applied the relativistic BS formalism to the magnetoexcitons in a graphene structures. In the regime of a strong magnetic field the electrons and holes are confined primarily to the lowest Landau Level (LLL), and the Coulomb energy is much smaller than the exciton cyclotron energy. The LLL approximation greatly simplify the calculations, but we may ask whether the magnetoexciton dispersion will be significantly affected by the contributions from the *infinity number* of Landau levels with indexes $n \geq 1$ neglected in the LLL approximation. Going beyond the LLL approximation is an ambitious task (see, e.g. (Shabad & Usov, 2006)) which will be left as a subject of our future research.

4. References

Berman, O. L., Lozovik, Y. E. & Gumbs, G. (2008). *Phys. Rev. B* Vol. 77: 155433.

Gorbar, E. V., Gusynin, V. P., Miransky, V. A. & Shovkovy, I. A. (2002). *Phys. Rev. B* Vol. 66: 045108.

Gusynin, V. P., Miransky, V. A. & Shovkovy, I. A. (1994). *Phys. Rev. Lett.* Vol. 73: 3499.

Gusynin, V. P., Miransky, V. A. & Shovkovy, I. A. (1995). *Phys. Rev. D* Vol. 52: 4718.

Haldane, F. D. M. (1988). *Phys. Rev. Lett.* Vol. 61: 2015.

Koinov, Z. G. (2008). *Phys. Rev. B* Vol. 77: 165333.

Koinov, Z. G. (2009). *Phys. Rev. B* Vol. 79: 073409.

Kosterlitz, J. M. & Thouless, D. J. (1973). *J. Phys. C* Vol. 6: 1181.

Kouzakov, K. A. & Studenikin, A. I. (2005). *Phys. Rev. C* Vol. 72: 015502.

Lozovik, Y. E. & Sokolik, A. A. (2008). *Pisma Zh. Eksp. Teor. Fiz.* Vol. 87: 61.

Min, H., Bistritzer, R., Su, J.-J. & MacDonald, A. H. (2008). *Phys. Rev. B* Vol. 78: 121401(R).

Nomura, K. & MacDonald, A. H. (2007). *Phys. Rev. Lett.* Vol. 98: 076602.

Novoselov, K. S., Geim, A. K., Morozov, S. V., Jiang, D., Katsnelson, M. I., Grigorieva, I. V., Dubonos, S. V. & Firsovl, A. A. (2005). *Nature (London)* Vol. 438: 197.

Novoselov, K. S., Geim, A. K., Morozov, S. V., Jiang, D., Zhang, Y., Dubonos, S. V., Grigorieva, I. V. & Firsov, A. A. (2004). *Science* Vol. 306: 666.

Salpeter, E. E. & Bethe, H. A. (1951). *Phys. Rev.* Vol. 84: 1232.
Semenoff, G. W. (1984). *Phys. Rev. Lett.* Vol. 53: 2449.
Shabad, A. E. & Usov, V. V. (2006). *Phys. Rev. D* Vol. 73: 125021.
Shevchenko, S. I. (1997). *Phys. Rev. B* Vol. 56: 10355.
Zhang, C.-H. & Jorlecar, Y. N. (2008). *Phys. Rev. B* Vol. 77: 233405.
Zhang, Y., Tan, Y.-W., Stormer, H. L. & Kim, P. (2005). *Nature* Vol. 438: 201.
Ziegler, K. (2006). *Phys. Rev. Lett.* Vol. 97: 266802.

11

Topological Effects, Index Theorem and Supersymmetry in Graphene

K.-S. Park
Department of Electrical and Computer Engineering, Pohang University of Science and Technology, San 31 Hyoja-Dong Nam-Gu, Pohang, Kyungbuk 790-784 Korea

1. Introduction

Carbon provides a fundamental material for all life and physical science. Carbon-based systems reveal a variety of structures with a great deal of physical properties. These physical properties result from the dimensionality of the structures among systems with carbon atoms. For a long time, in material science, both experimentalists and theorists have sought for the existence of a true two-dimensional (2D) material with the thickness of a single atom, or a membrane of atomic thickness. This 2D material was theoretically first studied on a monolayer of graphite by Wallace (1). Experimentally in the year 2004, a group led by A. K. Geim at the University of Manchester, U. K., realized such a 2D material under the name of graphene (2; 3). Graphene is composed of carbon atoms placed at the vertices of a two dimensional honeycomb lattice. It is regarded as a large molecule of carbon atoms which become strongly bound together on the sites of the honeycomb lattice. For each carbon atom on the lattice, three of the four outer electrons get strongly bond with its neighboring atoms by σ orbitals. The $2p_z$ orbital of the fourth electron produces a π bond with a neighboring carbon atom. The σ bonds form the covalent structure with a honeycomb geometry. The bond strength furnishes the flexibility and robustness for the lattice geometry. On the other hand, the π bonds generate the intrinsic electronic structure of graphene. Each π bond yields the half-filled electrons of p orbital to tunnel from a carbon atom to the neighboring one. Thus graphene should be regarded as a many body system on which electrons can get correlated from site to site, resulting in a rich collective behavior. The correlated behavior can be represented by quantum effects which can influence on graphene's electronic properties (2).

The electronic structure can be described by 2D massless relativistic fermions(4–7) in graphene. The massless fermions enable us to study topological effects on electronic properties of graphene. Topological effects are represented by the global properties of geometrical objects rather than their local ones. By varying the geometry, we can produce topologically different configurations such as a sphere or a torus on which the effective Dirac operators of massless fermions are well defined. Described by the Dirac fermions, graphene can have extraordinary properties of stability obtained in terms of geometry and topology of the underlined lattice. Apart from the robust structure of geometry, the topological properties can emerge due to long range quantum coherence in graphene. It follows that we can move electrons coherently through the whole graphene molecule, resulting in its detection of geometry or topology. This allows us to study a great deal of physical properties revealed by interplay between geometry and topology, and quantum effects(8; 9).

As quantum effects, there exist unconventional quantum Hall effects (QHE) which form a series of filling factors $\nu = \pm 2, \pm 6, \pm 10, \cdots$ as the four-fold degeneracy combined by spin and sublattice valley ones (6; 7). The energy dispersion shows a linear spectrum by the massless Dirac fermions with a Fermi velocity $v_F \approx 10^6 m/s$. The 4-fold degeneracy of the Landau level (LL) is lifted into 4 sublevels in the presence of an external high magnetic field. In the case of the tilted high magnetic field to the graphene plane, the spin degeneracy can be lifted at the firsr LL, resulting in the filling factor $\nu = 4$ QHE of monolayer graphene (10). Furthermore for bilayer graphene, the LL spectrum is composed of eightfold degenerate states at the zero energy and fourfold ones at finite energies under the high magnetic field. This can allows us to observe the quantum Hall plateaus at a seris of $\nu = \pm 4, \pm 8, \pm 12, \cdots$ (11–13). The charge carriers are chiral massive fermions which produce a parabolic energy band. The chiral fermions offer the unconventional integer QHE of the zero-LL anomaly which exhibits metallic behavior under the condition of low carrier densities and high magnetic fields in contrast to the conventional insulating phenomena (6; 7; 9–14).

Topological configurations can produce a crucial effect on the quantum states of a system. In particular, they can provide the possible quantum ground states which a system can have. This remarkable result is described in terms of the index theorem initiated by Atiyah and Singer (15). It gives the relationship between the analytic properties of the operator and the topological characteristic of the manifold upon which the operator is defined. The Dirac operator can be related to topological effects which is elegantly illuminated by the Atiyah-Singer index theorem in graphene (8). It leads to a topological invariant under deformations on a Dirac operator and plays an essential role in formulating supersymmetric quantum mechanics (SUSY QM) on the graphene sheet (9; 16).

In a theoretical sense, there has been at least the quantum mechanics (QM) of particles described by both fermionic and bosonic degrees of freedom. The SUSY QM may be hidden in the quantum mechanics of a particle acting on a group manifold which can be represented by a high degree of symmetry (17). As an example, a spin precessing in a magnetic field can have the hidden SUSY. In particular, under the uniform magnetic field the LL for an electron can be expressed by the spectrum of the SUSY oscillator which is composed of fermionic and bosonic ones. It is remarkable that this SUSY QM can possibly emerge in graphene with low carrier concentration and high mobility. The supersymmetry is built up over the Dolbeault complex due to the topological deformation on the lattice in a graphene system (9; 17; 18).

We exploit the G-index theorem and a high degree of symmetry to understand unusual quantum Hall effects of the $n = 0$ Landau level in graphene. The Dirac fermion results in SU(4) symmetry as a high degree of symmetry in the noninteracting Hamiltonian of the monolayer graphene. The high symmetries in graphene sheets can not couple to an external magnetic field. In the absence of the magnetic field the index theorem can provide a relation between the zero-energy state of the graphene sheet and the topological deformation of the compact lattice. Under the topological deformation the zero-energy states emerge naturally without the Zeeman splitting at the Fermi points in the graphene sheet. In the case of nonzero energy, the up-spin and down-spin states have the exact high symmetries of spin, forming the pseudospin singlet pairing. We describe the peculiar and unconventional quantum Hall effects of the $n = 0$ Landau level in monolayer graphene on the basis of the index theorem and the high degree of symmetry (9; 18).

This chapter is written as follows. We explain basic properties of graphene in section II. In section III, we discuss a path integral of coherent states in brief. In section IV, supersymmetry is introduced in graphene. In the following section, we investigate the Atiyah-Singer index theorem and topological properties. The G index theorem and deformation is covered. In section VI, SUSY QM and higher spin symmetry are described. Next we illuminate the low energy spectrum and unconventional quantum Hall effects in monolayer graphene. And finally we come to summary and conclusion.

2. Basic properties of graphene

Graphene is a molecule that is composed of carbon atoms placed on a two dimensional honeycomb lattice. The basic plaquette of the lattice has a hexagon and the atoms are located at the sites of the lattice. Electronic properties of graphene can be described by the tight binding model on which spinless electrons move from site to site along the links of the lattice without interaction each other. Under the tight-binding approximation graphene can be expressed by a simple Hamiltonian of coupled fermions on a hexagonal lattice (2; 19). The model Hamiltonian is a form given by

$$\mathrm{H} = -\frac{2v_{\mathrm{F}}}{3} \sum_{<i,j>} c_i^{\dagger} c_j, \tag{1}$$

where $< i,j >$ indicates nearest neighbors on the lattice. $c_i^{\dagger}$ and c_i are the creation and annihilation operators of the fermions located at site i with anticommuation relation $\{c_i, c_j^{\dagger}\} = \delta_{ij}$.

In order to calculate the spectrum of Hamiltonian (1), we account for a periodicity of honeycomb lattice which leads to a Fourier transformation. The periodic structure provides the energy eigenvalue problem for the Hamiltonian in a unit cell. The unit cell consists of two neighboring carbon atoms called A and B. They can be expressed by the three vectors $\vec{u}_i, \forall i = 1,2,3$. Under the Fourier transformation of $c(\vec{p}) = \sum_i e^{i\vec{p}\cdot\vec{u}_i} c_i$, the Hamiltonian is rewritten in terms of

$$\mathrm{H} = -\frac{2v_{\mathrm{F}}}{3} \int\int d^2p \left(c_A^{\dagger}(\vec{p}), c_B^{\dagger}(\vec{p})\right) \begin{pmatrix} 0 & \sum_{i=1}^{3} e^{i\vec{p}.\vec{u}_i} \\ \sum_{i=1}^{3} e^{-i\vec{p}.\vec{u}_i} & 0 \end{pmatrix} \begin{pmatrix} c_A(\vec{p}) \\ c_B(\vec{p}) \end{pmatrix}, \tag{2}$$

where $c_A(\vec{p})$ and $c_B(\vec{p})$ denote the Fourier transformed operators corresponding to the carbon atoms A and B, respectively.

Now it is easy to take the eigenvalue of the energy for electrons of graphene. The dispersion energy is given by (8; 9; 19)

$$E(p) = \pm\frac{2v_{\mathrm{F}}}{3}\sqrt{1 + 3\cos^2\frac{\sqrt{3}p_y}{2} + 4\cos\frac{3p_x}{2}\cos\frac{3p_y}{2}} \tag{3}$$

where the lattice distance between atoms becomes normalized to the unity. From the dispersion relation obtained above, graphene can have two independent Fermi points, $\vec{p} = K_{\pm} = \pm\frac{2\pi}{3}(1, \frac{1}{\sqrt{3}})$. We can expand it and then linearize it near the conical singularities of the Fermi points. Corresponding to the K_+ and K_- at the half-filling case, the Hamiltonian is expressed by the Dirac operators

$$\mathrm{H}^{\pm} = \pm v_{\mathrm{F}} \sum_{\mu=x,y} \gamma^{\mu} p_{\mu}, \tag{4}$$

where $p_{\mu} = -i\hbar\partial_{\mu}$ is the covariant momentum and the Dirac matrices γ^{μ} indicate the Pauli matrices $\gamma^{\mu} = \sigma^{\mu}$. Hence the low energy theory of graphene is described by means of free fermions.

The Hamiltonian can be written in the matrix form (8; 9)

$$\mathrm{H}^{\pm} = \begin{pmatrix} 0 & D_{\pm} \\ D_{\pm} & 0 \end{pmatrix}. \tag{5}$$

Here $D_{\pm}$ is a Dirac operator given by

$$D_{\pm} = \pm v_{\mathrm{F}}(\sigma^x p_x + \sigma^y p_y), = \mp 2i\hbar v_{\mathrm{F}} \begin{pmatrix} 0 & \partial_z \\ \partial_{\bar{z}} & 0 \end{pmatrix}. \tag{6}$$

Here we have expressed $D_{\pm}$ in terms of $\partial_z = \frac{1}{2}(\partial_x - i\partial_y)$ and $\partial_{\bar{z}} = \frac{1}{2}(\partial_x + i\partial_y)$. On the complex coordinates we can write

$$\nabla^2 = 4\partial_z\partial_{\bar{z}}. \tag{7}$$

So far, we have discussed the collective behavior of graphene electrons that can be governed by the Dirac equation. In particular, the velocity of the electrons is effectively 300 times smaller than the speed of light.

Next let us describe a curved graphene. On the curved surface of graphene the low energy physics can be illuminated by the Dirac equation that is defined on the corresponding curved manifold. The curvature generates an gauge field of magnetic flux going through the curved graphene. This yields to a picture about how to interact gauge fields with Dirac fermions. In order to have the way associated with the gauge fields, let us take into account a good method to include curvature to graphene. The simplest way is that we cut a $\frac{\pi}{3}$ piece of triangle from a graphene sheet and then glue the opposite ends of the lattice. This process results in a single pentagon at the apex of the generated cone while all the other plaquette keep a hexagon. Due to the minimal geometrical distortion, the honeycomb lattice has a positive curvature. The curvature can be obtained by calculating a circular tangent vector, V, around the apex by $\oint \mathrm{V} \cdot d\vec{r} = \frac{\pi}{3}$. The generation of a single pentagon gives rise to a dramatic effect on the spinor, resulting in deformation of the lattice. If the spinor is parallel transported around the apex by an angle 2π, it is forced at some point to make a jump from a site A to a site A while every site A takes only B neighborhoods or vice versa. This motion enables us to have the effect that the magnetic field gives on the wave function of a particle moving on a closed path. A full circulation provides accumulation of all phase factors to the particle wave function, and so generates the enclosed magnetic flux. This flux allows us to observe the Aharonov Bohm effect. Quantum mechanically, it can provide a discontinuity. Thus we have simultaneously to describe it through the wave function of the particle in the processes of being static or moving along the closed trajectory. This produces a vector potential term in the Hamiltonian that leaves the theory to be consistent.

By similar procedures made on the curved graphene that compensates the jump in the components of the spinor, we should take into account a nonabelian vector potential A in the effective Hamiltonian. Around the apex we can take the circulation of A along a path. It is expressed by $\oint_{\gamma} \mathrm{A} \cdot d\vec{r} = \frac{\pi}{2}\tau_2$, where τ_2 denotes the second Pauli matrix which couples the K_+ to the K_- components of the spinor. The effective gauge theory can be emerged due to the geometric deformation on geometric variants of graphene as topological effects. In next section we describe a path integral of the coherent states for geometrical and topological properties.

3. Path integral of coherent states

Let us construct a path integral of the coherent states on a group manifold. Following the Stone's approach (17), we discuss the coherent states on a general group G. Let us take $D(g)$, any $g \in G$, as an irreducible representation of the group. Assume that $|0>$ is some state in the space of representation. Then one can define $|g>$ by

$$|g>= D(g)|0> . \tag{8}$$

On the basis of the irreducible representation, Shur's lemma holds that

$$\frac{1}{V(G)} \int d[g]|g><g| = 1 \tag{9}$$

where $\frac{1}{V(G)}$ is the volume of the group manifold. In the expression of Eq. (9) $d[g]$ is the Haar measure on the group.
Let us compute a thermodynamic partition function

$$Z = \mathrm{Tr}(e^{-\beta \mathrm{H}}) \tag{10}$$

where β is the imaginary time. In the procedure of calculating the partition function, the trace is constrained to the representation space on which $D(g)$ acts. Dividing the Matsubara time-interval β into n parts, and using the Shur's lemma of Eq. (9), we can write down an iterated integral

$$\mathrm{Tr}(e^{-\beta \mathrm{H}}) = \mathrm{const.} \int (d[g]d[g^{'}]\cdots) < g|e^{-\beta \mathrm{H}/n}|g^{'} >< g^{'}|e^{-\beta \mathrm{H}/n} \cdots |g> . \tag{11}$$

where const. is the constant value taken on the representation space. When taking into account short time intervals, one can express $g^{'} \simeq g + \delta g$, and $\delta g \simeq O(\delta t)$, so that

$$\begin{aligned} &< g|e^{-\delta t \mathrm{H}}|g^{'} > \\ &\approx 1+ < g|\delta g > + < g|(-\delta t \mathrm{H})|g > +O(\delta t^2). \end{aligned} \tag{12}$$

On taking into consideration up to order $O(\delta t)$, we can write down the formal path-integral expression

$$\begin{aligned} &\mathrm{Tr}(e^{-\beta \mathrm{H}}) \\ &= \tfrac{1}{V(G)} \int d[g] \exp(\oint < g|\delta g > - \int_0^\beta dt < g|\mathrm{H}|g >). \end{aligned} \tag{13}$$

Here $d[g]$ is regarded as the path-integral measure given by the Haar measure at each time step.
The path-integral expression can be identified as a path integration over a quotient space of the group. In particular it is noted that a set of the $|g>$ can be different from only a phase so that the integrand is not sensitive to the phase factor. Now suppose that H is the subgroup of G, constructed from exponentiating a maximal commuting set of generator, i.e., a maximal torus. Then $|0>$ is expressed by an eigenstate of the generators of H which means a state of definite weight. The $|g>$ are represented by all phase multiples of one another in any one coset of G/H. And hence the coherent states can be described in terms of a bundle over G/H with the maximal torus as the gauge group, and the integration is made over the path in G/H.
Let us express the integrand in the coherent-state path integral without any choice of representatives. In order to make a natural procedure on the independent choice of the representatives, we define the projection operators as

$$P(g) = |g><g|. \tag{14}$$

They can be directly projected onto the physically distinct states since they do not have any phase ambiguity. In the integrand, the first term $\oint < g|\delta g >$ can be taken to be a gauge invariant form by using Stokes theorem

$$\oint_{\Gamma=\partial\Omega} < g|dg >= \int_{\Omega} d < g|dg > . \tag{15}$$

Making use of the identity form

$$d < g|dg >=< dg|dg >= -\mathrm{Tr}(dPPdP), \tag{16}$$

the first term yields

$$\oint_{\Gamma=\partial\Omega} < g|dg >= -\int_{\Omega} \mathrm{Tr}(dPPdP). \tag{17}$$

The second term in the integrand can be rewritten by

$$< g|\mathrm{H}|g >= \mathrm{Tr}(P(g)\mathrm{H}). \tag{18}$$

Hence combination of Eqs. (17) and (18) leads us to the path integral given by (17)

$$\mathrm{Tr}(e^{-\beta\mathrm{H}}) = \frac{1}{V(G/H)} \int d[g] \exp\left(-\int_{\Omega} \mathrm{Tr}(dPPdP) - \int_{\Gamma} \mathrm{Tr}(P(g)\mathrm{H})\right). \tag{19}$$

It is noted that the elements of G/H can only contribute to the integrand, so that the volume factor of the gauge group H gets out of the path integral.
There are certain subgroups corresponding to the Lie algebra $\mathcal{G}$ of a Lie group G. Let us describe some basic facts about semisimple Lie algebra in brief. Then it is known that the generators of the $\mathcal{G}$ can have a decomposition into a maximally commuting set $\mathcal{H} = \{H_i\}$, i.e., the Cartan subalgebra, and a set of ladder operators, E_α, one for each root vector $\alpha \in R$. The ladder operators are needed for complexifing the algebra to $\mathcal{G}^c$ which results from the group parameters to get the complex values. The H_i and the E_α hold that

$$[H_i, E_\alpha] = \alpha_i E_\alpha. \tag{20}$$

Assume that $|\lambda >$ is an eigenvector of the H_i with eigenvalues λ_i such as

$$H_i|\lambda >= \lambda_i|\lambda > . \tag{21}$$

Then

$$E_\alpha|\lambda >= |\lambda + \alpha > . \tag{22}$$

The roots may be classified into two sets by an arbitrary hyperplane on a root space. On one set the root objects are positive roots indicated by $\alpha \in R_+$ as increasing the weights while the others are negative roots, by $\alpha \in R_+$ as decreasing the weights. And thus we can take the greatest weight as a state which is annihilated by all $E_\alpha, \alpha \in R_+$.
This decomposition of the Lie algebra is made on some kinds of special subgroup. The Borel subgroups $B_\pm$ are constructed by exponentiating the algebras $\mathcal{B}_\pm$ which are spanned by the $E_\alpha, H_i; \alpha \in R_\pm, H_i \in \mathcal{H}^c$. Any $g \in G$ can be decomposed into the gaussian factors

$$g = \zeta_- h \zeta_+; \;\; \zeta_- \in Z_-, \zeta_+ \in Z_+, h \in H_+^c, \tag{23}$$

where the $Z_\pm$ are the groups obtained by exponentiating the $E_\alpha, \alpha \in R_\pm$. As an example, the gaussian factor ζ_- can be written as

$$\zeta_- = \exp(\sum_{\alpha \in R_-} z^\alpha E_\alpha). \tag{24}$$

When the representation matrix $D(g)$ is applied to a greatest weight, the factor ζ_+ is regarded as the identity. The z^α can be taken as complex coordinates for the coset space G^c/B_+. It follows that the set of physically distinct states can have an one-to-one correspondence to the G^c/B_+. For a complex manifold we can choose complex coordinates on the manifold with holomorphic functions. And constructed on non-greatest weight states, coherent states are described by nonholomorphic functions of $\bar{z}$. In the next section, after introducing supersymmetry on the complex manifold in brief, we will describe supersymmetric quantum mechanics in graphene (9; 16).

4. Supersymmetry in graphene

Let us use the methods of Witten to introduce fermionic creation operators $\psi^{x\dagger}$ and $\psi^{y\dagger}$ which correspond to the differential forms dx and dy (9; 16)

$$dx \leftrightarrow \psi^{x\dagger}|0>, dy \leftrightarrow \psi^{y\dagger}|0> . \tag{25}$$

In more detail $\psi^{x\dagger}$ performs the operation of exterior multiplication by dx while the adjoint, ψ_x, does that of interior multiplication by the vector dual to dx, say, ∂_x

$$\imath : \partial_x \to \psi_x, \mathcal{E} : dx \to \psi^{x\dagger}, \quad \imath : \partial_y \to \psi_y, \mathcal{E} : dy \to \psi^{y\dagger}. \tag{26}$$

The fermionic operators satisfy the anticommutation relations

$$\{\psi_\mu, \psi^{\nu\dagger}\} = \delta^\nu_\mu, \mu, \nu = x, y. \tag{27}$$

On the basis of these definitions let us set up

$$\frac{1}{2}(\psi_x - i\psi_y) = \psi_z = \frac{1}{2}\psi^{\bar{z}}, \quad \frac{1}{2}(\psi_x + i\psi_y) = \psi_{\bar{z}} = \frac{1}{2}\psi^z. \tag{28}$$

while holding the Hermitian conjugate relations.
Now let us describe a supersymmetry over the complex *Kähler* manifold. On the complex manifold we take two supercharges

$$Q_1 = \partial = \psi^{\dagger z}\partial_z, \; Q_1^\dagger = \delta = -\psi^{\bar{z}}\partial_{\bar{z}},$$
$$Q_2 = \bar{\partial} = \psi^{\dagger \bar{z}}\partial_{\bar{z}}, \; Q_2^\dagger = \bar{\delta} = -\psi^z\partial_z. \tag{29}$$

Making use of anticommutation relations for fermions, these supercharge operators allows us to express

$$\partial\delta + \delta\partial = Q_1Q_1^\dagger + Q_1^\dagger Q_1 = -\frac{1}{2}\nabla^2,$$
$$\bar{\partial}\bar{\delta} + \bar{\delta}\bar{\partial} = Q_2Q_2^\dagger + Q_2^\dagger Q_2 = -\frac{1}{2}\nabla^2. \tag{30}$$

Here it is easy to check that the cross terms such as

$$\partial\bar{\delta} + \bar{\delta}\partial = Q_1Q_2^\dagger + Q_2^\dagger Q_1 = 0, \tag{31}$$

do not have any contributions.
In terms of the two supercharge operators, the Dirac operators $D_{\pm}$ in graphene is given by

$$D_{\pm} = \mp i\hbar v_{\mathrm{F}}(Q_1 + Q_2), \equiv Q_{\pm}. \tag{32}$$

The Dirac operators can be described by means of the sum of the two supercharges which is the ordinary exterior derivative, d. And $Q_{\pm}$ become the supercharge of N = 1 supersymmetric quantum mechanics. It is not hard to check up that $D_{\pm} = Q_{\pm} + Q_{\pm}^{\dagger}$ is given by

$$Q_{\pm} + Q_{\pm}^{\dagger} = \mp 2i\hbar v_{\mathrm{F}} \begin{pmatrix} 0 & \partial_z \\ \partial_{\bar{z}} & 0 \end{pmatrix} \tag{33}$$

which are equivalent to Eq. (6). Furthermore the square of Eq. (33) can lead to

$$(Q_{\pm} + Q_{\pm}^{\dagger})^2 = -4(\hbar v_{\mathrm{F}})^2 \begin{pmatrix} \partial_z\partial_{\bar{z}} & 0 \\ 0 & \partial_{\bar{z}}\partial_z \end{pmatrix} = -\hbar^2 v_{\mathrm{F}}^2 \nabla^2, \tag{34}$$

where we have exploited Eq. (7). And hence in the sense of SUSY QM the Hamiltonian of graphene may be recapped in terms of the following form

$$\mathrm{H} \equiv 2(Q_{\pm} + Q_{\pm}^{\dagger})^2 = -2\hbar^2 v_{\mathrm{F}}^2 \nabla^2, \equiv -\frac{\hbar^2\nabla^2}{2m^*}, \tag{35}$$

provided that the mass, m^* were defined by $m^* = \frac{1}{4v_{\mathrm{F}}^2}$ in the last expression of the eq. (17). The Witten index is given by

$$\mathrm{Index(d)} = \mathrm{Tr}\left((-1)^F e^{-t\mathrm{H}}\right) \tag{36}$$

which accounts for the Euler number of the manifold as the exterior calculus of the de-Rham complex. In order to build up the supersymmety over the Dolbeault complex (22), we need one of the supercharges, $Q_2 = \bar{\partial}$. On the manifold of real dimension $2n$, the index of the Dolbeault complex is given by

$$\mathrm{Index}(\bar{\partial}) = \mathrm{Tr}\left((-1)^F e^{-t(Q_2+Q_2^{\dagger})^2}\right) \tag{37}$$

This index is more interested in the the SUSY QM over twisted Dolbeault complex which is associated with deformation of the topology of the lattice on a graphene system . And in the next section we build the Atiyah-Singer and G index theorem as well as the energy eigenvalues on the deformation of the compact manifold.

5. Index theorem in graphene

Let us describe the index theorem which gives an insight on the spectrum structure of certain operators such as the Dirac operators. In graphene this theorem enables us to have physical properties associated with the topology and geometry of the space in which the Dirac operators are defined. It provides the relationship between the analytic properties of the operator and the topological characteristics of the manifold.

5.1 The Atiyah-Singer index theorem in graphene

We illuminate the Atiyah-Singer index theorem by the method employed to the heat kernel expansion. The theorem furnishes a relation between zero eigenvalues of the Dirac operator of graphene and the total flux which goes through its surface. If the latter is connected to the genus of the surface through the Euler characteristic, we can find a close relation between the zero modes and the topology of the surface on a graphene system.

Let us start with a Dirac operator given by

$$\mathrm{K} = \begin{pmatrix} 0 & D^{\dagger} \\ D & 0 \end{pmatrix}. \tag{38}$$

Here D means an operator that maps a space M_{+} onto a space M_{-} while $D^{\dagger}$ is a map from M_{-} to M_{+}. If D is an $n \times m$ matrix, $D^{\dagger}$ becomes a $m \times n$ matrix. M_{+} and M_{-} are the space of n and m dimensional vectors, respectively. Because we are focusing on the zero modes of K such that the solutions of the equations $\mathrm{K}\Psi = 0$, let us define the number of different eigenstates of D with zero eigenvalue as η_{+} and the ones of $D^{\dagger}$ as η_{-}. As a bookkeeping of chirality, the chirality operator γ_5 is defined as

$$\gamma_5 = \begin{pmatrix} 1 & 0 \\ 0 & -1 \end{pmatrix}. \tag{39}$$

Its eigenstates can have eigenvalue $\pm$ provided that they act on $M_{\pm}$.

In order to calculate the number of zero eigenstates in which we are interested, we take into account the operator K^2 that has the same number of zero modes as K. The K^2 can be given by a diagonal form

$$\mathrm{K}^2 = \begin{pmatrix} D^{\dagger}D & 0 \\ 0 & DD^{\dagger} \end{pmatrix}. \tag{40}$$

It is claimed that the operators $D^{\dagger}D$ and $DD^{\dagger}$ get the same non-zero eigenvalues. To prove this statement, assume that $DD^{\dagger}\psi = \lambda\psi$ for eigenvalue $\lambda \neq 0$. Then it follows that

$$DD^{\dagger}\psi = \lambda\psi \longrightarrow D^{\dagger}D(D^{\dagger}\psi) = \lambda(D^{\dagger}\psi). \tag{41}$$

This means that the operator $D^{\dagger}D$ gives the same eigenvalue, λ, which corresponds to the eigenstate $D^{\dagger}\psi$. But it is not necessary to hold the case for $\lambda = 0$ when $D^{\dagger}\psi$ might be zero by itself.

Let us compute the trace of $\gamma_5 e^{-t\mathrm{K}^2}$ as followings:

$$\mathrm{Tr}(\gamma_5 e^{-t\mathrm{K}^2}) = \mathrm{Tr}(e^{-tD^{\dagger}D}) - \mathrm{Tr}(e^{-tDD^{\dagger}}) = \sum_{\lambda_{+}} e^{-t\lambda_{+}} - \sum_{\lambda_{-}} e^{-t\lambda_{-}}, \tag{42}$$

where λ_{+} and λ_{-} indicate the eigenvalues of the operators $D^{\dagger}D$ and $DD^{\dagger}$, respectively, and t is an arbitrary parameter. In the first step of the above procedures, γ_5 acts on the exponential so that it provides a $+1$ to the eigenvectors of $D^{\dagger}D$ when they are placed in M_{+}, and a -1 to the ones of $DD^{\dagger}$ when they belong to M_{-}. In the last step the trace is evaluated by a sum over all the eigenvalues of the corresponding operators. Every non-zero eigenvalue of $D^{\dagger}D$ is a one-to-one correspondence to an eigenvalue of $DD^{\dagger}$. Therefore all paired terms of non-zero eigenvalues cancel out each other. There are left over the zero eigenvalues of each operators, resulting in

$$\mathrm{Tr}(\gamma_5 e^{-t\mathrm{K}^2}) = \eta_{+} - \eta_{-}. \tag{43}$$

In general we cannot determine difference between the number of zero modes. It is seen that the above result is independent of t owing to the cancelation of the non-zero eigenvalue term. Actually we should evaluate Index(K). In order to calculate it practically, we take an alternative method of heat expansion for calculating $\mathrm{Tr}(\gamma_5 e^{-t\mathrm{K}^2})$. It says that for general $\hat{\Gamma}$ and $\hat{D}$ on a two dimensional compact manifold we can expand

$$\mathrm{Tr}(\hat{\Gamma}e^{-t\hat{D}}) = \frac{1}{4\pi t}\sum_{l\geq 0} t^{\frac{l}{2}} b_l(\hat{\Gamma},\hat{D}), \tag{44}$$

where Tr indicates the trace of matrices and the integration over coordinates of space. b_l denote expansion coefficients. For $\hat{\Gamma} = \gamma_5$ and $\hat{D} = \mathrm{K}^2$, we have to return to an expression that is t-independent. For this t-independence, the expansion coefficients should vanish for all l except for $l = 2$ under the condition that all the t contributions is canceled out each other. This allows us to determine the coefficient b_2 from the first order term in t in the series of expansion. For the evaluation, let us take D as $D = -ie^{\mu}_{\nu}\sigma^{\nu}(\nabla_{\mu} - ieA_{\mu})$. Here e^{μ}_{ν} indicates the zweibein of curved surface metric $g_{\mu\nu}$ that defines a local flat frame $\eta_{\alpha\beta} = e^{\mu}_{\alpha}e^{\nu}_{\beta}g_{\mu\nu}$ while σ_{μ} is the Pauli matrix. A_{μ} denotes a gauge field. It follows that

$$\mathrm{K}^2 = -g^{\mu\nu}\nabla_{\mu}\nabla_{\nu} + \frac{1}{4}[\gamma^{\mu},\gamma^{\nu}]F_{\mu\nu} - \frac{1}{4}R \tag{45}$$

where R is curvature, and $F_{\mu\nu} = \partial_{\mu}A_{\nu} - \partial_{\nu}A_{\mu}$ means the field strength. ∇_{μ} indicates a covariant derivative with respect to gauge and reparametrization transformation.
It is easy to see that the non-zero expansion coefficient b_2 is written by

$$b_2 = \mathrm{Tr}[\gamma_5(\frac{i}{4}[\gamma^{\mu},\gamma^{\nu}]F_{\mu\nu} - \frac{1}{4}R)] = 2\int\int \mathrm{B}\cdot d\mathrm{S}, \tag{46}$$

where B is the magnetic field given by $B_l = \frac{1}{2}\epsilon^{l\mu\nu}F_{\mu\nu}$. The integration has been over the whole surface. These two independent ways for calculating $\mathrm{Tr}(\gamma_5 e^{-t\mathrm{K}^2})$ allows u to arrive at the final formula of the index theorem (8)

$$\mathrm{Index(K)} = \nu_+ - \nu_- = \frac{1}{2\pi}\int\int \mathrm{B}\cdot d\mathrm{S}. \tag{47}$$

It states that the total flux which goes out of the surface is related to the number of zero modes of the K operator. The curvature doesn't contribute to the index formula because γ_5 is a traceless operator. Particularly no contribution of curvature is to show an intrinsic property of two dimensional surfaces. The index theorem leads to an integer number on the compact surfaces. Therefore $\int\int \mathrm{B}\cdot d\mathrm{S}$ produces the total magnetic monopole charge in discrete values inside the surface under the Dirac quantization condition of magnetic monopoles.
Using the index theorem, we observe the topological characteristics of graphene. Let us apply it to graphene and its geometric variants. In a certain configuration of graphene we have to account for the effective magnetic field on the graphene surface. Some plaquette deformations give rise to a specific circulation of the vector potential around a loop trajectory. Stokes's theorem helps us to have a relation between the circulation of the gauge potential around a loop γ_i and the flux of the corresponding magnetic field

$$\oint_{\gamma_i} \mathrm{A}\cdot d\vec{r} = \int\int_{S_i} \mathrm{B}\cdot d\mathrm{S}, \tag{48}$$

where S_i indicates the area. Hence going through the surface of graphene the total flux can be obtained from the fact that we can know the total number of deformations. It is necessary for us to have some information about the total number of plaguette deformations. The information is related to the topological properties of the surface through the Euler characteristics. Let us consider the Euler theorem in the following subsection.

5.2 Euler theorem

In general the Euler theorem gives rise to a relation between the structural information of a polyhedral lattice and its topological properties. There are a lot of proofs on this theorem. The most common methods are proofs on the basis of a reductions from the polyhedral lattice to the simpler one without changing its topological properties.

Let us consider a lattice placing on a compact surface with a certain genus g. Then we can compute the number of deformations in a lattice necessary to create such a surface by applying the Euler characteristic. Let V, E and F be the number of vertices, edges and faces of the lattice, respectively, and N_{end}, open ends. Then the Euler characteristic, χ is expressed by (8)

$$\chi = V - E + F = 2(1-g) - N_{end}. \tag{49}$$

The second step of Eq. (49) is satisfied by the Euler theorem. It is easily to check that a single cut in the surface can have a reduction of the genus by one and increase the number of open ends by two, say, $(g, N_{end}) \to (g-1, N_{end}+2)$, leaving the Euler characteristic χ preserved.

Let us apply the Euler theorem to the case of graphene molecules. There are three links on each vertex of graphene. Suppose that topological deformations such as pentagons or heptagons are present. Let us indicate the total number of pentagons, hexagons and heptagons by n_5, n_6 and n_7, respectively in the molecule. Then the total number of vertices is written by $V = (5n_5 + 6n_6 + 7n_7)/3$ when each k-gon has k vertices and each vertex takes three polygons. Similarly, the total number of edges is expressed by $E = (5n_5 + 6n_6 + 7n_7)/2$ provided that each edge has two polygons. The total number of faces is equal to the sum of different polygons, $F = n_5 + n_6 + n_7$. Combining these into the Euler characteristic, we see that

$$n_5 - n_7 = 6\chi = 12(1-g) - 6N_{end}. \tag{50}$$

This result reflects many facts. When equal numbers of pentagons and heptagons are inserted, they do not make any change about the topology of the surface in the case that they cancel out. On a flat graphene sheet we can put two pentagons and two heptagons on it without changing the curvature of the molecule away from these deformations. This is consistent with the effective gauge flux approaches where pentagons and heptagons give opposite flux contributions. On the other hand, it is known that nontrivial topologies necessarily provide an imbalance between pentagons and heptagons. The genus zero configurations result in an excess of pentagons while high genus surface has an excess of heptagons. Genus one surfaces do not have any pentagons or heptagons at all provided that they are equivalent to a flat sheet. It is seen that Eq. (50) recaps the known result of a sphere with $g = 0$ which leads to $\chi = 2$. This corresponds to a fact about $n_5 = 12$ and $n_7 = 0$ for the C_{60} fullerene. For a torus with $g = 1$, we can have $\chi = 0$ related to $n_5 = n_7 = 0$ in the case of the nanotubes. Thus no pentagons or heptagons may be required. If we account for the genus $g = 2$ surfaces, then we can obtain $\chi = -2$ where $n_5 = 0$ and $n_7 = 12$. In this situation equal numbers of pentagons and heptagons can be inserted without making any change of topology on the surface.

Suppose that K is a Dirac Hamiltonian H. Then we compute the Index(H). The Euler characteristic term allows us to compute the gauge field term in the Index(H). It can be given by including additionally the contributions from the surplus of pentagons or heptagons. Hence the total flux of the effective gauge field can lead to

$$\begin{aligned}\mathrm{Index(H)} &= \frac{1}{2\pi}\int\int_{S_i} \mathrm{B}\cdot d\mathrm{S} = \frac{1}{2\pi}\sum_{n_5-n_7}\oint_{\gamma_i} \mathrm{A}\cdot d\vec{r} \\ &= \frac{1}{2\pi}\frac{\pi}{2}(n_5-n_7) = 3(1-g) - \frac{3}{2}N_{end}. \end{aligned} \tag{51}$$

The total number of zero modes is equivalent to the sum contributed from each subsector of a Dirac operator. As a consequence, by adding the two contributions, we arrive at the index of the Dirac Hamiltonian that describes the graphene molecule. The Index(H) is expressed by (8)

$$\mathrm{Index(H)} = \nu_+ - \nu_- = 6(1-g) - 3N_{end} \tag{52}$$

which is consistent with the exact number of the zero modes if $\nu_+ = 0$ or $\nu_- = 0$. Therefore we have obtained the theorem which relates the number of zero modes existed in a certain graphene molecule to the topological characteristics of its surface.
This result provides the number of zero modes for the familiar cases of graphene molecules. For example, since a fullerene takes $g = 0$ and $N_{end} = 0$, it is expected that it has six zero modes which correspond to the two triplets of C_{60} and of similar large molecules. For the case of nanotubes, we have $g = 0$ and $N_{end} = 0$. This results in $\nu_+ - \nu_- = 0$ which is in agreement with previous theoretical and experimental results (20; 21). The index theorem gives rise to a surprising relation between the topology and the presence of magnetic flux which is effectively inserted in graphene molecules by geometrical deformations. The number of these deformations can be associated with the general topological characteristics of the lattice surface. They are related to the zero modes of a general graphene molecule with the genus and the number of open faces of its surface.

5.3 G index theorem

To build up supersymmetry over the Dolbeault complex which are associated with deformation on topology of the lattice (22), we need one of the supercharges, $Q_2 = \bar{\partial}$. The index theorem leads to a topological invariant under deformations on Q_2 and plays an essential role in formulating the SUSY QM over the twisted Dolbeault complex caused by the deformation in a graphene system.
Let us describe the general statement of the G-index theorem. Suppose that Q and $Q^\dagger$ are supercharges which have a map from a space of bosonic states to a space of fermionic states and vice versa. Further let us take a Lie group G generated by G_i. G_i satisfies the commutation relations with Q and $Q^\dagger$ such that (17) where

$$[G_i, G_j] = i f_{ij}^k G_k, \quad [Q, G_i] = 0, [Q^\dagger, G_i] = 0, \tag{53}$$

where f_{ij}^k are structure constants. G_i has also a commutation relation with the Hamiltonian of Eq. (5)

$$[G_i, \mathrm{H}] = 0. \tag{54}$$

A supercharacter can be expressed as a supertrace of the group elements

$$\Xi(e^{i\theta^i G_i}) = \mathrm{Tr}\left((-1)^F e^{i\theta^i G_i} e^{-t\mathrm{H}}\right). \tag{55}$$

The ordinary trace requires us to take a limit of $t \to 0$. But Eq. (55) does not have any dependence on t because the non-zero energy levels can not make contributions due to canceling in pairs between the bosonic and fermionic sectors. But the zero-energy levels can only give contributions. And these are not dependent on t.
The character is topologically invariant under deformations of the operators Q and $Q^\dagger$. On deformation let us consider the supersymmetry generator Q of the ordinary N = 1 SUSY QM on a manifold. In locally geodesic coordinates Q is expressed by

$$Q = \mathrm{d} = \psi^{\mu\dagger}\partial_\mu. \tag{56}$$

When deforming Q, it is changed into the new operator

$$Q_s = \mathrm{d} + s\iota_{\mathcal{K}} = \psi^{\mu\dagger}\partial_\mu + s\mathcal{K}^\mu\psi_\mu, \tag{57}$$

where $\mathcal{K}$ denotes a Killing vector field. And then the deformed Hamiltonian is given by

$$\begin{aligned} \mathrm{H}_s &\equiv 2(Q_s + Q_s^\dagger)^2, \\ &= -\partial^2 + s^2|\mathcal{K}|^2 - \frac{1}{2}s[\psi_\mu^\dagger, \psi_\nu^\dagger]\partial_\mu\mathcal{K}_\nu - \frac{1}{2}s[\psi_\mu, \psi_\nu]\partial_\nu\mathcal{K}_\mu. \end{aligned} \tag{58}$$

Now suppose that we decompose the deformed $N = 1$ SUSY operator into holomorphic and antiholomorphic sectors because of $\mathrm{d} = \partial + \bar{\partial}$. Then the G-index theorem can be associated with a high degree of symmetry on a graphene molecule. Let us take a Lie group G generated by G_i which satisfies the commutation relations with Q_2 and $Q_2^\dagger$ such that

$$[G_i, G_j] = if_{ij}^k G_k, \quad [Q_2, G_i] = 0, [Q_2^\dagger, G_i] = 0 \tag{59}$$

where f_{ij}^k mean structure constants. On graphene, there exists the Hamiltonian H such that

$$\mathrm{H} \equiv -2\hbar^2 v_\mathrm{F}^2(Q_2 + Q_2^\dagger)^2, \quad [G_i, \mathrm{H}] = 0. \tag{60}$$

On deforming Q_2, this can be changed into a new operator

$$Q_{2s} = \psi^{\dagger\bar{z}}\partial_{\bar{z}} + s\mathcal{K}^{\bar{z}}\psi_{\bar{z}}, \tag{61}$$

where s is a real parameter. $\mathcal{K}$ denotes a Killing vector field associated with the deformation. In the complex coordinates z and $\bar{z}$, the Killing vector field is given by $\mathcal{K} = -y\partial_x + x\partial_y = i(z\partial_z - \bar{z}\partial_{\bar{z}})$.
Under the deformation let us consider a graphene system corresponding to the K_+. Then we can obtain a deformed Hamiltonian $\mathrm{H}_{+,s}$ on the Dolbeault complex. For convenience, after replacing s by is, the deformed Hamiltonian is given by (9; 17)

$$\begin{aligned} \mathrm{H}_{+,is} &= -2\hbar^2 v_\mathrm{F}^2(Q_{+,2is} + Q_{+,2is}^\dagger)^2 \\ &= -2\hbar^2 v_\mathrm{F}^2\left(-2\partial_{\bar{z}z}^2 + |s|^2|z|^2\right) - 2\hbar^2 v_\mathrm{F}^2\left(s[\psi^{\dagger\bar{z}}, \psi_{\bar{z}}] + s(\bar{z}\partial_{\bar{z}} - z\partial_z)\right). \end{aligned} \tag{62}$$

This operator gives rise to eigenvalues (9; 17)

$$E_{+,nml} = -2\hbar^2 v_\mathrm{F}^2\left[|s|\left((n + \frac{1}{2}) + (m + \frac{1}{2}\right) + sl + s(\pm 1)\right]. \tag{63}$$

Depending on the choice of $s > 0$ or $s < 0$, we can consider two cases for zero eigenvalues satisfying the topological invariance imposed by the index theorem. First if $s > 0$, we should take -1 for the $[\psi^{\dagger\bar{z}}, \psi_{\bar{z}}]$. Now we get in the bosonic sector. We can have zero eigenvalues in the case of $l = 0, -1, -2, \cdots$. Second if $s < 0$, we have to choose $+1$ for $[\psi^{\dagger\bar{z}}, \psi_{\bar{z}}]$. And then we can get zero eigenvalues for $l = 0, 1, 2, \cdots$.
According to the choice of s, let us take two Hamiltonians $\mathrm{H}^{\uparrow}_{+,is} = \widehat{Q}^{\dagger}_{+,2is}\widehat{Q}_{+,2is}$ and $\mathrm{H}^{\downarrow}_{+,is} = \widehat{Q}_{+,2is}\widehat{Q}^{\dagger}_{+,2is}$ as two superpartners which imply the Z_2 grading over the Hilbert space. Here we have expressed $\widehat{Q}_{+,2is} = -i\sqrt{2}\hbar v_{\mathrm{F}}(Q_{+,2is} + Q^{\dagger}_{+,2is})$. Then we want to calculate the eigenvalues for the up-spin and down-spin components

$$\mathrm{H}^{\uparrow\downarrow}_{+,is}|\psi^{\uparrow\downarrow}_{+,nml}> = E^{\uparrow\downarrow}_{+,nml}|\psi^{\uparrow\downarrow}_{+,nml}>, \tag{64}$$

where $E^{\uparrow\downarrow}_{+,n+1ml} > E^{\uparrow\downarrow}_{+,nml} \geq E^{\uparrow\downarrow}_{+,000}$. Suppose that $E^{\uparrow}_{+,000}$ is zero. Then making use of the relations

$$\widehat{Q}_{+,i2s}\mathrm{H}^{\uparrow}_{+,is} = \widehat{Q}_{+,2is}\widehat{Q}^{\dagger}_{+,2is}\widehat{Q}_{+,2is} = \mathrm{H}^{\downarrow}_{+,is}\widehat{Q}_{+,2is}, \tag{65}$$

we can have

$$\begin{aligned}\mathrm{H}^{\downarrow}_{+,is}\widehat{Q}_{+,2is}|\psi^{\uparrow}_{+,nml}> &= \widehat{Q}_{+,2is}\mathrm{H}^{\uparrow}_{+,is}|\psi^{\uparrow}_{+,nml}> \\ = E^{\uparrow}_{+,nml}\widehat{Q}_{+,2is}|\psi^{\uparrow}_{+,nml}> .\end{aligned} \tag{66}$$

This means that if $E^{\uparrow}_{+,nml} \neq 0, \widehat{Q}_{+,2is}|\psi^{\uparrow}_{+,nml}>$ is an eigenstate of $\mathrm{H}^{\downarrow}_{+,is}$. And similarly

$$\begin{aligned}\mathrm{H}^{\uparrow}_{+,is}\widehat{Q}^{\dagger}_{+,2is}|\psi^{\downarrow}_{+,nml}> &= \widehat{Q}^{\dagger}_{+,2is}\mathrm{H}^{\downarrow}_{+,is}|\psi^{\downarrow}_{+,nml}> \\ = E^{\downarrow}_{+,nml}\widehat{Q}^{\dagger}_{+,2is}|\psi^{\downarrow}_{+,nml}> .\end{aligned} \tag{67}$$

If $E^{\downarrow}_{+,nml} \neq 0, \widehat{Q}^{\dagger}_{+,2is}|\psi^{\downarrow}_{+,nml}>$ is an eigenstate of $\mathrm{H}^{\uparrow}_{+,is}$. Therefore, for non-zero eigenvalues there exist the up-spin and down-spin eigenstates in pair. They form a supermultiplet connected by the supercharge $\widehat{Q}_{+,2is}$. The up-spin (down-spin) sector may be described as the bosonic (fermionic) sector at the K_+ point.
In the case of the zero eigenvalues for the down-spin sector, we should investigate both $E^{\downarrow}_{+,000} = 0$ and $E^{\downarrow}_{+,000} \neq 0$ separately. Let us assume that $E^{\downarrow}_{+,000} \neq 0$. Then the lowest eigenstate of $\mathrm{H}^{\downarrow}_{+,is}$ is expressed as $|\psi^{\downarrow}_{+,000}> \propto \widehat{Q}_{+,2is}|\psi^{\uparrow}_{+,1ml}>$. Hence, there exists a supermultiplet between the states $|\psi^{\uparrow}_{+,n+1ml}>$ and $|\psi^{\downarrow}_{+,nml}>$, which have the same energy $E^{\uparrow}_{+,n+1ml} = E^{\downarrow}_{+,nml}$, and we obtain energy eigenstates (9)

$$\begin{aligned}|\psi^{\downarrow}_{+,nml}> &= \frac{1}{\sqrt{|E^{\uparrow}_{+,n+1ml}|}}\widehat{Q}_{+,2is}|\psi^{\uparrow}_{+,n+1ml}>, \\ |\psi^{\uparrow}_{+,n+1ml}> &= \frac{1}{\sqrt{|E^{\downarrow}_{+,nml}|}}\widehat{Q}^{\dagger}_{+,2is}|\psi^{\downarrow}_{+,nml}>\end{aligned} \tag{68}$$

for $n \geq 0$. If $E^{\downarrow}_{+,000} = 0$ and $\widehat{Q}^{\dagger}_{+,2is}|\psi^{\uparrow}_{+,000}>= 0$, the relationships are written by (9)

$$|\psi^{\downarrow}_{+,nml}> = \frac{1}{\sqrt{|E^{\uparrow}_{+,nml}|}}\widehat{Q}_{+,2is}|\psi^{\uparrow}_{+,nml}>,$$
$$|\psi^{\uparrow}_{+,nml}> = \frac{1}{\sqrt{|E^{\downarrow}_{+,nml}|}}\widehat{Q}^{\dagger}_{+,2is}|\psi^{\downarrow}_{+,nml}> \tag{69}$$

for $n \geq 1$. Similarly, one can repeat eigenvalue problem for $H_{-,is}$, which corresponds to the K'_- point. As a relation between the up-spin and down-spin eigenstates, the bosonic (fermionic) sector is regarded as the down-spin (up-spin) sector at the K'_- point. And hence there exists the 4-fold degenerate energy spectrum.

6. Deformed energy eigenvalues and unconventional quantum Hall effect

On a sheet of graphene, let us consider the problem of magnetic field concentrated on a thin cylindrical shell of small, but finite radius $l_B = \sqrt{\frac{c\hbar}{eB}}$. The corresponding vector potential is given by $\vec{a} = (-y, x)/2l_B^2$ on the two dimensional plane of graphene. The problem in question is to compute the eigenvalues of the Dirac Hamiltonian in the field of a fractional magnetic flux on the graphene sheet (23; 24). Now under the fractional magnetic flux, the eigenvalues for $n = 0, l = 0, m = 0$ are given by

$$\sqrt{E^{\uparrow}_{+,000}} = \sqrt{E^{\downarrow}_{-,000}} = 0, \tag{70}$$

and

$$\sqrt{E^{\uparrow}_{+,n+100}} = \sqrt{E^{\downarrow}_{+,n00}} = \sqrt{E^{\downarrow}_{-,n+100}} = \sqrt{E^{\uparrow}_{-,n00}} = \pm\hbar w_{l_B}\sqrt{n+1} \tag{71}$$

for $n \geq 0, l = 0, m = 0$. Here $w_{l_B} \equiv \frac{\sqrt{2}v_F}{l_B}$. Equation (70) tells us that there is one zero-energy state only in the case of up-spin fermions but not in the case of down-spin fermions at the K_+ point. At K'_- point we can have one zero-mode state for down-spin fermions but not for up-spin fermions. The magnetic field direction at K_+ is opposite to that at K'_-. The zero-energy state may have the four-fold degeneracy emerging from electrons and holes (8). Since the LL of the zero-energy states becomes half-filled, no one would observe plateau at $\nu = 0$. But by index theorem, the flux quanta produce $4r$ $(r = 0, 1, 2, \cdots)$ zero-energy states. The $2r$ states of these are occupied. The flux quanta lift the $2r$ states to the Fermi energy. And then they can be removed by doping. The degeneracy between electrons and holes would be removed. We could observe the Hall plateau at $\nu = 0$ because holes are occupied before electrons. And hence we can describe an experimental observation of the Hall plateau emerging at $\nu = 0$. (10) On the basis of the index theorem, we compute the energy spectrum of the deformed Hamiltonian, $H_{+,is} = -2\hbar^2 v_F^2(Q_{+,2is} + Q^{\dagger}_{+,2is})^2$ at the K_+ point. The up-spin states of zero energy are $|0>, |1>, |2>, \cdots, |j^{\uparrow} - 1>$. They are degenerate in $|\psi^{\uparrow}_{+,0ml}>$. On the other hand, for down-spin states, we have to describe two cases. As the first case, assume that $j^{\downarrow} = 0$. Then, there do not exist any zero-energy states. So we may construct the supermultiplet given by Eq. (68). In the other case, if $j^{\downarrow} \neq 0$, the zero-energy states are given by $|0>, |1>, |2>, \cdots, |j^{\downarrow} - 1>$ as degenerate states of $|\psi^{\downarrow}_{+,0ml}>$. Therefore, these result in the $(j^{\uparrow} + j^{\downarrow})-$fold

degeneracy in the zero-energy state for fermions at the K_+ point. This degeneracy implies the exact correspondence between $j^\uparrow$ fermions and $j^\downarrow$ fermions under deformation. Similarly, we can investigate the energy spectrum of the zero-energy states at the K'_- point.
In order to generate the up-spin and down-spin states of zero energy the deformed superoperators are written, in terms of the original supercharges, by

$$\widehat{Q}_{+,2is} = \widehat{Q}^{\dagger j^\downarrow}_{+,2is} \widehat{Q}^{j^\uparrow}_{+,2is'}, \quad \widehat{Q}^\dagger_{+,2is} = \widehat{Q}^{\dagger j^\uparrow}_{+,2is} \widehat{Q}^{j^\downarrow}_{+,2is'} \tag{72}$$

where $j^\uparrow$ and $j^\downarrow$ are integers such as $j^\uparrow > j^\downarrow$. In terms of the $\widehat{Q}^\dagger_{+,2is}$ the state $|n>$ is given by

$$|n> = \frac{1}{\sqrt{n!}} (\widehat{Q}^\dagger_{+,2is})^n |0> . \tag{73}$$

Now, let us calculate the deformed eigenvalues by using Eq. (73) and solving the eigenvalue problems of Eq. (64) for the up-spin and down-spin components. In the bosonic sector, the deformed energy eigenvalues are expressed by

$$\sqrt{E^\uparrow_{+,000}} = \sqrt{E^\downarrow_{-,000}} = 0, \tag{74}$$

for $n = 0, l = 0, m = 0$. And we can have

$$\sqrt{E^\uparrow_{+,nml}} = \sqrt{E^\downarrow_{-,nml}} = \pm \hbar w_{l_B} \sqrt{\frac{(n + j^\uparrow - 1)!(n + j^\downarrow - 1)!}{[(n-1)!]^2}} \tag{75}$$

for $n \geq 1, l = -(j^\uparrow - 1), m = 2(j^\uparrow - 1)$. In the case of the fermionic sector, if $j^\downarrow = 0$, the deformed eigenvalues are

$$\sqrt{E^\downarrow_{+,nml}} = \sqrt{E^\uparrow_{-,nml}} = \pm \hbar w_{l_B} \sqrt{\frac{(n + j^\downarrow)!(n + j^\uparrow)!}{[(n)!]^2}} \tag{76}$$

for $n \geq 0, l = j^\downarrow, m = 2j^\downarrow$. If $j^\downarrow \neq 0$, the eigenvalues are given by

$$\sqrt{E^\downarrow_{+,000}} = \sqrt{E^\uparrow_{-,000}} = 0, \tag{77}$$

for $n = 0, l = 0, m = 0$. And we obtain

$$\sqrt{E^\uparrow_{+,n+1ml}} = \sqrt{E^\uparrow_{-,n+1ml}} = \pm \hbar w_{l_B} \sqrt{\frac{(n + j^\downarrow - 1)!(n + j^\uparrow - 1)!}{[(n-1)!]^2}} \tag{78}$$

for $n \geq 1, l = j^\downarrow - 1, m = 2(j^\downarrow - 1)$. We can check up that there exists $4(j^\uparrow + j^\downarrow)$-fold degeneracy in the zero-energy states and 4-fold one in all other states.
Among the energy spectrum given above, let us account for the special cases of $j^\uparrow = 1$, and $j^\downarrow = 0$ and $j^\uparrow = 2$, and $j^\downarrow = 0$. In the case of $j^\uparrow = 1$, and $j^\downarrow = 0$ it is not hard to check up that the energy eigenvalues are given by Eqs. (77) and (78). These results correspond to the spectrum of the monolayer graphene. For the case of $j^\uparrow = 2$, and $j^\downarrow = 0$, the energy spectra are expressed in terms of

$$\sqrt{E^\uparrow_{+,000}} = \sqrt{E^\downarrow_{-,000}} = 0, \tag{79}$$

for $n = 0, l = 0, m = 0$. And we can obtain (18)

$$\sqrt{E^{\uparrow}_{+,nml}} = \sqrt{E^{\downarrow}_{-,nml}} = \pm\hbar w_{l_B}\sqrt{n(n+1)} \tag{80}$$

for $n \geq 1, l = -1, m = 2$, while having

$$\sqrt{E^{\downarrow}_{+,nml}} = \sqrt{E^{\uparrow}_{-,nml}} == \pm\hbar w_{l_B}\sqrt{(n+1)(n+2)} \tag{81}$$

for $n \geq 0, l = 0, m = 0$. These energy spectra are eigenvalues of the bilayer graphene affected by the deformation, and are in agreement with the results in the literature (10; 12; 14). On the basis of the index theorem we have shown that there exist the $j^{\uparrow}$-fold and $j^{\downarrow}$-degeneracy in the zero-energy state at the K_+ point and similarly at the K'_-. And hence we can obtain the QHE characterized by (9)

$$\sigma_{xy} = \nu\frac{e^2}{h}, \nu = \pm 4(|n| + \tfrac{j^{\uparrow}+j^{\downarrow}}{2}). \tag{82}$$

7. Summary and conclusion

We presented the electronic properties of massless Dirac fermions characterized by geometry and topology on a graphene sheet in this chapter. Topological effects can be elegantly described by the Atiyah-Singer index theorem. It provides a topological invariant under deformations on the Dirac operator and plays an essential role in formulating supersymmetric quantum mechanics over twisted Dolbeault complex associated with the deformation on the topology of the lattice in a graphene system. Exploiting the G-index theorem and a high degree of symmetry, we explained deformed energy eigenvalues in graphene. The Dirac fermions result in SU(4) symmetry emerging out of both the pseudospin and spin as a high degree of symmetry in the noninteracting Hamiltonian of monolayer graphene. Under the topological deformation the zero-energy states emerge naturally without the Zeeman splitting at the Fermi points in the graphene sheet. Thus we observed an emergence of a higher degree of hidden symmetry under the topological deformation in graphene while the pseudospin is a good symmetry at the K and K' points in graphene. In the particular SU(2) of the pseudospin, the SU(2) is the exact spin symmetry of each Landau level. In the case of nonzero energy, the up-spin and down-spin states have the exact high symmetries of spin, forming the pseudospin singlet pairing. The pseudospin can play a key role on the physics of the $n = 0$ LL in the graphene sheet. The valley pseudospin degeneracy can lift only at the zeroth LL. The 4-fold degeneracy can be removed in the zero-energy states of monolayer graphene. If the mass terms were taken into account, the four-fold degeneracy can be removed in the zero-energy state of monolayer graphene. We can exploit this to understand the emergence of a Hall plateau at $n = 0$ in the experimental observations. But the four-fold degeneracy is not removed in the higher LLs. Including the Coulomb interaction, we can lift the degeneracy. The pseudospin symmetry $SU(2)$ is broken to $U(1) \times Z_2$. Therefore the total symmetry gives rise to $SU(2)_{spin} \times (U(1) \times Z)_{psuedospin}$ while the spin symmetry $SU(2)$ remains to be exact. Hence we understood the peculiar and unconventional quantum Hall effects of the $n = 0$ Landau level in monolayer graphene on the basis of the index theorem and the high degree of symmetry under the topological deformation without the Zeeman splitting. It would be very interesting and quite possible to apply the present approach to investigation of the composite Dirac fermions and fractional quantum Hall effects in graphene.

8. Acknowledgments

This work was supported by the National Research Laboratory Program (R0A-2005-001-10152-0), by Basic Science Research Program (2009-0083512), and by Priority Research Centers Program (2009-0094037) through the National Research Foundation of Korea funded by the Ministry of Education, Science and Technology, and by the Brain Korea 21 Project in 2010.

9. References

[1] P. R. Wallace, Phys. Rev. 71, 622 (1947).
[2] A. H. Castro Neto, et al., Rev. Mod. Phys. 81, 109 (2009).
[3] K. S. Novoselov *et al.*, Science 306, 66 (2004).
[4] G. W. Semenoff, Phys. Rev. Lett. 53, 2449 (1984).
[5] F. D. H. Haldane, Phys. Rev. Lett. 61, 2015 (1988).
[6] K. S. Novoselov *et al.*, Nature (London) 438, 197 (2005).
[7] Y. Zhang *et al.*, Nature (London)438, 201 (2005).
[8] J. K. Pachos and M. Stone, Int. J. Mod. Phys. B 21, 5399 (2007).
[9] K.-S. Park *et al.*, Int. J. Mod. Phys. B,23, 4801 (2009); K.-S. Park *et al.*, J. Korean Phys. Soc. 50, No 6 (2007).
[10] Y. Zhang *et al.*, Phys. Rev. Lett. 96, 136806 (2006).
[11] K. S. Novoselov *et al.*, Nature Physics 2, 177 (2006).
[12] B. E. Feldman *et al.*, Nature Phys. 5, 889 (2009).
[13] Y. Zhao, Phys. Rev. Lett. 104, 066801 (2010).
[14] E. McCann *et al.*, Phys. Rev. Lett. 96, 086805 (2006).
[15] M. F. Atiyah and I. M. Singer, Ann. Math. 87, 485 (1968).
[16] E. Witten, Nucl. Phys. B 185, 513 (1981); Nucl. Phys. B 202, 253 (1982); J. Differential Geometry 17, 661 (1982).
[17] M. Stone, Nucl. Phys. B 314, 557 (1989).
[18] M. Ezawa, Phys. Lett. B 372, 924 (2008).
[19] J. González *et al.*, Nucl. Phys. B 406, 77 (1993).
[20] S. Reich, C. Thomsen and P. Ordejon, Phys. Rev. B 65, 155411 (2002).
[21] R. Saito, M. Fujita, G. Dresselhuas, and M. S. Dresselhaus, Phys. Rev. B 46, 1804 (1992).
[22] T. Eguchi *et al.*, Phys. Rep. 66, 213 (1980).
[23] R. Jackiw and P. Rossi, Nucl. Phys. B 190, 681 (1981).
[24] S. Das Sarma and K. Yang, arXiv: 0906.2209v2 (2009).
[25] S. Kim, K.-S. Park, and K. S. Yi, J. Korean Phys. Soc. 49, S575 (2006).

Part 2

Graphene Nano-dots and Nanoribbons

12

Physics of Triangular Graphene

Motohiko Ezawa
Department of Applied Physics, The University of Tokyo
Japan

1. Introduction

Carbon forms many allotropic forms with different dimensionality: three-dimensional diamond, quasi-two-dimensional graphite(1; 2), one-dimensional nanotubes(3–10) and zero-dimensional fullerenes(11). These materials are well known to exhibit remarkable electronic properties depending on their geometry. There exists another form of carbon material, graphene, which is a monolayer graphite. Graphene can theoretically be considered the basic material for the other forms of carbon that can be obtained from it either by stacking (graphite), wrapping (nanotubes), or creation of topological defects (fullerenes). Furthermore, there are many derivatives in these graphene related materials. Their electrical properties, such as carrier type (particle or hole), can be continuously controlled by the application of external gate voltage(12).

Graphite is made out of stacks of graphene layers that are weakly coupled by van der Waals forces. A pencil is a typical graphite. When we use it, we are actually producing graphene stacks on a sheet of paper, which may well contain individual graphene layers. Nevertheless, it was only in 2004 when graphene was experimentally isolated(12). They produced high quality, large (up to 100 μm in size) graphene crystallites, which immediately triggered enormous theoretical and experimental studies. Moreover, the quality of the samples produced are so good that ballistic transport(12) and quantum Hall effects (QHE) have been observed(13; 14). For a recent review on graphene, see ref. (15).

Edge states of graphene show a very remarkable feature. There are three type of edges; zigzag edge, armchair edge and Klein edge(16). Graphene with a zigzag edge has the half-filled flat band at the zero-energy level and exhibits edge ferromagnetism(17). These edge states can be observed by STM(18; 19).

Among graphene derivatives, graphene nanoribbons(17; 20; 21) constitute a fascinating object due to a rich variety of band gaps, from metals to wide-gap semiconductors. In particular, the half-filled zero-energy states emerge in all zigzag nanoribbons and hence they are metallic. Such a peculiar band structure, as recognized by the pioneering work(17) in 1996 and revisited(21) in 2006, has motivated many researchers to investigate their electronic and magnetic properties. Now there are a profusion of papers on them, among which we cite some of early works(22–31). Important works on nanoribbons have been summarized into a map(32) as Essential Science Indicators by Tomson-Reuter in late 2008. Nanoribbons can be manufactured by patterning based on nanoelectronic lithography methods(26; 27; 33).

Another basic element of graphene derivatives is a graphene nanodisk(35; 36). It is a nanometer-scale disk-like material which has a closed edge. It is also referred to as nanoisland(37), nanoflake(38–40), nanofragment(41) or graphene quantum dot(42; 43).

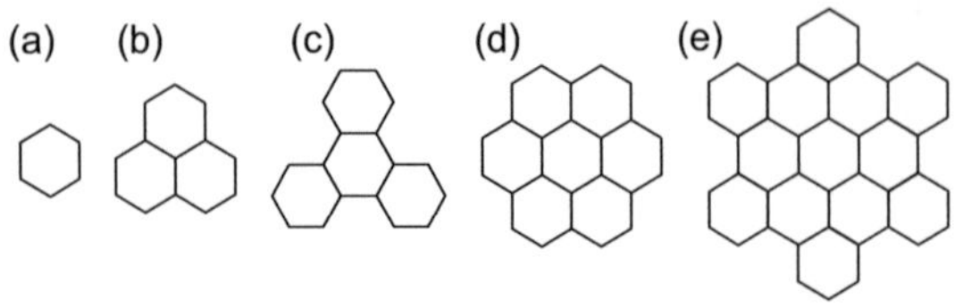

Fig. 1. Basic configurations of typical graphene nanodisks. (a) Benzene. (b) Trigonal zigzag nanodisk (phenalene). (c) Trigonal armchair nanodisk (triphenylene). (d) Hexagonal zigzag nanodisk (coronene). (e) Hexagonal armchair nanodisk (hexa benzocoronene)(34).

Nanoribbons and nanodisks correspond to quantum wires and quantum dots, respectively. They are candidates of future carbon-based nanoelectronics and spintronics alternative to silicon devices. A merit is that a nanoribbon-nanodisk complex can in principle be fabricated, embodying various functions, only by etching a graphene sheet.

There are many types of nanodisks, as displayed in Fig.1. It is possible to manufacture them by etching a graphene sheet by Ni nanoparticles(44). Among them, trigonal zigzag nanodisks are prominent in their electronic property because there exist half-filled zero-energy states in the non-interacting regime(36). This novel electric property was revealed first by the tight-binding model(36) in 2007, and then by first-principle calculations by other groups(37; 38; 45). Various remarkable properties of nanodisks have been investigated extensively in a series of works(36; 46–48).

Although there are yet only a few works on nanodisks, an experimental realization(44) must accelerate both experimental and theoritical studies on them. In this sense, the study of graphene nanodisk is very timely.

It has long been known(49–51) that the physics of electrons near the Fermi energy is described by the massless two-component Dirac equation or the Weyl equation in graphene. Graphene nanoribbons were successfully analyzed based on the Weyl equation in 2006(22). It is straightforward to generalize the method to investigate the structure of the zero-energy states in graphene nanodisks(48). Wave functions are explicitly constructed, which exhibits a texture of magnetic vortices peculiar to the representation of the trigonal symmetry group C_{3v}. We show the emergence of a vortex carrying the winding number 2. Such a vortex is highly unusual in all branches of physics. An analytic form of wave functions makes it possible to explore deep physics of graphene nanodisks.

The inclusion of Coulomb interactions provides the ground state with a finite spin proportional to the edge length according to the following argument. Hund's rule is well known in atomic physics, according to which the ground state of atoms has the maximum spin. Thus, assuming N-fold degenerate zero-energy states, the spin of the ground state is $N/2$. On the other hand, there exists Lieb's theorem applicable to flat-band ferromagnetism, according to which the sublattice imbalance $N = |L_A - L_B|$ per unit cell leads to the polarized ground state with spin $N/2$. Both theorems are relevant to graphene nanodisks, by assuring the emergence of the polarized ground state with spin $N/2$.

The nanodisk-spin system undergoes a quasi-phase transition between the quasiferromagnet and the parramagnet(47), which is a precursor of an actual phase transition. Graphene nanodisks can be interpleted as quantum dots made of graphene, as already noticed. In this point of view, it is natural to expect Coulomb blockades(46) and Kondo effects(47) by connecting a nanodisk with leads.

On the other hand, the study of spin-dependent transport phenomena has recently attracted much attention(52; 53). It has opened the way to the field of spintronics(54–57), literally spin electronics, where new device functionalities exploit both the charge and spin degrees of freedom. There are various approaches in this sphere. For instance, the use of a quantum-dot setup(58; 59) has been proposed, which can be operated either as a spin filter to produce spin-polarized currents or as a device to detect and manipulate spin states. However, there are difficulties of spintronics based on conventional materials, since spin directions are quickly relaxed and coherence is easily lost due to spin-orbit interactions. In graphene, spin relaxation length is as large as $2\mu m$ due to small spin-orbit interactions(60), as has been confirmed by spin precession measurement. Long spin relaxation length has motivated spintronics based on graphene(60–63) recently. Nevertheless, there exists a key issue how to generate and manipulate spin currents. Localized spins are necessary for this purpose. It can be solved by a new type of materials, graphene nanodisks. Some application to spintronics, such as a spin filter, a spin valve and a spin switch has been proposed(64–66).

This work is organized as follows. In Section 3, we summarize the basic nature of trigonal zigzag nanodisks. We introduce the size parameter N for trigonal zigzag nanodisks. There exists N-fold degenerate zero-energy states, implying that the zero-energy sector has the SU(N) symmetry. We classify them according to the representation of the trigonal symmetry group C_{3v}. In Section 4, we analyze these zero-energy states based on the Dirac theory of graphene. They are shown to be edge modes indexed by the edge momentum and grouped according to the representation of the trigonal symmetry group C_{3v}. Wave functions are explicitly constructed as analytic functions around the K point and as anti-analytic functions around the K′ point. By evaluating the probability density flow we find a texture of magnetic vortices perpendicular to the nanodisk plane. It is intriguing that a vortex with the winding number 2 emerges in the state belonging to the E (doublet) representation. In Section 5, we make an investigation of electron-electron interaction effects in the zero-energy sector. We derive explicitly the direct and exchange interactions, which break the SU(N) symmetry but not so strongly. We show that the spin stiffness is quite large due to large exchange interactions, which means that a nanodisk is a rigid ferromagnet. The system is well described by the infinite-range Heisenberg model within the SU(N) approximation, which is exactly solvable. We explore thermodynamical properties. Constructing the partition function, we calculate the specific heat, the entropy, the magnetization and the susceptibility. A sharp peak emerges at a certain temperature ($T = T_c$) in the specific heat, which we interpret as a quasi-phase transition between the quasiferromagnet and quasiparamagnet states. In Section 6, we argue that our analysis based on the zero-energy sector stands as it is even if the size of the nanodisk is large. In Section 8.0.0.3, we make an investigation of the spin current, the spin-filter effects. We first analyze how the spin of a nanodisk filters the spin of the current by assuming that the nanodisk is a rigid ferromagnet. However, the nanodisk is actually not a rigid ferromagnet but a quasiferromagnet. Hence an intriguing reaction phenomenon occurs: Namely, the spin of the nanodisk can be controlled by the spin of the current. We analyze the reaction to the spin of the nanodisk from the spin of electrons in the current. In Section 8, we propose a rich variety of spintronic devices made of nanodisks and leads, such as spin memory, spin amplifier, spin valve, spin-field-effect transistor, spin diode and spin switch. Graphene nanodisks could well be basic components of future nanoelectronic and spintronic devices.

2. Sample

It is possible to manufacture graphene nanoribbon and nanodisk by etching a graphene sheet. Here we review a method to produce crystallographically oriented cuts in single-layer

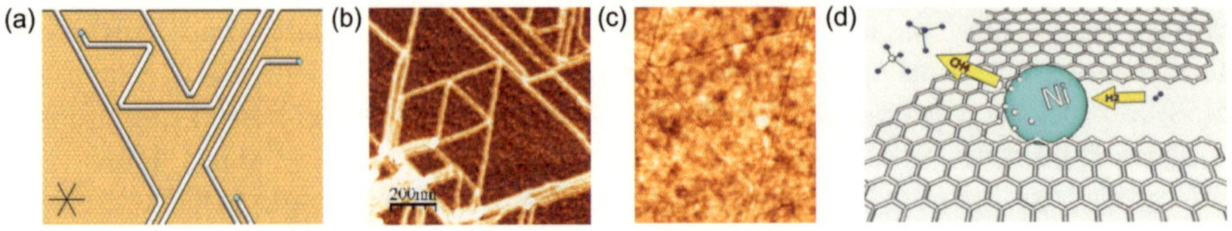

Fig. 2. (a) Key features of etching in SLG are chirality-preserving angles of 60 and 120, avoided crossing of trenches leaving $\sim$10 nm spacing between adjacent trenches and producing connected nanostructures, and trenches and nanoparticles with uniform width $<$ 10 nm. (b) AFM phase image of etched SLG with produced geometric nanostructures. The phase image obscures small details, making adjacent trenches appear to merge together. (c) AFM height image of equilateral triangle connected to three nanoribbons. (d) artoon of a Ni particle etching a graphene sheet (not to scale). Ni nanoparticles absorb carbon from graphene edges which then reacts with H2 to create methane. These figures are taken from the work of Campos et al. (44)

graphene (SLG) which effectively has all cuts oriented with the same edge-chirality based on the work of Campos et al. (44). By using this method nanoribbons, equilateral triangles (trigonal nanodisks), and other graphene nanostructures are fabricated, which could feature novel electronic behavior resulting from their specific edge orientations.

During the high-temperature etching stage, the Ni nanoparticles etch the graphene through catalytic hydrogenation of carbon, where carbon atoms from exposed graphene edges dissociate into the Ni nanoparticle, and then react with H_2 at the Ni surface. This process is summarized by the chemical reaction

$$\mathrm{Ni} + \mathrm{C}_{\mathrm{graphene}} + 2\mathrm{H}_2 \rightarrow \mathrm{Ni} + \mathrm{CH}_4, \tag{1}$$

where carbon from the graphene is hydrogenated into methane by the Ni nanoparticle catalyst. The same reaction can also be understood as the effective reverse of catalytic carbon nanotube (CNT) growth, and indeed it is found that the growth of CNTs competes with the desired etching results in graphene. This occurs because at high concentrations of carbon the Ni nanoparticles become supersaturated with carbon and can begin to expel carbon nanotubes. It is for this reason that extra precaution is necessary during preparation to avoid organic contaminants which can act as carbon sources and saturate the Ni nanoparticles. In addition, methane produced by the etching process itself can supersaturate the Ni nanoparticles if the amount of Ni on the substrate is too low.

The etching process produces a mosaic of clearly defined cuts across the SLG surface, as measured by atomic force microscopy [Fig.2(b,c)]. These continuous trenches left behind by individual nanoparticles run along straight lines, intermittently deflecting from their path or reflecting away from previously etched trenches. These deflections and reflections show a surprising regularity, with measured angles between any pair of trenches of either 60° or 120° [Fig. 2(a)]. Trenches forming edges at angles of 60° and 120° preserve the chirality of the edges, indicating that nearly all the cuts in our samples run along the same crystallographic orientation.

3. Electronic properties of nanodisks

In this section we summarize the basic nature of trigonal zigzag nanodisks(36). There exists N-fold degenerate zero-energy states, implying that the zero-energy sector has the SU(N) symmetry. All these states are classified into singlets (A representation) and doublets (E representation) according to the representation of the trigonal symmetry group C_{3v}. Wave functions are constructed explicitly by diagonalizing the Hamiltonian.

3.1 Classification of nanodisks

We first explore the electronic and magnetic properties of graphene nanodisks. It is a nanometer-scale disk-like graphene derivative which has a closed edge. There are many type of nanodisks, where typical examples are displayed in Fig.1. Graphene nanodisks are classified by the edge and shape. Typical edges are zigzag edge and armchair edge, while typical shapes are trigonal shape and hexagonal shape.

3.2 Energy Spectrum of nanodisks

We calculate the energy spectra of graphene derivatives based on the nearest-neighbor tight-binding model, which has been successfully applied to the studies of carbon nanotubes and graphene nanoribbons. The Hamiltonian is given by

$$H = \sum_i \varepsilon_i c_{i,\sigma}^{\dagger} c_{i,\sigma} + \sum_{\langle i,j \rangle} t_{ij} c_{i,\sigma}^{\dagger} c_{j,\sigma}, \tag{2}$$

where ε_i is the site energy, t_{ij} is the transfer energy, and $c_{i,\sigma}^{\dagger}$ is the creation operator of the π electron with spin σ at site i. The sum is taken over all nearest-neighboring sites $\langle i,j \rangle$. Owing to their homogeneous geometrical configuration, we may take constant values for these energies, $\varepsilon_i = \varepsilon_{\mathrm{F}}$ (Fermi energy) and $t_{ij} = t \approx 2.70$eV. Then, the diagonal term in (2) yields just a constant, $\varepsilon_{\mathrm{F}} N_{\mathrm{C}}$, where N_{C} is the number of carbon atoms in the system. The Hamiltonian (2) yields the Weyl electrons for graphene(12; 13; 67), as we review in Section 4.1. There exists one electron per one carbon and thus the band-filling factor is $1/2$. It is customarily chosen as the zero-energy level of the tight-binding Hamiltonian (2) at this point so that the energy spectrum is symmetric between the positive and negative energy states. Therefore, the system is metallic provided that there exists zero-energy states in the spectrum. It is understood that carbon atoms at edges are terminated by hydrogen atoms. We carry out the calculation together with this condition.

In analyzing a nanodisk containing N_{C} carbon atoms, the Hamiltonian (2) is reduced to an $N_{\mathrm{C}} \times N_{\mathrm{C}}$ matrix. It is possible to exactly diagonalize the Hamiltonian to determine the energy spectrum E_i together with its degeneracy g_i for each nanodisk. The density of state is given by

$$D(\varepsilon) = \sum_{i=1}^{N_{\mathrm{C}}} g_i \delta(\varepsilon - E_i). \tag{3}$$

We have explicitly diagonalized the Hamiltonian (2) for several nanodisks with trigonal zigzag shape, trigonal armchair shape and hexagonal zigzag shape. It is remarkable that there exist zero-energy states only in trigonal zigzag nanodisks. We have also checked explicitly the absence of the zero-energy state in a series of nanodisks with hexagonal armchair type. Actually there exist general arguments for the number of zero-energy states: See Section 3.4.2 with the use of the electron-hole symmetry and Section 3.4.4 with the use of the graph theory. We comment on the validity of the tight-binding model. Although the model may be viewed as too naive, it actually gives quite good results in the graphene system for the following reasons. First, the graphene is composed of a single kind of atoms, namely, carbons. It is a "super-clean material", as is the He liquid. Accuracy of the tight-binding model increases

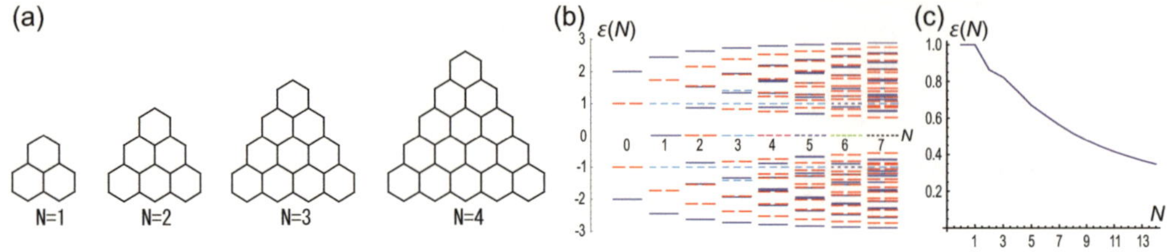

Fig. 3. (a) Geometric configuration of trigonal zigzag nanodisks. We define its size N by $N = N_{\text{ben}} - 1$ with N_{ben} the number of benzenes on one side of the trigon. Here, $N = 1, 2, 3, 4, 5$. The number of carbon atoms are related as $N_C = N^2 + 6N + 6$. (b) Density of states of the N-trigonal nanodisk for $N = 0, 1, 2, \cdots, 7$. The horizontal axis is the size N and the vertical axis is the energy $\varepsilon(N)$ in units of $t = 2.7$eV. Segments in each energy level indicate the degeneracy of the level. (c) The excitation gap ε as a function of the size N. It is approximately proportional to N^{-1}.

as the number of involved elements becomes lower. For instance, it is hard to determine the energy spectrum by the simple tight-binding model in transition metal oxides, since it contains many elements. Second, carbon is a light element, and there are only a few electrons in the shell. Especially in carbon there are only s and p orbitals. Furthermore, it is possible to neglect spin-orbit couplings since they are very small. Third, the graphene consists of the sp^2 orbital and the π orbital, where sp^2 orbitals contribute to make a rigid skeleton of the graphene structure, while π orbitals spread out in graphene and behave as free electrons. These two energy levels are well separated. Consequently, the simple tight-binding Hamiltonian presents a very good description of the graphene system by treating π electrons as free electrons.

3.3 Trigonal zigzag nanodisks

We shall investigate trigonal zigzag nanodisks in more details since they have zero-energy states. We define the size N of a nanodisk by $N = N_{\text{ben}} - 1$, where N_{ben} is the number of benzenes on one side of the trigon as in Fig.3(a). We show the density of state for $N = 1, 2, \cdots, 5$ in Fig.3(b).

It can be shown that the determinant associated with the Hamiltonian (2) has such a factor as

$$\det\left[\varepsilon I - H\left(N_C\right)\right] \propto \varepsilon^N, \tag{4}$$

implying N-fold degeneracy of the zero-energy states. Since there exist half-filled zero-energy states for $N \geq 1$, these nanodisks are "metallic". The gap energy between the first-excitation and the ground states decreases as the size becomes larger [Fig.3(c)].

3.4 Zero-energy sector

The size-N nanodisk has N-fold degenerate zero-energy states, where the gap energy is as large as a few eV. Hence it is a good approximation to investigate the electron-electron interaction physics only in the zero-energy sector, by projecting the system to the subspace made of those zero-energy states. The zero-energy sector consists of N orthonormal states $|f_\alpha\rangle$, $\alpha = 1, 2, \cdots, N$, together with the SU(N) symmetry. We can expand the wave function of the state $|f_\alpha\rangle$ as

$$f_\alpha(\boldsymbol{x}) = \sum_i \omega_i^\alpha \varphi_i(\boldsymbol{x}), \tag{5}$$

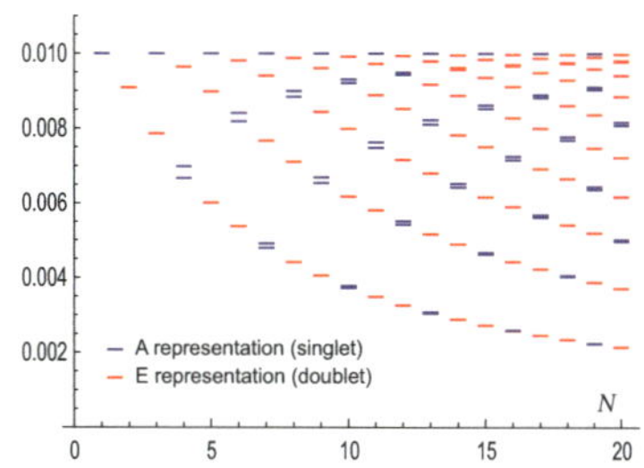

Fig. 4. Zero-energy states split into several nonzero-energy states by edge modifications. The horizontal axis is size N and the vertical axis is energy in units of $t = 2.7$eV. We take $\varepsilon_i = \varepsilon + \Delta\varepsilon$ for all edge carbons with $\Delta\varepsilon = 0.027$eV. States are grouped according to the representation of the trigonal symmetry C_{3V}. Doublet states (red) obey the E representation. Singlet states (blue) appear in pairs; the upper (lower) states obey the A_1 (A_2) representation.

where $\varphi_i(\boldsymbol{x})$ is the Wannier function localized at the site i. The operator c_i in the tight-binding Hamiltonian (2) annihilates an electron in the state described by the Wannier function $\varphi_i(\boldsymbol{x})$.

3.4.1 Trigonal symmetry

We are able to calculate the amplitude ω_i for zero-energy states in the trigonal zigzag nanodisk. In so doing it is necessary to select the N orthonormal states $|f_\alpha\rangle$ according to the representation theory of group. The symmetry group of the trigonal nanodisk is C_{3v}, which is generated by the $2\pi/3$ rotation $\mathfrak{c}_3$ and the mirror reflection $\sigma_\mathfrak{v}$. It has the representation $\{A_1$, A_2, $E\}$. The A_1 representation is invariant under the rotation $\mathfrak{c}_3$ and the mirror reflection $\sigma_\mathfrak{v}$. The A_2 representation is invariant under $\mathfrak{c}_3$ and antisymmetric under $\sigma_\mathfrak{v}$. The E representation acquires $\pm 2\pi/3$ phase shift under the $2\pi/3$ rotation. The A_1 and A_2 are 1-dimensional representations (singlets) and the E is a 2-dimensitional representation (doublet). These properties are summarized in the following character table:

C_{3v}	$\mathfrak{e}$	$2\mathfrak{c}_3$	$3\sigma_\mathfrak{v}$
A_1	1	1	1
A_2	1	1	-1
E	2	-1	0

(6)

The N orthonormal states $|f_\alpha\rangle$ together with the amplitude ω_i are constructed as follows. First, by requiring the mirror symmetry, we can explicitly construct the wave functions, all of which are found to be real. Next, we make appropriate linear combinations of these states to satisfy the trigonal symmetry C_{3v}. For this purpose, we resolve the degeneracy by applying C_{3v}-invariant perturbation to the size-N trigonal nanodisk. We note that all their wave functions are nonvanishing on edge sites and that they penetrate into the bulk with different penetration depth. Hence, if we apply an identical bias voltage to all edge carbons, we expect that the energy of the mode increases, as the penetration depth decreases. We take $\varepsilon_i = \varepsilon + \Delta\varepsilon$ for all edge carbons. We show how the zero-energy states split by taking $\Delta\varepsilon = 0.01t = 0.027$eV in Fig.4 for $N = 1, 2, \cdots, 20$. All zero-energy states are found to acquire positive energy and they become singlets or doublets. Singlet states obey the A_1 or A_2 representation, while appropriate linear combinatations of doublet states obey the E representation. There exists a nondegenerate state which has the highest energy for a nanodisk with N =odd, which is argued to belong A_2 representation. There are no such wave functions for nanodisks with N =even.

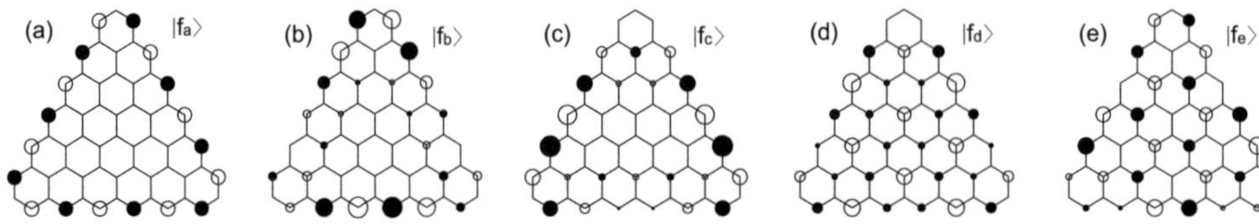

Fig. 5. (a)∼(e) The zero-energy states of the trigonal nanodisk with size $N = 5$. They are symmetric under the mirror reflection with respect to the vertical axis. The solid (open) circle denotes the positive (negative) amplitude ω_i. The amplitude is proportional to the radius of circle. Electrons are localized on edges in the state (a). When the site energy ε_i is increased at edges equally, the degeneracy is partially resolved, as illustrated in Fig.4. The state (a) has the highest energy and is a singlet in the A_2 representation of the symmetry group C_{v3}. The states (b) and (c) are degenerate; the states (d) and (e) are degenerate and have the lowest energy.

Let us explain the above scheme by taking an example of the nanodisk with size $N = 5$. First, we present 5 orthonormal states $|f_\alpha\rangle$ in Fig.5, where the solid (open) circle denotes the positive (negative) amplitude ω_i with $|\omega_i|$ being proportional to the radius of the circle. Second, the wave function $|f_a\rangle$ is entirely localized on edge sites as in Fig.5(a). Since it is invariant under the $2\pi/3$ rotation $\mathfrak{c}_3$ and antisymmetric under the mirror reflection $\sigma_\mathfrak{v}$, it is a singlet state obeying the A_2 representation. Third, we find two sets of states $(|f_b\rangle, |f_c\rangle)$ and $(|f_d\rangle, |f_e\rangle)$ are degenerate under the C_{3v}-invariant perturbation. Thus, we have identified 5 states in the spectrum of Fig.4. From the mirror symmetric states we construct the trigonal symmetric states by making a linear combination within each doublet, $|g_\pm\rangle = \frac{1}{\sqrt{2}}(|f_b\rangle \pm i|f_c\rangle)$, $|h_\pm\rangle = \frac{1}{\sqrt{2}}(|f_d\rangle \pm i|f_e\rangle)$. Each sets of complex wave functions, $(|g_+\rangle, |g_-\rangle)$ and $(|h_+\rangle, |h_-\rangle)$, transform properly under the E representation. We shall present an analytic approach to investigate these states based on the Dirac theory in Section 4.3. In the instance of $N = 5$, $(|g_+\rangle, |g_-\rangle)$ and $(|h_+\rangle, |h_-\rangle)$ correspond to $(|+k_2^+\rangle, |-k_2^+\rangle)$ and $(|+k_2^-\rangle, |-k_2^-\rangle)$, respectively, with $k_n^\pm$ being defined in (38).

3.4.2 Electron-hole symmetry

We have shown that the zero-energy sector of the trigonal zigzag nanodisk has the N-fold degeneracy by diagonalizing the Hamiltonian (2) explicitly. We wish to derive this fact from a general point of view. First, we appeal to the electron-hole symmetry of the tight-binding Hamiltonian (2). The Hamiltonian is invariant under the electron-hole transformation, $c_{\sigma,A} \to c^\dagger_{\sigma,A}$, $c_{\sigma,B} \to -c^\dagger_{\sigma,B}$. As a result, the band structure is symmetric between the positive-energy states and the negative-energy states, as is explicitly seen in Fig.3(b). The number of the positive-energy state ν_+ and the negative-energy state ν_- are equal $\nu_+ = \nu_-$. On the other hand, the number of the total states is equal to the number of carbon atoms, $N_C = N^2 + 6N + 6$. Hence, the number of the zero-enegy state η is given by $\eta = N_C - \nu_+ - \nu_-$. Now, the number of the positive or negative-energy state is less than 2 min $[N_A, N_B]$, $\nu_\pm \le 2\min[N_A, N_B]$, where N_A and N_B are the numbers of sites in sublattices A and B, respectively [see Fig.6]. We obtain $\eta \ge |N_A - N_B|$. It is easy to see $N_A = (N+1)(N+6)/2$, $N_B = (N+2)(N+3)/2$ for the trigonal zigzag nanodisk, and hence $\eta \ge N$ for the size-N nanodisk.

3.4.3 Bipartite lattice

We have derived the lower limit for the dimension of the zero-energy sector, $\eta \ge N$. It is a characteristic feature of the tight-binding Hamiltonian on the bipartite lattice that it can be

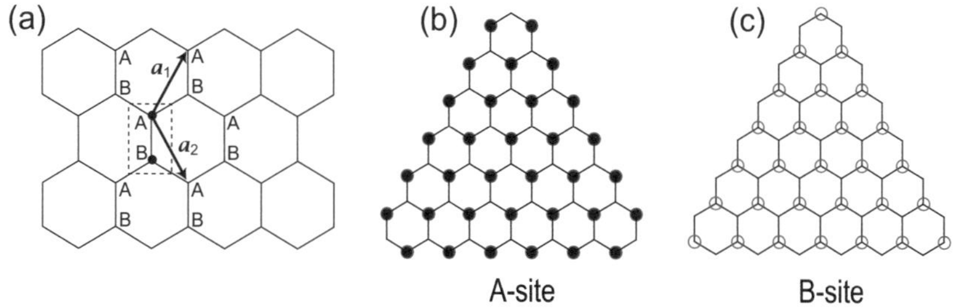

Fig. 6. (a) The two-dimensional honeycomb lattice is made of two trigonal sublattices generated by the two basis vectors $\boldsymbol{a}_1$ and $\boldsymbol{a}_2$ from the two base points A and B in primitive cell (dotted rectangle). (b,c) The A-sites and B-sites of a trigonal nanodisk. Electrons exist only in A sites.

represented in the form of

$$H = \begin{pmatrix} 0 & T_{AB} \\ T_{BA} & 0 \end{pmatrix}. \tag{7}$$

The wave function is decomposed into two parts describing the A sites and B sites, $|f\rangle = (f_A, f_B)^t$. The rank of the matrix (7) is $2N_B$ at the highest, but it is $2N_B$ for connected lattices. As a result the number of the zero-energy eigenvalues of the $(N_A + N_B)$-dimensional matrix (7) is given by $\eta = N_A - N_B = N$, where the last equality holds for the zigzag trigonal nanodisk with (**??**).

Furthermore, we are able to show that the wave function takes a nonzero value only on the A-site. The matrix T_{BA} is $N_B \times N_A$ dimensional. Let us choose N_A-dimensional eigenvectors satisfying $T_{BA} |f_A\rangle = 0$. The number of such eigenvectors is $N = N_A - N_B > 0$, as we have just shown. We now consider the $(N_A + N_B)$-dimensional vector, $|f_A\rangle = (f_A, 0)^t$. It is trivial to see that it is the eigenvector of H, $H |f_A\rangle = (0, T_{BA} f_A)^t = 0$. We have proved that there are N zero-energy eigenvalues whose wave functions vanish on all B sites. See Fig.5 and Fig.6.

3.4.4 Graph theory

There is another way to see the dimension of the zero-energy sector. The spectrum of the tight binding Hamiltonian of a honeycomb system can also be analyzed on the basis of a mathematically rigorous approach of the benzenoid graph theory(68). According to this graph theory, the number of zero-energy states is equal to the graph's nullity, $\eta = 2\alpha - N_C$, where N_C is the total number of sites and α is the maximum number of non-adjacent sites. Here, we have $\eta = N$, since $\alpha = N_A$ for the trigonal zigzag nanodisk [Fig.6].

3.5 Robustness against randomness and lattice defects

In actual application, however, it is important to discuss how stable the previous results are against lattice defects and randomness in transfer energy. We study three types of randomness: randomness in transfer energy, randomness in site energy and lattice defects. The modified Hamiltonian is

$$H_0 = \sum_i (\varepsilon_i + \delta\varepsilon_i)\, c^\dagger_{i,\sigma} c_{i,\sigma} + \sum_{\langle i,j \rangle} \left(t_{ij} + \delta t_{ij}\right) c^\dagger_{i,\sigma} c_{j,\sigma}, \tag{8}$$

where we take random values for $\delta\varepsilon_i$ and δt_{ij}.

First of all the total spin of the ground state is determined by Lieb's theorem although transfer randomness and lattice defects are included. (See Section 5.5 for Lieb's theorem.) The total

Fig. 7. Energy level with randomness. (a) Clean nanodisks. (b) Site energy randomness, $\delta\varepsilon_i/t \sim \pm 0.1$. (c) Transfer energy randomness, $\delta t_{ij}/t \sim \pm 0.1$. (d) Lattice defects with thre site at a corner. See Fig.8(a).

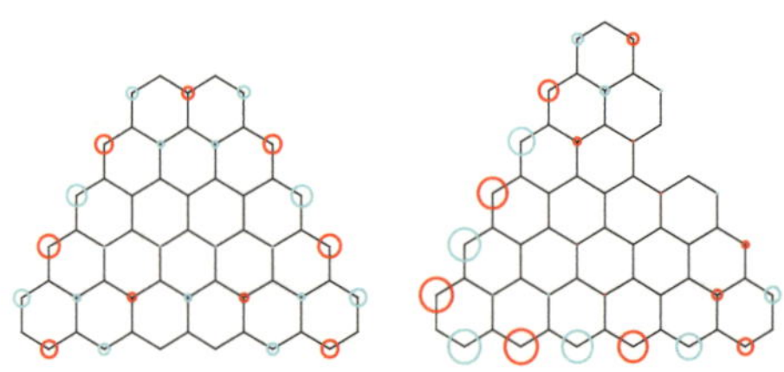

Fig. 8. Wave function with lattice defects. (a) Three atoms are absent at one corner. (b) One atom is absent at one edge. For both case the probability density drastically reduces near the lattice defects. The number of the zero-energy states reduces from N to $N-1$ for both cases.

spin is given by the difference of the A site and the B site, $S = \frac{1}{2}\left|N_A - N_B - \delta_A + \delta_B\right|$, with the number of lattice defects at A (B) site δ_A (δ_B), where N_A and N_B are number of A site and B site without lattice defects. The total spin does not change by introducing randomness in transfer energies. On the other hand, the number of the zero-energy states changes by the number of lattice defects. When site randomness is included, we can not resort to Lieb's theorem in the strict sense because the lattice is no longer bipertite. However, when the split due to the site randomness is smaller than that due to the Coulomb interaction, the total spin does not change.

We show the energy spectrum with the randomness in Fig.7, where we have taken rather large random values for $\delta\varepsilon_i$ and δt_{ij}: We have generated uniform random numbers in the region $|\delta\varepsilon_i|/t \leq \pm 0.1$ and $|\delta t_{ij}|/t \leq \pm 0.1$. The zero energy remains as it is even when we introduce lattice defects. However the zero-energy states split by the randomness in transfer energy. The wave functions also change by the lattice defect and the randomness in site and transfer energies. The changes are proportional to the site (transfer) modification $\delta\varepsilon_i$ (δt_{ij}), but slight in site and transfer randomness.

On the other hand, the wave functions drastically change by introducing lattice defects, as shown in Fig.8. We find the density reduces drastically near the lattice defects.

4. Dirac electrons in graphene

In this section we analyze the zero-energy states based on the Dirac theory of graphene(48). We explicitly construct wave functions as analytic functions around the K point and as anti-analytic functions around the K′ point. They are indexed by the edge momentum and grouped according to the representation of the trigonal symmetry group C_{3v}. By evaluating the probability density flow we reveal a novel texture of magnetic vortices perpendicular to the nanodisk plane. It is intriguing that a vortex with the winding number 2 emerges in the state belonging to the E (doublet) representation.

4.1 Weyl equations

Electrons obey the massless two-component Dirac equation, or the Weyl equation, in graphene(49–51). We start with a review how Dirac electrons arise from the one-particle electronic states in the honeycomb lattice [Fig.9]. We take the basis vectors $\boldsymbol{a}_1$ and $\boldsymbol{a}_2$ as $\boldsymbol{a}_1 = (1,0)\,a$, $\boldsymbol{a}_2 = \left(1/2, \sqrt{3}/2\right) a$, with a the lattice constant ($a \approx 2.46$Å). The honeycomb lattice has two different atoms per primitive cell, which we call the A and B sites. There are three B sites adjacent to an A site, which are specified by the three vectors $\boldsymbol{r}_i$ with $\boldsymbol{r}_1 = \left(0, -1/\sqrt{3}\right) a$, $\boldsymbol{r}_2 = \left(1/2, 1/\left(2\sqrt{3}\right)\right) a$, $\boldsymbol{r}_3 = \left(-1/2, 1/\left(2\sqrt{3}\right)\right) a$. The basis vectors $\boldsymbol{b}_i$ of the reciprocal lattice are given by solving the relations $\boldsymbol{b}_i \cdot \boldsymbol{a}_j = 2\pi\delta_{ij}$. The Brillouin zone is a hexagon in the reciprocal lattice with opposite sides identified.
We rewrite the tight-binding Hamiltonian (2) as

$$H = t\sum_{\boldsymbol{x},i}\left[c_{\mathrm{A}}^{\dagger}(\boldsymbol{x})\,c_{\mathrm{B}}(\boldsymbol{x}+\boldsymbol{r}_i) + c_{\mathrm{B}}^{\dagger}(\boldsymbol{x}+\boldsymbol{r}_i)\,c_{\mathrm{A}}(\boldsymbol{x})\right], \tag{9}$$

where c_S and $c_S^{\dagger}$ are the annihilation and creation operators for electrons localized at the S site (S =A,B). It describes a transfer of an electron between neighboring A and B sites without changing its spin. We introduce the Fourier transform of the electron annihilation operator $c_S(\boldsymbol{x})$,

$$c_S(\boldsymbol{x}) = \int \frac{d^2k}{2\pi} e^{i\boldsymbol{k}\cdot\boldsymbol{x}} c_S(\boldsymbol{k}). \tag{10}$$

The Hamiltonian (9) reads

$$H = t\int d^2k \left(c_{\mathrm{A}}^{\dagger}, c_{\mathrm{B}}^{\dagger}\right)\begin{pmatrix} 0 & f(\boldsymbol{k}) \\ f^*(\boldsymbol{k}) & 0 \end{pmatrix}\begin{pmatrix} c_{\mathrm{A}} \\ c_{\mathrm{B}} \end{pmatrix} \tag{11}$$

in the momentum space, where

$$f(\boldsymbol{k}) = e^{-ik_y a/\sqrt{3}} + 2e^{ik_y a/2\sqrt{3}} \cos\frac{k_x a}{2}. \tag{12}$$

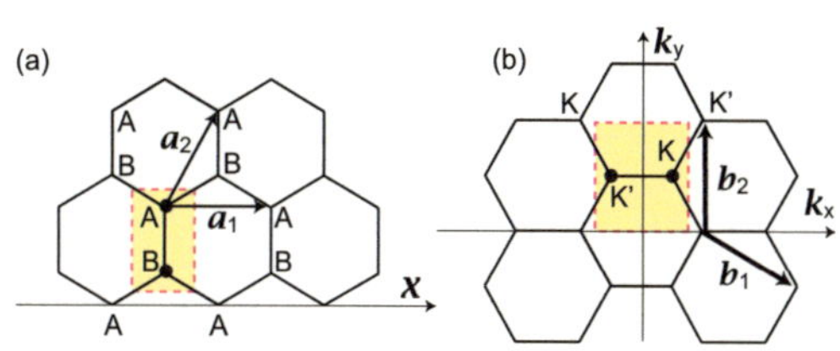

Fig. 9. (a) The two-dimensional honeycomb lattice is made of two triangular sublattices generated by the two basis vectors $\boldsymbol{a}_1$ and $\boldsymbol{a}_2$ from the base points A and B in a primitive cell (a dotted rectangle in yellow). (b) The reciplocal lattice is also a honeycomb lattice with the basis vectors $\boldsymbol{b}_1$ and $\boldsymbol{b}_2$. We take the Brillouin zone as indicated by a dotted rectangle in yellow. As two inequivalent points we take K and K′ as indicated.

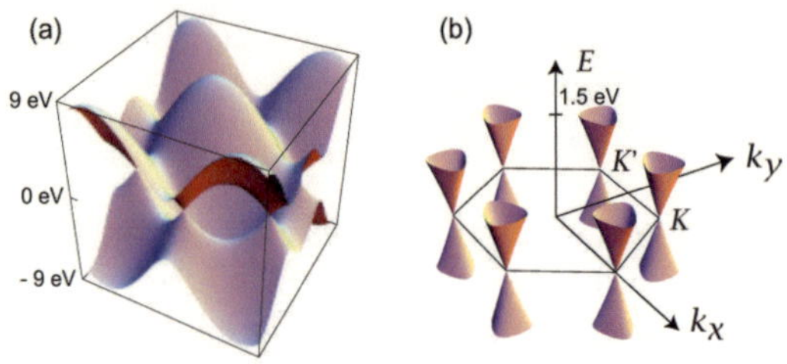

Fig. 10. (a) The overall valley structure of the dispersion relation (13) is illustrated. (b) A schematic diagram of the low-energy dispersion relation (13) near the Dirac points (K and K′ points) in the Brillouin zone. Only two Dirac cones are inequivalent to one another.

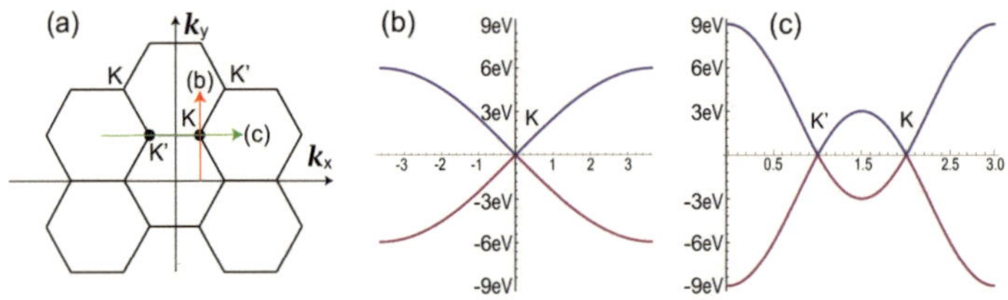

Fig. 11. Linear dispersion near the Brillouin zone corner. The linear region of the band is found to be dependent on the direction in the $\boldsymbol{k}$ space. In one of the direction (c) the linearity is good more than 3 eV.

The eigenvalues are found to be $\pm E(\boldsymbol{k})$ with

$$E(\boldsymbol{k}) = t\sqrt{1 + 4\cos\frac{\sqrt{3}k_y a}{2}\cos\frac{k_x a}{2} + 4\cos^2\frac{k_x a}{2}}. \tag{13}$$

The dispersion relation (13) implies that the Fermi level at $E(\boldsymbol{k}) = 0$ is reached by six corners of the first Brillouin zone, among which there are only two inequivalent points due to the periodicity of the reciprocal lattice. We take them as

$$\boldsymbol{K}^{+} = \left(2\pi/3, 2\pi/\sqrt{3}\right)/a, \qquad \boldsymbol{K}^{-} = \left(-2\pi/3, 2\pi/\sqrt{3}\right)/a, \tag{14}$$

and call them the K and K′ points [Fig.9(b) and Fig.10]. They endow graphene with a two-component structure, corresponding to two independent Fermi points. It is notable that the dispersion relation (13) has a valley structure around the K and K′ points.
The band dispersion (13) is linear in these Dirac valleys [Fig.10],

$$E_K(\boldsymbol{k}) = \hbar v_F |\boldsymbol{k} - \boldsymbol{K}^{\tau}| \qquad \text{for} \quad \boldsymbol{k} \simeq \boldsymbol{K}^{\tau}, \tag{15}$$

where $\tau = \pm$ is the valley index and $v_F = \sqrt{3}ta/(2\hbar)$ is the Fermi velocity. We show two sections of the band in Fig.11, where the linear region of the band is found to be dependent on the direction in the $\boldsymbol{k}$ space. In one of the section the linearity prevails more than 3 eV, as in Fig.11(c). This linear behavior has been confirmed experimentally(69) up to 3eV.
We may expand $f(\boldsymbol{x})$ as

$$f(\boldsymbol{x}) \simeq \frac{(-1)^{-1/3}\sqrt{3}}{2}\left[\tau\left(k_x - K^{\tau}\right) - i\left(k_y - K^{\tau}\right)\right] \tag{16}$$

for $\boldsymbol{k} \simeq \boldsymbol{K}^{\tau}$. It is convenient to introduce the reduced wave number $\hat{\boldsymbol{k}}$ by

$$\boldsymbol{k} = \boldsymbol{K}^{\tau} + \hat{\boldsymbol{k}}, \tag{17}$$

and rewrite the dispersion relation as

$$E(\hat{k}_x, \hat{k}_y) = \hbar v_{\mathrm{F}} \sqrt{\hat{k}_x^2 + \hat{k}_y^2} \tag{18}$$

near the K and K′ points.

It is clear from the dispersion relation illustrated in Fig.10 that the contribution to low-energy physics comes only from the regions around the two Fermi points, $\boldsymbol{k} \simeq \boldsymbol{K}^{\tau}$. We make the change of variable as in (17), and express the wave function $\psi_S^{\tau}(\boldsymbol{x})$ near the Fermi point $\boldsymbol{k} \simeq \boldsymbol{K}^{\tau}$ as

$$\psi_S^{\tau}(\boldsymbol{x}) = e^{i\boldsymbol{K}^{\tau}\cdot\boldsymbol{x}} \phi_S^{\tau}(\boldsymbol{x}), \tag{19}$$

where $\phi_S^{\tau}(\boldsymbol{x})$ is called the envelope function.

The dispersion relation (18) is that of 'relativistic' Dirac fermions. Indeed, we rewrite the Hamiltonian (11) as

$$\begin{aligned} H = \hbar v_{\mathrm{F}} \int d^2\hat{k} \left(c_{\mathrm{A}}^{\mathrm{K}\dagger}, c_{\mathrm{B}}^{\mathrm{K}\dagger} \right) \begin{pmatrix} 0 & \hat{k}_x - i\hat{k}_y \\ \hat{k}_x + i\hat{k}_y & 0 \end{pmatrix} \begin{pmatrix} c_{\mathrm{A}}^{\mathrm{K}} \\ c_{\mathrm{B}}^{\mathrm{K}} \end{pmatrix} \\ + \hbar v_{\mathrm{F}} \int d^2\hat{k} \left(c_{\mathrm{A}}^{\mathrm{K}'\dagger}, c_{\mathrm{B}}^{\mathrm{K}'\dagger} \right) \begin{pmatrix} 0 & -\hat{k}_x - i\hat{k}_y \\ -\hat{k}_x + i\hat{k}_y & 0 \end{pmatrix} \begin{pmatrix} c_{\mathrm{A}}^{\mathrm{K}'} \\ c_{\mathrm{B}}^{\mathrm{K}'} \end{pmatrix}, \end{aligned} \tag{20}$$

where c_S^{τ} is the annihilation operator for an electron at site S in the Dirac valley $\tau\,(= K, K')$. The quantum-mechanical Hamiltonian consists of two parts, $H = H^{\mathrm{K}} + H^{\mathrm{K}'}$, where

$$H^{\tau} = v_{\mathrm{F}}(\tau\sigma_x p_x + \sigma_y p_y) = v_{\mathrm{F}} \boldsymbol{\sigma} \cdot \boldsymbol{p}^{\tau}, \tag{21}$$

with $\boldsymbol{p}^{\tau} = (\tau p_x, p_y) = -i\hbar(\tau\partial_x, \partial_y)$. The Hamiltonian acts on the two-component envelope function,

$$\Phi^{\tau}(\boldsymbol{x}) = (\phi_{\mathrm{A}}^{\tau}(\boldsymbol{x}), \phi_{\mathrm{B}}^{\tau}(\boldsymbol{x}))^t. \tag{22}$$

The Weyl equations read

$$i\hbar\partial_t \Phi^{\tau}(\boldsymbol{x}) = v_{\mathrm{F}} \boldsymbol{\sigma} \cdot \boldsymbol{p}^{\tau} \Phi^{\tau}(\boldsymbol{x}). \tag{23}$$

Each Hamiltonian describes the two-component massless Dirac fermion, or the Weyl fermion. The symmetries of the system are as follows. First, we note that $H^{\mathrm{K}'} = \sigma_y H^{\mathrm{K}} \sigma_y$, where σ_y is the generator of the mirror symmetry. It transforms the K point into the K′ point. Next, $\sigma_z H^{\mathrm{K}} \sigma_z = -H^{\mathrm{K}}$, and $\sigma_z H^{\mathrm{K}'} \sigma_z = -H^{\mathrm{K}'}$, where σ_z is the generator of the electron-hole symmetry. It transforms the positive-energy state into the negative-energy state.

Eigenfunctions of the Hamiltonian (21) are readily obtained,

$$\Phi_{\pm}^{\tau}(\boldsymbol{x}) = e^{i\hat{\boldsymbol{k}}\cdot\boldsymbol{x}} u_{\pm}^{\tau}(\hat{\boldsymbol{k}}) \tag{24}$$

for the positive-energy state $\Phi_+^{\tau}(\boldsymbol{x})$ and the negative-energy state $\Phi_-^{\tau}(\boldsymbol{x})$, where

$$u_{\pm}^{\mathrm{K}}(\hat{\boldsymbol{k}}) = \frac{1}{\sqrt{2}} \left(1 \pm \frac{\boldsymbol{\sigma}\cdot\hat{\boldsymbol{k}}}{|\boldsymbol{k}|} \right) \begin{pmatrix} 1 \\ 0 \end{pmatrix}, \quad u_{\pm}^{\mathrm{K}'}(\hat{\boldsymbol{k}}) = \frac{1}{\sqrt{2}} \left(1 \pm \frac{\boldsymbol{\sigma}\cdot\hat{\boldsymbol{k}}'}{|\boldsymbol{k}'|} \right) \begin{pmatrix} 0 \\ 1 \end{pmatrix}, \tag{25}$$

with $\hat{\boldsymbol{k}}' = (-\hat{\boldsymbol{k}}_x, \hat{\boldsymbol{k}}_y)$. They transform as

$$\sigma_y u_\pm^{\mathrm{K}}(\hat{\boldsymbol{k}}) = i u_\pm^{\mathrm{K}'}(\hat{\boldsymbol{k}}), \quad \sigma_y u_\pm^{\mathrm{K}'}(\hat{\boldsymbol{k}}) = -i u_\pm^{\mathrm{K}}(\hat{\boldsymbol{k}}), \quad \sigma_z u_\pm^{\mathrm{K}}(\hat{\boldsymbol{k}}) = u_\mp^{\mathrm{K}}(\hat{\boldsymbol{k}}), \quad \sigma_z u_\pm^{\mathrm{K}'}(\hat{\boldsymbol{k}}) = -u_\mp^{\mathrm{K}'}(\hat{\boldsymbol{k}}), \tag{26}$$

under the mirror transformation and the electron-hole transformation.

The energy spectrum is symmetric between positive- and negative-energy states. There exists one electron per one carbon and the band-filling factor is 1/2 in graphene. Namely, all negative-energy states are filled up, as is a reminiscence of the Dirac sea.

4.2 Dirac electrons on zigzag edge

We analyze massless Dirac electrons on the zigzag edge based on the method due to Brey and Fertig(22). The Weyl equation holds for envelope functions satsifying

$$v_{\mathrm{F}} \begin{pmatrix} 0 & \tau p_x - i p_y \\ \tau p_x + i p_y & 0 \end{pmatrix} \begin{pmatrix} \phi_{\mathrm{A}}^{\mathrm{K}^\tau}(\boldsymbol{x}) \\ \phi_{\mathrm{B}}^{\mathrm{K}^\tau}(\boldsymbol{x}) \end{pmatrix} = E \begin{pmatrix} \phi_{\mathrm{A}}^{\mathrm{K}^\tau}(\boldsymbol{x}) \\ \phi_{\mathrm{B}}^{\mathrm{K}^\tau}(\boldsymbol{x}) \end{pmatrix}, \tag{27a}$$

where $p_x - i p_y = -2i\hbar\partial_z$, $p_x + i p_y = -2i\hbar\partial_{z^*}$ with $z = x + iy$ and $z^* = x - iy$. The Weyl equation is rewritten as

$$\partial_{z^*}\phi_{\mathrm{A}}^{\mathrm{K}}(\boldsymbol{x}) = i\frac{E}{2\hbar v_{\mathrm{F}}}\phi_{\mathrm{B}}^{\mathrm{K}}(\boldsymbol{x}), \qquad \partial_z\phi_{\mathrm{B}}^{\mathrm{K}}(\boldsymbol{x}) = i\frac{E}{2\hbar v_{\mathrm{F}}}\phi_{\mathrm{A}}^{\mathrm{K}}(\boldsymbol{x}), \tag{28a}$$

$$\partial_z\phi_{\mathrm{A}}^{\mathrm{K}'}(\boldsymbol{x}) = -i\frac{E}{2\hbar v_{\mathrm{F}}}\phi_{\mathrm{B}}^{\mathrm{K}'}(\boldsymbol{x}), \qquad \partial_{z^*}\phi_{\mathrm{B}}^{\mathrm{K}'}(\boldsymbol{x}) = -i\frac{E}{2\hbar v_{\mathrm{F}}}\phi_{\mathrm{A}}^{\mathrm{K}'}(\boldsymbol{x}). \tag{28b}$$

It is clear that, for the zero-energy state ($E = 0$), the envelope functions $\phi_{\mathrm{A}}^{\mathrm{K}}(\boldsymbol{x})$ and $\phi_{\mathrm{B}}^{\mathrm{K}'}(\boldsymbol{x})$ are analytic, while $\phi_{\mathrm{B}}^{\mathrm{K}}(\boldsymbol{x})$ and $\phi_{\mathrm{A}}^{\mathrm{K}'}(\boldsymbol{x})$ are anti-analytic.

We place a graphene sheet in the upper half plane ($y > 0$) with the edge at $y = 0$. Translational invariance in the x direction dictates the envelope functions are of the form $\phi_S^\tau(x, y) = e^{i\hat{k}_x x} f_S^\tau(y)$. Analyticity requirement allows us to write

$$\phi_{\mathrm{A}}^{\mathrm{K}}(\boldsymbol{x}) = C_{\mathrm{A}}^{\mathrm{K}} e^{i\hat{k}z}, \quad \phi_{\mathrm{B}}^{\mathrm{K}}(\boldsymbol{x}) = C_{\mathrm{B}}^{\mathrm{K}} e^{i\hat{k}z^*}, \quad \phi_{\mathrm{A}}^{\mathrm{K}'}(\boldsymbol{x}) = C_{\mathrm{A}}^{\mathrm{K}'} e^{i\hat{k}z^*}, \quad \phi_{\mathrm{B}}^{\mathrm{K}'}(\boldsymbol{x}) = C_{\mathrm{B}}^{\mathrm{K}'} e^{i\hat{k}z}, \tag{29a}$$

with C_S^τ being integration constant.

According to the tight-binding-model result, there are no electrons in the B site on edges with zero energy, and hence we require

$$\phi_{\mathrm{B}}^{\mathrm{K}}(y = 0) = \phi_{\mathrm{B}}^{\mathrm{K}'}(y = 0) = 0. \tag{30}$$

By avoiding divergence at $y \to \infty$, the resultant envelope functions are found to be

$$\phi_{\mathrm{A}}^{\mathrm{K}}(\boldsymbol{x}) = \sqrt{\hat{k}/L} e^{i\hat{k}z} = \sqrt{\hat{k}/L} e^{i\hat{k}x - \hat{k}y}, \quad \text{for} \quad \hat{k} > 0, \tag{31a}$$

$$\phi_{\mathrm{A}}^{\mathrm{K}'}(\boldsymbol{x}) = \sqrt{\hat{k}/L} e^{i\hat{k}z^*} = \sqrt{\hat{k}/L} e^{i\hat{k}x - |\hat{k}|y}, \quad \text{for} \quad \hat{k} < 0, \tag{31b}$$

with all other components being zero, where $2L$ is the size of the edge.

The wave number is a continuous parameter for an infinitely long graphene edge. According to the tight-binding-model result, as illustrated in Fig.13, the flat band emerges for

$$-\pi \le ak < -\frac{2\pi}{3} \quad \text{and} \quad \frac{2\pi}{3} < ak \le \pi, \tag{32}$$

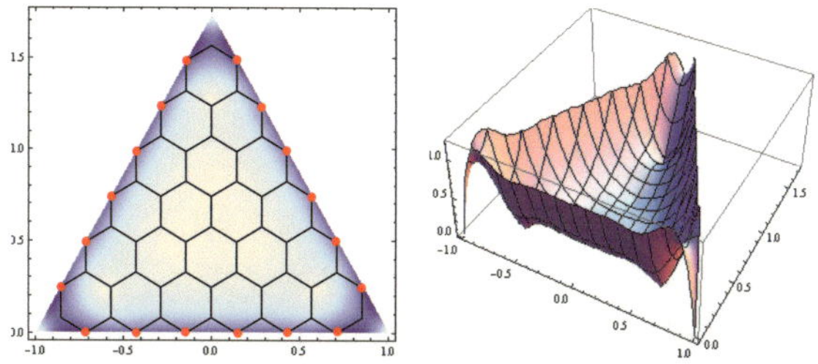

Fig. 12. The electron density is found to be localized along the edges of a trigonal nanodisk.

in terms of the original wave number k, or

$$-\frac{\pi}{3} \leq a\hat{k} < 0 \quad \text{and} \quad 0 < a\hat{k} \leq \frac{\pi}{3}, \tag{33}$$

in terms of the reduced wave number $\hat{k}$, around the K′ and K points, respectively. The boundary point $ak = -\pi$ of the K′ point is simultaneously the boundary point $ak = \pi$ of the K point in another Brillian zone, as illustrated in Fig.13: They are physically the same point and to be identified.

4.3 Dirac electrons in triangular Graphene

Our main purpose is to apply the above result to the analysis of the zero-energy sector of the trigonal zigzag nanodisk [Fig.12]. The envelope function of the trigonal zigzag nanodisk can be constructed by making a linear combination of envelope functions for three trigonal corners. We consider the trigonal region whose corners are located at $z_1 = (L, 0), z_2 = (-L, 0)$, $z_3 = \left(0, \sqrt{3}L\right)$. As the boundary conditions we impose $\phi(z_1) = \phi(z_2) = \phi(z_3) = 0$. The resultant envelope function is obtained around the K point ($\hat{k} > 0$) as

$$\phi(z) = e^{i\hat{k}z} - e^{i\hat{k}L} e^{i\hat{k}(z-L)\exp[-2\pi i/3]} - e^{-i\hat{k}L} e^{i\hat{k}(z+L)\exp[2\pi i/3]} \tag{34}$$

up to a normalization constant. The envelope function around the K′ point ($\hat{k} < 0$) is given by $\phi(z^*)$. We display the density of the wave function $|\phi(z)|^2$ in Fig.12.

The wave number is quantized for a finite edge such as in the trigonal nanodisk. We can determine it as follows. We focus on the wave function $\psi_A^\tau(\boldsymbol{x})$ at one of the A sites on an edge. There are N links along one edge of the size-N trigonal nanodisk. We can show that the correction terms are exactly cancelled out for these N links, and we obtain precisely $\Theta^\tau(a/2, N) = Nak$. On the other hand, the phase shift is π at the corner. The total phase shift is $3Nak + 3\pi$, when we encircle the nanodisk once. For the single-valueness of the wave function it is necessary that $3Nak + 3\pi = 2n\pi$, or

$$\hat{k}_n = \pm\frac{(2n+1)\pi}{3Na}, \qquad 0 \leq n \leq \frac{N-1}{2}, \tag{35}$$

with an integer n. The allowed region of the wave number is given by (32). The wave number is quantized as

$$k_n = \pm\left[\frac{2n+1}{3N} + \frac{2}{3}\right]\frac{\pi}{a}, \qquad 0 \leq n \leq \frac{N-1}{2}. \tag{36}$$

When N is even, there are $N/2$ states for $k_n > 0$ and $N/2$ states for $k_n < 0$. When N is odd, there are $(N-1)/2$ states for $k_n > 0$ and $(N-1)/2$ states for $k_n < 0$. Additionally, there seem

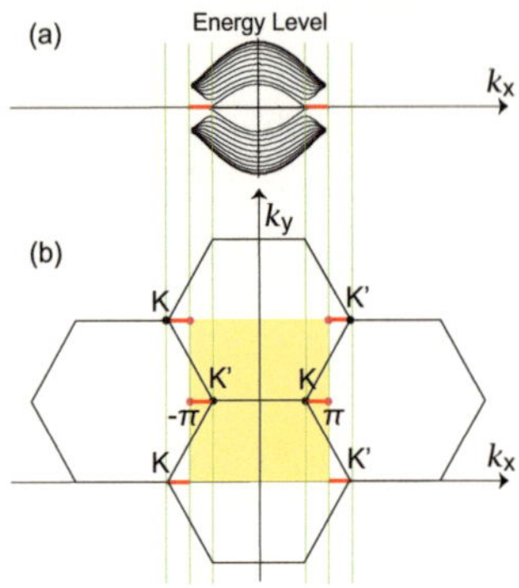

Fig. 13. (a) The energy spectrum in an infinitely long graphene edge. The flat band appears in the region indicated in red. (b) We take the Brillouin zone as indicated in yellow. The flat band appears in the region indicated in red. The boundary points at $\pm\pi$ are identified.

to appear two states with $k_n = \pm\pi/a$ at $n = (N-1)/2$. However, they are identified with one another, since they are located at the boundary of the Brillouin zone [13].
As we have reviewed in Section 3.4, the symmetry group of the trigonal nanodisk is C_{3v}, whose properties are summarized in the table (6). The mirror symmetry is equivalent to the exchange of the K and K′ points. With respect to the rotation there are three elements $\mathfrak{c}_3^0$, $\mathfrak{c}_3$, $\mathfrak{c}_3^2$, which correspond to 1, $e^{2\pi i/3}$, $e^{4\pi i/3}$. Accordingly, the phase shift of one edge is 0, $2\pi/3$, $4\pi/3$. They are determined by the conditions

$$Nak_n + \pi = 2n\pi, \quad Nak_n + \pi = 2n\pi + \frac{2\pi}{3}, \quad Nak_n + \pi = 2n\pi + \frac{4\pi}{3}, \tag{37a}$$

with an integer n, respectively. It follows that the state, indexed by the edge momentum k_n as in (36), are grouped according to the representation of the trigonal symmetry group C_{3v} as follows,

$$\left.\begin{array}{l} A_1\,(\text{singlet}): |k_n^0\rangle + |-k_n^0\rangle, \\ A_2\,(\text{singlet}): |k_n^0\rangle - |-k_n^0\rangle, \end{array}\right\} k_n^0 = \frac{6n+3}{3Na}\pi, \\ E\,(\text{doublet}): |k_n^{\pm}\rangle, \quad |-k_n^{\pm}\rangle, \; k_n^{\pm} = \frac{6n\pm 1}{3Na}\pi, \tag{38}$$

where the momentum k_n^{α} is subject to the condition (32). It follows that

$$\left\lfloor \frac{N+1}{3} \right\rfloor \leq n \leq \left\lfloor \frac{N}{2} \right\rfloor, \tag{39}$$

where $\lfloor a \rfloor$ denotes the maximum integer equal to or smaller than a, and $\lceil a \rceil$ denotes the minimum integer equal to or larger than a. The numbers of doublets (E-mode) and singlets (A_1-mode or A_2-mode) are given by

$$\left\lfloor \frac{N+1}{3} \right\rfloor, \quad N - 2\left\lfloor \frac{N+1}{3} \right\rfloor, \tag{40}$$

respectively.

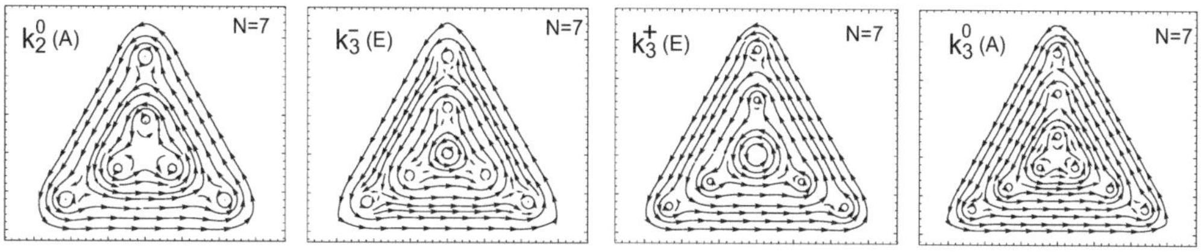

Fig. 14. Probability density flow for the state $|k_n^{(0,\pm)}\rangle$ in the nanodisk with $N=7$. The representation is indicated in the parenthesis. A vortex appears at the center of mass for the state belonging to the E (doublet) representation. It is interesting that the winding number is 2 in the state $|k_n^+\rangle$. A circulating current along the closed edge confines flux and generates a vortex texture respecting the trigonal symmetry.

4.3.1 Probability density flow (Berry connection)

To see the meaning of the wave number k_n^α more in detail, we have calculated the probability density flow

$$\mathcal{A}_i(x,y) = -i\phi^*(x,y)\partial_i\phi(x,y) \tag{41}$$

for various states, which we show for the case of $N=7$ in Fig.14. We observe clearly a texture of vortices: The number of vortices are $6,7,7,9$ for $|k_2^0\rangle$, $|k_3^-\rangle$, $|k_3^+\rangle$, $|k_3^0\rangle$, respectively. The vortex at the center of mass must have the winding number 2 in $|k_3^+\rangle$. In general, the total winding number N_{vortex} is calculated by

$$N_{\text{vortex}} = \frac{-i}{2\pi}\oint dx_i \frac{\phi^*(x,y)\partial_i\phi(x,y)}{|\phi(x,y)|^2} = N+m-1, \tag{42}$$

with $m = 0,1,2,\cdots,\lfloor(N-1)/2\rfloor$ in the size-N nanodisk, where the integration is made along the closed edge of a nanodisk. $N_{\text{vortex}} = 3n$ for k_n^0, $N_{\text{vortex}} = 3n+1$ for k_{n+1}^- and $N_{\text{vortex}} = 3n+2$ for k_n^+. The wave functions are classified in terms of modulo of the total winding number: the wave function belongs to the E-representation and has chiral edge mode for $N_{\text{vortex}} \equiv 1,2$ (mod 3), and belongs to the A-representation and has non-chiral edge mode for $N_{\text{vortex}} \equiv 0$ (mod 3). There are n vortices along the y-axis in the state $|k_n^\alpha\rangle$. The state $|k_n^\pm\rangle$, being the E-mode, has a vortex at the center of mass, where the winding number is 2 in the state $|k_n^+\rangle$. On the other hand, the state $|k_n^0\rangle$ does not have a vortex at the center of mass, and the combinations $|k_n^0\rangle \pm |-k_n^0\rangle$ belong to the A_1 and A_2 representations, respectively.

To demonstrate the above statement, we search for zero points of the wave function, where vortices appear. The probability density along the y-axis is given by

$$|\psi(y)|^2 = \left(-e^{-ky} + 2e^{k\left(y-\sqrt{3}Na/2\right)/2}\cos\frac{\sqrt{3}k\left(y+\sqrt{3}Na/2\right)}{2}\right)^2, \tag{43}$$

which is found to have n zeros for $k = k_n^{(\pm)}$. In particular we examine the emergence of a vortex at the center of mass, where we may expand $\psi(z)$ as

$$\psi(z) = C_0 + C_1 k\left(z - i\frac{Na}{2\sqrt{3}}\right) + C_2 k^2\left(z - i\frac{Na}{2\sqrt{3}}\right)^2 + \cdots. \tag{44}$$

The coefficient C_0 vanishes at

$$k = \frac{6n\pm 1}{3Na}\pi, \tag{45}$$

namely, for the state $|k_n^{\pm}\rangle$. Furthermore, C_1 vanishes at

$$k = \frac{6n+1}{3Na}\pi. \tag{46}$$

Hence we confirm that a vortex with the winding number 2 appears in the state $|k_n^{+}\rangle$. We can check that C_2 does not vanish where C_0 vanishes. Namely, there are no vortices with the winding number 3.

These vortices are present inside the nanodisk where the electron density is almost zero. The situation is similar to a flux confined within a solenoid or within a superconducting ring, where no electrons are present in the domain where the magnetic flux exists. The electric current confining the flux generates the Aharonov-Bohm phase (42) associated with vortices. Vortices make a texture respecting the trigonal symmetry C_{3v} within a nanodisk.

4.3.2 Magnetic vortices

Charged particles propagating along a closed edge generates magnetic field. The electromagnetic interaction is described in terms of the electromagnetic potential $\boldsymbol{A}$, which is introduced to the system by way of the Peierls substitution $\partial_j \rightarrow \partial_j + ieA_j/\hbar$. Here we consider the lowest order approximation, where $\phi_{\text{A}}^{\text{K}}(\boldsymbol{x})$ is not modified from (34) in the presence of the electromagnetic interaction. Then, from the Weyl equation (28), we derive

$$eA_i(\boldsymbol{x}) = -i\hbar\frac{\phi_{\text{A}}^{\text{K}*}(\boldsymbol{x})\partial_i\phi_{\text{A}}^{\text{K}}(\boldsymbol{x})}{|\phi_{\text{A}}^{\text{K}}(\boldsymbol{x})|^2} = \hbar\frac{\mathcal{A}_i(\boldsymbol{x})}{|\phi_{\text{A}}^{\text{K}}(\boldsymbol{x})|^2}. \tag{47}$$

The potential $A_i(\boldsymbol{x})$ exhibits the same texture of vortices as in Fig.14. The magnetic field is given by

$$B(x,y) = \nabla \times A(\boldsymbol{x}) = \frac{2\pi\hbar}{e}\sum_n \nu_n \delta\left(z - z_n\right), \tag{48}$$

where ν_n stands for winding number of the vortex at $z = z_n$. Hence a texture of vortices in the Berry connection leads to a texture of magnetic vortices. This δ-function type magnetic field would be smoothed out in a rigorous analysis of the coupled system of the Maxwell equation and the Weyl equation.

We have shown that the winding number of the vortex at the center of the nanodisk is $0, 1, 2$ in the state $|k_n^0\rangle, |k_n^-\rangle, |k_n^+\rangle$, respectively. By tuning the chemical potential any of them is made the ground state. As far as we are aware of, the vortex with the winding number 2 has never been found in all branches of physics. This is because two vortices with the winding number 1 have lower energy than one vortex with the winding number 2 in general. In the present case the disintegration of a vortex into two is prohibited by the trigonal symmetry.

As is well known, a single flux quantum has experimentally been observed in superconductor by using an electron-holographic interferometry(70). Then, in principle it is possible to observe a vortex texture in nanodisk as well. Furthermore, by attaching a superconductor film one may observe a disintegration of a vortex into two when the flux enters into the superconductor from the nanodisk. This would verify the winding number 2 of a vortex.

5. Electron-electron interactions

In this section we make an investigation of electron-electron interaction effects in the zero-energy sector. The spin stiffness is quite large due to large exchange interactions, which means that a nanodisk is a rigid ferromagnet. The system is well described by the infinite-range Heisenberg model, which is exactly solvable. Constructing the partition function, we explore thermodynamical properties(47; 65). A sharp peak emerges at a certain temperature in the specific heat, which we interpret as a quasi-phase transition between quasiferromagnetic and paramagnetic states.

5.1 Exchange interactions

Let us include the Coulomb interaction between electrons in the zero-energy sector. We take two states $|f_\alpha\rangle$ and $|f_\beta\rangle$, $\alpha \neq \beta$, each of which can accommodate two electrons with up and down spins at most. The two-state system is decomposed into the spin singlet χ_S and the spin triplet χ_T with the normalized wave functions,

$$f_{\alpha\beta}^{SS}(\boldsymbol{x},\boldsymbol{x}') = \frac{1}{\sqrt{2}}\Big(f_\alpha(\boldsymbol{x})f_\beta(\boldsymbol{x}') + f_\alpha(\boldsymbol{x}')f_\beta(\boldsymbol{x})\Big)\chi_S, \tag{49a}$$

$$f_{\alpha\beta}^{ST}(\boldsymbol{x},\boldsymbol{x}') = \frac{1}{\sqrt{2}}\Big(f_\alpha(\boldsymbol{x})f_\beta(\boldsymbol{x}') - f_\alpha(\boldsymbol{x}')f_\beta(\boldsymbol{x})\Big)\chi_T. \tag{49b}$$

The Coulomb energies are summarized as

$$\langle f_{\alpha\beta}^{SS}|H_C|f_{\alpha\beta}^{SS}\rangle = U_{\alpha\beta} + J_{\alpha\beta}, \qquad \langle f_{\alpha\beta}^{ST}|H_C|f_{\alpha\beta}^{ST}\rangle = U_{\alpha\beta} - J_{\alpha\beta}, \tag{50}$$

where $U_{\alpha\beta}$ and $J_{\alpha\beta}$ are the direct and exchange energies. In the many-state system the effective Hamiltonian is derived as

$$H_D = \sum_{\alpha\geq\beta} U_{\alpha\beta} n(\alpha)\, n(\beta) - \frac{1}{2}\sum_{\alpha>\beta} J_{\alpha\beta}[4\boldsymbol{S}(\alpha)\cdot\boldsymbol{S}(\beta) + n(\alpha)\, n(\beta)], \tag{51}$$

where $n(\alpha) = d_\sigma^\dagger(\alpha)d_\sigma(\alpha)$ is the number operator and $\boldsymbol{S}(\alpha) = \frac{1}{2}d_\sigma^\dagger(\alpha)\boldsymbol{\tau}_{\sigma\sigma'}d_{\sigma'}(\alpha)$ is the spin operator, with $d_\sigma(\alpha)$ the annihilation operator of electron with spin $\sigma =\uparrow,\downarrow$ in the zero-energy state $|f_\alpha\rangle$: $\boldsymbol{\tau}$ is the Pauli matrix. Note the existence of the on-state Coulomb term $U_{\alpha\alpha}n(\alpha)\, n(\alpha)$ in the effective Hamiltonian (51).

We expand $U_{\alpha\beta}$ and $J_{\alpha\beta}$ in terms of the Wannier functions with the use of (5),

$$U_{\alpha\beta} = \sum_S \omega_i^\alpha \omega_j^\alpha \omega_k^\beta \omega_l^\beta \int d^3x d^3y\; \varphi_i^*(\boldsymbol{x})\varphi_j(\boldsymbol{x})V(\boldsymbol{x}-\boldsymbol{y})\varphi_k^*(\boldsymbol{y})\varphi_l(\boldsymbol{y}), \tag{52a}$$

$$J_{\alpha\beta} = \sum_S \omega_i^\alpha \omega_j^\alpha \omega_k^\beta \omega_l^\beta \int d^3x d^3y\; \varphi_i^*(\boldsymbol{x})\varphi_j(\boldsymbol{y})V(\boldsymbol{x}-\boldsymbol{y})\varphi_k^*(\boldsymbol{y})\varphi_l(\boldsymbol{x}). \tag{52b}$$

The dominant contributions come from the on-site Coulomb terms with $i = j = k = l$ both for the direct and exchange energies. We thus obtain

$$U_{\alpha\beta} \simeq J_{\alpha\beta} \simeq U\sum_i (\omega_i^\alpha \omega_i^\beta)^2, \tag{53}$$

with

$$U \equiv \frac{e^2}{4\pi\varepsilon}\int d^3x d^3y\; \varphi_i^*(\boldsymbol{x})\varphi_i(\boldsymbol{x})\frac{1}{|\boldsymbol{x}-\boldsymbol{y}|}\varphi_i^*(\boldsymbol{y})\varphi_i(\boldsymbol{y}). \tag{54}$$

The Coulomb energy U is of the order of 1eV because the lattice spacing of the carbon atoms is $\sim$ 1Å in graphene. The nearest neighbor Coulomb interaction vanishes since no electrons exist in B sites: See Fig.6. It is straightforward to take into account higher order corrections but the effects are only to enhance the ferromagnetic order. The essential properties of the Coulomb interactions are well described only by taking the on-site terms.

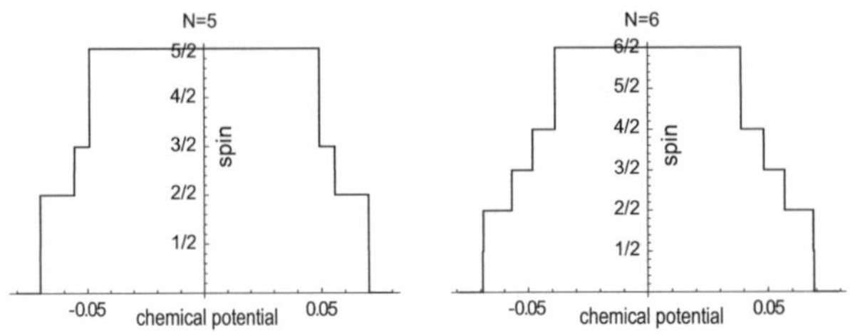

Fig. 15. The nanodisk spin as a function of the chemical potential. The vertical axis is the spin, where the highest value is $N/2$. The horizontal axis is the chemical potential μ. The nanodisk spin is controlled by changing μ.

5.2 Zero-energy splitting due to coulomb interactions

We have shown that the N-fold degeneracy in the zero-energy sector is resolved by Coulomb interactions. The Coulomb Hamiltonian has the trigonal symmetry C_{3v}, and the energy spectrum splits into different levels according to its representation. The energy split is approximately scaled by $U_{\alpha\alpha}$.

There exists additionally the spin-degeneracy in the noninteracting Hamiltonian: The total degeneracy is $2N$. The spin degeneracy is broken spontaneously with the introduction of Coulomb interactions. The splitting is symmetric with respect to the zero-energy level. At half-filling, electrons with the identical spin fill all energy levels under the Fermi energy. Then, the spin of the ground state is $N/2$, and it is a ferromagnet. We show the energy spectrum for $N = 5, 6$ in Fig.15.

It is interesting that we can control the total spin by changing the chemical potential μ. Let us assume that all spins are up polarized in the ground state. As μ increases from zero and reaches the next energy level, down-spin electrons fill the level. As a result, the ground-state spin is reduced by $1/2$ or 1 when the energy level is singlet or doublet, respectively. When μ increases higher than the highest energy level, the total spin becomes 0 and the ferromagnet disappears. The same results apply to the case when μ decreases. We show the total spin as a function of the chemical potential μ in Fig.15.

In our analysis the interaction effects are treated only within the subspace of zero-energy states. It is desirable to derive an effective low-energy Hamiltonian by integrating out the higher excitation levels in a renormalization procedure rather than just neglecting these levels. This has actually been done by functional-renormalization-group analysis recently(71). Our treatment neglecting the renormalizations by the higher levels are shown to be valid by functional-renormalization-group analysis. Furthermore, first-principle calculations of electronic structure of graphene nanodisk have been carried out, which are found to be consistent with our results(37; 38; 42). Result are shown in Fig.16.

5.3 SU(N) Approximation

As we have shown numerically, all $J_{\alpha\beta}$ are of the same order of magnitude for any pair of α and β, implying that the SU(N) symmetry is broken but not so strongly in the Hamiltonian (51). It is a good approximation to start with the exact SU(N) symmetry, where we set $J_{\alpha\beta} = U_{\alpha\beta} = J$. (We set the average of $J_{\alpha\beta}$ as J in the size-N nanodisk.) Then, the zero-energy sector is described by the Hamiltonian $H_D = H_S + H_U$, with

$$H_S = -J \sum_{\alpha \neq \beta} \boldsymbol{S}(\alpha) \cdot \boldsymbol{S}(\beta), \qquad H_U = \frac{J}{4} \sum_{\alpha \neq \beta} n(\alpha)\, n(\beta) + J \sum_{\alpha} n(\alpha). \tag{55}$$

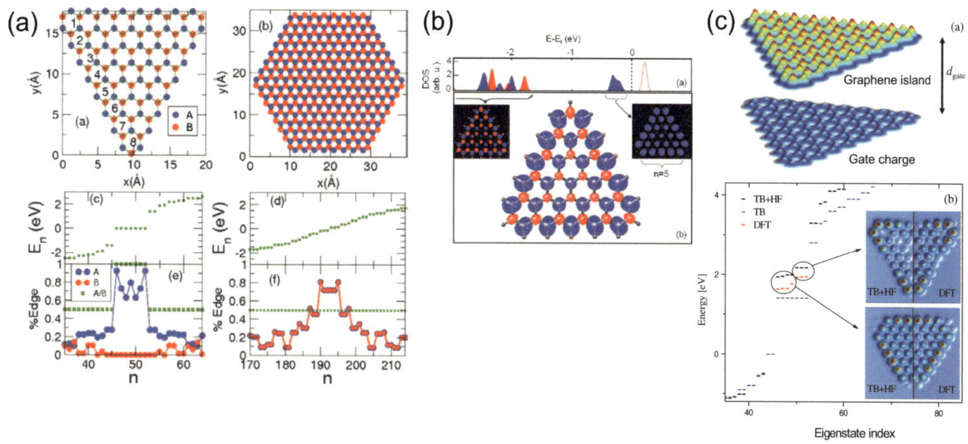

Fig. 16. The first-principle calculation of graphene nanodisdks. Fig. (a) is taken from the work of Fernández-Rossier et. al (37), Fig. (b) is taken from the work of Wang et. al (38) and Fig. (c) is taken form the work of Güçlü et al.(42).

The term H_S is known as the infinite-range Heisenberg model. We rewrite them as

$$H_S = -J\boldsymbol{S}_{\rm tot}^2 + \frac{3}{4}Jn_{\rm tot}, \qquad H_U = \frac{J}{4}(n_{\rm tot}^2 + 1), \tag{56}$$

where $n_{\rm tot} = \sum_\alpha n(\alpha)$ is the total electron number.
The ground states of nanodisks are half filled. We restrict the Hilbert space to the half-filling sector, where $n(\alpha) = n_\uparrow(\alpha) + n_\downarrow(\alpha) = 1$. The Hamiltonian (56) is reduced to the Heisenberg model,

$$H_S = -J\boldsymbol{S}_{\rm tot} \cdot \boldsymbol{S}_{\rm tot}, \tag{57}$$

where we have neglected an irrelevant constant term, $(3/4)JN$. This is exactly diagonalizable, $H_S|\Psi\rangle = E_s|\Psi\rangle$, with $E_s = -Js(s+1)$, where s takes values from $N/2$ down to $1/2$ or 0, depending on whether N is odd or even, $s = \frac{N}{2}, \frac{N}{2} - 1, \frac{N}{2} - 2, \cdots, \quad s \geq 0$. The Hilbert space is diagonalized, $\mathbb{H} = \otimes^N \mathbb{H}_{1/2} = \oplus^{g_N(s)} \mathbb{H}_s$, where $\mathcal{H}_s$ denotes the $(2s+1)$ dimensional Hilbert space associate with an irreducible representation of SU(2). The multiplicities $g_N(s)$ satisfies the recursion relation coming from the spin synthesizing rule, $g_N(s) = g_{N-1}(s - \frac{1}{2}) + g_{N-1}(s + \frac{1}{2})$. We solve this as $g_N(\frac{1}{2}N - q) = {}_NC_q - {}_NC_{q-1}$. The total degeneracy of the energy level E_s is $(2s+1)\, g_N(s)$. At half filling, the eigenstate of the Hamiltonian H_D is labeled as $|\Psi\rangle = |n_{\rm tot}, s, m\rangle$, where s is the total spin and m is its z-component.

5.4 Thermodynamical properties

We have a complete set of the eigenenergies together with their degeneracies. The partition function of the nanodisk with size N is exactly calculable(47). According to the standard procedure we can evaluate the specific heat $C(T)$, the entropy $S(T)$, the magnetization $\langle \boldsymbol{S}_{\rm tot}^2 \rangle$ and the susceptibility $\chi = \left(\langle S_{z,\rm tot}^2 \rangle - \langle S_{z,\rm tot} \rangle^2 \right) / (k_B T)$from this partition function. The ground-state value of the total spin is $S_g = \sqrt{(N/2)(N/2+1)}$. The entropy is given by $S(0) = k_B \log(N+1)$ at zero temperature. We display them in Fig.17 for size $N = 1, 2, 2^2, \cdots 2^{10}$.

There appear singularities in thermodynamical quantities as $N \to \infty$, which represent a phase transition at T_c between the ferromagnet and paramagnet states, $T_c = JN/(2k_B)$. For finite

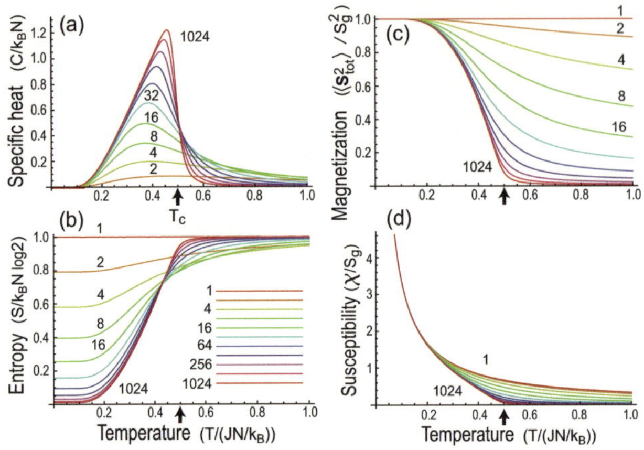

Fig. 17. Thermodynamical properties of the nanodisk-spin system. (a) The specific heat C in unit of $k_B N$. (b) The entropy S in unit of $k_B N \log 2$. (c) The magnetization $\langle \boldsymbol{S}_{\text{tot}}^2 \rangle$ in unit of S_g^2. (d) The susceptibility χ in unit of S_g. The size is $N = 1, 2, 2^2, \cdots 2^{10}$. The horizontal axis stands for the temperature T in unit of JN/k_B. The arrow represents the phase transition point T_c in the limit $N \to \infty$.

N, there are steep changes around T_c, though they are not singularities. It is not a phase transition. However, it would be reasonable to call it a quasi-phase transition between the quasiferromagnet and paramagnet states. Such a quasi-phase transition is manifest even in finite systems with $N = 100 \sim 1000$.

The specific heat and the magnetization take nonzero-values for $T > T_c$ [Fig.17(a),(c)], which is zero in the limit $N \to \infty$. The entropy for $T > T_c$ is lower than that of the paramagnet [Fig.17(b)]. These results indicate the existence of some correlations in the paramagnet state.

As shown in Fig.17(d), the susceptibility χ always shows the Curie-Weiss low $\chi \propto 1/T$ near $T = 0$, and exhibits also a behavior showing a quasi-phase transition at $T = T_c$. In the finite system, the expectation value of $S_{z,\text{tot}}$ is always zero because there is no spontaneous symmetry breakdown in the finite system, and the behavior is that of paramagnet.

5.5 Lieb's theorem

We make some remarkable observations on the spin of a nanodisk. Due to the exchange interaction in the Hamiltonian (51), all spins are spontaneously polarized into a single direction. Since the exchange energy is as large as the direct energy, which is the order of a few hundred meV, the spin stiffness $J_{\alpha\beta}$ is quite large, implying that nanodisks are almost rigid ferromagnets. We refer to the total spin of a nanodisk, $\boldsymbol{S}_{\text{tot}} = \sum_\alpha \boldsymbol{S}(\alpha)$, as the nanodisk spin. The size-N nanodisk spin is $\frac{1}{2}N$ in the ground state.

This is consistent with Lieb's theorem(72) valid for the Hubbard model. As far as only the on-site Coulomb interactions are taken into account, the electron-electron interaction is approximated by the Hubbard model,

$$H_U = U \sum_i d_{i\uparrow}^\dagger d_{i\uparrow} d_{i\downarrow}^\dagger d_{i\downarrow}, \tag{58}$$

where U is given by (54): It is estimated(73) that $U \approx t$. Lieb's theorem states that, in the case of repulsive electron-electron interactions ($U > 0$), a bipartite system at half-filling has the ground state whose total spin is $S_{\text{tot}} = \frac{1}{2}|N_A - N_B|$, where N_A and N_B are the numbers of sites in sublattices A and B, respectively [Fig.6]. They are given by (**??**), and hence $S_{\text{tot}} = \frac{1}{2}N$ for the size-N nanodisk.

6. Large size triangular graphene

In this appendix we argue that our analysis based on the zero-energy sector is essentially correct, even if the size N of the nanodisk is large and the band gap becomes very narrow. In this case, the DOS of the nanodisk is decomposed into the DOS due to the edge, which is δ-function like, and the DOS due to the bulk, which is almost continuous. The correction to the magnetization from the bulk part is shown to be of the order of $1/N$ with respect to the one from the edge part in the zero-temperature limit.

6.1 Density of states

It is possible to derive an analytical expression for the density of states of graphene per unit cell(74), which has the form

$$\rho(\varepsilon) = \frac{4}{\pi^2}\frac{|\varepsilon|}{t^2}\frac{1}{\sqrt{Z_0}}K\left(\sqrt{\frac{Z_1}{Z_0}}\right), \tag{59}$$

with

$$Z_0 = \left\{\begin{array}{ll} \left(1+\left|\frac{\varepsilon}{t}\right|\right)^2 - \left((\varepsilon/t)^2-1\right)^2/4; & -t \le \varepsilon \le t \\ 4\left|\varepsilon/t\right|; & -3t \le \varepsilon \le -t \vee t \le \varepsilon \le 3t \end{array}\right|, \tag{60a}$$

$$Z_1 = \left\{\begin{array}{ll} 4\left|\varepsilon/t\right|; & -t \le \varepsilon \le t \\ \left(1+\left|\frac{\varepsilon}{t}\right|\right)^2 - \left((\varepsilon/t)^2-1\right)^2/4; & -3t \le \varepsilon \le -t \vee t \le \varepsilon \le 3t \end{array}\right|, \tag{60b}$$

where $K(x)$ is the complete elliptic integral of the first kind. Thus, near the Dirac point we find

$$\rho(\varepsilon) = \frac{2A_c}{\pi v_{\text{F}}^2}|\varepsilon|, \tag{61}$$

where A_c is the unit cell area, $A_c = 3\sqrt{3}a^2/2$.

For large N nanodisks, the band gap decreases inversely proportional to the size. One may wonder if our analysis based on only the zero-energy sector is relevant. Indeed, the size of experimentally available nanodisks is as large as $N = 100 \sim 1000$. We would like to argue that our analysis of the zigzag trigonal nanodisk captures the basic nature of nanodisks even for such a large size system.

It is important to remark that each energy level is at most two-fold degenerate except for the states at $\varepsilon = 0$ and $\varepsilon = t$ irrespective to its size N, as illustrated in Fig.18. The degeneracy is governed by the representation (6) of the trigonal symmetry group C_{3v}. Hence, in the limit $N \to \infty$, the density of state (DOS) of the zigzag trigonal nanodisk reflects the following two aspects; one is that there appear van-Hove singularities at $\varepsilon = 0$ and $\varepsilon = t$. The other is that the DOS shows a continuous and dense structure, which has the same structure of the bulk graphene [Fig.18] except for the contributions from these singularities. Namely, near the Fermi energy, the DOS consists of that of the bulk graphene and an additional peak at the zero-energy states due to the edge states for $N \gg 1$, as illustrated in Fig.18. Hence, together with spin degrees of freedom, it behaves as

$$D(\varepsilon) = 2cN_c|\varepsilon| + 2N\delta(\varepsilon), \tag{62}$$

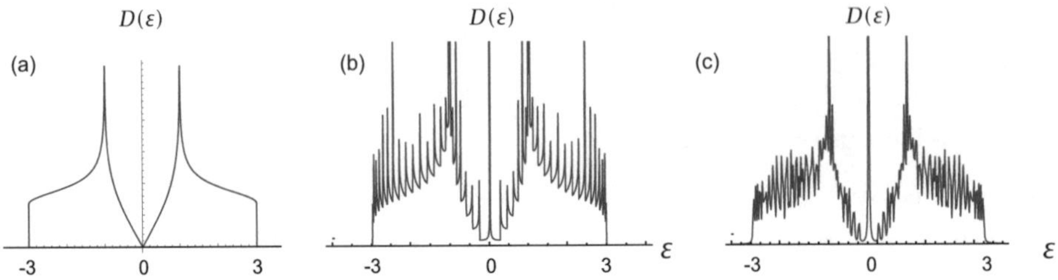

Fig. 18. The density of state of a (a) graphene, (b) graphene nanoribbon and (c) graphene nanodisk. The horizontal axis is the energy ε.

with $c = 2A_c/\pi v_F^2$. The linear term is due to the bulk states as given by (61), and the Dirac delta function term is due to the edge states. The important point is that the edge-state peak is clearly distinguished from the DOS due to the bulk part. It is enough to take into account only the zero-energy sector to analyze physics near the Fermi energy, since the contribution from the edge states is dominant.

6.2 Magnetism

For definiteness we explicitly calculate the magnetization of a nanodisk when its size is large. We start with the Hubbard Hamiltonian,

$$H = \sum_{ij\sigma} t_{ij} c_{i\sigma}^\dagger c_{i\sigma} + U \sum_i c_{i\uparrow}^\dagger c_{i\uparrow} c_{i\downarrow}^\dagger c_{i\downarrow} = \sum_{k\sigma} \varepsilon(k)\, c_{k\sigma}^\dagger c_{k\sigma} + U \sum_{kk'q} c_{k+q\uparrow}^\dagger c_{k'-q\downarrow}^\dagger c_{k'\downarrow} c_{k\uparrow}. \tag{63}$$

In the Hartree-Fock approximation the Hamiltonian is transformed into

$$H_{MF} = \sum \left(\varepsilon_k + \frac{U}{2N_C} \langle n \rangle + \sigma\Delta \right) c_{k\sigma}^\dagger c_{k\sigma} - N_C U \langle n_\uparrow \rangle \langle n_\downarrow \rangle,$$

where $N_C = N^2 + 6N + 6$,

$$\Delta = \frac{U}{2N_C} \langle m \rangle + \frac{h}{2}, \tag{64}$$

with the external magnetic field h and $\langle n_\sigma \rangle = \frac{1}{2}(\langle n \rangle + \sigma \langle m \rangle)$, or $\langle n \rangle = \langle n_\uparrow \rangle + \langle n_\downarrow \rangle$, $\langle m \rangle = \langle n_\uparrow \rangle - \langle n_\downarrow \rangle$. The statistical mean values $\langle n \rangle$ and $\langle m \rangle$ are determined self-consistently as

$$\langle n \rangle = \int d\varepsilon \rho(\varepsilon) \left[f(\varepsilon - \Delta) + f(\varepsilon + \Delta) \right], \tag{65}$$

$$\langle m \rangle = \int d\varepsilon \rho(\varepsilon) \left[f(\varepsilon - \Delta) - f(\varepsilon + \Delta) \right], \tag{66}$$

with the Fermi distribution function $f(x) = 1/(\exp[(x-\mu)/k_B T] + 1)$. We can rewrite (66) into

$$\langle m \rangle = \int d\varepsilon \rho(\varepsilon) \left[\tanh \frac{\beta(\varepsilon + \Delta)}{2} - \tanh \frac{\beta(\varepsilon - \Delta)}{2} \right]. \tag{67}$$

Substituting the formula (62) of the DOS into the Storner equation (66), we obtain

$$\langle m \rangle = N \tanh \frac{\beta\Delta}{2} + cN_c \left[\Delta^2 + \frac{1}{\beta^2} \left\{ \frac{\pi^2}{3} + 4\mathrm{Li}_2\left(-e^{-\beta\Delta}\right) \right\} \right], \tag{68}$$

with the dilogarithm function $\mathrm{Li}_2(x)$. It is difficult to solve this equation for $\langle m \rangle$ self-consistently at general temperature T. We examine two limits, $T \to 0$ and $T \to \infty$.

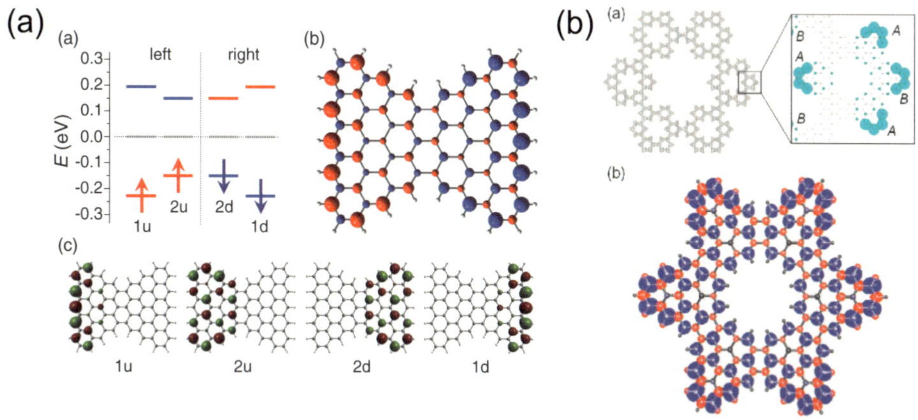

Fig. 19. (a) Clar goblet or bow-tie shaped graphene nanodisk. (b) Fractal sturucture of star of David graphene nanodisk. Fig (a) is taken from the work of Wang et. al (39), and Fig. (b) is taken from the work of Wang et al (38).

For the zero temperature ($T \to 0$) we obtain

$$\langle m \rangle = N + c\frac{U^2}{4N_C} \langle m \rangle^2 + \frac{cUh}{2} \langle m \rangle + O(h^2). \tag{69}$$

The magnetization $\langle m \rangle$ is determined by solving (69). Because $|\langle m \rangle| \leq N$, it follows that $\langle m \rangle = N + O(1)$. The contribution from the bulk gives a negligible correction to the total magnetization. Hence the magnetization is $\langle m \rangle = N$, and the ground state is fully poralized whenever $U \neq 0$. Ferromagnetism occurs irrespective of the strength of the Coulomb interaction.

The magnetization is propotional not to N_c but N. In this sence the ground state of nanodisk is not bulk ferromagnet but surface ferromagnet, which is consistent with the previous result. We next investigate the high temperatuer limit ($T \to \infty$). Using the Taylor expansion of the dilogarithm function,

$$\mathrm{Li}_2\left(-e^{-\beta\Delta}\right) = -\frac{\pi^2}{12} + \beta\Delta \log 2 - \frac{\beta^2\Delta^2}{4} + \frac{\beta^3\Delta^3}{24} + \cdots, \tag{70}$$

we find

$$\langle m \rangle = N \tanh \frac{\beta\Delta}{2} + cN_c \left[\frac{4\Delta}{\beta} \log 2 + \frac{\beta\Delta^3}{6} + \cdots\right]. \tag{71}$$

The leading term is the second term, and hence the main contribution comes from the bulk. The solution is only $\langle m \rangle = 0$ for which $\Delta = 0$. There is no magnetization at high temperature.

7. Complex structure of nanodisk

7.1 Bowtie-shaped graphene nanodisk

A bowtie-shaped graphene nanodisk is proposed by Wang et. al(39). A bowties-shaped graphene nanodisk is composed of two trigonal graphene nanodisks sharing one hexagon: See Fig.19. Lieb's theorem predicts the ground state with $S = 0$. On the other hand, there are degenerate zero-energy states predicted from the graph theory(68). To satisfy the spin-singlet ground state, the ground electronic configuration exhibits antiferromagnetic ordering.

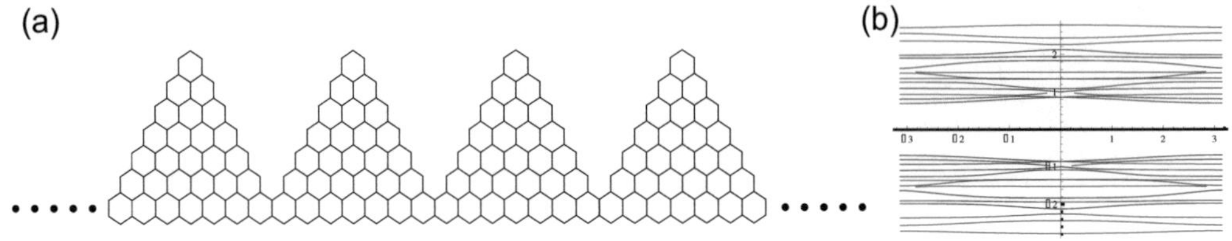

Fig. 20. (a) Illustration of nanodisk array. (b) Band structure.

7.2 "Star of David" graphene nanodisk

A "Star of David" shape graphene nanodisk is proposed by Wang et. al(38). The structure is fractal and generated by repeatedly overlapping two triangles in opposite directions and removing the overlap portion: See Fig.19. The structure has a fractal dimension,

$$\frac{\log 1/6}{\log 1/3} = 1.62. \tag{72}$$

The total spin increases exponentially with the fractal level q as $S_q = S_0 2^q$, where S_0 is the spin of the initial graphene triangle(38). The increase of the total spin is due to the increase of the boundary length, a hallmark of fractal structures.

7.3 Nanodisk array

We investigate nanodisk arrays, which are materials that nanodisks are connected in one-dimention. We show an example of trigonal zigzag nanodisk array in Fig.20(a). We show the corresponding band structure in Fig.20(b). It is intriguing that there are N-fold degenerate perfect flat band in the nanodisk with size N. This fact is also confirmed by the Leib theorem. Each nanodisk has spin N/2 and make ferromagnetic coupling between two nanodisks. In the same way we can make two-dimensional nanodisk arrays.

7.4 Nanomechanical switching

We can construct a nanomechanical switching contacting two graphene trigonal corners. We assume the angle between two corners is θ. The conductance is determined by the overlap integral of pi electrons between two corners, which is given by

$$\left|\langle p_z \cos\theta + p_y \sin\theta \, | p_z \rangle\right| = \cos^2\theta. \tag{73}$$

When the two planes are parallel ($\theta = 0$), the overlap takes the maximum value and pi-electrons can go through the contact. This is the on state. When the two planes are orthogonal ($\theta = \pi/2$), the overlap takes the minimum value and pi-electrons can not go through the contact. This is the off state. The angle is changed by nanomechanically. In this sence this system acts as a nanomechanical switch. This system could detect the angle very sensitively and be useful for detect nanomechanical oscillations.

8. Spintronic devices

We anticipate graphene derivatives to be the key elements of future nanoelectronics and spintronics. First of all, graphene is common material and ecological. In graphene, spin relaxation length is as large as $2\mu m$ due to small spin-orbit interactions. Long spin relaxation length has motivated spintronics based on graphene recently. Nevertheless, there exists a key issue yet to be resolved, that is, how to generate and manipulate spin currents. Graphene

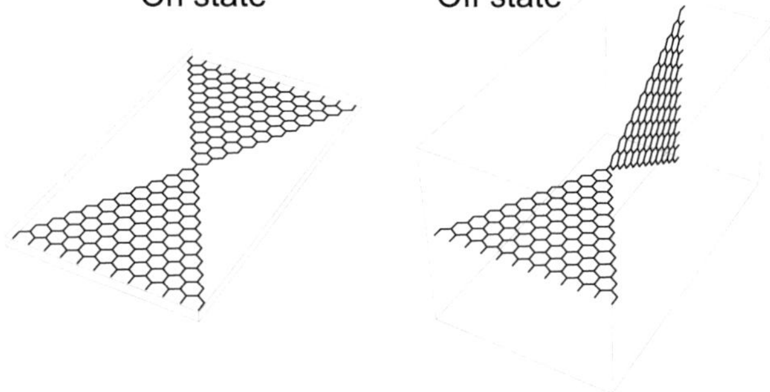

Fig. 21. Nanomechanical switching. (a) On state. (b) Off state.

nanodisks with size $N = 10 \sim 100$ would be ideal for this purpose, where localized spins are stored. In the rest of this section we shall propose a rich variety of spintronic devices made of nanodisks and leads, such as spin memory, spin valve, spin-field-effect transistor, spin diode and spin switch.

8.0.0.1 Basic properties

We summarize the relation between the spin properties of a nanodisk and an incoming electric current. First of all, being a quasiferromagnet, the nanodisk has a definite polarization. With respect to the incoming electric current there are three cases. (1) The polarized current, where all electrons have a definite polarization, rotates the nanodisk spin to that of the incoming current. (2) The unpolarized current, where the polarization of each electron is completely random, does not induce any effective magnetic field. Hence it is filtered so that the outgoing current is polarized to that of the nanodisk. (3) The partially polarized current, where the polarization of each electron is at random but the averaged polarization has a definite direction, induces a net effective magnetic field. Hence it rotates the nanodisk spin to that of the incoming current, and then is filtered so that the outgoing current is completely polarized to the averaged polarization of the incoming current. Furthermore, it is possible to control the nanodisk spin externally by applying magnetic field. Then the outgoing current has the same polarization as that of the nanodisk, irrespective of the type of incoming current. Using these properties we propose some applications of graphene nanodisks for spintronic devices.

8.0.0.2 Spin memory

The first example is a spin memory(58). For a good memory device three conditions are necessary: (i) It keeps a long life time information; (ii) Information stored in the memory can be read out without changing the information stored; (iii) It is possible to change the information arbitrarily.

First, since the life time of the nanodisk quasiferromagnet is very long compared to its size, we may use the nanodisk spin as an information. Next, we can read-out this information by applying a spin-unpolarized current. The outgoing current from a nanodisk is spin-polarized to the direction of the nanodisk spin. Thus we can obtain the information of the nanodisk spin by observing the outgoing current. Finally, the direction of the nanodisk spin can be controlled by applying a spin-polarized current into the nanodisk. Thus, the nanodisk spin satisfies the conditions as a memory device. The important point is that the size is of the order of nanometer, and it is suitable as a nanodevice.

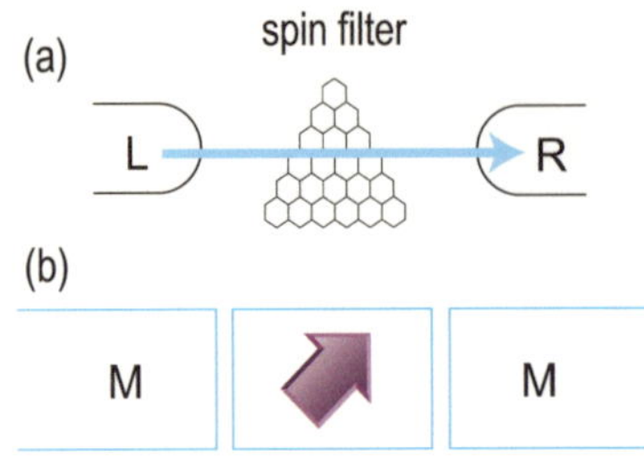

Fig. 22. (a) An electron tunnels from the left lead to the nanodisk and then to the right lead. Only electrons with the same spin direction as the nanodisk spin can pass through the nanodisk freely. As a result, when we apply a spin-unpolarized current to the nanodisk, the outgoing current is spin polarized to the direction of the nanodisk spin. Consequently, this system acts as a spin filter. (b) The system is a reminiscence of a metal-ferromagnet-metal junction.

8.0.0.3 Spin filter

We make an investigation of the spin-filter effects(64). We first analyze how spin of a nanodisk filters spin of the current by assuming that the nanodisk is a rigid ferromagnet. However, since the nanodisk is not a rigid ferromagnet but a quasiferromagnet, its spin can be controlled by the spin of the current. We analyze the reaction to the spin of the nanodisk from the spin of electrons in the current.

We go on to investigate applications of nanodisks to spintronics. The basic one is a spin filter(75). We consider a lead-nanodisk-lead system, where an electron makes a tunnelling from the left lead to the nanodisk and then to the right lead. This system is a reminiscence of a metal-ferromagnet-metal junction (Fig.22). If electrons in the lead has the same spin direction as the nanodisk spin, they can pass through the nanodisk freely. However, those with the opposite direction feel a large Coulomb barrier and are blocked (Pauli blockade)(64). As a result, when we apply a spin-unpolarized current to the nanodisk, the outgoing current is spin polarized to the direction of the nanodisk spin. Consequently, this system acts as a spin filter.

8.0.0.4 Spin valve

A nanodisk can be used as a spin valve, inducing the giant magnetoresistance effect(76–78). We set up a system composed of two nanodisks sequentially connected with leads [Fig.23]. We apply external magnetic field, and control the spin direction of the first nanodisk to be $|\theta\rangle = \cos\frac{\theta}{2}|\uparrow\rangle + \sin\frac{\theta}{2}|\downarrow\rangle$, and that of the second nanodisk to be $|0\rangle = |\uparrow\rangle$. We inject an unpolarized-spin current to the first nanodisk. The spin of the lead between the two nanodisks is polarized into the direction of $|\theta\rangle$. Subsequently the current is filtered to the up-spin one by the second nanodisk. The outgoing current from the second nanodisk is

$$I_{\uparrow}^{\text{out}} = I\cos\frac{\theta}{2}. \tag{74}$$

We can control the magnitude of the up-polarized current from 0 to I by rotating the external magnetic field. The system act as a spin valve.

8.0.0.5 Spin switch

We consider a chain of nanodisks and leads connected sequentially (Fig.22). Without external magnetic field, nanodisk spins are oriented randomly due to thermal fluctuations, and a

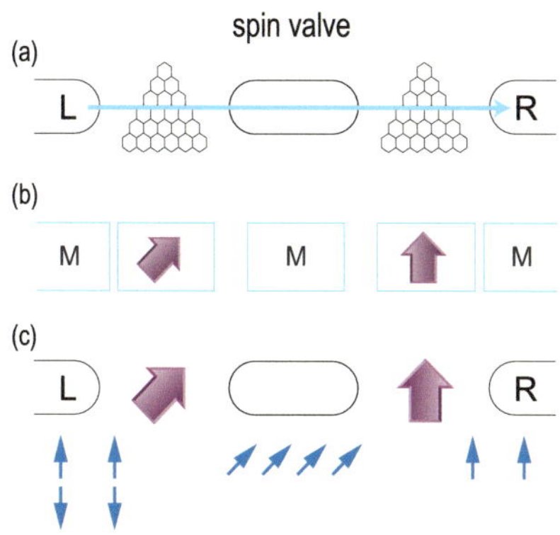

Fig. 23. Illustration of spin valve. (a) The spin valve is made of two nanodisks with the same size, which are connected with leads. (b) The system is analogous to the one made of metals and ferromagnets. (c) Applying external magnetic field, we control the spin direction of the first nanodisk to be $|\theta\rangle$, and that of the second nanodisk to be $|0\rangle = |\uparrow\rangle$. The incomming current is unpolarized, but the outgoing current is polarized, $I_{\uparrow}^{\text{out}} = I\cos\frac{\theta}{2}$, $I_{\downarrow}^{\text{out}} = 0$. Its magnitude can be controlled continuously. This acts as a spin valve.

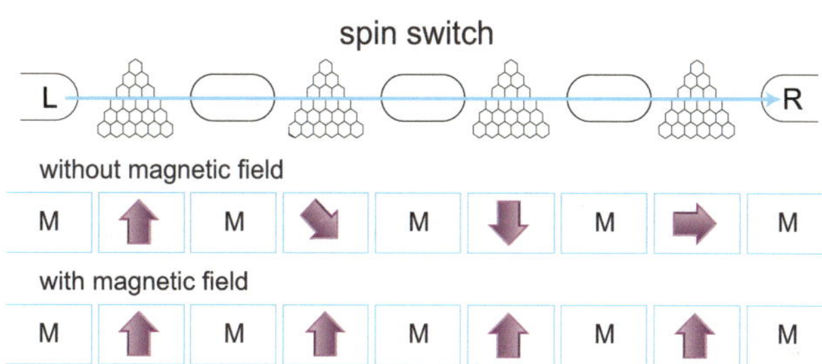

Fig. 24. A chain of nonodisks and leads acts as a spin switch. Without external magnetic field, nanodisk spins are oriented randomly due to thermal fluctuations, and a current cannot go through the chain. However, as soon as a uniform magnetic field is applied to all nanodisks, the direction of all nanodisk spins become identical and a current can go through.

current cannot go through the chain. However, when and only when a uniform magnetic field is applied to all nanodisks, the direction of all nanodisk spins become identical and a current can go through. Thus the system acts as a spin switch, showing a giant magnetoresistance effect. The advantage of this system is that a detailed control of magnetic field is not necessary in each nanodisk.

8.0.0.6 Spin-field-effect transistor

We again set up a system composed of two nanodisks sequentially connected with leads [Fig.25]. We now apply the same external magnetic field to both these nanodisks, and fix their spin direction to be up, $|0\rangle = |\uparrow\rangle$. As an additional setting, we use a lead acting as a spin rotator with the spin-rotation angle θ. The outgoing current from the second nanodisk is $I_{\uparrow}^{\text{out}} = I\cos\frac{\theta}{2}$. It is possible to tune the angle θ by applying an electric field. Hence we can control the magnitude of the up-polarized current. The system acts as a spin-field-effect transistor(79).

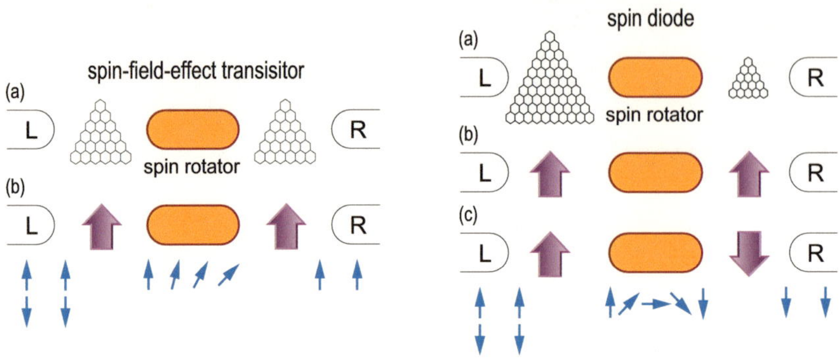

Fig. 25. (Left) Illustration of spin-field-effect transistor made of two nanodisks with the same size. Two nanodisks are connected with a rotator. We set the spin direction of the two nanodisks to be up by magnetic field. The incomming current is unpolarized, but the outgoing current is polarized and given by $I_{\uparrow}^{\text{out}} = I \cos \frac{\theta}{2}$, $I_{\downarrow}^{\text{out}} = 0$. The up-spin current is rotated by the angle θ within the central lead ascting as a rotator. Illustration of spin diode made of two nanodisks with different size. By controlling the bias voltage $\Delta\mu$, the current flows from the left lead to the right lead ($\Delta\mu > 0$), or in the opposite way ($\Delta\mu < 0$). The incoming current is unpolarized, which is made polarized by the first nanodisk. The electron spin in the central lead is rotated by the Rashba-type interaction. (Right) Illustration of spin diode made of two nanodisks with different size. By controlling the bias voltage $\Delta\mu$, the current flows from the left lead to the right lead ($\Delta\mu > 0$), or in the opposite way ($\Delta\mu < 0$). The incoming current is unpolarized, which is made polarized by the first nanodisk. The electron spin in the central lead is rotated by the Rashba-type interaction.

8.0.0.7 Spin diode

The third example is a spin diode [Fig.25]. We set up a system composed of two nanodisks sequentially connected with leads, where two nanodisks have different sizes. The left nanodisk is assumed to be larger than the right nanodisk. Then the relaxation time of the left nanodisk $\tau_{\text{L}}(\equiv \tau_{\text{filter}}^{\text{L}})$ is larger than that of the right nanodisk $\tau_{\text{R}}(\equiv \tau_{\text{filter}}^{\text{R}})$, $\tau_{\text{L}} > \tau_{\text{R}}$. Second, the applied magnetic field is assumed to be so small that the nanodisk spin can be controlled by a polarized current. For definiteness we take the direction of the magnetic field to be up. Third, the lead is assumed to have the Rashba-type interaction(80),

$$H_{\text{R}} = \frac{\lambda}{\hbar} \left(p_x \sigma^y - p_y \sigma^x \right). \tag{75}$$

Spins make precession while they pass through the lead. The spin-rotation angle is given(55) by

$$\Delta\theta = \frac{2\lambda m^* L}{\hbar}, \tag{76}$$

where m^* is the electron effective mass in the lead and L is the length of the lead. We can control $\Delta\theta$ by changing the coupling strength λ externally by applying electric field(81).
When no currents enter the nanodisk, the direction of two nanodisk spins is identical due to the tiny external magnetic field, which is up. When we inject the current in this state, the net outgoing current is very small,

$$I^{\text{out}} = \cos \Delta\theta \simeq 0 \qquad \text{for} \quad \Delta\theta \approx \pi. \tag{77}$$

This is the "off" state of the spin diode.
Let us inject an unpolarized pulse square current to the system, starting at $t = t_i$ and finishing at $t = t_f$,

$$I_\sigma(t) = I^{\text{in}}\theta(t - t_i)\theta(t_f - t), \tag{78}$$

where σ denotes the spin. The system become the "on" state by the pulse. When the bias voltage is such that $\Delta\mu > 0$, the current flows into the left nanodisk and then into the right nanodisk. The left nanodisk acts as a spin filter. The current in the central lead is initially up-polarized but is rotated by the angle $\Delta\theta$ due to the Rashba-type coupling effect. Then it enters the right nanodisk. This rotation angle acts as the initial phase for the second nanodisk. On the other hand, when $\Delta\mu < 0$, the current enters the right nanodisk and goes out from the left nanodisk. Since the relaxation time is τ_{L}, the total spin-dependent charge is given by the above formulas but with the replacement of τ_{R} by τ_{L}. Because the sizes of two nanodisks are different, these two currents behave in a different way.
The important result is that the system acts as a rectifier so that the up-spin current flows from the left to the right, or the large nanodisk to the small nanodisk. We may call it "spin diode".

9. Discussions

The physics of graphene related materials is currently one of the most active and attractive research areas in condensed matter physics. Among these graphene derivatives nanoribbons and nanodisks are mainly focused. They correspond to wires and quantum dots, respectively. There are a profusion of papers on nanoribbons, but there are yet only a few works on nanodisks. This may be because nanodisks are difficult to manufacture. However, nanodisks are experimentally isolated recently by the Ni etching techniques. An experimental realization must accelerate both experimental and theoritical studies on nanodisks.
In this paper we have reported the results on the electronic and magnetic properties of nanodisks. The trigonal zigzag nanodisk has N-fold degenerate zero-energy states when its size is N. The low-energy physics near the Fermi energy is well described by this zero-energy sector. Wave functions are explicitly constructed and classified according to the symmetry group C_{3v}. The emergence of a quasi-phase transition has been found between the quasi-ferromagnet and the paramagnet as a function of temperature even for samples with $N \approx 100$. We have also studied nanodisk-lead systems. In the intermediate coupling regime, as the chemical potential increases, a salient series of Coulomb blockade peaks develops in the conductance, reflecting the energy spectrum of nanodisk with a broken SU(N) symmetry. An appropriate size to observe Coulomb blockades is $N = 2 \sim 10$. In the strong coupling regime, by investigating the spin-spin correlation, we present some indications of many-spin Kondo effects. An appropriate size to observe Kondo effects is $N = 2 \sim 4$.

10. Acknowledgements

I am very much grateful to Professors N. Nagaosa, T. Ando, S. Tarucha, Y. Takada, H. Tsunetsugu, A. Oshiyama, M. Ueda, S. Hasegawa, P. Kim, P.J. Herrero, B. Nikolic and C. Honerkamp for fruitful discussions on the subject and reading through the manuscript. This work was supported in part by Grants-in-Aid for Scientific Research from the Ministry of Education, Science, Sports and Culture 20840011, No. 20940011 and No. 22740196.

11. References

[1] M. S. Dresselhaus, G. Dresselhaus, K. Sugihara, I. L. Spain and H. A. Goldberg, *Graphite Fibers and Filaments*, Springer Series in Material Science Vol 5 (Springer-Verlag, Berlin, 1988).

[2] J. W. McClure, Phys.Rev. 104, 666 (1956).

[3] S. Iijima, Nature (London) 354, 56 (1991).

[4] N. Hamada, S. I. Sawada, and A. Oshiyama, Phys. Rev. Lett. 68, 1579 (1992).

[5] R. Saito, M. Fujita, G. Dresselhaus, and M. S. Dresselhaus, Appl. Phys. Lett. 60, 18 (1992).

[6] J. W. Mintmire, B. I. Dunlap, and C. T. White, Phys. Rev. Lett. 68, 631 (1992).

[7] R. Saito, G. Dresselhaus, and M. S. Dresselhaus, *Physical Properties of Carbon Nanotubes*, Imperial College Press, 1998, London.

[8] J. W. G. Wildöer, L. C. Venema, A. G. Rinzler, R. E. Smalley, and C. Dekker, Nature (London) 391, 59 (1998).

[9] A. Jorio, R. Saito, J. H. Hafner, C. M. Lieber, M. Hunter, T. McClure, G. Dresselhaus, and M. S. Dresselhaus, Phys. Rev. Lett. 86, 1118 (2001).

[10] M. Ouyang, J. L. Huang, C. M. Lieber, Ann. Rev. Phys. Chem 53, 201 (2002).

[11] H. W. Kroto, J. R. Heath, S. C. O'Brien, R. F. Curl, R. E. Smalley, Nature (London) 318, 167 (1985).

[12] K.S. Novoselov, A. K. Geim, S. V. Morozov, D. Jiang, Y. Zhang, S. V. Dubonos, I. V. Grigorieva, and A. A. Firsov, Science 306, 666 (2004).

[13] K.S. Novoselov, A. K. Geim1, S. V. Morozov, D. Jiang1, M. I. Katsnelson, I. V. Grigorieva, S. V. Dubonos and A. A. Firsov, Nature 438, 197 (2005).

[14] Y. Zhang, Yan-Wen Tan, Horst L. Stormer and Philip Kim, Nature 438, 201 (2005).

[15] A. H. Castro Neto, F. Guinea, N. M. R. Peres, K. S. Novoselov and A. K. Geim, Rev. Mod. Phys. 81, 109 (2009).

[16] D. J. Klein, Chem. Phys. Lett. 217, 261 (1994).

[17] M. Fujita, K. Wakabayashi, K. Nakada, and K. Kusakabe, J. Phys. Soc. Jpn. 65, 1920 (1996).

[18] Y. Niimi, T. Matsui, H. Kambara, K. Tagami, M. Tsukada, and H. Fukuyama. Appl. Surf. Sci. 241, 43 (2005). Y. Kobayashi, K. Fukui, T. Enoki and K. Kusakabe, Phys. Rev. B, 73, 125415 (2006).

[19] O. E. Andersson, B. L. V. Prasad, H. Sato, T. Enoki, Y. Hishiyama, Y. Kaburagi, M. Yoshikawa and S. Bandow, Phys. Rev. B, 58, 16387 (1998).

[20] K. Kusakabe and M. Maruyama, Phys. Rev. B, 67, 092406 (2003).

[21] M. Ezawa, Physical Review B, 73, 045432 (2006).

[22] L. Brey, and H. A. Fertig, Phys. Rev. B, 73, 235411 (2006).

[23] F. Muñoz-Rojas, D. Jacob, J. Fernández-Rossier, and J. J. Palacios, Phys. Rev. B, 74, 195417 (2006).

[24] Y. -W Son, M. L. Cohen, and S. G. Louie, Phys. Rev. Lett., 97, 216803 (2006).

[25] V. Barone, O. Hod, and G. E. Scuseria, Nano Lett., 6, 2748 (2006).

[26] M. Y. Han, B. Oezyilmaz, Y. Zhang, and P. Kim, Phys. Rev. Lett., 98, 206805 (2007).

[27] Z. Chen, Y. -M. Lin, M. J. Rooks, and P. Avouris, Physica E, 40, 228 (2007).

[28] Z. Xu and Q. -S. Zheng, Appl. Phys. Lett. 90, 223115 (2007).

[29] B. Özyilmaz, P. Jarillo-Herrero, D. Efetov and P. Kim, Appl. Phys. Lett. 91, 192107 (2007).
[30] Y.W. Son, M.L. Cohen and S.G. LouieSon, Nature 444, 347 (2006).
[31] O.Hod, V. Barone, J.E. Peralta and G.E. Scuseria, Nano Lett. 7, 2295 (2007).
[32] See http://sciencewatch.com/ana/st/graphene/.
[33] C. Berger, Z. Song, X. Li, X. Wu, N. Brown, C. Naud, D. Mayou, T. Li, J. Hass, A.N. Marchenkov, E. H. Conrad, P.N. First and W. A. de Heer, Science 312, 119 (2006).
[34] H. J. Räder, A. Rouhanipour, A. M. Talarico, V. Palermo, P. Samorì, and K. Müllen, Nature materials 5, 276 (2006).
[35] M. Ezawa, Physica Status Solidi (c) 4, No.2, 489 (2007).
[36] M. Ezawa, Phys. Rev. B 76, 245415 (2007): M. Ezawa, Physica E 40, 1421-1423 (2008).
[37] J. Fernández-Rossier, and J. J. Palacios, Phys. Rev. Lett. 99, 177204 (2007).
[38] W. L. Wang, S. Meng, and E. Kaxiras, Nano Lett 8, 241 (2008).
[39] W. L. Wang, Oleg V. Yazyev, S. Meng, and E. Kaxiras, Phys. Rev. Lett. 102, 157201 (2009).
[40] J. Akola, H. P. Heiskanen, and M. Manninen, Phys. Rev. B 77, 193410 (2008).
[41] O. V. Yazyev, Rep. Prog. Phys 73, 5 (2010).
[42] A. D. Güçlü, P. Potasz, O. Voznyy, M. Korkusinski, and P. Hawrylak, Phys. Rev. Lett. 103, 246805 (2009).
[43] P. Potasz, A. D. Güçlü and P. Hawrylak, Phys. Rev. B 81, 033403 (2010).
[44] L. C. Campos, V. R. Manfrinato, J. D. Sanchez-Yamagishi, J. Kong and P. Jarillo-Herrero, Nano Lett., 9, 2600 (2009).
[45] O. Hod, V. Barone, and G. E. Scuseria, Phys. Rev. B 77, 035411 (2008).
[46] M. Ezawa, Phys. Rev. B 77, 155411 (2008).
[47] M. Ezawa, Phys. Rev. B 79, 241407(R) (2009).
[48] M. Ezawa, Phys. Rev. B 81, 201402(R) (2010).
[49] J.C. Slonczewski and P.R. Weiss, Phys. Rev. 109, 272 (1958).
[50] G. W. Semenoff, Phys. Rev. Lett. 53, 2449 (1984).
[51] H. Ajiki and T. Ando, J. Phys. Soc. Jpn., 62, 1255 (1993); T. Ando, Y. Zheng and H. Suzuura, Microelectronic Engineering, 63, 167 (2002).
[52] S. Murakami, N. Nagaosa, S.-C. Zhang, Science 301, 1348 (2003).
[53] Y. Ohno, D. K. Young, B. Beschoten, F. Matsukura, H. Ohno and D. D. Awschalom, Nature 402, 790 (1999).
[54] See, e.g., Realizing Controllable Quantum States: Mesoscopic Superconductivity and Spintronics, edited by H. Takayanagi and J. Nitta, World Scientific, Singapore, 2005, and references therein.
[55] I. Zǔtić, J. Fabian and S. Das Sarma, Rev. Mod. Phys., 76, (2004), and references therin.
[56] S. A. Wolf, D. D. Awschalom, R. A. Buhrman, J. M. Daughton, S. von Molnár, M.L.Roukes, A. Y. Chtchelkanova and D. M. Treger, Science 294, 1488 (2001), and references therin.
[57] G. A. Prinz, Science 282, 1660 (1998), and references therin.
[58] P. Recher, E. V. Sukhorukov and D. Loss, Phys. Rev. Lett., 85, 1962 (2000).
[59] J. A. Folk, R. M. Potok, C. M. Marcus and V. Umansky, Science 299, 679 (2003).
[60] N. Tombros, C. Jozsa, M. Popinciuc, H. T. Jonkman, B. J. van Wees, Nature 448, 571 (2007).

[61] V. M. Karpan, G. Giovannetti, P. A. Khomyakov, M. Talanana, A. A. Starikov, M. Zwierzycki, J. van den Brink, G. Brocks, P. J. Kelly, Phys Rev Lett 99, 176602 (2007).
[62] L.Brey, H.A.Fertig, Phys. Rev. B 76, 205435 (2007).
[63] M. Wimmer, N. Adagideli, S. Berber, D. Tománek, and K. Richter, Phys. Rev. Lett., 100, 177207 (2008).
[64] M. Ezawa, Eur. Phys. J. B 67, 543 (2009).
[65] M. Ezawa, New J. Phys., 11, 095005 (2009).
[66] M. Ezawa, Physica E 42, 703-706 (2010).
[67] K.S. Novoselov, E. McCann, S. V. Morozov, V. I. Fal'ko, M. I. Katsnelson, U. Zeitler, D. Jiang, F. Schendini and A. K. Geim, Nature Phys. 2, 177 (2006).
[68] S. Fajtlowicz, P. E. John, and H. Sachs, Croat. Chem. Acta 78, 195. (2005).
[69] S.Y. Zhou, G.-H. Gweon, J. Graf, A. V. Fedorov, C. D. Spataru, R. D. Diehl, Y. Kopelevich, D.-H. Lee, S.G. Louie, A. Lanzara, Nature Phys. 2, 595 (2006).
[70] T. Matsuda, S. Hasegawa, M. Igarashi, T. Kobayashi, M. Naito, H. Kajiyama, J. Endo, N. Osakabe and A. Tonomura, Phys. Rev. Lett. 62, 2519 (1989).
[71] M. Kinza, J. Ortlo, and C. Honerkamp, cond-mat/arXiv:1002.2513
[72] E. H. Lieb, Phys. Rev. Lett. 62, 1201 (1989).
[73] L. Pisani, J.A. Chan, B. Montanari and N.M. Harrison, Phys. Rev. B 75, 64418 (2007).
[74] J. P. Hobson and W. A. Nierenberg, Phys. Rev. 89, 662 (1953).
[75] T. Koga, J. Nitta, H. Takayanagi and S. Datta, Phys. Rev. Lett. 88, 126601 (2002).
[76] M. N. Baibich, J. M. Brot, A. Fert, N. V. Dau and F. Petroff, Phys. Rev. Lett. 61, 2472 (1988).
[77] G. Binasch, P. Grunberg, F. Saurenbach and W. Zinn, Phys. Rev. B 39, 4828 (1989).
[78] Y. M. Lee, J. Hayakawa, S. Ikeda, F. Matsukura and H. Ohno, Appl. Phys. Lett. 90, 212507 (2007).
[79] S. Datta and B. Das, Appl. Phys. Lett. 56, 665 (1990).
[80] E. I. Rashba, Fiz. Tverd. Tela (Leningrad) 2, 1224 (1960) [Sov. Phys. Solid State 2, 1109 (1960)]; Y. A. Bychkov and E. I. Rashba, J. Phys. C 17, 6039 (1984).
[81] J. Nitta, T. Akazaki, and H. Takayanagi, Phys. Rev. Lett. 78, 1335 (1997).

13

Graphene Nano-Flakes and Nano-Dots: Theory, Experiment and Applications

Ian Snook[1] and Amanda Barnard[2]
[1]*Applied Physics, School of Applied Sciences, RMIT University,*
[2]*Virtual Nanoscience Laboratory, CSIRO Materials Science & Engineering,*
[1,2]*Australia*

1. Introduction

The experimental isolation of graphene (Novoselov et al., 2004) was one of the greatest discoveries in materials physics for many years. The time between the prediction of the electronic structure of graphene (Wallace, 1947) and its experimental isolation was nearly 60 years (Novoselov et al., 2004) but, by contrast, since this experimental discovery 6 years ago there has been an enormous amount of theoretical and experimental interest in graphene membranes. This is because graphene has proved to have exceptional properties including very fast electron transport, room temperature quantum hall effect, the highest mechanical strength and greatest thermal conductivity yet measured (Castro Neto et al., 2009; Abergel et al., 2010). In particular its fascinating electrical properties have lead to the speculation that graphene may one day replace silicon as the material of choice for most electronic applications (van Norden, 2006).

From a physical perspective graphene is a large 2-D sheet or membrane of sp^2 bonded carbon atoms which we will term graphene membranes (GMBs), see figure 1. The basic structure of GMBs is that of fused heaxagonal rings of aromatic benzene so GMBs may be regarded as giant aromatic molecules. In chemistry they are often termed polycyclic aromatic hydrocarbons (PAH's) if the edges are hydrogenated (Wu et al., 2007; Zhi and Mullen, 2008). It may also be thought of as honeycomb lattice since, at least ideally, GMBs are large, flat structures consisting of hexagonal rings of carbon atoms with a nearest neighbour distance of 1.42 Å and bond angles of 120° (as shown in figure 1) just like the structure of a layer of honeycomb. As we shall see in section 4 another useful way of looking at these structures for some purposes is as two interpenetrating trigonal lattices, labeled A- and B-. But the important point here is that all of these descriptions and metaphors are referring to the same thing.

Despite having many exceptional properties GMBs have one very severe limitation from the point of view of electronics applications; they have no band-gap and a vanishingly small density of states at the Fermi level i.e. it is a semi-metal (Castro Neto et al., 2009; Abergel et al., 2010), (see section 4). Several methods have been suggested to induce a band gap in graphene and thus overcome this fundamental limitation. One method is to cut a GMB into nanoribbons (GNRs, see figure 1b) which may produce a band-gap, depending on the width (Castro Neto et al., 2009; Abergel et al., 2010). This observation has subsequently generated enormous research activity on GNRs. The reason for the appearance of a gap is

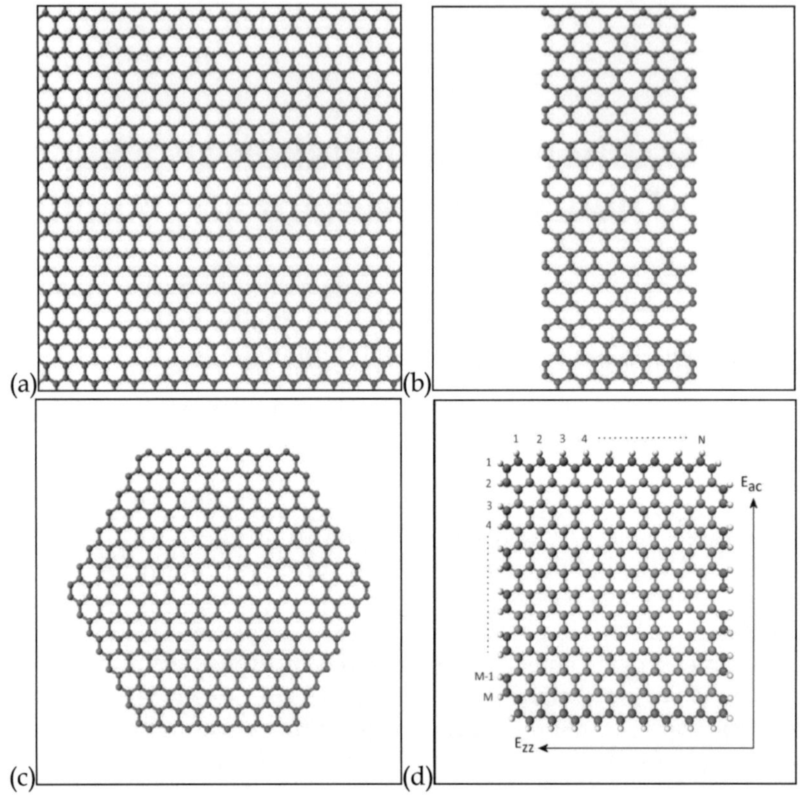

Fig. 1. The three basic type of graphene, (a) an infinite graphene sheet or membrane, 2-D graphene (b) a graphene nanoribbon, 1-D graphene, (c) a graphene nanoflake or graphene nanodot, 0-D graphene and (d) shows a GNF with a zigzag edge labeled by E_{zz}, an armchair edge labeled by E_{ac}.

that along with the finite width edge states begin to play a significant role in modifying the electronic properties since a significant number of atoms reside at the edges. Therefore, GNRs may be termed 1-D graphene and we may look at their production as adding one extra degree of engineering freedom to graphene as the fraction of edge atoms and hence edge states may be systematically controlled. There are two major types of idealized edge states, armchair and zigzag as can be seen on the sides and tops respectively of the structure represented in figure 1d. It has been shown that GNRs can be metals, semiconductors, half-metals, feromomagnets and antiferomagnets depending on their edge structures, width, shape and chemical termination (Castro Neto et al., 2009; Abergel et al., 2010; Son, et al, 2006a; Son et al., 2006b; Hod et al., 2007).

However, this critical dependence of their properties on their edge states also leads to new limitations in terms of fabrication (for use in electronic devices) because it is difficult to consistently and reliably produce GNRs with edge states of a particular type. Another limitation is that the cutting of GMBs into GNRs leads to electron mobility degradation and loss of performance in devices.

The zero-dimensional (0-D) form of graphene, which may be called graphene nano-flakes (GNFs) or graphene nano-dots (GNDs) (see figure 1c) also exists but has been much less extensively studied than GMBs or GNRs. This is surprising given that preliminary studies suggest they show some very interesting properties which differ from those of 2-D and 1-D graphene and have great potential for a variety of applications principally as electronic and magnetic devices. These potential applications arise because GNFs not only have edge states, but also corner states, and may also be cut into a much larger variety of different shapes (see figure 2). The addition of these structural features may be regarded as giving GNFs yet another degree of engineering freedom over GMRs and twice as many degrees of freedom than GMBs. Examples of corner states can be seen where edge states meet in figure 1c and in figure 2 we see some examples of different shapes which are possible with GNFs, and the different types of corners that can be introduced.

Furthermore, in contrast to GNRs, GNFs can potentially range in size from molecular to semi-infinite 2-D structures, and consequently their electronic structures will vary from having discrete molecular levels to being band-like as their dimensions are made larger. This leads to the potential of spanning the range of electronic and magnetic properties from molecular-like to 2-D like by using GNFs of different dimensions. Of course, just as we see in GMBs and GNRs, GNFs also have their own inherent limitations, as we shall see in later sections.

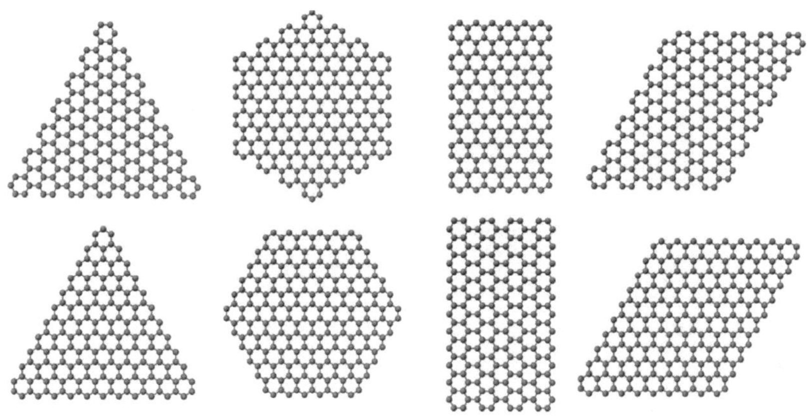

Fig. 2. This shows a variety of possible GNF shapes which might be produced with different geometries and corners, but always terminated with either zigzag or armchair edges.

In this chapter we give an outline of how GNF structures may be created, give a theoretical description of the structure and its stabilty, the electronic structure and basic characteristics. We then outline some of the potential applications of these nano-flakes in the areas of electronic and magnetic devices. Where there exists a large literature which is particularly the case for GMBs and GNRs but also in some areas on GNFs (particularly in the area of synthesis) we have attempted to reference comprehensive and up-to-date reviews in order to keep this chapter within reasonable limits.

2. The production of GNFs

As stated in the introduction graphenes consist of fused, combinations of sp^2 bonded carbon hexagons referred to as rings. Thus, at the one end of the size spectrum of graphenes we have a benzene radical or if terminated by hydrogen atoms we have benzene and at the other extreme an infinite 2-D graphene membrane. These limiting cases are usually treated by chemists or materials engineers, respectively, with materials physics spanning the sizes in between. Depending of the perspective, there are two distinct classes of methods that have been generally used to make GNFs, partly because of this wide range of sizes of GNFs, and these are usually classified as bottom-up and top-down.

2.1 Bottom-up production of GNFs

As suggested above, these are essentially chemical method of production based on the chemistry of aromatic molecules. Small molecular units are "fused" to form large aromatic hydrocarbons by a large variety of chemical reactions some of which are shown in figure 3 below (Wu et al., 2007; Zhi and Mullen, 2008).

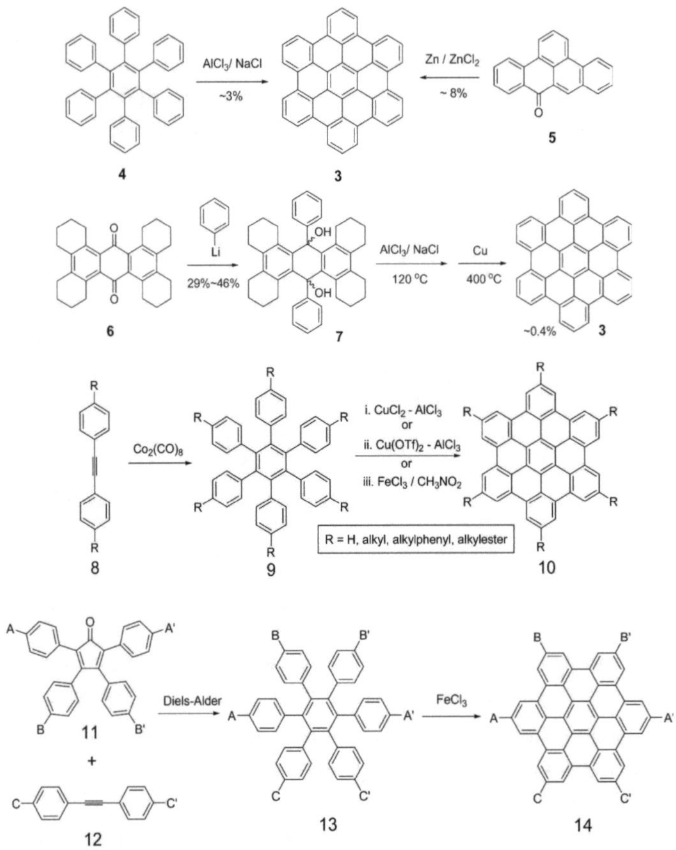

Fig. 3. Examples of various chemical reactions used to produce GNFs (Wu, et al., 2007).

The largest such structure produced by these methods reported to date is a 222 ring GNF. A variety of terminations of the benzene ring edges may be made by hydrogen, alkyl groups, etc (Wu et al., 2007; Zhi and Mullen, 2008).

Chemical methods, as well as producing nanoflakes can also be combined with other techniques to assemble these units into films. This is most commonly done in solution by mechanical extrusion, zone casting/refining, aligned substrate, magnetic field alignment and thermal annealing (Wu et al., 2007; Zhi and Mullen, 2008) but this can also be in vacuum by the use of soft-landing mass spectroscopy (Rader et al., 2006) as illustrated in the schematic diagram in figure 4.

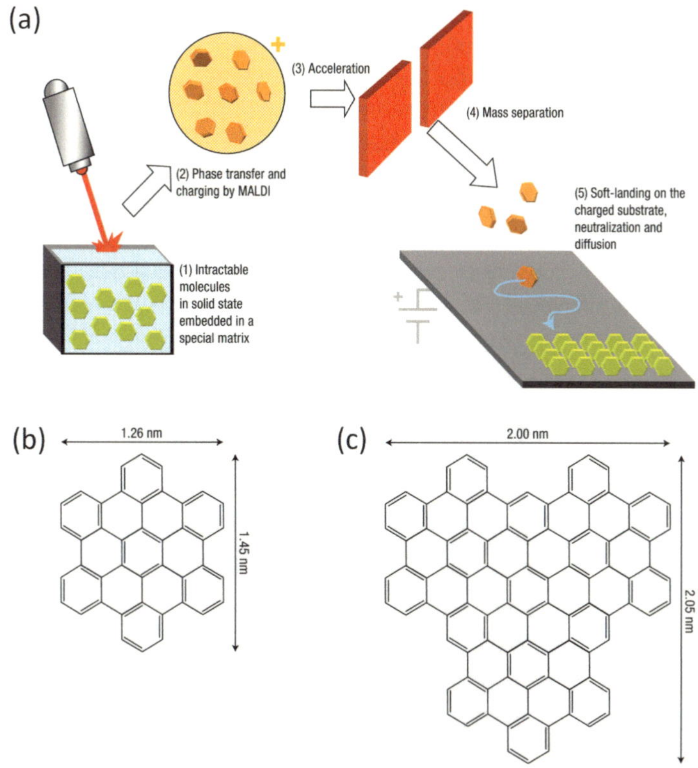

Fig. 4. Schematic diagram of the soft-landing set-up used with typical molecules shown as are the van der Waals dimensions of these molecule obtained by molecular modelling (Rader et al. , 2006).

2.2 Top-down production of GNFs

These methods start with a large piece of graphene sheet or graphene related material such as graphene oxide and cut GNFs directly from the sheets. These methods usually produce graphene islands in a membrane of graphene, which are effectively non-free standing GNFs that are often quite large. Such structures are then usually referred to as graphene quantum dots (GQDs).

Naturally, this method requires that one first produce large sheets (GMBs) and this has been done by a variety of methods, micromechanical cleavable of a graphite single crystal (Novoselov et al., 2004) by chemical means e.g. by starting from graphite oxide (Park & Ruoff, 2006) or by chemically "unzipping" carbon nanotubes (CNTs) (Terrones, 2009).

Presumably some or all of these techniques could be adapted to produce graphene GNFs. A possible adaption of the "unzipping" methods may be to directly "unroll" small fullerenes such as bucky-balls, nanohorns or small pieces of CNTs to produce free-standing GNFs. This does not seem to have been reported at the time of writing, but if such methods prove to be feasible it would open-up the possibility of transferring GNFs to a variety of substrates rather than just making them on graphene.

Graphene layers have also been produced on surfaces, either by removal of layers from a SiC crystal surface (Hass et al., 2008) or by chemical vapour deposition (Obraztov, 2009)

These methods might also be adaptable to the production of GNFs especially in light of recent work which showed that dome-shaped carbon nanoislands may be produced on the (111) surface of Ir (Lacovig et al., 2009).

Once sheets of 2-D graphene are produced, GNFs have to be "cut" from them which has currently been done by a variety of methods; combined e-beam lithography and plasma etching (Berger et al., 2006; Schedin et al., 2007; Stampfer et al., 2008; Neubeck el al., 2010), chemical stripping (Li et al., 2008), scanning tunneling microscope lithography (Tapaszto et al., 2008) and atomic force microscope lithography (Neubecket al., 2010), hydrocarbon lithography (Meyer et al., 2008) and catalytic cutting by atoms (Datta et al., 2008; Ci et al., 2008; Ci et al. 2009; Campos et al., 2009).

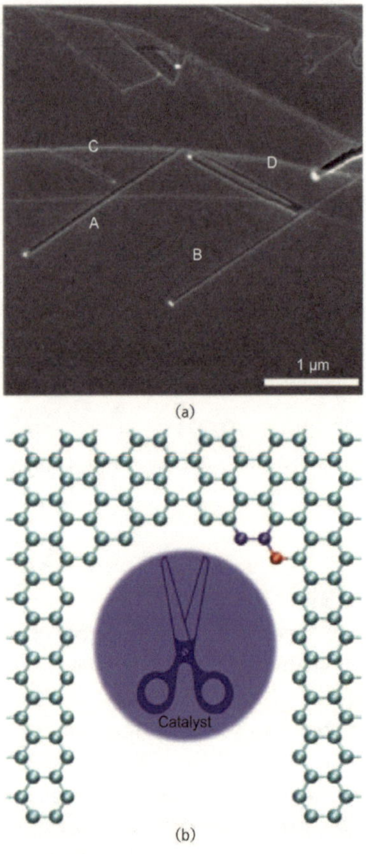

Fig. 5. An illustration of catalytic "cutting"of graphene. From Chi, L., Xu, Z., Wang, L., Gao, G., Ding, F., Kelly, F, Yakobso, I. & Ajayan, P (2008). Controlled nanocutting of graphene. *Nano Research,* Vol. 1 (2008) pp. 116-122

Standard e-beam lithography methods seem to be limited to producing features of greater than 10's of nanometers but scanning tunneling microscope lithography can produce much smaller features with pre-determined crystallographic orientations (Tapaszto et al., 2008).

Catalytic cutting seems particularly well suited to the production of GNFs as it can produce a variety of "cuts" giving very clean edges with definite edge geometries. This method was first developed using Fe atoms for multi-layered structures (Datta et al., 2008) and later adapted (Ci et al., 2008; Ci et al., 2009; Campos et al., 2009) to produce single layer nanostructures (see figure 5). Being able to produce definite edge geometries is very important in determining the properties of GNFs so these techniques look very promising.

2.3 Functionalization

Graphene is hydrophobic and is therefore difficult to solubilize or disperse in most liquids, which limits their easy processability by many traditional methods. Thus, chemists have spent considerable effort in functionalizing graphenes so that they may be more readily solubilized or dispersed, especially in water. The bottom-up approach outlined in section 2.1 has the advantage that GNFs produced in this manner may be constructed from organic molecules which have solubilizing groups already attached before the GNF is synthesized. However, methods have been developed to make larger GNFs (as well as GMBs and GNRs) more soluble in a variety of solvents such as water and tetrahydrofurane (Loh et al., 2010). One study demonstrated that during the chemical production of graphenes one can avoid using graphene oxide (Salzmann et al., 2010). This is important as most chemical methods of producing GMBs are harsh, and often leave many defects and unwanted chemical groups attached. Groups such as carboxyl epoxy and hydroxyl are commonly used to solubilize in water and long alkyl chains make GNFs soluble in many organic solvents.

Unfortunately, many of these methods produce heterogeneous, non-stoichiometric functionalized products which contain defects. As a consequence of this is that, of course, their properties are modified by the attached chemical groups themselves. All these effects usually lead to a degradation of the properties of the graphenes produced in this way, but in some cases they may lead to the new and interesting behavior. For example, the optical properties of GNFs may be tailored by use of different attached groups, and this may open up the possibility of using graphene in optical applications (Loh et al., 2010).

2.4 Production of GNFs by confinement

As an alternative to producing regions of a material with actual physical boundaries as discussed in section 2.2, one of the standard methods of producing semiconducting nanodots is to confine a region of a bulk material by using an electrostatic potential (a gate potential) to produce a confined structure. This has been used for such system as GaAs. The confinement of electrons in such regions of graphene by use of potentials is not straightforward as the relativistic nature of the electronic excitations leads to Klein tunneling where these electrons can tunnel unimpeded through such barriers (Castro Neto et al., 2009; Abergel et al., 2010). Methods other than the straightforward use of "cutting" have been suggested to overcome this problem. This includes, creating a confinement potential within the graphene layer to trap an electron for a finite time, the introduction of a finite gap by coupling with the substrate or by using a bilayer, or the use of a magnetic confinement potential, the review by Abergel et al. gives an excellent and comprehensive summary and explanation of such methods (Abergel et al., 2010).

3. Structure and stability of GNFs in practice

In the introduction we discussed the idealized structures of GNFs, however, in practice real structures may differ considerable from these idealizations. In this section we will discuss some more realistic features of the structure of GNFs.

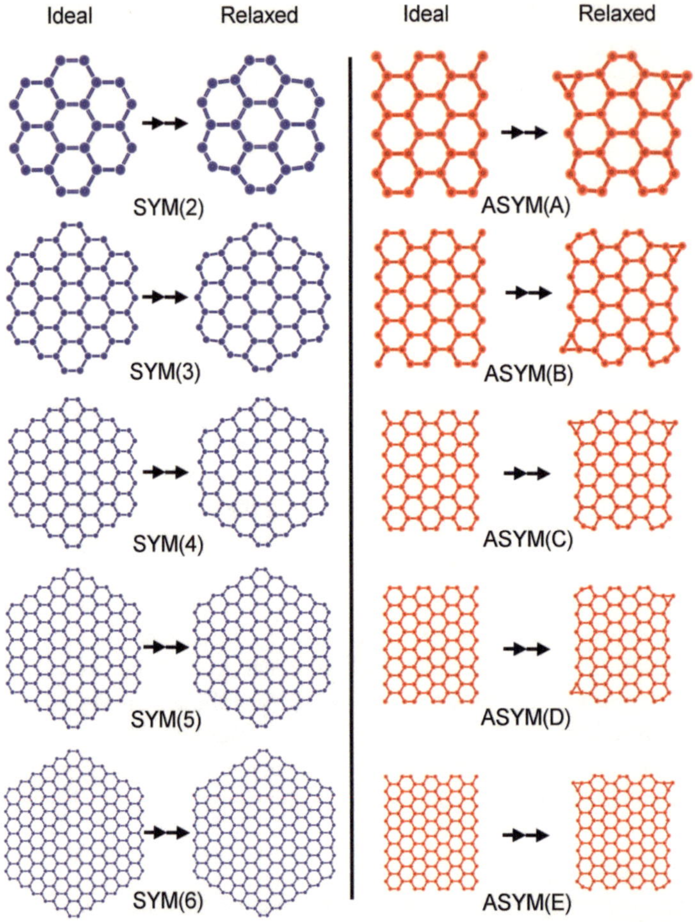

Fig. 6. The initial (ideal) structures on the left and final relaxed structures on the right for all symmetric (SYM) and asymmetric (ASYM) structures obtained by DFT (Barnard & Snook, 2008).

3.1 Edge reconstructions

The first factor one must consider when studying realistic GNFs is that, if the GNF is un-terminated then reconstructions at edges will occur, in order to break aromaticity, and lower the total energy of the flake. This is evident on the right-most structures in figure 6 where the edges of the relaxed structures of the asymmetric GNFs obtained by density functional theory (DFT) calculations (Barnard & Snook, 2008) show several types of reconstructions.

As we shall outline in section 4 the edges of GNFs (and of GNRs) play a very important role in determining their properties, and in particular their electronic and magnetic structure. As stated in the Introduction there are two basic types of edges in graphenes zigzag and armchair edges as illustrated in figures 1, 2 and 6. Note that the atoms on opposite zigzag edges belong to different A- and B- sub-lattices. However, other types of edge state are possible, if less numerous, such as reconstructed zigzags and armchairs (Gass et al., 2008; Koskinen et al., 2008; Koskinen et al., 2009) and those arising from unconventional "cuttings"(Wu & X.C. Zeng, 2008), as well as corner states which are formed where two edges meet. Reconstructed corners can also be seen in the relaxed structure on the right hand side of figure 6. In fact as we mentioned in section 1, and shall see in section 4, GNFs offer the potential to tailor the edge states by cutting different shaped GNFs which is much more versatile than in GNRs and can lead to interesting advantages in using GNFs in electronic and magnetic applications.

3.2 Defects, adsorption, impurities and doping

In addition to edges and corners defects are another deviation from structural ideality. It was also recognized early on that defects can occur in GMBs, GNRs and GNFs and that they may have an important effect on the properties of these materials (Gass et al., 2008; Kraheninikov et al., 2009). In many cases the effect of defects may be detrimental as they often degrade the properties of graphene but in some cases may be useful, as they may introduce new properties such as magnetism, or may even increase the mechanical strength (Kraheninikov et al., 2009). Some of the earliest types of defects identified were point defects and Stone-Wales (SW) defects (Ma et al., 2009) consisting of two pentagons and two heptagons. Several types of defects have been subsequently predicted and/or observed in graphenes some of which are shown schematically in figure 7 (Carr & Lusk, 2010). One of the most interesting recent developments is the suggestion that an extended one-dimensional defect made of paired pentagons and octagons may function as a conducting nano-wire which would be of use in making graphene electronics (Carr & Lusk, 2010). Defects may also be of use in creating spintronic devices (see section 4), making microfluidic channels, forming porous membranes and offering active sites where controlled chemical reactions could take place. There is also theoretical and experimental evidence that SW-type represent preferred sites for atomic and molecular attachment and may be of benefit in modifying the properties of pure graphene structures. This may very well be a way of increasing the capacity of graphene to adsorb hydrogen and, thus, lead to a solution to the long standing question of providing a viable way of storing hydrogen to be used as a fuel. Defects may be deliberately produced by electron or ion beam irradiation as well as by their inadvertent introduction as a consequence of the fabrication processes, or through interaction with the substrate and with the environment.

Other related "defects" are adsorbed atoms and molecules (which, as mentioned above, may "decorate" defect sites) and dopants. There is a growing literature on defects, doping and adsorption in graphene but most of this is concerned with GMBs and GNRs and not with GNFs so there is a need and an opportunity for new research in this area. In particular, since GNFs are of finite size, the properties of a defect can dominate the structure, depending on the size and defect density. If the GNF is small, a high fraction of atoms will either participate in (or be in the vicinity of) a defect, and this may have a consequence for the thermal/chemical stability and structural integrity. In addition to this, while the opportunities for traditional defect patterning of GNFs (as per example in figure 7) may be limited by the finite size, new opportunities are introduced when one considers combinations of defects, edges and corners.

One may also regard the attachment of functional groups to the periphery (or to defect sites) of graphenes as an "impurity", particularly when they are non-functional, as they have the potential to quell the properties of pure graphene structures as mentioned in section 2.3. It should also be highlighted that adsorbates, defects and the attachment of functional groups to graphenes may change their structure and/or lead to non-planarity of the system. This will be particularly severe for small GNFs as is illustrated in figure 7d.

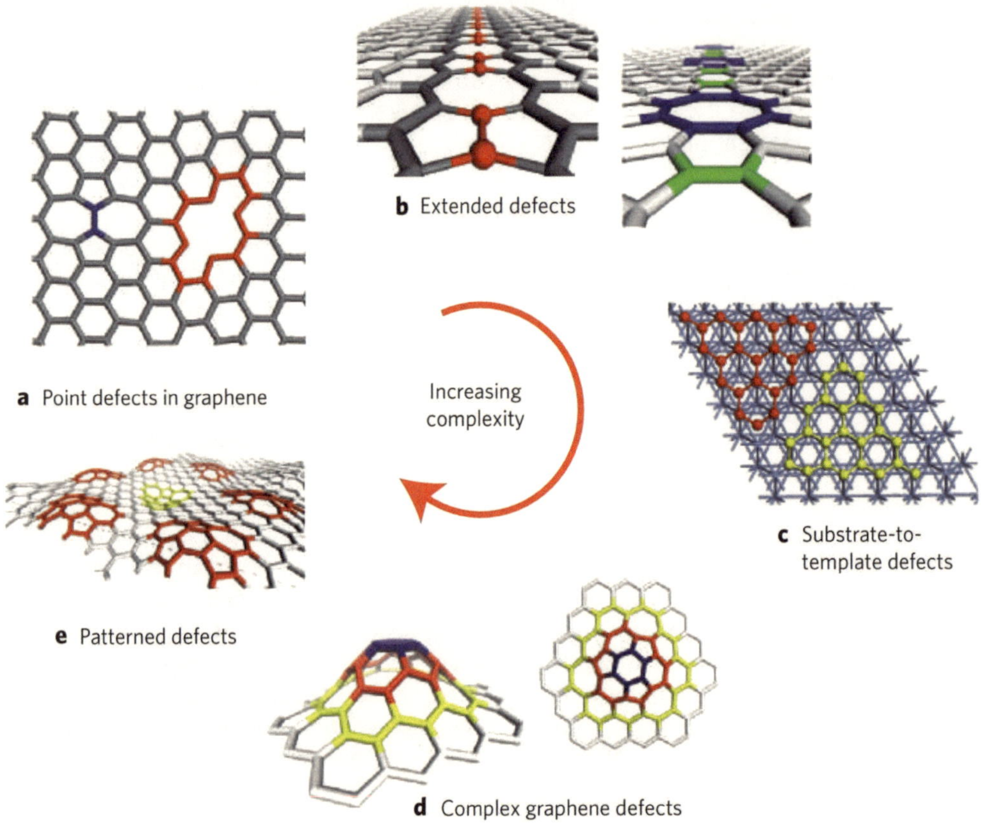

Fig. 7. A representation of defects which may occur in graphenes calculated by means of DFT (Carr & Lusk, 2010)

3.3 Vibrational states and ripples

When researching the properties and applications of graphenes, one must also remember that the structures outlined above are static, or represent a statistical average, and that vibrational states occur around these values. These states have been well studied in GMBs and in GNRs (Ferrari et al, 2006; Ni et al., 2008) but do not appear to have been extensively studied for GNFs. This is a problem that needs to be addressed as these vibrational states lead to Raman and/or infrared spectra which are very useful in identifying the structure of graphenes. The characteristic Raman G-band of graphene has been extensively used to differentiate single, double and multi-layer graphenes (Ferrari et al, 2006; Ni et al., 2008) as has the low frequency breathing modes of CNTs. These are considered signatures of these materials.

Another importance feature of vibrational effects is their contribution to the thermal conductivity of graphene. This is, possible, more important for GMBs as, because of their large thermal conductivity, they have been suggested as being useful for use as heat sinks. Nevertheless, the importance for GNFs remains largely unexplored.

Finally at finite temperature (T), not only do normal modes occur, but graphene structures develop ripples (Meyer et al., 2007; Castro Neto et al., 2009; Abergel et al., 2010). This was first seen in GMBs, and is another departure from ideality. Now as has been shown theoretically (Mermin, 1968) that a truly 2-D crystal with harmonic forces is unstable for T > 0K, so this rippling is presumably the mechanism by which this instability is overcome. Although these structures are still composed of a single layer of C atoms, planarity is no longer preserved, and the structures are no longer mathematically 2-D. It has also been shown by ab initio molecular dynamics calculations that similar large scale vibrational modes occur in GNFs (Snook et al., 2005; Barnard & Snook, 2008) as shown in figure 8. More recently similar modes have been predicted theoretically in boron nitride nanoflakes (Topsakal et al., 2009) and boron nitride nanoribbons (Kan et al., 2008).

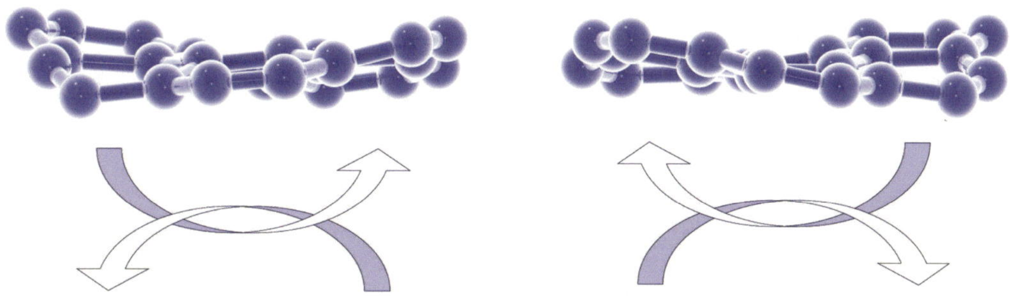

Fig. 8. Examples of the saddle-like distortions of the symmetric C_{24} flake, annealed at 900K. A dynamical rocking between these structures is observed following equilibration (Barnard & Snook, 2008)

4. Electronic structure and magnetic behavior of GNFs

One of the initial and, indeed, continuing interests with graphene membranes is in their fascinating electronic properties which we will only briefly outline here as there exists a large and growing literature on this topic (Castro Neto et al., 2009; Abergel et al., 2010). In fact because this literature is so vast and growing we will only give a brief outline of some of the fundamental aspects of electronic structure and their consequences for the particular case of GNFs.

4.1 The electronic structure of GMBs

As mentioned in section 1 the 2-D hexagonal lattice structure of GMBs may be looked on as two triangular lattices A- and B- , the reason for this is shown in figure 9. Wallace used a tight binding (or Huckel) model (Wallace, 1947) to show that the highest electronic states could be described by occupied π and unoccupied π* bands which meet only at the K and K* points in reciprocal space on the Fermi surface as shown in figure 11. This is why no band gap exists and as there is also a vanishing density of states at these points this makes 2-D graphene a semi-metal (or zero-bandgap semiconductor). The elementary excited electronic states show a linear dispersion relationship similar to massless, chiral, relativistic particles and are now termed "Dirac Fermions". This leads to the electron intrinsic mobility being much higher than in silicon (Castro Neto et al., 2009), and suggests that faster electronics can be made from graphene (than from Si). However, the zero-gap means that GMBs cannot be used for the construction of current types of Si-based electronic devices.

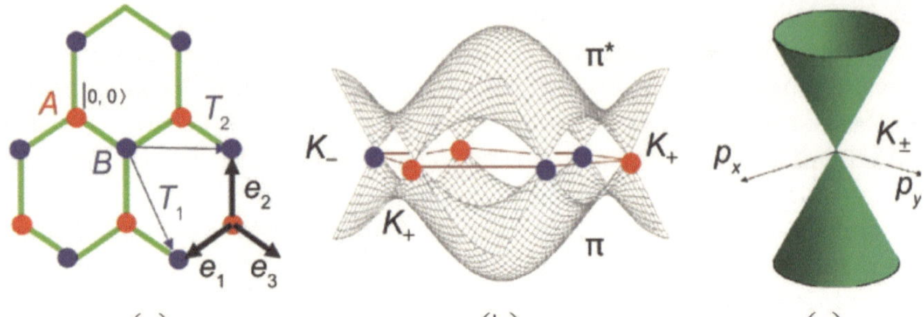

Fig. 9. (a) the graphene lattice (b) the π and π* bands and (c) the dispersion relation close to the $K_\pm$ points of the first Brillouin zone, a "Dirac cone" (from Cresti, A., Nemec, N., Biel, B., Neibler, G., Triozon, F., Cunerberti, G. & Roche, S. (2008). Charge transport in disordered graphene-based low dimensional materials. *Nano Research,* Vol. 1 (2008) pp. 361-394)

4.2 The effect of edges on the electronic structure of graphene

As previously outlined the simplest way to modify the electronic structure of GMBs is the introduction of edges and corners which has been most extensively studied in GNRs (Son, et al, 2006a; Son et al., 2006b; Hod et al., 2007; Castro Neto et al., 2009; Abergel et al., 2010). Although the properties of GNRs may be tailored using their edge states, edge states are actually even more important in GNFs than in GNRs as the fraction of atoms in edges can be made very large. In contrast, although corner states are unique to GNFs, and are likely to be invaluable for a variety of purposes, far fewer atoms reside at or near corners, and even in complicated shapes the corner-to-edge ratio is very low. Therefore, it is important to understand the effect of edges on electronic states of graphene nanostructures if we are to engineer designer GNFs in the near future.

As we know there are two basic types of edge structures, zigzag (ZZ) and armchair (AC) as displayed in figures 1 and 2, and that these edges contain dangling ("unsatisfied") bonds.

This introduces a mixture of sp^2 and sp hybridization into the basic sp^2 graphene lattice unless the edges are chemically bonded to non-carbon atoms or functional groups (see figure 1d). Therefore, both un-terminated and terminated edge structures can alter the basic electronic structure of graphene and play a crucial role in determining the electronic and magnetic properties of finite structures. The effect of edge states on electronic and magnetic behaviour has been most extensively studied for GNRs, so at this stage we will selectively summarize some findings of relevance to GNFs (Son, et al, 2006a; Son et al., 2006b; Hod et al., 2007; Castro Neto et al., 2009; Abergel et al., 2010),

- GNRs with either ZZ edges or AC edges have a finite band gap and are semiconducting although the origin of the gap is different in each case.
- It has also been found theoretically that AC edged GNRs can be metallic but this has been suggested to be due to the limitations of the tight binding model used in these studies to describe their electronic structure.
- ZZ edges introduce localized states as is illustrated in figure 12. As opposite edges of such a GNR belong to different sublattices so the spin ordering is different on each edge.
- ZZ edges can create ferromagnetic (FM) or antiferromagnetic (AFM) phases.
- There are no localized states at AC edges which is illustrated in figure 12.

- Large magnetic moments occur at ZZ edges but there are none for AC edges.
- Functionalization ("passivation") of edges can significantly change the properties of GNRs with either AC or ZZ edges and in the latter case this can generate a fully spin polarized state.
- In an applied electric field ZZ terminated GNRs can produce a ½ metallic state.

An interesting consequence of the effect of edges on the spin state of GNRs is that ZZ terminated GNRs may be useful in the field of spintronics, whereas AC edges may not.

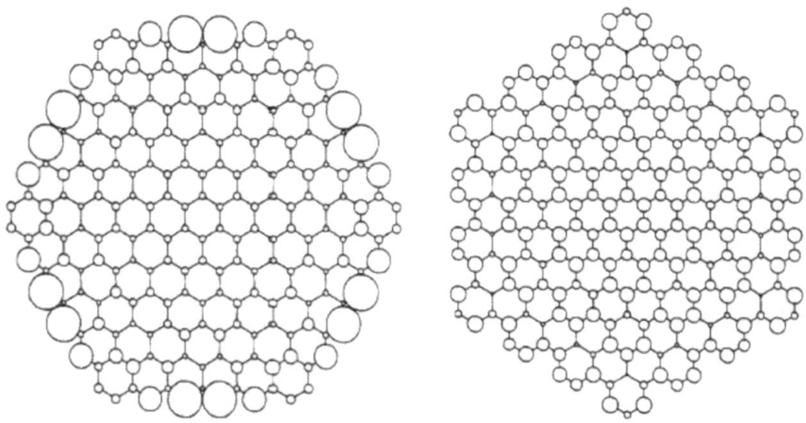

Fig. 10. The spatial distribution of populations of the HOMO level for GNFs having on the left zigzag edges and on the right having armchair edges (adapted from Stein & Brown, 1987).

4.3 The transition from discrete to band-like electronic states in GNFs

Small GNFs have discrete, molecular electronic energy levels, so the energy of the highest occupied molecular orbital (HOMO) takes the place of the Fermi Energy (E_f) and the difference in the energies of the HOMO and that of the lowest unoccupied molecular orbital (LUMO) corresponds to the energy gap (E_g). GNFs only have a continuous band structure when their dimensions are very large (see section 5) and a representation of this transition is shown in figure 11 (Chen & Tao, 2009). Along with this basic change in the nature of the electronic level, there is also a consequent change in electrical character from insulator to semiconductor, and then to a small or zero energy gap material (see figure 11).

Furthermore, for small GNFs the details of this discrete electronic structure can vary strongly with their shape and dimensions as illustrated in figure 12 and can be heavily influenced by termination i.e. passivation or functionalization.

4.4 Some general comments concerning the electronic states of GNFs

There have been a small but growing number of studies of the electronic states of GNFs using a variety of techniques such as tight binding models (TBMs), Hubbard models (HM), density functional theory (DFT) and the Dirac equation (DE). In these studies a variety of different shapes have been investigated including squares, rectangles, circles, triangles, disks, hexagons, polygons and non-symmetric shapes.

As outlined in section 4.2 it has been found that the electronic state and magnetic properties of GNFs are strongly influenced by the presence of edges and for small GNFs the effect is more pronounced than GMBs and GNRs.

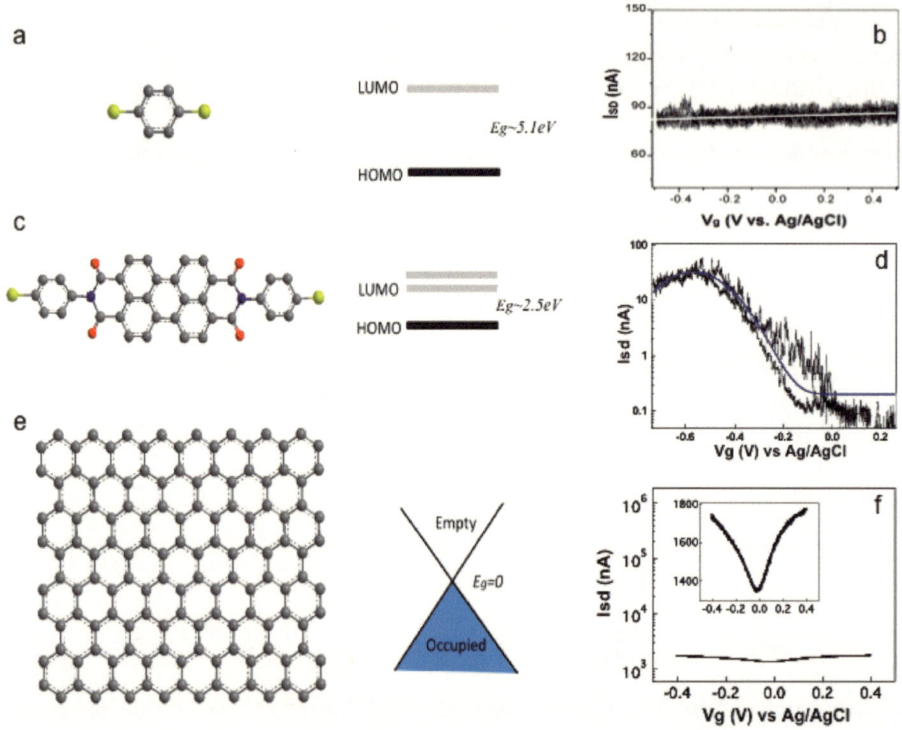

Fig. 11. Molecular structures, energy diagrams, and electrochemical gate effects of benzenedithiol, PTCDI, and a graphene sheet: (a) Benzenedithiol (containing a single benzene ring) has a large LUMO-HOMO gap (5.1 eV) and is "insulating" with a weak gate effect (b). (c) The LUMO-HOMO gap of PTCDI (containing seven rings) is ~2.5 eV, and the molecule is "semiconducting" with a large gate effect (d). (e) Graphene (containing a large number of rings) has a zero energy gap between the conduction and valence bands and shows semimetallic behavior with a weak gate effect (f)." (Chen & Tao, 2009).

4.5 Some interesting results for simple geometric shapes

4.5.1 Squares and rectangles

It has been shown for square and rectangular shaped GNFs (GNRs of finite length) that the band gap may be readily manipulated by altering the width of the GNF (Jing et al., 2007). The smallest rectangular GNFs consisting of a chain of n- fused single benzene rings (acenes) have been shown to have an AFM ground state for $n > 7$ (Jing & Sheng, 2008).

Interestingly a study GNFs with AC edges showed that unlike infinite GNRs they can be metallic (Shemella et al., 2007) and half-metallic states have also been predicted by DFT calculations with ZZ edges in the presence of an electric field (Hod et al., 2008; Zheng & Duley, 2008b).

The effect of functionalization and doping rectangular GNFs by the H, N, O, F, V atoms and by groups such as -OH, -CH_3 has also been studied by means of DFT calculations which showed that the electronic and magnetic properties of GNFs may be readily manipulated by these means (Zheng & Duley, 2008a; Sahin & Senger, 2008; Berashevich & Chakraborty, 2009; Olivi-Tran, 2010). This leads to the expectation that they may be extremely useful in areas such as spintropics, sensors and transistors.

4.5.2 Nanodisks

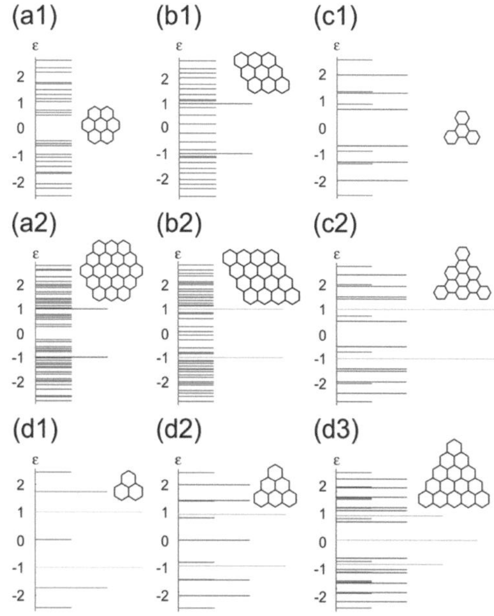

Fig. 12. The electronic energy level diagrams for some GNFs whose shapes are shown beside the energy level diagram (Ezawa, 2008).

The electronic structure of a variety of compact structures based on highly symmetrical shapes such as large PAHs, triangular, hexagonal and other polygonal shapes, which are often termed nanodisks, have been studied by several groups (Stein & Brown, 1987; Banerjee & Bhattacharyya, 2008; Ezawa, 2007; Ezawa, 2008; Heiskanen et al., 2008; Kuc & Heine, 2010). The results show that their electronic and magnetic structure may be varied considerably by changing the shape and size of the nanodisk and also by functionalizing them, once again suggesting that they may be extremely useful in fields such as spintronics. An example of the change in the electronic energy levels of some small graphene nanodisks with size and shape changes are shown in figure 12.

4.5.3 Some results for more complex shapes

The interesting magnetic properties of graphene arise from the properties of edge states which suggests the possibility to engineer materials with very interesting magnetic properties by using shapes other than the simple geometric ones mentioned above. Some research has begun into investigating the properties of more complicated shapes (Wang et al., 2008; Yazyev et al. , 2008 ; Yu et al. 2008) and because of the variety of shapes that GNFs can exist in they are ideal candidates for such studies.

A complication arises, however, as the relative energies of the various possible phases (AFM, FM, metal, semiconducting, ½ metal, non-magnetic) have to be worked out in order to find the most stable state. This is usually done by means of extensive and time consuming calculations so to try to circumvent this problem there have been various methods suggested to find simple rules to predict these stabilities (Yu et al. 2008; Dias, 2008; Potasz et al. 2010; Wang et al., 2009; Fernandez & Palacios, 2007). These methods are based on considering the underlying geometry and topology of the graphene structures e.g. the fact that graphene may be pictured as two sublattices A- and B- as pointed out in the Introduction. Such methods should prove of great value in helping design GNF structures with the optimum magnetic behavior for applications such as spintropics (Trauzettel et al., 2007; Tombros, et al., 2007; Fal'ko, 2007; Rycerz et al, 2007). This prospect is enhanced by work that shows that the magnetic behavior of GNFs is very robust to disorder (Bhowmick & Shenoy, 2008; Wimmer et al., 2010).

5. Graphene and graphane nanoflakes

We have already introduced the simplest GNF; benzene, which is a planar six membered ring of C atoms each being terminated by a single H atom (monohydride). However, there is another stable six membered ring of C atoms; cyclohexane, which is a non-planar structure in which the C atoms are terminated by two H atoms (dihydride). If a GMB is completely hydrogenated on all carbon atoms on both sides of the sheet then the resulting structure, called graphane, is non-planar and consists entirely of sp^3 bonded carbon atoms and two sigma bonded hydrogen atoms per C atom. This structure was predicted theoretically by means of DFT calculations (Sofo et al., 2007) and discovered experimentally (Ryu et al., 2008; Elias et al., 2009) only quite recently. As graphane is an insulator it may be an ideal material to use in combination with graphene to form hybrid graphene/graphane integrated electronic devices and detectors. However, there remains the problem of how to "connect" these two materials together in an integrated hybrid device. Two recent theoretical studies have suggested ways that this might be achieved.

In one study (Barnard & Snook, 2010a), tight binding density functional theory (DFTB) was used to show that if a graphene nanoflake were anionically charged by adding excess electrons (for example by means of an electron beam) then beyond a certain level of added charge regions of the GNF would transform into hydrogen-less graphane. These results are shown in figure 13, where we can see that the graphene/graphane ratio, and the graphane distribution is dependent on the level of charge and the shape of the flake, but in each case this produces graphene/graphane boundaries and leaves the graphane regions ripe for functionalization by atoms or functional groups. Further work revealed that the charge-induced graphene to graphane can be predicted if the mass and the structure of the flake is known (or can be estimated) and identified by way of the shift in the Fermi level (Barnard & Snook, 2010 b). If this can be realized experimentally then it will enable structures consisting of co-existing regions of graphene and graphane to be produced spontaneously, and in the absence of volatile or expensive chemical environments.

As a follow-up to simulating "nanoroads" of graphene in graphane sheets (long strips of graphene in a graphane sheet) as an alternative to GNRs Yakobson et al. used DFT and DFTB methods to simulate graphene nanodots of various sizes embedded in a graphane sheet (Singh et al., 2010) as shown in figure 16 below. In this study it was found that the size n, shape and stability of the simulated dots were governed by the aromaticity of the dot and the interface between the two regions. Sizes corresponding to stable embedded aromatic molecules such as benzene and coronene were predicted to be stable and for larger dots hexagonal clusters are favoured i.e. structures congruent with the lattice hexagons which have armchair edges. Clusters had large band gaps of about $15/\sqrt{n}$ eV with the size dependence being characteristic of confined Dirac fermions (see figure 14). The largest dots have an electronic spectrum of GMBs (see section 4). The authors also simulated some arrays of dots and found that the band gaps of these structures are similar to that of isolated dots.

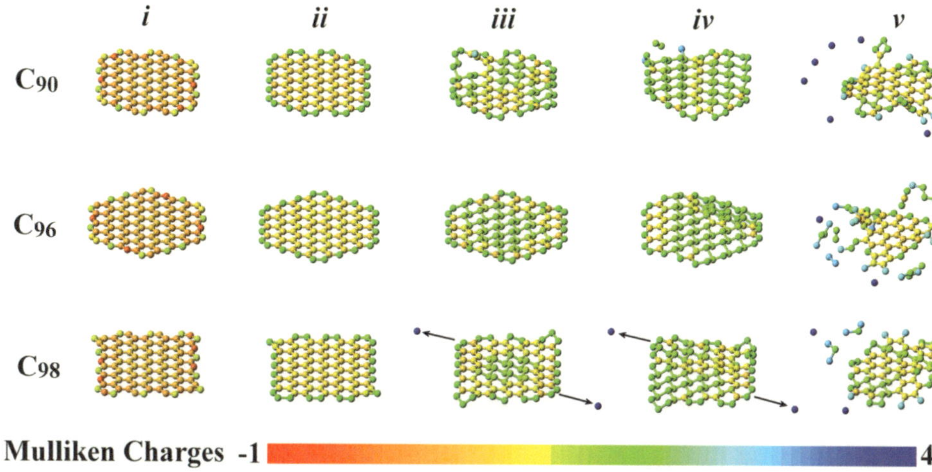

Fig. 13. Structure and distribution of charges in C_{90} C_{96} and C_{98} anionic graphene nanoflakes at different stages of electron injection (Barnard & Snook, 2009).

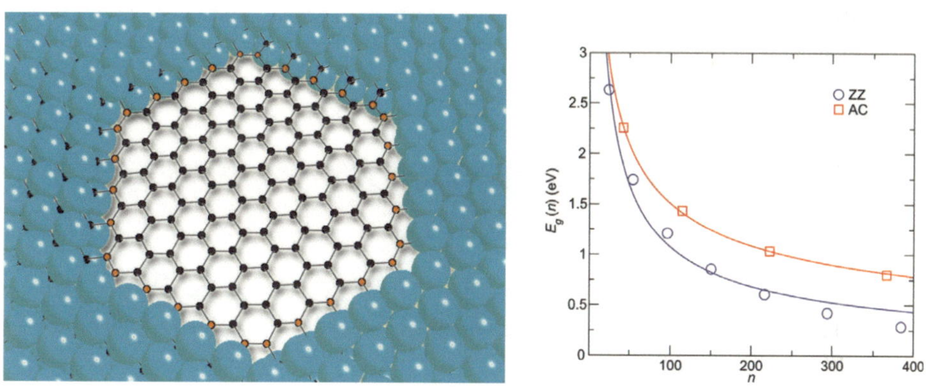

Fig. 14. Hexagonal GNF formed by removing atoms from a graphane sheet and the energy gap E_g as a function of the size of the GNF as a function of the size n (adapted from Singh et al, 2010).

6. Potential uses of graphene/graphane nanoflakes

Potential applications have been mentioned previously in this chapter particularly in section 4 but here we will review how some of these ideas have resulted in real device prototypes being made. Since GNFs may range in size from molecular to mesoscopic they have the potential to bridge the gap between molecular electronics and nano-electronics, and to integrate with existing technologies. Here we will give just two examples which will illustrate the extreme limits.

An example of the former class of devices is the use of coronenes to make a graphene field-effect transistor (FET) a "circuit diagram" of which is shown in figure 15. This device consists of a coronene molecule with wires to the source and drain electrodes consisting of linker molecules which behaved electrically as a FET and thus this is a molecular graphene transistor.

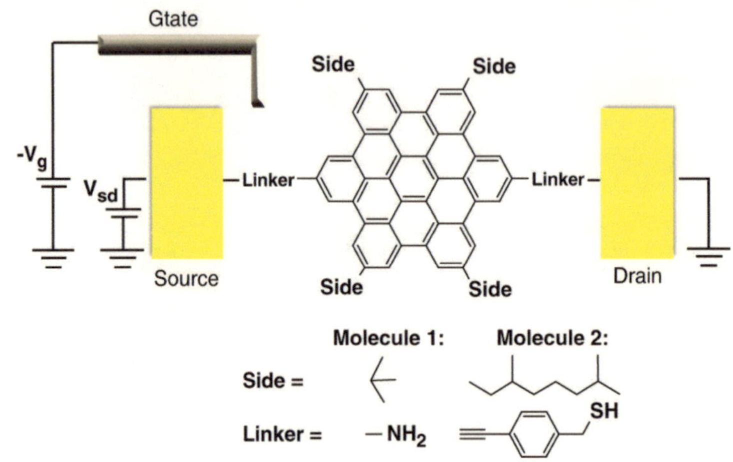

Fig. 15. Single HBC molecule FET (Diez-Perez et al., 2010).

At the other end of the size scale there has been considerable interest in building devices such as single electron transistors (Ponomarenko et al. 2008 and Stampfer et al., 2008). Figure 16 shows an SEM image and an artistic view of just such a transistor which was cut from GNRs to create a GQD by use of a combination of electron beam lithography and reactive plasma etching.

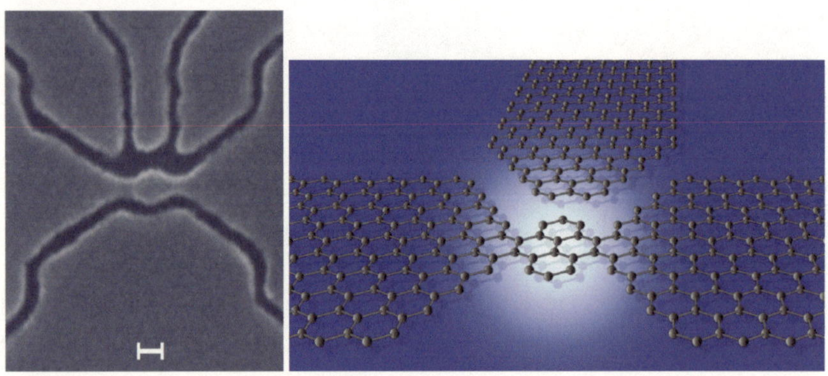

Fig. 16. An SEM image and an artistic impression of a single electron transistor (Novoselov & Geim, http://www.condmat.physics.manchester.ac.uk/pictures/).

Currently the features of devices such as these are around 10 nm and so the GQD is quite large but it has been projected that dimensions should be able to be reduced to 1 nm in the future (Ponenmarkenov et al., 2008).
The work outlined above, and that of others (see for example Stampfer et al., 2008a and Stampfer et al., 2008b) shows that it is indeed possible to fabricate working electronic devices from graphene which contain GNFs and GQDs and this is set to become a growing area of research in the future.

7. Some other areas of future interest

Previous sections of this chapter have outlined what we belief have been the major areas of interest in GNFs up until the present and those that we believe will continue to be of major interest in the future. However, there are several other areas which we feel deserve more attention in light of their potential for yielding new scientific discoveries and technological applications.

7.1 The role of GNFs in the synthesis of CNTs

CNTs continue to be of great importance particularly for many applications and occur in two forms (either semiconducting or metallic) which have quite different properties and uses. However, most syntheses produce a mixture of these different types of CNTs and it is very hard to separate them especially on the large scale needed for use in industrial applications. Recently it has been shown that GNFs may be directly converted to CNTs which involves GNFs containing between 60 and 100 atoms. This leads to the possibility of understanding at least one mechanism of formation of CNTs and it suggests that it may be possible to use this type of route to directly synthesis CNTs with specific chirality. This is potentially a very interesting area for increased activity combining scientific insight with practical applications.

7.2 The possible production of other nano-carbon flakes

The main thrust of research involving materials consisting of large aromatic carbon rings has been those where the rings are fused i.e. structures of the type shown in figures 1 and 2. However, the bottom-up methods which have been used to produce GNFs are also capable of producing nanoflakes of a different type.

For example it is already possible to combine small pieces of GNFs with various chemical groups (Wu et al., 2007; Zhi and Mullen, 2008) to produce interesting properties both similar and dissimilar to GNFs depending upon the mixing and configurations. An example might be to "hard-wire" together several different GNFs in much the same way that the single coronene molecule shown in figure 17 is hard-wired into a transistor circuit.

It may also be possible to make NFs of other aromatic molecules containing other atoms as well as carbon and from other planar systems containing no carbon atoms at all, such as boron nitride sheets. This could lead to even more interesting and versatile NF properties and applications.

7.3 Multiple layers and stacks of GNFs

In this review we have ignored few layered GNFs and large stacks of GNFs both of which have been the subject of theoretical and experimental interest. Such structures can have properties which differ from those of single GNFs, and of graphite nanocrystals, often in a beneficial way (Jackel et al., 2006; Abergel et al., 2010). An example of this is that bilayer GNDs have a different band gap than single layer GNDs and electrons in such structures may be confined with potential barriers in contrast to single layer GNFs (Abergel et al, 2010).

7.4 Optical properties

It has been shown experimentally that GMBs, graphene oxide and GNFs (Gokus, et al., 2009; Luo et al., 2009; Sun eta al., 2008; Loh, et al., 2010) have interesting optical properties when functionalized. This leads to the interesting possibility of developing graphenes with optical properties which may be tuned by changing the functional groups (Loh et al., 2010). Indeed it has been suggested that graphene composites may be used commercially for optical applications such as saturable absorbers for laser applications before they are used for electronic ones. Thus it may well be a promising area to investigate the optical properties of GNFs both theoretically and experimentally. In fact it has already been shown theoretically that GNFs of different shapes can have a significant second hyperpolarizability which may be of use in non-linear optics applications (Nagai et al., 2010).

8. Conclusions

As we can see from the descriptions here, and throughout this book, graphene has already proved to be an extremely exciting new material which has many fascinating properties. Much of the research on graphene has been concerned with 2-D graphene (graphene membranes, GMBs) and 1-D graphene (graphene nanoribbons, GNRs), because of their interesting and potentially very useful properties have been relatively simple to isolate and address. Indeed it has already been shown that it is feasible to make working devices (albeit only at the laboratory level) using GMBs and GNRs for applications such as transistors, molecule detectors and flexible electrodes.

Less well studied is the 0-D form of graphene (graphene nanoflakes, GNFs or graphene nanodots, GNDs) which presents a greater degree of complexity, but offers a greater potential for flexibility and selectivity (both literally and figuratively). In this chapter we have attempted to show that GNFs share many useful features in common with these 2-D and 1-D forms of graphene, but also have unique additional properties such as interesting electronic and magnetic states and also many potential applications in electronics and detectors. Many of these additional features arise because GNFs may be engineered to form many more shapes than GMBs and GNRs and, as has been pointed out, GNFs can also span the dimensional scale from molecular to 2-D enabling potentially great versatility in many applications. A very promising area of future applications appears to lie in the area of spintronics and perhaps quantum computing as the magnetic properties of GNFs are not only comparable with those of GNRs which have previously been suggested for these applications but they may be manipulated and altered in a wide variety of ways.

Aside from the currently known features of GNFs, readers will undoubtedly be left with a range of questions. Far from being a failure, this serves to highlight that there is obviously great potential to further explore and exploit many of their properties, and so much more to be learned. The most obvious of these is their functional and optical properties which hold much promise for making optical devices.

Furthermore, GNFs also offer many challenges to refine existing scientific knowledge of their preparation and properties including, their improved synthesis and assembly by bottom-up and top-down methods, the refinement of confinement methods for GNDs and the exploration of their properties such as their vibrational spectra and magnetic behavior.

9. References

Abergel, D.; Apalkov, V.; Berashevich, J.; Zieler, K. & Chakraborty, T. (2010). Properties of graphene: a theoretical perspective. *Advances in Physics,* Vol. 59 (4) (July 2010) pp.261-482

Banerjee, S. & Bhattacharyya, D. (2008) Electronic properties of nano-graphene sheets calculated by quantum chemical DFT. *Computational Materials Science,* Vol. 44 (March, 2008) pp. 41-45

Barnard, A. & Snook, I. (2008). Thermal stability of graphene edge structure and graphene nanoflakes. *Journal of Chemical Physics, Vol.* 128 (9) (March 2008) pp. 094707-1 - 094707-7

Barnard, A. & Snook, I. (2010). Transformation of graphene into graphane in the absence of hydrogen. *Carbon, Vol.* 48 (November 2009) pp. 981-986

Barnard, A. & Snook, I. (2010). Size- and Shape- dependence of the graphene to graphane transformation in the absence of hydrogen. *Journal of Materials Chemistry*, Vol. 20 (2010) pp. 10459-10464

Barone, V.; Hod, O. & Scuseria, G.(2006). Electronic structure and stability of semiconducting graphene nanoribbons. *Nano Letters*, Vol. 6(12) (November 2006) pp. 2748-2754

Berashevich, J. & Chakraborty, T. (2009). Impurity-induced spin gap asymmetry in nanoscale graphene. *Physical Review B*, Vol. 80 (September, 2009) pp. 115430 - 1-6

Berger, C.; Song, Z.; Li, X.; Wu, X.; Brown, N.; Naud, C.; Mayou, D.; Li, T.; Hass, J.; Marchenkov, A.; Conrad, E.; First, P. & de Heer, W. (2006). Electronic confinement and coherence in patterned epitaxial graphene. *Science*, Vol. 312 (May 2006) pp. 1191-1196

Bhowmick, S. & Shenoy, V. (2008). Edge magnetism of single layer graphene nanostructures. *The Journal of Chemical Physics*, Vol. 128 (June, 2008) pp. 244717 -1 - 244717-7

Campos, L., Manfrinato, V., Sanchez-Yamagishi, J., Kong, J. & Jarillo-Herrero, P. (2009). Anisotropic etching and nanoribbons formation in single-layer graphene. *Nano Letters*, Vol. 9(7) (2009 June) pp. 2600-2604

Carr, L. & Lusk, M. (2010). Graphene gets designer defects. *Nature Nanotechnology* , 5 (May 2010) pp. 316-317

Castro Neto, A.; Guinea, F.; Peres, N.; Novoselov, K. & Geim, A. (2009). The electronic properties of graphene. *Reviews of Modern Physics*, Vol. 81 (January 2009) pp. 109-162

Chen, F. & Tao, N. (2009). Electron transport in single molecules: From Benzene to Graphene. *Accounts of Chemical Research*, Vol. 42 (2009) 429-438

Ci. L.; Xu, Z.; Wang, L.; Gao, W..; Ding, F.; Kelly, K.; Yakobson, B. and Ajayan, P. (2008). Controlled Nanocutting of Graphene. *Nano Research*, Vol. 1 (2008) pp. 116-122

Ci, L.; Song, L.; Jariwala, D.; Elias, A.; Gao, W.; Terrones, M. & Ajayan, P. (2009). Graphene shape control by multistage cutting and transfer. *Advanced Materials*, Vol. 21 (2009) pp. 4487-4491

Chuvilin, A.; Kaiser, U.; Bichoutskaia, E.; Besley, N. & Khlobystov, A. (2010). Direct transformation of graphene to fullerene. *Nature Chemistry*, Vol. 2 (June 2010) pp. 450-453

Cresti, A.; Nemec, N.; Biel, B.; Niebler, G.; Triozon, F.; Cuniberti, G. & Roche, S. (2008). Charge transport in disordered graphene-based low dimensional materials. *Nano Research*, Vol. 1 (2008) pp. 361-394

Datta, S.; Strachan, D.; Khamis, S. and Johnson, A. (2008). Crystallographic etching of few-layer graphene. *Nano Letters*,Vol. 8 (7) (June 2008) pp.1912-1915

Dias, J. (2008). Resonance-theoretical calculations of the ground state spin density of the pπ-system of edge atoms on graphene nanodots and nanoribbons. *Chemical Physics Letters*, Vol. 467 (November, 2008) pp. 200-203;

Diez-Perez, I.; Li, Z.; Hihath, J.; Li, J.; Zhang, C.; Yang, X; Zang, L.; Dai, Y.; Feng, X.; Mullen, K. and Tao, N. (2010). Gate-controlled electron transport in coronenes as a bottom-up approach towards graphene transistors. *Nature Communications*, Vol. 1 (June, 2010) pp. 1-5

Elias, D.; Nair, R.; Mohiuddin, T.; Morozov S.; Blake, P.; Halsall, M., Ferrari, A.; Boukhvalov, D.; Katsnelson, M.; Geim, A. & Novoselov, K. (2009). Control of graphene's properties by reversible hydrogenation: Evidence for graphane. *Science*, Vol. 323 (January 2009) pp. 610-613

Ezawa, M. (2007). Metallic graphene nanodisks: Electronic and magnetic properties. *Physical Review B*, Vol. 76 (December, 2007) pp. 245415 -1 – 6

Ezawa, M. (2008). Graphene Nanoribbon and Graphene Nanodisk. *Physica E*, Vol. 40 (September, 2008) pp. 1421-1423

Fal'ko, V. (2007). Quantum information on chicken wire. *Nature Physics*, Vol. 3 (March, 2007) pp. 151-152

Fernandez-Rossier, J. & Palacios,J. (2007). Magnetism in graphene nanoislands. *Physical Review Letters*, Vol. 99 (October, 2007) pp. 177204 – 1- 4

Ferrari, A.; Meyer, J.; Scardaci, V.; Casiraghi, C.'; Lazzeri, M.; Mauri, F.; Piscanec, S.; Jiang, D.; Novoselov, K.; Roth, S. & Geim, A. (2006). Raman spectrum of graphene and graphene layers. *Physical Review Letters*, Vol. 97 (18) (November 2006) pp. 187401-1 - 187401-4

Gass, M.; Bangert, U.; Bleloch, A.; Wang, P.; Nair, R. & Geim, A. (2008). Free-standing graphene at atomic resolution. *Nature Nanotechnology*, Vol. 3 (November, 2008) pp. 676 - 681

Gokus, T.; Nair, R.; Bonetti, A.; Bohmler, M.; Lombardo, A.; Novoselov, K.; Geim, A.; Ferrari, A. & Hartschuh, A. (2009). Making graphene luminescent by oxygen plasma treatment. *ACS Nano*, Vol. 3 (12) (2009) pp. 3963-3968

Hass, J.; de Heer, W. & Conrad, E. (2008). The growth and morphology of epitaxial multilayer graphene. *Journal of Physics: Condensed Matter, Vol.* 20 (July 2008) pp. 323202-1 – 323202-27

Heiskanen, H.; Manninen, M. & Akola, J. (2008). Electronic structure of triangular, hexagonal and round graphene flakes near the Fermi level. *New Journal of Physics*, Vol. 10 (October, 2008) pp. 103015 – 1-14

Hod, O.; Barone, V.; Peralta, J. & Scuseria, G. (2007). Enhanced half-metallicity in edge-oxidized graphene nanoribbons. *Nano Letters, Vol.* 7(8) (July, 2006) pp. 2295-2299

Hod, O.; Barone, V. & Scuseria, G. (2008); Half-metallic graphene nanodots: A comprehensive first principles theoretical study. Physical Review B, Vol. 77 (January, 2008) pp. 035411 1-6

Jackel, F., Watson, M., Mullen, K. & Rabe, J. (2006). Tunneling through nanographene stacks. *Physiical Review B*, Vol. 73 (January 2006) pp. 045423 - 1-6

Jiang, D.; Sumpte, D. & Dai, S. (2007). First principles study of magnetism in nanographenes. *Journal of Chemical Physics* , Vol. 127 (September, 2007) pp. 124703 – 1-5

Jing, D. & Sheng, D., (2008). Electronic structure of higher acenes. *Journal of Physical Chemistry A*, Vol. 112 (October, 2008) pp. 332-335

Kan, E.-J.; Wu. X.; Li, Z.; Zeng. X.; Yang, J. & Hou, J. (2008). Half-metallicity in hybrid BCN nanoribbons. *Journal of Chemical Physics*, Vol. 129 (August 2008) pp. 084712-1 - 084712-5

Koskinen, P.; Malola, S. & Hakkinen, H. (2009). Self-passivating edge reconstructions of graphene. *Physical Review Letters*, Vol. 101 (September, 2009) pp. 115502-1 -4

Koskinen, P.; Malola, S. & Hakkinen, H. (2009). Evidence for graphene edges beyond zigzag and armchair. *Physical Review B*, Vol. 80 (2009) pp. 073401-1 073401-3

Krasheninnikov, A.; Berber, S. & Tomanek,D. (2009). Computational Studies of Defects in Nanoscale Carbon Materials. CECAM-HQ-EPFL, Lausanne, Switzerland. May 2009.
http://ww.cecam.org/workshop-0-313.html MSU Home Page:
http://nanotube.msu.edu/dnc09/

Kuc, A. & Heine, T. & Seifert, G. (2010). Structural and electronic properties of graphene nanoflakes. *Physical Review B*, Vol. 81 (February, 2010) pp. 085430 - 1- 7

Lacovig, P.; M. Pozzo, M; Alfe, D.; Vilmercati, P.; Baraldi,A. & Lizzit, S. (2009). Growth of Dome-Shaped Nanoislands on Ir (111): The Intermediate between Carbidic Clusters and Quasi-Free-standing Graphene. *Physical Review Letters*, Vol. 103 (October 2009) 166101-1 - 166101-4

Li, X.; Wang, X.; Zhang, L.; Lee, S. & Dai, H. (2008). Chemically derived, ultrasmooth graphene nanoribbons semiconductors . *Science*, Vol. 319 (February 2008) pp. 1229-1232

Libisch, F.; Stampfer, C. & Burgdorfer, J. (2009). Graphene quantum dots: Beyond a Dirac billiard. *Physical Review B*, Vol. 79 (2009) 115423-1 - 115423-6

Loh, K.; Bao, Q.; Ang, P. & Yang, J. (2010). The chemistry of graphene. *Journal of Materials Chemistry*, Vol. 20 (January 2010) pp. 2277-2289

Luo, Z.; Vora, P.; Mele, E.; Johnson, A. & Kikkawa, J. (2009). Photoluminescence and band gap modulation in graphene oxide. *Applied Physics Letters*, Vol. 94 (March 2009) pp. 11909 1-3

Ma, J.; Alfe, D.; Michaelides, A. and Wang, E. (2009). Stone-Wales defects in graphene and other planar sp^2 - bonded materials. *Physical Review B*, Vol. 80 (July, 2009) pp. 033407 - 1 - 4

Mermin, N. (1968). Crystalline order in two dimensions. *Physical Review*, Vol. 176 (1) (December, 1968) pp. 250-254

Meyer, J; Geim, A.; Katsnelson, M.; Novoselov, K.; Booth, T. & Roth, S. (2007). The structure of suspended graphene sheets. *Nature*, Vol. 446 (March 2007) pp. 60 - 63

Meyer, J.; Girit, C.; Crommie, M. & Zettl, A. (2008). Hydrocabon lithography on graphene membranes. *Applied Physics Letters*, 92 (March, 2008) pp. 123110 -1 - 123110-3

Nagai, H.; Nakano, M.; Yoneda, K.; Kishi, R.; Takahashi, H.; Shimizu, A.; Kubo, T.; Kamada, K.; Ohta, K.; Botek, E.; & Champagne, B. (2010) . Signature of multiradical character in second hyperpolarizabilities of rectangular graphene nanoflakes. *Chemical Physics Letters*, Vol. 489 (March, 2010) pp. 212–218

Neubeck, S. ; Ponomarenko, L. ; Freitag, F.; Giesbers, A. ; Zeitler, U.; Morozov, S.; Blake, P.;

Geim, A. and Novoselov, K. (2010). From One Electron to One Hole: Quasiparticle Counting in Graphene Quantum Dots Determined by Electrochemical and Plasma Etching, *Small*, Vol. 6 (14) (June, 2010) pp. 1469–1473

Ni, Z.; Wang, Y.; Yu, T. & Shen, Z. (2008). Raman spectroscopy and imaging of graphene. *Nano Research*, Vol. 1 (2008) pp. 273-291

Novoselov, K.; Geim, A.; Morozov, S.; Jiang, D.; Zhang, Y., Dubonos, S.; Grigorieva, I. & Firsov, A. (2004). Electric Field Effect in Atomically Thin Carbon Films. *Science*, Vol. 306 (October, 2004) pp. 666-669

Obraztsov, A. (2009). Making graphene on a large scale. *Nature Nanotechnology*, Vol. 4, (April, 2009) pp. 212-213

Olivi-Tran, N. (2010). Pure, Si and sp^3 – C- doped graphene nanoflakes: A numerical study of density of states. *Physica B*, Vol. 405 (2010), pp. 2749-2752

Park, S. & Ruoff, R. (2009). Chemical methods for the production of graphenes. *Nature Nanotechnology*, Vol. 4 (March, 2009) pp. 217-224

Ponomarenko, L.; Schedin, F.; Katsnelson, M.; Yang, R.; Hill, E.; Novoselov, K. & Geim, A. (2008). Chaotic Dirac billiard in graphene quantum dots. *Science*, Vol. 320 (April , 2008) pp. 356 – 358

Potasz, P.; Guclu, A. & Hawrylak, P. (2010). Zero-energy states in triangular and trapezoidal graphene structures. *Physical Review B*, Vol. 81 (January, 2010) pp. 033403 – 1- 4

Rader, H.; Rouhanipour, A.; Talarico, A.; Palermo, V. ; Samori, P. & Mullen, K. (2006). Processing of giant graphene molecules by soft-landing mass spectroscopy. *Nature Materials*, Vol. 5 (April, 2006) pp. 276 – 280

Rycerz, A.; Tworzdlo, J. & Beenakker, C. (2007). Valley filter and valley valve in graphene. *Nature Physics*, Vol. 3 (March, 2007) pp. 172-175

Ryu, S.; Han, M.; Maultzsch, J.; Heinz, T.; Kim, P.; Steigerwald, M. & Brus, L. (2008). Reversible basal plane hydrogenation of graphene. *Nano Letters*,Vol. 8 (12) (2008) pp. 4597-4602

Sahin, H. & Senger, R. (2008). First-principles calculations of spin-dependent conductance of graphene nanoflakes. *Physical Review B*, Vol. 78 (November, 2008) pp. 205423 1- 8

Salzmann, C., Nicolosi, V. and Green, M. (2010). Edge-carboxylated graphene nanoflakes from nitric acid oxidized arc-discharge materials. *Journal of Materials Chemistry*, Vol. 20 (2010) pp. 314-319

Schedin, F.; Geim, A.; Morozov, S.; Hill, E.; Blake, P.; Katsnelson, M. & Novoselov, K. (2007). Detection of individual gas molecules adsorbed on graphene. *Nature Materials*, Vol. 6 (September, 2007) pp. 652 -655

Shemella, P.; Zhang, Y. & Mailman, M.; Ajayan, P. & Nayak, S. (2007). Energy gaps in zero-dimensional graphene nanoribbons. *Applied Physics Letters*, Vol. 91 (July, 2007) pp. 042101 – 1- 3

Singh, A., Penev, E. & Yakobson, B. (2010). Vacancy clusters in graphane as quantum dots. *ACS Nano*, Vol. 4 (6) (May, 2010) pp. 3510-3514

Snook, I.; Barnard, A.; Russo, S.; Springal, R. & Srbinovsky, J. (2005). Simulating nano-carbon materials. *Molecular Simulation, Vol.* 31 (6-7) (May, 2005) pp. 495 – 504

Sofo, J.; Chaudhari, A. & Barber, G. (2007). Graphane: A two-dimensional hydrocarbon. *Physical Review*, Vol. 75 (April , 2007) pp. 153401-1 – 153401-4

Son, Y.; Cohen, M. & Louie, S. (2006). Half-metallic graphene nanoribbons. *Nature*, Vol. 444 (November, 2006) pp. 347-349

Son, Y.; Cohen, M. & Louie, S. (2006). Energy gaps in graphene nanoribbons. *Physical Review Letters*, Vol. 97 (November, 2006) pp. 216803-1 - 216803-4

Stampfer, C; Schurtenberger, E.; Molitor, F.; Guttinger, J.; Ihn, T. & Ensslin, K. (2008). Tunable Graphene single electron transistor. *Applied Physics Letters*, Vol. 8 (8) (July, 2008) pp. 2378-2383

Stampfer, C; Guttinger, J.; Molitor, F.; Graf. D.; Ihn, T. & Ensslin, K. (2008). Tunable Coulomb blockage in nanostructured graphene. *Nano Letters*, Vol. 92 (2008) pp. 012102- 1- 012102-3

Stein, S. and Brown, R. (1987). π-Electron Properties of Large Condensed Polyaromatic Hydrocarbons. *Journal of the American Chemical Society*, Vol. 109 (1987) pp. 3721-3729

Sun, X.; Liu, Z.; Welsher, K.; Robinson, J.; Goodwin, A.; Zaric, S. & Dai, H. (2008). Nano-graphene oxide for cellular imaging and drug delivery. *Nano Research*,Vol. 1 (2008) pp. 203-212

Tapaszto, L.; Dobrik, G.; Lambin, P. & Biro, L. Tailoring the atomic structure of graphene nanoribbons by scanning tunneling microscope lithography. *Nature Nanotechnolgy*, Vol. 3 (July, 2008) pp. 397-401

Terrones, M. (2009). Nanotubes unzipped. *Nature*, 458 (April 2009) pp. 845-846

Tombros, N.; Jozsa, C.; Popinciuc, M.; Jonkman. H. & van Wees, B. (2007). Electronic spin transport and spin precession in single graphene layers at room temperature. *Nature*, Vol. 448 (August, 2007) pp. 571-574

Topsakal, M; Aktürk, E. & Ciraci, S. (2009). First-principles study of two- and one-dimensional honeycomb structures of boron nitride. *Physical Review B*, Vol. 79 (March, 2009) 115442- 1- 115442-11

Trauzettel, B.; Bulaev, D.; Loss, D. & Burkard, G. (2007). Spin qubits in graphene quantum dots. *Nature Physics*, Vol. 3 (March, 2007) pp. 192 - 196

Van Noorden, R. (2006). Moving towards a graphene world. *Nature*, Vol. 442 (July, 2006) pp. 228-229

Wallace, P. (1947). The Band Theory of Graphite. *The Physical Review*, Vol. 71(9) (May, 1947) pp. 622- 634

Wang, W.; Meng, S. & Kaxiras, E. (2008). Graphene nanoflakes with large spin. *Nano Letters*, Vol. 8 (1) (December, 2007) pp. 241-245

Wang, W.; Yazyev, O. ; Meng, S. & Kaxiras, E. (2009). Topological frustration in graphene nanoflakes: Magnetic order and spin logic devices; *Physical Review Letters*, Vol. 102 (April, 2009) pp. 157201 1 – 4

Wimmer, M.; Akhmerov. A.& Guinea, F. (2010). Robustness of edge states in graphene quantum dots. *Physical Review B*, Vol. 82 (July, 2010) pp. 045409-1 - 045409- 9

Wu, J.; Pisula, W. & Mullen, K. (2007). Graphenes as potential material for electronics. *Chemical Reviews*, Vol. 107(3) (October, 2007) pp. 718-747

Wu, X. & Zeng, X. (2008). Sawtooth-like graphene nanoribbons. *Nano Research*, 1(2008) pp. 40-45

Yazyev, O.; Wang, W.; Meng, S. & Kaxiras, E. (2008). Comment on Graphene nanoflakes with large spin: Broken symmetry states. *Nano Letters*, Vol. 8 (2) (January, 2008) pp. 766

Yu, D.; Lupton, E.; Gao, H.; Zhang, C. & Liu, F. (2008). A unified geometric rule for designing nanomagnetism in graphene. *Nano Research*, Vol. 1 (2008) pp. 497-501

Zheng, H. & Duley, W. (2008). First-principles study of edge chemical modifications in graphene nanodots. *Physical Review B*, Vol. 78 (July, 2008) pp. 045421 -1 – 5

Zheng, H. & Duley, W. (2008). Field effects on the electronic and spin properties of undoped and doped graphene nanodots. *Physical Review B*, Vol. 78 (October, 2008) pp. 155118 -1 – 7

Zhi, L. & Mullens, K. (2008). A bottom-up approach from molecular nanographenes to unconventional materials. *Journal of Materials Chemistry*, Vol. 18 (February, 2008) pp. 1472-1484

14

Finite-Size Effects in Graphene Nanostructures

Antonio Monari[1,2] and Stefano Evangelisti[1]
[1]*Laboratoire de Chimie et Physique Quantiques. Université de Toulouse et CNRS.*
[2] *Equipe de Chimie et Biochimie Théoriques, SRSMC, CNRS, Nancy-Université*
France

1. Introduction

Since its characterization in 2004 [Novoselov et al. (2004; 2005)] graphene has emerged as a material showing somehow rather unusual and particular characteristics [Castro Neto et al (2009); Geim (2009); Katselnson et al. (2006)]. In October 2010 A. Geim and K. Novoselov have been awarded the Physics Nobel Prize for the characterization of graphene. Much of this particular behavior is due to its nature of a pure two dimensional (2-D) materials made up of Carbon atoms in a honeycomb lattice. Moreover graphene, can be thought not only as the constituting material of the well known graphite (made up of a series of graphene stacks kept together by weak van der Waals bonds), but also as the constituent of other carbon-based nanomaterial. For instance the well known carbon nanotubes [Charlier et al. (2007)] can be considered as a rolled-up graphene sheet where edges have been joined. Fullerene [Andreoni (2000)] on the other hand can be seen as graphene with the presence of pentagonal defects responsible for the positive curvature that leads to the spherical geometry. Despite its importance, and despite the fact that graphene sheets could be produced every time someone uses a graphite-pencil, graphene was isolated and characterized only very recently, due to the technically difficulties in evidencing its formation. Moreover, even if theoretical considerations on the electronic structures of a single graphite sheets date back to the work of Pauling in 1950s [Pauling (1972)], the existence of a purely 2-D systems was considered as impossible until recently. However, soon afterwards its characterization graphene has open the way to an impressive mole of experimental and theoretical studies (see for instance Castro Neto et al (2009) and reference therein), in particular because of its remarkable electronic properties that make graphene a very important material to be exploited in the field of molecular and nano-electronics. Technical applications of graphene, or graphene-based devices, can go from single molecule detection [Schedin et al. (2007)], to field effect transistors [Novoselov et al. (2004)] and quantum information processing [Trauzettel et al. (2007)]. One should also cite that the robustness of the σ skeleton of graphene makes it one of the strongest materials ever tested [Lee et al. (2008)], and this fact obviously suggest the possible use of carbon-fiber reinforcements in novel composite materials. Although our work will not focus mainly on the properties of infinite graphene sheets, let us cite that graphene also proved to be a zero gap semi-metal, characterized by a very high electron mobility. Moreover, graphene is able to show unusual linearly dispersing electronic excitations: the electronic excitations in the vicinity of the Fermi level remind those of massless Dirac fermions. Graphene, therefore, allowed predictions of quantum electrodynamics to be tested in a solid-state system. It is also noteworthy to cite that many of these findings have been anticipated in the framework of independent electrons band theory by Wallace back in 1947 [Wallace (1947)]. Since then, even if theoretical works have become much more accurate and have benefit of the possibility to compare to experiments, the role of electron-electron correlation in graphene has been still an object of intense debate [Castro Neto et al (2009)]. Another very interesting problems, strongly related to electronic properties and to the Dirac fermions, concerns the influence and effects produced by disorder in the physics of electrons in graphene as well as its transport properties. Some of the parameters developed in the following of this Chapter, and in particular the localization tensor, could be of seminal importance in elucidating these effects.

If, as we have briefly recalled, infinite graphene presents absolutely remarkable properties, also finite-size systems (nano ribbons or nano islands) show a behavior that needs a careful investigation [Berger et al. (2006)]. In particular the geometry of the islands, and especially the form of edges, can module the properties of the system in a very impressive way [Castro Neto et al (2009)]. We would like to stress that in the near past the control over the nano-islands geometry was rather poor, due to the fact that graphene was produced by a top-down approaches by stripping away sheets from graphite. Today, on the other hand, with the emergence of bottom-up approaches, based on controlled deposition on metallic surfaces (for instance Ruthenium), the control of the final system geometry has significantly improved, and it is now possible to produce nano-islands of different size and forms with atomically defined edges [Fernandéz-Rossier & Palacios (2007)], ranging from the quasi-one dimensional ribbons to the zero-dimension dots. One of the first and better known and studied finite-size effect on graphene is the emergence of edge-states in the case of nano-ribbons [Son et al. (2004)]. From a physical points of view this means that the electrons close to the Fermi level tend to preferentially occupy regions of space that are close to the borders of the ribbon. This feature has been observed experimentally and has been predicted with high level *ab initio* computations for instance by Hod et al. (2008; 2007). It is also known that the behavior of the nano-ribbon strongly depends on the type of edges, giving rise to the two well known classes of "zig-zag" and "armchair" edges. More complex structures, such as hexagonal, rhomboidal and triangular ones, will show an even more complex behavior and can also give rise to open shell, high multiplicity ground states [Fernandéz-Rossier & Palacios (2007); Yu et al. (2008); Ezawa (2007)]. This could result in the possibility of rationally design advanced nano-magnets, also exploiting the connection of different multiplicity sub-units (see for instance the works of Yu et al. (2008); Trinquier et al. (2010)). In the same way, as suggest by Fürst et al (2009) the possibility to create antidot graphene lattices, in which graphene can be artificially periodically perforated to create a precise arrangement of holes, appears very promising to further control and module electronic properties. All this attempts can be seen as steps toward the production of metametarials [Pendry et al. (1996)], i.e. materials that derive their properties from their artificial, man-made, periodic small-scale structure. Anyway it is noteworthy to cite also all the attempts made to enhance physically properties of graphene-like structures by their interaction with atoms or small molecules, for instance following the procedure of the spin doping.

1.1 Magnetic properties of graphene nano-systems

An important feature to rationalize the relation between structure and electronic properties, and more generally, to study finite size effects on graphene nano-structures, is to recognize that the honeycomb graphene is composed by a bipartite lattice [Fernandéz-Rossier & Palacios (2007)], with two compenetrating triangular sublattices A and B. Each atom constituting the graphene can be associated to one of the two sublattices, one can therefore speak of A and B graphene atoms or centers. Moreover, each A atom will only have B nearest neighborings and viceversa. So for instance, zig-zag edges are composed of atoms that all belong to the same sublattices, while the same does not hold for armchair edges. Moreover in order to rationalize the emergence of magnetic properties in graphene nanosystems one can recall the Lieb's theorem [Lieb (1989)]. This theorem, also known as the theorem of itinerant magnetism and usually applied within the framework of the Hubbard one-orbital model, is indeed intended to predict the total spin of the ground state in bipartite lattices. In particular one can see that an imbalance on the number of atom in one sublattice will yeld a magnetic ground state with spin

$$S = \frac{|N_A - N_B|}{2} \qquad (1)$$

where N_A and N_B are the numbers of atoms of A and B type respectively. Moreover, in a Hückel type approximation, $|N_A - N_B|$ will also be the number of eigenvalues equals to zero. It is also important to note that the magnetization originating from localized edge-states give also rise to a high density of states at the Fermi level which in turn can determine a spin polarization instability. Moreover the relation between the unbalanced number of atoms and the ground state spin implies that two centers will be ferromagnetically coupled if they belong to the same sublattice and antiferromagnetically coupled if they do not [Yu et al. (2008)].
If now we will consider triangular type nanoislands (as will be evident in the following sections) one can see that the two sublattices are unbalanced, i.e. $N_A \neq N_B$. For this reason, the fundamental ground state will be a magnetic one, as was confirmed for instance by Fernandéz-Rossier & Palacios (2007). When, on the other hand, one deals with hexagonal coronene-like structures $N_A = N_B$, the ground state will be a closed shell singlet. In particular, in the latter case no eigenvalue of the Hückel Hamiltonian will be equal to zero. In reality, since even if a gap still exist at the Fermi level, increasing the size of the hexagonal island will diminish the difference between the highest occupied and the lowest unoccupied molecular orbitals (HOMO, LUMO), thus favoring an open-shell solution. Indeed a quantum phase transition has been predicted for hexagonal structures leading to a compensated ferrimagnet [Yu et al. (2008)].
In the following, we will rationalize the relation between magnetic and electronic properties of graphene nanostructures, presenting also a comparison between the simple Hückel, tight-binding, Hamiltonian and more sophisticated multireference *ab initio* ones.

1.2 Modern theory of conductivity: Polarizability and localizability

Graphene and graphene nano-ribbons are known to exhibit a very particular and in some cases exotic electronic properties [Castro Neto et al (2009)]. Graphene, for instance, is very well known for being a zero-gap semiconductor, characterized by an infinite electron mobility. Finite graphene nanoribbons, or Graphene Nano Islands (GNI), on the other hand show electronic properties that can vary drastically with the geometrical structure and shape [Castro Neto et al (2009); Fernandéz-Rossier & Palacios (2007); Yu et al. (2008)]. For these reasons, a study of electronic properties of GNI in the framework of the modern theory of conductivity [Resta (1998); Resta & Sorella (1999)] can be of valuable importance in order to elucidate the characteristic of these systems, as well as their correlation with the shape, as was recently pointed out by Evangelisti et al. (2010).
For readers' convenience, we report here a brief review of the so called modern theory of conductivity, explicitating the most relevant quantities appearing in this formalism to

discriminate between a metal and an insulator.
Modern theory of conductivity first emerged in a well established way with the seminal works of Kohn in 1964 [Kohn (1964)]. In this work, for the first time, the character of an insulating state was related not only to the few electronic states close to the Fermi level, as it was customary on the "classic" theory, but indeed to the ground state as a whole. To use Kohn's words, *"insulator behavior does not appear to depend on the notion of a filled band, but rather reflects a certain type of organization of the electrons"*. Going even further, Kohn affirms that in the insulating state the electrons *"so organize themselves as to satisfy a many electron localization condition"*.
Therefore, one can understand that, in this framework, the emergence of a metallic or insulator behavior is no more related to the opening or closing of a gap at Fermi level but indeed to the emergence of electron localization or delocalization [Resta & Sorella (1999); Kohn (1964)].
Different authors contributed to the development of the theory in the last years, and in particular the localization condition invoked by Kohn was given a quantitative indicator by means of the localization tensor, as defined by Resta and Sorella [Resta (1998); Resta & Sorella (1999); Aoki & Imamura (1995)]. Although this theory was first derived within Periodic Boundary Condition, its extension to the case of finite systems is possible, and is clearly much more appropriate for the systems we want to investigate here [Resta (2005; 2006)].
Suppose we consider an electronic wavefunction $|\Psi>$, and indicate by $\hat{\mathbf{r}}_i$ the position operator of the i-th electron (the position operator being a vector quantity of Cartesian components x, y, z). Then $\hat{r_\beta}$ will be one Cartesian component of the total position operator

$$\hat{\mathbf{r}} = \sum_{i=1}^{n} \mathbf{r}_i \tag{2}$$

It will be possible to define the localization tensor (or localizability) of the state associated to $|\Psi>$ as the cumulant of the second-order momentum with respect of the operator $\hat{\mathbf{r}}$ i.e. the quadratic fluctuation of the position

$$< r_\beta r_\gamma >_c = \frac{1}{n}(< \Psi|\hat{r}_\beta \hat{r}_\gamma|\Psi > - < \Psi|\hat{r}_\beta|\Psi >< \Psi|\hat{r}_\gamma|\Psi >) \tag{3}$$

Notice that the $\frac{1}{n}$ factor has been introduced in order to eliminate the dependence on the number of electrons n in the case of identical *non interacting* subsystems. In the following, for the sake of simplicity of notation, the locality-tensor components $< r_\beta r_\gamma >_c$ will be simply indicated as $\rho_{\beta\gamma}$ and the definition of Localizability will be used in analogy with polarizability. As shown by Resta [Resta & Sorella (1999)]in the case of a metallic system the locality ρ will diverge, while for insulators it will converge to a finite value. We will also remind that the locality has been directly related to the macroscopic conductivity $\sigma(\omega)$ by Souza et al. (2000)

$$\rho_{\alpha\beta} = \delta_{\alpha\beta} \frac{V}{n} \frac{\hbar}{\pi e^2} \int_0^\infty \frac{d\omega}{\omega} \Re\sigma(\omega) \tag{4}$$

An equivalent, in the case of a complete expansion space, and in some instances more convenient expression of the localization tensor can be obtained by invoking the sum-over-states or spectral resolution formalism as:

$$\rho_{\alpha\beta} =< \Psi_0|(r_\beta - < r_\beta >)|(r_\gamma - < r_\gamma >)|\Psi_0 >= \sum_{k>0} < \Psi_0|r_\beta|\Psi_k >< \Psi_k|r_\gamma|\Psi_0 > \tag{5}$$

Here the vectors $|\Psi_k>$ are the (excited) eigenvectors of the system while $|\Psi_0>$ represents the ground state. In this formalism the relation between polarizability and locality becomes much more evident if one recall the spectral resolution form of the (static) polarizability:

$$\alpha_{\beta\gamma} = 2\sum_{k>0}\left(\frac{<\Psi_0|r_\beta|\Psi_k><\Psi_k|r_\gamma|\Psi_0>}{E_k - E_0}\right) \tag{6}$$

where E_k are the eigenvalues corresponding to the Ψ_k eigenvector, E_0 being the ground state energy. We will also recall that for conducting systems too the polarizability will diverge, implying, from a more physical point of view, and infinite possibility of deformation of the electronic cloud of the systems.
Hence, the metallic or insulator characteristic of a system can be investigated by three criteria:

1. Zero energy gap.
2. Infinite per-atom polarizability.
3. Infinite fluctuation of the position operator (Localizability).

Anyway, it has to be pointed out that while the energy gap relates the metallic behavior of a system only to the states close to the Fermi level the polarizability and the Localizability are able to-take into account the properties of the whole system as it is clear from the spectral-resolution formalism.
We would also like to remind that, strictly speaking, since we are dealing with finite systems, it is clearly impossible to obtain a true metallic behavior. However, the divergent or convergent behavior of the Localizability (and of the per-atom - or Specific - Polarizability) can be clearly evidenced. Therefore, one can reasonably speak of a sort of metalicity, or at least of the presence of "precursors" of metallic characters.

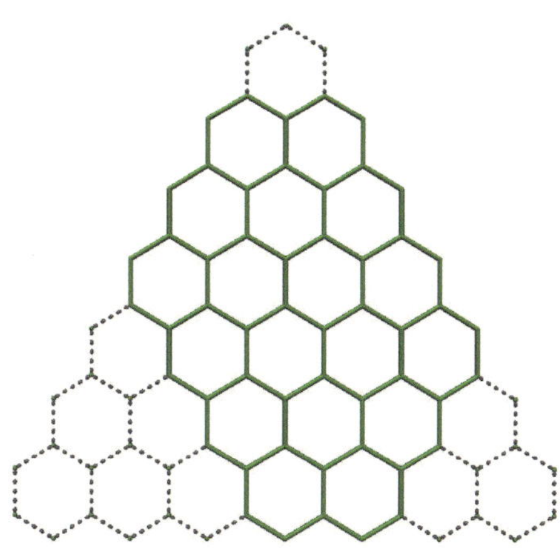

Fig. 1. The (7|3, 2, 1) HGNI. Note that the triangles eliminated from the original hypertriangulene structure have been represented with dots

The previous formalism has already been applied to the study of metal insulator transition in different systems, both at *ab initio* [Vetere et al. (2008)] and tight binding level [Evangelisti et al. (2010); Monari et al. (2008); Bendazzoli et al. (2010)] and it has been shown that the equivalence of the previous cited criteria does not always hold. For instance, disordered [Bendazzoli et al.

(2010)] systems are gapless while being insulator, and graphene nano-ribbons [Evangelisti et al. (2010)] can show a divergent polarizability and a convergent localizability. For this reason, modern theory of conductivity proves its uselfullness in giving the possibility of a deeper characterization of the metallic and insulator states.

2. Hexagonal Graphene Nanoislands

In this section, a possible classification of GNI is proposed and illustrated. We will limit ourselves to Hexagonal GNI (or HGNI), which are defined as the convex structures that have a (*a priori* irregular) hexagonal shape. Because of the hexagonal structure of the building block of graphene, the hexagonal motifs representing HGNI will have angles of 60, 120 or 180 degrees. In order to establish a classification of HGNI, we will start from a hypertriangulene having a side containing Λ hexagons. The most general HGNI that can be derived from it well be obtained by deleting three triangular patterns out of the triangulene vertices (see Figure 1). Each one of these three triangles will be uniquely identified by the number of elementary hexagons of its side, λ_i, with $i = 1, 2, 3$. Notice that, in some cases, some of the λ_i can be zero. The resulting HGNI will then be indicated by $(\Lambda|\lambda_1, \lambda_2, \lambda_3)$. In order to avoid any ambiguity, we will assume

$$\lambda_1 \geq \lambda_2 \geq \lambda_3$$

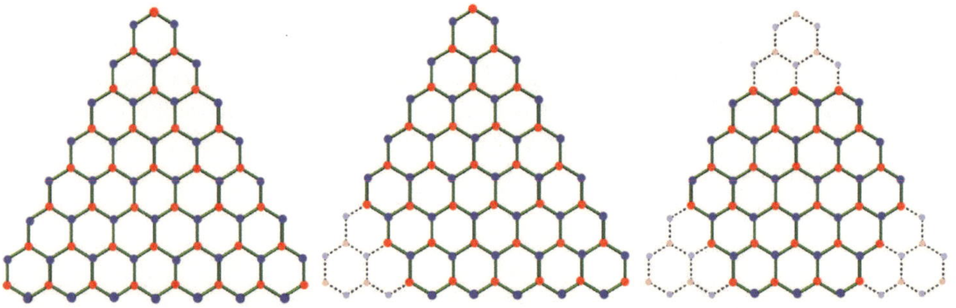

Fig. 2. The A "lateral" (dark blue) and B "apical" (dark red) sublattice sites of various HGNI. Notice in light blue and light red the eliminated lateral and apical sites, respectively, when $\lambda \neq 0$. Left the $(7|0,0,0)$ triangulene, with a lateral-apical sites difference equal to $\Lambda - 1 = 6$. Center the $(7|2,0,0)$ characterized by a lateral-apical sites difference of $\Lambda - \lambda_1 - 1 = 4$. Right the $(7|2,2,2)$ coronene that exhibits a lateral-apical sites difference of $\Lambda - \sum_i \lambda_i - 1 = 0$

Notice that, in our notation, the small triangles are not allowed to overlap. This implies that the sum of two different λ's must be smaller than (or at most equal to) Λ. Because of the ordering we have assumed, this condition is automatically fulfilled if we have

$$\lambda_1 + \lambda_2 \leq \Lambda$$

As we said in the Introduction, Graphene is a bipartite lattice. Therefore, HGNI also are obviously bipartite lattices. This implies that, at the Hückel level, the number of zero-energy orbitals is given by the difference between the two types of lattices.

Let us consider a triangulene structure. With our previous notation, it will be described as $(\Lambda|\,0,0,0)$. The lattice sites will be called "lateral" (blue in Figure 2) or "apical" (red in Figure 2). In particular, lateral sites will be defined as the Carbon atoms on one of the three edges in the $(\Lambda|\,0,0,0)$ triangulene, while the three terminal atoms will be apical.

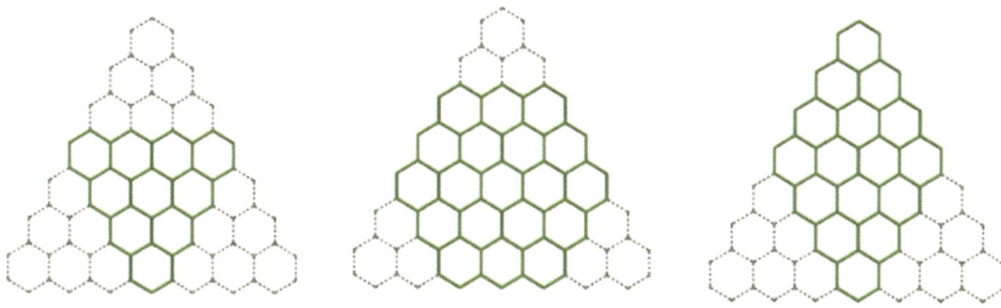

Fig. 3. HGNI obtained from the $(7|0,0,0)$ triangulene. Left the $(4|0,0,0)$ triangulene. Center the $(7|2,2,2)$ coronene. Right the $(7|2,2,0)$ pyrene. Eliminated triangles represented with dots

If one looks at Figure 2, it is straightforward to note that the difference between A and B sites is equal to the difference between lateral and apical sites. On the other hand it is also evident that the lateral sites outnumber the apical ones by $\Lambda - 1$.
When a triangular portion is cut at a triangulene vertex, the difference between the lateral and apical sites is reduced by λ. As a general result, we obtain that the difference between the lateral and apical sites (referred to the original triangulene structure) is given by

$$\Lambda - \sum_{i=1}^{3} \lambda_i - 1$$

The modulus of this quantity gives the number of zero-energy orbitals of the system if treated at the Hückel level, n_0.
It is useful to consider some special cases (see Figure 3):

- Triangulenes: They are described by the pattern $(\Lambda|\,0,0,0)$, and have $n_0 = \Lambda - 1$. They can also be obtained by cutting three identical triangular portions having side λ out an original triangulene having side equal to $2\lambda + 1$: $(2\lambda+1|\lambda,\lambda,\lambda) = (\lambda+1|\,0,0,0)$.
- Coronenes: They are characterized by the pattern $(3\lambda+1|\lambda,\lambda,\lambda)$, which implies that $n_0 = 0$.
- Pyrenes: The are characterized by the pattern $(2\lambda+1|\lambda,\lambda,0)$, which also implies that $n_0 = 0$.

Finally, we notice that HGNI having a D_{3h} symmetry are obtained if and only if the three λ_i are equal: $(\Lambda|\lambda,\lambda,\lambda)$, which gives $n_0 = |\Lambda - 3\lambda - 1|$ It is possible to start from a triangular structure and, by *adding* hexagons on the sides, to pass through a series of D_{3h} shapes, finally ending up with a larger triangulene:

$(2\lambda+1|\lambda,\lambda,\lambda)$
$(2\lambda+1|\lambda-1,\lambda-1,\lambda-1)$
$(2\lambda+1|\lambda-2,\lambda-2,\lambda-2)$
...
$(2\lambda+1|0,0,0)$

If λ is an integer multiple of 3, the series will pass from the coronene $(2\lambda+1|\frac{2}{3}\lambda,\frac{2}{3}\lambda,\frac{2}{3}\lambda)$.

3. Computational details

In this Section, we will recall the computational details and strategies used in the present work. In particular the details concerning both the *ab initio* and the *tight binding* approach will be discussed.

3.1 Tight binding

All the computation at the tight binding (Hückel in the language of Chemists) level have been performed by using a specific home-made code [Evangelisti et al. (2010)]. In particular, the diagonalization of the Hückel Hamiltonian, in order to get the eigenvectors and eigenvalues, have been performed by using high efficiency BLAS and Lapack [lapack (1999)] subroutines. The Hückel Hamiltonian has been constructed starting from the connectivity of each carbon atoms in each HGNI, defining

$$\langle i|\hat{H}|i\rangle = \alpha = 0 \tag{7}$$

and

$$\langle i|\hat{H}|j\rangle = \beta\gamma_{ij} \tag{8}$$

where γ_{ij} is 1 if i and j are different sites connected in the HGNI skeleton, and 0 otherwise. We remind that $-\beta$ is sometimes called the hopping integral t in physical literature. In HGNI, due to a more complex connectivity, the Hückel Hamiltonian matrix will not be a tridiagonal one, as it is customary in model linear systems (polyacetilene).

As will be discussed in the next section for some systems by diagonalization of the Hamiltonian we obtained some strictly zero eigenvalues, in that case we considered the system as an open shell one, and each zero eigenvector was considered as a half filled orbital. Polarizability and Localizability have been computed by direct application of equations 4 and 5, anyway we would like to stress that since we are dealing also with open shell systems it was necessary to compute separately the contribution coming from spin up and spin down (α and β) electrons.

3.2 Ab initio

All the *ab initio* calculations have been performed by using the MOLPRO [Knowles & Werner (2002)] Quantum-Chemistry package. The nanostructures investigated at *ab initio* level have at least a D_{3h} symmetry. Coronene islands are more symmetric, having a D_{6h} point-group symmetry. However, most *ab initio* codes (and MOLPRO is one of these) are able to treat only abelian subgroups. Because of this symmetry constraint, calculations have been performed in C_{2v} and D_{2h} instead of D_{3h} and D_{6h}, respectively. All HGNI edges have been saturated with Hydrogens atoms.

In *ab initio* calculations, all the angles concerning connected atoms have been fixed to the ideal value of 120 degrees. The internuclear C-C distance between connected atoms was fixed at 1.4 Å, while the corresponding C-H distance was fixed at 1.0 Å.

The atomic basis set is the minimal STO-3G [Hehre et al. (1969)] contracted gaussian basis for both Carbon and Hydrogen. Although this basis set is certainly too small to give quantitatively reliable results, it is known to reproduce correctly the qualitative behavior of many organic and inorganic systems. Its reduced size, on the other hand, permits to investigate at a correlated *ab initio* level systems whose size is relatively large.

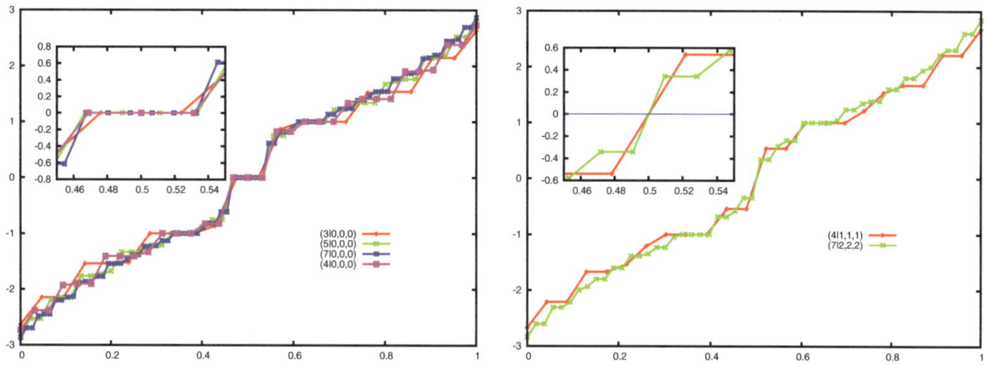

Fig. 4. The tight binding energy spectrum of some triangulene (left panel) and coronene (right panel) structures. In the boxes see a zoom of the Fermi level

The systems have been studied at Complete Active Space Self Consistent Field (CAS-SCF) level [Roos et al. (1980)], and Multi-Reference Perturbation-Theory level by using the n-electrons valence perturbation theory (NEVPT2) formalism [Angeli et al. (2001; 2002; 2007)]. In all the CAS-SCF calculations, the active space has been selected by choosing the orbitals that are strictly degenerated at Hückel level. The orbitals have been optimized at RHF level for the high-spin wavefunction. These same orbitals have been then frozen for the CAS-SCF calculations on the other spin multiplicities. In this way, our calculations are actually of CAS-CI type. The NEVPT2 formalism has then be applied to these CAS-CI wavefunctions in order to recover the dynamic correlation.

4. Results and discussion

The different behavior of Coronenes and triangulene finite size graphene structure will be illustrated in this Section. In particular we will first consider the tight binding approximation, before switching to an *ab initio* approach that allows us to better analyze the magnetic behavior of such systems.

4.1 Tight binding

Even if the tight-binding approach represents a crude approximation of a chemical systems, it can be important to sketch out some general tendencies, and to elucidate the behavior of different classes of compounds. This is particularly true since it allows the treatment of very large systems due to its extremely reduced computational cost.
The tight binding (Hückel) calculations of the different HGNI show, first of all, a quite different behavior between the triangulene and coronene classes. This difference is another confirmation of the very important role played by the finite size effects in graphene nano-islands. The energy spectra of some Triangulenes are shown in the left panel of Figure 4, while Coronenes are shown in the right panel (Notice that the spectra have been normalized with respect to the number of carbon atoms, and therefore with respect to the number of eigenvalues for graphical reasons). If the tail of the spectra looks similar between the two classes of compounds and between islands of different size, a remarkable difference can be seen at the Fermi level. As expected, [Fernandéz-Rossier & Palacios (2007); Yu et al. (2008)] Coronenes show a gap at the Fermi level, while Triangulenes shows the presence of some degenerate energy eigenvalues at the Fermi level. As can be seen on Table 1, where

system	n_C	N_f	α_x	ρ_x
(4|1,1,1)	24	0	0.37321	0.48330
(7|2,2,2)	54	0	0.62249	0.59056
(3|0,0,0)	22	2	0.40466	0.48802
(5|0,0,0)	46	4	0.53029	0.55306
(7|0,0,0)	78	6	0.64423	0.59801
(4|0,0,0)	33	3	0.46865	0.52370
(7|1,1,1)	69	3	0.71761	0.61768

Table 1. Localizability (ρ) and Specific Polarizability (α), in arbitrary units, for the tight-binding approximation; n_C indicates the number of Carbon atoms; n_0 represents the number of states at the Fermi level, i.e. having an eigenvalue exactly zero.

the number of zero level is reported, the Lieb theorem is exactly respected, as well as our empirical formula. Moreover the Lieb theorem appears also to be respected for intermediate structures like the (7|1,1,1) island, which present three zero-energy levels again as predicted from our formula. The magnetic levels arise from orbitals concentrated at the edge of the structure (edge states). This will induce a concentration of the spin density at the border of the island, as can be seen in Figure 5, where the Hückel spin density for the (7|0,0,0) triangulene has been shown. The spin density has been obtained by using Hückel eigenvectors and considering an occupation equal to one of the degenerate zero-energy eigenvalues. For graphical reasons, an arbitrary totally symmetric gaussian function has been placed on each Carbon center.

Finally if one enlarges the island the effect imposed by the different edges, and by the unbalancement of apical and lateral sites becomes less important. The HGNI energy spectrum is expected to converge toward the infinite graphene structure. In Figure 6 we report the spectra of the (121|40,40,40) coronene and of the (95|0,0,0) triangulene (both structures having about 10,000 carbon atoms). The two spectra are nearly superposable, even if a slight difference still remains in the close vicinity of the Fermi level. Indeed even if coronene will present a gap, while triangulene will be gapless, the energy difference between occupied and unoccupied orbitals in the first system will become smaller and smaller with the system size, tending to the gapless spectrum of infinite graphene. This fact can be considered as the origin of the spin instability of large coronene systems, for which the ground state will be an open shell one instead of a closed shell. In Table 1 we also report the values of the specific (i.e. per atom) polarizability and Localizability of the different classes of compounds. We can see that for small size HGNI the two parameters behave in a similar way with the

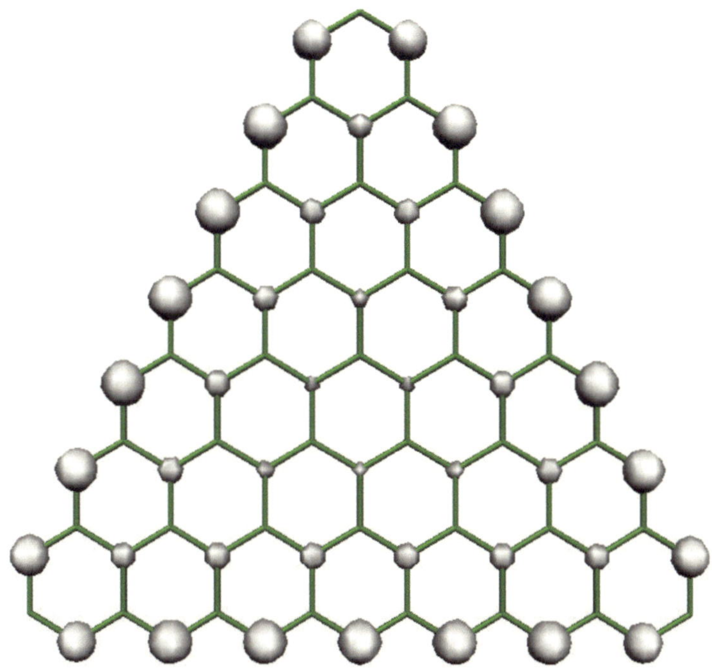

Fig. 5. The tight binding spin density of the (7|0,0,0) triangulene (septuplet state)

system size, and within the different classes of compounds although polarizability appears to have a slightly higher value than Localizability. In Figure 7 we report the behavior of specific polarizability and Localizability for triangulenes of different sizes up to the (100|0,0,0) HGNI. If Localizability appears to converge to a value close to 1, specific polarizability diverges linearly (implying an overall quadratic divergence of the polarizability). These results, which shows a qualitatively different behavior of the two indicators, can be considered to be in coherence with the observation that extended graphene is a semiconductor with infinite electronic mobility.

4.2 Ab initio

In Table 2, we report CAS-SCF and NEVPT2 computations on different graphene nano-island. In particular, as stated in Computational Details, we included in the active space the orbitals that give rise to the degenerate zero-energy manifold at tight-binding level. This choice assures to obtain a Multi-Configurational wavefunction that is able to correctly represent the physics of the problem, without presenting an explosive size of the configuration space. As it was already evident at Hückel level, the splitting of the different levels follows quite remarkably the Lieb theorem for a Hubbard bipartite lattice. Indeed, in the case of Triangulenes, the electronic state with multiplicity coherent with the Lieb theorem ($n_0 + 1$, where n_0 is the difference between apical and lateral sites) is the ground state. The same is true for the (7|1,1,1) intermediate system. In a similar way, Coronenes, at least for the size of the ones investigated here, show always a singlet ground state.

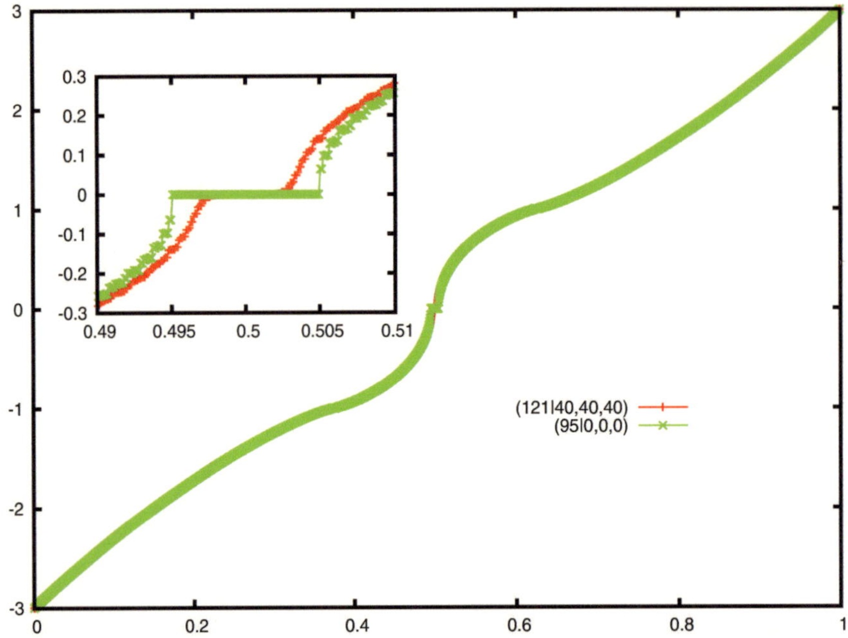

Fig. 6. The tight binding energy spectrum of (121|40,40,40) coronene and (95|0,0,0) triangulene both tending to the infinite graphene spectrum

The same effects can be seen on the left and right panel of the Figure 8, where we report the energy of the different multiplicity state, with respect to the highest multiplicity ground state, of different odd Triangulenes at CAS-SCF and NEVPT2 level.

As far as the effect of the size of the island is concerned, one can easily see that the energy splitting is large in larger structures. Therefore, one can assume that enlarging the structure will tend to favor the ferromagnetic coupling of the unpaired electrons. Rather interestingly, on the other hand, the inclusion of the dynamic correlation (at NEVPT2 level) determines a very strong lowering of the ferromagnetic coupling, up to a factor of one half in the case of odd Triangulenes. Probably this means that the relaxation of the other π double occupied orbitals induce a rather important stabilization of the low spin states. We would like to stress that, to such a magnitude, this phenomenon is quite uncommon in the magnetic spectrum of organic and inorganic magnetic materials, thus suggesting a peculiar role played by the graphene-like electronic structure.

5. Conclusion

We have reported a tight binding and *ab initio* study of graphene triangular and hexagonal nano structures. In particular, we proposed a general classification of these structures based on an index Λ and three indices λ_i. We showed how Coronene and Triangulene classes of HGNI can be easily identified within this classification via an opportune choice of the indices. By using the Lieb theorem and considering that graphene is actually a bipartite lattices, we showed how it is possible to guess the magnetic properties of these nano-structures from the values of Λ and λ_i. This heuristic findings have been confirmed also by high level *ab initio* computations. We also showed how the magnetic properties of Triangulenes structures are

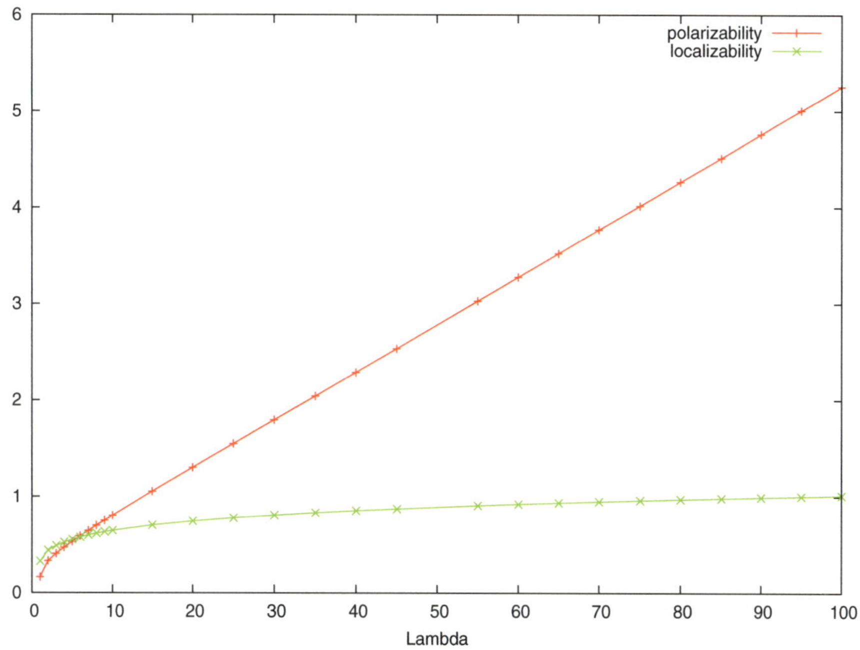

Fig. 7. The specific polarizability and Localizability of different Triangulenes $(\Lambda|0,0,0)$ as a function of Λ

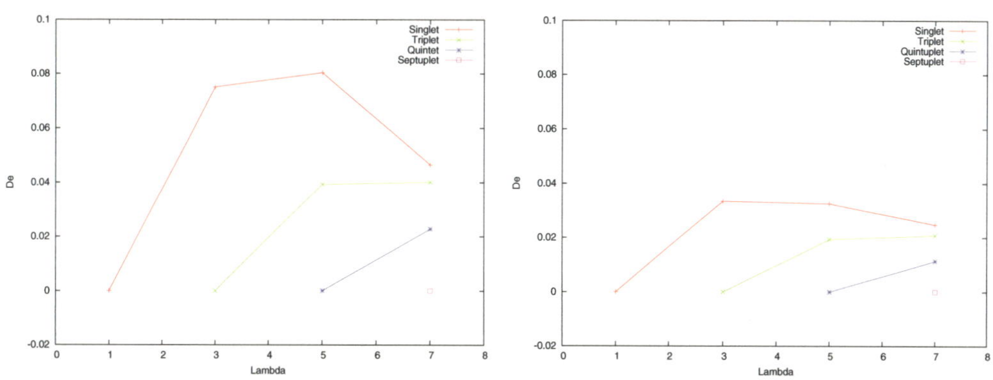

Fig. 8. The CAS-SCF (left panel) and NEVPT2 (right panel) magnetic spectrum for the odd Triangulenes with respect to the highest multiplicity ground state. Energy in hartree

given by edge states and orbitals, giving rise to a concentration of the spin density at the border. However, we underlined how the particular nature of graphene induces a specific behavior of the magnetic spectrum. In particular, the ferromagnetic coupling is very strongly reduced by the inclusion of dynamical correlation.

The behavior of graphene nano-islands has also been studied in the framework of the modern theory of conductivity, in particular by using the localization tensor (Localizability) and the polarizability. We showed how, in the limit of an extended graphene sheet, the two indicators of a metal-insulator transition behave in a qualitatively different way: while polarizability diverges Localizability shows a convergence. This fact can be ascribed to the nature of

system	n_C	n_H	S	E_{CAS-CI}	E_{NEVPT2}
(4|1,1,1)	24	12	1	0.00000000	0.00000000
(7|2,2,2)	54	18	1	0.00000000	0.00000000
(3|0,0,0)	22	12	3	0.00000000	0.00000000
			1	0.07504935	0.03331113
(5|0,0,0)	46	18	5	0.00000000	0.00000000
			3	0.03924909	0.01935507
			1	0.08033144	0.03242700
(7|0,0,0)	78	24	7	0.00000000	0.00000000
			5	0.02271725	0.01135881
			3	0.04009214	0.02078559
			1	0.04647274	0.02467302
(4|0,0,0)	33	15	4	0.00000000	0.00000000
			2	0.03632651	0.02469043
(7|1,1,1)	69	21	4	0.00000000	0.00000000
			2	0.00466476	0.00624823

Table 2. Relative Energies (hartree) of nanoislands species, computed at *ab initio* level:

extended graphene, that is known to behave like a gapless semi conductor with infinite electron mobility. These results also confirm how the use of modern theory of conductivity can be of valuable importance in the study of exotic materials.

6. Acknowledgments

The authors would like to kindly acknoweldge support from the University of Toulouse and Nancy and from the French CNRS, also under the PICS action 4263. Support from European Union under the Cost in Chemistry Actions D37 (GRIDCHEM) and CM0702 is also gratefully acknoweldged.

7. References

K. S. Novoselov, A. k. Geim, S. V. Morozov, D. Jang, Y. Zhang, S. V. Dubonos, I. V. Grigorieva, A. A. Firsov"Electric Field Effect in Atomically Thin Carbon Films" Science 306 666 (2004).

K. S. Novoselov, A. K. Geim, S. V. Morozov, D. Jang, M. I. Katselnov, I. V. Grigorieva, S. V. Dubonov, A. A. Firsov "Two-dimensional gas of massless Dirac fermions in graphene" Nature 438 197 (2005).

A. H. Castro Neto, F. Guinea, N. M. R. Peres, K. S. Novoselov, A. K. Geim, "The electronic property of graphene" Rev. Mod. Phys. 81 109 (2009).
A. K. Geim "Graphene: Status and Prospects" Science 324 1530 (2009).
M. I. Katselnson, K. S. Novoselov, A. K. Geim "Chiral tunnelling and the Klein paradox in graphene" Nat. Phys. 2 620 (2006).
J. C. Charlier, X. Blase, S. Roche "Electronic and transport properties of nanotubes" Rev. Mod. Phys. 79, 677 (2007)
W. Andreoni "The physics of Fullerene-Based and Fullerene Related Materials" Springer Berlin (2000).
L. Pauling "The nature of chemical bond" Cornell University Press, Ithaca, NY (1972).
F. Schedin, A. K. Geim, S. V. Morozov E. W. Hill, P. Blake, M. I. Katselson, K. S. Novoselov "Detection of individual gas molecules adsorbed on graphene" Nat. Mater. 6 652 (2007)
B. Trauzettel, D. V. Bulaev, D. Loss, G. Burkard "Spin qubits in graphene quantum dots" Nat. Phys. 3 192 (2007).
C. Lee, X. Wei, J. W. Kysar, J. Hone "Measurement of the Elastic Properties and Intrinsic Strength of Monolayer Graphene" Science 321 385 (2008).
P. R. Wallace "The Band Theory of Graphite" Phys. Rev. 71, 622 (1947).
C. Berger, Z. Song, X. Li, X. Wu, N. Brown, C. Naud, D. Mayou, T. Li, J. Hass, A. N. Marchenkov, E. H. Conrad, P. N. First, W. A. de Heer "Electronic Confinement and Coherence in Patterned Epitaxial Graphene" Science 312 1191 (2006).
J. Fernandéz Rossier, J. J. Palacios "Magnetism in Graphene Nanoislands" Phys. Rev. Lett. 99, 177204 (2007).
Y.-W. Son, M. L. Cohen, S. G. Louie "Half-metallic graphene nanoribbons" Nature 444, 347 (2006).
O. Hod, V. Barone, G. E. Scuseria "Half-metallic graphene nanodots: A comprehensive first-principles theoretical study" Phys. Rev. B 77, 035411 (2008).
D. Yu, E. M. Lupton, H. J. Gao, C. Zhang, F. Liu. "A unified rule for designing nanomagnetism in graphene" Nano Res. 1 497 (2008).
G. Trinquier, N. Suaud, J.-P. Malrieu, "Theoretical study of high-spin polycyclic hydrocarbons" Chem. Europ. J. 16 8762 (2010).
M. Ezawa "Metallic graphene nanodisk: Electronic and magnetic properties" Phys. Rev. B 76, 245415 (2007).
J. A. Fürst, J. G. Pederson, C. Flindt, N. A. Mortensen, M. Brandbyge, T. G. Pedersen, A.-P. Jauho "Electronic Properties of graphene antidot lattice" J. of Phys. 11 095020 (2009).
J. B. Pendry, A. J. Holden, W. J. Stewart I. Youngs "Extremely Low Frequency Plasmons in Metallic Mesostructures" Phys. Rev. Lett. 76 4773 (1996).
O. Hod, J. E. Peralta, G. E. Scuseria Phys. Rev. B "Edge effects in finite elongated graphene nanoribbons" 76, 233401 (2007).
E. H. Lieb "Two theorems on the Hubbard model" Phys. Rev. Lett. 62 1201 (1989)
R. Resta, "Quantum-Mechanical Position Operator in Extended Systems" Phys. Rev. Lett. 80, 1800 (1998).
R. Resta and Sorella, "Electron Localization in the Insulating State" Phys. Rev. Lett. 82, 370 (1999).
S. Evangelisti, G. L. Bendazzoli and A. Monari "Electron localizability and polarizability in tight-binding graphene nanostructures", Theor. Chem. Acc. 126, 257 (2010)
W. Kohn, "Theory of the Insulating State" Phys. Rev. 133, A171 (1964)

Y.Aoki and A. Imamura, "An analytical Hückeltype approach to the relationship between Peierls instability in polyenes and interchain interaction" J. Chem. Phys., 103, 9726 (1995).

R. Resta, "Electron Localization in the Quantum Hall Regime", Phys. Rev. Lett. 95, 196805 (2005).

R. Resta, J. Chem. Phys. "Kohns theory of the insulating state: A quantum-chemistry viewpoint" 124, 104104 (2006).

I. Souza, T. Wilkens, and R. M. Martin, "Polarization and localization in insulators: Generating function approach" Phys. Rev. B 62, 1666 (2000).

V. Vetere, A. Monari, G.L. Bendazzoli, S. Evangelisti, and B. Paulus, "Full configuration interaction study of the metal-insulator transition in model systems: Li_N linear chains (N = 2,4,6,8)" J. Chem. Phys. 128, 024701 (2008).

A. Monari, G. L. Bendazzoli, S. Evangelisti, "The metal-insulator transition in dimerized Hückel chains" J. Chem. Phys. 129, 134104 (2008)

G. l. Bendazzoli, S. Evangelisti, A. Monari, R. Resta, "Kohns localization in the insulating state: One-dimensional lattices, crystalline versus disordered" J. Chem. Phys. 133, 064703 (2010)

http://www.netlib.org/lapack/

Molpro quantum chemistry package, by P. Knowles and H. J. Werner http://www.molpro.net

W.J. Hehre, R.F. Stewart and J.A. Pople, "SelfConsistent MolecularOrbital Methods. I. Use of Gaussian Expansions of SlaterType Atomic Orbitals" J. Chem. Phys. 51 2657 (1969).

B. O. Roos, P. R.Taylor, P. E. M. Sieghbahn "A complete active space SCF method (CASSCF) using a density matrix formulated super-CI approach" Chem. Phys. 48, 157 (1980).

C. Angeli, R. Cimiraglia, S. Evangelisti, T. Leininger, and J.-P. Malrieu, "Introduction of n-electron valence states for multireference perturbation theory" J. Chem. Phys., 114, 10252, (2001).

C. Angeli, R. Cimiraglia, J.-P. Malrieu, "n-electron valence state perturbation theory: A spinless formulation and an efficient implementation of the strongly contracted and of the partially contracted variants" J. Chem. Phys., 117, 9138, (2002)

C. Angeli, M. Pastore, R. Cimiraglia, Theor. Chem. Acc., "New perspectives in multireference perturbation theory: the n -electron valence state approach" 117, 743 (2007).

15

Quantum Computation with Graphene Nanostructure

Zhi-Rong Lin, Guo-Ping Guo, Tao Tu, Qiong Ma and Guang-Can Guo
Key Laboratory of Quantum Information, University of Science and Technology of China, Chinese Academy of Sciences, Hefei, 230026
P.R.China

1. Introduction

Quantum computers which can efficiently simulate complex quantum systems and solve certain classes of hard mathematical problems are of great interest and importance. However, practical implementation of a large-scale quantum computer represents a formidable challenge. One significant obstacle is combining good access to the quantum system with high degree isolation from the environment in a scalable system.

The spin qubit proposal is a promising approach to address the central issues (Loss et al., 1998). The electron spin is a two-level system which is a natural candidate for realization of a quantum bit. Spin qubits in semiconductor nanostructures can be accessed and scaled easily. After the original spin-qubit proposal, there has been enormous research effort in implementing the spin-based information processing and the major breakthroughs in basic proof-of-principle experiments have been achieved in GaAs/AlGaAs quantum dots. First, the single-shot measurement of an individual electron spin has been realized (Elzerman et al., 2004). Second, the demonstrations of the $\sqrt{SWAP}$-gate on two-electron spin states (Petta et al., 2005) and single spin rotations (Koppens et al., 2006; Nowack et al., 2007; Pioro-Ladrière et al., 2008) suffice to universal quantum operations of spin qubits. By now all parts of the original Loss-DiVincenzo proposal have been demonstrated in the proof-of-principle experiments. To build a scalable quantum computer requires that the gate error should be smaller than the threshold and the decoherence time should be 10^4 times longer than the operation time. However because of interaction with the nuclei environment in the host GaAs/AlGaAs material via both lattice-mediated spin-orbit interactions and hyperfine interactions, the decoherence times are not long enough compared to the single qubit operations. Because of the completely eliminated hyperfine interactions in graphene, there is great potential for electron spin qubits in a nuclear-spin-free quantum world (Trauzettel et al., 2007). It is highly desirable to propose an efficient architecture made with graphene nanostructures to implement the quantum information processing (QIP) (Pedersen et al., 2008).

Owing to the special band structure of graphene (Castro Neto et al., 2009), its low-energy quasi-particles behave as Dirac fermions, and the Klein tunneling and Chiral effect make it non-trivial to form well controllable quantum dots in graphene. There are several ways in which one could localize electrons (holes) in graphene; by using suitable transverse states in Graphene nanoribbon (GNR) (Trauzettel et al., 2007; Silvestrov et al., 2007), by electrical confinement in bilayer graphene (Peeters et al., 2007), or by using the topological structure

(Wang et al., 2007). In the chapter, we propose two alternative approaches that the localized states can exist in the zigzag region of a GNR with a sequence of Z-shaped structures (Guo et al., 2009) or substrate modulated graphene quantum dot (Ma et al., 2009). The localized electron (hole) spin states can be used, as the physical qubit. For the GNR quantum dot chain, the interaction between qubits is found to be of the always-on Heisenberg form. Moreover, for a practical quantum computer to operate, it is essential to properly tailor the disturbing environment. An important technique for doing this is the use of quantum bang-bang (BB) control strategy and the decoherence-free subspaces (DFS) encoding method, both of which are traditionally discussed in the context of atomic, molecular and optical setup (Morton et al., 2006; Kwiat et al., 2000; Kielpinski et al., 2001; Viola et al., 2001; Zhang et al., 2004). In this chapter, these ideas will be introduced to construct an effective quantum information circuit in new graphene nanostructure below.

2. Electron localization in graphene quantum dots

2.1 Quantum dots on a graphene nanoribbon

It is difficult to form a conventional-type quantum dot inside an infinite graphene because of Klein tunneling which wound induce charge transmission through the interface of p-n junctions. Trauzettel, Bulaev, Loss and Burkard firstly introduced a method to overcome such difficulties and form spin qubits in quantum dots based on GNR with armchair boundaries in 2007 (Trauzettel et al., 2007). For the semiconducting armchair boundary conditions, both sublattices of graphene's hexagonal structure will be terminated equally on both side which result in the emergence of a gap and destruction of the valley degeneracy. The charge carriers can be confined on the quantum dot regions between two barrier regions in which electric potential can be tuning by applying a appropriate local gate voltage. Since all the bound states are non-degenerate in valley space, the spin qubits are proposed in this graphene nanostructure. The two spin qubits are coupled via an exchange coupling and the exchange coupling is controlled by the tunnel barrier between the dots. In combination with single spin operations, universal quantum gates can be achieved. The interest idea of the proposal is non-local electron spins in any two of dots can be coupled with the others being decoupled by detuning. Therefore long-distance quantum gates in graphene quantum dots are feasible.

2.2 Substrate modulated graphene quantum dot

In this section, we concentrate on a new method to use gapped graphene as barrier to confine electrons in gapless graphene and form a good quantum dot, which can be realized on an oxygen-terminated SiO_2 substrate partly H-passivated. Further, this method can be upgraded to form two-dimensional quantum dot arrays. We systematically investigate two-dimensional system and find that the coupling strength between neighboring dots can be uniquely anisotropic. The ability to achieve more complex and scalable pattern in our proposal suffice to design a large-scale quantum computer in principle.

Recently, it was discussed that the electronic energy spectrum of the monolayer graphene depended strongly on the surface characteristic of the substrate (Zhou et al., 2007; Shemella et al., 2009). For example, if a single layer graphene is deposited onto a SiO_2 surface, a finite energy gap will open between conduction and valence bands for an oxygen-terminated surface, but close when the oxygen atoms on the substrate are passivated with hydrogen atoms. The confinement can be achieved by a gapless nanoscale graphene regions connecting with gapped regions, which serve as barriers, as shown in Fig 1. The devices are realizable in experiment. In an oxygen terminated SiO_2 substrate, we use protective stuff to cover the

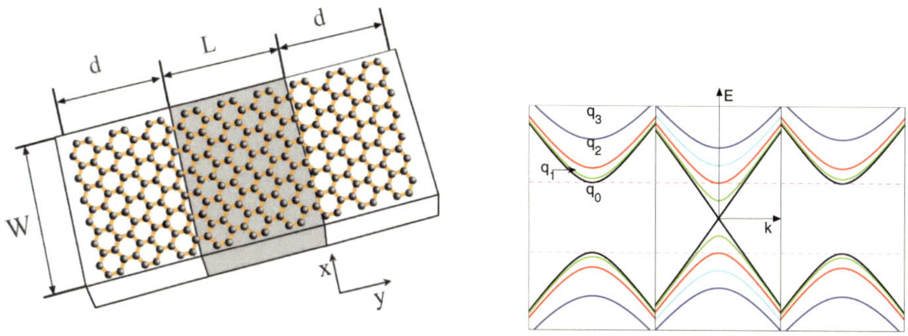

(a) (b)

Fig. 1. (a) Schematic of a single quantum dot. A GNR is deposited on a SiO_2 substrate. The dark part is the dot region, which is fully hydrogen-passivated and gapless. The light part is the barrier region, which is non hydrogen-passivated or slightly hydrogen-passivated and gapped. (b) Energy bands of this system.

barrier regions and the dot region is fully exposed to hydrogen atoms atmosphere. Then a single layer graphene is deposited on this substrate and the bound states exit in the hydrogen-passivated regions. Compared to the approach of forming quantum dots on GNR with semiconducting armchair boundary conditions (Trauzettel et al., 2007), the realization of quantum dot will not depend much on the boundary conditions. Since ferromagnetic insulators deposited on graphene can induce ferromagnetic correlations in graphene, we consider adding ferromagnetic insulator such as EuO upon the two gapped graphene barriers. The induced exchange interaction is estimated to achieve 5 meV by using EuO (Haugen et al., 2008).

The electron waves in graphene system are usually described by four component spinor envelop wavefunction $\Psi = (\psi_A^{(K)}, \psi_B^{(K)}, -\psi_A^{(K')}, -\psi_B^{(K')})$. The behaviors of low energy electron can be described by 4×4 Dirac equation for massless or massive particles, which can be written as (Recher et al., 2009): in the dot region (where $0 \le y \le L$),

$$-i\hbar v_F \begin{pmatrix} \sigma_x\partial_x + \sigma_y\partial_y & 0 \\ 0 & -\sigma_x\partial_x + \sigma_y\partial_y \end{pmatrix} \Psi = E\Psi, \tag{1}$$

and in the barrier region (where $y < 0$ or $y > L$),

$$-i\hbar v_F \begin{pmatrix} \sigma_x\partial_x + \sigma_y\partial_y & 0 \\ 0 & -\sigma_x\partial_x + \sigma_y\partial_y \end{pmatrix} \Psi + \Delta \begin{pmatrix} \sigma_z & 0 \\ 0 & \sigma_z \end{pmatrix} \Psi - \eta V_\sigma \Psi = E\Psi, \tag{2}$$

where $\hbar$ is the Planck constant, $v_F \approx 10^6$ m/s is the Fermi velocity, $\sigma_x, \sigma_y, \sigma_z$ are Pauli matrices acting on two-spinor states related to the two triangular sublattices of graphene, $\eta = \pm 1$ stands for the two spin indexes (spin up and spin down). 2Δ is the gap induced by the substrate, $2V_\sigma$ is the spin splitting energy due to the correlation with ferromagnetic contacts. We consider metallic armchair shaped GNR with the quantized transverse momentum $q_n = n\frac{\pi}{W}$ and the wave vectors in the y direction should satisfy different conditions as

$$E = \pm\sqrt{(\hbar v_F q_n)^2 + (\hbar v_F k)^2}, \tag{3}$$

in the dot and

$$E = \pm\sqrt{(\hbar v_F q_n)^2 + (\hbar v_F k')^2 + \Delta^2} - \eta V_\sigma, \tag{4}$$

in the barriers, where k is the wave vector in the dot and k' in the ferromagnetic barrier with $\pm$ signs referring to conduction band $(+)$ and valence band $(-)$ respectively. The bound state requires that k' is a pure imaginary, which means the bound state energy should satisfy

$$|E| \geqslant \hbar v_F |q_n|, \ |E + \eta V_\sigma| < \sqrt{(\hbar v_F q_n)^2 + \Delta^2}. \tag{5}$$

The energy levels of the bound states can be obtained by matching the wavefunctions at $y = 0$ and $y = L$. We use $1/L$ as the unit of q_n and the characteristic energy $\hbar v_F/L$ as the energy unit. In Fig. 2, we show the energy spectrum as a function of the substrate induced interaction Δ for different transverse momentums (q_n) where V_σ is assumed to be 5 meV (Haugen et al., 2008). As shown in Fig. 2, when Δ increases, the number of bound states is increasing at the same time, which can be deduced from Eq. 5.

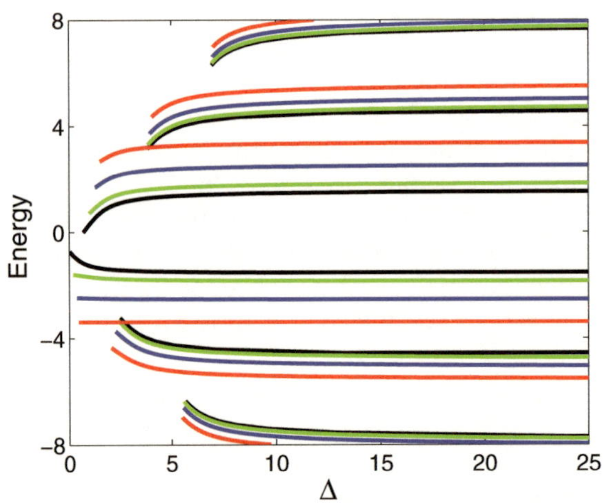

Fig. 2. Bound-state energy levels of a substrate modulated graphene quantum dot with $V_\sigma = 5meV$ for $q_n = 0, 1, 2, 3$ and $\eta = 1$, black: $q_n = 0$, green: $q_n = 1$, blue: $q_n = 2$, red: $q_n = 3$. Both axe labels are in the characteristic energy unit of $\hbar v_F/L$. The length L and width W of dot are assumed to be 100 nm and 300 nm respectively.

Our approach can be easily developed to more complex and scalable pattern. Here we study a two-dimensional quantum dot array as shown in Fig. 3a. Instead of forming transverse modes of nanoribbon, the wavefunctions must match between dot and barrier along both x and y directions. The energy spectrum of this system has been plotted in Fig. 3b. The neighboring dots are coupled by the exchange coupling J. We obtain J by calculating the exchange coupling $J = 4t^2/U$ according to Hubbard approximation. Here U is onsite Coulomb energy and $t = \varepsilon \int \varphi_1^\dagger(\vec{r_1})\varphi_2(\vec{r_2})$ is tunnelling matrix element between two neighboring quantum dots, where ε is the single-particle bound state energy and $\varphi_1(\vec{r})$ and $\varphi_2(\vec{r})$ are the wavefunctions of two neighboring or next-nearest neighboring dots. We estimate $U \approx 10$ meV for the dot size $L \approx 30$ nm and the characteristic energy unit of this system $\hbar v_F/L$ is about 22 meV. Fig. 4 shows the nearest coupling strength (J_1) and the next-nearest coupling strength (J_2) of the ground state versus the opened gap Δ in the barrier region (when $d = 3L$), from which we

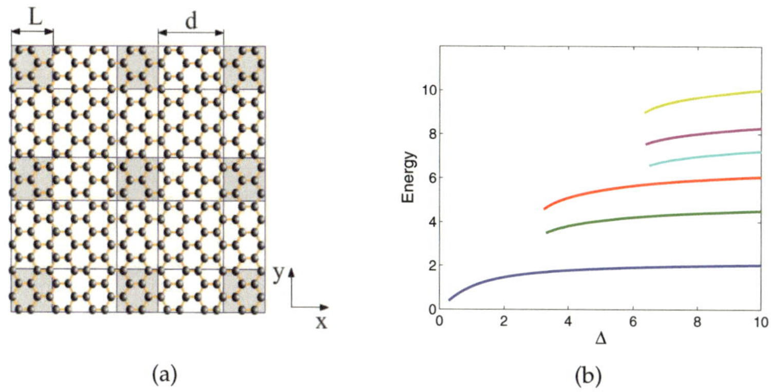

Fig. 3. (a) Schematic drawing of two-dimensional quantum dot arrays. The dot regions (dark regions) are defined by the gapped barrier regions (light regions). (b) The bound-state energy levels of this two-dimensional system with $L = 30$ nm and $d = 3L$. Both axe labels are in the characteristic energy unit of $\hbar v_F / L$.

find that the nearest coupling strength decreases sharply when the opened gap increases. We also find that the nearest coupling both along the x direction and y direction are the same for the ground state, and coupling strengths between the nearest and next-nearest neighboring dots are anisotropic for excited states.

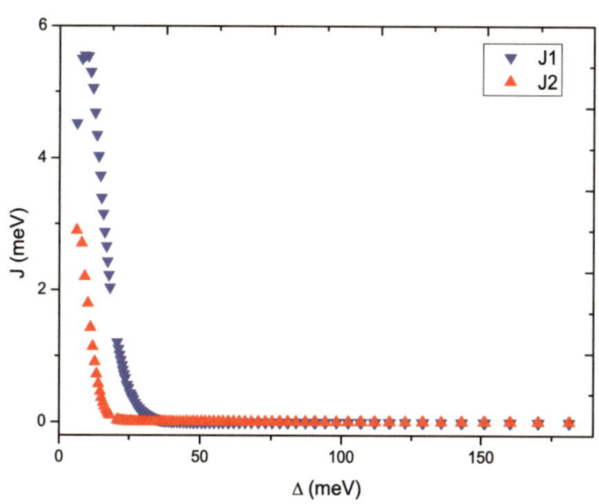

Fig. 4. Coupling strength of the nearest (J_1) and next-nearest (J_2) dots as a function of the opened gap Δ in the barrier region for the ground state.

2.3 Graphene nanoribbon quantum dot chain

In this section, we introduce a graphene quantum system in which the proposed GNR consists of an array of Z-shaped structures and each Z-shaped structure includes a central region with zigzag edge connecting to two regions with armchair edge, as presented in Fig. 5d.

Using the π orbital tight-binding(TB) approximation, the density of states and the spectrum of the zigzag region in this GNR system can be obtained by the direct diagonalization with periodic boundary conditions. We find that there are several localized states with electron-hole symmetry around the zero energy point, as shown in Fig.5a. Considering higher-order hopping terms, we calculated the DOS and the spectrum using a third nearest-neighbor TB model instead of nearest-neighbor TB model with the second and third neighbor interaction energies $\gamma_1 = -0.12$eV and $\gamma_2 = -0.068$eV (Reich et al., 2002; Son et al., 2006). As shown in Fig. 5a and Fig. 6, higher-order hopping terms destroy the electron-hole symmetry, but don't destroy the confined states in each zigzag region. Hence we can choose to get one localized electron or hole in the zigzag region by adjusting the Fermi level through the individual top gates. Furthermore, in the calculations we find that the spectrum of GNR depends very much on the nature of their geometry. There are no localized states in the zigzag regions when the width of this GNR $N = 3m - 1$ or $N = 2m$ (unit cells), as shown in Fig. 6. Here m is an integer.
We now discuss two coupled Z-shaped quantum dots, which are connected with armchair GNR. Fig. 5b shows the spatial distribution of local probability density of a typical GNR with two Z-shaped structures, and $N = 7$, $L = 4$, $D = 6$ for $E_0 = \pm 0.3$ eV discrete states. As shown in Fig. 5c, each zigzag region (quantum dot) confines one electron and these two electrons are coupled by the exchange interaction J_1. J_1 can be obtained by calculating the exchange interaction $J_1 = 4t^2/U$. Obviously, the exchange coupling J_1 is determined by the geometry of the nanoribbon. For each N and L, J_1 depends on the distance between two neighboring dots D (unit cells), as shown in Fig. 7.
The spin of the localized charge carrier is used as the physical qubit and the GNR with a sequence of Z-shaped structures forms an one-dimensional spin qubit chain as shown in Fig. 5d. In this chapter, we neglect the magnetic effect of GNR edge (Son et al., 2009). The neighboring qubits in this chain have an always-on Heisenberg interaction $H = J_1 \vec{S_1} \cdot \vec{S_2}$. Here $\vec{S_1}$ and $\vec{S_2}$ are the spin operators of the neighboring localized charge carrier. For a sequence of Z-shaped structure GNR with parameters as $N = 7$, $L = 4$, $D = 18$, the Hamiltonian of the system can be expressed as

$$H_I = \sum_{\langle i,j \rangle} J_{i,j}(\sigma_i^x \otimes \sigma_j^x + \sigma_i^y \otimes \sigma_j^y + \sigma_i^z \otimes \sigma_j^z), \tag{6}$$

where $\sigma_{i,j}^{x,y,z}$ are the spin Pauli operators of the localized charge carrier in the quantum dots, $\langle i, j \rangle$ represent two nearest neighboring dots.

3. Quantum-noise control in graphene nanoribbon quantum dot chain

The main challenges for solid-state QIP are achieving the high accuracy of gate operation and tailoring the disturbing environment. In our one-dimensional spin qubit chain, the quantum information is disturbed by the charge noise, the nuclear spins in substrate and inherent spin-spin interaction. Therefore achieving noise control is indispensable. A variety of strategies have been devised to meet this challenge, no single method can suppress the complex noise and decoherence. Rather, constructing a reliable QIP scheme depends crucially on the errors that happen. First, to avoid the spin qubits to entangle with the environment, we can apply a BB operation $U_z = exp(-i\sigma_z \pi/2)$ to each quantum dot region. Such rotation operations can be implemented through the electrically driven single-electron spin resonance by localized a.c. electric field pulses if ferromagnetic strips are integrated on top of the graphene quantum dots, which has been successfully realized in GaAs/AlGaAs quantum dot

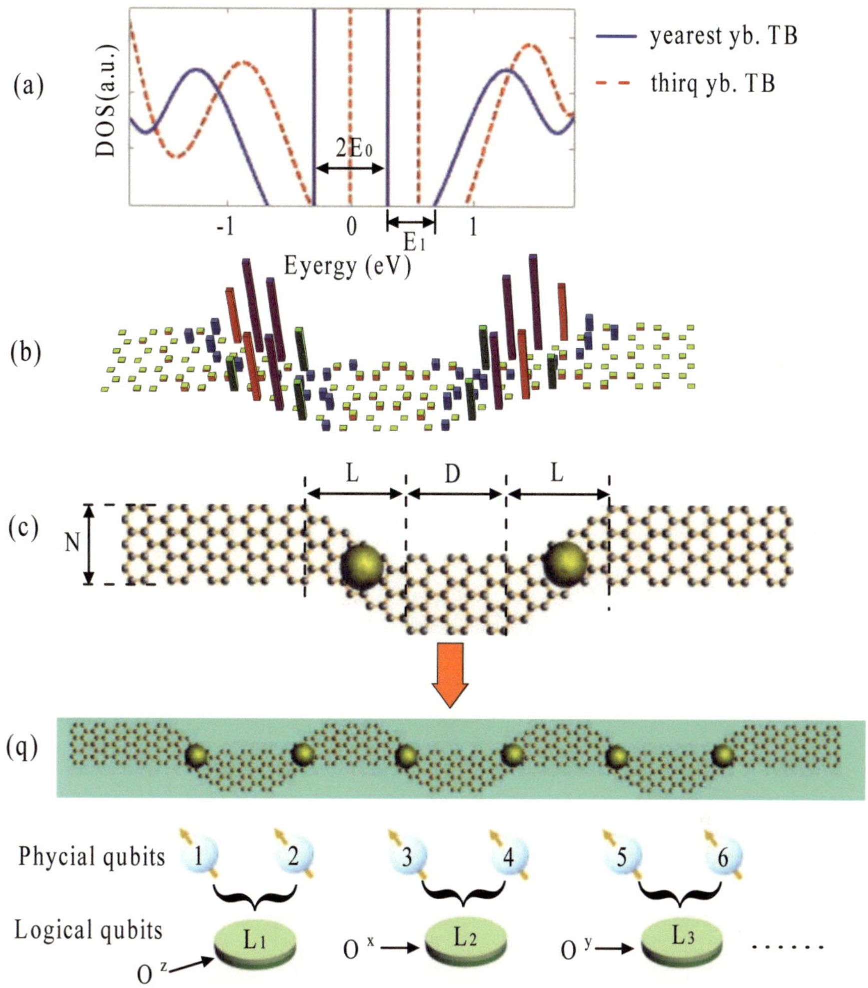

Fig. 5. Schematic illustration of the proposed architecture of GNR quantum dot chain. (a) The density of states of the GNR are calculated by nearest neighbor TB approximation and third nearest-neighbor TB approximation with the second and third neighbor interaction energies $\gamma_1 = -0.12$eV and $\gamma_2 = -0.068$eV. (b) The spatial distribution of local probability density of GNR with two coupled quantum dots, and $N = 7$, $L = 4$, $D = 6$ for ground states. (c) Two coupled graphene quantum dots in which each dot is filled with a single electron. The physical qubit is encoded into the spin of the confined electron. (d) The proposed periodic architecture with three logical qubits as a unit for quantum computation. Physical qubits 1 and 2 form logical qubit L_1; physical qubits 3 and 4 form logical qubit L_2; physical qubits 5 and 6 form logical qubit L_3. The G^z , G^x, G^y are the BB operation sets of L_1, L_2 and L_3 respectively. The cyan ribbon indicates the micromagnet integrated on top of the GNR structure to apply an slanting magnetic field. The GNR and micromagnet are isolated by an insulating layer. Each zigzag region has a nearby gate. The nearby gates are not outlined for clarity.

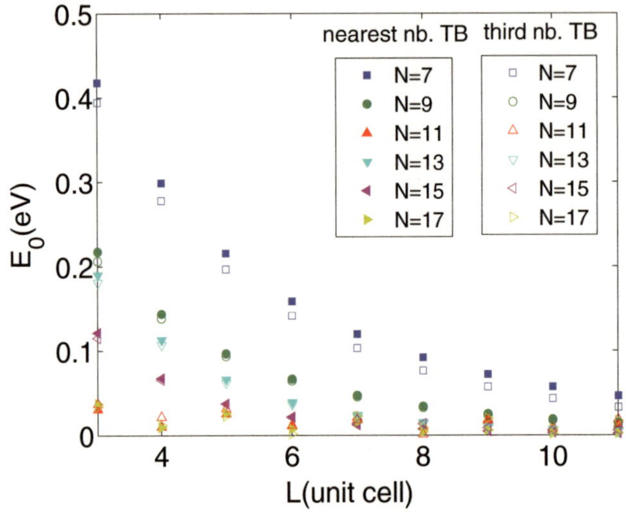

Fig. 6. The ground-state energy E_0 of the quantum dot in the GNR chain versus length of quantum dot region L (unit cells) for different width of the nanoribbon N (unit cells).

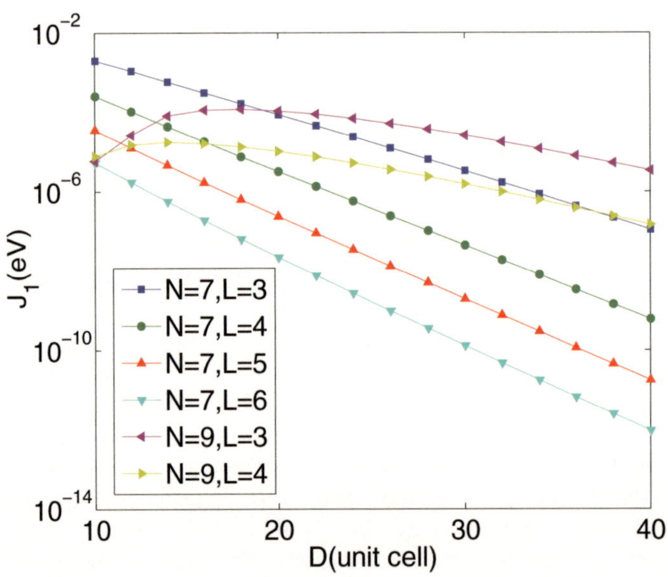

Fig. 7. The coupling energy J_1 of two nearest neighboring dots in GNR quantum dot chain versus the distance between two dots D (unit cells) for different size of quantum dot region.

experiment recently (Pioro-Ladrière et al., 2008). Next, to counteract the phase decoherence, we can use DFS encoding (Duan et al., 1997; Lidar et al., 1998; Benjamin et al., 2003). For a simply DFS encoding, two physical qubits can encode a logical qubit:

$$|0\rangle_L = |\uparrow_1\downarrow_2\rangle, |1\rangle_L = |\downarrow_1\uparrow_2\rangle. \tag{7}$$

As shown in Fig. 5c, localized electron spins in the two neighboring zigzag regions can be used to form a logical qubit.

Furthermore, a special encoding method and a nonsynchronous BB pulse operations are exploited to overcome the untunable spin-spin interactions between two neighboring physical qubits (Zhang et al., 2004). We proposed a GNR quantum dot chain architecture, which forms a periodic structure $L_1L_2L_3L_1L_2L_3\cdots$ with three logical qubits as a unit, as shown in Fig. 5d. L_1 represents a logical qubit encoded as Eq. (7). L_2 is a logical qubit encoded as

$$|0\rangle_{L2} = \frac{1}{2}(|\uparrow\rangle_3 + |\downarrow\rangle_3)(|\uparrow\rangle_4 - |\downarrow\rangle_4), \tag{8}$$

$$|1\rangle_{L2} = \frac{1}{2}(|\uparrow\rangle_3 - |\downarrow\rangle_3)(|\uparrow\rangle_4 + |\downarrow\rangle_4). \tag{9}$$

And L_3 is a logical qubit encoded as

$$|0\rangle_{L3} = \frac{1}{2}(|\uparrow\rangle_5 + i|\downarrow\rangle_5)(|\uparrow\rangle_6 - i|\downarrow\rangle_6), \tag{10}$$

$$|1\rangle_{L3} = \frac{1}{2}(|\uparrow\rangle_5 - i|\downarrow\rangle_5)(|\uparrow\rangle_6 + i|\downarrow\rangle_6). \tag{11}$$

With this periodic architecture, we need to apply nonsynchronous BB pulse operations respectively to L_1, L_2, L_3 from the operation sets $G^z = \{I, U_z, R_z\}$, $G^x = \{I, U_x, R_x\}$, $G^y = \{I, U_y, R_y\}$, where $U_z = -\sigma_1^z \otimes \sigma_2^z$, $R_z = -iI_1^z \otimes \sigma_2^z$, $U_x = -\sigma_1^x \otimes \sigma_2^x$, $R_x = -iI_1 \otimes \sigma_2^x$, $U_y = -\sigma_1^y \otimes \sigma_2^y$, and $R_y = -iI_1 \otimes \sigma_2^y$. Thus we obtain a quantum computation system with entirely decoupled logical qubits.

4. Universal quantum gates in GNR quantum dot chain

In this section, we discuss the scheme to perform universal set of quantum gates on encoded qubits. Since arbitrary single-qubit rotations can be constructed from the two elementary logic operations $\bar{X}$ and $\bar{Z}$, we show how to implement the two gate operations in the GNR quantum dot chain. For logical qubit L_1, $\bar{X} = \frac{1}{2}(\sigma_1^x \otimes \sigma_2^x + \sigma_1^y \otimes \sigma_2^y)$, $\bar{Z} = \frac{1}{2}(\sigma_1^z - \sigma_2^z)$. $\bar{X}$ gate can be easily achieved by adjusting the BB pulses of both qubits 1 and 2 to be synchronous. The operation time is $\Delta t = \hbar\pi/4J = 0.2$ ns, for $N = 7$, $L = 4$, $D = 18$. $\bar{Z}$ gate can be achieved by localized pluses on the two physical qubits respectively. The operation time of $\bar{Z}$ gate can be nanosecond scale when a slanting magnetic field with large field gradient is applied onto each quantum dot region (Pioro-Ladrière et al., 2008). The fidelity of the $\bar{X}$ gate is limited by fluctuations in the qubit splitting J caused by charge noise, such as $1/f$ noise. The accuracy of the $\bar{Z}$ gate is dominated by spin dephasing due to the nuclear field fluctuations. The fidelity of the $\bar{Z}$ gate can be very high due to the small nuclear field in graphene system.

Combined with the arbitrary single-qubit rotation, two-qubit CNOT gate on any two logical qubits is required to complete our universal set of quantum gates. For example, we construct CNOT gate between two neighboring logical qubits, L_1 and L_2. It is shown that CNOT gate can be implemented by W gate $W = e^{i\theta\bar{Z}\otimes\bar{Z}}$ conjugating Hadamard operation. It has been

known that W gate is equivalent to a controlled rotation about the z axis: $W = e^{i\theta\bar{Z}\otimes\bar{Z}} = |0\rangle\langle 0| \otimes I + |1\rangle\langle 1| \otimes e^{2i\theta\bar{Z}}$ (Bremner et al., 2002). By performing Hadamard transformation to the two physical qubits of the second logical qubit L_2 and changing the BB control pulse to be the same with L_1, we can recouple the two neighboring logical qubits and implement W gate of logical qubits of L_1 and L_2.
The spin decoherence time of graphene quantum dot has been predicted to be more than 10 μs in the nature carbon material (Trauzettel et al., 2007; Fischer et al., 2009). This decoherence time is 4 orders longer than the gate operation time of the present scheme and the gate error might meet the required threshold in principle. This combined DFS and BB control method is a useful approach to offer the possibility for coherent controlling spin qubits on graphene.

5. Conclusion

To conclude, we have discussed the potential to implement spin-based quantum computation on graphene nanostructures. Several approaches have been introduced to achieved quantum confinement of charge carriers. To overcome the dependence on the boundary conditions of GNR, we proposed a method to form quantum dots in substrate modulated graphene. We presented the theoretical proposals for forming logical qubit encoding in a DFS subspace and achieving noise control by BB control strategy in a GNR quantum dot chain with always-on Heisenberg interaction. Furthermore, universal set of quantum gates on encoded qubits has been achieved by a sequence of pulse control. Recently, the experimental breakthroughs in few electrons or holes graphene quantum dots (Neubeck et al., 2010) open an avenue for realization of spin qubit in graphene nanostructure.

6. Acknowledgments

This work was supported by the National Science Foundation (ECS-0601478), the National Basic Research Program of China (Grants No. 2011CBA00200), and the National Natural Science Foundation of China (Grants 10804104, 10874163, and 10934006).

7. References

Loss D. & DiVincenzo D. P. (1998). Quantum computation with quantum dots, *Physical Review A*, Vol. 57, Issue 1, pp. 120-126.

Elzerman J. M.; Hanson R.; Willems Van Beveren L. H.; Witkamp B.; Vandersypen L. M. & Kouwenhoven L. P. (2004). Single-shot read-out of an individual electron spin in a quantum dot, *Nature*, Vol. 430, No. 6998, pp. 431-435.

Petta J. R.; Johnson A. C.; Taylor J. M.; Laird E. A.; Yacoby A.; Lukin M. D.; Marcus C. M.; Hanson M. P. & Gossard A. C. (2005) Coherent manipulation of coupled electron spins in semiconductor quantum dots, *Science*, Vol. 309, No. 5744, pp. 2180-2184.

Koppens F. H.; Buizert C.; Tielrooij K. J.; Vink I. T.; Nowack K. C.; Meunier T.; Kouwenhoven L. P. & Vandersypen L. M. (2006) Driven coherent oscillations of a single electron spin in a quantum dot, *Nature*, Vol. 442, No. 7104, pp. 766-771.

Nowack K. C.; Koppens F. H. L.; Nazarov Yu. V. & Vandersypen L. M. K. (2007) Coherent control of a single electron spin with electric fields , *Science*, Vol. 318, No. 5855, pp. 1430-1433.

Pioro-Ladrière M.; Obata T.; Tokura Y.; Shin Y. S.; Kubo T.; Yoshida K.; Taniyama T. & Tarucha S. (2008) Electrically driven single-electron spin resonance in a slanting Zeeman field, *Nature Physics*, Vol. 4, Issue 10, pp. 776-779.

Trauzettel B.; Bulaev D. V.; Loss D. & Burkard G. (2007) Spin qubits in graphene quantum dots, *Nature Physics*, Vol. 3, Issue 3, pp. 192-196.

Pedersen T. G.; Flindt C.; Pedersen J.; Mortensen N. A.; Jauho A. P. & Pedersen K. (2008) Graphene antidot lattices: designed defects and spin qubits, *Physical Review Letters*, Vol. 100, Issue 13, pp. 136804.

Castro Neto A. H.; Guinea F.; Peres N. M. R.; Novoselov K. S. & Geim A. K. (2009) The electronic properties of graphene, *Reviews of Modern Physics*, Vol. 81, Issue 1, pp. 109-162.

Silvestrov P. G. & Efetov K. B. (2007) Quantum dots in graphene, *Physical Review Letters*, Vol. 98, Issue 1, pp. 016802.

Pereira J. M. Jr.; Vasilopoulos P.; & Peeters F. M. (2007) Tunable quantum dots in bilayer graphene, *Nano Lett.*, Vol. 7, Issue 4, pp. 946-949.

Wang Z. F.; Shi Q. W.; Li Q. X.; Wang X. P.; Hou J. G.; Zheng H. X.; Yao Y. & Chen J. (2007) *Applied Physics Letters*, Vol. 91, Issue 5, pp. 053109.

Guo G. P.; Lin Z. R.; Tu T.; Cao G.; Li X. P. & Guo G. C. (2009) Quantum computation with graphene nanoribbon, *New Journal of Physics*, Vol. 11, pp. 123005.

Ma Q.; Lin Z. R.; Tu T.; Guo G. C. & Guo G. P.(2009) Substrate modulated graphene quantum dot, *arxiv*, 0910.4708.

Morton J. J. L.; Tyryshkin A. M.; Ardavan A.; Benjamin S. C.; Porfyrakis K.; Lyon S. A. & Briggs G. A. D. (2006) Bang-bang control of fullerene qubits using ultrafast phase gates, *Nature Physics*, Vol. 2, Issue 1, pp. 40-43.

Kwiat P. G.; Berglund A. J.; Altepeter J. B. & White A. G. (2000) Experimental Verification of Decoherence-Free Subspaces, *Science*, Vol. 290, No. 5491, pp. 498-501.

Kielpinski D.; Meyer V.; Rowe M. A.; Sackett C. A.; Itano W. M.; Monroe C. & Wineland D. J. (2001) A Decoherence-Free Quantum Memory Using Trapped Ions, *Science*, Vol. 291, No. 5506, pp. 1013-1015.

Viola L.; Fortunato E. M.; Pravia M. A.; Knill E.; Laflamme R. & Cory D. G. (2001) Experimental realization of noiseless subsystems for quantum information processing, *Science*, Vol. 293, No. 5537, pp. 2059-2063.

Zhang Y.; Zhou Z. W.; Yu B. & Guo G. C. (2004) Concatenating dynamical decoupling with decoherence-free subspaces for quantum computation, *Physical Review A*, Vol. 69, Issue 4, pp. 042315.

Zhou S. Y.; Gweon G.-H.; Fedorov A. V.; First P. N.; de Heer W. A.; Lee D.-H.; Guinea F.; Castro Neto A. H. & Lanzara A. (2007) Substrate-induced bandgap opening in epitaxial graphene, *Nature Materials*, Vol. 6, Issue 10, pp. 770-775.

Shemella P. & Nayak S. K. (2009) Electronic structure and band-gap modulation of graphene via substrate surface chemistry, *Applied Physics Letters*, Vol. 94, Issue 3, pp. 032101.

Recher P.; Nilsson J.; Burkard G. & Trauzettel B. (2009) Bound states and magnetic field induced valley splitting in gate-tunable graphene quantum dots *Physics Reiview B*, Vol. 79, Issue 8, pp. 085407.

Haugen H.; Huertas-Hernando D. & Brataas A. (2008) Spin transport in proximity-induced ferromagnetic graphene, *Physics Reiview B*, Vol. 77, Issue 11, pp. 115406.

Reich S.; Maultzsch J. & Thomsen C. (2002) Tight-binding description of graphene, *Physics Reiview B*, Vol. 66, Issue 3, pp. 035412.

Son Y. W.; Cohen M. L. & Louie S. G. (2006) Energy gaps in graphene nanoribbons, *Physics Reiview Letters*, Vol. 97, Issue 21, pp. 216803.

Muñoz-Rojas F.; Fernández-Rossier J. & Palacios J. J. (2009) Giant magnetoresistance in ultrasmall graphene based devices, *Physics Reiview Letters*, Vol. 102, Issue 13, pp. 136810.

Duan L. M. & Guo G. C. (1997) Preserving coherence in quantum computation by pairing quantum bits, *Physics Reiview Letters*, Vol. 79, Issue 10, pp. 1953-1956.

Lidar D. A.; Chuang I. L. & Whaley K. B. (1998) Decoherence-free subspaces for quantum computation, *Physics Reiview Letters*, Vol. 81, Issue 12, pp. 2594-2597.

Benjamin S. C. & Bose S. (2003) Quantum computing with an always-On Heisenberg interaction, *Physics Reiview Letters*, Vol. 90, Issue 24, pp. 247901.

Bremner M. J.; Dawson C. M.; Dodd J. L.; Gilchrist A.; Harrow A. W.; Mortimer D.; Nielsen M. A. & Osborne T. J. (2002) Practical scheme for quantum computation with any two-Qubit entangling gate, *Physics Reiview Letters*, Vol. 89, Issue 24, pp. 247902.

Fischer J.; Trauzettel B. & Loss D. (2009) Hyperfine interaction and electron-spin decoherence in graphene and carbon nanotube quantum dots, *Physics Reiview B*, Vol. 80, Issue 15, pp. 155401.

Neubeck, S.; Ponomarenko, L. A.; Freitag, F.; Giesbers, A. J. M.; Zeitler, U.; Morozov, S. V.; Blake, P.; Geim, A. K. & Novoselov, K. S. (2010). From One Electron to One Hole: Quasiparticle Counting in Graphene Quantum Dots Determined by Electrochemical and Plasma Etching, *Small*, Vol. 6, Issue 14, pp. 1469-1473.

16

Graphene Nanoribbons: Geometric, Electronic, and Magnetic Properties

Erjun Kan, Zhenyu Li and Jinlong Yang
University of Science and Technology of China, China

1. Introduction

Graphene nanoribbons (GNRs) have one-dimensional structures with hexagonal two-dimensional carbon lattices, which are stripes of graphene. Their structures and their electronic and magnetic properties have been intensively studied both experimentally and theoretically. Due to their various edge structures, GNRs present different electronic properties ranging from normal semiconductors to spin-polarized half metals, which opens the possibility of GNRs as electric devices.

In this chapter, the geometric, electronic, and magnetic properties of GNRs are discussed. First, the electronic and magnetic properties of pristine GNRs are understood with their special structures. We emphasize the importance of one-dimensional quantum confinement effect and edge states. Secondly, since GNRs have large surface-volume ratio and special edge states, their properties can be modified by many methods, such as doping and adsorption. The electronic property and its response to modulation are described in detail. Finally, the experimental realizations of GNRs are introduced, which provide substantial bases to theoretical prediction of GNRs' electronic and magnetic properties. Possible future research directions are also discussed.

2. Magnetic and electronic properties of pristine GNRs

After the successful isolation of graphene, its amazing properties make it become a rising star of current materials research. However, as we know, graphene is a zero-gap semiconductor. To extend the real applications, an energy gap is needed, which enables the basic electric logic states: on and off. Besides, carbon-based magnetic materials are very important, which have small spin scattering and large potential to be immigrated into future electric devices. Due to the modern technology, etching or patterning graphene in some special direction has been realized. When graphene is etched or patterned along one specific direction, a novel quasi one-dimensional (1D) structure is obtained, which is a strip of graphene, referred as graphene nanoribbon (GNR). There are some critical questions about such GNRs: How do they organize the structures? Do they have energy gaps? And is there any magnetic state in these GNRs?

2.1 Geometric structures of pristine GNRs

The typical width of GNR is of nanometers. Different with 2D graphene, the termination in one direction introduces important quantum confinement effect (QCE), which endows GNR various attractive properties. Depends on the termination style, normally, nonchiral GNR can be divided into two kinds: Armchair and Zigzag (Fig. 1 shows the structures of Armchair and Zigzag GNRs). Adopting the standard convention, the width of armchair GNRs is classified by the number of dimer lines (N_a) across the ribbons. Likewise, the one of zigzag GNRs is classified by the number of zigzag chains (N_z) across the ribbons. Perpendicular to the direction of defined width, GNRs repeat their geometric structures, and form one-dimensional periodic structures.

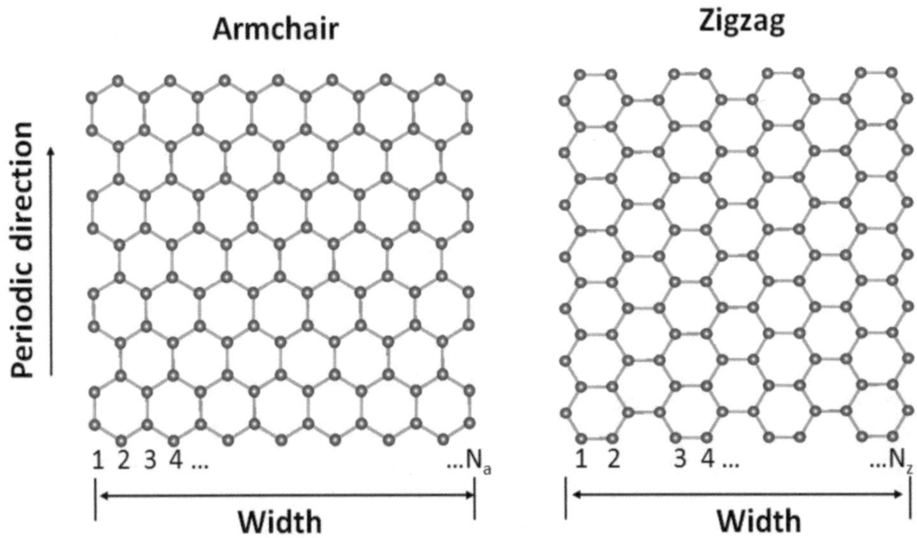

Fig. 1. Structure of armchair and zigzag nanoribbons

Since GNRs are stripes of graphene, edge atoms are not saturated. Active edge states become an important factor to determine the edge structures. For armchair GNRs, there is no any edge rescontructions, and the planar patterns are kept. While for zigzag GNRs, it is unexpectedly found that the zigzag edge is metastable, and reconstructions spontaneously take place at high temperature. Some special structures have been proposed as possible reconstructed patterns (Koskinen et al., 2008), however the detailed reconstructions still need to be further studied. To keep the planar structures of zigzag GNRs, hydrogen atoms are introduced to saturate the edge atoms of GNRs. Besides, in most of the reported structures in experiments, GNRs are found no reconstructions. Thus, most theoretical research has taken the saturated GNRs as the starting point. In the following context, edge atoms of GNRs are saturated by hydrogen atoms without special notation.

2.2 Electronic properties of Armchair GNRs

Many theoretical studies have been devoted into investigating the electronic properties of Armchair GNRs, such as tight-binding calculations, density functional theory (DFT) calculations, and many-electron green's function approach within GW approximation. Among those methods, DFT calculations adopt parameter-free self consistent field calculations, and their reliability has been broadly proved in solid state field and nano-scale systems. Thus, most of the theoretical investigations have been carried out with DFT calculations. However, it is well established that DFT calculations underestimate band gaps.

Other methods, such GW approximation, have been adopted to correct DFT calculations, and get the reliable band gaps. In the following discussions, both DFT and GW results are included.

As shown in the Fig. 1, armchair GNRs are defined by the number of dimer lines (N_a) across the ribbons. Their electronic structures have been carefully investigated by DFT calculations (Son et al. 2006a). Their results show that all armchair GNRs are semiconductors with energy gaps, which decrease as a function of increasing ribbon widths. No magnetism has been found in armchair GNRs. As seen from Fig. 2., the energy gaps as a function of ribbon width are well separated into three different kinds: N_a = 3p, 3p+1, 3p+2 (p is integer). Moreover, the gap size hierarchy is well separated. N_a = 3p+1 categories has the largest energy gap, while the N_a = 3p+2 series is the smallest one.

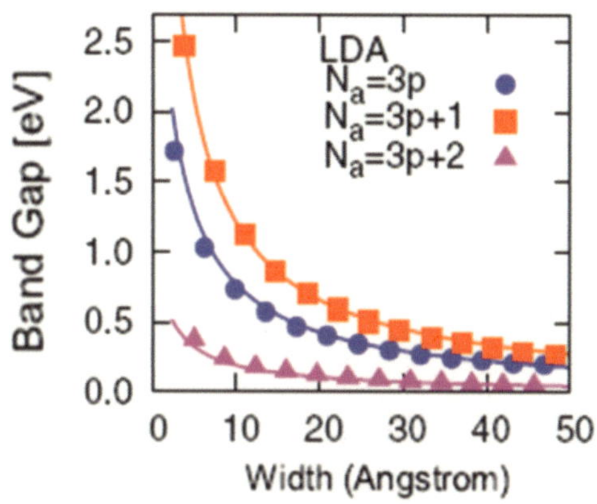

Fig. 2. The variation of band gaps of armchair GNRs with DFT calculations (Son et al. 2006a).

Armchair GNRs show semiconducting behaviours with a direct energy gap. The determining factor comes from the quantum confinement effect (QCE), which can be characterized by [energy gaps] versus $[\text{width}]^{-1}$. Besides the QCE, son et al. has pointed that the edge effects play an important role to force the armchair GNRs to be semiconductors (Son et al. 2006a). As discussed in the previous parts, the edge carbon atoms of armchair GNRs are passivated by hydrogen atoms, which leads to the bonding of carbon atoms at the edges different with other carbon atoms. As a consequence, the bond lengths of carbon atoms at the edges are shorter than that in the middle of ribbons, and open the energy gaps of armchair GNRs.

Although armchair GNRs have three typical families (corresponding to Na = 3p, 3p+1, 3p+2, respectively) with distinguished energy gaps, they have similar band shapes. As an example, the band structure of armchair GNR with N_a = 13 is shown in Fig. 3. (Sun et al. 2008), there are four important subbands which dominate the electronic behaviors of armchair GNRs. These subbands are constituted by the π bonds of carbon atoms, and have different shapes, which provide the possibility of external modulations, such as strain.

On the other hand, although DFT calculations can provide the correct qualitative pictures, the accurate energy gaps of armchair GNRs are needed. To solve the band-gap problems, Li et al. have performed first-principles calculations using many-electron Green's function approach within the GW approximation, which is treated as one of the most accurate methods to predict the energy gaps. It is clear that GW calculations give the same hierarchy

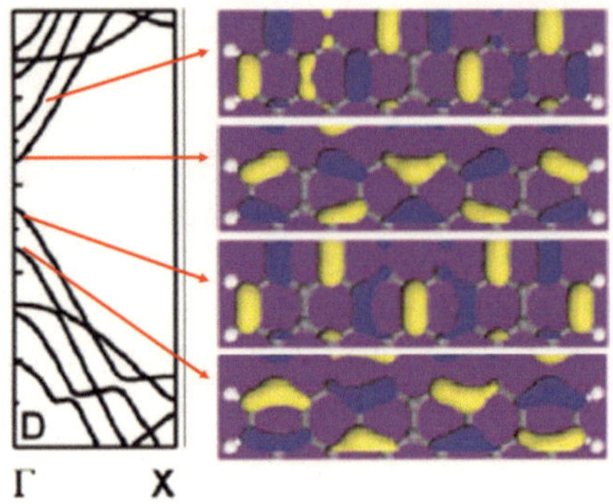

Fig. 3. Band structure and the space charge distribution of armchair GNRs (N_a = 13) (Sun et al. 2008).

as those obtained in DFT calculations. But the GW corrections to the band gaps of armchair GNRs are significant. The corrections are from 0.5 to 3 eV for the GNRs with width from 1.6 to 0.4 nm (Li et al. 2007). Thus, GW calculations show that armchair GNRs can have large energy gaps with limited widths.

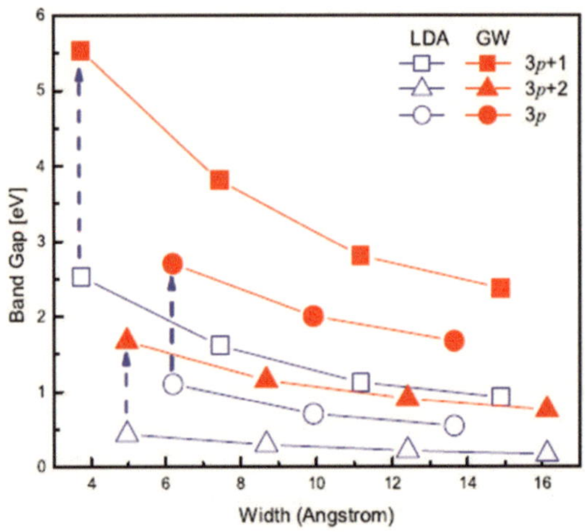

Fig. 4. The variation of band gaps of armchair GNRs with GW calculations (Li et al. 2007).

2.3 Electronic and magnetic properties of zigzag GNRs

For nanoribbons with zigzag shaped edges, without considering spin states, DFT calculations have shown that a set of doubly degenerate flat edge-state bands at Fermi level (E_F) (Son et al. 2006a; Wu et al. 2009), which give rise to a very large density of states (DOS) at E_F. This DOS peak at E_F is half filled, which therefore provides Stoner instability leading to magnetic states.

By inclusion of the spin degrees of freedom within DFT methods, the zigzag GNRs have been predicted to have a magnetic insulating ground state with ferromagnetic ordering at each zigzag edge, and antiparallel spin orientation between the two edges (Son et al. 2006a; Wu et al. 2009). Different with armchair GNRs, the zigzag GNRs have the same hierarchy of band gap-width relationship.

In Fig. 5., the spin-density plot clearly shows that spin moments are mainly distributed at the edge carbon atoms. Compared with the nonspin-polarized solutions, spin-polarized edge states are more favoured, and the total energy difference between these states increases with the width of GNRs. For example, the energy difference is 20 meV per edge carbon atoms for N_z= 8, while becomes 24 meV for N_z = 16 (Son et al. 2006a; Wu et al. 2009). Moreover, the spin-polarized states are further stabilized by ferromagnetic coupling at the edge, while antiferromagnetic coupling between the two edges. However, the energy difference between ferromagnetic and antiferromagnetic coupling is very small, and decreases as width of ribbons increases. When the width is significantly larger than the decay length of the spin-polarized edge states, antiferromagnetic and ferromagnetic states almost have the same stability(Son et al. 2006a; Wu et al. 2009).

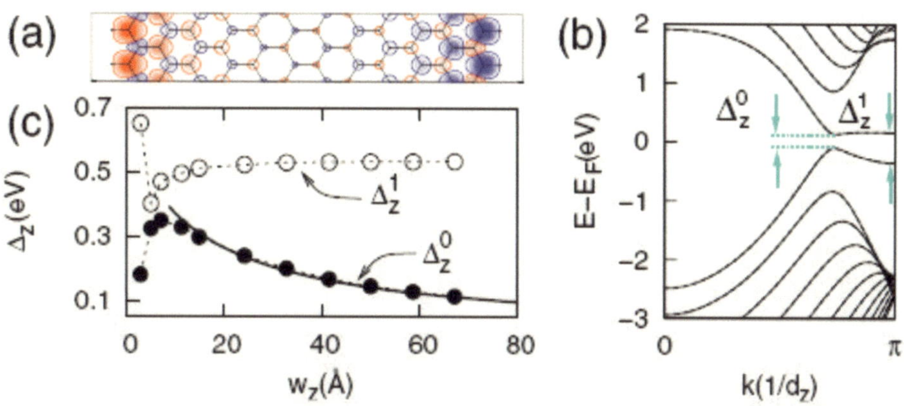

Fig. 5. (a) Spin density for zigzag GNRs with N_z = 12, red and blue mean different spin direction. (b) The band gaps of zigzag GNRs with N_z = 12, spin-up channel is degenerate with spin-down channel in all energy bands. (c) The variation of band gap and energy splitting as function of the width of zigzag GNRs. (Son et al. 2006a).

GW approach has large effect on such localized states in zigzag GNRs. Li et al's results show that the magnitudes of the corrections to the DFT energy gaps in zigzag GNRs are similar to those in AGNRs (Li et al., 2007). As plotted in Fig. 6., the corrections enlarge the band gap by roughly 1 eV for the studied ribbons. On the other hand, the spin-polarized edge states are kept.

Based on the above results, it is clear that DFT method can correctly predict the qualitative properties of zigzag GNRs, namely, spin-polarized edge states and semiconducting with direct energy gaps. However, GW approach and similar methods can provide better quantitative results. In this sense, DFT calculations as a simple method will provide reliable physical pictures in most cases, and satisfy our requirements in studying GNRs.

3. Magnetic and electronic properties of GNRs under external modulations

As discussed above, both armchair and zigzag GNRs have direct band gaps. Compared with graphene, which is a zero-gap semiconductor, the insulating behaviour of GNRs is very important. It is well known that two distinguished states, metallic and insulating states, are necessary to develop reliable electric devices. Thus, for the potential applications based on graphene, GNRs greatly extend the functionality of graphene.

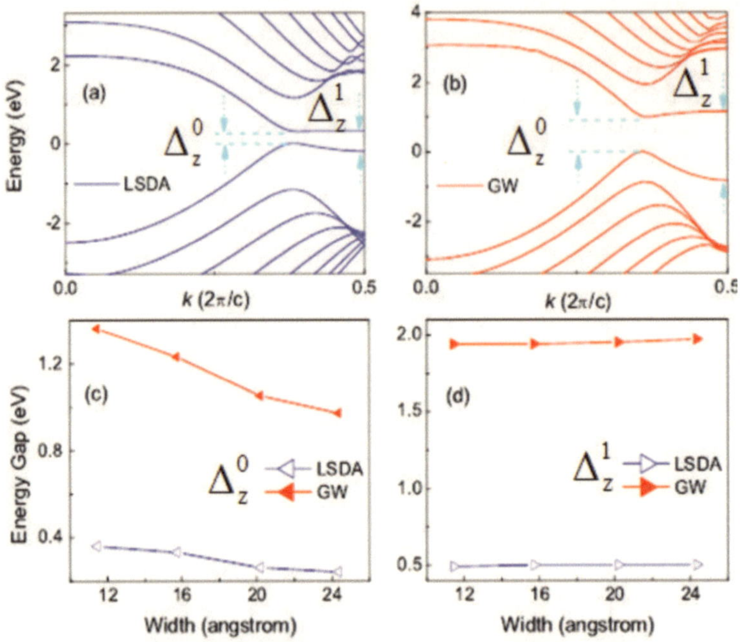

Fig. 6. Calculated band structure of zigzag GNRs with N_z =12 for (a) DFT calculations and (b) GW calculations, respectively. (c) and (d) represent the variation of direct band gap and energy gap at the zone boundary with the width of zigzag GNRs (Li et al. 2007).

On the other hand, because of the existence of edges in GNRs, there are many possibilities of external modulations to modify the electronic and magnetic properties of GNRs, such as applying electric field, strain, and edge modifications. Besides, the special geometric and electronic structures of GNRs provide the potential possibilities of producing graphene-based spintronics. As we know, zigzag GNRs have the spontaneous spin-polarized edge states. Thus, how to manipulate the electronic and magnetic properties of GNRs becomes a very interesting topic.

In the following part, we will discuss the reported external modulations on the armchair and zigzag GNRs. Since the pristine armchair GNRs are non spin-polarized materials, we discuss the reported results of armchair GNRs under external modulations first. While for zigzag GNRs, the situations are more complicated and interesting because of the spin-polarized solutions.

3.1 Armchair GNRs

3.1.1 Applying strain

Armchair GNRs show semiconducting behaviours with a direct energy gap, and are well separated into three categories. To build future nano devices based on GNRs, the capability to control GNRs' electronic properties is highly desirable. One of the possible and effective ways is to apply external strain. Thus, how are the geometric structures and the related electronic structures deformed under external strain becomes interesting problems.

Sun et al. have carried out comprehensive DFT studies to investigate those problems (Sun et al. 2008). To clearly indicate such effect, the deformation of armchair GNRs can be represented by the strain (ε), defined as $\varepsilon = (r-r_0)/r_0$, where r and r_0 are the deformed and initial equilibrium lattice constants (r_0= 4.287 Å) along the periodic direction of armchair GNRs. As we know, the structures of armchair GNRs are deformed to release the strain energy by applying external strain. In order to describe the vibration of C-C bonds in GNRs, four representative C-C bonds are selected in armchair GNRs with N_a = 13 (sketched in Fig. 7.). It is clear that the C-C separations change almost linearly with the strain, and the deformation leads to the largest change in the inner C-C bond length (a_1). As reported in the pristine armchair GNRs, the C-C bonds (a_3 and a_4) are shorted because of the edge effect. Therefore, the stronger C-C bonds (a_3 and a_4) have smaller response under the same external strain tensor. Here we should note all armchair GNRs have the same tendency, which represents the general physical picture of armchair GNRs under external strain.

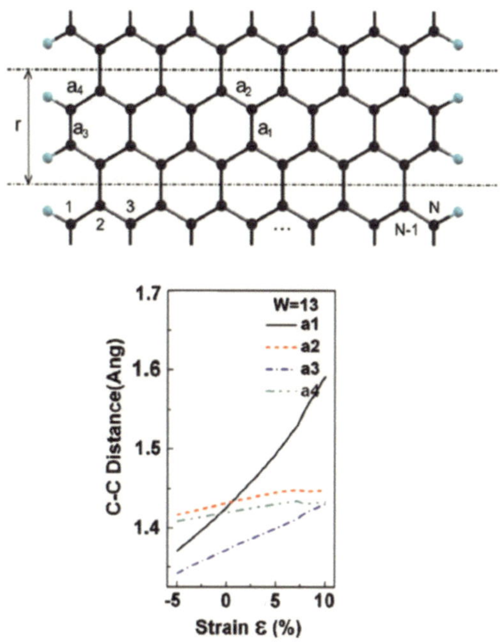

Fig. 7. The top one is the schematic of armchair GNRs with N_a = 13, four representative C-C bonds are shown, while the bottom one is the vibrations of C-C bong lengths under external strain (Sun et al. 2008).

Next, let us turn to the energy gap modification. Because armchair GNRs have three different categories, Sun et al. have selected N_a = 12, 13, 14 as the representative ones to investigate this issue. As shown in Fig. 8., the calculated maximal energy gaps for the armchair GNRs with N_a = 12, 13, 14 appear at ε= 5.0%, -0.8%, and 9.5%, respectively. While the minimal energy gaps occur at ε= -4.5%, 7.3%, and 1.3%, respectively. The shapes of the calculated curves display a zigzag feature for N_a = 12, 13, 14, which indicate that the energy gap is sensitive to the external strain.

To extend the studied widths, the variations of the energy gaps of three families of structures with different widths are plotted in Fig. 8. Sun et al. have summarized several features: (1) The minimal energy gaps of all deformed armchair GNRs are of the order of several meV. (2) The zigzag feature is observed in all deformed armchair GNRs (Sun et al. 2008).

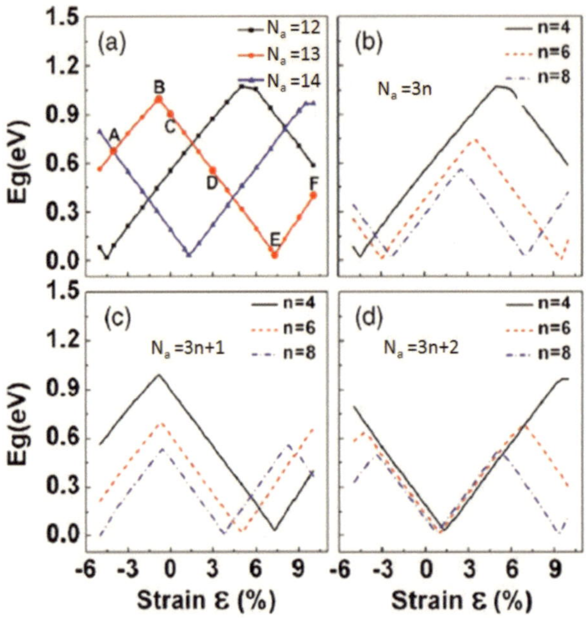

Fig. 8. (a) Variation of the energy gaps as a function of strain for the armchair GNRs with N_a = 12, 13, 14. (b), (c), and (d) are variation of energy gaps for three different categories (Sun et al. 2008).

3.1.2 Chemical adsorption

Armchair GNRs are non-spin-polarized semiconductors, therefore, how to produce armchair GNRs-based spintronics becomes an interesting question. Similar with carbon nanotubes, chemical adsorption of transition-metal (TM) atoms may be a possible solution. Kan et al. have proposed that armchair GNRs can be adsorbed by one-dimensional TM chains, leading to the desired magnetism (Kan et al. 2007a).

During their studies, they selected N_a= 10, 11, 12 to represent the three different categories, and titanium chains are taken as the source of magnetism. To investigate the adsorptive behaviours, three different concentrations of adsorption have been considered, namely, one, two, and four Ti atoms per two unit cell. As shown in Fig. 9, Ti atoms are found to prefer the hollow sites near the edges, and form zigzag atomic chain (Kan et al. 2007a). The same tendency is observed for all three categories of GNRs. Their results indicate that Ti atoms will spontaneously produce one-dimensional atomic chains on GNRs under thermal equilibrium.

Since atomic Ti chains can form stable adsorption on GNRs, there are two natural questions: Do the hybrid structures have magnetic signals? And do they have special electronic structures? To answer these questions, Kan et al. have performed comprehensive theoretical studies within the DFT framework. The following results show all the hybrid structures (three different concentrations) are spin polarized with ferromagnetic coupling. For example, in the case of four Ti atoms per two unit cell, ferromagnetic states are lower in energy than antiferromagnetic states by 0.16, 0.16, and 0.14 eV for N_a = 10, 11, and 12, respectively. The large energy difference of different magnetic coupling ensures the survival of ferromagnetism under high temperature.

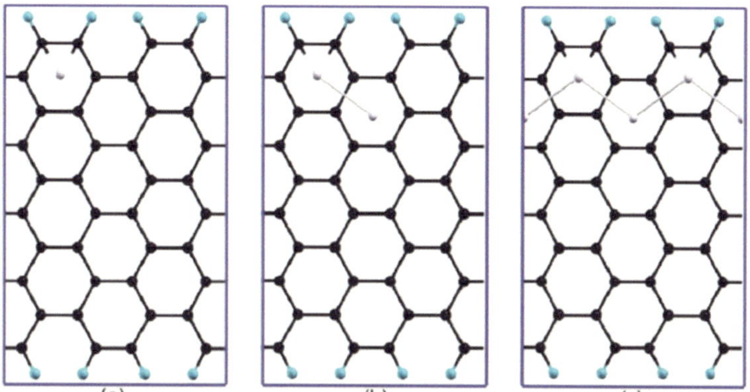

Fig. 9. (a) The geometric structures of armchair GNRs adsorbed by Ti chains under three concentrations, namely, (a) one Ti atom per two unit cell, (b) two Ti atoms per two unit cell, and (c) four Ti atoms per two unit cell. Black balls are C atoms, sapphire balls are H atoms, and white balls are Ti atoms (Kan et al. 2007a).

In Fig. 10., the calculated band structures of the hybrid structures are plotted. It is clear that their electronic structures show spin-selective characters. Especially, for N_a =10, and 11, the band structures show perfect half-metallic behaviours.

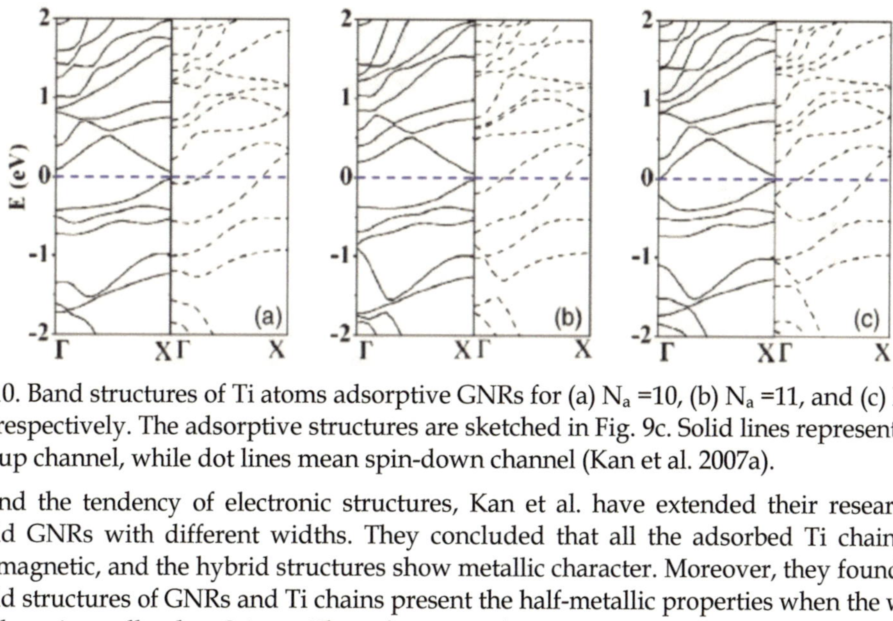

Fig. 10. Band structures of Ti atoms adsorptive GNRs for (a) N_a =10, (b) N_a =11, and (c) N_a =12, respectively. The adsorptive structures are sketched in Fig. 9c. Solid lines represent spin-up channel, while dot lines mean spin-down channel (Kan et al. 2007a).

To find the tendency of electronic structures, Kan et al. have extended their research in hybrid GNRs with different widths. They concluded that all the adsorbed Ti chains are ferromagnetic, and the hybrid structures show metallic character. Moreover, they found that hybrid structures of GNRs and Ti chains present the half-metallic properties when the width of ribbons is smaller than 2.1 nm. Thus, this research paves a new way to explore spintronics at the nanometer scale based on GNRs.

3.2 Zigzag GNRs

3.2.1 Applying electric field

It is now well known that zigzag GNRs are semiconductors with two localized electronic edge states. These two states are ferromagnetically ordered at each edge, and antiferromagnetically coupled each other, which means the total spin of zigzag GNRs is zero. Since these states are edge states, the effects of external transverse fields are expected to be significant.

Son et al. have investigated the aforementioned problems by performing DFT calculations. In their research, the external transverse fields are simulated by a periodic saw-tooth-type potential, which is perpendicular to the direction of the ribbons edge. Taking N_z =16 as a representative model, Son found that the valence and conduction bands associated with one spin orientation becomes closer, and eventually close their gap under strong enough external electric fields. Whereas band gaps of the other spin orientation are widen. As shown in Fig. 11, spin-degenerate band structures become spin-selective, and eventually changed into half metals by applying external electric fields (Son et al, 2006b).

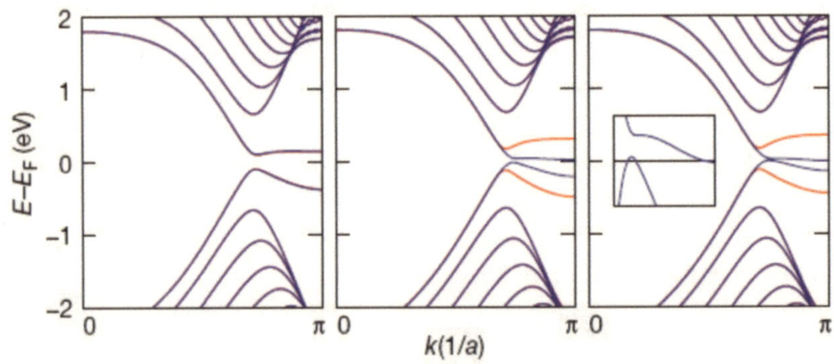

Fig. 11. From left to right, the spin-resolved band structures of zigzag GNRs with N_z = 16 under electric fields 0.0, 0.05 and 0.1 V/Å, respectively. The red and blue lines denote bands of different spin channels (Son et al, 2006b).

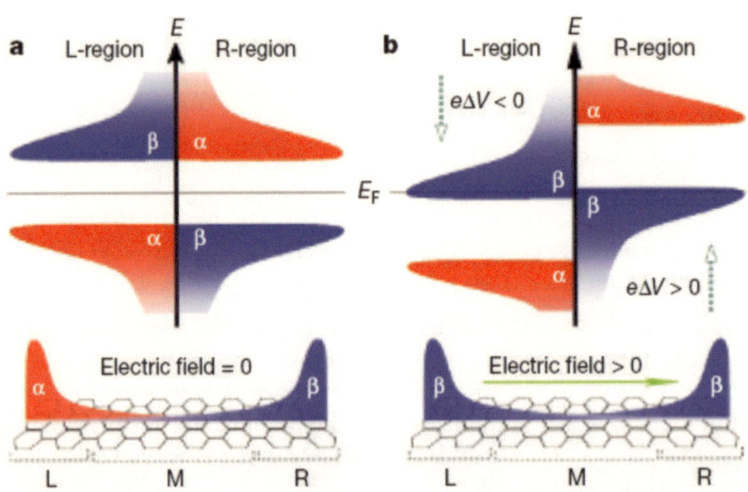

Fig. 12. (a) Schematic density-of-states diagram of the zigzag GNRs without external electric field. L-region (R-region) means the left (right) side of GNRs, and α and β are two spin channels. (b) Density-of-states diagram under applied electric field (Son et al, 2006b).

The half-metallic property originates from the fact that the applied electric fields induce energy-level shifts of edge states. Based on the analysis of density of states, Son et al. have used a simple model to explain the appearance of half-metallic property. As shown in Fig. 12, the spin-polarized states localized at two edges (left and right edges) are well spatially separated and degenerate in energy without applying electric fields. Since both occupied and unoccupied states of one special spin channel are located at the opposite sides of the zigzag GNRs, the effect of external electric field on them is different, namely, raising the energy-level of occupied, while lowering the one of unoccupied. Consequently, the energy gaps of one spin channel will be closed by external electric field, while leaving the other ones insulating. Thus, the half-metallicity comes from the relative movement in energy of edge states under electrostatic potential (Son et al, 2006b).

However, for any potential applications, the strength of critical electric field is very important. As we know, the semilocal DFT theory can only predict the qualitative band gaps. Thus, how strong does the electric field need to induce the half-metallicity becomes a mysterious. To solve the problems, Kan et al. have performed hybrid density functional calculations (B3LYP), which is viewed as one of the most accurate methods to get the quantitative results (Kan et al, 2007b). Their B3LYP calculations reproduce the half-metallicity under electric fields, but the critical electric fields are much higher than those from normal DFT calculations.

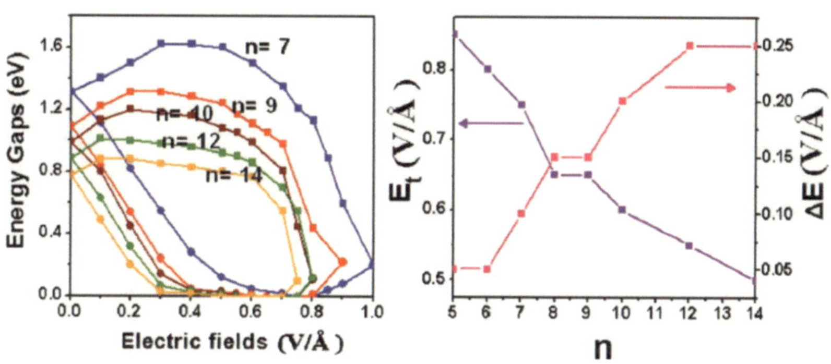

Fig. 12. Left: Band gaps of zigzag GNRs against external electric fields, n means the width of ribbons. The line with squares are spin-up channel, while filled circles are spin-down one. Right: The critical electric fields (E_t) to achieve half-metallicity and the range of electric field strength to keep half-metallicity (Kan et al, 2007 b).

As shown in Fig. 12, the critical electric fields decrease with the increasing of ribbon widths. However, it is still too high for applications. On the other hand, the electric field range at which zigzag GNRs remain half-metallic increases with the ribbon width. All these results indicate that the half-metallicity induced by electric fields is robust, and independent on the theoretical methods adopted. Thus, the only problem is how to overcome the obstacle of high critical electric field in applications.

3.2.2 Edge modifications

Since spin-polarized edge states dominate the electronic and magnetic properties of zigzag GNRs, the edge modifications play an important role. To achieve the amazing half-metallicity, Kan et al. have proposed that edge modifications by chemical groups can overcome the obstacle of high electric field (Kan et al, 2008a). The designs are neat and simple as sketched in Fig. 13.

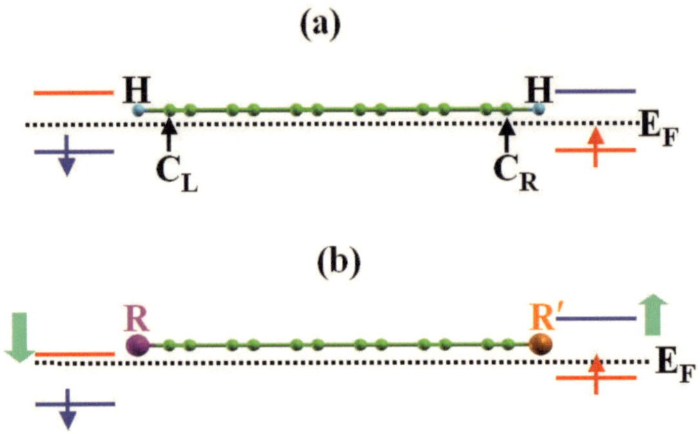

Fig. 13. (a) Structures of the H-saturated zigzag GNRs and schematic energy diagram for the edge states. (b) Schematic edge-state energy diagram of zigzag GNRs with different chemical modifications at two edges. (Kan et al, 2008a).

As we know, when zigzag GNRs are symmetrically saturated by hydrogen atoms at two edges, the spin-polarized edge states are degenerated in energy. One edge is occupied by spin-down electrons, while the other one is occupied by spin-up electrons. However, when the two edges are anti-symmetrically modified with different chemical groups, as shown in Fig. 13b, the induced potential shifts at two edges are different, which breaks the degenerancy of energy level of two edge states. Therefore, the modified ribbons become spin selective. Once the induced potential difference is large enough, the energy gap of one spin channel is closed, and zigzag GNRs are expected to become half metals.

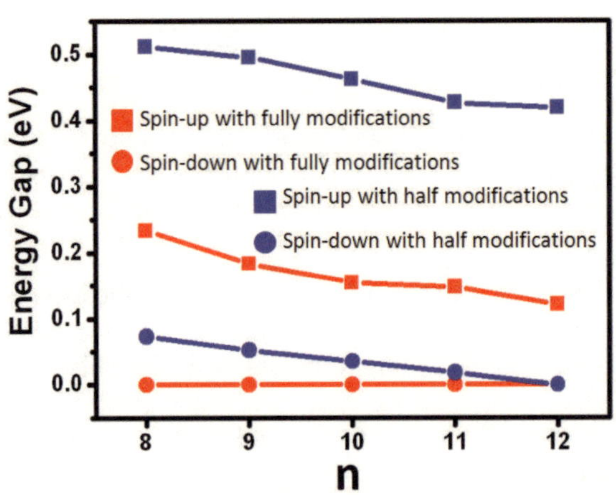

Fig. 14. The band gaps of zigzag GNRs with edge modifications. (Kan et al, 2008a).

To confirm the above predictions, Kan et al. performed DFT calculations to study the effect of different chemical groups (Kan et al, 2008a). NO_2 and CH_3 are selected as the typical groups achieving the half-metallicity. In order to study the effect of chemical modifications, two concentrations are considered, namely, fully and half modifications. As the definition, fully modifications mean that all the carbon atoms at one edge are modified by NO_2, while the carbon atoms at the other edge are saturated by CH_3. For half modifications, the selected chemical groups and hydrogen atoms alternatively saturate the carbon atoms at two edges. As plotted in Fig. 14, the calculated band gaps show perfect spin-selective behaviours. For both fully and half modifications, half-metallicity can be achieved with enough width.

Similar with Kan et al's works, Wu et al. reported that half-metallicity can be obtained by other chemical groups, such as OH groups at one edge, while NO_2 or SO_2 at the other edge (Wu et al, 2010). They have studied many chemical groups, and given many possibilities to future research. Thus, edge-modification has been viewed as a powerful tool to tail the electronic structures of GNRs.

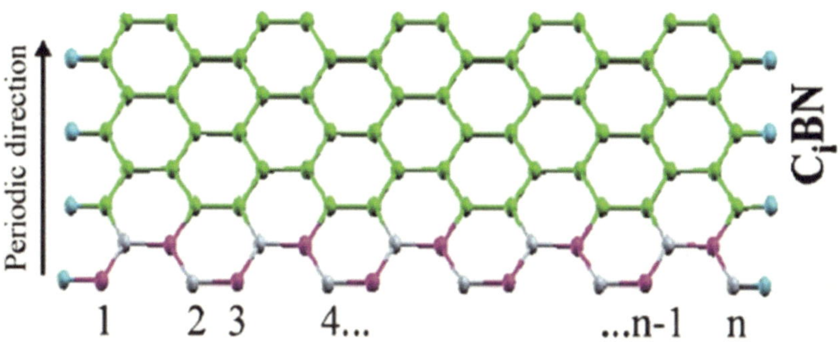

Fig. 15. The atomic structures of hybrid BCN nanoribbons. Green, pink, gray, and sapphire balls denote carbon, boron, nitrogen, and hydrogen atoms, respectively (Kan et al, 2008b).

3.2.3 Chemical doping

Since half-metallicity can be achieved with anti-symmetrical potential at two edges, doping boron and nitrogen atoms in zigzag GNRs becomes one possible way. Kan et al. has proposed one interesting model, as shown in Fig. 15 (Kan et al, 2008b). BN chains are integrated into zigzag GNRs, and each BN chain is well separated by *n* carbon chains ($n = 1, 2, 3$). As we know, boron and nitrogen atoms provide holes and electrons when they are separately doped into zigzag GNRs. Therefore, the effect on the edge states of boron and nitrogen atoms definitely are different, which may induce half-metallicity.

To confirm their proposal, Kan et al. carefully studied the electronic structures of such hybrid BCN nanoribbons with DFT methods (Kan et al, 2008b). As the direct index, spin-resolved energy gaps are sketched in Fig. 16. There are three significant characters: firstly, all studied hybrid BCN nanoribbons, namely, C_1BN, C_2BN, and C_3BN, are spin-selective materials when N_z is larger than 8. Secondly, all the studied BCN nanoribbons become half metals with enough width of ribbons. Thirdly, the critical width realizing half-metallicity is different. For C_1BN, the critical width is $N_z = 12$, while 10 and 8 for C_2BN, and C_3BN, respectively. In another words, the critical width decreases with the increasing of intervened carbon chains. These results mean that half-metallicity in hybrid BCN ribbons can be easily realized by controlling the doping methods of BN chains.

Similar with Kan et al's results, Dutta et al. have proposed that doping boron and nitrogen atoms into zigzag GNRs can lead to half-metallicity (Dutta et al, 2009). Slightly different with Kan et al's idea, Dutta et al. suggested that doped BN chains are perpendicular to the periodic direction. However, the electronic behaviours of doped BN chains are similar with the reported results of Kan et al.

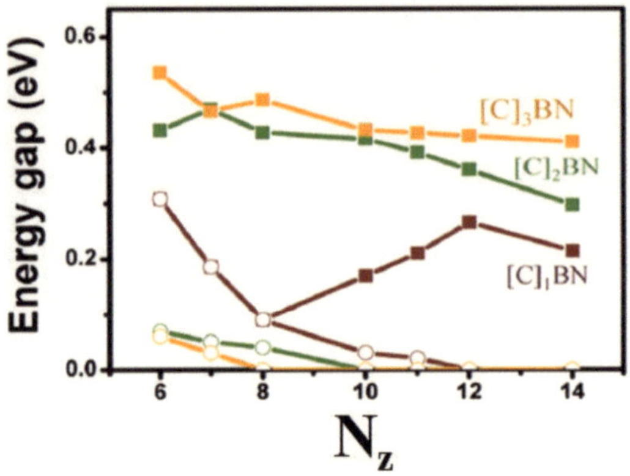

Fig. 16. The spin-resolved band gaps of hybrid BCN nanoribbons (Kan et al, 2008b).

3.2.4 Chemical adsorptions

Although the half-metallicity can be obtained by many methods, such as the electric field, edge modifications, and doping, we should keep in mind that the whole nanoribbons have no net magnetic moments. In applications of nano devices and spintronics, large net magnetic moments are sometimes desirable. On the other hand, transition-metal free magnetism is also interesting because of the weak spin-orbit coupling. Thus, the next important question is how to get large magnetic moments in such nanoribbons without magnetic atoms. To handle this problem, Kan et al. have theoretically suggested that zigzag nanoribbons have a large magnetic moment by adsorbing main-group atoms. In their studies, they used boron, carbon, and nitrogen atoms as adatoms (Kan et al, 2010).

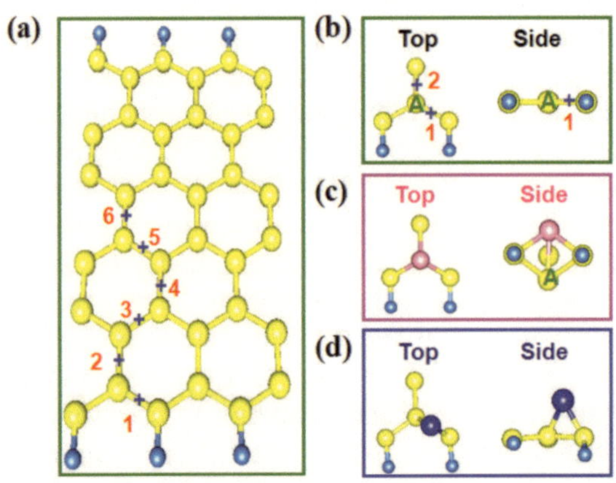

Fig. 17. (a) The possible adsorptive sites of zigzag GNRs. (b) The edge structures of pristine zigzag GNRs. The adsorptive structures for (c) boron (carbon) and (d) nitrogen adatoms (Kan et al, 2010)

As we know, zigzag GNRs are antiferromagnetic coupling between two edges, leading to net magnetic moments as zero. By adsorbing atoms, the bonding of pristine zigzag GNRs is broken, which may result in local states. Thus, the first question is how these adatoms bond with the zigzag GNRs. Using DFT calculations, Kan et al. found that the preferred adsorptive sites are near the edges. As shown in Fig. 17, they took zigzag GNRs with $N_z = 6$

as an example, and considered all possible adsorptive sites. For boron and carbon, the adatoms prefer the site labelled as "A" in the figures, and push the original carbon atoms out of the plane of ribbons, while for the nitrogen atom, it prefers the site 1, which gives the lowest energy. The different adsorptive behaviours can be due to the strong tendency of sp^3 hybridization in boron and carbon atoms.

The next question is whether adatoms can induce large magnetic moments. Luckily, the following studies give a positive answer. As seen in Fig. 18, the plotted spin density clearly shows that antiferromagnetic distributions of pristine have been switched into ferromagnetic distributions after carbon adsorption, which leads to a net magnetic moment of 2 μ_B per carbon adatom. While for boron and nitrogen adatoms, they have the same tendency with carbon adatoms, but the inducing magnetic moments are 1 μ_B per adatom.

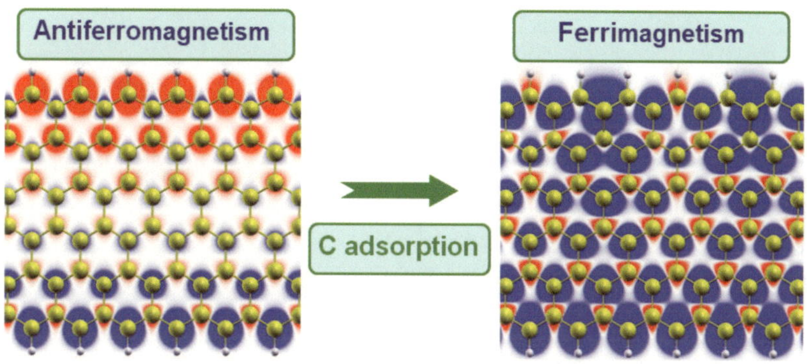

Fig. 18. Spin density of zigzag GNRs before and after adsorption. Red means spin-up density, while the blue means spin-down density (Kan et al, 2010).

Although they have answered the question whether adatoms can induce magnetic moments, there are still two critical questions: Whether do the magnetic moments survive with multi-adatoms? And whether all the zigzag GNRs have the same situations? To solve these questions, Kan et al. investigated the case with two adatoms. They found that two adatoms prefer the ferromagnetic coupling, and give a sum of induced magnetic moments of individual adatoms. By extending the studied model to other nanoribbons, they also found that all zigzag GNRs have the same results. Thus, their studies present an important way to manipulate the magnetism of zigzag GNRs.

4. Experimental realizations of GNRs

Although theoretical works have proposed and predicted many interesting properties of GNRs, the experimental realizations of GNRs are still not easy. According to the fabricated methods, we can divide them into chemical and physical realizations. Normally, chemical methods are assembling the small molecules into GNRs, while physical methods are tailing the large graphene or nanotube into GNRs.

Many groups have reported the chemical fabrication of GNRs with small molecules. Yang et al. reported that the 1,4-diiodo-2,3,5,6-tetraphenylbenzene and 4-bro-mophenylboronic acid with other catalyst can produce GNRs with lengths of up to 12 nm (Yang et al, 2008). Recently, Cai et al. have reported atomically precise bottom-up fabrication of GNRs (Cai et al, 2010). In their research, they used surface-assisted coupling of molecular precursors into linear polyphenylenes and their subsequent cyclodehydrogenation. Therefore, the topology, width and edge periphery of the GNRs products are defined by the structure of the precursor monomers, which can be designed to a wide range of different GNRs (Cai et al, 2010). However, accurate controlling the edge structures, width, and lengths of produced GNRs is still not easy for chemists.

Fig. 19. Representation of the gradual unzipping of one wall of a carbon nanotube to form a nanoribbon (Kosynkin et al, 2009)

In this sense, cutting graphitic materials (Li et al, 2009) to form GNRs provides an alternative solution to partially overcome the obstacle. However, how to open the strong C-C bonding is a big challenge. As shown in Fig. 19, Kosynkin et al. have creatively proposed that using solution-based oxidative process to break the C-C bonds of carbon nanotubes. Their experimental results show a nearly 100% yield of nanoribbon structures (Kosynkin et al, 2009). On the other hand, Jiao et al. have proposed an approach to making GNRs by unzipping multiwalled carbon nanotubes by plasma etching of nanotubes partly embedded in a polymer film (Jiao et al, 2009), which is a completely physical method.

5. Summary

In this chapter, the geometric, electronic, and magnetic properties of GNRs are carefully discussed in both experiment and theory. Theoretical studies have shown that there are energy gaps in GNRs, due to the edge effects and quantum confinement effect. These results have well confirmed by experimental observations. However, the amazing magnetic property predicted in zigzag GNRs has never been observed in experiment. Several factors, such as carriers concentrations, edge structures, and thermal dynamic under limited temperature, are responsible for this situation. Thus, one very imporant question is how to make the magnetism more robust.

On the other hand, the experimantal realization of GNRs are far away from the satisfaction. Since both edge structures, width, and length of GNRs are difficult to control, finding an effective way to overcome these problems becomes more and more important, both for acadamic research and industry applications.

6. References

Cai, J.; Ruffieux, P.; Jaafar, R.; Bieri, M.; Braun, T.; Blankenburg, S.; Muoth, M.; Seitsonen, A.; Saleh, M.; Feng, X.; Mullen, K. & Fasel, R. (2010). Atomically precise bottom-up fabrication of graphene nanoribbons. *Nature,* Vol. 466, No.22, 470-474, ISSN: 0028-0836

Dutta, S.; Manna, A. & Pati, S. (2009). Intrinsic half-metallicity in modified graphene nanoribbons. *Physical Review Letters,* Vol. 102, No. 9, 096601, ISSN: 0031-9007

Jiao, L.; Zhang, L.; Wang, X.; Diankov, G. & Dai, H. (2009). Narrow graphene nanoribbons from carbon nanotubes. *Nature,* Vol. 458, No.16, 878-880, ISSN: 0028-0836

Kan, E; Xiang, H.; Yang, J. & Hou, J. (2007a). Electronic structures of atomic Ti chains on graphene nanoribbons : A first-principles study. *The Journal of Chemical Physics,* Vol. 127, No. 16, 164706, ISSN: 0021-9606

Kan, E; Li, Z.; Yang, J. & Hou, J. (2007b). Will zigzag graphene nanoribbon turn to half metal under electric field ? *Applied Physics Letters,* Vol. 91, No. 24, 243116, ISSN: 0003-6951

Kan, E; Li, Z.; Yang, J. & Hou, J. (2008a). Half-metallicity in edge-modified zigzag graphene nanoribbons. *Journal of the American Chemical Society,* Vol. 130, No. 13, 4224, ISSN: 0002-7863

Kan, E; Wu, X.; Li, Z.; Zeng, X.; Yang, J. & Hou, J. (2008b). Half-metallicity in hybird BCN nanoribbons. *The Journal of Chemical Physics,* Vol. 129, No. 8, 084712, ISSN: 0021-9606

Kan, E.; Xiang, H.; Wu, F; Lee, C.; Whangbo, M. & Yang, J. (2010). Ferrimagnetism in zigzag graphene nanoribbons induced by main-group adatoms. *Applied Physics Letters,* Vol. 96, No. 10, 102503, ISSN: 0003-6951

Koskinen, P; Malola, S. & Hakkinen, H. (2008). Self-passivating edge reconstructions of graphene. *Physical Review Letters,* Vol. 101, No.11, 115502, ISSN: 0031-9007

Kosynkin, D.; Higginbotham, A.; Sinitskii, A.; Lomeda, J.; Dimiev, A.; Price, B. & Tour, J. (2009). Longitudinal unzipping of carbon nanotubes to form graphene nanoribbons. *Nature,* Vol. 458, No.16, 872-876, ISSN: 0028-0836

Li, Y.; Park, C.; Son, Y; Cohen, M. & Louie, S. (2007). Quasiparticle energies and band gaps in graphene nanoribbons. *Physical Review Letters,* Vol. 99, No.18, 186801, ISSN: 0031-9007

Li, Z; Zhang, W; Luo, Y; Yang J; Hou JG. (2009) How Graphene Is Cut upon Oxidation? *Journal of the American Chemical Society,* Vol. 131, No. 18, 6320, ISSN: 0002-7863

Son, Y; Cohen, M. & Louie, S. (2006a). Energy gaps in graphene nanoribbons. *Physical Review Letters,* Vol. 97, No.21, 216803, ISSN: 0031-9007

Son, Y; Cohen, M. & Louie, S. (2006b). Half-metallic graphene nanoribbons. *Nature,* Vol. 444, No.16, 347-349, ISSN: 0028-0836

Sun , L; Li, Q.; Ren, H.; Su, H.; Shi, Q & Yang, J. (2008). Strain effect on electronic structures of graphene nanoribbons : A first-principles study. *The Journal of Chemical Physics,* Vol. 129, No. 7, 074704, ISSN: 0021-9606

Wu, F; Kan, E.; Xiang, H.; Wei, S.; Whangbo, M. & Yang, J. (2009). Magnetic states of zigzag graphene nanoribbons from first principles . *Applied Physics Letters,* Vol. 94, No. 22, 223105, ISSN: 0003-6951

Wu, M; Wu, X. & Zeng, X. (2010). Exploration of half metallicity in edge-modified graphene nanoribbons. *Journal of Physical Chemistry C*, Vol. 114, No. 9, 3938, ISSN: 1932-7447

Yang, X.; Dou, X.; Rouhanipour, A.; Zhi, L.; Rader, H. & Muller, K. (2008). Two-dimensional graphene nanoribbons. *Journal of the American Chemical Society*, Vol. 130, No. 13, 4216, ISSN: 0002-7863

17

Mechanical and Electronic Properties of Graphene Nanostructures

Ricardo Faccio, Luciana Fernández-Werner, Helena Pardo, Cecilia Goyenola, Pablo A. Denis and Álvaro W. Mombrú
Centro NanoMat, DETEMA, Polo Tecnológico de Pando, Facultad de Química, Cryssmat-Lab, DETEMA, Facultad de Química, CINQUIFIMA, Espacio Interdisciplinario, Universidad de la República, Facultad de Química, Uruguay

1. Introduction

Quite recently a new carbon nanostructure, called graphite nanoribbon (GNR), has emerged, taking the attention of the scientific community because of its promising use in spintronics. It is manly attributed to the work of Son et al. (Son et al. 2006 a; Son et al. 2006 b), who predicted that in-plane electric field, perpendicular to the periodic axis, induces a half-metal state in zigzag nanoribbons (ZGNR). This state corresponds to a one spin flavour with metallic behaviour, while the opposite spin flavour experiences an increase in the energy gap. Apart from the interesting dependence of the electronic structure upon an electric field, this is a promising material for future spintronic devices, since it could work as a perfect spin filter. Very recently Campos-Delgado et al. (Campos-Delgado et al. 2008) reported a chemical vapour deposition route (CVD) for the bulk production of long, thin, and highly crystalline graphene ribbons (less than 20-30 μm in length), with widths from 20 to 300 nm and small thicknesses (2 to 40 layers). In addition, the bottom up synthesis of these nanostructures may be feasible as noted by Hoheisel and collaborators (Hoheisel et al. 2010). This experimental advance further increases the expectations for the use of these materials in high-tech devices. In parallel there is an increased interest in the physical properties of carbon nanostructures in general, due to their outstanding mechanical and electronic properties. Recently, Lee et al. (Lee et al. 2008) measured the mechanical properties of a single graphene layer, demonstrating that graphene is the hardest material known, since the elastic modulus reaches a value of 1.0 TPa. Besides, many efforts have been dedicated to study the electronic properties of graphene, because creating a gap could allow the use of graphene in field effect transistors. Many mechanisms have been proposed with that purpose: nano-pattering, creating quantum dots, using multilayer, doping with heteroatoms such as sulphur (Denis et al. 2009), covalent functionalization (Bekyarova et al. 2009) and applying mechanical stress (Pereira et al. 2009; Gui et al. 2008). Recently Gui (Gui et al. 2008) proposed that graphene under a symmetrical strain distribution always leads to a zero band-gap semiconductor, and the pseudogap decreases linearly with the strain strength in the elastic linear regime. However, asymmetrical strain induces an opening of band gaps at the Fermi level. The evidence presented above clearly points that it is important to know how the electronic properties of ZGNR depend on stress, in order to predict its performance in future devices (e.g. gates). The bibliography shows a few works related with the study of strain in graphene nanoribbons (Sun et al. 2008; Xu et al. 2009; Su et al. 2008), neither of them informing the Young's modulus of GNR. The main conclusions from those works indicates that there is no important variation of the electronic properties of zigzag nanoribbons upon stress-strain effects (i.e. energy gaps and local magnetic moments), while there is no information regarding the mechanical properties of this nanostructure. Recently our group (Faccio et al. 2009) presented the first systematic determination of the Young's modulus, Poisson's ratio and calculated Shear modulus for graphene nanoribbons.

1.1 Theoretical background

The Young's modulus can be considered a measure of the stiffness of a solid. It is widely used for isotropic and continuous media, in which, the elastic parameters together defines the mechanical properties of the material. These parameters are: the Young's modulus E^{3D}, the shear modulus G and the Poisson's ratio υ. These three parameters are related by simple expressions that allows the calculation of one of them if two are known. For a three dimensional (3D) solid they can be defined as:

$$E^{3D} = \frac{\sigma_x}{\varepsilon_x} = \frac{1}{V_0}\left(\frac{\partial^2 E_S}{\partial \varepsilon_x^2}\right)_{E_0} \tag{1.1}$$

$$G^{3D} = \frac{\sigma_{xy}}{\varepsilon_{xy}} = \frac{1}{V_0}\left(\frac{\partial^2 E_S}{\partial \varepsilon_{xy}^2}\right)_{E_0} \tag{1.2}$$

$$\upsilon = -\frac{\varepsilon_y}{\varepsilon_x} = -\frac{\varepsilon_z}{\varepsilon_x} \tag{1.3}$$

$$G^{3D} = \frac{E^{3D}}{2(1+\upsilon)} \tag{1.4}$$

The Young's modulus represents the ratio between the uniaxial stress σ_x and the uniaxial strain ε_x, and can be easily calculated in terms of the strain energy E_S, the tensile strain ε_x, and the equilibrium volume V_0. The shear modulus is also defined in terms of the shear components of the strain and stress tensors. Finally, the Poisson's ratio υ relates the axial strain ε_x with the transversal strain ε_y or ε_z.

In the case of graphene it has more sense to define the in-plane stiffness instead of the classical 3D Young's modulus, because of the reduced dimensionality of this material. For this reason in graphite the elastic properties can be considered independent of the interlayer distance between graphene layers, c_0=3.35 Å, and the constants can be described as follows:

$$E^{2D} = \frac{1}{A_0}\left(\frac{\partial^2 E_S}{\partial \varepsilon_x^2}\right)_{E_0} = E^{3D} c_0 \tag{2.1}$$

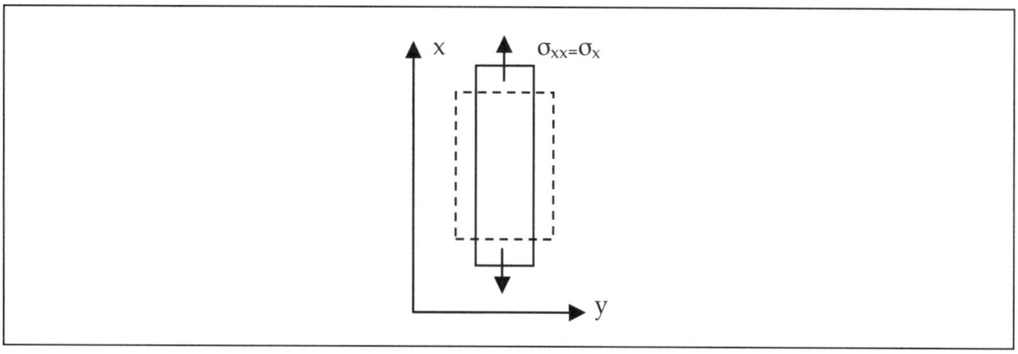

Fig. 1. Lateral contraction produced by an axial stress

$$\upsilon = -\frac{\varepsilon_y}{\varepsilon_x} = -\frac{l_0}{w_0}\frac{\partial w}{\partial l} \tag{2.2}$$

where A_0 corresponds to the equilibrium reference area of the 2D material, with length l_0 and width w_0. The in-plane stiffness of graphite is obtained considering an axial load applied over graphene. The value obtained in this case is E^{3D}=1.02(3) TPa (Blakslee et al. 1970). It allows us to obtain E^{2D}= 3.41(9) TPaÅ. This value is almost the same as that the one obtained experimentally for graphene, E^{2D}=3.42(30) TPaÅ (Lee et al. 2008) using nano indentation with atomic force microscope. This result is in agreement with those reported by Kudin et al. (Kudin et al. 2001) and Van Lier et al. (Lier et al. 2000). Using *Ab initio* methods they reported Young's modulus of E^{3D}= 1.02 TPaÅ (Kudin et al. 2001) and 1.11 TPaÅ (Lier et al. 2000). The Poisson's ratio is unambiguously defined in terms of the transversal ratio over the longitudinal variation with a value of υ= 0.149.

Some reports based on another theoretical approach; using Brenner potentials, show wide scattered results, with strong dependence on the equilibrium adjustment yield used in the calculation (Reddy et al. 2006). The results obtained by this method correspond to Young's modulus of E^{3D}=0.7 TPa and 1.11 TPa, with Poisson's ratio of υ=0.40 and 0.25,with the use of modified and non-modified potential respectively. Many representative results, based on Reddy et al. (Reddy et al. 2006), are presented in Table 1.

Single-walled carbon nanotubes (SWCNT) are an example of a one dimensional system described in terms of 2D property E^{2D}, since two parameters must be informed, the tube length (L) and the tube radius (r), in order to gain independence of size effects. Several expressions have been published for their mechanical properties in terms of multidimensional Young's modulus as: E^{3D}, E^{2D}, etc (Wu et al. 2008, Bogár et al. 2005). The values reported show a wide variation on experimental E^{nD}'s values, up to an order of magnitude of difference. This is mainly due to the difficulty in determining the precise structure of nanotubes under study, the presence of defects, chirality, etc. Recently, Wu et al. (Wu et al. 2008) used a combined optical characterization of individual SWCNT, coupled with magnetic actuation technique, to measure the Young's modulus of nanotubes with known chirality. The Young's modulus obtained was E^{3D}=0.97(16) TPa, assuming a wall thickness of c=3.4 Å corresponding to the interlayer spacing in graphite. No dependence on the nanotube's chiral index within the experimental accuracy was found. This result agrees quite well with theory, in particular with the values reported by Bogár et al. (Bogár et al. 2005). Employing an all electron DFT method, they reported E^{2D} for different tubes radius,

Reference	E^{3D} (TPa)	υ	Remarks
		Graphene	
Kudin et al. 2001	1.02	0.149	Graphene (*Ab initio*)
Lier et al. 2000	1.11	-	Graphene (*Ab initio*)
Reddy et al. 2006	1.012	0.245	Graphene (Brenner**)
Reddy et al. 2006	0.669	0.416	Graphene (Brenner*)
Arroyo et al. 2004	0.694	0.412	Graphene (Brenner)
Reddy et al. 2005	1.11	0.45	Graphene (Truss model)
Faccio et al. 2009	0.96	0.17	Graphene (*Ab initio*)
		Carbon Nanotubes	
Zhang et al. 2002	0.694	-	SWNT (Brenner)
Lu et al. 1997	0.97	0.28	SWNT (Empirical model)
Shen et al. 2004	0.213–2.08	0.16	SWNT (MM)
Yu et al. 2000	0.32–1.47	-	SWNT (Experiments)
Sammalkorpi et al. 2004	0.7	-	SWNT (MD)
Yoon et al. 2005	1.0	0.25	DWNT (Vibrations)
Wu et al. 2008	0.81–1.13	-	SWNT (Experiments)
Bogár et al. 2005(***)	0.8-1.05	-	SWNTS (*Ab initio*)
Bogár et al. 2005	1.05	-	SWNT (5,5) - (*Ab initio*)
Faccio et al. 2009	1.01	-	SWNT (5,5) - (*Ab initio*)

(*) Minimized potential and (**) Non-minimized Potential

(***) This result was converted to E^{3D} for comparison purposes, using: $E^{3D}=E^{2D}c_0$.

Table 1. Representative results for different carbon nanostructures.

that ranges from r=1.32 Å to 4.11 Å. They concluded that there is no dependence between the Young's modulus and the chirality of the nanotube.

1.2 Electronic and geometrical structure of graphene nanoribbons

In graphene nanoribbons, the presence of different types of boundary shapes, called edges, modifies the electronic structure of the material. The major effects are observed at the Fermi level, displaying unusual magnetic and transport features (Palacio et al. 2005). There are two typical shapes for graphite edges, with quite opposite consequences, zigzag and armchair. The zigzag edges present electronic localized states at the boundaries, corresponding to non-bonding states that appear at the Fermi level as a large peak in the density of states. It gives rise to important effects on the magnetic and transport properties of nanographite. These localized states are totally absent in the case of armchair edges. For these reasons the more interesting case correspond to nanographite with zigzag edges.

A graphite nanoribbon consists of strips of graphene; it is created by cutting the sheet along two parallel lines. The direction of these lines defines the type of edge, while the distance between them defines its width. This kind of structures was first introduced by Fujita et al. (Fujita et al. 1996; Nakada et al. 1996; Wakabayashi et al. 1999) who studied its electronic structure using tight-binding models over graphite. They found that the typical diamagnetic ordering of graphene is disrupted because of the presence of zigzag edges, as mentioned earlier, that breaks the delocalized π-bands creating an instability that is resolved via electronic spin polarization; see Figure 2(b). In order to study size and edge effects in carbon nanoribbons Fujita & Wakabayashi (Palacio et al. 2005) proposed a simple model that consists in N dimmers that establish the nanoribbon's width; the edges are terminated with

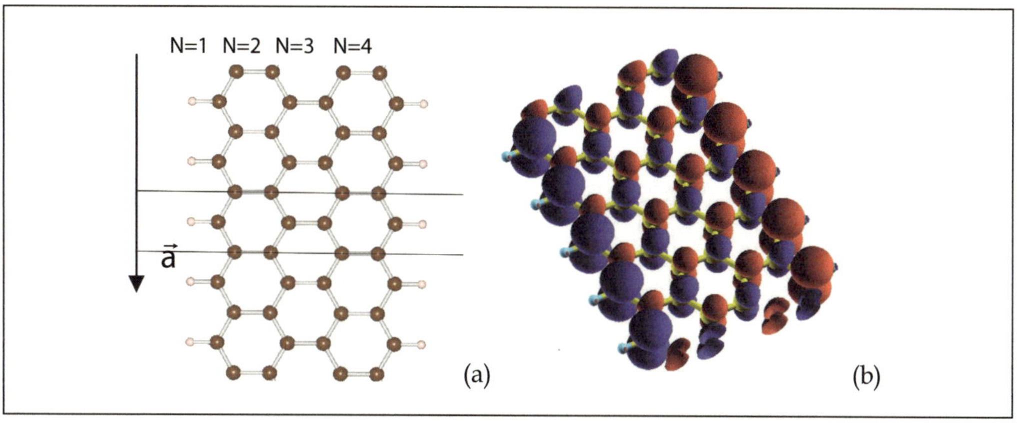

Fig. 2. (a) Graphite nanoribbon with N=4, displaying its smallest unit cell; the arrow shows the periodic direction $\vec{a}$. (b) Spin density map, showing the antiferromagnetic arrangement between opposite edges.

hydrogen atoms, while the translational invariance of graphene is kept along the direction perpendicular to the width, see Figure 2(a). Regarding the electronic structure, the non-magnetic solution has many states at the Fermi level, which produces a strong instability that can be resolved by spin polarization or geometrical distortion. Due to the non-bonding character of the zigzag localized edge states, the geometrical reconstruction is unlikely to happen (Miyamoto et al. 1999) and another mechanism should be explored for this purpose. The other alternative mechanism consists in a spin polarization of the electronic density, in particular with an antiparallel arrangement between near carbon atoms, with a zero total magnetization. The local magnetic moments increase at the edges; they couple oppositely with their near neighbours carbon atoms, and damp inwards the ribbons. Since opposite carbon atoms at the edges always belong to different sublattices the net magnetization is zero. Another important consequence of the spin polarization of the electronic density is the aperture of a gap in ZGNR, yielding a Slater insulator (Pisani et al. 2007). The opening of the gap is related with the ZGNR width, since it is a consequence of the interaction between edges. For this reason wider ribbons, with longer distances between opposite edges, recovers the graphene geometry with a gap equal to zero. The tendency observed corresponds to an exponential decay of the energy gaps when increasing the nanoribbon's width (N). Table 2 shows the results for N=4, 5, 6, 7, 8, 9 and 10.

Width (N)	Energy Gap (eV)
4	0.63
5	0.59
6	0.54
7	0.50
8	0.46
9	0.43
10	0.40

Table 2. Energy gaps for different zigzag graphene nanoribbons

The antiferromagnetic arrangement is always the ground state, having lower energy than the nonmagnetic and ferromagnetic solution. The ferromagnetic solution refers to a local ferrimagnetic arrangement, but with equal polarization between opposite carbon atoms at the edges, with a non-zero magnetic moment. The electronic structure of N=4 ZGNR is shown in Figure 3, where the band structure is presented together with the corresponding density of states.

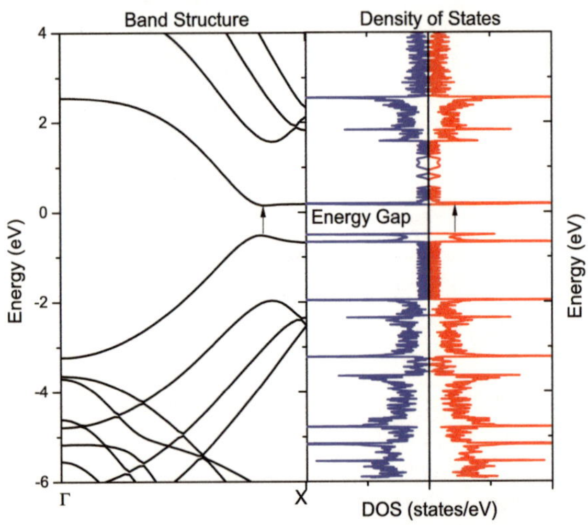

Fig. 3. Band structure along the periodical path Γ-X, and the corresponding total density of states for the N=4 zigzag graphite nanoribbon.

Recently the scientific community has revitalized the attention in GNR not only from the theoretical standpoint, but for the promising of its technological applications. It is manly attributed to the work of Son et al. (Son et al. 2006 a; Son et al. 2006 b) who predicted that in-plane electric field, perpendicular to the periodical axis, induces a half-metal state in ZGNR. This state corresponds to a one spin flavour with metallic behaviour; while the opposite spin flavour experiences an increase in the energy gap. Apart of the interesting dependence of the electronic structure upon an electric field, this is a promising material for future spintronic devices, since it could work as a perfect spin filter. Figure 4 shows the case of N=4 ZGNR where different electric fields were applied obtaining a half-metal state for an E=0.8 eV/Å.

2. Methods

2.1 Density functional theory

There exist many methodologies for obtaining the electronic structure of many-particle systems. "Density Functional Theory (DFT)" (Hohenberg et al. 1964; Kohn et al. 1965) is a way to study the electronic structure of many-particle systems with a relative low computational cost, since the description of a quantum system is based in functionals that depends on the electronic density instead of the many-particle wave functions. It is the corner stone of the theory, since the complexity of solving the many-particle wavefunction decreases and we only must deal with a set of almost single-electron-wavefunctions instead. In this way the Hamiltonian of the many-particle system, called Kohn-Sham Hamiltonian, can be described as:

$$E[\rho]=T_0[\rho]+V_{ee}[\rho]+V_{xc}[\rho]+V_{ext}[\rho] \tag{3.1}$$

where T_0 corresponds to the kinetic energy functional, V_{ee} is the inter-electronic repulsion, V_{ext} is the external potential coming from the nucleus arrangement, and finally V_{xc} is the exchange and correlation potential. While T_0, V_H and V_{ext} can be easily incorporated in the Hamiltonian, since they are well know because of its universal expressions, the V_{xc} is not known in an exact form and must be estimated or approximated.

$$E[\rho]=T_0[\rho]+V_{ee}[\rho]+V_{xc}[\rho]+V_{ext}[\rho] \tag{3.2}$$

The exchange and correlation effects must be incorporated in the equations because the universal functionals T_0 and V_{ee} do not take into account the fermionic behaviour of the electronic wavefunction, and fail to describe instantaneously electron-electron interactions, and it is attributed to the mean field form of the handled expressions. This justifies the incorporation of an additional term into the Hamiltonian, called the exchange-correlation potential V_{xc}, with the purpose to deal with the limitations early mentioned.
The incorporation of the V_{xc} term represents an additional problem, since there is no knowledge about the exact form of the functional and approximations should be made in order to estimate its value.

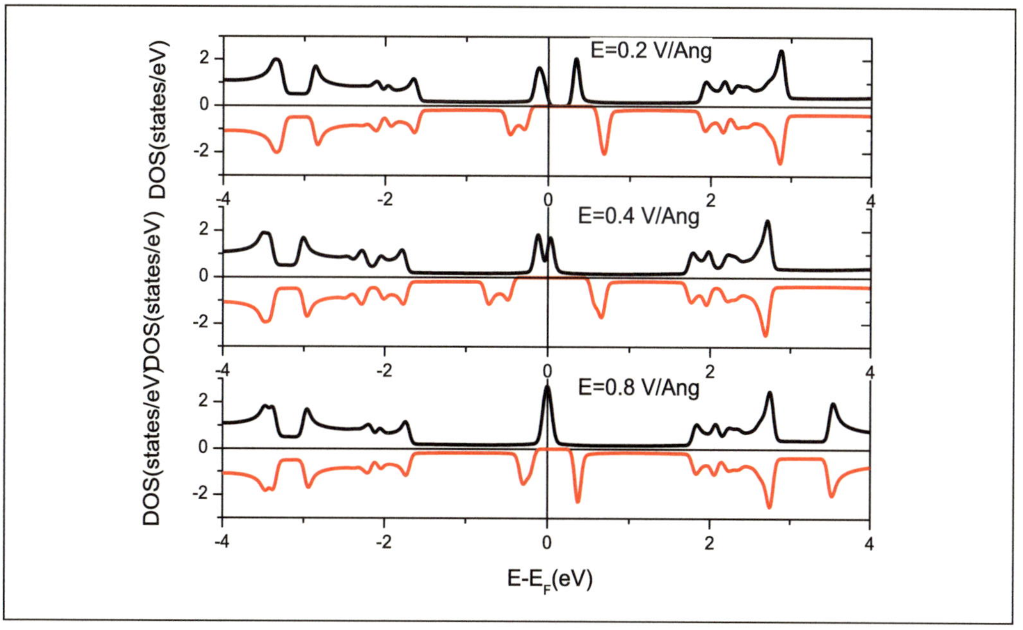

Fig. 4. Density of states for N=4 ZGNR under different applied electric fields (a) 0.2 V/Å, (b) 0.4 V/Å and (c) 0.8 V/Å.

2.2 Exchange-correlation functionals and dispersive interactions

The last section described the philosophy behind DFT and the origin of the exchange-correlation potential (xc-potential). The first and most widely used xc-potential is the so called "Local Density Approximation (LDA)" introduced by W. Kohn and L. J. Sham. The model assumes that the exchange-correlation energy of a particular density ρ(r) can be obtained with the help of the results obtained for an interacting homogeneous electron gas. The total density of the material is divided in infinitesimal portions of constant density $\delta\rho(r)$ each. Then the exchange correlation energy is added by an amount equal to the one corresponding to a uniform interacting electron gas of density $\rho_{EG}(r) = \delta\rho(r)$ that occupies the same piece of volume (Perdew et al. 1981; Ceperley et al. 1980). This model clearly neglects the fact that true densities are not homogeneous, but the results obtained are surprisingly good either for slowly varying or strongly varying densities either. The next step points towards the description of non homogeneous electron density, building a V_{xc} that accounts for the density of the neighbour pieces of volume. This is the Generalized Gradient Approximation (GGA), and because there is no a unique way to incorporate the density gradient in it, there exist many versions of this functional. While GGA gives better description of geometrical and electronic issues than LDA under particular circumstances (Konahoff et al. 2002), both systematically underestimate electronic band gaps, and generally tends to delocalise the electronic density. A possible solution to these problems consist in building hybrid xc-potentials that allow us to join the advantages of DFT and Hartree-Fock formalism together in one functional. Probably, the most widely used hybrid functional is called B3LYP (Becke et al. 1993; Lee et al. 1988; Vosko et al. 1980; Stephens et al. 1994). It is based on the original idea from Becke (Becke et al. 1993), where the functional is assembled with the correlation term suggested by Lee, Yang and Parr (LYP) (Lee et al. 1988), and the exchange term is a weighted sum of three terms that includes the Hartree-Fock term. The three empirical parameters were determined by fitting the predicted values to a set of atomization energies, ionization potentials, proton affinities, and total atomic energies (Stephens et al. 1994). Since B3LYP includes in some amount the Hartree-Fock exact-exchange this functional can be considerer as a non-local, and its implementation at the moment is done for localized basis set codes. This is main the reason why LDA and GGA are still widely used, including for Van der Waals materials, and the selection depends on the particular interest over the system under study.

While LDA, GGA and B3LYP functionals perform reasonably well describing metallic, covalent and ionic bonds; Van der Waals interaction is still a missing aspect within DFT implementations. This kind of binding is explained in terms of dynamical dipole-dipole correlations, which is inherently a non-local interaction and therefore neglected within LDA/GGA schemes. Recently there have been serious efforts to solve this problem with significant results (Khon et al. 1998; Rydberg et al. 2003; Dion et al. 2004). Román-Pérez (Román-Pérez et al. 2009) proposed an efficient implementation of the Van der Waals functional proposed by Dion (Dion et al. 2004) in the SIESTA code (Ordejón et al. 1996; Soler et al. 2002), with really promising results. They found an overhead in total computational cost, in comparison with semilocal functionals, very moderate for medium and large systems. For these reasons LDA and GGA exchange-correlation functionals are still used for describing Van der Waals solids such as graphite. The use of one over the other depends on the physical property to be assessed. LDA is widely used for graphite simulations since it gives good geometrical parameters and energy differences, in particular the calculated interlayer distance between graphene seems to be in good agreement with experiment (Tournus et al. 2005; Charlier et al. 1994). At first, would seem that LDA describes quite well the Van der Waals interactions in graphite, but what really happens is a fortuitous cancellation of errors (Tournus et al. 2005). A detailed analysis shows that the electron density determined by the LDA method is not adequate, offering a poor description of the

topology of the electron density in particular in the direction perpendicular to graphene (Strocov et al. 2000).
Moreover GGA xc-potential offers better geometrical reconstruction in the graphene plane, with better C-C distances (Konstantinova et al. 2006; Zhang et al. 2007) and better in-plane phonon frequencies (Wirtz et al. 2004). Furthermore GGA reproduces qualitatively the corrugation of the crystal potential (Strocov et al. 2000), but the energy involved in the $\pi \cdots \pi$ interactions is totally underestimated, leading to an increase of the interlayer distance of c_{GGA}=8.20 Å which is about 20% higher than the corresponding experimental value c_{GGA}=6.70 Å (Trucano et al. 1975). This is the main reason why LDA is preferred over GGA, when $\pi \cdots \pi$ interactions are involved. Thus the results of these approximations can be interpreted as the upper and lower bound for the interaction strength.

2.3 Basis Sets and DFT codes for *Ab initio* simulations

Ab initio methods can be selected in terms of the basis set used to generate the KS orbitals. The more representative methods correspond to plane wave expansion and localized basis sets. Each method has its advantages and pitfalls. The plane-wave methods are the most accurate within DFT. They are asymptotically complete allowing for systematic convergence and offering for spatially unbiased wave-functions. The main disadvantage is the computational cost, since many wave-functions per atom are needed for a good accuracy. Additionally the methods led to very high total energies, thus energy differences are very small and thus sensitive to the converged basis set. On the other hand the localized basis sets are in general very efficient, since the number of basis functions per atom is very small, reducing the computational cost. The physical interpretation of many properties is straightforward, since the basis sets are based on atomic like functions. In difference to plane waves methods localized basis set shows systematic lack for convergence and previous knowledge is needed for an accurate treatment of the basis sets.

2.4 Simulation of carbon nanoribbons

The theoretical study of the uniaxial stress on different ZGNR is based on the First Principles - Density Functional Theory (Hohenberg et al. 1964, Kohn et al. 1965) which we successfully used to study, bulk graphene, thioepoxidated SWCNTs and sulfur doped graphene (Denis et al. 2009; Faccio et al. 2008; Denis et al. 2008). The simulations are performed using the SIESTA code (Ordejón et al. 1996; Sánchez-Portal et al. 1997; Soler et al. 2002) which adopts a linear combination of numerical localized atomic-orbital basis sets for the description of valence electrons and norm-conserving non-local pseudopotentials for the atomic core. The pseudopotentials were constructed using the Trouiller and Martins scheme (Troullier et al. 1991) which describes the interaction between the valence electrons and atomic core. We selected a split-valence double-ζ basis set with polarization orbitals for all the carbon atoms. The extension of the orbitals is determined by cutoff radii of 4.994 a.u. and 6.254 a.u. for s and p channels respectively, as obtained from an energy shift of 50 meV due to the localization. The total energy was calculated within the Perdew-Burke-Ernzerhof (PBE) form of the generalized gradient approximation GGA xc-potential (Perdew et al. 1996 (a), 1997 (b)). The real-space grid used to represent the charge density and wavefunctions was the equivalent of that obtained from a plane-wave cutoff of 230 Ry. The atomic positions were fully relaxed in all the cases using a conjugate-gradient algorithm (Press et al. 1986) until all forces were smaller than 10 meV/Å was reached. A Monkhorst Pack grid

(Monkhorst et al. 1976) of 300x2x2 supercell, defined in terms of the actual supercell, was selected to obtain a mesh of 600 k-points in the full Brillouin Zone. All these parameters allow the convergence of the total energy. In order to validate our methodology we calculated the Young's modulus of (5,5) SWCNT, for which the literature shows several results from *Ab initio* methods (see Table 1). The smallest unit cell contains a total of 20 carbon atoms. With the purpose to study the dependence on the number of carbon atoms, we simulated the case 40 carbon atoms per unit cell. The Young's modulus obtained are E^{3D}=1.03(2) and E^{3D}=1.01(3) TPa, for 20 and 40 carbon atoms in the unit cell respectively. The results are consistent within the uncertainty which was estimated from the variance obtained from the adjustment of the second order fitting of the energy upon unitary deformation. Therefore one can conclude that the results are not affected by the number of supercells used, in particular along the periodic direction. Additionally the results are in good agreement with the reported in the bibliography see Table 1, in particular with an excellent agreement with those from Bogár et al. (Bogár et al. 2005). Regarding geometry in graphene nanoribbons, we can distinguish two C-C bond orientations: the bond perpendicular to the crystalline periodic direction d(|) and the bond diagonal to the normal direction d(/). The bond distances differ from the inner part of the ribbon (bulk) respecting the atoms at the edge. In the case of bulk C-C distances we found d(|)bulk= 1.44 Å and d(/)bulk=1.44 Å, while at the edge of the ribbon we found d(|)edge= 1.46 Å, and d(/)edge=1.43 Å. This result agrees with the tendency observed by Pisani et al. (Pisani et al. 2007), where the perpendicular bond elongates at the edge, contracting the corresponding diagonal bond at the edge. It promotes an increase of the zigzag C-C-C angle from 120° at the bulk to 121.9° at the edge. This trend is observed for the whole un-stressed studied ribbons. For all of these reasons, we can unequivocally conclude that our methodology is valid.

3. Results and discussion

The selected ZGNR for simulation correspond to N=4, 5, 6, 7, 8, 9 and 10. Since the code handled was designed for three dimensional materials, we designed special unit cells. All the cells were orthogonal, with the GNR placed in the ab plane, and oriented with the periodic direction along the *a* axis, see Figure 1 for the ZGNR N=4 sketch. In order to avoid interference between symmetry images, vacuum regions of 15 Å were added along *b* and *c* directions. In the case of the smallest unit cell, the *a* axis value for every cell is approximately a_0= 2.495 Å, with a total number of atoms of 2N+2. With the purpose of increase the number of degrees of freedom in each case, the cells were expanded in four units along the *a* axis ($a=4a_0$), it allows us to multiply by four the number of atoms inside the supercells according to 8N+8. The total number of atoms in each case is: 40, 48, 56, 64, 72, 80 and 88. The stress-strain curves are obtained applying different stress to the GNR, allowing full atomic relaxation together with full unit cell parameters optimization, until the desired stress tensor is reached. Since we are considering uniaxial strain only, the Voigt tensor has only one non-zero component: $[\sigma_x, \sigma_y, \sigma_z, \sigma_{xy}, \sigma_{xz}, \sigma_{yz}]\rightarrow[\sigma_x, 0, 0, 0, 0, 0]$. The selected stress components of the Voigt tensor allow us to establish strains in the range of $\varepsilon_x= \pm0.020$ for the whole series, which assures a linear stress regime (Liu et al. 2007, Khare et al. 2007). It corresponds to a quadratic dependence of the total energy upon the strain. The most important features of the data treatment are presented in Figure 5 for N=10 ZGNR.

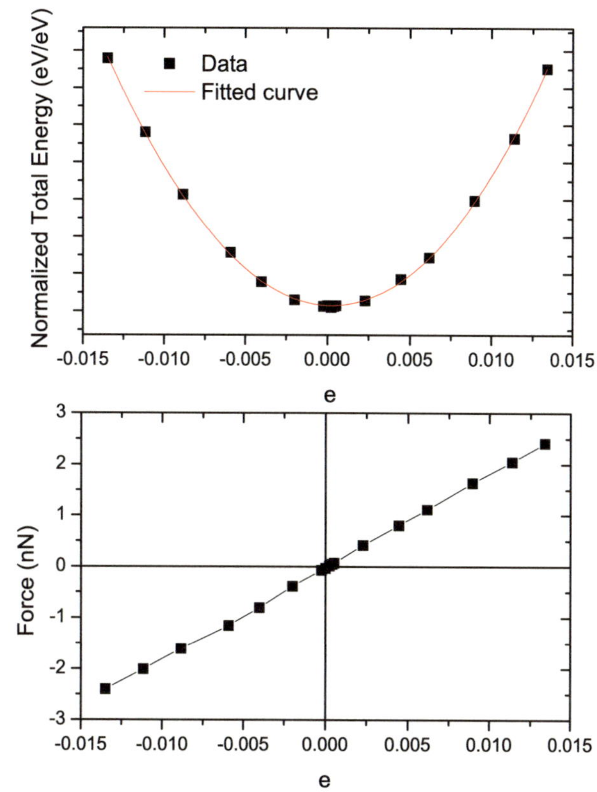

Fig. 5. Normalized total energy versus strain, and the corresponding force versus strain for N=10 ZGNR (Faccio et al. 2009).

While the second derivative of the total energy is easily obtained, the reference surface is ambiguously defined, with a dependence of the results upon the surface selection. In particular, the problem arises with the selection of the GNR's width, since it is a surface of pruned edges. In our case we have selected two different ways of determine the reference width of GNR: the shortest C-C width (d_A) and the longest C-C width (d_B). A sketch of these distances is presented in Figure 6(a). It is clear that neither of them are the best selection, and it becomes a problem when we want to compare these results in the N-infinity limit, corresponding to graphene. For this reason all the results are presented, together with the results for graphene. Figure 6(b) shows the variation of the E^{2D} upon the GNR's width N. The same results are presented in Table 3. To check the reliability of the calculations, the case of N=∞ (graphene) was studied. In this case we take a rectangular supercell with 32 carbon atoms. Each periodic crystalline axis were oriented along the zigzag and armchair directions, selecting a *c* value of 20 Å in order to avoid interference between images. The stress was applied along the zigzag axis. The obtained Young's modulus $E^{3D}=E^{2D}/c_0=$ 0.964(9) TPa agrees quite well with early reported values (see Table 1), as well as the Poisson's ratio υ= 0.17, that matches with the one reported by Kudin et al. (Kudin et al. 2001) υ=0.149 and Liu et al. (Liu et al. 2007) υ=0.186. It is another point that helps us to validate our methodology.

N	d_A	d_B	$E^{2D}{}_A$	$E^{2D}{}_B$
	Å		TPaÅ	
4	05.80	07.19	5.04	4.07
5	08.66	09.35	4.21	3.90
6	10.12	11.51	4.27	3.76
7	12.98	13.68	3.88	3.68
8	14.45	15.83	4.08	3.72
9	17.30	18.00	3.84	3.69
10	18.77	20.16	3.91	3.64
∞	-		3.23	

Table 3. Final E^{2D}'s Young modulus obtained from the different GNR width (d_i's)

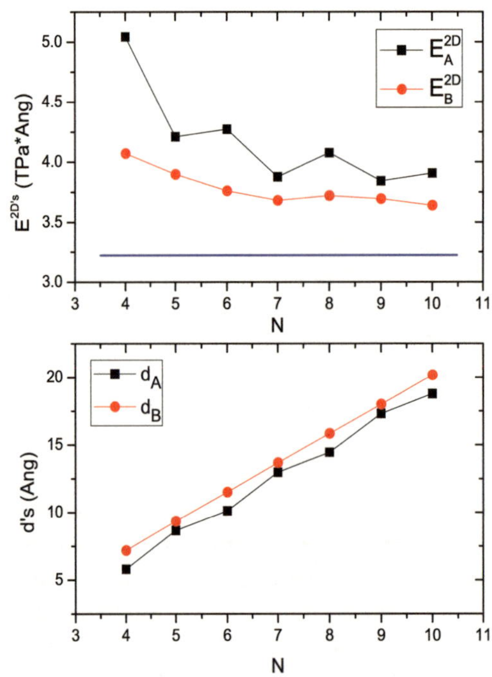

Fig. 6. The N=5 ZGNR sketching the distances: d_A(square-black) and d_B (circle-red); and the E^{2D}'s Young modulus according to the different distances considered in the model which are expressed in terms of the ribbon's width (N). The horizontal blue line corresponds to the graphene results.

The results show $E^{2D}{}_A$ and $E^{2D}{}_B$ decreases while N increases, always having a Young's modulus higher than the graphene one. We can argue that ZGNR are harder than graphene. This tendency is the opposite of the case for carbon nanotubes, and the reason can be easily explained in terms of graphene bending. The curvature of CNTs softens the rolled-up graphene sheet because of the lost of overlapping between of the sp^2 orbitals, with a pronounced effect for smaller tubes (Bogár et al. 2005). In the case of GNR the sheet is always plane, with a perfect sp^2 overlapping and strong stiffness. At first, this result would not be expected, but the response could be understood qualitatively in terms of two opposing effects: the curvature of graphene and geometrical edge reconstruction. The highest curvature corresponds with the lowest the orbital overlap, and hence with the lowest hardness. Furthermore our results indicate that the energy necessary to deform the ribbons (strain energy), expressed as energy per atom, is lower when more carbon atoms are involved, thus fewer atoms hardness of the material. The origin of this effect lies in the geometrical reconstruction of the C-C bonds positioned at the edge. As was mentioned in the introduction, the diagonal C-C distances of GNT at the edges contracts ~ 0.02 Å at the same time that the zigzag C-C-C angle increases ~2°, orientating the stronger C-C diagonal bonds more parallel to the periodic direction of the nanostructure and hardening the bonds. This effect is more evident in the case of thin GNR since there are few C-C bulk bonds, and as the GNR width increases, the bulk bonds prevails diluting the effect of the harder C-C bonds at the edge. In the case of nanotubes the relaxation effect on the edge does not exist, and therefore the curvature effect prevails.

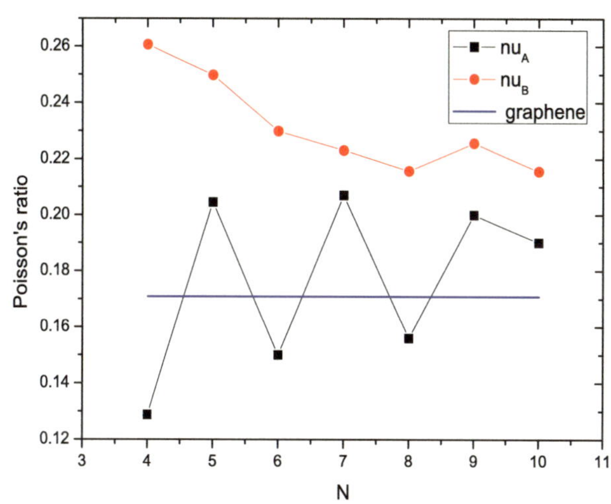

Fig. 7. The υ's dependence upon the GNR's width (N). The horizontal blue line corresponds to graphene results.

The Poisson's ratio presents a similar tendency to the one observed for the Young's modulus. The results are shown in Figure 7 and Table 4, where the (i=A & B) values are presented together with the value for graphene. The tendency between υ and N corresponds to a damped oscillation in the case of υ_A, while the dependency is smoother for the case of υ_B. In an extrapolated limit the infinite widths the ratios υ_i are: υ_A= 0.18 and υ_B= 0.22.

As was expressed in section 2, the shear modulus (G^{3D}), can be obtained employing equation 1.4 and taken the calculated values for Young's modulus and the Poisson's ratio. For graphene we obtained G^{3D}= 0.408 TPa. This value agrees with G^{3D}=0.384 TPa reported by Reddy et al. (Reddy et al. 2006), but differs in almost two times with those reported for Sakhaee-Pour (Sakhaee-Pour et al. 2009). Employing a force field method for finite graphene sheets, with different edge terminations, he obtained G values that range from 0.21 to 0.23 TPa. It is important to note that the corresponding Poisson's ratio reported by Sakhaee-Pour was obtained using equation 1.4. However our results are more similar to those reported for

N	υ_A	υ_B	G_A	G_B
4	0.129	0.261	0.667	0.482
5	0.204	0.250	0.522	0.466
6	0.150	0.230	0.555	0.456
7	0.207	0.223	0.480	0.449
8	0.156	0.216	0.526	0.457
9	0.200	0.226	0.478	0.450
10	0.190	0.216	0.490	0.447
∞	0.179		0.408	

Table 4. Poisson's ratio and estimated shear modulus for d_A and d_B models

SWCNT (Li et al. 2003; Tserpes et al. 2005) for which there have been reported G^{3D} values ranging from 0.250 to 0.485 TPa. This is a valid reference for our results, since in this case the mechanical load involves only a single graphene layer. This is the main reason why shear modulus of SWCNT are higher than MWCNT, since in this last case there exist sliding effect between nanotubes that reduces the shear modulus. In one hand, this discrepancy can be attributed to the different nature of the methods used for the simulation. On the other hand our results were estimated for two independent parameters, assuming equation 1.4 being valid. Regarding the dependence of shear modulus upon ribbon width, see Figure 8 and Table 4, what we found is a similar dependence to Young's modulus vs. N. This is an expected result since equation 1.4 is dominated by its numerator, corresponding to the Young's modulus, while the denominator remains almost constant, since the Poisson's ratio remains almost constant. Further simulations, including shear deformation, should be done in order to shed more light on this subject.

Regarding electronic structure features of GNR we found no significant dependence of its properties upon strain. These results agree to those early reported (see Table 1.), whereas for the case of zigzag GNRs it has been found a small variation of energy gaps and local magnetic moments, with no variation in the ordering of the occupied-bands. In our case the energy gaps increases in δE_{gap}=0.02 eV for a positive strain of ε= 0.02, and reduces in δE_{gap}= -0.02 eV for compressive strain ε= -0.02. These results are valid for all the studied GNR's widths. Similar results are obtained for local magnetic moments at the carbon edges, in all the cases the variation are in the order of ± 3% for the same strain range studied.

4. Conclusions

In summary, the electronic and mechanical properties of stressed ZGNR were calculated using *Ab initio* Density Functional Theory. The proposed models allowed us to obtain the corresponding Young's modulus, shear modulus and Poisson's ratio for ZGNR with different width. In all the cases the GNR present higher constants than graphene, but they approximate to this value when the GNR's width is increased. All of this indicates that the narrowest GNR could be one of the strongest materials. This effect could be explained in terms of the hardness of the C-C bonds positioned at the edges of the GNR, due to observed geometrical reconstruction. This property could lead to important consequences regarding the structure of the edge of this nanostructure because chemical substitution, the appearance of defects, and chemical doping could soft or stiff the edges. All these possibilities could lead to an important variation of the mechanical properties of GNR, in particular for the case of shorter GNR of low dimensional systems. It would be interesting to simulate the presence

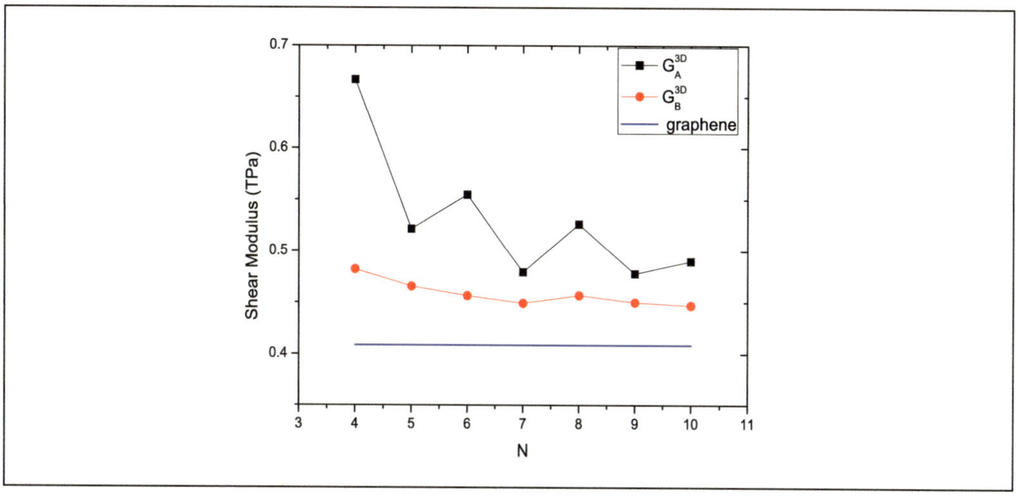

Fig. 8. Shear modulus G3D for GNR indicating the estimated value for graphene.

of strong donating and strong acceptor groups as functional groups substituting the presence of the single H atoms. Regarding the mechanical properties it has been observed a first order dependency of stress upon strain in the region from ε= -0.02 to ε= +0.02. A non-linear dependence is found for higher strain. Electronic structure features are not sensitive to strain in this linear elastic regime, being an additional promise for the using of carbon nanostructures in nano-electronic devices in the near future.

5. Acknowledgments

The authors thank the PEDECIBA Química, CSIC and Agencia Nacional de Investigación e Innovación (ANII) -Uruguayan organizations- for financial support.

6. References

Arroyo, M. & Belytschko, T. (2004). Finite crystal elasticity of carbon nanotubes based on the exponential Cauchy-Born rule. *Phys. Rev. B*, 69, 115415.

Becke, A. D. (1993). Density-functional thermochemistry. III. The role of exact exchange. *J. Chem. Phys.*, 98, 5648.

Bekyarova, E.; Itkis, M. E.; Ramesh, P.; Berger, C.; Sprinkle, M.; de Heer, W. A.; Haddon, R. C. (2009). Chemical Modification of Epitaxial Graphene: Spontaneous Grafting of Aryl Groups. *J. Am. Chem. Soc.*, 131(4), 1336-1337.

Blakslee, O. L. (1970). *J. Appl. Phys.*, 41, 3373.

Bogár, F.; Mintmire, J. W.; Bartha, F.; Mezo, T. & Van Alsenoy, C. (2005). Density-functional study of the mechanical and electronic properties of narrow carbon nanotubes under axial stress. *Phys. Rev. B*, 72, 085452.

Campos-Delgado, J. ; Romo-Herrera, J. M. ; Jia, X. T. ; Cullen, D. A. ; Muramatsu, H. ; Kim, Y. A. ; Hayashi, T. ; Ren, Z. F. ; Smith, D. J. ; Okuno, Y. ; Ohba TKanoh, H. ; Kaneko, K.; Endo, M. ; Terrones, H. ; Dresselhaus, M. S. &Terrones, M. (2008). Bulk production of a new form of sp(2) carbon: Crystalline graphene nanoribbons. *Nano Letters*, 8 (9), 2773-2778.

Ceperley, D. M. and Alder, B. J. (1980). Ground State of the Electron Gas by a Stochastic Method. *Phys. Rev. Lett.*, 45, 566-569.

Charlier, J.-C.; Gonze, X. & Michenaud, J.-P. (1994). First-principles study of the stacking effect on the electronic properties of graphite(s). *Carbon*, 32, 289-299.

Denis, P. A.; Faccio, R. (2008). Theoretical characterization of thioepoxidated single wall carbon nanotubes. *Chem. Phys Lett.*, 460, 486-491.

Denis, P. A.; Faccio, R.; Mombrú, A. W. (2009). Is It Possible to Dope Single-Walled Carbon Nanotubes and Graphene with Sulfur? *ChemPhysChem*, 10, 715-722.

Dion, M.; Rydberg, H.; Schröder, E.; Langreth, D. C. & Lundqvist, B. I. (2004). Van der Waals Density Functional for General Geometries. *Phys. Rev. Lett.*, 92, 246401.

Faccio, R.; Pardo, H.; Denis, P.A.; Yoshikawa-Oreiras, R.; Araújo-Moreira, F. M.; Verisimo-Alves, M. & Mombrú, A. W. (2008). Magnetism induced by single carbon vacancies in a three-dimensional graphitic network. *Phys. Rev. B.*, 77, 035416.

Faccio, R.; Denis, P.; Pardo, H.; Goyenola, C. & Mombrú, A. W. (2009). Mechanical properties of graphene nanorribons. *J. Phys: Cond. Matt.*, 21(8), 285304.

Friesecke, G. & Theil, F. (2002). Validity and failure of the Cauchy-Born hypothesis in a two-dimensional mass-spring lattice. *J. Nonlinear Sci.*, 12, 445-478.

Fujita, M.; Wakabayashi, K.; Nakada, K. & Kusakabe, K. (1996). Peculiar Localized State at Zigzag Graphite Edge. *J. Phys. Soc. Jpn.*, 65, 1920-1923.

Gui, G.; Li, J. & Zhong, J. (2008). Band structure engineering of graphene by strain: First-principles calculations. *Phys. Rev. B*, 78, 075435.

Hoheisel, T. N.; Schrettl, S.; Szilluweit, R. & Frauenrath, H. (2010). Nanostructured Carbonaceous Materials from Molecular Precursors. *Angew. Chem. Int. Ed.*, 49, 6496 - 6515

Hohenberg, P. & Kohn, W. (1964). Inhomogeneous Electron Gas. *Phys. Rev.*, 136, B864-B871.

Khare, R.; Mielke, S. L.; Paci, J. T.; Zhang, S.; Ballarini, R.; Schatz, G. C. & Belytschko, T. (2007). Coupled quantum mechanical/molecular mechanical modeling of the fracture of defective carbon nanotubes and graphene sheets. *Phys. Rev. B*, 75, 075412.

Kohn, W. & Sham, L. J. (1965). Self-Consistent Equations Including Exchange and Correlation Effects. *Phys. Rev.*, 140, A1133-A1138.

Khon, W.; Meir, Y. & Makarov, D. E. (1998). van der Waals Energies in Density Functional Theory. *Phys. Rev. Lett.*, 80, 4153-4156.

Kohanoff, J. and Gidopoulos, N. I. (2002). Density Functional Theory: Basics, New Trends and Applications - The Handbook of Molecular Physics and Quantum Chemistry, Ed. S. Wilson, Molecular Electronic Structure, John Wiley & Sons, ISBN 0471623741, Chichester.

Konstantinova, E.; Dantas, S. O. & Barone, P. M. V. B. (2006). Electronic and elastic properties of two-dimensional carbon planes. *Phys. Rev. B*, 74, 035417.

Kudin, K. N.; Scuseria, G. E. & Yakobson, B. I. (2001). C_2F, BN, and C nanoshell elasticity from *Ab initio* computations. *Phys. Rev. B*, 64, 235406.

Lee, C.; Yang, W. & Parr, R. G. (1988). Development of the Colle-Salvetti correlation-energy formula into a functional of the electron density. *Phys. Rev. B*, 37, 785-789.

Lee, C.; Wei, X.; Kysar, J. W.; Hone, J. (2008). Measurement of the Elastic Properties and Intrinsic Strength of Monolayer Graphene. *Science*, 321, 385-388.

Li, C.; Chou, T.-W. (2003). A structural mechanics approach for the analysis of carbon nanotubes. Int. J. Solid Struct., 40, 2487-2499.

Lier, G. V.; Alsenoy, C. V.; Doren, V. V. & Greelings, P. (2000). *Ab initio* Study of the Elastic Properties of Single-Walled Carbon Nanotubes and Graphene". *Chem. Phys. Lett.*, 326, 181.

Liu, F.; Ming, P. M. & Li, J. (2007). *Ab initio* calculation of ideal strength and phonon instability of graphene under tension. *Phys. Rev. B*, 76, 064120.

Lu, J. P. (1997). Elastic Properties of Carbon Nanotubes and Nanoropes. *Phys. Rev. Lett.*, 79, 1297-1300.

Miyamoto, Y.; Nakada, K. & Fujita, M. (1999). First-principles study of edge states of H-terminated graphitic ribbons. *Phys. Rev. B*, 59, 9859.

Monkhorst, H. J. & Pack, J. D. (1976). Special points for Brillouin-zone integrations. *Phys. Rev. B*, 13, 5188-5192.

Nakada, K.; Fujita, M.; Dresselhaus, G. & Dresselhaus, M. S. (1996). Edge state in graphene ribbons: Nanometer size effect and edge shape dependence. *Phys. Rev. B*, 54, 17954.

Ordejón, P.; Artacho, E. & Soler, J.M. (1996). Self-consistent order-N density-functional calculations for very large systems. *Phys. Rev. B*, 53, R10441

Palacio, F. & Makarova, T. (2005). *Carbon-based Magnetism,* ISBN 13:978.0.444.51947-4, Elsevier. B. V., Amsterdam, The Netherlands.

Perdew, J. P. and Zunger, A. (1981). Self-interaction correction to density-functional approximations for many-electron systems. *Phys. Rev. B*, 23, 5048-5079.

Perdew, J. P.; Burke, K.; and M. Ernzerhof, (1996). Generalized Gradient Approximation Made Simple. Phys. Rev. Lett., 77, 3865;

Perdew, J. P.; Burke, K.; and M. Ernzerhof (1997). Generalized Gradient Approximation Made Simple [Phys. Rev. Lett. 77, 3865 (1996)]. Phys. Rev. Lett., 78, 1396-1396.

Pereira, V. M.; Castro Neto, A. H. & Peres, N. M. R. (2009). Tight-binding approach to uniaxial strain in graphene. *Phys. Rev. B*, 80, 045401.

Press, W. H.; Flannery, B. P.; Teukolsky, S. A.; Vetterling, W. T. (1986). *New Numerical Recipes*, Cambridge University Press, New York.

Pisani, L.; Chan, J.A.; Montanari, B. & Harrison, N.M. (2007). Electronic structure and magnetic properties of graphitic ribbons. *Phys. Rev. B*, 75, 064418.

Reddy, C. D.; Rajendran, S. & Liew, K. M. (2005). Equivalent continuum modeling of graphene sheet. *Int. J. Nanosci.*, 4, 631-636.

Reddy, C. D.; Rajendran, S. & Liew, K. M. (2006).Equilibrium configuration and elastic properties of finite graphene. *Nanotechnology*, 17, 864-870.Xu, Z. (2009). Graphene Nano-Ribbons Under Tension. *Journal of Computational and Theoretical Nanoscience*, 6, 1-3.

Román-Pérez, G. & Soler, J. M. (2009). Efficient Implementation of a van der Waals Density Functional: Application to Double-Wall Carbon Nanotubes. *Phys. Rev. Lett.*, 103, 096102.

Rydberg, H.; Dion, M.; Jacobson, N.; Schröder, E.; Hyldgaard, P.; Simak, S. I.; Langreth, D. C. & Lundqvist, B. I. (2003). Van der Waals Density Functional for Layered Structures. *Phys. Rev. Lett.*, 91, 126402.

Sakhaee-Pour, A. (2009). Elastic properties of single-layered graphene sheet. *Solid State Communications*, 149, 91-95.

Sammalkorpi, M.; Krasheninnikov, A.; Kuronen, A.; Nordlund, K. & Kaski, K. (2004) Mechanical properties of carbon nanotubes with vacancies and related defects. Phys. Rev. B 70, 245416.

Sánchez-Portal, D.; Ordejón, P.; Artacho, E.; Soler, J.M. (1997). Density-functional method for very large systems with LCAO basis sets. *Int. J. Quantum Chem.*, 65, 453-461.

Shen, L. & Li, J. (2004). Equilibrium structure and strain energy of single-walled carbon nanotubes. *Phys. Rev. B*, 69, 045414.

Soler, J. M.; Artacho, E.; Gale, J.; D. García, A.; Junquera, J.; Ordejón, P.; Sánchez-Portal, D. (2002). The SIESTA method for *Ab initio* order-N materials simulation. *J. Phys.: Condens. Matter*, 14, 2745-.

Stephens, P. J.; Devlin, F. J.; Chabalowski, C. F. & Frisch, M. J.(1994). *Ab initio* Calculation of Vibrational Absorption and Circular Dichroism Spectra Using Density Functional Force Fields. *J. Phys. Chem.*, 98, 11623-11627.

Strocov, V. N.; Blaha, P.; Starnberg, H. I.; Rohlfing, M.; Claessen, R.; Debever, J.-M. & Themlin, J.-M. (2000). Three-dimensional unoccupied band structure of graphite: Very-low-energy electron diffraction and band calculations. *Phys. Rev. B*, 61, 7, 4994-5001.

(a) Son, Y.-W.; Cohen, M. L. & Louie, S. G. (2006). Half-metallic graphene nanoribbons. *Nature* (London) 444, 347-349.

(b) Son, Y.-W.; Cohen, M. L. & Louie, S. G. (2006). Energy Gaps in Graphene. *Phys. Rev. Lett.* 97, 216803.

Su, W. S.; Wu, B. R. & Leung, T. C. (2008). The Deformation Effect on the Electronic Structure of the Graphite Nanoribbon Arrays, *cond-mat*/0810.4582

Sun, L.; Li, Q.; Ren, H.; Su, H.; Shi, Q. W. & Yang, J. (2008). Strain effect on electronic structures of graphene nanoribbons: A first-principles study. *J. Chem. Phys.*, 129, 074704.

Tournus, F. ; Charlier, J. C. ; Mélinon, P. (2005). Mutual orientation of two C60 molecules : an *Ab initio* study. *J. Chem. Phys.*, 122, 094315.

Troullier, N.; Martins, J.L. (1991).Efficient pseudopotentials for plane-wave calculations. *Phys. Rev. B*, 43, 1993-2006.

Trucano, P. & Chen, R. (1975). Structure of graphite by neutron diffraction. *Nature (London), 258*, 136-137.

Tserpes, K. I.; Papanikos, P. (2005). Composites Part B: Engineering, 36(5), 468-477.

Vosko, S. H.; Wilk, L. & Nusair, M. (1980). Accurate spin-dependent electron liquid correlation energies for local spin density calculations: a critical analysis". *Can. J. Phys.*, 58(8), 1200-1211.

Wakabayashi, K.; Fujita, M.; Ajiki, H. & Sigrist, M. (1999). Electronic and magnetic properties of nanographite ribbons. *Phys. Rev. B*, 59, 8271.

Wirtz, L. & Rubio, A. (2004). The phonon dispersion of graphite revisited. *Solid State Communications*, 131, 141-152.

Wu, Y.; Huang, M.; Wang, F.; Henry Huang, X. M.; Rosenblatt, S.; Huang, L.; Yan, H.; O'Brien, S. P.; Hone, J. & Heinz, T. F. (2008). Determination of the Young's Modulus of Structurally Defined Carbon Nanotubes. *Nano Letters*, 8(12), 4158–4161.

Yoon, J.; Ru, C. Q. & Mioduchowski, A. (2005). Terahertz Vibration of Short Carbon Nanotubes Modeled as Timoshenko Beams. *J. Appl. Mech.*, 72(1), 10-17.

Yu, M. F.; Files, B. S.; Arepalli, S. & Ruoff, R. S. (2000). Tensile Loading of Ropes of Single Wall Carbon Nanotubes and their Mechanical Properties. *Phys. Rev. Lett.*, 84, 5552-5555.

Zhang, P.; Jiang, H.; Huang, Y.; Geubelle, P. H. & Hwang, K. C. (2002). The elastic modulus of single-wall carbon nanotubes: a continuum analysis incorporating interatomic potentials. Int. J. Solids Struct. 39, 3893-3906.

Zhang, Y.; Talapatra, S.; Kar, S.; Vajtai, R.; Nayak; S. K. and Ajayan, P. M. (2007). First-Principles Study of Defect-Induced Magnetism in Carbon. *Phys. Rev. Lett.*, 99, 107201.

18

Fracture Behaviors of Graphene Sheets and Carbon Nanotubes

J. R. Xiao[1*] and J. W. Gillespie Jr.[1,2,3**]
[1]*Center for Composite Materials,*
[2]*Department of Materials Science and Engineering,*
[3]*Department of Civil and Structural Engineering,*
University of Delaware,
USA

1. Introduction

Graphene sheet is a one-atom-thick planar sheet of sp2-bonded carbon atoms that are densely packed in a hexagonal crystal lattice. Graphene is the basic structural element of some carbon allotropes including graphite, carbon nanotubes and fullerenes. Carbon nanotubes (CNTs) (Iijima, 1991) exhibit exceptional physical properties: small size, low density, high stiffness, high strength and excellent electronic and thermal properties (Dresselhaus et al., 1996; Wong et al., 1997; Che et al., 2000; Yakobson and Avouris, 2001; Thostenson et al., 2001). These exceptional mechanical and physical properties along with low weight of CNTs and recent improvements in their synthesis and purification techniques make CNTs excellent candidates for use in tailoring properties of composites (Wagner et al., 1998; Cadek et al., 2002; Dalton et al., 2003).

Molecular mechanics/dynamics and *ab initio* methods are suitable for studying the mechanical properties and fracture behaviors of grapheme sheet and nanotubes but are limited in scale and are computationally expensive. Recently, continuum mechanics based models for CNTs have been developed using the harmonic energy potential (Li and Chou, 2003; Chang and Gao, 2003). These models reduce computational cost significantly, but can only be used to investigate elastic properties, such as the Young's modulus or Poisson's ratio. To fully predict the stress-strain relationship and failure mechanism of CNTs, these methods are not sufficient. The Brenner potential function (Brenner, 1990) is considered more accurate and versatile. It can handle changes in atom hybridization and bonds with atoms other than carbon. A continuum mechanics approach directly incorporating the Brenner potential function has been developed by Huang's group (Zhang et al., 2002; Jiang et al., 2003) to model elastic properties and stress-strain relationships of carbon nanotubes based on a modified Cauchy-Born rule. Based on the modified Morse potential function (Belytschko et al., 2001), which is simpler than the Brenner potential, the authors have developed models for perfect and defective CNTs (Xiao et al., 2005; Xiao et al., 2009). The developments have the ability to predict the ultimate stress and other mechanical properties, including nanotube's nonlinear stress-strain relationship. The analytical model (Xiao et al., 2005) has been extended to solve mechanical responses of defect-free single- and multi-walled CNTs under internal and external pressure loadings (Xiao et al., 2006; Xiao and Gillespie, 2006a) as well as aligned nanotube-composites (Xiao and Gillespie, 2006b). The analytical model not only provides simple closed-form solutions but also presents a better insight of the role of the atomic networks.

The present work discusses the atomistic based finite bond element model and its application to study the effects of Stone-Wales (5-7-7-5) defects on mechanical properties and fracture behaviors of graphene sheets and carbon nanotubes. The element formulation includes eight degrees of freedom reducing computational cost compared to the 12 degrees of freedom used in other FE type models. The coefficients of the elements are determined based on the analytical molecular structural mechanics model developed by the authors. The model uses the modified Morse potential to predict the Young's modulus and stress-strain relationship of perfect and defective nanotubes and graphene sheets. Research involving the Stone-Wales defect preformed by Tserpes, and Papanikos (2007) using the Finite Element (FE) based model does not include the deformation of the original nanotube structure around the nucleation site, which may not be true in general as atoms redistribute to minimize energy. The authors (Xiao et al., 2009) proposed a simple way to simulate the formation of a Stone-Wales (5-7-7-5) defect using an interaction mechanics method to calculate the deformations caused by the formation of a Stone-Wales.

The variation of ultimate stress, strain at failure, and Young's modulus values of carbon nanotubes and graphene sheets has been examined as a function of the distance between two defects aligned in the axial and hoop directions has been studied. The fracture failure patterns will be presented and discussed. The mechanical properties as a function of the number of defects in the hoop direction are also studied. It is found that the moduli are sensitive to the tube lengths when the total tube length is used to compute the overall effective axial strain. If one uses a characteristic length based on the size of the local perturbation in the deformation fieldcreated by the defect to define the strain, a size independent modulus can be obtained for the defective region. The diameter of the affected region (2nm) from a single defect is defined as the defective length and is used for all different tube lengths examined in the present study. The effects of defect density on mechanical properties of tubes of any lengths are also discussed. Progressive failure in the post failure region is investigated and a few computational examples are discussed. The predictions compare favorably to the corresponding published results from experiments and numerical calculations (i.e tight binding or molecular dynamics) for armchair and zigzag carbon nanotubes.

2. An atomistic based finite bond element model

A single-walled carbon nanotube can be viewed as a hollow cylinder rolled from a graphene sheet, composed of carbon hexagons. The diameter of the nanotube can be calculated as $d = \sqrt{3(n_1^2 + n_2^2 + n_1 n_2)} / \pi$, where a = 0.142 nm is the C-C bond length, and the pair of integers (n_1, n_2) are indices to represent its helicity such as armchair ($n_1 = n_2$) and zigzag ($n_2 = 0$) nanotubes.

There are several different potential functions available (Brenner, 1990; Abell, 1985; Tersoff, 1988) for describing C-C bond interaction other than simple harmonic functions. Among them, the modified Morse potential function (Belytschko et al., 2001) is simple and is used in the present study. The modified Morse potential function was correlated to the Brenner potential function for strains below 10%. In this paper, a new term $E_{torsion}$ is added to consider the bond energy due to angle variation of bond twisting $\Delta\phi$ which has been found to be negligible for defect-free tubes (Li and Chou, 2003) and was neglected in earlier studies (Chang and Gao, 2003; Xiao et al., 2005; Xiao et al., 2006; Xiao and Gillespie, 2006a; Xiao and Gillespie, 2006b). However, this term may play a more important role in defective tubes. The energy potential function is given as follow:

$$E = E_{stretch} + E_{angle} + E_{torsion} \tag{1}$$

where $E_{stretch}$ is the bond energy due to bond stretch Δr, and E_{angle} is the bond energy due to bond angle variation $\Delta\theta$, and

$$E_{torsion} = \frac{1}{2} k_{\phi} (\Delta\phi)^2 \tag{2}$$

The parameters associated with the terms $E_{stretch}$ and E_{angle} can be seen in Ref (Belytschko et al., 2001). The force constant associated with the term (2) is taken as (Li and Chou, 2003; Cornell et al., 1995; Jorgensen and Severance, 1990) $k_{\phi} = 0.278 \quad nN - nm / rad^2$.
The stretch force, the angle-variation moment and the torsional moment can be obtained from differentiations of Eq. (1) as functions of bond stretch, bond angle variation and torsion angle variation, respectively:

$$F(\Delta r) = 2\beta D_e \left(1 - e^{-\beta\Delta r}\right) e^{-\beta\Delta r} \tag{3a}$$

$$M(\Delta\theta) = k_{\theta} \Delta\theta \left[1 + 3k_{sextic} (\Delta\theta)^4\right] \tag{3b}$$

$$M_{\phi}(\Delta\phi) = k_{\phi} \Delta\phi \tag{3c}$$

A constant bond torsional stiffness is implied by Eq. (3c). The stretch stiffness and the angle-variation stiffness can be further obtained from differentiations of Eqs. (3a-3b) as functions of bond stretch and bond angle variation, respectively:

$$k_r(\Delta r) = 2\beta D_e^2 \left(1 - 2e^{-\beta\Delta r}\right) e^{-\beta\Delta r} \tag{4}$$

$$k_{\theta}(\Delta\theta) = k_{\theta} \left[1 + 15k_{sextic} (\Delta\theta)^4\right] \tag{5}$$

Analytical solutions for predicting nonlinear mechanical behaviors of defect-free SWCNTs have been investigated by using an effective "stick-spiral" model based on a unit cell approach (Xiao et al., 2005). The effective "stick-spiral" model uses a stick with Eq. (3a) to model the force-stretch relationship of the C-C bond and a spiral spring with Eq. (3b) to model the angle bending moment resulting from an angular variation of bond angle. The stick is assumed to have an infinite bending stiffness and finite torsional stiffness. One can include the torsional term by using the stick with Eq. (3c).
For a defective nanotube with loss of local symmetry, the unit cell approach becomes inappropriate, and a generalized molecular mechanics (MM) model (Sears and Batra, 2004) or a finite element (FE) type model (Li and Chou, 2003) are needed to include the entire molecular structure system of the defective nanotube in order to consider the effect of

defects on its mechanical response. In this paper, we propose a new FE type model for nanotubes based on the effective "stick-spiral" model. Bond elements are developed for simulating deformation modes of the chemical bonds.

Typically the FE type model (Li and Chou, 2003) uses a beam element (Kanchi 1993) with sectional stretch stiffness for the force-stretch relationship of the C-C bond, sectional flexural rigidity for the angle bending moment (Eq. 3b), and sectional torsional stiffness for the torsional moment (Eq. 3c). Each element has 12 degrees of freedom. The major differences between the FE type model and the effective "stick-spiral" model are the assumptions made with respect to the bending stiffness (flexible vs rigid). The bond element used in the present paper has infinite bending stiffness (the stick) with finite bending stiffness of the two end joints (the spirals) indicated by the square box shown in Fig. 1. Each of the three molecular deformation modes (stretching, angle variation, and angle torsion) are represented by tension, bending and torsion of a bond element with 8 degrees of freedom $u_e = [u_{xi} \quad u_{yi} \quad u_{zi} \quad \phi_i \quad u_{xj} \quad u_{yj} \quad u_{zj} \quad \phi_j]$. The element can be stretched (pure tension) and torqued (pure torsion) along its axial direction and bent by relative transverse displacement without angle changes at the two ends. The bond angle variation of the bond element can be associated with the relative transverse displacement (e) between the two ends as $d\theta = e/a$ (Fig. 1c). The present approach, designated the finite bond element model, is expected to give the same solution as the "stick-spiral" model (Xiao et al., 2005; Chang and Gao, 2003) for defect-free CNTs. The stick-spiral model with infinite bending stiffness represents the true physical deformation modes and is able to predict both in-plane stiffness (Young' modulus) and Poisson's ratio of CNTs accurately.

For the bond element defined in Fig. 1 in a three-dimensional space, the elemental equilibrium equation can be established for every bond element. The final system of equations with appropriate boundary conditions imposed can be solved by the displacement-control Newton-Raphson method. A MATLAB program has been written based on finite bond element method for SWCNTs subjected to tensile loadings.

3. Validation on defect-free SWCNTs

The calculations based on our FE-based finite bond element model will be validated by comparing with the results obtained from the analytical model (Xiao et al., 2005) for defect-free CNTs. The initial equilibrium state of the atomistic system of a CNT is created based on the *ab initio* calculations (Ye et al., 2001) where the approximate expressions of angle α and β of armchair nanotubes are taken as $\alpha \approx 2\pi/3$ and $\beta = \pi - \arccos[0.5\cos(\pi/2n_1)]$, respectively. The angle α and β for zigzag nanotubes as $\alpha \approx 2\pi/3$ and $\beta = \arccos[0.25\text{-}0.75\cos(\pi/n_1)]$. The displacement boundary conditions for the simulations are shown in Figure 2. The atoms on the bottom edge of the tube are fixed in the axial (z) direction. One atom is fully fixed in all three (x-, y- and z) directions to prevent rigid body rotations. The other end of the CNT was axially displaced incrementally to introduce load into the tube.

The interatomic force per atom (f) is calculated for all atoms along the end of the nanotube where the displacement is prescribed. For defect-free nanotubes, the force is identical for all atoms on the end. The axial strain of the CNT is computed as $\varepsilon = \Delta L/L_0$, where L_0 is the initial length of the CNT. The force-strain relationship of nanotubes is predicted using the procedures outlined above. Fig. 3 shows the computed force-strain relationships for

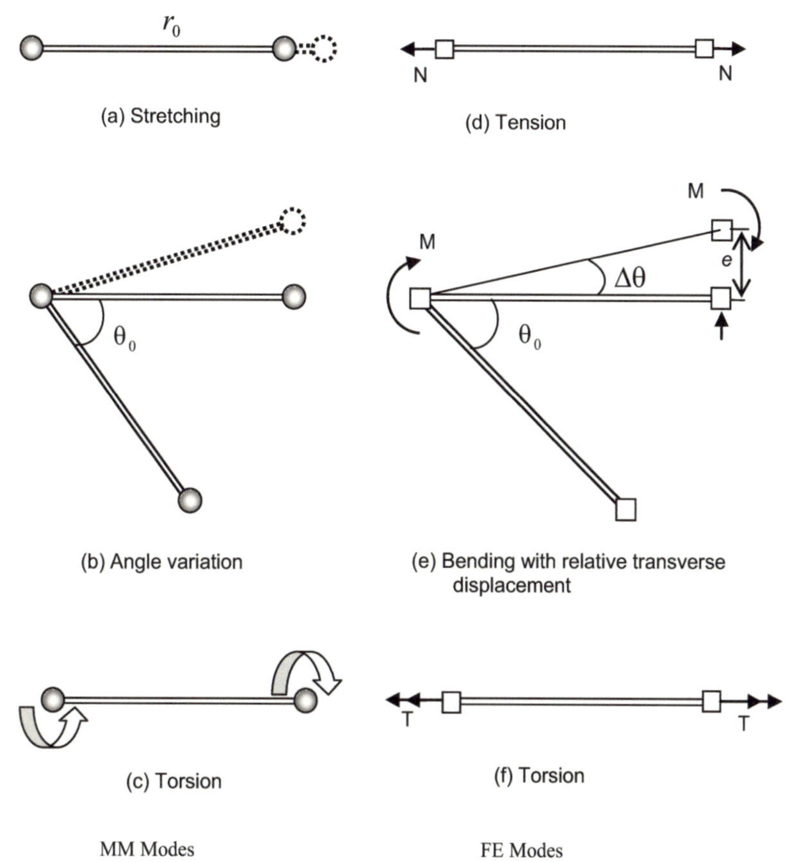

Fig. 1. Molecular mechanics modes are compared with finite bond element method

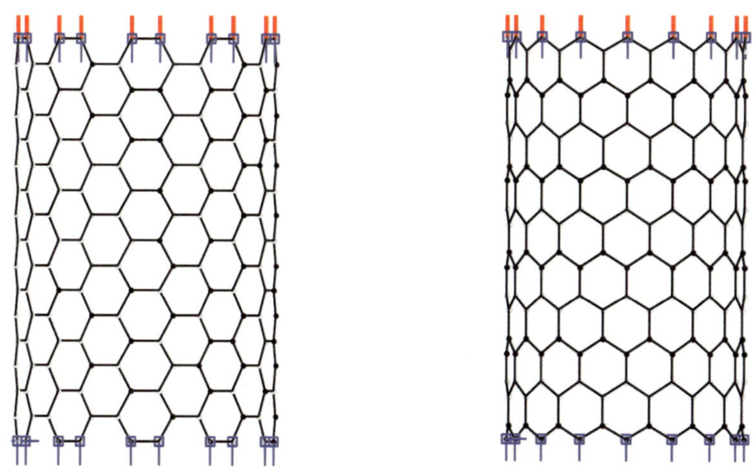

a (10, 10) nanotube   a (16, 0) nanotube

Fig. 2. Atomistic armchair and zigzag nanotubes (blue bar denotes constrained displacement and red bar indicates prescribed displacement) armchair and zigzag nanotubes. Only two different types of nanotubes (i.e. a (12, 12) armchair and a (20, 0) zigzag) are presented for illustration purpose.

The numerical procedure is able to give the post failure behavior (i.e. beyond the inflection point where the maximum of the interatomic force occurs) as shown in Fig. 3. Similar predictions have been reported in the study (Jiang et al., 2003). However, the accuracy of the predicted post failure response (dash lines) is limited to the strains shown in Fig. 3 (vertical solid line) since the simple interatomic potential function used is not capable of describing the behavior of the nanotube after the bonds are broken where the formation of new bonds, rehybridization and structural transformations may occur.

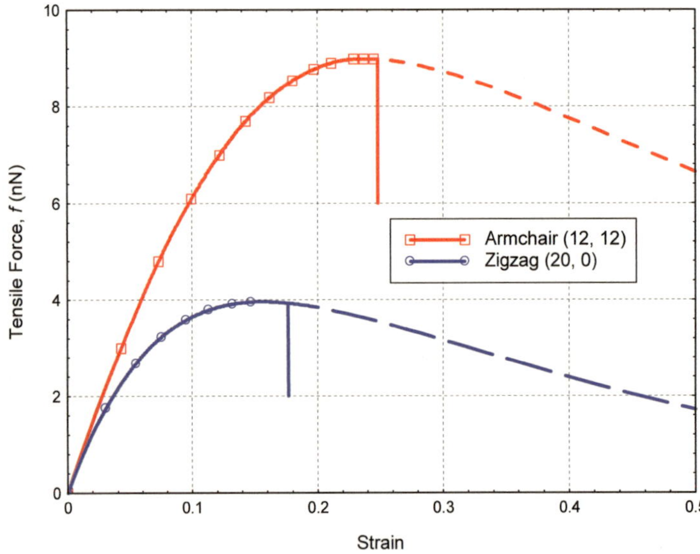

Fig. 3. Tensile force-strain curves for armchair and zigzag nanotubes

From the experimental (Yu et al., 2000) and theoretical (Belytschko et al., 2001) studies on the tensile behavior of nanotubes, it was found that the stress exhibits a sudden drop to zero when stress reaches the tensile strength and the fracture is brittle. The predicted maximum load corresponding to the inflection point is taken as the tensile capacity of the defect free nanotube in the present study which is highlighted in Fig. 3.

It should be noted that the present method does not require a tube thickness to be defined. However, in order to compare the results with published data, conventional moduli and strengths can be calculated using an assumed wall thickness for CNTs (e.g. t = 0.34 nm is used in the literature). The total force carried by the CNTs can be given as $F = 2n_1 f$, and then the stress can be computed as $\sigma = F/\pi dt$ and the Young's modulus as $E = \frac{F/\pi dt}{\Delta L/L_0}$, respectively. It is not surprising that the predicted tensile behaviors are almost identical to those of the analytical molecular mechanics model (Xiao et al., 2005) because the coefficients used in the present atomistic based FE model are the same as those used in the analytical model. The negligible difference (less than 0.5%) is attributed partially to the effects of the torsional mode and numerical round-off error. Consequently, the present study also confirmed that the torsional term in the potential energy is negligible when the defect free tubes are subjected to tensile loading which justifies the assumptions made in other studies.

The size-dependent feature of in-plane stiffness and Poisson's ratio can also be captured by the present model. The Young's moduli for both armchair and zigzag nanotubes decrease with decreasing tube diameter and approach the predicted graphite value (383 J/m^2, corresponding to a Young's modulus of 1.13 TPa with thickness of 0.34nm) when the tube diameter is increased. The Poisson's ratios for both armchair and zigzag tubes decrease with increasing tube diameter and approach the limit value of 0.20 for graphene sheet, which agrees with both molecular dynamics simulations and experimental values. It should be noted that the similar FE based model (Li and Chou 2003) predicted a very low value (<0.08) for the Poisson's ratios of graphene sheets and nanotubes. Consequently, the present formulation overcomes this limitation in previous work. In the context of the present study, accurate modeling of Poisson interactions in the prediction of defect formation is important as shown in the next section.

The predicted nonlinear behaviors of nanotubes are very similar to results calculated from molecular dynamics (Belytschko et al., 2001). The predicted strengths and failure strains are significantly higher than the experimental values (11 ~ 63 GPa for strength and 10% ~ 13% for failure strain) (Yu et al., 2000). This difference can be partially explained by the presence of defects that reduce stiffness and strength. The mechanical behavior of defective nanotubes is studied in the following section.

4. Progressive failure of defective SWCNTs

4.1 Stone-wales defect formation

Various types of defects exist in CNTs, such as vacancies and topological defects (5-7-7-5). The Stone–Wales 5-7-7-5 defect involves the 90° rotation of a carbon bond with a new configuration as shown in Fig. 4. The effect of the SW defect on the configuration is believed to be local and limited to atoms in the neighborhood of the defect. Atoms far away from the defect undergo uniform deformation and their geometry configurations are the same as the defect-free tubes. For atoms near the defect, an interaction (similar to contact) mechanics concept is used to determine their initial equilibrium positions, which is possible by treating the defect formation as the result of interaction between two deformable 'bodies' as shown in Fig. 5. 'Inserting' the body *b* of a zigzag unit into the body *a* with an armchair hole as

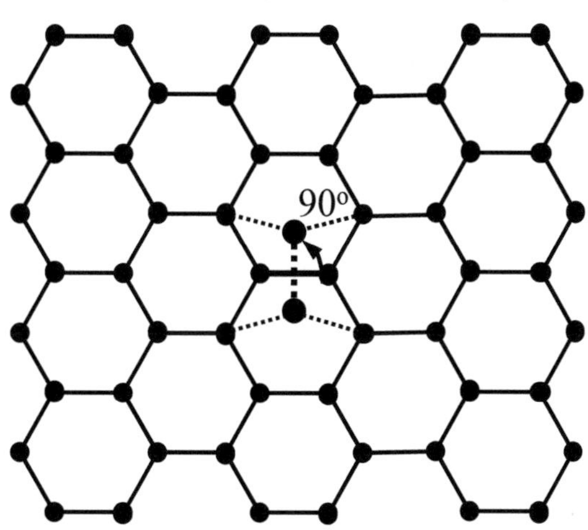

Fig. 4. SW defect generated by rotating the C–C bond

defined in Fig. 5 and forcing them to be coincident at four specific locations will cause re-configuration of the system within a localized region because of the geometrical distortions that exist at the four locations. Assuming the same in-plane properties for each body (governed by the same potential function), self-equilibrating 'residual' or internal forces will be generated between disturbed atoms if we assume that there is no net atomic force between atoms at equilibrium state of defect-free region (reference state). The resulting configuration corresponds to the minimized energy state of a defective graphene sheet and CNT.

The finite bond element method is used for describing atomistic force-displacement interaction for both bodies to study how the two regions conform to each other. The two deformable bodies Ω^e ($e = a, b$) are shown in Fig. 5. The interaction sites (contact 'surface') consists of four atoms (1, 2, 3 and 4) that exist in both bodies with prescribed displacements along Γ_u^e and tractions applied on Γ_t^e ($e = a, b$). If the initial gap along the interaction surface of the two bodies is denoted by g (= (u_{xi} u_{yi} u_{zi}), i = 1,2,3,4) the interaction condition can be described in the form of inequalities as:

$$F_c \le 0 \qquad g \ge 0 \tag{8a}$$

$$g\, F_c = 0 \tag{8b}$$

where F_c (F_{xi} F_{yi} F_{zi}) is the interaction force vector of the four interfacial atoms. For each body, displacement and force fields are determined by the molecular mechanics and can be written in the following condensed matrix form:

$$\begin{bmatrix} K_a & 0 \\ 0 & K_b \end{bmatrix} \begin{Bmatrix} u_a \\ u_b \end{Bmatrix} = \begin{Bmatrix} F_t - F_c \\ F_c \end{Bmatrix} \tag{9}$$

in which u_a and u_b are components of the displacement vector (same size as the gap vector g) of the four atoms in the body a and b, respectively. F_t is the applied external force vector. K_a is the condensed stiffness matrix of the body a with the displacements u_a as the basic unknowns and contents of stiffness contributions from all elements of the body a. K_b is the condensed stiffness matrix of the body b. By introducing the following transformation:

$$\begin{Bmatrix} u_a \\ u_b \end{Bmatrix} = \begin{bmatrix} I & I \\ 0 & I \end{bmatrix} \begin{Bmatrix} u_a \\ g \end{Bmatrix} \tag{10}$$

where [I] is the identity matrix. Eq. (9) can be easily transformed to:

$$Kg = F + F_c \tag{11}$$

in which

$$K = K_a - K_a \left[K_a + K_b\right]^{-1} K_a \tag{12a}$$

$$F = F_t - K_a \left[K_a + K_b\right]^{-1} F_t \tag{12b}$$

So that the force F_c, which is required to close the gap to achieve coincident atom locations, can be calculated from Eq. (11) for the initial ($F = 0$) and/or loaded states as:

$$F_c = -Kg - F \tag{13}$$

The gap function is generated by aligning nodes 1 and 1′ of two bodies in Fig. 5. The boundary condition applied to the system matrix in Eqs. 11 and 12 is that node 1 is fixed. With the force F_c known one can determine the equilibrium positions of atoms in the defective system which corresponds to the minimized system energy. It should be noted that the proposed analysis involves a nonlinear procedure since the modified Morse potential energy function is used. The formation energy can be calculated as $\int F_c dg \approx \sum_i 0.5F_c^i(dg^i + dg^{i+1})$ or simply calculated as $0.5F_c\,g$ without significant loss of accuracy.

The formation energy of a single SW defect occurring at the center of nanotubes is slightly different for various tube diameters examined. A typical value of 2.86eV for a (12, 12) tube agrees with the range of published values (2.5~3.5eV) (Nardelli et al., 1998; Pan et al., 2000). Generally the larger diameter requires the higher formation energy with a plateau value of 2.97eV for a graphene sheet based on the present model as shown in Fig. 6. The calculation of the defect formation energy is more complicated than the approximate method presented in this work. However, we have compared our simple approach to predict the defect formation energy and found reasonable accuracy with published molecular dynamic simulations. The formation of SW defect causes elongation of the local tube structure along the axial direction, but also shrinking along the hoop direction. Fig. 7 shows the affected area and configuration due to the formation of the SW defect in a graphene sheet and the (12, 12) CNT. The shrinking can be seen from the side view of the defective tube as shown in Fig. 7.

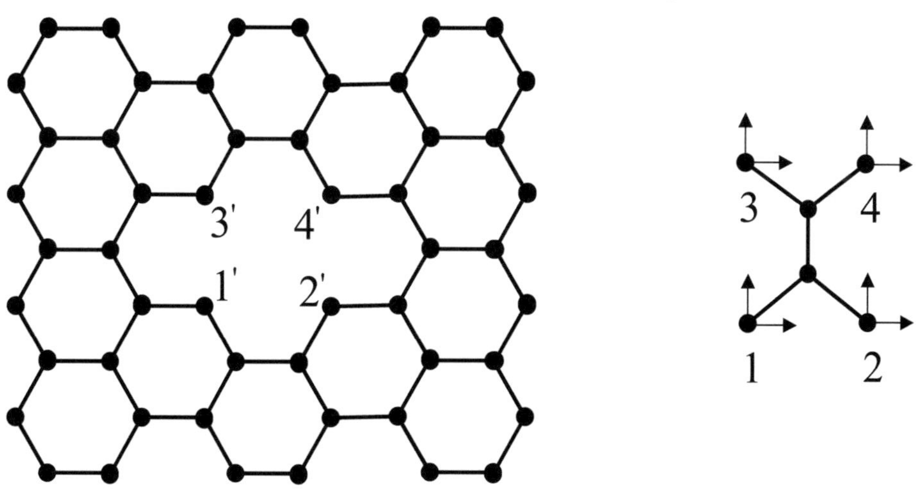

(a) Ideal armchair configuration with a hole (b) A zigzag unit

Fig. 5. Modeling of atom position rearrangements due to SW defect formation

Based on our simulations, the affected region in a tube/sheet is localized and covers four neighboring rings (hexagons) around the defect. The affected area is slightly sensitive to the tube diameter when the tube diameter is smaller than 2 nm and becomes insensitive for larger diameters. The molecular structure outside the region remains undisturbed. All chemical bonds within the affected region are preloaded by the presence of the SW defect.

With the application of external tensile force on the tube ends, the defect region becomes the weakest point and will serve as a damage nucleation site for progressive failure as given in the following section.

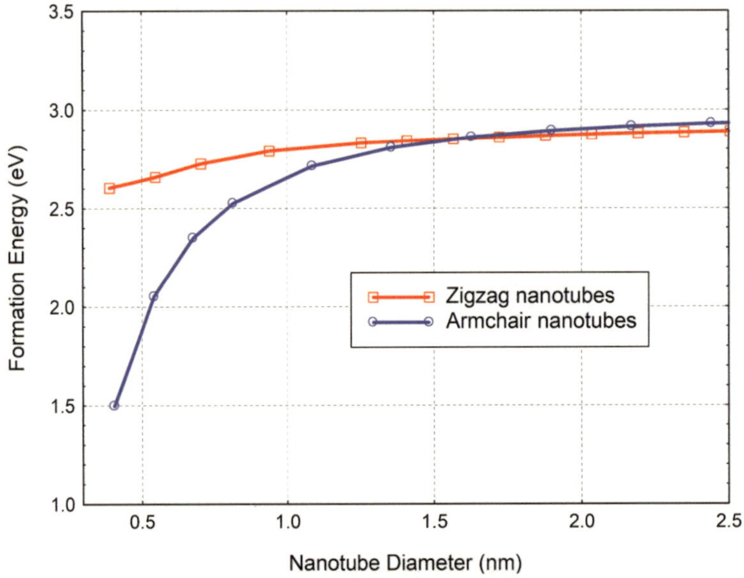

Fig. 6. Formation energy of SW defect vs nanotube diameter

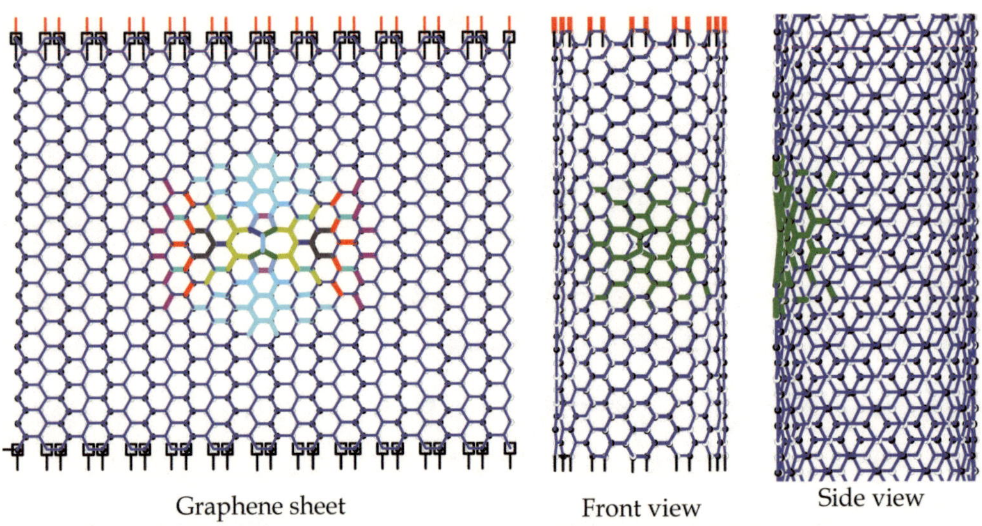

Fig. 7. Formation of SW defect in a graphene sheet and a (12, 12) SWCNT

4.2 Progressive failure analysis

After the defect formation in a graphene sheet or nanotube, one can apply further load (tension) onto the defective sheet and tube as shown in Fig. 7, where the load and boundary conditions are applied in the same fashion as described above. The tube length is taken as 6 nm for all simulations.

The bond breaking criterion brings an important issue in simulating fracture of solids. Very often, a cut-off distance r_{cf} is used in atomistic simulations. Such distance-based criterion is adopted in the present study. In our analytical study (Xiao et al., 2005), the inflection point (corresponding to a r_{cf} = 0.168 nm) was used. In the present study the effect of the cut-off distances on the fracture and failure of defective graphene sheets and nanotubes will be investigated.

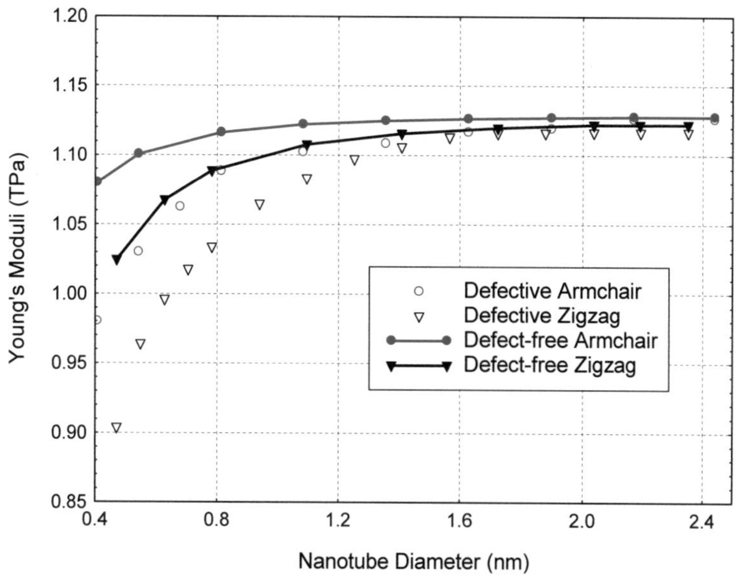

Fig. 8. Young's moduli of carbon nanotubes vs tube diameter

Different cut-off distances have been suggested (Xiao et al., 2005; Lu and Bhattacharya, 2005; Huhtala et al., 2004; Xia et al., 2002; Dumitrica et al., 2003). It has been found that there is no significant change in the critical bond-breaking force with values ranging from 0.17 nm to 0.19 nm. Clearly, the cut-off distance used in our analytical study (Xiao et al., 2005) is close to this range. A value of r_{cf} = 0.175 nm is adopted in the present study unless otherwise stated. As shown in the study (Belytschko, et al., 2003), the force field shape of the modified Morse potential function is essentially the same as that of the Brenner potential function before the inflection point (i.e. the maximum of the interatomic force) and totally different after the inflection point. Belytschko, et al. (2003) reported that the fracture is essentially independent of the separation energy and depends primarily on the inflection point of the interatomic potential, i.e. the shape of the potential function after the inflection point is not important to fracture behavior. Consequently, any cut-off distances beyond the inflection point will not affect the fracture response but will significantly increase the computational cost.

Fig. 8 shows the calculated Young's modulus of defective nanotubes (open points) compared with defect-free tubes (solid lines). It can be seen that the Young's moduli are size dependent for both defective and defect-free tubes, and the Young's moduli of defective tubes are reduced because of the presence of SW defect, particularly for small diameter tubes (10% reduction for diameters ranging from 0.4-0.6 nm). Such reduction becomes insignificant for larger diameter (> 2nm) tubes simply because the ratio of the affected region (almost constant) over the whole area along the hoop direction decreases as the tube diameter increases.

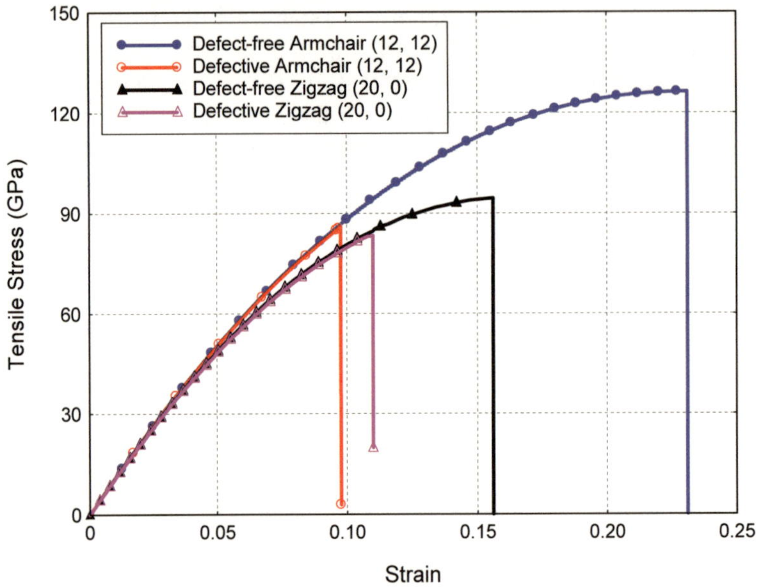

Fig. 9. Tensile force-strain curves of carbon nanotubes

Fig. 9 shows the calculated stress-strain relationships for armchair and zigzag nanotubes with and without 5-7-7-5 (SW) defect at the center. Again, only two different types of nanotubes ((12, 12) armchair and (20, 0) zigzag with diameter around 1.6 nm) are presented. As can be seen from Fig. 9, the predicted tensile strength (85.9 GPa) of defective armchair nanotubes is much less than that (126.2 GPa) of defect-free armchair nanotubes (reduced 32%). Whereas, the effect of the SW defect on the failure strength of zigzag tubes is less significant for armchair tubes because of nanotube chirality. The failure strength of defective (20, 0) zigzag tube is 83.3 GPa, which is 12% reduced from the pristine value of 94.5 GPa. The predicted failure strains in the present study are 9.8% for the defective (12, 12) armchair nanotube (23.1% for the defect-free), and 11.0% for the defective (20, 0) zigzag nanotube (15.6% for the defect-free). All these predictions agree well with the MD results (Belytschko, et al., 2003). It should be noted that the predicted strengths are still significantly higher than the experimental values (11 ~ 63 GPa) of Yu et al. (2000) though the predicted failure strains are comparable to the measured results (10% ~ 13%) (Yu et al., 2000). This issue can be partially explained by the present of multiple SW defects (Lu, B. Bhattacharya, 2005; Dumitrica et al., 2003) as well as other types of defects (Xiao and Hou, 2006).

Tubes are found to exhibit brittle behavior at fracture. Once the tube deformation reaches a critical level (corresponding to the bond cutoff distance), atomic bonds break successively and lead to a complete fracture with little strain applied. The effect of the cutoff distance on the progressive failure has been examined with three different cutoff values (0.168, 0.175 and 0.185 nm) used. It is found that the ultimate strengths and failure strains are almost identical (difference is less than 0.1%) for all three values examined and the failure is brittle no matter what value is used for the cutoff distance. However, the computing time with r_{cf} = 0.185 nm is much longer than those with other two smaller values. The effect of the torsional term on the mechanical behaviors of defective tubes was also assessed. The present study shows that the torsional term in the potential energy still plays a minor role when the defective tubes are subjected to tensile loading.

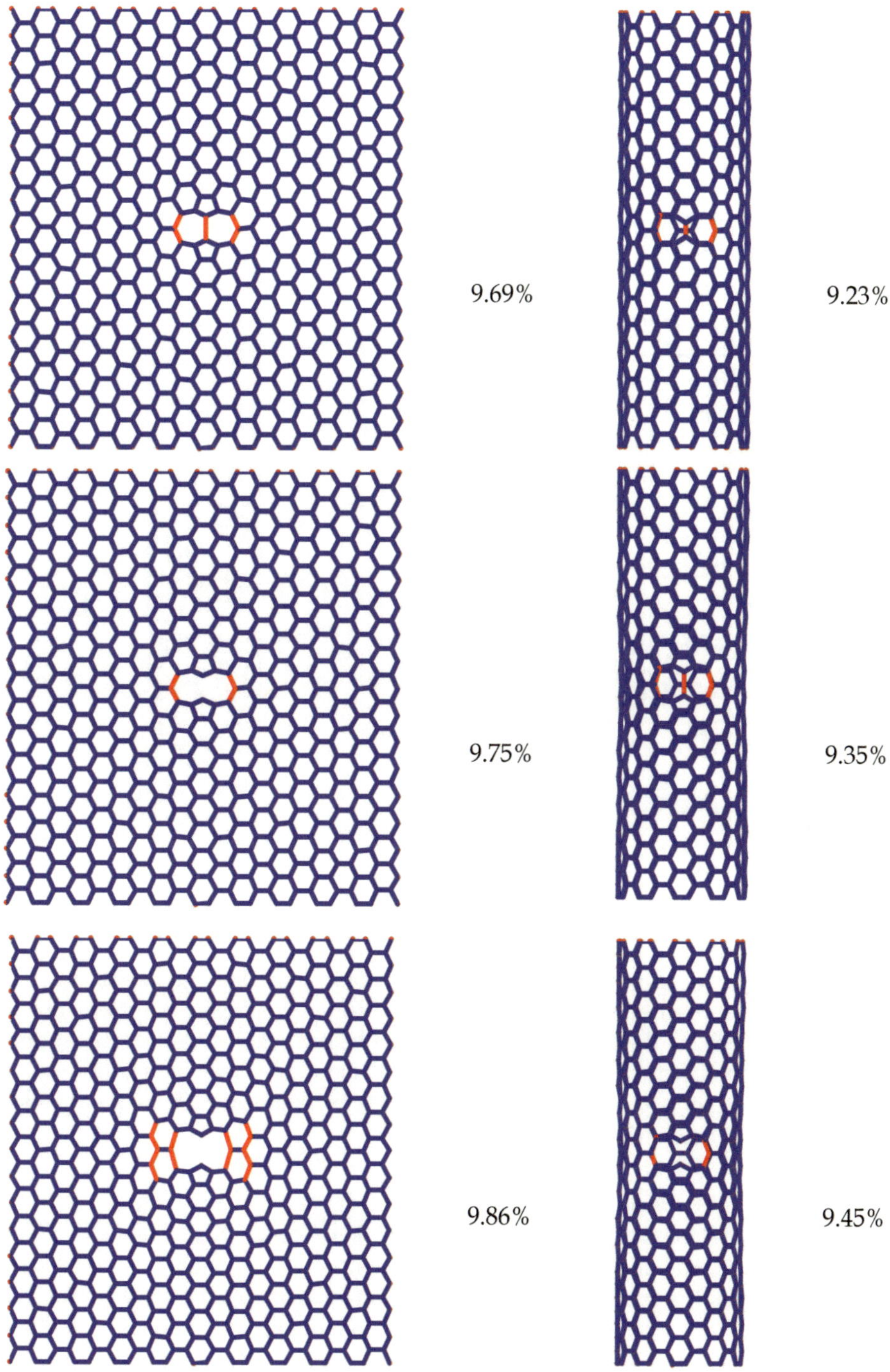
9.69%
9.23%
9.75%
9.35%
9.86%
9.45%

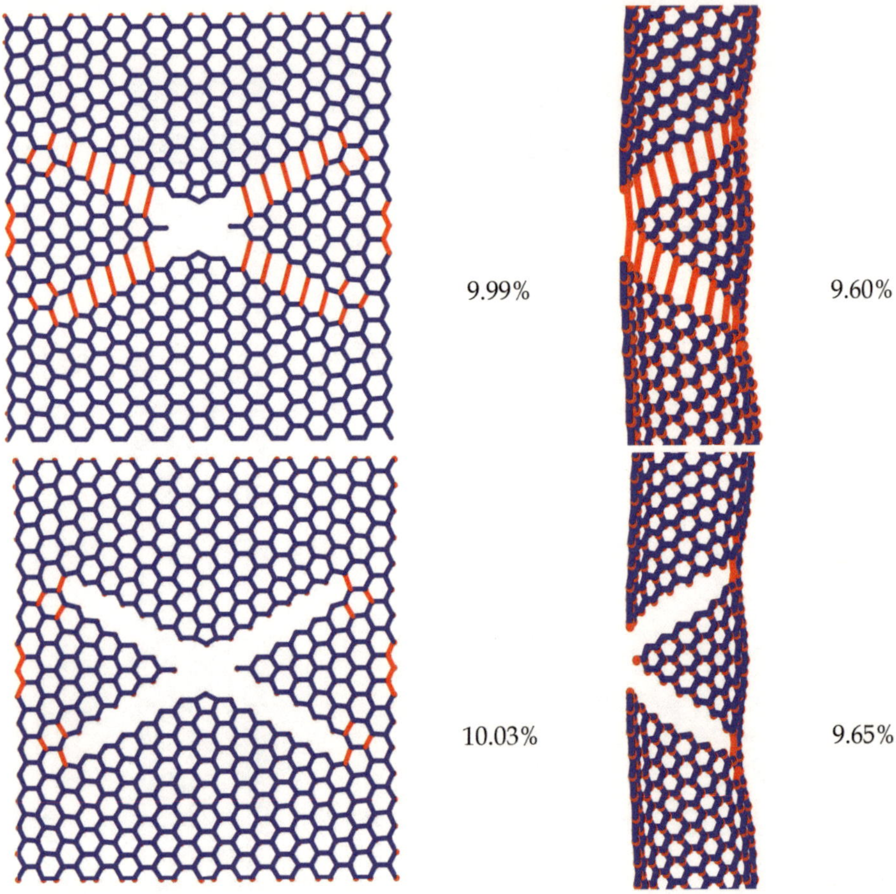

(a) a grahpene sheet (b) a (10, 10) armchair nanotube

Fig. 10. Fractured profiles at different strains of a defective graphene sheet and a defective nanotube under tension

Fig. 10 shows the crack evolution (elimination of failed bond elements based on the cutoff distance failure criterion) at different strain levels for a graphene sheet and an (10, 10) armchair CNT. Similar failure patterns can be seen between the graphene sheet and CNT which exhibits diagonal crack paths. The failure pattern in the CNT wraps around the tube in the $\pi/4$ direction along its circumference, which is similar to MD simulations (Belytschko, et al., 2003; Lu and Bhattacharya, 2005).

4.3 Effects of multiple stone-wales defects

The present study also considers CNTs with multiple defects along its axial direction and hoop direction, respectively. The effect of the distance between adjacent defects and the number of defects on the mechanical behaviors of defective graphene sheets and CNTs will be examined. The developed MATLAB program of the finite bond element model has been further improved to be able to form several Stone-Wales defects randomly at multiple points in the CNT structure. The present method for creating multiple defects on a nanotube is similar to creating a single defect. Once the first defect is created at a given location, any deformations of the tube are taken into account for the next defect. The defects are therefore created at specific locations sequentially, not simultaneously. The effect of randomly distributed multiple defects on mechanical properties of nanotubes has been studied by Lu and Bhattacharya [17] using MD. Predictions on mechanical properties of CNTs with multiple defects based on continuum mechanics based models have not been well studied. The present study provides understanding of the effects of multiple defects on the Young's modulus, ultimate strength, and strain at failure of the defective CNTs. After defect formations in a graphene sheet or nanotube, one can apply further load (tension) onto the defective sheet and tube as shown in Fig. 11, where the load and boundary conditions are applied in the same fashion as described above. Different tube lengths are examined.

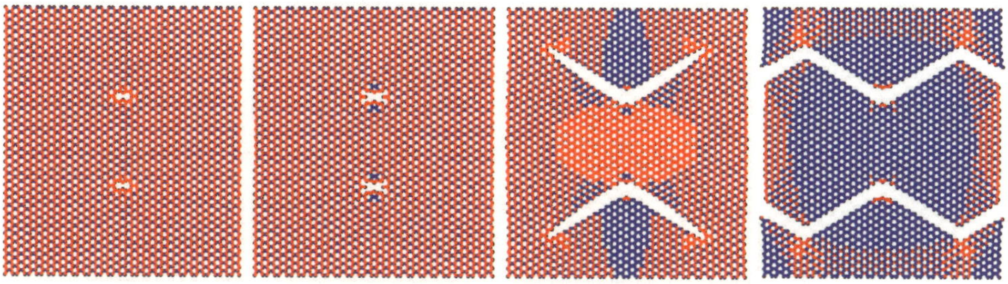

(a) Graphene sheet

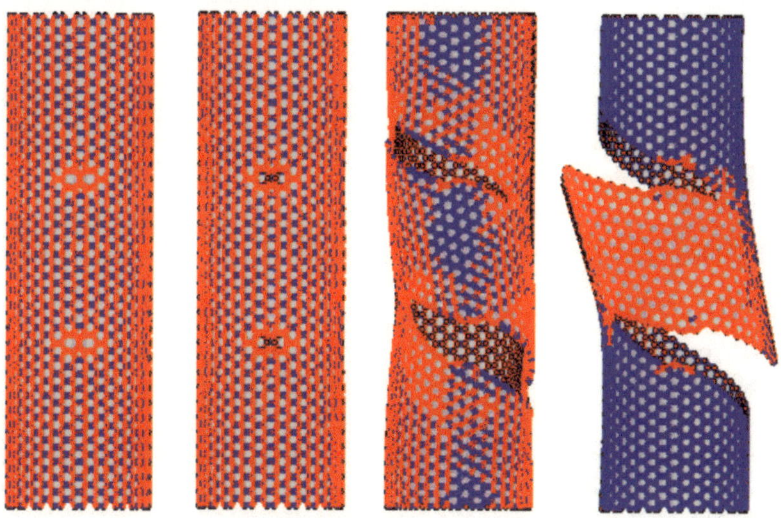

(b) CNT

Fig. 11. Fracture pattern of a graphene sheet and tube with two defects along its axial direction

4.3.1 Effects on young's modulus

From our simulations, it has been found that graphene sheets and CNTs with multiple defects along its axial direction (with separation distance larger than 2nm) show similar moduli and strengths as those with single defect, because the determining factor is the weakest cross-section. It should be noted that there is no defect interactions when the separation distance is greater than 2nm. The present study focuses on the effects of two defects located at the same axial location but separated by a prescribed angle around the circumference. We look at a (20, 20) CNT with a diameter of 2.71 nm. Based on our previous study (Xiao et al., 2005), the Young's moduli of nanotubes with a diameter larger than 2nm are insensitive to their diameters. Fig. 12 compares the calculated Young's Moduli of the CNT with different angles (in degree) between the two defects for 4 different tube lengths. The effect that the difference in angle between defects has on the modulus is negligible for angles > 70 degree (around 2 nm separation). Similar studies carried out on (10,10), (17,0), and (35,0) tubes of varying lengths yield similar results. It is also noticed that the shorter the tube length is, the lower Young's Modulus, because the ratio of the affected area over the total length is larger for a short tube length. This is a consequence of how axial strain of the tube is defined. In our study we have adopted the definition commonly used in experimental studies.

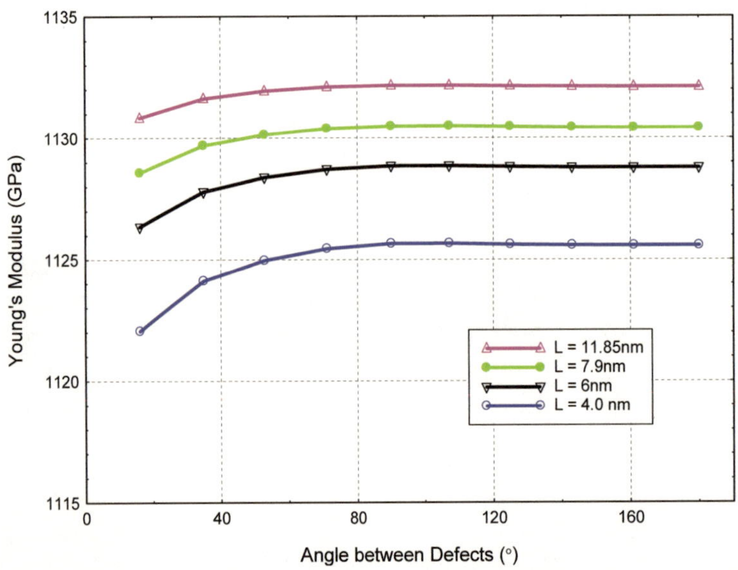

Fig. 12. Young's modulus vs. angle between two defects for (20,20) CNTs with different lengths (Young's modulus calculated using total tube length)

To better understand these local effects on modulus one can separate the defective tubes into two different regions, i.e. the affected region which is about 2nm long (correlates to the failure pattern of a single defect as shown in Xiao, et al., (2009), and unaffected region (total length minus 2nm). In the unaffected region the Young's modulus is the same as the defect-free tubes reported previously (Xiao et al., 2005). In the affected region, local strains are defined as $\varepsilon = \Delta L_D / L_{D0}$, where L_{D0} is the initial length of the affected region (i.e. 2nm) and ΔL_D is the change in length of the affected region. Consequently, two different moduli can be defined, i.e. (a) apparent modulus E_{total} based on the total length as shown in Fig. 12, and (b) defective modulus E_{local} based on the local affected length as shown in Fig. 13.

In Fig. 12, a plateau modulus is reached for separation angles greater than 72 degrees for all tube lengths. The magnitude of this plateau modulus increases as the tube length increases. For infinitely long tubes, the plateau modulus approaches the modulus of the defect free tube. For angles less than 72 degrees, modulus decreases from the associated plateau level due to defect interactions (i.e. distance between defects is less than 2nm affected zone).

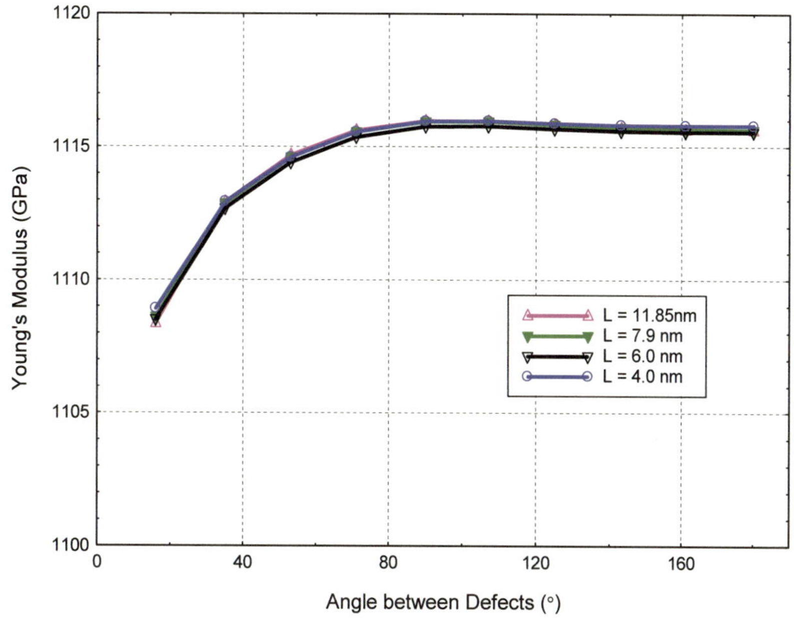

Fig. 13. Young's modulus vs. angle between two defects for (20,20) CNTs with different lengths (Young's modulus calculated using local length of 2nm)

In Fig. 13, it can be seen that E_{local} also exhibits a plateau modulus but is insensitive to the tube length. Results for all tube lengths collapse to a single curve. For angles less than 72 degrees, we see the same drop-off in modulus due to defect interaction. It should be noted that one can use different defective lengths other than 2nm to obtain different defective moduli E_{local}. We chose 2nm as the size of the affected area based on the study on single defect formation (Xiao, et al., 2009) as the defective length.

The relationship between Young's modulus and defect angle in Figs. 12 and 13 indicates that two defects start to interact when their distance (hoop direction) is smaller than 2 nm (about 72 degree defect angle). In Fig. 14, we study defect interactions by increasing the number of defects at a given axial location to look at the effects of defect number on the Young's Modulus. In this case we have considered multiple defects that are spaced at uniform angular increments around the circumference (e.g. 4 defects correspond to a separation angle of 90 degrees). Fig. 14 illustrated the difference between our two moduli definitions (E_{total} and E_{local}) computed from original tube lengths and local defective length (2 nm) for the four- and eight-defect cases. The local modulus which is insensitive to overall tube length shows that the modulus drops from the defect-free value as the number of defects increases (4 and 8 defects reduce the modulus by 3 and 6%, respectively). For the case of apparent modulus (E_{total}) at a given tube length, an increase in the number of defects in the hoop direction reduces the modulus. As the tube length increases, the modulus monontonically increases and approaches the defect-free level. As explained above, this is a consequence of defining strain as change in axial length over initial length. Local deformations around the defect are identical but contribute less to the overall deformation as length increases. One concludes from Fig. 14 that for a given defect pattern, one may take the local modulus (red lines) as the lower bound of the apparent modulus (blue lines) and the defect-free modulus (black line) as the upper bound as the tube length varies.

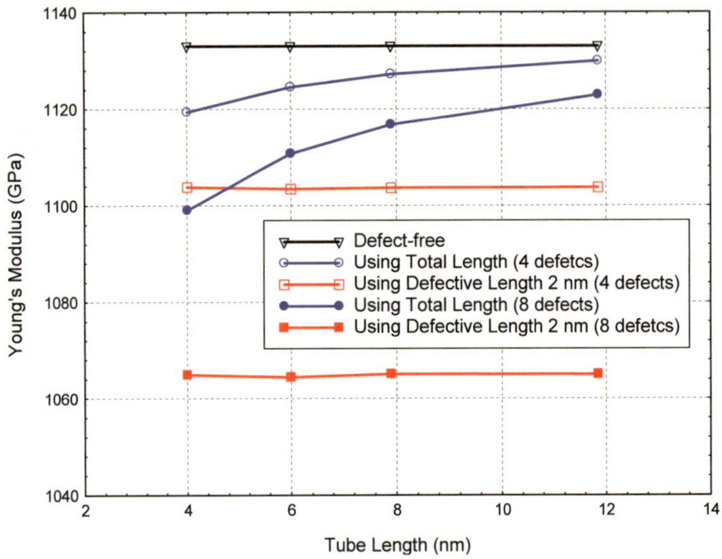

Fig. 14. Young's moduli of (20,20) CNTs with different lengths (4 or 8 defects uniformly distributed along the hoop direction)

The computed moduli of tubes with higher number of defects ranging from 1 to 20 along its hoop direction are presented in Fig. 15. Similar trends are observed where modulus reductions increase with increasing number of defects that are interacting. The local modulus provides a lower bound and the apparent modulus increases with tube length. The results converge to the defect-free result as the number of defects approach zero and the tube length increases. At these higher levels of defects, modulus reductions are quite significant (e.g. local modulus decreases by 30% for the case of 20 circumferential defects).

One may now generalize the present results to study nanotubes with defects uniformly distributed along both the hoop and axial directions of the tube. Consider each combination of defect number and tube length in Fig. 15 as a unit cell taken from a longer tube with a periodic defect pattern along the tube length. For instance, the calculated Young's modulus of the 6mn long tube with 8 defects uniformly distributed along its hoop direction, as shown in Fig. 15, can be treated as that of any longer tubes with the 8 hoop defects distributed every 6nm along its axial direction. Consequently, one can define defect density as the number of defects per unit surface area within the unit cell of interest. This approach is, based on the assumption that defects exist with equal probability along the entire length of the tube. By following this approach, modulus as a function of defect density is calculated as shown in Fig. 16. Results are now independent of tube length (i.e. local and apparent moduli are equivalent for a given defect density). As the defect density approaches zero, the modulus approaches the defect-free value as expected. Interestingly a linear relationship is observed with modulus decreasing with increasing defect density. A family of curves are presented in Fig. 16 that illustrate the effects of defect interaction for a given defect density. As the separation angle decreases (i.e. interaction increases), the negative slope of the modulus reduction versus defect density increases. The modulus results presented in Fig. 16 are expected to apply to nanotubes of larger diameter than considered in this study (>2.71 nm) including graphene sheets. The effects of defects on nanotube strength are considered next.

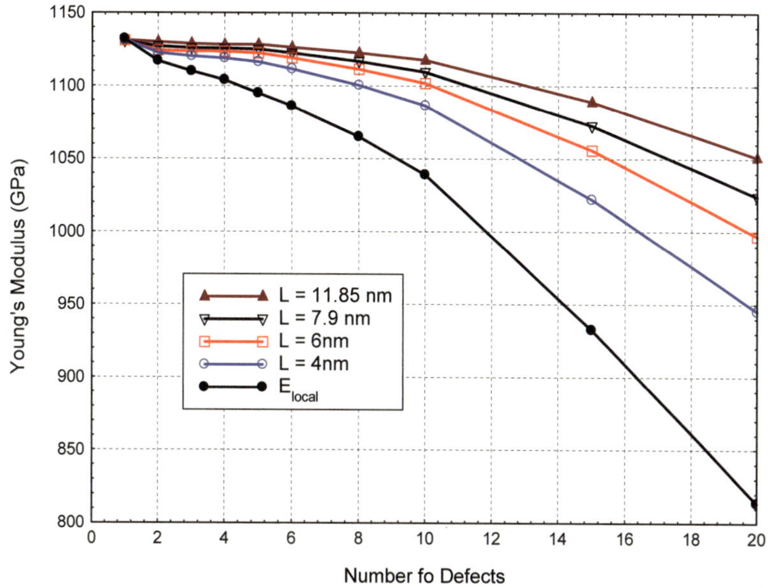

Fig. 15. Young's modulus vs. number of defects for (20,20) CNTs with different lengths

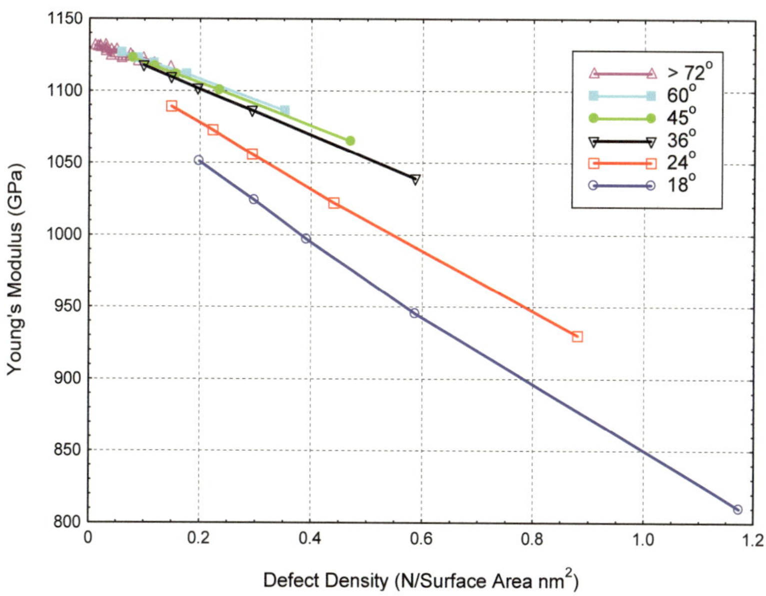

Fig. 16. Young's modulus vs. defect density for a (20,20) CNT

4.3.2 Effects on ultimate strength

The effects of isolated defects (i.e. no interactions) on strength are first examined for a 4nm long (10,10) nanotube. The ultimate strengths for a defect free tube, a tube with one defect and a tube with two defects along the axial direction at different hoop locations are 124 GPa, 84.8 GPa, and 84.2 GPa respectively. In contrast to the modulus results discussed above, single defects significantly reduce failure strength (approximately 30%). Despite having two defects along the axial direction, failure strength is effectively the same as a single defective tube, because the determining factor for ultimate strength is the weakest cross-section. These results illustrate that multiple defects that are sufficiently separated such that there are no interactions between them have comparable strength.

To study the effects of multiple interacting defects, let us focus on tubes with defects distributed around their hoop direction. Similar to the Young's modulus study, simulations were conducting using (10,10), (20,20), (17,0), and (35,0) nanotubes with different lengths of 4, and 6 nanometers. It should be noted that a (10,10) tube has about the same diameter as a (17,0) tube, and the same is true for (20,20) and (35,0). In each simulation, the tube was loaded quasi-statically in tension under prescribed displacement until failure. For failure of a bond to occur, the individual strain of the bond had to reach 18.5%. This value is consistent with published data as an inflection point for a C-C bond. The simulations predict that once a single initial bond has broken, the stress would drop abruptly under displacement controlled loading. The basic nanotube structure will still be intact and have some post-failure properties. In our simulations, the ultimate strength is defined as the maximum stress the tube reaches. Also, the strain at failure is the strain that corresponds to this ultimate stress.

In Fig. 17, we consider the armchair cases with two defects at the same axial location as a function of separation angle. At angles greater than approximately 70 degrees a strength plateau is established equal to the ultimate strength of a single defect mentioned above. When the angle drops below 70 degrees (0.83 nm distance for a (20, 20) CNT, which is about the same size of the defective area), the defects begin to interact and a dramatic reduction in strength is predicted. For the case of 18 degree separation angle, the strength has dropped 58% of the defect-free ultimate strength. Results from the (10, 10) CNT simulation follow the same trend. In contrast to the modulus results, it should be noted that the computed strengths are not sensitive to tube length.

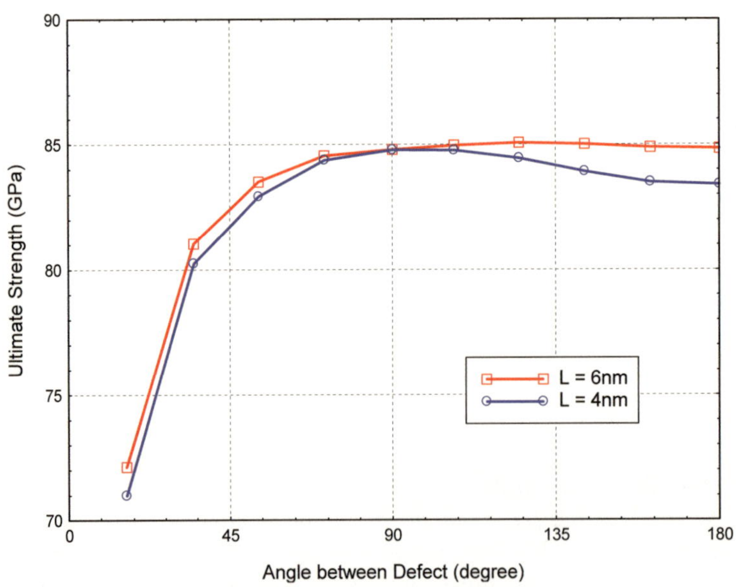

Fig. 17. Ultimate strength vs. angle between two defects for (20, 20) CNTs with different lengths

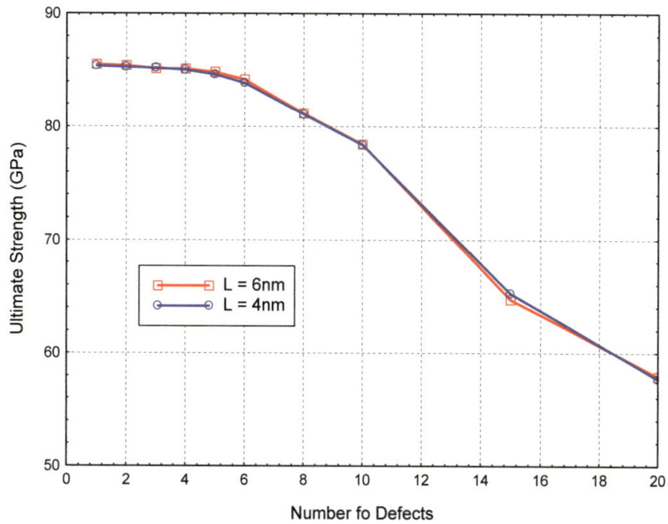

Fig. 18. Ultimate strength vs. Number of defects for (20,20) CNTs with different lengths

The effect of multiple defects at the same axial location on the ultimate strength is considered next. In the armchair configuration, increasing the amount of defects present around the circumference decreases the ultimate strength as shown in Fig. 18 from the plateau strength corresponding to a single defect. At approximately 4-6 defects, the separation angle drops below 70 degrees and significant interactions between multiple defects develop. In the case of multiple defects, the strength drops to the same level as the results for 2 defects separated by the same angle. For example, 20 defects have a separation angle of 18 degrees and one observes the same strength reduction as shown in Fig. 18 (i.e. 58% of the defect-free tube). One concludes that the strength of a tube with multiple defects is determined by the degree of interaction indicated by the separation distance.

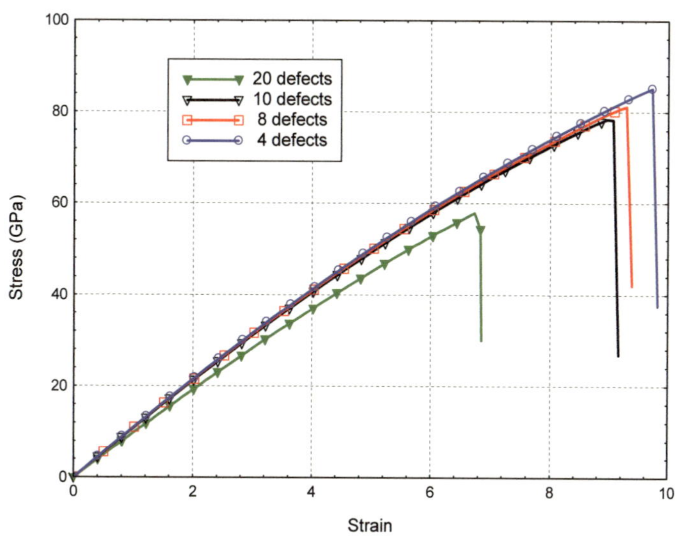

Fig. 19. Stress-strain curves of a 6nm (20, 20) CNT with multiple defects equally spaced around the circumference at the middle section

Fig. 19 gives the computed stress-strain relationships of a 6nm (20, 20) tube with different number of defects spaced uniformly around the circumference at the same axial location (about the middle section of the tube). Based on the discussion above, the initial modulus decreases as the number density of defects increases. In addition, the strength and strain to failure decreases as the separation angle between defects decreases. In this figure, the 20 defect curve has both the lowest modulus and lowest ultimate strength and strain to failure. The failure patterns of the 6nm (20, 20) tube with 4 and 8 defects equally spaced around the circumference direction under tension loading are shown in Fig. 20. It can be seen that the failure of the tube with 8 defects (Fig. 20b) is extremely localized with no diagonal crack propagation as seen in the tube with less hoop defects (see Fig. 11 and Fig. 20a).

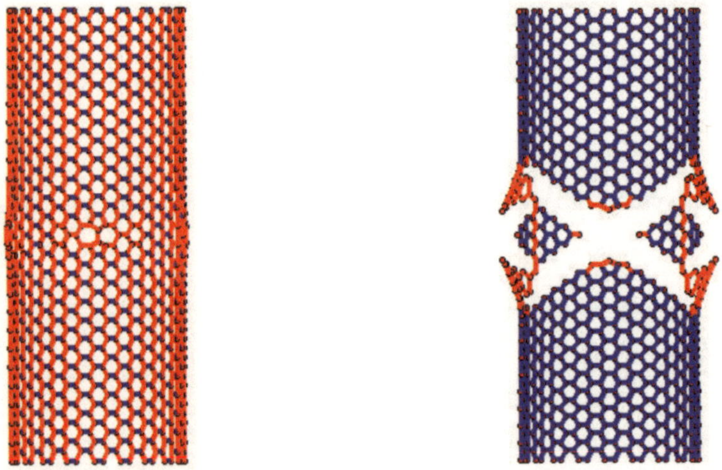

(a) With 4 defects

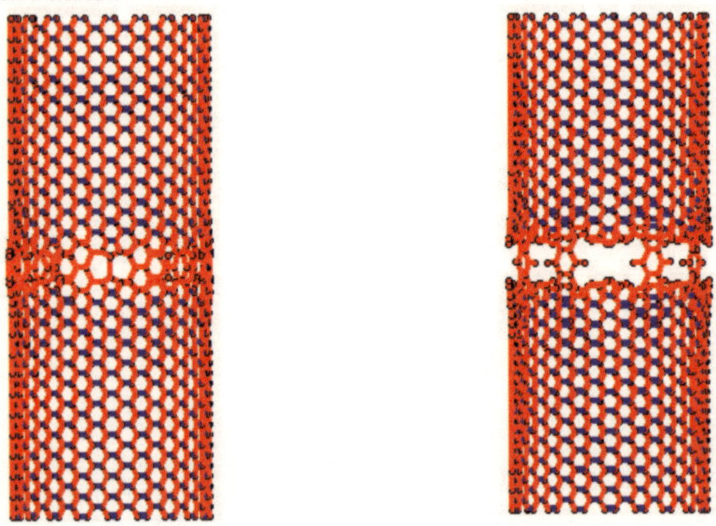

(b) With 8 defects

Fig. 20. Failure patterns of a 6nm (20, 20) CNT with multiple defects equally spaced around the circumference at the middle section

5. Conclusions

By incorporating the modified Morse potential function into a novel atomistic finite bond element molecular structural mechanics model, the mechanical responses of graphene sheet and single-walled nanotubes under tension conditions are investigated. The finite bond element has 8 degrees of freedom (DOF) which is used to simulate the interatomic connections in CNTs. Compared to other similar finite element approaches (12DOFs) the computational cost of the present method is reduced. The coefficients in the finite bond element model are taken from the analytical molecular mechanics model (Xiao et al., 2005). The present approach is capable of predicting Young's moduli, Poisson's ratios, and stress-strain relationships of graphene sheets and nanotubes with or without a SW defect. An interaction mechanics approach is introduced to model the formation of a 5-7-7-5 SW defect in CNTs which reasonably captures the physical phenomena in terms of reconfiguration, local deformation and formation energy. Consequently, effects of the SW defect on the Young's moduli, fracture and progressive failure of defective CNTs have been investigated. Using the present approach, it is feasible to model multiple defects and their interaction in both SWCNT and MWCNT since the present approach is much simpler and computationally efficient than the classical molecular dynamics model.

A methodology to create multiple defects in a CNT at given locations has been implemented into our MATLAB code. Consequently, the effect and interaction of multiple defects on a SWCNT were studied. The resulting simulations were able to predict the Young's modulus, ultimate strength, and strain at failure. The influence of single and multiple defects on mechanical properties were studied. In the case of Young's modulus, increasing the number of defects along the hoop direction can change its properties dramatically, particularly when the defect distance is smaller than 2 nm (defect interaction). It is also found that moduli are sensitive to the tube lengths when the total tube length is used to compute the strain. This finding has major implications for comparing experimental data from tests on different tube lengths. A local defective length (2 nm) is introduced to separate the overall deformation into two different regions: defective and defect-free. By doing so, a size independent modulus has been obtained for the defective region, which can be treated as the minimum modulus for a given defective tube of any length larger than 2nm.

In addition, results were generalized to consider defect density (number of defects per unit surface area). Stiffness reductions were found to be linearly related to defect density (number of defects per unit surface area) and to become independent of tube length. It was found that further reductions occur when the defects interact with each other (typically when the distance between defects is smaller than 2 nm). The effects of single and multiple defects on ultimate strength were considered. Significant strength reductions are predicted to occur by the presence of single defects. Further strength reductions are predicted for cases where adjacent defects are interacting. In this case the ultimate strength is governed more by the separation distance than the defect density. Finally the overall stress-strain response has been predicted as a function of number of defects. Results including the effects of defects more closely match experimental data reported in the literature. It should be noted that this present model has a more simple approach than quantum or the classical molecular dynamics model and is less computationally expensive.

6. References

Abell GC. (1985). Empirical chemical pseudopotential theory of molecular and metallic bonding. Physical Review B, Vol. 31, pp.6184-6196.

Belytschko T.; Xiao SP.; Schatz GC.; & Ruoff RS. (2002). Atomistic simulations of nanotube fracture. Physical Review B, Vol. 65, pp.235430.

Brenner DW. (1990). Empirical potential for hydrocarbons for use in simulating the chemical vapor-deposition of diamond films. Physical Review B, Vol. 42, pp.9458–9471.

Cadek M.; Coleman JN.; Barron V, et al., (2002). Morphological and mechanical properties of carbon-nanotube-reinforced semicrystalline and amorphous polymer composites. Appl. Phys. Lett.; Vol. 81, pp.5123-5125

Chang T.; & Gao H. (2003). Size-dependent elastic properties of a single-walled carbon nanotube via a molecular mechanics modelnanotubes. J. the Mech. Phys. Solids, Vol.51, pp.1059 -1074.

Che J, Cagin T, & Goddard WA, (2000). Thermal conductivity of carbon nanotubes. Nanotechnology, Vol. 11, pp. 65–69.

Cornell WD.; Cieplak P, Bayly CI, et al., (1995). A second generation force field for the simulation of proteins, nucleic acids, and organic molecules. American Chem. Soc.. Vol.117, pp.5179-5197.

Dalton AB.; Collins S.; Munoz E.; et al., (2003). Super-tough carbon-nanotube fibres - These extraordinary composite fibres can be woven into electronic textiles. Nature (London). Vol. 423, pp.703-703.

Dresselhaus MS.; Dresselhaus G.; & Eklund PC, (1996). *Science of fullerenes and carbon nanotubes,* Academic Press, San Diego.

Dumitrica T.; Belytschko T.; Yakobson BI, (2003). Bond-breaking bifurcation states in carbon nanotube fracture. J. Chem. Phys. Vol.118(21), pp.9485–9488.

Kanchi MB. (1993). Matrix Methods of Structural Analysis, John Willey & Sons,.

Huhtala M, Krasheninnikov AV, Aittoniemi J, et al., (2004). Improved mechanical load transfer between shells of multiwalled carbon nanotubes. Physical Review B, vol.70(4), pp.045404

Huxtable ST, Cahill DG, Shenogin S, et al., (2003). Interfacial heat flow in carbon nanotube suspensions. Nat. Mater. Vol. 2, pp.731-734.

Iijima S. (1991). Helical microtubes of graphitic carbon. Nature, Vol. 354, pp. 56–58.

Jiang H, Feng XQ, Huang Y, et al., (2004). The effect of nanotube radius on the constitutive model for carbon nanotubes. Comput. Methods Appl. Mech. Engrg. Vol.193, pp.3419-3429.

Jiang H, Zhang P, Liu B, Huang Y, et al., (2203). The effect of nanotube radius on the constitutive model for carbon nanotubes. Computational Material Science. Vol. 28, pp.429-442.

Jorgensen WL. & Severance DL, (1990). Aromatic aromatic interactions - free-energy profiles for the benzene dimer in water, chloroform, and liquid benzene. American Chem. Soc. Vol.112, pp.4768-4774.

Li CY. & Chou TW, (2003). A structural mechanics approach for the analysis of carbon nanotubes. Int. J. Solids Struct. Vol.40, pp.2487–2499.

Lu Q & Bhattacharya B, (2005). Fracture resistance of zigzag single walled carbon nanotubes. Nanotechnology. Vol.16, pp.555-566.

Lu Q & Bhattacharya B, (2005). The role of atomistic simulations in probing the small-scale aspects of fracture - a case study on a single-walled carbon nanotube. Eng. Fract. Mech. Vol.72, pp.2037–2071.

Nardelli MB, Yakobson BI, & Bernholc J, (1998). Mechanism of strain release in carbon nanotubes. Physical Review B, Vol.57, pp.R4277

Pan C, Yang WS, & Yang J, (2000). Formation energies of topological defects in carbon nanotubes. Physical Review B, Vol.62, pp.12652.

Sears A, Batra RC, (2004). Macroscopic properties of carbon nanotubes from molecular-mechanics simulations. Physical Review B, Vol. 69, pp.235406.

Stone AJ, Wales DJ, (1986). Theoretical studies of icosahedral c60 and some related species. Chem. Phys. Lett. Vol.128, pp.501-503.

Tersoff J, (1988). Empirical interatomic potential for carbon, with applications to amorphous-carbon.Phys. Rev. Let. Vol.61, pp.2872–2879.

Thostenson ET, Ren Z, Chou TW. (2001). Advances in the science and technology of carbon nanotubes and their composites: a review. Composites Sci. Tech.Vol.61, pp.1899–1912.

Tserpes KI, Papanikos P, (2007). The effect of Stone-Wales defect on the tensile behavior and fracture of single-walled carbon nanotubes. Comp. Struct. Vol. 79, pp.581-589.

Wagner HD, Lourie O, Feldmann Y, & Tenne R, (1998). Stress-induced fragmentation of multiwall carbon nanotubes in a polymer matrix. Appl Phys Lett. Vol. ;72, pp.188-190.

Wong EW, Sheehan PE, & Lieber CM, (1997). Nanobeam mechanics: elasticity, strength, and toughness of nanorods and nanotubes. Science. Vol. 277, pp.1971–1975.

Xia Y, Zhao M, Ma Y, et al., (2002). Tensile strength of single-walled carbon nanotubes with defects under hydrostatic pressure. Physical Review B, Vol. 65, pp.155415.

Xiao JR, Gama BA, & Gillespie JW, (2005). An analytical molecular structural mechanics model for the mechanical properties of carbon nanotubes. Int. J. Solids Struct.;vol. 42, pp.3075-3092.

Xiao JR, Lopatnikov S, Gama BA, & Gillespie JW, (2006) Nanomechanics on the deformation of single- and multi-walled carbon nanotubes under radial pressure. Materials Science and Engineering A. Vol. 416, pp.192-204.

Xiao JR & Gillespie JW. (2006a). Nonlinear deformation and progressive failure of multiwalled carbon nanotubes under internal radial pressure Physical Review B, Vol. 74, pp.155404

Xiao JR & Gillespie JW. (2006b). Nanomechanics of single-walled carbon nanotubes as composite reinforcement. Polymer Eng. Sci.; Vol. 46, pp.1051-1059.

Xiao SP & Hou WY. (2006). Studies of nanotube-based resonant oscillators through multiscale modeling and simulation. Physical Review B, Vol. 73, pp.115406.

Yakobson BI & Avouris P. (2001). In *Carbon Nanotubes. Topics in Applied Physics*, Dresselhaus, M.S., Dresselhaus, G., Avouris, P. (Eds.), Springer Verlag, Berlin-Heidelberg, Germany. V. 80, p.287.

Ye LH, Liu BG, & Wang DS, (2001). *Ab initio* molecular dynamics study on small carbon nanotubes. Chinese Phys. Lett., Vol. 18, pp.1496–1499.

Yu MF, Files BS, Arepalli S, & Ruoff RS, (2000). Strength and breaking mechanism of multiwalled carbon nanotubes under tensile load. Phys. Rev. Lett.; vol. 84, pp.5552-5555.

Zhang P, Huang Y, Geubelle PH, et al., (2002). The elastic modulus of single-wall carbon nanotubes: a continuum analysis incorporating interatomic potentials. International Journal of Solids Structures, Vol. 39, pp.3893–3906.

Zhang S, Mielke SL, Khare R, et al., (2005). Mechanics of defects in carbon nanotubes: Atomistic and multiscale simulations. Physical Review B, Vol. 71, pp.115403

Part 3

Magnetic Properties

19

Orbital Magnetism of Graphenes

Mikito Koshino
Department of Physics, Tohoku University
Sendai, 980-8578
Japan

1. Introduction

The magnetism of conventional metal is composed of two different contributions, the Pauli paramagnetism due to the spin magnetic moment, and the Landau diamagnetism due to the orbital motion of electrons. In a free electron system, the magnitude of the spin component is larger than the orbital component so that the system exhibits paramagnetism in total. In a condensed matter system, on the other hand, the orbital magnetism sensitively depends on the detail of the band structure, and sometimes largely deviates from the conventional Landau diamagnetism. Particularly, narrow gap materials such as graphite(1–3) or bismuth (4–6) possess a strong orbital diamagnetism which overcomes the spin paramagnetism.
Graphene monolayer (7–9) is an extreme case, in which the conduction and valance bands stick together with zero gap and a linear dispersion analogous to massless Dirac fermions. (1; 10–13) Accordingly, the orbital susceptibility has a strong singularity at band touching point (Dirac point), which has a Fermi energy dependence at zero temperature as (1; 14–25)

$$\chi(\varepsilon_F) \propto \delta(\varepsilon_F). \tag{1}$$

This anomalous behavior can be understood as a zero-mass limit of conventional magnetism for massive Dirac electron, in which the pseudo-spin magnetic moment associated valleys (K and K') gives an essential contribution (25). Monolayer graphene exhibits a non-trivial response also to non-uniform magnetic fields. Owing its scale-less electronic structure, graphene works as a magnetic mirror, where the response current creates a mirror magnetic field which mimics any external field distributions. (24)
The orbital magnetism was also studied for graphene-based materials, such as graphite intercalation compounds,(14–17) and multilayer graphenes. (26–28) Those systems have strong diamagnetism as well, while the delta-function singularity is strongly modified by the electronic coupling between different graphene layers. Bilayer graphene (29–32) has a zero-gap structure with a finite mass in contrast to massless band in monolayer (33–41). This leads to a less singular, logarithmic peak of the susceptibility.(15; 21) In graphene stacks with more than three layers, the Hamiltonian and thus susceptibility can be decomposed into contributions from sub-systems equivalent to monolayer or bilayer graphene.(21)
In this chapter, we review those anomalous properties of the orbital magnetism in graphene and related materials. In Sec. 2, we discuss the susceptibility of the monolayer graphene. We describe the origin of the delta function singularity at the Dirac point, and also a peculiar response to non-uniform magnetic field. We argue the orbital magnetism of bilayer graphene

in Sec. 3, and of general graphene stacks in Sec. 4. In Sec. 5, we extend the analysis to a three-dimensional Dirac system, which is known to give an effective model for bismuth.

2. Monolayer graphene

2.1 Magnetic susceptibility of graphene

Electronic states of graphene in the vicinity of K and K' points in the Brillouin zone are well described by the effective mass approximation. (1; 10–13) Let $|A\rangle$ and $|B\rangle$ be the Bloch functions at the K point, corresponding to A and B sublattices, respectively. In a basis $(|A\rangle, |B\rangle)$, the Hamiltonian for the monolayer graphene around the K point becomes

$$\mathcal{H}^K = \begin{pmatrix} \Delta & v\pi_- \\ v\pi_+ & -\Delta \end{pmatrix}, \tag{2}$$

where $v \approx 1 \times 10^6$ m/s is the band velocity (8; 9), $\pi_\pm = \pi_x \pm i\pi_y$, and $\boldsymbol{\pi} = -i\hbar\boldsymbol{\nabla} + (e/c)\mathbf{A}$ with vector potential $\mathbf{A}$ giving external magnetic field $\mathbf{B} = \boldsymbol{\nabla} \times \mathbf{A}$. In the following, we neglect the spin Zeeman energy because the spin splitting is much smaller than Landau-level separations. The Hamiltonian at the K' point is obtained by exchanging $\pi_\pm$ in Eq. (2).

The diagonal terms $\pm\Delta$ represent the potential asymmetry between A and B sites, which opens an energy gap at the Dirac point. Although A and B are intrinsically symmetric in usual graphene, the asymmetry can arise in a sample placed on a certain substrate material, where the interaction between the graphene and the substrate lattice produces different potentials between A and B. (42; 43) Theoretically, the singular behavior in ideal graphene with vanishing gap is intuitively understood by taking the limit $\Delta \to 0$, as will be shown below. We can safely assume $\Delta \geq 0$ without loss of generality. The energy band at $B = 0$ is given by

$$\varepsilon_s(p) = s\sqrt{v^2p^2 + \Delta^2}, \quad (s = \pm 1) \tag{3}$$

with electron momentum $\mathbf{p} = (p_x, p_y)$ and $p = \sqrt{p_x^2 + p_y^2}$. The density of states is

$$D(\varepsilon) = \frac{g_v g_s |\varepsilon|}{2\pi\hbar^2 v^2}\theta(|\varepsilon| - \Delta), \tag{4}$$

where $g_s = 2$ and $g_v = 2$ represent the degrees of freedom associated with spin and valley, respectively, and $\theta(t)$ is a step function, defined by

$$\theta(t) = \begin{cases} 1 \ (t > 0); \\ 0 \ (t < 0). \end{cases} \tag{5}$$

The Landau-level spectrum can be found using the relation $\pi_+ = (\sqrt{2}\hbar/l_B)a^\dagger$ and $\pi_- = (\sqrt{2}\hbar/l_B)a$, where $l_B = \sqrt{c\hbar/(eB)}$ is magnetic length and $a^\dagger$ and a are raising and lowering operators for usual Landau-level wave functions, respectively. The eigenfunction of the Hamiltonian at K point is written as $(c_1\phi_{n-1}, c_2\phi_n)$ with integers $n \geq 0$, where ϕ_n is the usual Landau-level wave function and ϕ_n of $n < 0$ is regarded as 0. The Hamiltonian matrix for (c_1, c_2) becomes

$$H^K = \begin{pmatrix} \Delta & \hbar\omega_B\sqrt{n} \\ \hbar\omega_B\sqrt{n} & -\Delta \end{pmatrix}. \tag{6}$$

For $n = 0$, the eigenvector $(c_1, c_2) = (1, 0)$, corresponding to the eigenvalue Δ, is not associated with any real eigenstates since the first component of wavefunction ϕ_{-1} is zero. Similarly, the eigenfunction of K' point is written as $(c_1\phi_n, c_2\phi_{n-1})$, and the eigenvector $(0,1)$ of $n = 0$, corresponding to the eigenvalue $-\Delta$, is not a real state. From these arguments, we obtain the eigen energies,

$$\varepsilon_n^K = \mathrm{sgn}_-(n)\sqrt{(\hbar\omega_B)^2|n| + \Delta^2},$$
$$\varepsilon_n^{K'} = \mathrm{sgn}_+(n)\sqrt{(\hbar\omega_B)^2|n| + \Delta^2} \tag{7}$$

with $n = 0, \pm 1, \pm 2, \cdots$, $\hbar\omega_B = \sqrt{2}\hbar v/l_B$, $l_B = \sqrt{\hbar/eB}$, and

$$\mathrm{sgn}_\pm(n) = \begin{cases} +1 \ (n > 0); \\ \pm 1 \ (n = 0); \\ -1 \ (n < 0). \end{cases} \tag{8}$$

The Landau levels of $n \neq 0$ are doubly degenerate between the K and K' valleys, while those of $n = 0$ are not. Figure 1(a) shows an example of energy levels at $\hbar\omega_B = 2\Delta$.
The thermodynamical potential at temperature T becomes

$$\Omega = -\frac{1}{\beta}\frac{g_v g_s}{2\pi l_B^2}\sum_{s=\pm}\sum_{n=0}^{\infty}\varphi\left[\varepsilon_s((\hbar\omega_B)^2 n)\right]\left(1 - \frac{\delta_{n0}}{2}\right), \tag{9}$$

where $\beta = 1/k_B T$, $\varepsilon_s(x) = s\sqrt{x + \Delta^2}$, $\varphi(\varepsilon) = \log\left[1 + e^{-\beta(\varepsilon-\zeta)}\right]$ with ζ being the chemical potential, and $g_s = 2$ and $g_v = 2$ represent the degrees of freedom associated with spin and valley, respectively. The magnetic susceptibility is defined by

$$\chi = -\left(\frac{\partial^2\Omega}{\partial B^2}\right)_\zeta\bigg|_{B=0}. \tag{10}$$

In weak magnetic field, using the Euler-Maclaurin formula, the summation in n in Eq. (9) can be written as an integral in continuous variable x and a residual term proportional to B^2. At zero temperature, we have (25)

$$\chi(\varepsilon_F) = -g_v g_s \frac{e^2 v^2}{6\pi c^2}\frac{1}{2\Delta}\theta(\Delta - |\varepsilon_F|). \tag{11}$$

In the limit of $\Delta \to 0$, this approaches

$$\chi(\varepsilon_F) = -g_v g_s \frac{e^2 v^2}{6\pi c^2}\delta(\varepsilon_F). \tag{12}$$

The susceptibility of Eq. (11) and the density of states of Eq. (4) are shown in Fig. 1 (b). The susceptibility is not zero in the gap, because the completely filled valence band gives a constant diamagnetic susceptibility. When the Fermi energy enters the conduction band, the susceptibility jumps downs to zero, resulting in zero total magnetism.
Because the Hamiltonian is equivalent to that of a Dirac electron with a nonzero mass, the magnetic susceptibility around the band edge should correspond to that of a conventional electron. This is clearly illustrated by the effective Hamiltonian expanded in the vicinity of

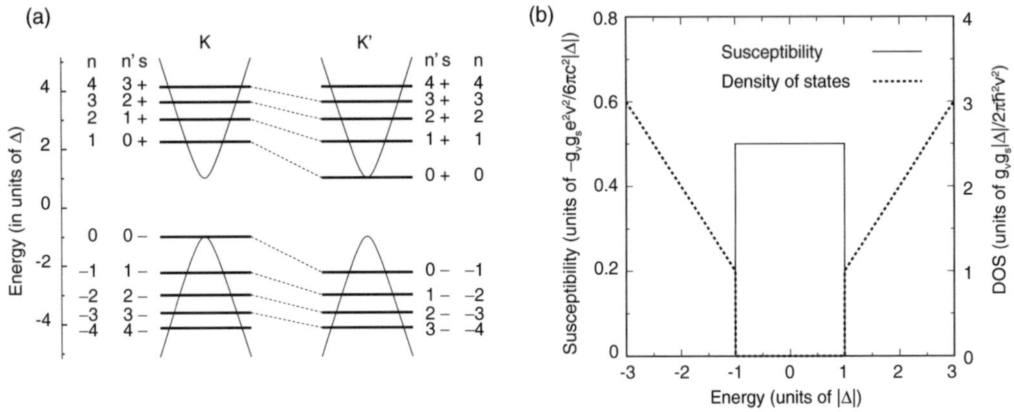

Fig. 1. (a) Landau-level energies of gapped monolayer graphene for $\hbar\omega_B = 2\Delta$. (b) Orbital susceptibility (solid) and density of states (dashed) of monolayer graphene with asymmetric potential Δ. For susceptibility, the upward direction represents negative (i.e., diamagnetic).

$\mathbf{k} = 0$.(25) For the conduction band, the effective Hamiltonian for the A site near the band bottom ($\varepsilon = \Delta$) is written apart from the constant energy as

$$\mathcal{H}^K \approx \frac{v^2}{2\Delta}\pi_-\pi_+ = \frac{\pi^2}{2m^*} + \frac{1}{2}g^*\mu_B B, \tag{13}$$

$$\mathcal{H}^{K'} \approx \frac{v^2}{2\Delta}\pi_+\pi_- = \frac{\pi^2}{2m^*} - \frac{1}{2}g^*\mu_B B, \tag{14}$$

where $\mu_B = e\hbar/(2mc)$ is the Bohr magneton with m being the free electron mass, and we defined $m^* = \Delta/v^2$, $g^* = 2m/m^*$. For instance, the g factor is estimated at $g^* \sim 60$ at $\Delta = 0.1$ eV, and diverges as $\propto \Delta^{-1}$ as the gap decreases. The last term in each Hamiltonian can be regarded as the pseudo-spin Zeeman term, where the different valleys K and K' serve as pseudo-spin up ($\xi = +1$) and down ($\xi = -1$), respectively. This agrees with the Zeeman energy expected for an intrinsic magnetic moment, that originates from the self-rotation of the wave packet in Bloch electron. (44; 45)

Obviously, the pseudo-spin Zeeman term gives the Pauli paramagnetism and the first term containing π^2 gives the Landau diamagnetism in the usual form as

$$\chi_P(\varepsilon) = \left(\frac{g^*}{2}\right)^2 \mu_B^2 D(\varepsilon), \tag{15}$$

$$\chi_L(\varepsilon) = -\frac{1}{3}\left(\frac{m}{m^*}\right)^2 \mu_B^2 D(\varepsilon), \tag{16}$$

with density of states $D(\varepsilon) = g_v g_s m^*/(2\pi\hbar^2)\,\theta(\varepsilon)$. The total susceptibility $\chi_P + \chi_L$ actually agrees with the amount of the jump at the conduction band bottom in χ of Eq. (11). Because $g = 2m/m^*$ in the present case, we have $\chi_L = -\chi_P/3 \propto 1/m^*$ as in the free electron, giving the paramagnetic susceptibility in total. Therefore the susceptibility exhibits a discrete jump toward the paramagnetic direction when the Fermi energy moves off the Dirac point.

In the original Hamiltonian, the Landau-level energies in Eq. (7) can be rewritten as

$$\varepsilon_{\xi,s,n'} = s\sqrt{(\hbar\omega_B)^2\left(n' + \frac{1}{2} + \frac{\xi s}{2}\right) + \Delta^2}$$
$$(n' = 0, 1, 2, \cdots). \tag{17}$$

Figure 1(a) shows energy levels for $\hbar\omega_B = 2\Delta$ and the relationship between the different labeling schemes of Eqs. (7) and (17). For the conduction band, the levels of the same n' with opposite pseudo-spins $\xi = \pm 1$ share the same Landau level function labeled by n' on the A site, on which the states near the conduction-band bottom ($\varepsilon = \Delta$) have most of the amplitude. For the valence band, similarly, n' describes the index of the Landau-level function at the B site.

2.2 Response to non-uniform magnetic field

We extend the argument to spatially modulated magnetic fields. The graphene in a non-uniform magnetic field was studied in the context of the electron confinement, (46–48) the peculiar band structures in superlattice, (49; 50) transport,(51) and the quantum Hall effect.(52; 53) In the following, we introduce a linear response theory to general field distributions. (24) We consider an isotropic 2D system, under a magnetic field given by $B(\mathbf{r}) = [\nabla \times \mathbf{A}(\mathbf{r})]_z$ with vector potential $\mathbf{A}(\mathbf{r})$. Here $\mathbf{r} = (x, y)$ denotes 2D position on the graphene. We define $\mathbf{j}(\mathbf{r}) = (j_x, j_y)$ as the 2D electric current density induced in the system. Within the linear response, the Fourier-transforms of $\mathbf{j}(\mathbf{r})$ and $\mathbf{A}(\mathbf{r})$ satisfy

$$j_\mu(\mathbf{q}) = \sum_\nu K_{\mu\nu}(\mathbf{q}) A_\nu(\mathbf{q}), \tag{18}$$

with response function $K_{\mu\nu}$. The gauge invariance for $\mathbf{A}$ requires $\sum_\nu K_{\mu\nu}(\mathbf{q})\, q_\nu = 0$, and the continuous equation, $\nabla \cdot \mathbf{j}(\mathbf{r}) = 0$, imposes another constraint $\sum_\mu q_\mu K_{\mu\nu}(\mathbf{q}) = 0$. To meet both requirements, tensor $K_{\mu\nu}$ needs to be in the form,

$$K_{\mu\nu}(\mathbf{q}) = K(q)\left(\delta_{\mu\nu} - \frac{q_\mu q_\nu}{q^2}\right). \tag{19}$$

On the other hand, because $\nabla \cdot \mathbf{j}(\mathbf{r}) = 0$, we can express $\mathbf{j}(\mathbf{r})$ as $j_x = c\,\partial m/\partial y$, $j_y = -c\,\partial m/\partial x$, with $m(\mathbf{r})$ being the local magnetic moment perpendicular to the layer. In the linear response, its Fourier transform is written as

$$m(\mathbf{q}) = \chi(q) B(\mathbf{q}), \tag{20}$$

with the magnetic susceptibility $\chi(q)$. Equations (18) and (20) are complementary, and both response functions $\chi(q)$ and $K(q)$ are related by

$$\chi(q) = \frac{1}{cq^2} K(q). \tag{21}$$

We apply the above formulation to graphene without gap, i.e., $\Delta = 0$. The eigenstates at zero magnetic field are labeled by $(s, \mathbf{k})$ with $s = +1$ and -1 being the conduction and valence bands, respectively, and wave vector $\mathbf{k}$. The eigen energy is given by $\varepsilon_{s\mathbf{k}} = s\hbar v k$, and the corresponding wavefunction is $\psi_{s\mathbf{k}}(\mathbf{r}) = e^{i\mathbf{k}\cdot\mathbf{r}}\mathbf{F}_{s\mathbf{k}}/\sqrt{S}$ with $\mathbf{F}_{s\mathbf{k}} = \frac{1}{\sqrt{2}}\begin{pmatrix} e^{i\theta} \\ s \end{pmatrix}$, where k and θ are defined by $(k_x, k_y) = k(\cos\theta, \sin\theta)$ and S is the system area.

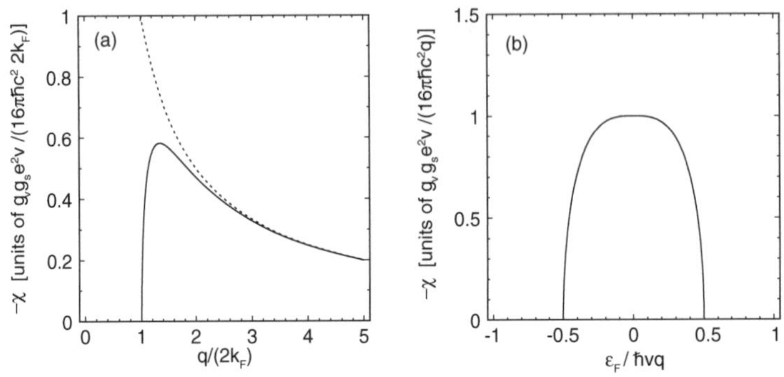

Fig. 2. Magnetic susceptibility $\chi(q\,;\varepsilon_F)$ in graphene plotted against (a) q with fixed ε_F, and (b) ε_F with fixed q. Dotted line in (a) represents the undoped case, Eq. (24).

The local current density at $\mathbf{r}_0$ is calculated as the expectation value of current-density operator $\hat{\mathbf{j}}(\mathbf{r}_0) = ev\,\boldsymbol{\sigma}\,\delta(\mathbf{r}-\mathbf{r}_0)$ over the occupied states. In the first order perturbation in $\mathbf{A}$, we have

$$K_{\mu\nu}(\mathbf{q}) = -\frac{g_v g_s e^2}{c}\frac{1}{S}\sum_{ss'\mathbf{k}}\frac{f(\varepsilon_{s\mathbf{k}}) - f(\varepsilon_{s'\mathbf{k}+\mathbf{q}})}{\varepsilon_{s\mathbf{k}} - \varepsilon_{s'\mathbf{k}+\mathbf{q}}}(\mathbf{F}^\dagger_{s\mathbf{k}}\, v_\nu\, \mathbf{F}_{s'\mathbf{k}+\mathbf{q}})\,(\mathbf{F}^\dagger_{s'\mathbf{k}+\mathbf{q}}\, v_\mu\, \mathbf{F}_{s\mathbf{k}}), \tag{22}$$

where $v_\mu = v\sigma_\mu$.
At the zero temperature, this can be explicitly calculated as

$$\chi(q\,;\varepsilon_F) = -\frac{g_v g_s e^2 v}{16\hbar c^2}\frac{1}{q}\theta(q-2k_F)\left[1+\frac{2}{\pi}\frac{2k_F}{q}\sqrt{1-\left(\frac{2k_F}{q}\right)^2} - \frac{2}{\pi}\sin^{-1}\frac{2k_F}{q}\right], \tag{23}$$

where $k_F = |\varepsilon_F|/(\hbar v)$ is the Fermi wave number. Significantly, χ vanishes in range $q < 2k_F$, i.e., no current is induced when the external field is smooth enough compared to the Fermi wavelength. At $\varepsilon_F = 0$, particularly, we have

$$\chi(q;0) = -\frac{g_v g_s e^2 v}{16\hbar c^2}\frac{1}{q}. \tag{24}$$

We plot $\chi(q\,;\varepsilon_F)$ against (a) q and (b) ε_F in Fig. 2. As a function of q, the susceptibility suddenly starts from zero at $q = 2k_F$, and rapidly approaches the universal curve (24). As a function of ε_F with fixed q, it is nonzero only in a finite region satisfying $|\varepsilon_F| < \hbar vq/2$, and its integral over ε_F becomes constant $-g_v g_s e^2 v^2/(6\pi c^2)$. Thus, in the limit of $q \to 0$ it goes to the previous result Eq. (12). The susceptibility of the carbon nanotube to a uniform field perpendicular to the axis has the equivalent expression of Eq. (24) where q is replaced by $2\pi/L$ with tube circumference L. (54; 55)
The undoped graphene ($\varepsilon_F = 0$) has a special property in which the counter magnetic field induced by the response current has a spacial distribution similar to that of external magnetic field, regardless of its length scale (24). Let us assume that a sinusoidal external field $B(\mathbf{r}) = B\cos qx$ is applied to undoped graphene. With the susceptibility Eq. (24), the response current is calculated as $j_y(\mathbf{r}) = -[g_v g_s e^2 vB/(16\hbar c)]\sin qx$. The current induces a counter magnetic field which reduces the original field. The z component of the induced field on graphene becomes

$$B_{\rm ind}(\mathbf{r}) = -\alpha_g B(\mathbf{r}), \quad \alpha_g = \frac{2\pi g_v g_s e^2 v}{16\hbar c^2} \approx 4\times 10^{-5}. \tag{25}$$

Because the ratio is independent of q, Eq. (25) is actually valid for any external field $B(\mathbf{r})$, i.e., the magnetic field on the graphene is always reduced by the same factor $1-\alpha_g$. This property holds whenever $\chi(q)$ is written as C/q with constant C.
The argument of the magnetic field screening can be extended in the three dimensional field distribution. Let us suppose a situation when a certain magnetic object is placed above the undoped graphene ($z>0$), which produces an external magnetic field $\mathbf{B}(\boldsymbol{\rho})$ in 3D space $\boldsymbol{\rho}=(x,y,z)$. Then, the followings are concluded: (i) On the other side of the graphene ($z<0$), the induced field becomes $-\alpha_g\mathbf{B}(\boldsymbol{\rho})$, i.e., the external field is screened by the factor $1-\alpha_g$. (ii) On the same side ($z>0$), the induced field is given by $\alpha_g R_z[\mathbf{B}(x,y,-z)]$, where R_z is the vector inversion with respect to $z=0$. Namely, this is equivalent to a field of the mirror image of the original object reflected with respect to $z=0$, and reduced by α_g.
Using the property of magnetic mirroring, we can easily estimate the magnetic repulsive force which works between undoped graphene and magnetic objects. If we put on the top of graphene the permanent magnet with surface magnetic flux σ_m, the repulsive force per unit area is given by $2\pi\alpha_g\sigma_m^2$, which is a force between a sheet with magnetic monopole density σ_m and its mirror image with $\alpha_g\sigma_m$. At σ_m which amounts to the surface flux 1T (e.g., neodymium magnet),(56) this is as large as 0.16 gram weight /cm^2, which is surprisingly large as a force generated by a thin film only one atom thick.
The $1/q$ dependence of $\chi(q)$, which is responsible for peculiar diamagnetic responses argued above, is a characteristic property common to general k-linear Hamiltonian. This can be shown using the scaling argument as follows. We consider an isotropic two-dimensional system, in which Hamiltonian $\mathcal{H}$ contains only terms linear in π_x and π_y. The velocity operator $v_\mu = \partial\mathcal{H}/\partial\pi_\mu$ is then a constant matrix independent of $\boldsymbol{\pi}$. Similarly to graphene, the eigen energy and the wavefunction are written as $\varepsilon_{n,\mathbf{k}}$ and $\psi_{n,\mathbf{k}}(\mathbf{r}) = e^{i\mathbf{k}\cdot\mathbf{r}}\mathbf{F}_{n,\mathbf{k}}/\sqrt{S}$, respectively, with the subband index n and the wavenumber $\mathbf{k}$. The response function $K_{\mu\nu}(\mathbf{q})$ can then be written in the same form as Eq. (22), with index s replaced with n.
If we change the energy and wave number scales by an arbitrary factor α as

$$\varepsilon = \alpha\tilde{\varepsilon}, \quad k_i = \alpha\tilde{k}_i, \tag{26}$$

then the Hamiltonian becomes formally identical under that transformation, since the coefficients of k-linear terms in the Hamiltonian remain unchanged. The eigen energy and eigen function obey $\varepsilon_{n,\tilde{\mathbf{k}}} = \varepsilon_{n,\mathbf{k}}/\alpha$ and $\mathbf{F}_{n,\tilde{\mathbf{k}}} = \mathbf{F}_{n,\mathbf{k}}$. Using them, we can show that the response function $K_{\mu\nu}(\mathbf{q})$ scales at zero temperature and at zero Fermi energy as

$$K_{\mu\nu}\left(\frac{\mathbf{q}}{\alpha}\right) = \frac{K_{\mu\nu}(\mathbf{q})}{\alpha}, \tag{27}$$

which allows us to write as $K(q) = Cq$ with certain number C. With Eq. (21), the susceptibility χ becomes

$$\chi(q) = \frac{C}{cq}. \tag{28}$$

Similar scaling argument also applies to $\chi(\varepsilon_F)$ at $q=0$, leading to $\chi(\varepsilon_F) \propto \delta(\varepsilon_F)$ in any k-linear Hamiltonian.(26)

3. Bilayer graphene

Bilayer graphene is a pair of graphene layers arranged in AB (Bernal) stacking and includes A_1 and B_1 atoms on layer 1 and A_2 and B_2 on layer 2.(33–41) The schematics of the AB stacked

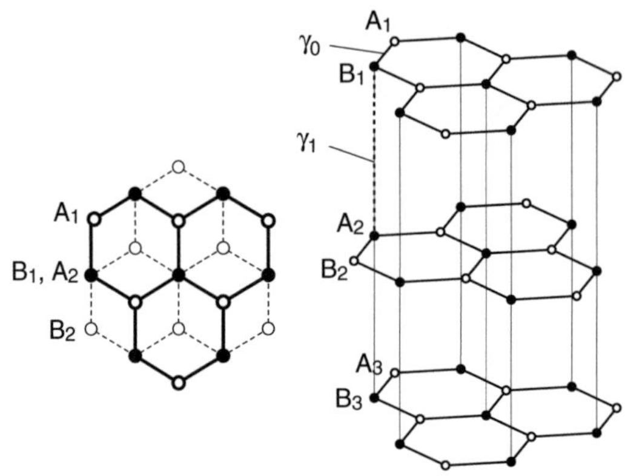

Fig. 3. Atomic structure of AB (Bernal)-stacked multilayer graphene

graphite is illustrated in Figure 3, and the only layer 1 and 2 exist in bilayer graphene. The Hamiltonian at the K point for the basis $(|A_1\rangle, |B_1\rangle, |A_2\rangle, |B_2\rangle)$ is given by

$$\mathcal{H}^K = \begin{pmatrix} \Delta & v\pi_- & 0 & 0 \\ v\pi_+ & \Delta & \gamma_1 & 0 \\ 0 & \gamma_1 & -\Delta & v\pi_- \\ 0 & 0 & v\pi_+ & -\Delta \end{pmatrix}, \tag{29}$$

and that for K' point is obtained by exchanging $\pi_\pm$. The parameter $\gamma_1 \approx 0.39$ eV (57) represents vertical coupling between B_1 and A_2 (33; 34; 60), and Δ describes potential asymmetry between layer 1 and 2 (not A and B sites), which gives rise to an energy gap. (33–37; 39; 60; 61) Experimentally the potential asymmetry can be induced by applying an electric field perpendicular to the layer, (30–32; 62; 63) and the energy gap as large as 0.2 eV was actually observed in spectroscopic measurements. (30; 62; 63) Several interlayer coupling parameters other than γ_1 were introduced for the description of the band structure of bulk graphite, (58; 59) while they do not change the qualitative feature of the low-energy spectrum (59) and will be neglected in the following arguments.
The energy band at $B = 0$ is given by (37)

$$\varepsilon_{s\mu}(p) = s\left(\frac{\gamma_1^2}{2} + v^2p^2 + \Delta^2 + \mu\left[\frac{\gamma_1^4}{4} + v^2p^2(\gamma_1^2 + 4\Delta^2)\right]^{1/2}\right)^{1/2}, \tag{30}$$

with $s = \pm 1$ and $\mu = \pm 1$. The index $\mu = +1$ and -1 give a pair of bands further and closer to zero energy, respectively, and $s = +1$ and -1 in each pair represent the electron and hole branches, respectively. The band-edge energies corresponding to $p = 0$ are given by $|\varepsilon| = \varepsilon_\pm$ for $\mu = \pm 1$, where

$$\varepsilon_+ = \sqrt{\gamma_1^2 + \Delta^2}, \quad \varepsilon_- = |\Delta|. \tag{31}$$

For $\mu = -1$, the band minimum becomes

$$\varepsilon_0 = \frac{\gamma_1|\Delta|}{\sqrt{\gamma_1^2 + 4\Delta^2}}, \tag{32}$$

which corresponds to an off-center momentum.(34) The density of states diverges here as $D(\varepsilon) \propto (\varepsilon - \varepsilon_0)^{-1/2}$. The energy bands and the density of states with several Δ's are plotted

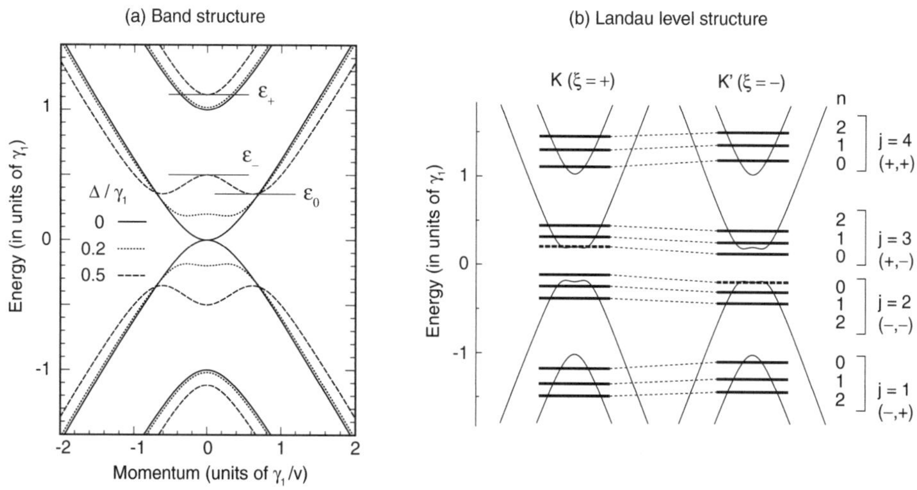

Fig. 4. (a) Band structure of bilayer graphenes with the asymmetry gap $\Delta/\gamma_1 = 0, 0.2$, and 0.5. Horizontal lines indicate the energies of ε_0, ε_- and ε_+ for $\Delta/\gamma_1 = 0.5$. (b) Landau-level energies of bilayer graphene given by Eq. (39) with $\Delta = 0.2\gamma_1$ and $\hbar\omega_B = 0.5\gamma_1$. At each valley, an energy level indicated as broken bar represents the Landau level which originally belongs to $n = -1$ at opposite valley. The quantum number (s, μ) is indicated below j.

in Figs. 5 (a) and (b), respectively. Vertical lines in (a) indicate the energies of ε_0, ε_-, and ε_+ for $\Delta = 0.5\gamma_1$.

In a magnetic field, the eigenfunction of the Hamiltonian at the K point is written as $(c_1\phi_{n-1}, c_2\phi_n, c_3\phi_n, c_4\phi_{n+1})$ with integer $n \geq -1$. For $n \geq 1$, the Hamiltonian matrix for (c_1, c_2, c_3, c_4) becomes(25; 34; 64)

$$H^K_{n\geq 1} = \begin{pmatrix} \Delta & \hbar\omega_B\sqrt{n} & 0 & 0 \\ \hbar\omega_B\sqrt{n} & \Delta & \gamma_1 & 0 \\ 0 & \gamma_1 & -\Delta & \hbar\omega_B\sqrt{n+1} \\ 0 & 0 & \hbar\omega_B\sqrt{n+1} & -\Delta \end{pmatrix}, \quad (33)$$

For $n = 0$, the first component does not actually exist because $\phi_{-1} = 0$. The matrix for (c_2, c_3, c_4) becomes

$$H^K_0 = \begin{pmatrix} \Delta & \gamma_1 & 0 \\ \gamma_1 & -\Delta & \hbar\omega_B \\ 0 & \hbar\omega_B & -\Delta \end{pmatrix}. \quad (34)$$

For $n = -1$, only the component c_4 survives and the Hamiltonian is

$$H^K_{-1} = -\Delta. \quad (35)$$

For the K' point, the eigenfunction is written as $(c_1\phi_{n+1}, c_2\phi_n, c_3\phi_n, c_4\phi_{n-1})$. For $n \geq 1$, the Hamiltonian matrix for (c_1, c_2, c_3, c_4) is

$$H^{K'}_{n\geq 1} = \begin{pmatrix} \Delta & \hbar\omega_B\sqrt{n+1} & 0 & 0 \\ \hbar\omega_B\sqrt{n+1} & \Delta & \gamma_1 & 0 \\ 0 & \gamma_1 & -\Delta & \hbar\omega_B\sqrt{n} \\ 0 & 0 & \hbar\omega_B\sqrt{n} & -\Delta \end{pmatrix}. \quad (36)$$

For $n = 0$, the matrix for (c_1, c_2, c_3) becomes

$$H_0^{K'} = \begin{pmatrix} \Delta & \hbar\omega_B & 0 \\ \hbar\omega_B & \Delta & \gamma_1 \\ 0 & \gamma_1 & -\Delta \end{pmatrix}, \tag{37}$$

and for $n = -1$, that for c_1 is

$$H_{-1}^{K'} = \Delta. \tag{38}$$

If we extend the definition of the matrix of Eq. (33) to $n = 0$, its three eigenvalues agree with those of H_0^K and the rest with that of $H_{-1}^{K'}$. Similarly, the matrix of Eq. (36) with $n = 0$ gives eigenvalues of $H_0^{K'}$ and H_{-1}^K. Thus we can use Eqs. (33) and (36) with $n \geq 0$ to produce the full spectrum. By introducing the pseudo-spin variable $\xi = \pm 1$, the Hamiltonian is combined into a single expression,

$$H_n^{\xi} = \begin{pmatrix} \Delta & \sqrt{x_{n-}} & 0 & 0 \\ \sqrt{x_{n-}} & \Delta & \gamma_1 & 0 \\ 0 & \gamma_1 & -\Delta & \sqrt{x_{n+}} \\ 0 & 0 & \sqrt{x_{n+}} & -\Delta \end{pmatrix}, \tag{39}$$

with

$$x_{n\pm} = x_n \pm \frac{1}{2}\xi\delta, \tag{40}$$

and

$$x_n = \left(n + \frac{1}{2}\right)\delta, \quad \delta = (\hbar\omega_B)^2. \tag{41}$$

We write the eigenvalues of H_n^{ξ} as

$$\varepsilon_j(x_n, \xi\delta) \quad (j = 1, 2, 3, 4), \tag{42}$$

in the ascending order in energy ($j = 1$ and 2 for valence bands and $j = 3$ and 4 for the conduction bands). The second argument in $\varepsilon_j(x_n, \xi\delta)$ represents the dependence on B which are not included in x_n. Figure 4 shows the example of the Landau-level spectrum at $\Delta/\gamma_1 = 0.2$ and $\hbar\omega_B/\gamma_1 = 0.5$, where the thick dashed lines represent the Landau level which originally belongs to $n = -1$ at opposite valleys. The correspondence between quantum numbers j and (s, μ) are indicated in the figure.
The thermodynamic potential becomes

$$\begin{aligned} \Omega &= -\frac{1}{\beta}\frac{g_s}{2\pi l_B^2}\sum_{\xi,j}\sum_{n=0}^{\infty}\varphi\left[\varepsilon_j(x_n, \xi\delta)\right] \\ &= -\frac{1}{\beta}\frac{g_s}{4\pi\hbar^2 v^2}\sum_{\xi,j}\left[\int_0^{\infty}\varphi\left[\varepsilon_j(x, \xi\delta)\right]dx + \frac{\delta^2}{24}\frac{\partial\varphi[\varepsilon_j(x,0)]}{\partial x}\bigg|_{x=0}\right] + O(\delta^3), \end{aligned} \tag{43}$$

where we used the Euler-Maclaurin formula in the second equation. The first term in the bracket can be transformed by changing the integral variable from x to ε as

$$\frac{1}{\beta}\int_0^{\infty}\varphi\left[\varepsilon_j(x, \xi\delta)\right]dx = \int_{-\infty}^{\infty} f(\varepsilon)\, n_j(\varepsilon, \xi\delta) d\varepsilon, \tag{44}$$

where we used $\varphi'(\varepsilon) = -\beta f(\varepsilon)$ with $f(\varepsilon)$ being the Fermi distribution function, and defined

$$n_j(\varepsilon, \xi\delta) \equiv s_j(\varepsilon, \xi\delta) x_j(\varepsilon, \xi\delta), \tag{45}$$

where $x_j(\varepsilon, \xi\delta)$ is a real and positive solution of $\varepsilon = \varepsilon_j(x, \xi\delta)$ and

$$s_j(\varepsilon, \xi\delta) \equiv \mathrm{sgn}\Big(\frac{\partial x_j(\varepsilon, \xi\delta)}{\partial \varepsilon}\Big). \tag{46}$$

If there are more than one solution of x_j, we regard n_j as their sum. The quantity $n_j(\varepsilon, \xi\delta)/(4\pi\hbar^2 v^2)$ represents the electron density below ε for the conduction band and the hole density above ε for the valence band. By expanding

$$n_j(\varepsilon, \xi\delta) = n_j^{(0)}(\varepsilon) + n_j^{(1)}(\varepsilon)\, \xi\delta + \frac{1}{2} n_j^{(2)}(\varepsilon)\, \delta^2 + \cdots, \tag{47}$$

we can further expand Ω of Eq. (43) in terms of $\delta \propto B$. We have (25)

$$\chi(\varepsilon) = g_s g_v \frac{e^2 v^2}{\pi c^2} \sum_j \Big[\int_{-\infty}^{\varepsilon} n_j^{(2)}(\varepsilon') d\varepsilon' - \frac{1}{12} \theta[\varepsilon - \varepsilon_j(0,0)] \frac{\partial \varepsilon_j(x,0)}{\partial x}\Big|_{x=0} \Big]. \tag{48}$$

For the Hamiltonian of Eq. (39), the eigenequation $\det(\varepsilon - H_n^{\xi}) = 0$ can be solved for x ($\equiv x_n$) as

$$x_{\pm} = \varepsilon^2 + \Delta^2 \pm \frac{1}{2}\sqrt{(4\varepsilon\Delta - \xi\delta)^2 + 4\gamma_1^2(\varepsilon^2 - \Delta^2)}, \tag{49}$$

which gives $x_j(\varepsilon, \xi\delta)$ when being real and positive. At symmetric bilayer, $\Delta = 0$, the susceptibility can be explicitly calculated as

$$\chi(\varepsilon) = -\frac{g_v g_s}{4\pi c^2} \frac{e^2 \gamma^2}{\hbar^2} \frac{\theta(\gamma_1 - |\varepsilon_F|)}{\gamma_1} \left(-\ln \frac{|\varepsilon_F|}{\gamma_1} - \frac{1}{3} \right), \tag{50}$$

agreeing with the previous results (15; 21). The susceptibility diverges logarithmically toward $\varepsilon_F = 0$, becomes slightly positive for $|\varepsilon_F| \lesssim \gamma_1$, and vanishes for $|\varepsilon_F| > \gamma_1$ where the higher subband enters. The integration of χ in Eq. (50) over the Fermi energy becomes $-(g_v g_s/3\pi)(e^2\gamma^2/\hbar^2)$ independent of γ_1, which is exactly twice as large as that of the monolayer graphene Eq. (12).

For finite asymmetry Δ, we can analytically argue several important properties as follows. (25) Let us first consider the case $\varepsilon > \varepsilon_+$, where two conduction bands are occupied by electrons. In this case $x_\pm$ are both real and positive and we have $x_1 = x_2 = 0$, $x_3 = x_+$, and $x_4 = x_-$. Then, we have

$$\sum_j n_j(\varepsilon, \xi\delta) = x_+ + x_- = 2(\varepsilon^2 + \Delta^2), \tag{51}$$

independent of $\xi\delta$. Therefore, $\sum_j n_j^{(2)}(\varepsilon)$ identically vanishes, resulting in susceptibility independent of energy in the region $\varepsilon > \varepsilon_+$ regardless of the value of Δ. The same is true for $\varepsilon < -\varepsilon_+$. Because $\chi = 0$ for $\varepsilon = \pm\infty$, i.e., in the case of empty or filled band, we can conclude that the susceptibility vanishes for $\varepsilon > \varepsilon_+$ and $\varepsilon < -\varepsilon_+$ independent of interlayer interaction γ_1 and asymmetry Δ.

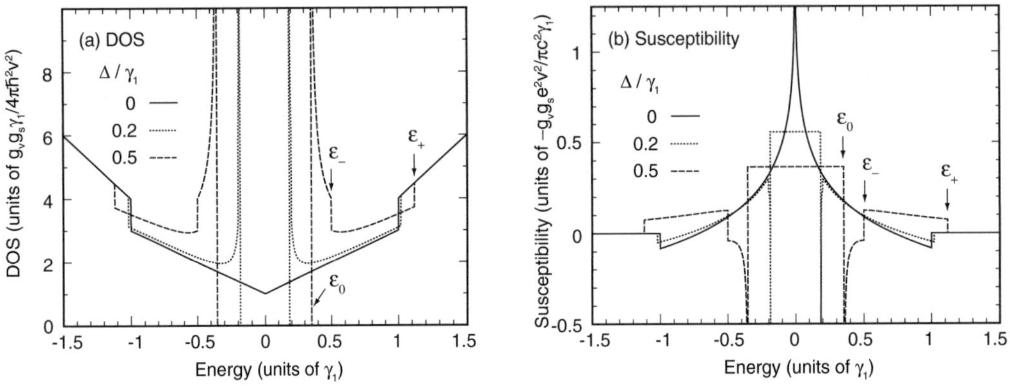

Fig. 5. (a) Density of states, and (b) magnetic susceptibility of bilayer graphenes with the asymmetry gap $\Delta/\gamma_1 = 0, 0.2$, and 0.5. Arrows indicate the energies of ε_0, ε_- and ε_+ for $\Delta/\gamma_1 = 0.5$. In (b), upward direction represents negative (i.e., diamagnetic) susceptibility.

In the energy region $-\varepsilon_- < \varepsilon < -\varepsilon_0$ near the top of the valence band, both x_+ and x_- are real and positive, giving the states at outer and inner equi-energy circle of the band $j = 2$, respectively. Then we have

$$n_2^{(2)}(\varepsilon) = \frac{\partial^2}{\partial\delta^2}(-x_+ + x_-)\Big|_{\delta=0} = \frac{\gamma_1^2(\Delta^2-\varepsilon^2)}{2[(4\Delta^2+\gamma_1^2)(\varepsilon^2-\varepsilon_0^2)]^{3/2}}. \quad (52)$$

When the energy approaches to $-\varepsilon_0$ from the negative side, the integral of $n_2^{(2)}(\varepsilon)$, thus the susceptibility, diverges in positive direction as $\propto (\varepsilon+\varepsilon_0)^{-1/2}$ in the same manner as the density of states. The same divergence occurs at the bottom of the conduction band, $+\varepsilon_0$, because of the electron-hole symmetry.
In the vicinity of the bottom of the excited conduction band, $\varepsilon = \varepsilon_+$, we have

$$n_4^{(2)}(\varepsilon) = \frac{\partial^2}{\partial\delta^2}x_-\theta(x_-)\Big|_{\delta=0} = \left[\frac{\partial^2 x_-}{\partial\delta^2}\theta(x_-) + \left(\frac{\partial x_-}{\partial\delta}\right)^2\delta(x_-)\right]_{\delta=0}, \quad (53)$$

where we used $x_-\delta(x_-) = 0$ and $x_-\delta'(x_-) = -\delta(x_-)$. Using Eq. (48), we find that the susceptibility makes a discrete jump at ε_+ as

$$\chi(\varepsilon_+ + 0) - \chi(\varepsilon_+ - 0) = g_v g_s \frac{e^2v^2}{\pi c^2}\left(\frac{\Delta^2\sqrt{\Delta^2+\gamma_1^2}}{\gamma_1^2(2\Delta^2+\gamma_1^2)} - \frac{2\Delta^2+\gamma_1^2}{12\gamma_1^2\sqrt{\Delta^2+\gamma_1^2}}\right), \quad (54)$$

where the first term in the bracket comes from the integral of the delta function in Eq. (53) and the second term from the step function in Eq. (48).
Near ε_+, the eigenstates are given primarily by the dimer states composed of $|B_1\rangle$ and $|A_2\rangle$. The effective Hamiltonian is described by the second order in interband interaction with the conduction-band bottom $|A_1\rangle$ and the valence-band top $|B_2\rangle$, where each process gives a term $\propto \pi_+\pi_-$ or $\propto \pi_-\pi_+$. In symmetric bilayer with $\Delta = 0$, the terms $\pi_+\pi_-$ and $\pi_-\pi_+$ have the same coefficient and the pseudo-spin Zeeman term identically vanishes. When Δ becomes nonzero, the two coefficients shift from each other linearly in Δ because of the band-gap

opening, leading to a nonzero Zeeman term. The resulting effective Hamiltonian is given by Eqs. (13) and (14) with

$$m^* = \frac{\gamma_1^2\sqrt{\Delta^2+\gamma_1^2}}{2v^2(2\Delta^2+\gamma_1^2)}, \quad g^* = \frac{4\Delta\sqrt{\Delta^2+\gamma_1^2}}{2\Delta^2+\gamma_1^2}\frac{m}{m^*}. \tag{55}$$

In the region $\Delta \lesssim \gamma_1$, the typical magnitude of the effective mass m^* is of the order of $\sim \gamma_1/(2v^2) \approx 0.035m$.

The susceptibility is written as Pauli and Landau magnetism in Eqs. (15) and (16), respectively, which together give a susceptibility jump of Eq. (54). The paramagnetic component χ_P is zero at $\Delta = 0$ and monotonically increases as Δ becomes larger. At $g^* = (2/\sqrt{3})(m/m^*)$ or $\Delta \approx 0.34\gamma_1$, χ_P exceeds χ_L and the susceptibility step changes from diamagnetic to paramagnetic. In the limit $\Delta \to \infty$, we have $g^* = 2m/m^*$ as in the monolayer. This is to be expected, because the bilayer graphene in this limit can be regarded as a pair of independent monolayer graphenes, where interlayer coupling γ_1 opens an energy gap at each Dirac point. Similar argument also applies to the behavior around ε_-.

Figure 5 (c) plots the susceptibility for $\Delta = 0$, 0.2, and 0.5. In accordance with the above analytic consideration, we actually observe that the susceptibility vanishes in the region $\varepsilon > \varepsilon_+$ and $\varepsilon < -\varepsilon_+$ and that the susceptibility step at $\varepsilon = \varepsilon_+$ changes from diamagnetic to paramagnetic with increasing Δ. We also see that the susceptibility for $\Delta \neq 0$ diverges in the paramagnetic direction at $\varepsilon = \pm\varepsilon_0$.

4. Multilayer graphenes

We consider a multilayer graphene composed of N layers of a carbon hexagonal network, which are arranged in the AB (Bernal) stacking. A unit cell contains A_j and B_j atoms on the layer $j = 1, \cdots, N$. If the basis is taken as $|A_1\rangle, |B_1\rangle$; $|A_2\rangle, |B_2\rangle$; $\cdots$; $|A_N\rangle, |B_N\rangle$, the Hamiltonian for the multilayer graphene around the K point becomes (26; 34–36; 40; 41)

$$\mathcal{H} = \begin{pmatrix} H_0 & V & & & \\ V^\dagger & H_0 & V^\dagger & & \\ & V & H_0 & V & \\ & & \ddots & \ddots & \ddots \end{pmatrix}, \tag{56}$$

with

$$H_0 = \begin{pmatrix} 0 & v\pi_- \\ v\pi_+ & 0 \end{pmatrix}, \quad V = \begin{pmatrix} 0 & 0 \\ \gamma_1 & 0 \end{pmatrix}, \tag{57}$$

where $\pi_\pm = \pi_x \pm i\pi_y$ with $\boldsymbol{\pi} = -i\hbar\nabla + e\mathbf{A}$ and the vector potential $\mathbf{A}$. Here we do not consider the interlayer potential asymmetry. The effective Hamiltonian for K' is obtained by exchanging π_+ and π_-.

Let us define functions

$$f_m(j) = \frac{2}{\sqrt{N+1}}\sin\left(\frac{\pi}{2}j\right)\cos\left[\frac{m\pi}{2(N+1)}j\right], \tag{58}$$

$$g_m(j) = \frac{2}{\sqrt{N+1}}\cos\left(\frac{\pi}{2}j\right)\sin\left[\frac{m\pi}{2(N+1)}j\right]. \tag{59}$$

Here $j = 1, 2, \cdots, N$ is the layer index, and m is the subsystem index which ranges as

$$m = \begin{cases} 1, 3, 5, \cdots, N-1, & N = \text{even} \\ 0, 2, 4, \cdots, N-1, & N = \text{odd} \end{cases} \tag{60}$$

Obviously $f_m(j)$ is zero at even j, while $g_m(j)$ is zero at odd j. We construct the basis by assigning $f_m(j), g_m(j)$ to each cite as

$$\begin{aligned} |\phi_m^{(\text{A,odd})}\rangle &= f_m(1)|A_1\rangle + f_m(3)|A_3\rangle + \cdots, \\ |\phi_m^{(\text{B,odd})}\rangle &= f_m(1)|B_1\rangle + f_m(3)|B_3\rangle + \cdots, \\ |\phi_m^{(\text{A,even})}\rangle &= g_m(2)|A_2\rangle + g_m(4)|A_4\rangle + \cdots, \\ |\phi_m^{(\text{B,even})}\rangle &= g_m(2)|B_2\rangle + g_m(4)|B_4\rangle + \cdots. \end{aligned} \tag{61}$$

The superscript such as (A, odd) indicates that the wavefunction has an amplitude only on $|A_j\rangle$ with odd j's.
In the basis of Eq. (61), the Hamiltonian (56) is decomposed into independent blocks labeled by m. (26) The subspace of $m = 0$ is special in that $g_m(j)$ is identically zero, so that only two bases $\{|\phi_0^{(\text{A,odd})}\rangle, |\phi_0^{(\text{B,odd})}\rangle\}$ survives in Eq. (61). The Hamiltonian matrix projected on this subspace is written as

$$\mathcal{H}_0 = \begin{pmatrix} 0 & v\pi_- \\ v\pi_+ & 0 \end{pmatrix}, \tag{62}$$

which is equivalent to the Hamiltonian of the monolayer graphene. Note that $m = 0$ only exists when the total layer number N is odd as shown in Eq. (60). For $m \neq 0$, the projected matrix on $\{|\phi_m^{(\text{A,odd})}\rangle, |\phi_m^{(\text{B,odd})}\rangle, |\phi_m^{(\text{A,even})}\rangle, |\phi_m^{(\text{B,even})}\rangle\}$ becomes

$$\mathcal{H}_m = \begin{pmatrix} 0 & v\pi_- & 0 & 0 \\ v\pi_+ & 0 & \lambda\gamma_1 & 0 \\ 0 & \lambda\gamma_1 & 0 & v\pi_- \\ 0 & 0 & v\pi_+ & 0 \end{pmatrix}, \tag{63}$$

where $\lambda \equiv \lambda_m$, is defined by

$$\lambda_m = 2\cos\kappa_m, \quad \kappa_m = \frac{\pi}{2} - \frac{m\pi}{2(N+1)}. \tag{64}$$

Eq. (63) is equivalent to the Hamiltonian of a bilayer graphene, except that interlayer-coupling parameters γ_1 are multiplied by λ. An odd-layered graphene with $N = 2M + 1$ is composed of one monolayer-type and M bilayer-type subsystems, while even-layered graphene with $N = 2M$ is composed of M bilayers but no monolayer. Figure 6 shows the low-energy band structures in the multilayer graphenes with $N = 3, 4$, and 5.
The eigenstate of a finite-layered graphene can be regarded as a part of a standing wave in 3D limit, which is a superposition of opposite traveling waves with $\pm k_z$. The quantity $\kappa(= \kappa_m)$ in our representation corresponds to the 3D wave number via $\kappa = |k_z|d$, with d being the distance between adjacent layers. Thus the monolayer-type subsystem $\kappa = \pi/2$ is related to H-point in the 3D Brillouin zone and $\kappa = 0$ is to K-point.
Using the decomposition of the Hamiltonian, the magnetization of the N-layered graphene can be written as a summation over each sub-Hamiltonian.(26) The contribution from

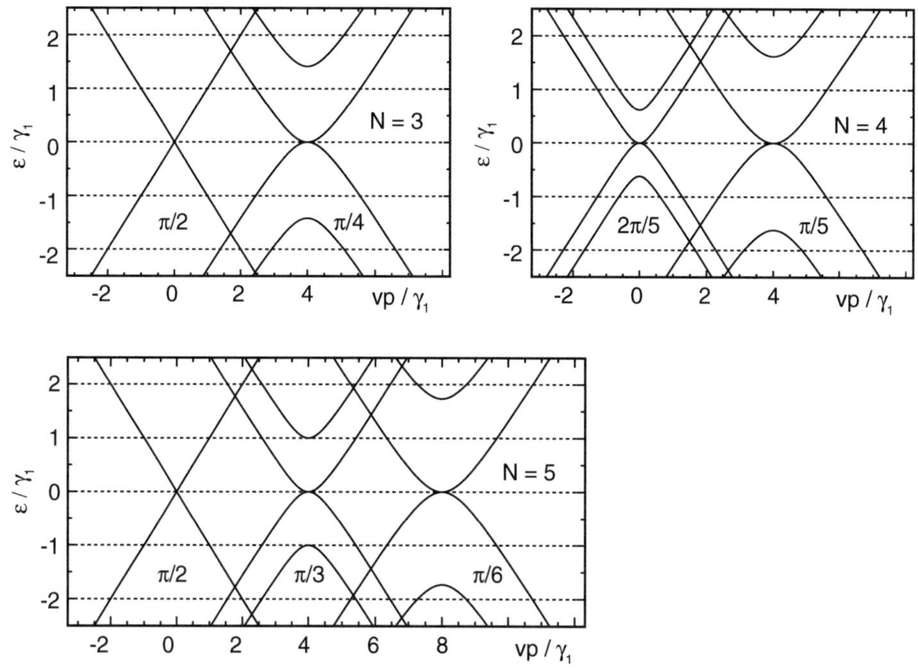

Fig. 6. Band structures of multilayer graphenes of $N = 3, 4$, and 5. The energy bands are shown separately for each of subsystems with a horizontal shift. The number represents κ.

$m = 0$ is exactly equivalent to the susceptibility of a monolayer graphene, Eq. (12), which is proportional to delta function of ε_F. Thus, the odd-layer graphene always has a large diamagnetic peak at zero energy. The susceptibility of a bilayer-like graphene becomes Eq. (50) with γ_1 replaced with $\lambda\gamma_1$. Figure 7 shows $\chi(\varepsilon_F)$ of graphenes with layer number from $N = 1$ to 5, where we include a constant energy broadening Γ as a phenomenological disorder effect. We can see that odd-layered graphenes exhibit a particularly large peak, which mainly comes from the monolayer-type band.

5. Three dimensional Dirac system

The results of graphene can be directly extended to three-dimensional Dirac Hamiltonian, which approximately describes the electronic structure of bismuth. (4–6; 65) The Hamiltonian is given by

$$\mathcal{H} = \begin{pmatrix} \Delta & 0 & v\pi_z & v\pi_- \\ 0 & \Delta & v\pi_+ & -v\pi_z \\ v\pi_z & v\pi_- & -\Delta & 0 \\ v\pi_+ & -v\pi_z & 0 & -\Delta \end{pmatrix}, \tag{65}$$

where four bases correspond to two orbital and two spin degrees of freedom. The anisotropy of the velocity present in bismuth is ignored for simplicity, as we focus on the parallel argument to its 2D counterpart. The orbital susceptibility was previously calculated for realistic Hamiltonian retaining the anisotropy and other factors. (6)

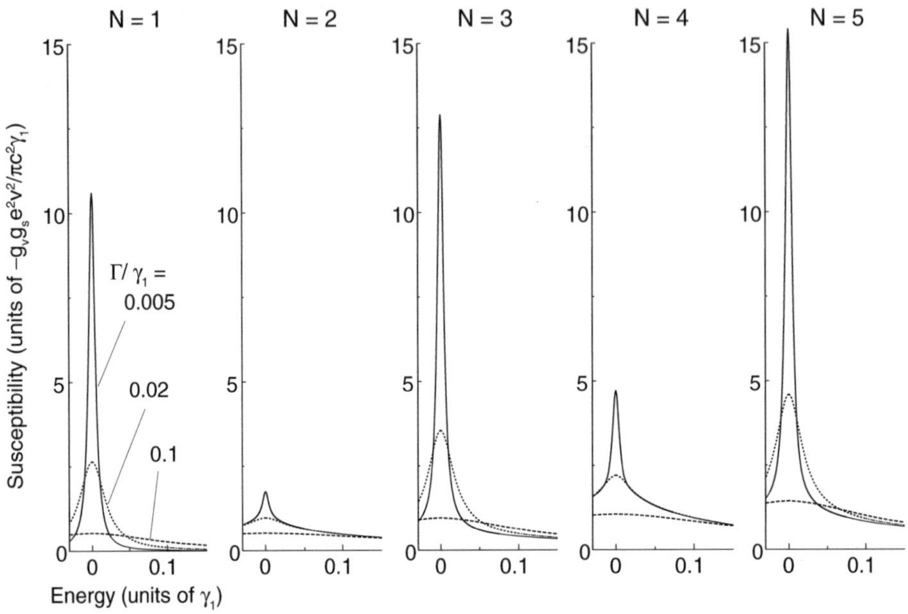

Fig. 7. Susceptibility of multilayer graphenes with layer numbers $N = 1$ to 5, plotted against the Fermi energy. Results shown for several disorder strengths specified by constant scattering rate Γ.

The density of states at zero magnetic field is

$$D(\varepsilon) = \frac{g_v g_s}{\pi^2 \hbar^3 v^3} |\varepsilon| \sqrt{\varepsilon^2 - \Delta^2}\, \theta(\varepsilon^2 - \Delta^2), \tag{66}$$

where g_v is the valley degeneracy allowing the presence of different k points described by the above Hamiltonian in the first Brillouin zone. The Landau levels in a uniform magnetic field in z direction are given by (25)

$$\varepsilon_{s,n,\sigma} = s\sqrt{(\hbar\omega_B)^2\left(n + \frac{1}{2} + \frac{\sigma}{2}\right) + v^2 p_z^2 + \Delta^2} \quad (n = 0, 1, 2, \cdots), \tag{67}$$

with $\hbar\omega_B = \sqrt{2}\hbar v/l_B$, $s = \pm 1$, and $\sigma = \pm 1$. This is equivalent to the two-dimensional Dirac system, Eq. (17), when the term Δ^2 is replaced with $\Delta^2 + v^2 p_z^2$. The susceptibility $\chi(\varepsilon)$ is calculated by integrating Eq. (11) in p_z as

$$\begin{aligned} \chi(\varepsilon) &= -\frac{g_v g_s e^2 v^2}{6\pi c^2} \int \frac{dp_z}{2\pi\hbar} \frac{\theta(\Delta^2 + v^2 p_z^2 - \varepsilon^2)}{2\sqrt{\Delta^2 + v^2 p_z^2}} \\ &= -\frac{g_v g_s e^2 v}{12\pi^2 \hbar c^2} \begin{cases} \log \dfrac{2\varepsilon_c}{|\Delta|} & (|\varepsilon| < |\Delta|); \\ \log \dfrac{2\varepsilon_c}{|\varepsilon| + \sqrt{\varepsilon^2 - \Delta^2}} & (|\varepsilon| > |\Delta|), \end{cases} \end{aligned} \tag{68}$$

where ε_c is a cut-off energy. In the limit of $\Delta \to 0$, the susceptibility at zero energy logarithmically diverges.

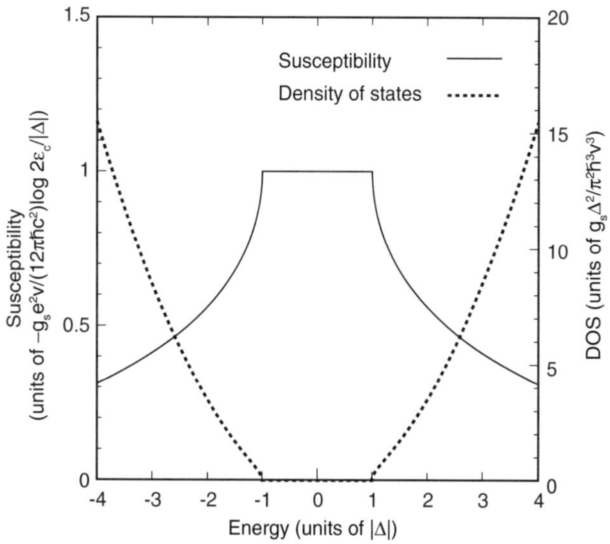

Fig. 8. Orbital susceptibility and density of states of three-dimensional Dirac electron.

At an energy ε just above the band bottom $|\Delta|$, we obtain the paramagnetic contribution

$$\chi(\varepsilon) - \chi(0) \approx \frac{2}{3}\left(\frac{m}{m^*}\right)^2 D(\varepsilon)\mu_B^2, \tag{69}$$

where $D(\varepsilon) = (g_s g_v/4\pi^2)(2m^*/\hbar^2)^{3/2}\sqrt{\varepsilon}$ with $m^* = \Delta/v^2$. This is nothing but the magnetic susceptibility, dominated by the Pauli paramagnetism, of a three-dimensional metal with mass m^* and g factor $g^* = 2m/m^*$. Figure 8 shows the susceptibility and the density of states in the present system. The singular decrease of the susceptibility at the band edges is understood in terms of the appearance of the dominant spin paramagnetism inside the band.

We note that in bismuth the index σ in Eq. (67) represents real spin, while it was valley pseudo-spin in Eq. (17) for graphene. The Pauli component included in Eq. (69) thus describes the real spin paramagnetism enhanced by the strong spin-orbit coupling, apart from the bare electron paramagnetism.

6. Conclusion

We have reviewed the orbital magnetism of graphene and related materials. Graphene monolayer has a delta-function singularity at Dirac point, which can be understood in zero-mass limit of massive Dirac electrons. There the Pauli paramagnetism associated with valley degree of freedom gives an essential contribution in addition to the conventional Landau diamagnetism. The graphene multilayers also exhibits strong diamagnetism, while the interlayer coupling significantly modifies the magnetic singularity at the band touching point. The argument can be extended for a three-dimensional Dirac system, which explains the strong diamagnetism of bismuth in a parallel fashion to that for graphene. The susceptibility in graphene related systems is expected to be observed by employing the experimental techniques used for two-dimensional electron systems in semiconductor.(66–71)

7. Acknowledgments

The materials presented here are based on close collaborations with Tsuneya Ando, Yasunori Arimura and Edward McCann. This work was supported in part by Grant-in-Aid for Scientific Research on Priority Area "Carbon Nanotube Nanoelectronics," by Grant-in-Aid for Scientific Research, by Global Center of Excellence Program at Tokyo Tech "Nanoscience and Quantum Physics" from Ministry of Education, Culture, Sports, Science and Technology Japan, and by EPSRC First Grant EP/E063519/1 and by JST-EPSRC Japan-UK Cooperative Programme Grant EP/H025804/1.

8. References

[1] J. W. McClure, Phys. Rev. 104, 666 (1956).
[2] J. W. McClure, Phys. Rev. 119, 606 (1960).
[3] M. P. Sharma, L. G. Johnson, and J. W. McClure, Phys. Rev. B 9, 2467 (1974).
[4] P. A. Wolff, J. Phys. Chem. Solids 25, 1057 (1964).
[5] H. Fukuyama and R. Kubo, J. Phys. Soc. Jpn. 27, 604 (1969).
[6] H. Fukuyama and R. Kubo, J. Phys. Soc. Jpn. 28, 570 (1970).
[7] K. S. Novoselov, A. K. Geim, S. V. Morozov, D. Jiang, Y. Zhang, S. V. Dubonos, I. V. Grigorieva, and A. A. Firsov, Science 306, 666 (2004).
[8] K. S. Novoselov, A. K. Geim, S. V. Morozov, D. Jiang, M. I. Katsnelson, I. V. Grigorieva, S. V. Dubonos, and A. A. Firsov, Nature 438, 197 (2005).
[9] Y. Zhang, Y.-W. Tan, H. L. Stormer, and P. Kim, Nature 438, 201 (2005).
[10] J. C. Slonczewski and P. R. Weiss, Phys. Rev. 109, 272 (1958).
[11] D. P. DiVincenzo and E. J. Mele, Phys. Rev. B 29, 1685 (1984).
[12] G. W. Semenoff, Phys. Rev. Lett. 53, 2449 (1984).
[13] T. Ando, J. Phys. Soc. Jpn. 74, 777 (2005). Theory of electronic states and transport in carbon nanotubes
[14] S. A. Safran and F. J. DiSalvo, Phys. Rev. B 20, 4889 (1979).
[15] S. A. Safran, Phys. Rev. B 30, 421 (1984).
[16] J. Blinowski and C. Rigaux, J. Phys. (Paris) 45, 545 (1984).
[17] R. Saito and H. Kamimura, Phys. Rev. B 33, 7218 (1986).
[18] S. G. Sharapov, V. P. Gusynin, and H. Beck, Phys. Rev. B 69, 075104 (2004). Magnetic oscillations in planar systems with the Dirac-like spectrum of quasiparticle excitations
[19] H. Fukuyama, J. Phys. Soc. Jpn. 76, 043711 (2007).
[20] M. Nakamura, Phys. Rev. B 76, 113301 (2007).
[21] M. Koshino and T. Ando, Phys. Rev. B 75, 235333 (2007).
[22] A. Ghosal, P. Goswami, and S. Chakravarty, Phys. Rev. B 75, 115123 (2007).
[23] A. Kobayashi, Y. Suzumura, and H. Fukuyama, J. Phys. Soc. Jpn. 77, 064718 (2008).
[24] M. Koshino, Y. Arimura, and T. Ando, Phys. Rev. Lett. 102, 177203 (2009).
[25] M. Koshino and T. Ando, Phys. Rev. B 81, 195431 (2010).
[26] M. Koshino and T. Ando, Phys. Rev. B 76, 085425 (2007).
[27] M. Nakamura and L. Hirasawa, Phys. Rev. B 77, 045429 (2008).

[28] A. H. Castro Neto, F. Guinea, N. M. Peres, K. S. Novoselov, and A. K. Geim, Rev. Mod. Phys. 81, 109 (2009).

[29] K. S. Novoselov, E. McCann, S. V. Morozov, V. I. Falko, M. I. Katsnelson, U. Zeitler, D. Jiang, F. Schedin, and A. K. Geim, Nature Phys. 2, 177 (2006).

[30] T. Ohta, A. Bostwick, T. Seyller, K. Horn, and E. Rotenberg, Science 313, 951 (2006)

[31] E. V. Castro, K. S. Novoselov, S. V. Morozov, N. M. R. Peres, J. M. B. Lopes dos Santos, J. Nilsson, F. Guinea, A. K. Geim, and A. H. Castro Neto, Phys. Rev. Lett. 99, 216802 (2007).

[32] J. B. Oostinga, H. B. Heersche, X.-L. Liu, A. F. Morpurgo, and L. M. K. Vandersypen, Nat. Mat. 7, 151 (2008).

[33] E. McCann and V. I. Falko, Phys. Rev. Lett. 96, 086805 (2006).

[34] F. Guinea, A. H. Castro Neto, and N. M. R. Peres, Phys. Rev. B 73, 245426 (2006).

[35] C. L. Lu, C. P. Chang, Y. C. Huang, J. M. Lu, C. C. Hwang, and M. F. Lin, J. Phys.: Cond. Matt. 18, 5849 (2006).

[36] C. L. Lu, C. P. Chang, Y. C. Huang, R. B. Chen, and M. L. Lin, Phys. Rev. B 73, 144427 (2006).

[37] E. McCann, Phys. Rev. B 74, 161403 (2006).

[38] M. Koshino and T. Ando, Phys. Rev. B 73, 245403 (2006).

[39] J. Nilsson, A. H. Castro Neto, N. M. R. Peres, and F. Guinea, Phys. Rev. B 73, 214418 (2006).

[40] B. Partoens and F. M. Peeters, Phys. Rev. B 74, 075404 (2006).

[41] B. Partoens and F. M. Peeters, Phys. Rev. B 75, 193402 (2007).

[42] S. Y. Zhou, G.-H. Gweon, A. V. Fedorov, P. N. First, W. A. de Heer, D.-H. Lee, F. Guinea, A. H. Castro Neto, and A. Lanzara, Nature Mater. 6, 770 (2007).

[43] S. Y. Zhou, D. A. Siegel, A. V. Fedorov, F. El Gabaly, A. K. Schmid, A. H. Castro Neto, D.-H. Lee, and A. Lanzara, Nature Mater. 7, 259 (2008).

[44] M.-C. Chang and Q. Niu, Phys. Rev. B 53, 7010 (1996).

[45] D. Xiao, W. Yao, and Q. Niu, Phys. Rev. Lett. 99, 236809 (2007).

[46] N. M. R. Peres, F. Guinea, and A. H. Castro Neto, Phys. Rev. B 73, 125411 (2006).

[47] A. De Martino, L. DellAnna, and R. Egger, Phys. Rev. Lett. 98, 066802 (2007).

[48] M. R. Masir, P. Vasilopoulos, A. Matulis, and F. M. Peeters, Phys. Rev. B 77, 235443 (2008).

[49] C.-H. Park, L. Yang, Y.-W. Son, M. L. Cohen, and S. G. Louie, Phys. Rev. Lett. 101, 126804 (2008).

[50] C.-H. Park, Y.-W. Son, L. Yang, M. L. Cohen, and S. G. Louie, Nano Lett. 8, 2920 (2008).

[51] Y.-X. Li, J. Phys.: Condens. Matter 22, 015302 (2010).

[52] J. S. Park, K. Sasaki, R. Saito, W. Izumida, M. Kalbac, H. Farhat, G. Dresselhaus, and M. S. Dresselhaus, Phys. Rev. B 80, 081402 (2009).

[53] I. Snyman, Phys. Rev. B 80, 054303 (2009).

[54] H. Ajiki and T. Ando, J. Phys. Soc. Jpn. 62, 2470 (1993); *ibid.* 63, 4267 (1994); *ibid.* 64, 4382 (1995).

[55] M. Yamamoto, M. Koshino, and T. Ando J. Phys. Soc. Jpn. 77, 084705 (2008).

[56] See, for example, J.M.D. Coey: *Rare-earth Iron Permanent Magnets*, (Clarendon Press, 1996)

[57] A. Misu, E. Mendez, and M. S. Dresselhaus, J. Phys. Soc. Jpn. 47, 199 (1979).
[58] M. S. Dresselhaus and G. Dresselhaus, Adv. Phys. 51, 1 (2002).
[59] M. Koshino and T. Ando, Solid State Commun. 149, 1123 (2009).
[60] T. Ando and M. Koshino, J. Phys. Soc. Jpn. 78, 034709 (2009).
[61] T. Ando and M. Koshino, J. Phys. Soc. Jpn. 78, 104716 (2009).
[62] Y. Zhang, T.-T. Tang, C. Girit, Z. Hao, M. C. Martin, A. Zettl, M. F. Crommie, Y. R. Shen, and F. Wang, Nature 459, 820 (2009).
[63] K. F. Mak, C. H. Lui, J. Shan, and T. F. Heinz, Phys. Rev. Lett. 102, 256405 (2009).
[64] M. Koshino and E. McCann, Phys. Rev. B 81, 115315 (2010).
[65] Y. Fuseya, M. Ogata, and H. Fukuyama, Phys. Rev. Lett. 102, 066601 (2009).
[66] H. L. Stormer, T. Haavasoja, V. Narayanamurti, A. C. Gossard, and W. Wiegmann, J. Vac. Sci. Technol. B 1, 423 (1983).
[67] J. P. Eisenstein, H. L. Stormer, V. Narayanamurti, A. Y. Cho, A. C. Gossard, and C. W. Tu, Phys. Rev. Lett. 55, 875 (1985).
[68] I. Meinel, D. Grundler, S. Bargst adt-Franke, C. Heyn, and D. Heitmann, Appl. Phys. Lett. 70, 3305 (1997).
[69] A. Potts, R. Shepherd, W. G. Herrenden-Harker, M. Elliott, C. L. Jones, A. Usher, G. A. C. Jones, D. A. Ritchie, E. H. Linfield, and M. Grimshaw, J. Phys. C 8, 5189 (1996).
[70] S. A. J. Wiegers, M. Specht, L. P. Levy, M. Y. Simmons, D. A. Ritchie, A. Cavanna, B. Etienne, G. Martinez, and P. Wyder, Phys. Rev. Lett. 79, 3238 (1997).
[71] M. Zhu, A. Usher, A. J. Matthews, A. Potts, M. Elliott, W. G. Herrenden-Harker, D. A. Ritchie, and M. Y. Simmons, Phys. Rev. B 67, 155329 (2003).

Part 4

Transport Properties

20

Electronic and Transport Properties of Defected Graphene Nanoribbons

Narjes Gorjizadeh[1], Yoshiyuki Kawazoe[1] and Amir A. Farajian[2]
[1]Tohoku University,
[2]Wright State University,
[1]Japan
[2]U.S.A.

1. Introduction

Graphene nanoribbons, quasi-one-dimensional structures of carbon, are fascinating materials. These structures can be constructed as strips of graphene sheet, the two dimentional honeycomb lattice of carbon with sp^2 hybridization. Geometrically two main types of slice can be cut from a graphene sheet, with zigzag edge and armchair edge (Niimi05; Kobayashi05). The edge geometry is the key parameter which determines the electronic properties of the nanoribbons. Although the two-dimensional graphene is a zero band-gap semi-metal, electronic structure of nanoribbons depend on their edge geomtery (Saito92; Klein94; Fujita96; Son06a). A simple tight-binding model with one orbital per atom predicts that zigzag nanoribbons are metallic. But density functional calculations shows that all graphene nanoribbons are semiconductors at their ground state with band gaps which depend on their width and edge geometry, closing at infinite width, i. e. infinite graphene. Moreover, the electronic structure of graphene nanoribbons can be modified by chemical functionalization, such as functionalization by various atomic sepcies or by functional groups (Maruyama04a; Gunlycke07; Hod07; Gorjizadeh08). A large variety of electronic and magnetic properties, such as semiconducting with a wide range of band gap, metallic, ferromagnetic, antiferromagnetic, half-metallic, half-semiconducting, can be obtained by chamical modifications of the nanoribbons. Modification of the edge or using an adsorbate or substitution of carbons of the nanoribbon with an appropriate host are different options of functionalizations of these materials. These properties, along with the ballistic electronic transport, and the quantum Hall effect (Novoselov05a; Zhang05) and high carrier mobility (Novoselov05a) cause these quasi-1D materials to be promising candidates for nanoelectronics applications (Novoselov04b; Son06b; Obradovic06; Li08; Hod09; Zhu10). Various junctions can be constructed by connecting nanoribbons of different widths and types with perfect atomic interface, and electronic device can be integrated on them by selective chemical funtionalization on a single nanoribbon sheet (Huang07; Yan07; Gorjizadeh08).

In order to achieve their potential for these applications it is essential to have a better understanding of the electronic structure of graphene nanoribbons and have ability to control them. From a practical point of view, when nanoribbons are fabricated experimentally, they will have some structural defects. Vacancy and adatom–vacancy defects are among the most probable ones (Hashimoto04). These defects should be taken into account in practical aspects of the electronic transport in nanodevices based on graphene nanoribbons. In this chapter we will discuss the effects of defects in quantum conductance of grahene nanoribbons. In next section we will review the geometry of armchair and zigzag nanoribbons. Dependence of their electronic properties on the width and edge shape will be discussed. Then, an overview of nonequilibrium Green's function formalism (NEGF) for calculation of quantum conductance will be presented. After that the results of conductance calculation of armchair and zigzag nanoribbons with some structural defects such as vacancies and adatom-vacancies will be discussed using NEGF techniques combined with a four-orbitals-per-atom tight-binding method. We will then show dependence of the conductance on the width of nanoribbons and the position of the defects and their distance from the edge of the nanoribbons.

2. Geometry and electronic structure of graphene nanoribbons

There are two types of nanoribbons, based on their edge shapes, called zigzag and armchair edges (Saito92). The width of the nanoribbons is labled by an integer, N, which stands for the number of carbon dimers counted from one edge toward the other edge of armchair nanoribbons and the number of zigzag lines for zigzag nanoribbons. Figure 1 shows the two types of nanoribbons with their width indices. The dotted rectangle in this figure represents the unit cell of the nanoribbons. Nanoribbon is like an unrolled carbon nanotube (CNTs). The difference between carbon nanotubes and nanoribbons is the periodic boundary condition. Instead of a seamless tube, nanoribbon has two open boundaries. Therefore, the periodic boundary condition of nanotube in circumferential direction changes to open boundary condition (Möbius boundary condition) (Wakabayashi03). The electronic properties of nanoribbons are affected by these open boundaries. In carbon nanotubes, the circumferential periodic boundary condition imposes its electronic properties, making it metallic or semiconducting (Saito92). The electronic properties of nanoribbons, on the other hand, are affected by their open boundaries, as well as their edge geometry. Intrinsically there are dangling bonds at the edges, whose linear combinations form some of the eigenstates near the Fermi energy, and determine the properties of nanoribbons.

The earliest theoretical studies of graphene nanoribbons, using a simple tight-binding method with one π-orbital per atom, predicts that 1/3 of the armchair nanoribbons, whose width index N satisfies N=3M-1 (M is an integer), are metallic (Fujita96), and the other 2/3 are semiconductor with band gaps depending on their width, while all zigzag nanoribbons are metallic. This behaviour is similar to characteristics of carbon nanotubes. A characteristic peak is also predicted in the density of states (DOS) of zigzag nanoribbons near Fermi energy (Niimi06; Sasaki06). This peak is a nano-size effect and decreases by increasing the width of nanoribbon. By considering a four-orbitals-per atom tight-binding model, and optimizing the structure of nanoribbons, or simply consider a larger hopping energy for the C-C bonds of the edge in the one-orbital per atom model, the 1/3 rule for armchair nanoribbons does not apply any more, and all the armchair nanoribbons show an energy gap around the Fermi energy. This gap changes with the width of the nanoribbon, which is the effect of the open boundary.

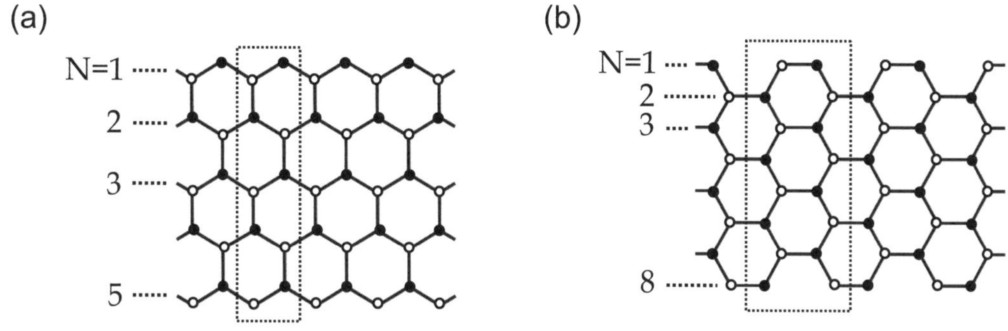

Fig. 1. Graphene nanoribbons with zigzag (a) and armchair (b) edges. The black circles denote sublattice A, while the white circles refer to sublattice B. The dotted rectangle represents the unit cell of the nanoribbon. The width of the nanoribbon is specified by N.

First principle studies based on spin polarized density functional theory (DFT) predicts that all graphene nanoribbons are semiconductors at their ground state with band gaps which depend on their width and edge geometry, closing at infinite width, i. e. infinite graphene (Pisani07). Meanwhile zigzag nanoribbons have localized edge states which are ferromagnetically ordered, but with opposite spin orientation at the two edges which makes them antiferromagnetically coupled. The magnetism in zigzag nanoribbon, a pure carbon system, which arises from π-orbitals of carbon localized at the edge is specially notable (Wakabayashi99).

There are two types of basic site in primitive cell of graphene and the atomic sites of graphene are divided into two sublattices, called A and B. In zigzag nanoribbons the edge atoms belong to the same sublattice, while the two edges carry atoms of two different sublattices. In these nanoribbons, the A-type and B-type atoms are placed alternatively from one edge toward the other one, as shown in Fig. 1(a). The spin in the same sublattice atoms of zigzag nanoribbon is localized in one direction, and in the opposite of the other sublattice. Hence, the total spin of the zigzag nanoribbons is zero, and the edges of the nanoribbons are antiferromagnetically ordered in the ground state. In case of armchair nanoribbons, the edge of the nanoribbon contains both A- and B-type atoms, as dimers, which sit along the edge, as depicted in Fig. 1(b). The black circles in this figure illustrate sublattice A and the white circles refer to sublattice B.

Another important matter in the structure of nanoribbons is the C-C bond lengths with respect to their distance from the edge. There are two types of C-C bonds in a graphene nanoribbon, those which are parallel to the axis of nanoribbon, where we call them type 1, and those which make an 30° or 60 ° angle with the axis, and we call them type 2. The change of C-C bond length with the number of atomic layers distant from edge are shown in Fig 2 for armchair nanoribbons N = 8, and N = 14. The 0th layer implies the edge, as shown in the inset. In nanoribbon N=8 all C-C bonds are affected by the edge and distort from 1.42 Å, which is the optimized bond length of graphene. But, for N=14, the effect of edge can be seen only till second atomic layer distant from the edge. From the third layer, the bonds are 1.42 Å, the same as graphene. The bond length of the edge, which is the triple bond, is 1.26 Å or 1.27 Å depending on the width of nanoribbon. Patterns of other structures with N > 14 are the same as N = 14, with edge bond length of 1.26 Å and the bonds of the bulk are equal to 1.42 Å.

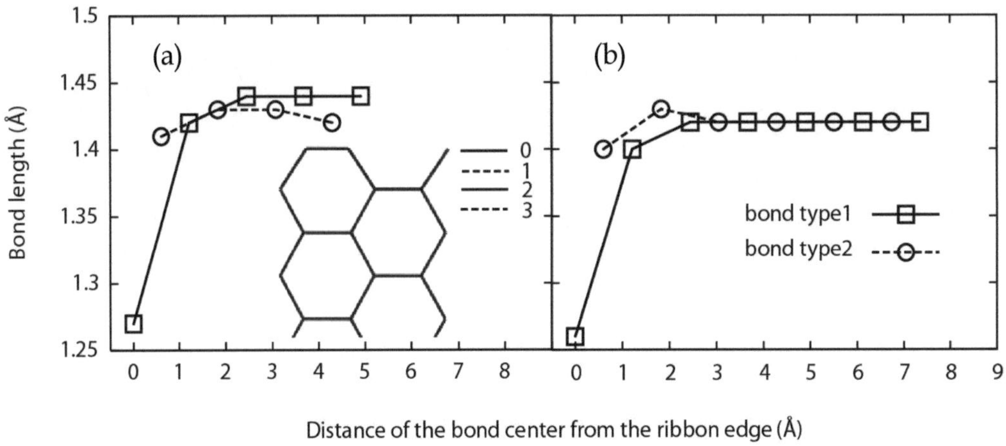

Fig. 2. C-C bond length for different layers of carbon counted as their distance from edge of armchair nanoribbons N=8 (a) and N=14 (b).

3. Tight-binding and NEGF

We use the nonequilibrium Green's function (NEGF) technique to calculate quantum conductance of nanoribbons (Datta95). The electronic structure and forces are obtained through a four-orbitals-per-atom tight-binding model based on semiempirical parameters developed by Xu et al. for carbon (Xu92). This parameterization has been used in tight binding studies of carbon systems, such as fullerenes and nanotubes (Esfarjani98; Ozaki00; Farajian01). We divide the system under study to three parts, as sketched in Fig 3: left lead, conductor, and right lead. According to Landauer formula (Landauer70), the conductance through the conductor region is:

$$C = \frac{2e^2}{h} T \,, \tag{1}$$

where the coefficient $2e^2/h$ is the conductance quantum in ballistic conductor, and T is the transmission function, and can be expressed in NEGF formalism as follows (Pastawski91,Tian98):

$$T = Tr(\Gamma_L G^r \Gamma_R G^a) \,. \tag{2}$$

Here, *Tr* indicates the transmission per energy channel between the leads, Γ_L and Γ_R are the coupling functions to the left and right leads, and G^r and G^a are the retarded and advanced Green's functions of the conductor (Torelli00), respectively, given by:

$$G^r = (EI - H - \Sigma^r)^{-1} \,, \tag{3}$$

$$G^a = [G^r]^\dagger \,. \tag{4}$$

The advanced Green's function is the Hermitian conjugate of the retarded Green's function. The Green's function describes the dynamics of the electron inside the system. Here, *H* is the

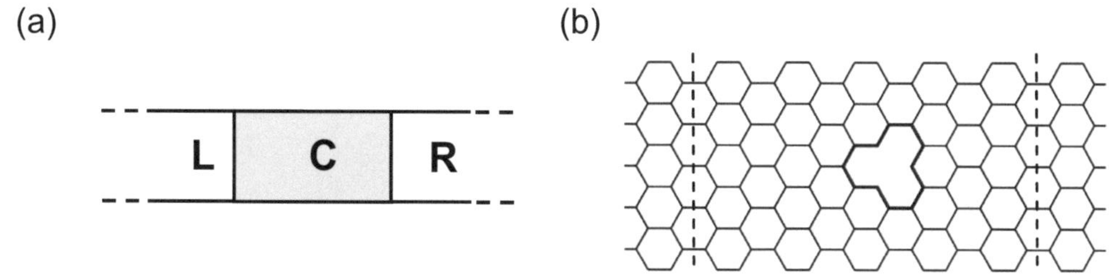

Fig. 3. (a) Schematic picture of a conductor (C) sandwiched between left (L) and right (R) contacts. (b) An armchair nanoribbon N=11 including a vacancy defect as an example of the L-C-R system.

Hamiltonian of conductor part, E is the (complex) energy of conducting electron ($E=\varepsilon+i\,\eta$), with η arbitrarily small, and I is the identity matrix. Σ^r is the retarded self-energy which contains the self-energy of the Left and Right leads, $\Sigma^r=\Sigma^r_L+\Sigma^r_R$, where Σ^r_L and Σ^r_R are obtained from surface Green's functions of the leads as follows (Munoz87):

$$\Sigma_L^r = H_{LC}^{\dagger} G_L^0 H_{LC} \tag{5}$$

$$\Sigma_R^r = H_{CR} G_R^0 H_{CR}^{\dagger} \tag{6}$$

where $G^0{}_L$ and $G^0{}_R$ are the surface Green's functions of the semi-infinite Left and Right leads, respectively; and H_{LC} and H_{CR} represent the coupling matrices to the leads. In fact, the semi-infinite leads are mapped into complex self-energies. The imaginary part of the self-energy implies that the eigenstates are not confined leak from the boundaries. The surface Green's function, which includes the effect of the semi-infinite lead projected at its surface layer, is calculated by López Sancho algorithm (Lopez Sancho84).
The Hamiltonian of the whole system can be written in tight-binding block matrix as follows:

$$H = \begin{pmatrix} H_L & H_{LC} & 0 \\ H_{LC}^{\dagger} & H_C & H_{CR} \\ 0 & H_{CR}^{\dagger} & H_R \end{pmatrix}. \tag{7}$$

where H_C is the Hamiltonian of the conductor part, which is used in the Green' function in eq. 3. The coupling functions of the leads, Γ_L and Γ_R, are identified by

$$\Gamma_{L(R)} = i[\Sigma_{L(R)}^r - \Sigma_{L(R)}^a], \tag{8}$$

where Σ^a is the Hermitian conjugate of Σ^r.
In eq. 3, the Hamiltonian, the Green's function, and the self-energies are all represented by matrices, using the tight-binding basis. In the tight-binding expression for carbon used in this work, there are four orbitals representing the 2*s* and 2*p* orbitals of carbon. Hence, for a system with N carbon atoms, the basis set contains $4N$ elements, and the matrices are $4N\times 4N$. In this formalism DOS is equal to to $-Im[Tr(G)]/\pi$, where *Im* inducates the maginary part.

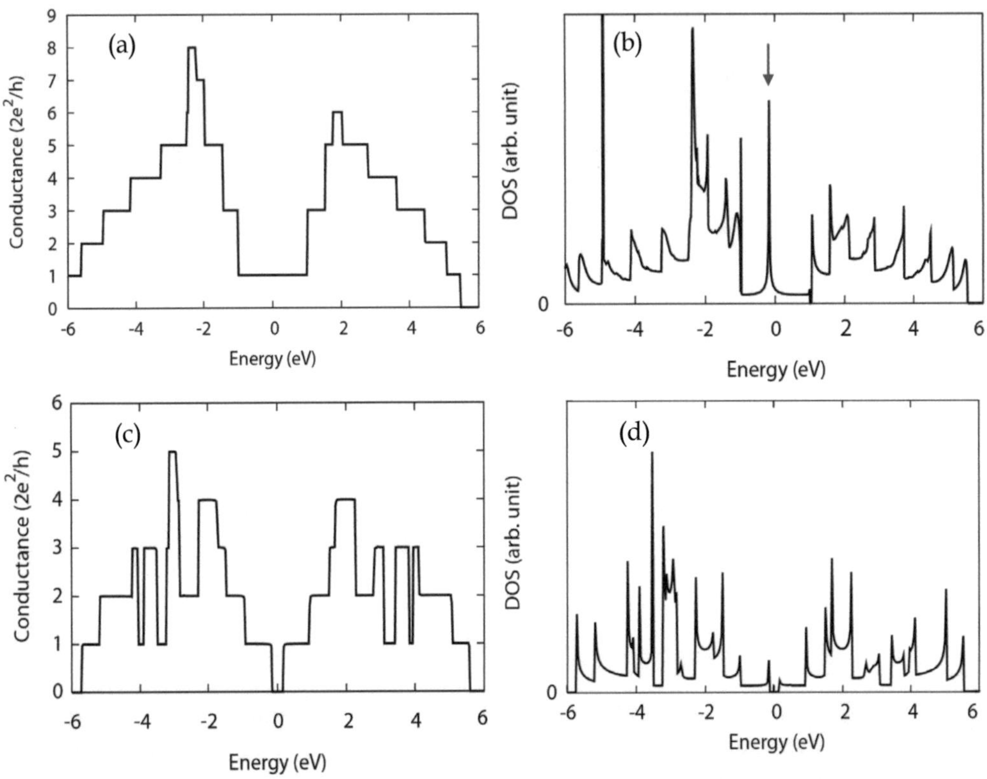

Fig. 4. (a) Conductance and (b) DOS of a zigzag nanoribbon N=6. (c) Conductance and (d) DOS of an armchair nanoribbon N=8. The Fermi energy is set to zero. The sharp peak specified by an arrow in (a) refers to edge state of the zigzag nanoribbons.

In order to calculate the conductance, the first step is to optimize the geometry of the structure, using the tight-binding description parameterized by Xu et al. (Xu92), to obtain the Hamiltonian description of the system. Then, we calculate the Green's function of the conductor, which includes the effect of the leads through their self-energies. The conductance is then obtained through eq. 1, 2. We do not consider the spin of electrons in the calculations of this chapter, and take part the charge of the carriers in the transport properties. The edges in all structures are pure carbon and are not hydrogen saturated. Figure 4 illustrates the conductance and DOS of a zigzag N=6 and an armchair nanoribbon N=8, versus the energy of carrier. The Fermi energy of the structures is set to zero. Zigzag nanoribbons with width N=6 has metallic behaviour within tight-binding approximation. The sharp peak at the energy 0.18 eV in the DOS, specified by an arrow in Fig. 4(b), refers to localized edge states of zigzag nanoribbons. These edge states correspond to the non-bonding π orbitals. The existence of these edge states has been confirmed by experimental observations also (Niimi06). The edge states are the characteristics of zigzag nanoribbons and do not appear in the armchair edges. The conductance and DOS of the armchair nanoribbons shows an energy gap around Fermi energy. This energy gap depends on the width of the nanoribbons and decreases by increasing the width. In graphene, the in-plane bonds are the σ bonds with sp^2 hybridization, which does not contribute in conduction.

These are the π bonds, perpendicular to nanoribbon surface, which are responsible for the conduction. The change in bond length results in a change in overlap of Pz orbitals, i.e. the π bonds, and leads to changing the conduction property. For those nanoribbons whose π pattern is distorted more from the graphene sheet, the energy gap is larger. Regardless of appropriate boundary condition, it seems that the typical bond length of 1.42Å is the most suitable for π orbitals to contribute in conduction. It is worth mentioning that saturation of the edge by hydrogen reduces the gap, but does not remove it. The energy gap of armchair nanoribbons N=8, for instance, decreases from 0.45 eV to 0.27 eV by hydrogen saturation of the edge.

4. Quantum conductance of nanoribbons with structural defects

For calculating the conductance of nanoribbons with defects we consider five unit cells of the nanoribbon as the conductor part, which is sandwiched between two semi-infinite contacts, as sketched in Fig 3(b). The contacts are defectless graphene nanoribbons with the same width and edge geometry of the middle junction.

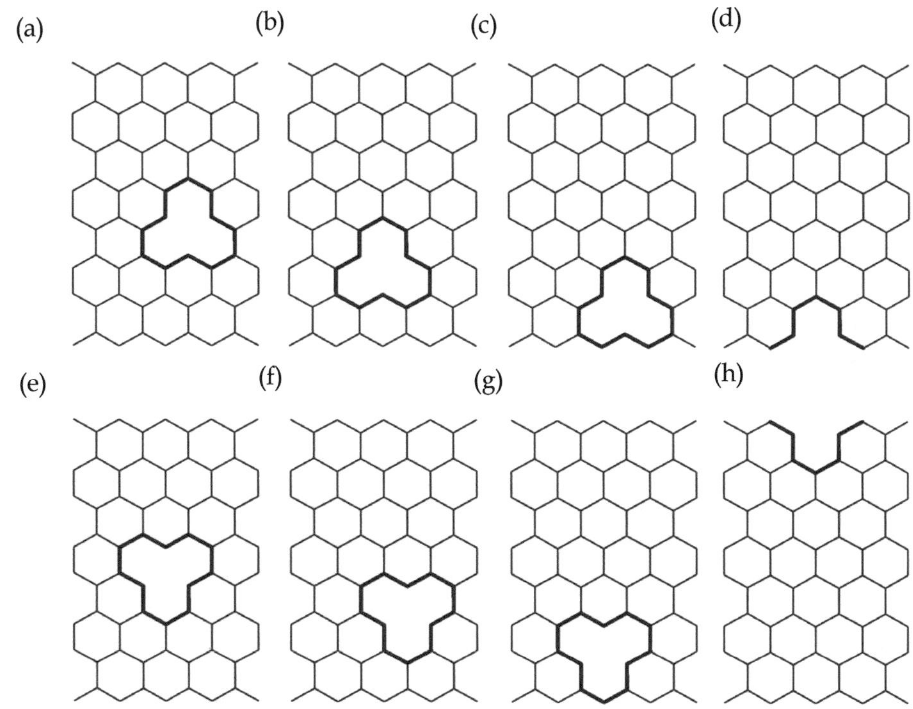

Fig. 5. Zigzag nanoribbons N=8 with vacancies of type B (top) and type A (bottom) at different positions toward the edge.

In order to obtain the Hamiltonian of the middle junction, i.e conductor, the nanoribbon is sandwiched between five unit cells of nanoribbons with the same width from left and right and the structure is optimized, such that the maximum force on each atom is less than 0.01 eV/Å (Gorjizadeh09). After the optimization, the five extra unit cells on each of the two sides of the nanoribbons are eliminated, and the five unit cells of the nanoribbon left in the

middle are used as the junction part for calculations of conductance. Using this method, the extra unit cells which were eliminated, omit the effects of open boundary along the axis, such that the junction part represents a portion of an infinite nanoribbon. The geometries of the semi-infinite left and right contacts are also extracted from the relaxed structures, obtained the same way as the junction part. This way, we are able to simulate the effect of a single defect in the junction part, without having periodic boundary condition.
The vacancy defect can appear in the site of an atom of type A or type B in the nanoribbon. The position and type of vacancy, A or B, versus the edge of the nanoribbon is the matter of considering in the calculations.

4.1 Defects in zigzag nanoribbons

We consider a zigzag nanorirbbon N=8 with five unit cells in the middle and calculate the conductance when there is a vacancy of type A and B at different positions versus edge, as illustrated in Fig. 5. First we consider the structure of Fig. 5(a), a vacancy of type B in the middle of the nanoribbon. Figure 6 depicts the calculated conductance of nanoribbon with this defect, compared with the conductance of a perfect (defectless) nanoribbon. It can be noticed from this figure that number of conducting channels decreases in the defected nanoribbon. By moving the position of this vacancy from center toward the edge which has the same type of atom, according to Fig. 5(a-d), the conductance near the Fermi energy is changed (Fig. 6(b)). The behaviour of the conductance clearly depends on the place of the vacancy versus the edge. When the vacancy is in the middle of the nanoribbon, the conductance at Fermi energy decreases to zero. By moving the vacancy from the center toward the edge, the conductance near the Fermi energy increases until it reaches the conductance of a perfect nanoribbon, when the vacancy is at the edge. The energy for which the decrease of conductance is the strongest is slightly below the Fermi energy, around -0.18 eV. On the other hand, the vacancy of type A in the center of the nanoribbon, as in Fig. 5(e), even opens a gap around the energy -0.18 eV (Fig. 6(c)). Moving the place of this vacancy toward the same edge, according to Fig. 5(e-g), will result in the same trend as vacancy B, as depicted in Fig. 6(d). As mentioned in section 2, all zigzag nanoribbons have a typical peak in their DOS near and slightly below the Fermi energy which is due the edge states. Our results show that a vacancy defect inside the nanoribbon affects the edge states of the nanoribbon. By removing one atom from the lattice of the nanoribbon, three sp^2 bonds are broken, creating three dangling bonds in the neighboring carbon atoms. These dangling bonds tend to spread toward the neighboring carbon atoms of the nanoribbon in order to overlap with electronic orbitals of their neighbors and lower their energy. Hence, the electronic wave functions near the vacancy are affected. In other words, some localized states appear in the vicinity of the vacancy and affect the π bonds of the nanoribbon. Therefore, the effect of the vacancy on the π bands (from -1to 1 eV in Fig. 6(b)), which are responsible for conduction near the Fermi energy, is significant. If the localized vacancy states are in the middle of the nanoribbon, their effect on the whole electronic wave function will be maximum. As the width of the nanoribbon is small, on the order of a few Å, the effect of the localized states of the vacancy are spread until both nanoribbon edges. Once the vacancy is moved toward one edge, the effect of the localized vacancy states at and near the other edge will decrease, due to increase of the distance between the vacancy and the other edge of the nanoribbon. Therefore, the number of less affected conducting channels of the nanoribbon increases compared to when the vacancy is in the middle of nanoribbon.

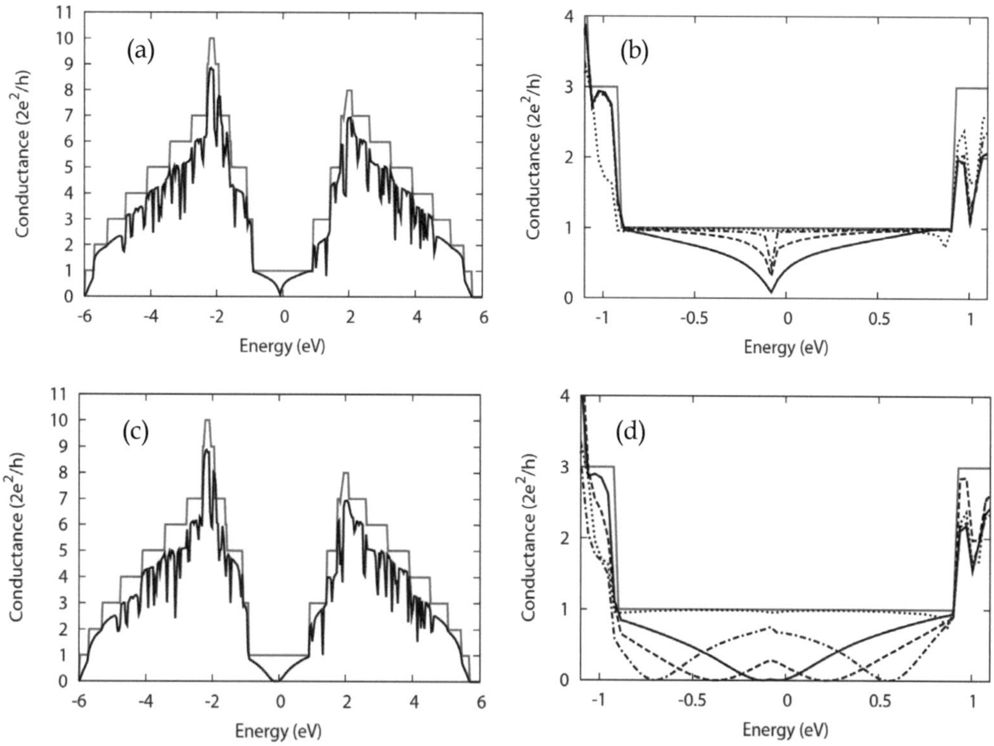

Fig. 6. Conductance of a zigzag nanoribbon N=8 with (a) vacancy defect of Fig. 5(a) compared with the defectless nanoribbon sketched in gray, (b) vacancy defects of type B at different positions from the edge. The solid line is defect of Fig. 5(a), dashed line is defect of Fig. 5(b), dashed-dotted line is defect of Fig. 5(c), and dotted line is defect of Fig. 5(d). Conductance of the same zigzag nanoribbon N=8 with (c) vacancy defect of Fig. 5(e) compared with the defectless nanoribbon sketched in gray. (d) Vacancy defects of type A at different positions from the edge. The solid line is defect of Fig. 5(e), dashed line is defect of Fig. 5(f), dashed-dotted line is defect of Fig. 5(g), and dotted line is defect of Fig. 5(h).

The same situation happens for zigzag nanoribbons with other widths, such as N=6 and N=4. However, the gap opening around the energy -0.18 eV when the vacancy is in the middle of the nanoribbon, is larger when the width of the nanoribbon is smaller. This gap is equal to 1.0 eV for the nanoribbon N=4, while it is 0.35 eV for the nanoribbon N=8. Therefore, by increasing the width of nanoribbon, this effect will diminish.

Next, we place an adatom at a nearest neighbor position of vacancy. An adatom-vacancy defect of this kind is more stable than one adatom and one vacancy defect (Hashimoto04). A carbon adatom is not stable on graphene. It will diffuse on graphene with a diffusion barrier of 0.14 eV (Lehtinen03; Hashimoto04) It appears mostly in the vicinity of the vacancies, because this combination of vacancy and neighboring adatom has been predicted to be metastable and long lived (Hashimoto04; Krasheninnikov01; Lu04).

An adatom can be attached on top of an A-type or a B-type site. Figure 7(a-b) shows schematic of the adatom-vacancy defect, when adatom is in site A or B at a nearest neighboring site of a vacancy type A in zigzag nanoribbon N=6. The C-C bond between the

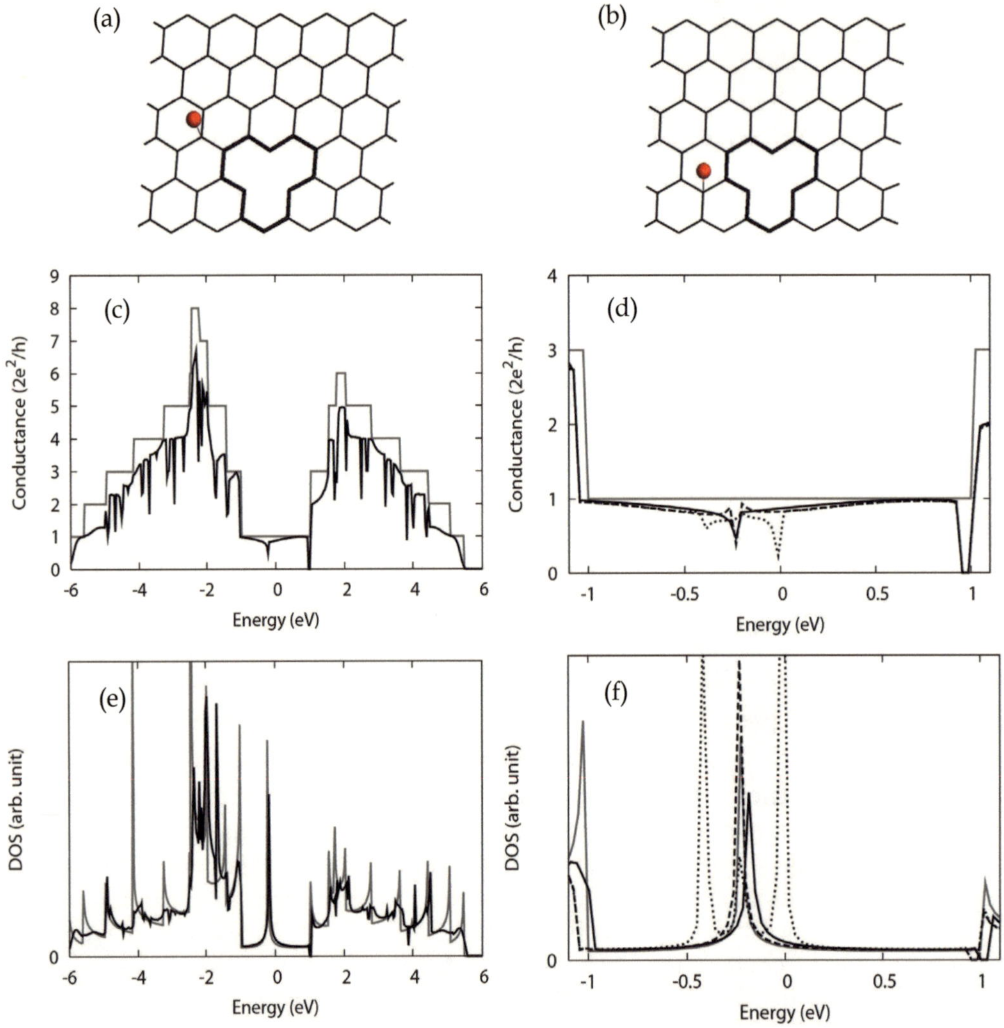

Fig. 7. Schematic of a zigzag nanoribbon N=6 with adatom-vacancy defect, when adatom is in (a) site A or, (b) site B, in vicinity of a vacancy type A. (c) Conductance of the zigzag nanoribbon with vacancy type A. (d) Compariosn of the conductance of plot (c) with adatom-vacancy defects of (a) and (b). The solid line is the conductance of vacancy type A, dahsed line is that of adatom-vacancy type A-A (Fig. 5(a)), and dotted line is that of adatom-vacancy type B-A. Two peaks appear near and at the two sides of the original peak of the nanoribbon, in case of adatom-vacancy type B-A, indicating that localized states appear at these energies.

adatom and the carbon atom of the nanoribbon is 2.04 Åwhich indicates that the adatom is physisorbed on the nanoribbon. Conductance of the zigzag nanoribbon N=6 with the vacancy is sketched in Fig. 7.c, while the conductance of adatom-vacancy defects of type A-A and B-A are shown in Fig. 7.d. The results of DOS in Fig. 7. e-f show that two peaks appear near and at the two sides of the original peak of the nanoribbon, in case of adatom-vacancy type B-A, indicating that localized states appear at these energies which are located at energy intervals -0.18 eV and +0.21 eV away from the original peak. These two energies show a drastic drop, instead of the drop at the energy of the original peak at -0.18 eV. However, the DOS and conductance of adatom of type A near vacancy of type A does not change compared to that of vacancy A defect. Therefore, the adatom does not change the transport property of the zigzag nanoribbon when it is attached to a site of the same type as vacancy, while it splits the original peak of the DOS at -0.18 eV and decreases the conductance when it is attached to a site of different type site compared to the vacancy. The same results were obtained for zigzag nanoribbons of other widths, i. e. N=4 and 8.

4.2 Defects in armchair nanoribbons

We consider the same defects of the vacancy and adatom-vacancy in the structures of armchair nanoribbons N=8, 12, and 16, and compare the conductance characteristics with those of perfect structures. Armchair nanoribbons are semiconducting. Figure 8(a-b) shows the conductance of an armchair nanoribbon N=8 with a vacancy type B in the middle of the nanoribbon, and its comparison with those of structures with one and two atom vacancy at the edge. Similar to the zigzag nanoribbons, in armchair nanoribbons also localized states appear when there is vacancy in their structures, and the conductance decreases due to defects compared with the perfect nanoribbon. But the energy gap is not affected by the defects. This could be expected as the gap arises owing to the lack of conducting channels (bands), for that particular energy interval, within the two semi-infinite defectless contacts.The vacancy states are localized near the vacancy within the junction part, which attaches the two semi-infinite graphene contacts. These localized states cannot generate extra bands within the contacts, and therefore the gap remains upon introducing the vacancy.

Figure 8(c-d) depicts the conductance and DOS of adatom-vacancy defects. The conductance of adatom-vacancy is the same as that of vacancy defect, indicating that adatom does not affect the transport properties of the armchair nanoribbon, both for the adatom and vacancy of the same type and different types. The same results are obtained for the nanoribbons N=12 and 16. This effect is attributed to the lack of edge state in armchair nanoribbon. The edge state in zigzag nanoribbon produces a peak in DOS at -0.18 eV. This peak is split into two by different sites of adatom-vacancy defects, as is shown in Fig. 7(f). With the absence of edge-state peak in armchair nanoribbons, the split does not occur and the conductance of adatom-vacancy defect is similar to that of vacancy.

5. Conclusion

We study the effect of defects such as vacancy and adatom-vacancy on the conductance of graphene nanoribbons. Our results show that localized states appear when there is vacancy inside the nanoribbon, which affects its conductance. The drop of conductance depends on the place of the vacancy. When the vacancy is at the middle of the nanoribbon, its effect on conductance at energies near Fermi energy is maximum. For zigzag nanoribbons, there is a drastic drop of conductance at the energy -0.18 eV which is the energy of the original peak in the density of states (DOS) of the nanoribbon. A gap opens around this energy, whose width depends on the width of the nanoribbon: The thinner the nanoribbon, the wider the gap. By moving the vacancy from the middle of the nanoribbon toward the edge, the decrease of

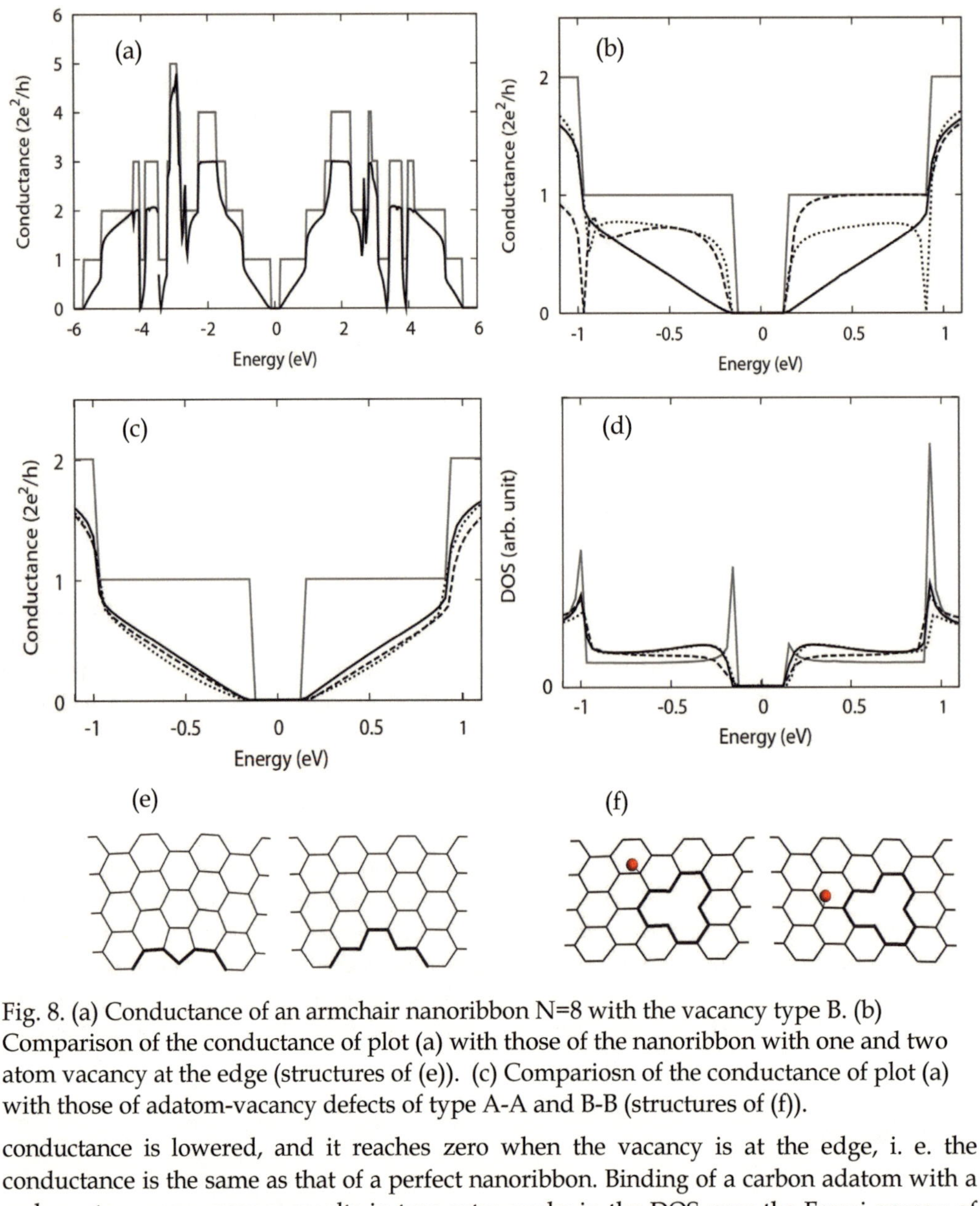

Fig. 8. (a) Conductance of an armchair nanoribbon N=8 with the vacancy type B. (b) Comparison of the conductance of plot (a) with those of the nanoribbon with one and two atom vacancy at the edge (structures of (e)). (c) Compariosn of the conductance of plot (a) with those of adatom-vacancy defects of type A-A and B-B (structures of (f)).

conductance is lowered, and it reaches zero when the vacancy is at the edge, i. e. the conductance is the same as that of a perfect nanoribbon. Binding of a carbon adatom with a carbon atom near vacancy results in two extra peaks in the DOS near the Fermi energy of zigzag nanoribbons. This binding makes localized states at energies of the new peaks and results in the decrease of conductance at these energies. These effects are very sensitive to the adatom and vacancy types and their relative distances. The peak splitting in the DOS of zigzag nanoribbons occurs when adatom and vacancy are of different types. In the case of thin armchair nanoribbons, which are semiconducting, the presence of vacancy and adatom-vacancy defects decreases the number of conducting channels, but the gap does not change due to these defects. The vacancy and adatom-vacancy defects have the same effect on conductance owing to the absence of edge state near the Fermi energy of armchair nanoribbons. We neglected the spin of electrons in all the calculations in this chapter, and considered the electronic conductance through tight-binding formalism.

6. Acknowledgments

The authors would like to express their sincere thanks to the crew of the Center for Computational Materials Science, Institute for Materials Research, for their support of the Hitachi SR11000 (model K2) supercomputer system and also Global COE program of Tohoku University for financial support.

7. References

[Datta, 1995] Datta, S. (1995). *Electronic transport in mesoscopic systems,* Cambridge University Press, ISBN, 0-521-41604-3, Cambridge, UK

[Esfarjani et al., 1998] Esfarjani, K.; Hashi, Y.; Onoe, J.; Takeuchi, K. & Kawazoe, Y. Vibrational modes and IR analysis of neutral photopolymerized C_{60} dimers. *Physical Review B,* Vol. 57, No. 1, (1998), pp. 223-229

[Farajian & Mikami, 2001] Farajian, A. A. & Mikamai, M. Electronic and mechanical properties of C_{60}-doped nanotubes. *Journal of Physics: Condensed Matter,* Vol. 13, (3 September 2001), pp. 8049-8059

[Fujita et al., 1996] Fujita, M.; Wakabayashi, K.; Nakada, K. & Kusakabe, K. Peculiar localized state at zigzag graphite edge. *Journal of the Physical Society of Japan,* Vol. 65, No. 7, (1996), pp. 1920–1923

[Gorjizadeh et al., 2008] Gorjizadeh, N.; Farajian, A.A.; Esfarjani, K. & Kawazoe, Y. Spin and band-gap engineering in doped graphene nanoribbons. *Physical Review B,* Vol. 78, (2008), Article ID 155427, 6 pages

[Gorjizadeh et al., 2009] Gorjizadeh, N.; Farajian, A. A. & Kawazoe, Y. Effects of defects on conductance of graphene nanoribbons. *Nanotechnology* Vol. 20, (2009), Article ID 015201, 6 pages

[Gunlycke et al., 2007] Gunlycke, D.; Li, J.; Mintmire, J. W. & White, C. T. Altering low-bias transport in zigzag edge graphene nanostrips with edge chemistry. *Applied Physics Letters,* Vol. 91, No. 11, (2007), Article ID 112108, 3 pages

[Hashimoto et al., 2004] Hashimoto, A.; Suenaga, K.; Gloter, A.; Urika; K. & Iijima, S. Direct evidence for atomic defects in graphene layers. *Nature,* Vol. 430, No.7002, (19 August 2004), pp. 870–873

[Hod et al., 2007] Hod, O.; Barone, V.; Peralta, J. E. & Suseria, G. E. Enhanced half-metallicity in edge oxidized zigzag graphene nanoribbons *Nano Letters,* Vol. 7, No. 8, (2007), pp. 2295–2299

[Hod & Scuseria, 2009] Hod O. & Scuseria, G. E. Electromechanical properties of suspended graphene nanoribbons. *Nano Letters,* Vol. 9, No. 7, (2009), pp. 2619–2622

[Huang et al., 2007] Huang, B.; Yan, Q.; Zhou, G.; Wu, J.; Gu, B.-L.; Duan, W. & Liu, F. Making a field effect transistor on a single graphene nanoribbon by selective doping *Applied Physics Letters,* Vol. 91, No. 25, (2007), Article ID 253122, 3 pages

[Klein, 1994] Klein, D. J. Graphitic polymer strips with edge states. *Chemical Physics Letters,* Vol. 217, No. 3, (1994), pp. 261–265

[Kobayashi et al., 2005] Kobayashi, Y.; Fukui, K.-I.; Enoki, T.; Kusakabe, K. & Y. Kaburagi,. Observation of zigzag and armchair edges of graphite using scanning tunneling microscopy and spectroscopy. *Physical Review B*, Vol. 71, No. 19, (2005), Article ID 193406, 4 pages

[Krasheninnikov et al., 2001] Krasheninnikov, A. V.; Nordlund, K.; Sirviö, M.; Salonen, E. & Keinonen, J. Formation of ion-irradiation-induced atomic-scale defects on walls of carbon nanotubes. *Physical Review B,* Vol. 63, No. 24, (2001), Article ID 245405, 6 pages

[Landauer, 1970] Landauer, R. Electrical resistance of disordered one-dimensional lattices. Philosophical Magazine, Vol. 21, No. 172, (1970), pp. 863-872

[Lehtinen et al., 2003] Lehtinen, P. O.; Foster, A.S.; Ayuela, A.; Krasheninnikov, A.; Nordlund, K. & Nieminen, R. M. Magnetic properties and diffusion of adatoms on graphene sheet. *Physical Review Letters,* Vol. 91, No. 1, (4 July 2003), Article ID 17202, 4 pages

[Li et al.,2008] Li, Z.; Qian, H.; Wu, J.; Gu, B.-L. & Duan, W. Role of symmetry in the transport properties of graphene nanoribbons under bias. *Physical Review Letters,* Vol. 100, No. 20, (2008), Article ID 206802, 4 pages

[Lopez Sancho et al., 1984] Lopez Sancho, M. P.; Lopez Sancho, J. M. & Rubio, J. Quick iterative scheme for the calculation of transfer matrices: application to Mo (100). *Journal of Physics F: Metal Physics,* Vol. 14, No. 5, (1984), pp. 1205-1216

[Lu & Pan, 2004] Lu, A. J. & Pan, B.C. Nature of single vacancy in achiral carbon nanotubes. *Physical Review Letters,* Vol. 92, No. 10, (2004), Article ID 105504, 4 pages

[Munoz et al., 1987] Munoz, M. C.; Velasco, V. R. & Garcia-Moliner, F. Tight binding models for non ideal semiconductor interfaces. *Progress in Surface Science,* Vol. 26, No. 1-4, (1987), pp. 117-133

[Maruyama et al., 2004] Maruyama, M.; Kusakabe, K.; Tsuneyuki, S.; Akagi, K.; Yoshimoto, Y. & Yamauchi, J. Magnetic properties of nanographite with modified zigzag edges. *Journal of Physics and Chemistry of Solids,* Vol. 65, No. 2-3, (2004), pp. 119–122

[Niimi et al., 2005] Niimi, Y.; Matsui, T.; Kambara, H.; Tagami, K.; Tsukada, M. & Fukuyama, H. Scanning tunneling microscopy and spectroscopy studies of graphite edges. *Applied Surface Science,* Vol. 241, No. 1-2, (2005), pp. 43–48

[Niimi et al., 2006] Niimi, Y.; Matsui, T.; Kambara, H.; Tagami, K.; Tsukada, M. & H. Fukuyama. Scanning tunneling microscopy and spectroscopy of the electronic local density of states of graphite surfaces near monoatomic step edges. *Physical Review B*, Vol. 73, No. 8, (2006), Article ID 085421, 8 pages

[Novoselov et al., 2004] Novoselov, K. S.; Geim, A. K.; Morozov, S. V.; Jiang, D.; Zhang, Y.; Dubonos, S. V.; Grigoreiva, I. V. & Firsov A. A. Electric field effect in atomically thin carbon films. *Science,* Vol. 306, No. 5696, (2004), pp. 666–669

[Novoselov et al., 2005] Novoselov, K. S.; Geim, A. K.; Morozov S. V. et al. Two dimensional gas of massless Dirac fermions in graphene. *Nature,* Vol. 438, No. 7065, (2005), pp. 197–200

[Obradovic et al., 2006] Obradovic, B.; Kotlyar, R.; Heinz, F. et al. Analysis of graphene nanoribbons as a channel material for field-effect transistors. *Applied Physics Letters*, Vol. 88, No. 14, (2006), Article ID 142102, 3 pages

[Ozaki et al., 2000] Ozaki, T.; Iwasa, Y.& Mitani, T. Stiffness of single-walled carbon nanotubes under large strain. *Physical Review Letters*, Vol. 84, No. 8, (2000), pp. 1712-1715

[Pastawski , 1991] Pastawski, H. M. Classical and quantum transport from generalized Landuer-Büttiker equations. *Physical Review B*, Vol. 44, No. 12, (1991), pp. 6329-6339

[Pisani et al., 2007] Pisani, L.; Chan, J. A.; Montanari, B. & Harrison, N. M. Electronic structure and magnetic properties of graphitic ribbons. *Physical Review B*, Vol. 75, No. 6, (2007), Article ID 064418, 9 pages

[Saito et al., 1992] Saito, R.; Fujita, M.; Dresselhaus, G. & Dresselhaus, M. S. Electronic structure of chiral graphene tubules. *Applied Physics Letters*, Vol. 60, No. 18, (1992), pp. 2204–2206

[Sasaki et al., 2006] Sasaki, K.; Murakami, S. & R. Saito. Stabilization mechanism of edge states in graphene. *Applied Physics Letters*, Vol. 88, No. 11, (2006), Article ID 113110, 3 pages

[Sols et al., 2007] Sols, F.; Guinea, F. & Neto, A. H. C. Coulomb blockade in graphene nanoribbons. *Physical Review Letters*, Vol. 99, No. 16, (2007), Article ID 166803, 4 pages

[Son et al., 2006] Son, Y.-W.; Cohen, M. L. & Louie, S. G. Energy gaps in graphene nanoribbons. *Physical Review Letters*, Vol. 97, No. 21, (2006), Article ID 216803, 4 pages

[Son et al., 2006] Son, Y.-W.; Cohen, M. L. & Louie, S. G. Half-metallic graphene nanoribbons. *Nature*, Vol. 444, No. 7117, (2006), pp. 347–349

[Tian et al., 1998] Tian, W.; Datta, S.; Hong, S.; Reifenberger, R.; Henderson, J. I. & Kubiak, C. P. Conductance spectra of molecular wires. *The Journal of Chemical Physics*, Vol. 109, No. 7, (15 August 1998), pp. 2874-2882

[Torelli & Mitas, 2000] Torelli, T. & Mitas, L. Electron Correlation in C_{4N+2} Carbon Rings: Aromatic versus Dimerized Structures. *Physical Review Letters*, Vol. 85, No. 8, (2000), pp. 1702-1705

[Wakabayashi et al., 1999] Wakabayashi, K.; Fujita, M.; Ajiki, H. & Sigrist, M. Electronic and magnetic properties of nanographite ribbons. *Physical Review B*, Vol. 59, No. 12, (1999), pp. 8271–8282

[Wakabayashi & Harigaya, 2003] Wakabayashi, K. & Harigaya, K. Magnetic Structure of Nano-Graphite Möbius Ribbon. *Journal of the Physical Society of Japan*, Vol. 72, No. 5, (May 2003), pp. 998-1001

[Wu et al., 2010] Wu, M. H.; Pei, Y. & Zeng, X. C. Planar tetracoordinate carbon strips in edge decorated graphene nanoribbon *Journal of the American Chemical Society*, Vol. 132, No. 16, (2010), pp. 5554–5555

[Xu et al., 1992] Xu, C. H.; Wang, C. Z.; Chan, C. T. & Ho, K. M. A transferable tight-binding potential for carbon. *Journal of Physics: Condensed Matter*, Vol. 4, No.28, (13 July 1992), pp. 6047-6054

[Yan et al., 2007] Yan, Q.; Huang, B.; Yu, J.; Zheng, F.; Zang, J.; Wu, J.; Gu, B.-L.; Liu, F. & Duan, W. Intrinsic current-voltage characteristics of graphene nanoribbons transistors and effect of edge doping. *Nano Letters,* Vol. 7, No. 6, (2007), pp. 1469-1473

[Zhang et al., 2005] Zhang, Y.; Tan, Y.-W.; Stormer, H. L. & Kim, P. Experimental observation of the quantum Hall effect and Berry's phase in graphene. *Nature,* Vol. 438, No. 7065, (2005), pp. 201–204

[Zhu et al., 2010] Zhu, L.; Wang, J.; Zhang, T.; Ma, L., Lim, C. W.; Ding, F. & Zeng, X. C. Mechanically robust tri-wing graphene nanoribbons with tunable electronic and magnetic properties. *Nan Letters,* Vol. 10, No. 2, (2010), pp. 494-498

21

Propagation of Electron Waves in Monolayer Graphene and Optical Simulations with Negative-Zero-Positive Index Metamaterials

Xi Chen[1] and Li-Gang Wang[2]
[1] *Department of Physics, Shanghai University and Departamento de Química Física, UPV-EHU*
[2] *Department of Physics, Zhejiang University and Department of Physics, The Chinese University of Hong Kong*
[1]*China and Spain*
[2]*China and HongKong*

1. Introduction

The analogies between phenomena occurring in two different physical systems open a route to find new effects or to translate solution on techniques or devices, and quite often help to understand both systems better (Dragoman & Dragoman, 2004). In particular, electronic analogues of many optical behaviors such as reflection, refraction (Gaylord & Brennan, 1989), focusing (Sivan et al., 1990; Spector et al., 1990; van Houten et al., 1988), collimation (Molenkamp et al., 1990), and interference (Ji et al., 2003; Yacoby et al., 1994) have been achieved in two-dimensional electron gas (2DEG) enabling the systems as a basic platform to study foundation problems in quantum mechanics as well as quantum information processing. The close relation between optics and electronics results from the fact that the electrons act as wave due to the ballistic transport properties of a highly mobility 2DEG created in semiconductor heterostructures (Palevski et al., 1989). As a result, there is a growing interest in the design and development of devices based on electron wave propagation, which has given rise to a research field described as electron wave optics (Datta, 1996; Gaylord et al., 1991).

Over the past six years, monolayer graphene has attracted much attention (Beenakker, 2008; Castro Neto et al., 2009) since the graphitic sheet of one-atom thickness has been fabricated experimentally by A. K. Geim *et al.* in 2004 (Novoselov et al., 2004). The valence electron dynamics in such a truly two-dimensional (2D) material is governed by a massless Dirac equation. Thus graphene has many unique electronic and transport properties (Beenakker, 2008; Castro Neto et al., 2009), including half-integer and unconventional quantum Hall effect (Zhang et al., 2005), observation of minimum conductivity (Novoselov et al., 2005), and Klein tunneling (Katsnelson et al., 2006).

The great progress on graphene has brought the analogy between ballistic electrons and light propagations to a new level, and has lead to the Dirac electron wave optics. In the regard, various electron optics phenomena, such as focusing (Cheianov et al., 2007), collimation (Park et al., 2008), subwavelength optics (Darancet et al., 2009), Bragg reflection (Ghosh et al., 2009), and Goos-Hänchen effect (GH) (Beenakker et al., 2009; Zhao et al., 2010) provide different ways to control the electronic wave propagations in different graphene-based electron devices. One of the most important discoveries is that Cheianov *et al.* (Cheianov et al., 2007) have recently demonstrated the negative refraction at the *n-p* graphene interface when the incidence angle is less than the critical angle for total reflection and proposed the electron focusing effect as a "perfect lens" in metamaterial. However, the propagations of electron waves become quite different, when the evanescent waves are considered in total reflection. Zhao and Yelin (Zhao et al., 2010) have once studied that the electron beam will experience the so-called GH shift at the graphene interface, when the incidence angle is larger than the critical angle. C. W. J. Beenakker *et al.* (Beenakker et al., 2009) have further found that the GH effect at a *n-p* interface in graphene doubles the degeneracy of the lowest propagating mode. In addition, following the seminal paper on the negative refraction and electron focusing, Garcia-Pomar *et al.* (Pomar et al., 2008) have also proposed an *n-p-n* graphene transistor to realize the valley beam splitter or collimation. So the manipulation of electron beam propagation can benefit from all these optical-like phenomena by applying an external electric or magnetic field to alter the flowing of electrons (Wang and Liu, 2010).

From a somewhat different but relevant perspective, a growing interest has appeared regarding the simulations of Dirac equation and relativistic effects by many controllable physical setups, for instance, cold atom in optical lattice (Zhu et al., 2007), spin-1/2 particles in single trapped ion (Lamata et al., 2007), ultrarelativistic atom with internal energy levels in a tripod configuration (Juzeliunas et al., 2008), and light in fiber Bragg gratings (Longhi, 2010). It is worthwhile to point out that R. Gerritsma *et al.* (Gerritsma et al., 2010) have lately implemented for the first time a quantum optical simulation of a tunable relativistic quantum mechanical system. In the optical field, the Dirac point (DP) in photonic crystals for the Bloch states is also of significance and interest, based on the similarity of the photonic bands of the two-dimensional photonic crystals (2DPCs) with the electronic bands of solids. Consequently, the conical diffraction (Peleg et al., 2007), "pseudodiffusive" scaling (Sepkhanov et al., 2007), and photon's Zitterbewegung (Zhang, 2008) have been found near the DP in photonic crystals. As we know that, compared to solids, atomic and quantum optical systems, pure optical systems offer clean and easy controlled way to test theoretical predictions. Stimulated by the realization of the DP with a double-cone structure for optical field in the negative-zero-positive index metamaterial (NZPIM) (Wang et al., 2009 , a;b), the pseudodiffusive property, Zitterbewegung effect, Bragg-like reflection, and the unique GH effect (Chen et al., 2009 , b) inside the NZPIM slab are extensively investigated. Taking into account the close relation between Klein paradox and negative refraction (Guney & Meyer, 2009), the novel phenomena in the NZPIM will definitely motivate the further work to simulate many exotic phenomena in graphene with relatively simple optical system.

This chapter presents a review on the propagation of electron waves in monolayer graphene and optical simulations with NZPIM. The chapter is organized as follow. In Sec. 2, the transmission gap (corresponding to Bragg-like reflection) and GH effect are discussed in single graphene barrier. Similar to the transmission gap, the zero-averaged wavenumber gap associated with new DP in monlayer graphene superlattice is also studied in Sec. 3 by the transfer matrix method. In Sec. 4, the optical simulations of the relevant phenomena mentioned above are made by the NZPIM with the DP. Finally, we make brief conclusion in Sec. 5.

2. Transmission gap, Bragg-like reflection and Goos-Hänchen effect in monolayer graphene barrier

As we know, one of the most interesting phenomena is the perfect transmission, in particular, for normal incidence, through arbitrarily high and wide graphene barriers, referred to as Klein tunneling (Katsnelson et al., 2006). Therefore, such important property may lead to the design of various graphene-based device. Until recently, the transport properties of massless Dirac fermions, including Klein tunneling and perfect transmission, have been extensively studied in single graphene barriers (Katsnelson et al., 2006), n-p-n junctions (Cheianov et al., 2007), graphene-based double barriers (Perira et al., 2007) and even graphene superlattice (Bai et al., 2007; Barbier et al., 2008; 2010; Bliokh et al., 2009). In the meanwhile, inhomogeneous magnetic fields on the nanometer scale have been suggested to circumvent the Klein tunneling and produce confined graphene-based structures (Anna et al., 2009; Martino et al., 2007). It was found that the angular range of the transmission through monolayer and bilayer graphene with magnetic barrier structures can be efficiently controlled and gives the possibility to construct a direction-dependent wave vector filter (Masir et al., 2008 , a;b).
In the following section, we shall firstly investigate the transmission properties of Dirac-like electron waves in single monolayer graphene barrier, when the ballistic electrons are obliquely incident on the monolayer graphene barrier. It is shown that the transmission at non-zero incidence angle has a gap, which can be considered as Bragg-like phenomenon. This controllable transmission gap is quite different from the perfectly transparent for the angles close to the normal incidence (Katsnelson et al., 2006) and does result from the evanescent waves in two cases of classical motion and Klein tunneling due to the effect of parallel wave vector. Based on the tunable transmission gap (Chen & Tao, 2009), GH shifts for Dirac fermions in transmission through a two-dimensional (2D) monolayer graphene barrier can be enhanced by transmission resonances, and can also be negative as well as positive. So these lateral shifts, associated with the transmission gap and Bragg-like reflection, lead to the significant difference from the conventional GH shift in total reflection.

2.1 Transmission and reflection in the graphene barrier

We consider the incident electron wave propagates with Fermi energy E at angle ϕ with respective to the x axis through a single 2D graphene barrier, as shown in Fig. 1, where V_0 and d are the height and width of the potential barrier, respectively. From the view point of its electronic properties, graphene is a two-dimensional zero-gap semiconductor with the linear dispersion relation, $E = \hbar k_F v_F$, thus the low-energy quasi-particles are formally described by the Dirac-like Hamiltonian (Katsnelson et al., 2006), $\hat{H}_0 = -i\hbar v_F \sigma \nabla$, $v_F \approx 10^6 m \cdot s^{-1}$ is the Fermi velocity, k_F is the Fermi wave vector, and $\sigma = (\sigma_x, \sigma_y)$ are the Pauli matrices. So the wave function of the incident electrons is assumed to be

$$\Psi_{in}(x,y) = \begin{pmatrix} 1 \\ se^{i\phi} \end{pmatrix} e^{i(k_x x + k_y y)}, \tag{1}$$

the wave function of the transmitted electrons can be expressed as follows,

$$\Psi_t(x,y) = t \begin{pmatrix} 1 \\ se^{i\phi} \end{pmatrix} e^{i(k_x x + k_y y)}, \tag{2}$$

where $s = sgn(E)$, $k_x = k_F \cos\phi$ and $k_y = k_F \sin\phi$ are the perpendicular and parallel wave vector components outside the barrier. According to the boundary conditions, transmission coefficient is determined by

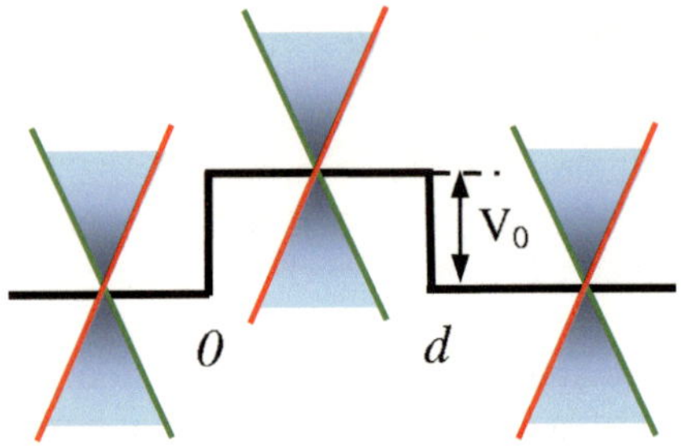

Fig. 1. Schematic diagram for a single monolayer graphene barrier with height V_0 and width d.

$$t \equiv \frac{1}{f} e^{i\varphi} = \frac{1}{\cos(q_x d) - i(ss' \sec\phi \sec\theta - \tan\phi \tan\theta)\sin(q_x d)}, \tag{3}$$

where $s' = sgn(E - V_0)$, $q_x = (k_F'^2 - k_y^2)^{1/2}$, $k_F' = |E - V_0|/\hbar v_F$, and $\theta = \arctan(k_y/q_x)$. The critical angle ϕ_c for total reflection can be defined by

$$\phi_c = \arcsin\left(\frac{|V_0 - E|}{E}\right), \tag{4}$$

so that when $\phi > \phi_c$, the wave function in the propagating case becomes evanescent wave by replacing q_x by $i\kappa$, where $\kappa = (k_y^2 - k_F'^2)^{1/2}$. Next, we will discuss the transmission in two cases of Klein tunneling ($E < V_0$) and classical motion ($E > V_0$) to show the unique transmission properties of electron waves in monolayer graphene barrier.

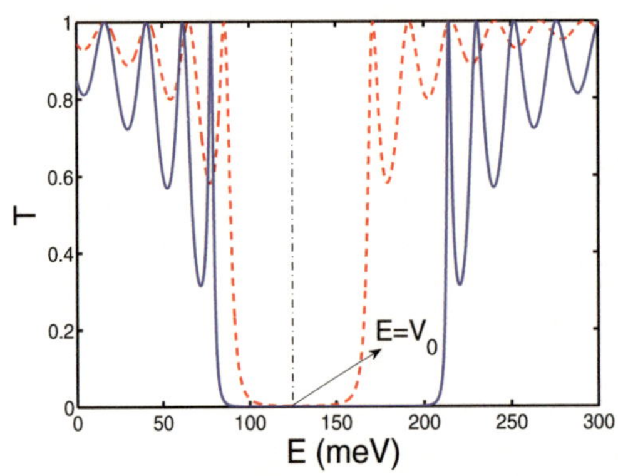

Fig. 2. (Color online) Transmission gap as the function of incident energy E, where $d = 80$ nm, $V_0 = 120$ meV, solid and dashed curves correspond to $\phi = 25°$ and $\phi = 15°$, respectively.

Case 1: Klein tunneling ($ss' = -1$). The transmission probability T can be given by Eq. (3),

$$T \equiv |t|^2 = \left[\cos^2(q_x d) + \frac{(k_y^2 + k_F k_F')^2}{k_x^2 q_x^2} \sin^2(q_x d)\right]^{-1}. \tag{5}$$

It is clear that the angular-dependent transmission probability T becomes equal to 1 under the resonance condition, $q_x d = N\pi$, ($N = 0, \pm 1, ...$). However, when the angle of incidence satisfies $\phi > \phi_c'$, where the critical angle (4) tends to $\phi_c' = \sin^{-1}(V_0/E - 1)$, with the

necessary condition $E < V_0 < 2E$, the tunneling of the Dirac-like electron through the monolayer graphene barrier occurs, thus the transmission probability damps exponentially in the following form: $T \approx 4k_x^2\kappa^2 e^{-2\kappa d}/[k_x^2\kappa^2 + (k_y^2 + k_F k_F^{'})^2]$.
Case 2: Classical motion ($ss' = 1$). The transmission probability can be rewritten as

$$T = \left[\cos^2(q_x d) + \frac{(k_y^2 - k_F k_F^{'})^2}{k_x^2 q_x^2}\sin^2(q_x d)\right]^{-1}. \tag{6}$$

Similarly, when the incidence angle ϕ is less than the critical angle, $\phi_c^{''} = \sin^{-1}(1 - V_0/E)$ given by the critital angle (4), the transmission probability T in this case depends periodically on the width d of barrier. On the contrary, when $\phi > \phi_c^{''}$, wave vector q_x becomes imaginary number, thus the transmission probability tends to $T \approx 4k_x^2\kappa^2 e^{-2\kappa d}/[k_x^2\kappa^2 + (k_y^2 - k_F k_F^{'})^2]$.
Figure 2 shows that the transmission as the function of incidence energy E has a gap, where $d = 80nm$, $V_0 = 120meV$, solid and dashed curves correspond to $\phi = 25°$ and $\phi = 10°$. The energy region of the transmission gap, $V_0 - \hbar v_F k_y < E < V_0 + \hbar v_F k_y$, since $q_x^2 < 0$. Thus the width of transmission gap is given by

$$\Delta E = 2\hbar v_F k_y, \tag{7}$$

which depends strongly on the parallel wave vector k_y, and can be controlled by the incidence angle, as shown in Fig. 2, where the center of transmission gap is $E = V_0$. It is clear that the transmission gap becomes narrower with the decrease of the incidence angle, and even vanishes at normal incidence. Since the transmission gap results from the evanescent wave in two cases of Klein tunneling and classical motion, the incidence angle, the height and width of potential barrier play important roles in the transmission gap. Especially, the transmission gap will become deeper with increasing the barrier width, due to the decrease of the the decay factor $\exp(-2\kappa d)$ in the evanescent case. Here we shall point out that the transmission gap can be tunable by changing the gate-voltage, which can realize the electron wave energy filter by the shift of transmission gap.
The tunable transmission gap implies Bragg-like reflection, which usually happens only in the single potential barrier, instead of multiple barriers. It turns out that Dirac-like electrons can perfectly reflected by the graphene barrier in the region of energy. This perfect reflection is similar to but different from the Bragg reflection in magnetic barrier in graphene (Ghosh et al., 2009). Actually, this Bragg-like phenomenon discussed here can also be applied to select electron energy by the reflection window.
Next, we will study the total conductance (G) and Fano factor (F) in the single graphene barrier. Using the Büttiker formula (Datta, 1996), the total conductance of the system at zero temperature is given by

$$G = G_0 \int_{-\pi/2}^{\pi/2} T(E_F, E_F \sin\phi)\cos\phi d\phi, \tag{8}$$

with the Fermi energy E_F and the units of conductance $G_0 = (2e^2/\hbar)(\ell/\pi\hbar v_F)$, where ℓ is the length of the structure along the y direction. Furthermore, we can also study the Fano factor (F) (Tworzydlo et al., 2006), which is given by

$$F = \frac{\int_{-\pi/2}^{\pi/2} T(1-T)\cos\phi d\phi}{\int_{-\pi/2}^{\pi/2} T\cos\phi d\phi}. \tag{9}$$

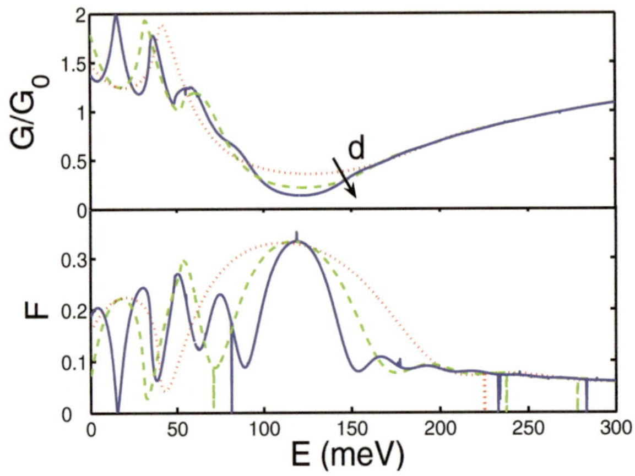

Fig. 3. (Color online) Conductance G/G_0 and Fano factor F as the function of incident energy, where $d = 80$ nm (blue solid line), $d = 50$ nm (green dashed line), $d = 30$ nm (red dotted line), and other parameters are the same as in Fig. 2.

Figure 3 illustrates the dependence of the conductance and Fano factor on the incidence energy E, where the physical parameters are the same as in Fig. 2. Obviously, it is shown that the visible kinks of the conductance due to transmission resonance are closely related to the quasi-bound state. More importantly, all conductance curves indicate a pronounced forbidden region, which is the region of almost zero conductance corresponding to the transmission gap. In addition, the behavior of Fano factor resembles Bragg reflection. Also the interesting point is that Fano factor will approach the maximum value 1/3 in the transmission gap. In a word, the transmission gap has great effect on the electron transport in monolayer graphene, but also provides a novel phenomenon to design various electron wave devices.

2.2 Quantum Goos-Hänchen shift

As mentioned above, Cheianov *et al.* (Cheianov et al., 2007) have found the negative refraction and electron focusing in graphene p-n junction, when the incidence angle is less than the critical angle for total reflection. However, when the incidence angle is larger than the critical angle, the quantum GH effect in total reflection has been investigated (Beenakker et al., 2009; Zhao et al., 2010) at a p-n interface in graphene, which is analogous to the lateral shift of the light beam totally reflected from a dielectric interface (Goos & Hänchen, 1947; 1949). Here we would like to discuss the negative and positive GH shifts for Dirac fermions in transmission through a 2D monolayer graphene barrier, based on the tunable transmission gap.
To calculate the lateral shifts, we consider the incidence beam as

$$\Psi_{in}(x,y) = \int_{\infty}^{\infty} dk_y A(k_y) e^{i(k_x x + k_y y)} \begin{pmatrix} 1 \\ se^{i\phi} \end{pmatrix}, \tag{10}$$

where the angular-spectrum distribution $A(k_y)$ is sharpen distribution around k_{y0}, which can be simply assumed to be Gaussian function $A(k_y) = w_y \exp[-(w_y^2/2)(k_y - k_{y0})^2$ with $w_y = w \sec\phi_0$, and the half width of the incident beam at waist w. Accordingly, the transmitted beam can also expressed as

$$\Psi_t(x,y) = \int_{\infty}^{\infty} dk_y t(k_y) A(k_y) e^{i(k_x x + k_y y)} \begin{pmatrix} 1 \\ se^{i\phi} \end{pmatrix}, \tag{11}$$

For the well-collimated beam, that is, $\delta\phi = \lambda_f/(\pi w) \ll 1$, the lateral shifts can be defined, according to the stationary phase method (Beenakker et al., 2009; Zhao et al., 2010), as

$$s_t = -\frac{\partial \varphi}{\partial k_{y0}}, \tag{12}$$

where the subscript 0 denotes the values taken at $k_y = k_{y0}$ corresponding to the central incidence angle $\phi = \phi_0$. It is noted that the lateral shifts of up and down components are both the same as s_t in transmission at the interface $x = d$, whereas on the total reflection the shifts of up and down components are different due to the phase in the spinor wave function, and the GH shift is thus defined as their average value in the literature (Beenakker et al., 2009).

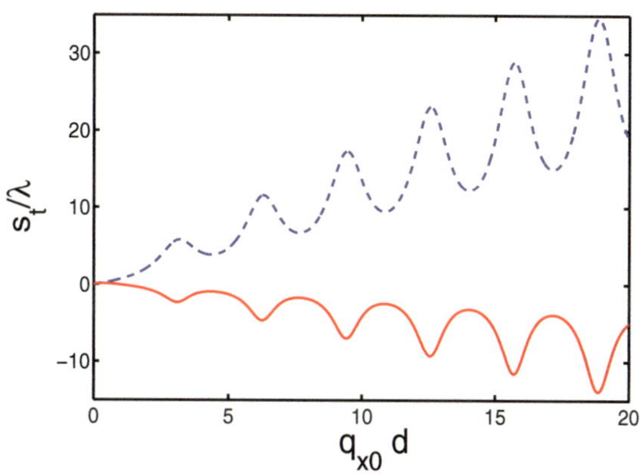

Fig. 4. (Color online) Dependence of the lateral shifts in the propagating case on the barrier's width d, where $\phi_0 = 20°$, $V_0 = 120$ meV, d is re-scaled to $q_{x0}d$, $E = 80$ meV (solid line), and $E = 200$ meV (dashed line).

Figure 4 shows the most impressive behavior that in the propagating case the lateral shift can be negative for Klein tunneling, $E < V_0$, and also be enhanced by the transmission resonances, whereas for classical motion, $E > V_0$, the shift is always large and positive. On the contrary, when the incidence angle ϕ_0 is larger than the critical angle ϕ_c, the lateral shifts become in the order of Fermi wavelength due to the evanescent wave, which is similar to those in total reflection at a single graphene interface (Beenakker et al., 2009; Zhao et al., 2010). Instead of the enhancement by the transmission resonances shown in Fig. 4, Fig. 5 illustrates that the lateral shifts for Klein tunneling and classical motion saturate respectively to negative and positive constants with increasing the barrier's width in the evanescent case, where (a) $\phi_0 = 40°$ and (b) 70° (which are both larger than the critical angle for total reflection).

Similarly, we shall discuss the properties of the lateral shifts in two cases of Klein tunneling and classical motion.

Case 1: Klein tunneling ($ss' = -1$). When the incidence angle ϕ_0 is less then the critical angle ϕ_c, where $\phi'_c = \sin^{-1}(V_0/E - 1)$, the lateral shifts are given by

$$s_t = \frac{d \tan\phi_0}{f_0^2}\left\{\left[2 + \left(\frac{k_0^2}{k_{x0}^2} + \frac{k_0^2}{q_{x0}^2}\right)\right]\frac{\sin(2q_{x0}d)}{2q_{x0}d} - \frac{k_0^2}{q_{x0}^2}\right\}, \tag{13}$$

where $k_0 = (k_F k'_F + k_{y0}^2)^{1/2}$. Obviously, lateral shifts obtained above can be positive as well as negative because of $|\sin(2q_{x0}d)/(2q_{x0}d)| \leq 1$. So for the thin barrier $d \to 0$, that is, the lateral shifts can be positive, while the lateral shifts become negative for an enough thick barrier. It is also interesting that the negative lateral shifts can be enhanced by the transmission resonances. The exotic behaviors of negative and positive shifts are analogous to those of the transmitted light beam though a left-handed metamaterial slab (Chen et al., 2009 , b).

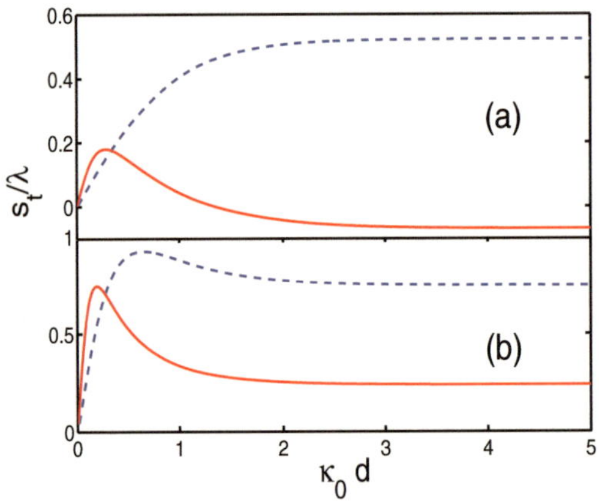

Fig. 5. (Color online) Dependence of the lateral shifts in the evanescent case on the barrier's width d, where (a) $\phi_0 = 40°$ and (b) $\phi_0 = 70°$, d is re-scaled to $\kappa_0 d$, $E = 80$ meV (solid line), and $E = 200$ meV (dashed line).

On the contrary, when $\phi_0 > \phi'_c$, the lateral shifts become

$$s_t = \frac{d \tan\phi_0}{f_0^2}\left\{\left[2 + \left(\frac{k_0^2}{\kappa_0^2} - \frac{k_0^2}{k_{x0}^2}\right)\right]\frac{\sinh(2\kappa_0 d)}{2\kappa_0 d} + \frac{k_0^2}{\kappa_0^2}\right\}. \tag{14}$$

In the limit of opaque barrier, $\kappa_0 d \to \infty$, the lateral shifts tend to the constants as follows,

$$s_t = \frac{\tan\phi_0}{\kappa_0}\frac{2k_{x0}^2\kappa_0^2 - k_0^2(k_{x0}^2 - \kappa_0^2)}{k_{x0}^2\kappa_0^2 - k_0^4}, \tag{15}$$

which are proportional to $1/\kappa_0$, and imply that the shifts in the evanescent case are in the same order of electron wavelength as GH shifts in a single graphene interface (Beenakker et al., 2009; Zhao et al., 2010). More interestingly, the saturated lateral shifts are negative when the incidence angle satisfies $\phi'_c < \phi_0 < \phi^*$, where the critical angle is defined by $\phi^* = \arcsin\sqrt{\sin\phi'_c}$. But the shifts in this case will become positive when $\phi_0 > \phi^*$. The sign change of the lateral shifts described by Fig. 5 (b) appears at the incidence angle $\phi_0 = \phi^*$, which is similar to the result of the quantum GH effect in graphene, taking the pseudospin degree into account (Beenakker et al., 2009).

Case 2: classical motion ($ss' = 1$). When the incidence angle is less than the critical angle for total reflection, $\phi_0 < \phi''_c$, the lateral shifts can be written as

$$s_t = \frac{d \tan\phi_0}{f_0^2}\left\{\left[2 - \left(\frac{k_0'^2}{k_{x0}^2} + \frac{k_0'^2}{q_{x0}^2}\right)\right]\frac{\sin(2q_{x0} d)}{2q_{x0} d} + \frac{k_0'^2}{q_{x0}^2}\right\}, \tag{16}$$

where $k'_0 = (k_F k'_F - k_{y0}^2)^{1/2}$. Similarly, the lateral shifts for classical motion also depend periodically on the barrier's width, and can be enhanced by the transmission resonances.

However, these lateral shifts in classical motion are always positive. When $\phi_0 > \phi_c''$, the shifts in the evanescent case will become

$$s_t = \frac{d \tan \phi_0}{f_0^2} \left\{ \left[2 - \left(\frac{k_0'^2}{\kappa_0^2} - \frac{k_0'^2}{k_{x0}^2} \right) \right] \frac{\sinh(2\kappa_0 d)}{2\kappa_0 d} - \frac{k_0'^2}{\kappa_0^2} \right\}. \tag{17}$$

Then the lateral shifts in the limit, $\kappa_0 d \to \infty$, are given by

$$s_t = \frac{\tan \phi_0}{\kappa_0} \frac{2k_{x0}^2 \kappa_0^2 + k_0'^2 (k_{x0}^2 - \kappa_0^2)}{k_{x0}^2 \kappa_0^2 - k_0'^4}, \tag{18}$$

which are always positive constants.

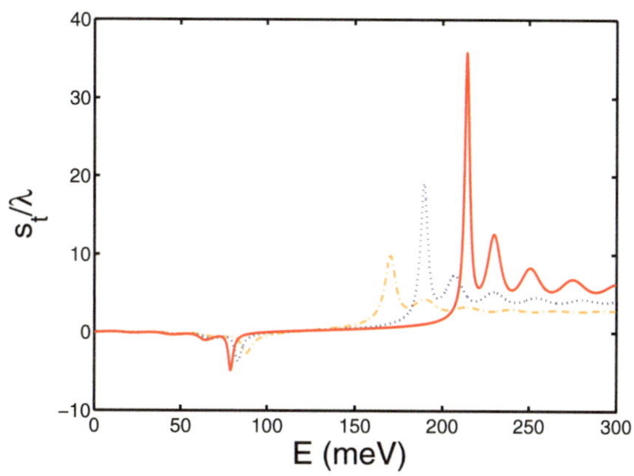

Fig. 6. (Color online) The lateral shifts as the function of incident energy E, where $d = 80$ nm, and $V_0 = 120$ meV, $\phi_0 = 25°$ (red solid line), $\phi_0 = 20°$ (blue dotted line), and 15° (orange dot-dashed line).

Based on the properties in two cases of Klein tunneling and classical motion, the lateral shifts as the function of incidence energy E are shown in Fig. 6. It is shown that the lateral shifts are closely related to the transmission gap $\Delta E = 2\hbar k_y v_F$, as descried in last subsection. Figure 6 indicates that the lateral shifts change the sign near the DP $E = V_0$, and can also be enhanced by the transmission resonances near the boundaries of energy gap. Actually, the sign change of the shifts does result from the fact that the DP $E = V_0$ means the transition between Klein tunneling ($E < V_0$) and classical motion ($E > V_0$), which correspond to the negative and positive group velocities, respectively.

Obviously, the lateral shifts considered here are quite different from the GH shifts in total reflection. The lateral shifts can be enhanced by the transmission resonances, thus can be easily modulated by various parameters such as potential heights and incidence angles. Actually, the periodical dependence of negative and positive lateral shifts on the gap provides an efficient way to modulate the spatial position in a fixed graphene barrier, which is useful for the manipulation of electron beam propagation in graphene (Wang and Liu, 2010).

3. Electronic Band gaps and transport properties inside graphene superlattices

Most recently, there have been a number of interesting theoretical investigations on the graphene supperlatices with periodic potential structures, which can be generated by different methods. As a matter of fact, it is well known that the superlattice are very successful in controlling the electronic structures of many convectional semiconductor material [For review, see Ref. (Tsu, 2005)]. The peculiar properties of localized superlattice states including such as the effect of the Bragg confinement of electron in coupled superlattice and the formation of above-barrier quasi-bound states at the junction of superlattices are of significance for both the fundamental research and various devices (Steslicka et al., 2002). In this section, we will present the result on a new DP which is exactly located at the energy corresponds to the zero-averaged wavenumber inside the one-dimensional (1D) periodic potentials. It is emphasized here that the gap for the zero-averaged wavenumber is quite different from the Bragg gap, which is analogous to the case of the 1DPCs containing left-handed and right-handed materials (Bliokh et al., 2009; Wang & Zhu, 2010 , b).

3.1 Transfer Matrix method for mono-layer graphene superlattices

First of all, we will develop the transfer matrix method for such system in this subsection.
The Hamiltonian of a low-energy electron moving inside a mono-layer graphene in the presence of the electrostatic potential $V(x)$, which only depends on the coordinate x, is given by

$$\hat{H} = v_F \sigma \cdot \hat{p} + V(x)\hat{I}, \tag{19}$$

where $\hat{p} = (p_x, p_y) = (-i\hbar\frac{\partial}{\partial x}, -i\hbar\frac{\partial}{\partial y})$ is the momentum operator with two components, $\hat{I}$ is a 2×2 unit matrix. This Hamiltonian acts on a state expressed by a two-component pseudospinor $\Psi = (\tilde{\psi}_A, \tilde{\psi}_B)^T$, where $\tilde{\psi}_A$ and $\tilde{\psi}_B$ are the smooth enveloping functions for two triangular sublattices in graphene. Due to the translation invariance in the y direction, the wave functions $\tilde{\psi}_{A,B}(x,y)$ can be written as $\tilde{\psi}_{A,B}(x,y) = \psi_{A,B}(x)e^{ik_y y}$. Therefore, from Eq. (19), we obtain

$$\frac{d\psi_A}{dx} - k_y\psi_A = ik\psi_B, \tag{20}$$

$$\frac{d\psi_B}{dx} + k_y\psi_B = ik\psi_A, \tag{21}$$

where $k = [E - V(x)]/\hbar v_F$ is the wavevector inside the potential $V(x)$. Obviously, when $E < V(x)$, the wavevector inside the barrier is opposite to the direction of the electron's velocity, which is relevant to the negative refraction in the graphene (Cheianov et al., 2007).
In what follows, we assume that the potential $V(x)$ is comprised of periodic potentials of square barriers as shown in Fig. 7. Inside the j th potential, $V_j(x)$ is a constant, therefore, from Eqs. (20) and (21), we can obtain

$$\frac{d^2\psi_A}{dx^2} + (k_j^2 - k_y^2)\psi_A = 0, \tag{22}$$

$$\frac{d^2\psi_B}{dx^2} + (k_j^2 - k_y^2)\psi_B = 0. \tag{23}$$

Here the subscript "j" denotes the quantities in the j th potential. The solutions of Eqs. (22) and (23) are the following forms: $\psi_A(x) = ae^{iq_j x} + be^{-iq_j x}$ and $\psi_B(x) = ce^{iq_j x} + de^{-iq_j x}$, where q_j =sign$(k_j)\sqrt{k_j^2 - k_y^2}$ is the x component of the wavevector inside the j th potential V_j for $k_j^2 > k_y^2$, otherwise $q_j = i\sqrt{k_y^2 - k_j^2}$; and a (c) and b (d) are the amplitudes of the forward and backward propagating spinor components. Following the calculations in the literature (Wang

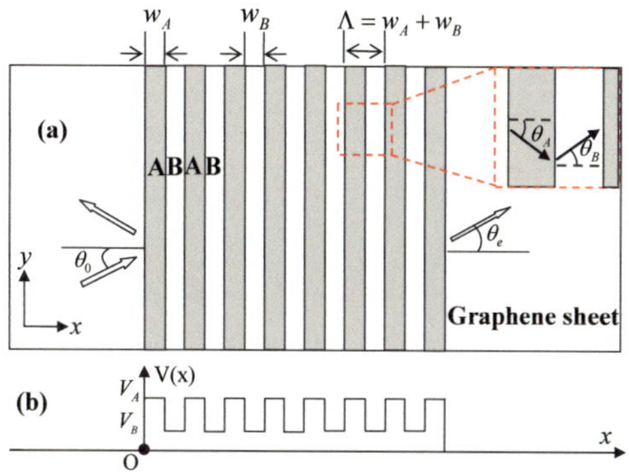

Fig. 7. (Color online) (a) Schematic representation of the finite periodic potentials of square barriers in $x-y$ plane. Grey regions denote the electrodes to apply the periodic potentials on the graphene, where θ_0 (θ_e) denotes the incidence (exit) angles of the carriers passing through the graphene superlattice. In the inset, θ_A (θ_B) denotes the angles of the carriers in the barriers A and B for the cases with $V_B < E < V_A$. (b) The whole profiles of the periodic potentials applied on the monolayer graphene.

& Zhu, 2010 , b), we obtain the relation between $\binom{\psi_A(x_{j-1})}{\psi_B(x_{j-1})}$ and $\binom{\psi_A(x_{j-1}+\Delta x)}{\psi_B(x_{j-1}+\Delta x)}$ can be finally written as:

$$\begin{pmatrix}\psi_A(x_{j-1}+\Delta x)\\ \psi_B(x_{j-1}+\Delta x)\end{pmatrix} = M_j(\Delta x, E, k_y)\begin{pmatrix}\psi_A(x_{j-1})\\ \psi_B(x_{j-1})\end{pmatrix}, \tag{24}$$

where the matrix M_j is given by

$$M_j(\Delta x, E, k_y) = \begin{pmatrix}\frac{\cos(q_j\Delta x-\theta_j)}{\cos\theta_j} & i\frac{\sin(q_j\Delta x)}{\cos\theta_j}\\ i\frac{\sin(q_j\Delta x)}{\cos\theta_j} & \frac{\cos(q_j\Delta x+\theta_j)}{\cos\theta_j}\end{pmatrix}. \tag{25}$$

It is easily to verify the equality: $\det[M_j] = 1$. Here we would like to point out that in the case of $E = V_j$, the transfer materix (25) should be recalculated with the similar method and it is given by

$$M_j(\Delta x, E, k_y) = \begin{pmatrix}\exp(k_y\Delta x) & 0\\ 0 & \exp(-k_y\Delta x)\end{pmatrix}. \tag{26}$$

With help of the above equations (25) and (26), we manage to build up the boundary condition in order to obtain the transmission and reflection coefficients. As shown in Fig. 7, we assume that the Dirac fermions of E is incident from the region $x \leq 0$ at any incidence angle θ_0. In this region, the wave function is the superposition of the incident and reflected waves, so we have the following equation:

$$\begin{pmatrix}\psi_A(x_e)\\ \psi_B(x_e)\end{pmatrix} = \mathbf{X}\begin{pmatrix}\psi_A(0)\\ \psi_B(0)\end{pmatrix}, \tag{27}$$

with

$$\mathbf{X}=\begin{pmatrix}x_{11} & x_{12}\\ x_{21} & x_{22}\end{pmatrix} = \sum_{j=1}^{N} M_j(w_j, E, k_y). \tag{28}$$

After some algebraic calculations, we find the reflection and transmission coefficients given by (Wang et al., 2010; Wang & Chen, 2010)

$$r(E,k_y) = \frac{(x_{22}e^{i\theta_0} - x_{11}e^{i\theta_e}) - x_{12}e^{i(\theta_e+\theta_0)} + x_{21}}{(x_{22}e^{-i\theta_0} + x_{11}e^{i\theta_e}) - x_{12}e^{i(\theta_e-\theta_0)} - x_{21}}, \tag{29}$$

$$t(E,k_y) = \frac{2\cos\theta_0}{(x_{22}e^{-i\theta_0} + x_{11}e^{i\theta_e}) - x_{12}e^{i(\theta_e-\theta_0)} - x_{21}}, \tag{30}$$

where we have used the property of $\det[\mathbf{X}] = 1$. In the following subsection, we will discuss the properties of the electronic band structure and transmission for the graphene-based periodic potentials of square barriers.

3.2 New Dirac point and zero-averaged wavenumber gap

In this section, we would like to discuss some unique properties of the band structures in the graphene-based periodic-potential systems by using the above transfer matrix method.

3.2.1 Infinite periodic structure

Firstly, let us investigate the electron's bandgap for an infinite periodic structure $(AB)^N$, where the periodic number N tends to infinity. The magnitude and width of the potential A (B) are with the electrostatic potential $V_{A(B)}$ and width $w_{A(B)}$, as shown in Fig. 7. According to the Bloch's theorem, the electronic dispersion at any incident angle follows the relation

$$2\cos[\beta_x\Lambda] = 2\cos[q_A w_A + q_B w_B] + \frac{[2\cos(\theta_A - \theta_B) - 2]}{\cos\theta_A \cos\theta_B}\sin(q_A w_A)\sin(q_B w_B). \tag{31}$$

Here $\Lambda = w_A + w_B$ is the length of the unit cell. When the incident energy of the electron satisfies $V_B < E < V_A$, we have $\theta_A < 0$, $q_A < 0$, $\theta_B > 0$, and $q_B > 0$ for the propagating modes. The angles for θ_A and θ_B are schematically shown in the inset of Fig. 7 (a). Then if $q_A w_A = -q_B w_B$, the above equation (31) becomes

$$\cos[\beta_x\Lambda] = 1 + \frac{[1-\cos(2\theta_A)]}{\cos^2\theta_A}|\sin(q_A w_A)|^2. \tag{32}$$

This equation indicates that, when $q_A w_A = -q_B w_B \neq m\pi$ and $\theta_A \neq 0$, there is no real solution for β_x, i.e., existing a bandgap; Additionally, this bandgap will be close at normal incident case ($\theta_A = 0$) from Eq. (32). Therefore, the location of the touching point of the bands is exactly given by $q_A w_A = -q_B w_B$ at $\theta_A = 0$, i. e., $k_A w_A = -k_B w_B$, or $(V_A - E)w_A = (E - V_B)w_B$, which is consistent with the condition $q_A w_A = \pm\Omega_B$ at $\theta_A = 0$ for the location of the touch point of the band in the limiting case of a periodic δ-barrier structures, called as the Kronig-Penney model (Barbier et al., 2009) or the Dirac comb (Arovas et al., 2010) with $w_B \to 0$ with $V_B w_B \to \pm\hbar v_F \Omega_B$ finite (Ω_B is a dimensionless positive quantity). Therefore the above condition for the periodic δ-barrier structures with $w_B \to 0$ and $V_B w_B \to \pm\hbar v_F \Omega_B$ is simplified into $k_A w_A = \pm\Omega_B$.

Figure 8 shows clearly that a band gap opens exactly at energy $E = 25$meV under the inclined incident angles (i.e., $k_y \neq 0$), where the condition $q_A w_A = -q_B w_B \neq m\pi$ is satisfied. At the case of normal incidence ($\theta_A = \theta_B = 0$), the upper and lower bands linearly touch together and form a new double-cone DP. The location of the new DP is governed by the condition, $k_A w_A = -k_B w_B$, or $(V_A - E)w_A = (E - V_B)w_B$. For the graphene-based periodic-barrier structure with $V_A \neq 0$ and $V_B = 0$, the distribution of the periodic potentials as an example is

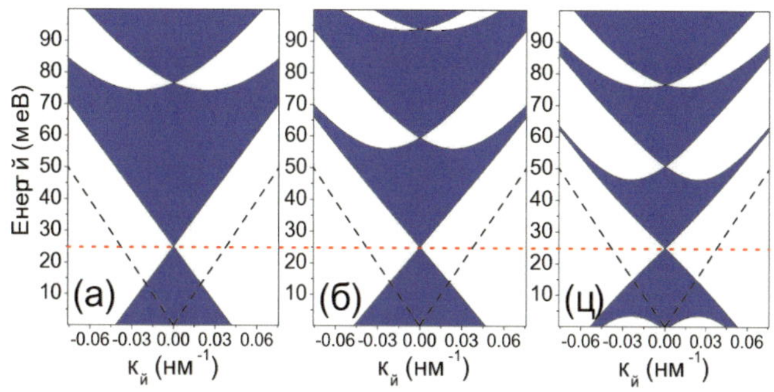

Fig. 8. (Color online) Electronic band structures for (a) $w_A = w_B =$ 20nm, (b) $w_A = w_B = 30$ nm and (c) $w_A = w_B = 40$ nm, with $V_A = 50$ meV and $V_B = 0$. The dashed lines denote the "light cones" of the incident electrons, and the dot line denotes the location of the new DP.

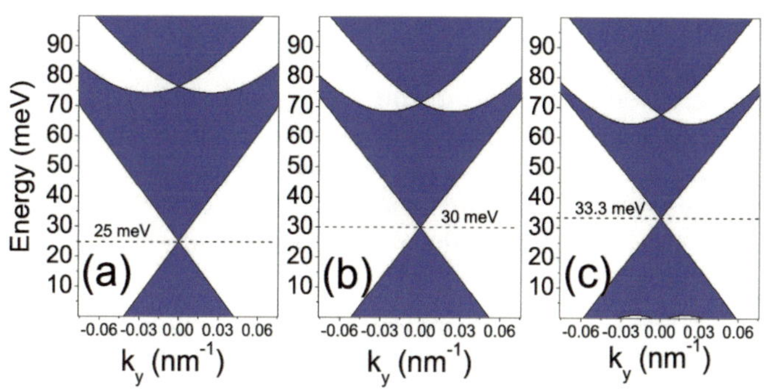

Fig. 9. (Color online) Electronic band structures for (a) $w_A/w_B = 1$, (b) $w_A/w_B = 3/2$, and (c) $w_A/w_B = 2$, with $V_A = 50$ meV, $V_B = 0$ and $w_B = 20$ nm in all cases. The dashed lines denote the locations of the new DP.

shown in Fig. 7 (b), and in this case the new DP is exactly located at $E = V_A/(1 + w_B/w_A)$. It turns out that the location of the new DP has nothing to do with the lattice constants; and the position of the opened gap associated with the new DP is not only independent of the lattice constants but also is weakly dependent on the incident angles. Figure 9 further illustrates that the locations of both the new DP and the opened gap are determined by the ratio of w_A/w_B for the cases with the fixed heights V_A of the potentials. From the above discussions, we find that the volume-averaged wavenumber at the energy of the new DP is zero, therefore such an opened gap associated with the new DP may be called as *the zero-averaged wavenumber gap*, which is very similar to that in the 1DPCs containing the left-handed metamaterials (Li et al., 2003), where the so-called zero (volume) averaged index gap is independent of the lattice constant but only dependent on the ratio of the thicknesses of the right- and left-handed materials. For a special case of the graphene-based periodic-potential structures with $V_A = -V_B$ and $w_A = w_B$, the usual DP is located at the energy $E = 0$ with $k_y = 0$. This result is the same as the discussion in Ref. (Barbier et al., 2010), and it is also similar to that for the cos-type modulated electric-potential structure in Refs. (Brey & Fertig, 2009; Ho et al., 2009). As a matter of fact, the properties of these novel zero-averaged wavenumber states are similar to those of the zero-energy states in the previous work (Brey & Fertig, 2009), and semimetallic properties are induced due to the effect of the modulated electric potential (Ho et al., 2009).

3.2.2 Finite periodic-barrier systems

Next, we turn to the study on the transmission, conductance and Fano factor in the finite periodic-barrier system. To illustrate the robust transmission gap, we plot the transmission as a function of incident energy in Fig. 9 for the finite structures, for example, $(AB)^{25}$, where (a) different lattice constants with a fixed ratio $w_A/w_B = 1$ and an incident angle $\theta_0 = 10^\circ$ and (b) different incident angles with the fixed lattice parameters $w_A = w_B = 30$nm. Compared to the novel zero-averaged wavenumber gap, the other bandgap structures, that is, Bragg gap, are not only dependent on the lattice constants but also strongly dependent on different angles. Further calculations can demonstrate that the higher opened gap is destroyed by strong disorder, but the zero-averaged wavenumber gap survives. The robustness of the zero-averaged wavenumber gap comes from the fact that the zero-averaged wavenumber solution remains invariant under disorder. It should be emphasized again that the position of the zero-averaged wavenumber gap near the new DP is insensitive to both the incidence angles and the disorder. Moreover, we can also consider the effect of a defect barrier on the property of the electron's transport inside the zero-averaged wavenumber gap. One can compare the two cases of the defect modes respectively occurring inside the zero-averaged wavenumber gap and the higher bandgaps. As discussed in Ref. (Wang & Zhu, 2010 , b), the location of the defect mode inside the zero-averaged wavenumber gap is very weakly dependent on the incident angle but the defect mode in the higher bandgap is strongly sensitive to the incident angle.

Accordingly, the zero-averaged wavenumber gap has great effect on the electronic properties of graphene. Figure 10 shows the electronic conductance and its Fano factor as a function of energy inside the finite graphene-based superlattices. It is clearly seen that the conductance becomes minimal at the new DP of $E = 25$meV, and the corresponding Fano factor has a value of $1/3$. Near this new DP the conductance is almost decreasing linearly for the energy below the DP and increasing for the energy above the DP. More interestingly, the Fano factor may become larger than $1/3$ for the higher gap (Bragg gap) due to the fact that the higher band gap is highly shifted to the higher energy at the non-zero incidence angles. All the results can be applicable to the gapped graphene superlattice, as we discussed in Ref. (Wang & Chen, 2010). From all above discussions, we can draw the conclusion that the physical meaning of the zero-averaged wavenumber gap in the graphene superlattice is very similar to the transmission gap in the monolayer graphene, as mentioned in last section. The transmission properties do result from the Klein tunneling in graphene, which is analogous to the negative refraction in the left-handed metamaterials. These phenomena can be applied to predict novel effect in electron wave optics, thus design various electron waves devices. Following the optical analogy in 1DPCs containing left-handed metamaterials (Wang & Zhu, 2010 , a), the one thing that we can do is to investigate the negative and positive GH shifts in the graphene superlattice with the zero-averaged wavenumber gap, where the condition for the extra Dirac points in the graphene-based superlattices is the same as that for the band-crossing effect in 1DPCs consisted of left-handed material and right-handed material. More interesting optical analogies or simulations of the phenomena in graphene will be discussed in the next

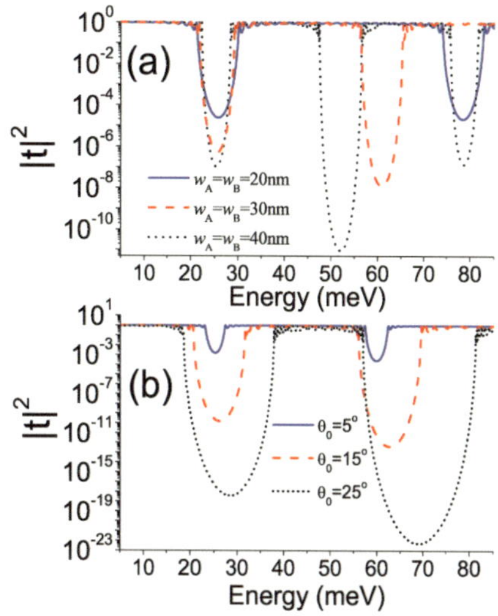

Fig. 10. (Color online) Transmission probabilities of the finite periodic-potential structure $(AB)^{25}$ under (a) different lattice constants with a fixed ratio $w_A/w_B = 1$ and $\theta_0 = 10°$ and (b) different incidence angles with the fixed lattice parameters $w_A = w_B = 30$ nm.

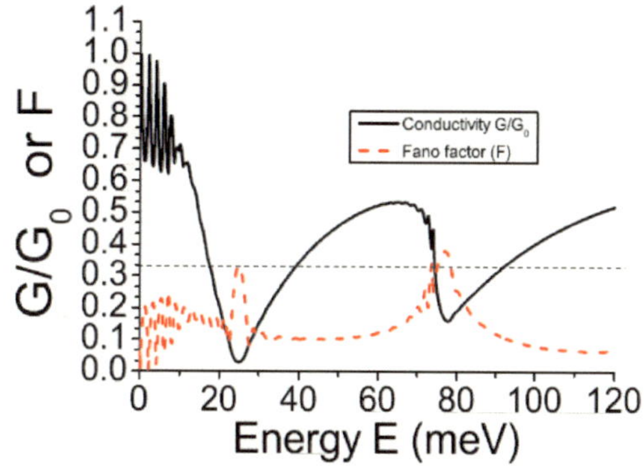

Fig. 11. (Color online) The dependence of the electronic conductance and Fano factor on the energy in the graphene superlattices with the periodic-barrier structures $(AB)^{25}$, where the parameters are $w_A = w_B = 20$ nm, $V_A = 50$ meV and $V_B = 0$ meV.

section, according to the link between Klein tunneling in graphene and negative refraction in metamaterial.

4. Optical simulations with negative-zero-positive index metamaterial

Compare to solids, optical systems offer clean and easy controlled way to test theoretical predictions. The experimental test in electronic systems is usually hindered by the difficulty to maintain system homogeneity. Our central study is to emphasize that the DP with the double-cone structure for the light field and its applications can be realized in a homogenous negative-zero-positive index (NZPI) medium, instead of the 2DPCs (Zhang, 2008).

4.1 Dirac point with double cones in optics

It is well known that that Maxwell's equations for light field reduce to Helmholtz equation, which could be written as $\wp E_z(x,y,\omega) + k^2(\omega)E_z(x,y,\omega) = 0$, with a wavenumber k and $\wp = (\partial^2/\partial x^2 + \partial^2/\partial y^2)$ in two dimensional case for an homogenous material when the polarization of the light field is in the z direction. In general, the Helmholtz equation is written as the Dirac equation,

$$\begin{bmatrix} 0 & -i(\frac{\partial}{\partial x} - i\frac{\partial}{\partial y}) \\ -i(\frac{\partial}{\partial x} + i\frac{\partial}{\partial y}) & 0 \end{bmatrix} \Psi = k(\omega)\Psi, \quad (33)$$

where $\Psi = \begin{pmatrix} E_{z1}(x,y,\omega) \\ E_{z2}(x,y,\omega) \end{pmatrix}$ are the eigenfunctions of the electric fields with the same $k(\omega)$. It is amazing that when the index of the homogenous optical medium varies from negative to zero and then to positive with frequency (Wang et al., 2009 , a), $k(\omega_D) = 0$ and the higher-order terms in the series of $k(\omega) = k(\omega_D) + (\omega - \omega_D)/v_D + \beta(\omega - \omega_D)^2$ can be neglected, where $\omega_D > 0$ is the frequency of the DP (corresponding wavelength is $\lambda_D = 2\pi c/\omega_D$) and group velocity $v = (d\omega/dk)|_{\omega=\omega_D}$. In this case, the homogenous material with linear dispersion

$$k(\omega) = (\omega - \omega_D)/v_D, \quad (34)$$

is called as NZPI media. In this case, it is seen from the transmission spectral at point $L = 40\lambda_D$ that two bands touch each other forming a double-cone structure, see Fig. 12. Thus, the light transport near the DP obeys the massless Dirac equation as follows:

$$\begin{bmatrix} 0 & -i(\frac{\partial}{\partial x} - i\frac{\partial}{\partial y}) \\ -i(\frac{\partial}{\partial x} + i\frac{\partial}{\partial y}) & 0 \end{bmatrix} \Psi = \left(\frac{\omega - \omega_D}{v_D}\right)\Psi, \quad (35)$$

In fact, these two eigenfunctions near DP correspond to the two eigenvalues, $\pm|\omega - \omega_D|/v_D$ of $k(\omega)$, which can be respectively realized by NZPIM in the cases of $\omega > \omega_D$ and $\omega < \omega_D$. So the combination of these eigenfunctions for the same $k(\omega)$ will result in the same properties of transmission and reflection as those in 1DPCs, as discussed below. For example, we take the Drude model as the parameters for both the relative permittivity and permeability of the NZPIM (Wang et al., 2009 , a;b): $\varepsilon_1(\omega) = 1 - \omega_{ep}^2/(\omega^2 + i\gamma_e\omega)$, $\mu_1(\omega) = 1 - \omega_{mp}^2/(\omega^2 + i\gamma_m\omega)$, where ω_{ep}^2 and ω_{mp}^2 are the electronic and magnetic plasma frequencies, and γ_e and γ_m are the damping rates relating to the absorption of the material. Here we can assume $\gamma_e = \gamma_e = \gamma \ll \omega_{ep}^2, \omega_{mp}^2$. It is important that when $\omega_{ep} = \omega_{mp} = \omega_D$ and $\gamma = 0$ (no loss), then both $\varepsilon_1(\omega_D)$ and $\mu_1(\omega_D)$ may be zero simultaneously. In this case, we find $k(w_D) \approx 0$ and $v_D \simeq c/2$, where c is the light speed in vacuum (Wang et al., 2009 , b). So far, we find out the condition to have the Dirac dispersion for light field in the homogenous media, and also provide the candidate material to implement it. Clearly, the difference from the DP in 2DPCs is that the eigenfunctions of the light fields in our system are not the Bloch states in periodic structures but the electromagnetic fields itself in the homogenous NZPIM.

4.2 Pseudodiffusive property and Zitterbewegung effect

As discussed in the above subsection, for NZPI media, we have the DP with a double-cone structure. When frequency ω is close to ω_D, owing to $k^2 = k_x^2 + k_y^2 \to 0$, k_x becomes an imaginary number for real k_y, so that the field along the x direction between the interval

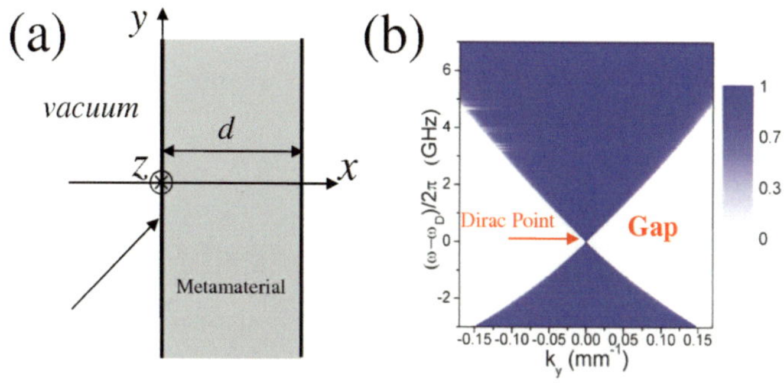

Fig. 12. (Color online) (a) Schematic of a NZPIM slab. (b) Distribution of spectral transmittance for a slab with $L = 40\lambda_D$. Blue area denotes the high transmission and white area denotes the prohibition of light. The parameters of the NZPIM are $\omega_{ep} = \omega_{mp} = \omega_D = 10 \times 2\pi$ GHz and $\gamma = 10^{-5}$ GHz.

between 0 and L have the following relation: $t(L, k_y) = E(L)/E(0) = \exp(-|k_y|L)$. Then the total energy transmittance is

$$T_{total}(L, k_y) = \int_{-\infty}^{\infty} |t(L, k_y)|^2 dk_y = \frac{1}{L}, \tag{36}$$

which implies that the propagation of light field at ω_D has the $1/L$ scaling law a main characteristic of diffusion phenomenon, which has been studied for electron transport in graphene (Katsnelson, 2006) and light transport in photonic crystals (Sepkhanov et al., 2007). Consider the $1/L$ scaling law in a semi-infinite NZPI media, as indicated in Fig. 12 (a). Such a structure may reduce but cannot completely eliminate the non-ideal interface effect $x = 0$. As a simple proof, ε_1 and μ_1 are assumed to be real. The transmission coefficient at $x = L$ is $t(k_y, \omega) = \alpha \exp(ik_x L)$, where $\alpha = 2(qq_m)^{1/2}/(q + q_m)$ is determined by the boundary condition, $q_m = k_{x1}/(\mu_1 k_0)$, $q_0 = (k_0^2 - k_y^2)^{1/2}/k_0$ for $k_0 > k_y$ and otherwise $q_0 = i(k_y^2 - k_0^2)^{1/2}/k_0$, where $k_0 = \omega/c$, k_{x1} is the x-component wave number in the metamaterial. Near the DP , we have $t(k_y, \omega) = \alpha \exp(-|k_y|L)$. Usually, α depends on k_y. For the large distance L, this function decreases quickly with increasing k_y. In this sense we assume that α does't depend on k_y (when an ideal interface is considered). Then the total transmittance is $|\alpha|^2/L$, which is different from Eq. (36) only by a value α due to the interface. Therefore the light transport near DP, namely $\omega = \omega_D$, is proportional to $1/L$ inside the NZPIM.

Figure 12 (b) shows the transmission spectrum at $L = 40\lambda_D$ insider the NZPIM, as shown in Fig. 12 (a). We see that both the upper and lower passbands touch at $\omega_D = 2\pi \times 10$ GHz and nearby the dispersion is linear. With increasing the distance L, the touch at ω_D is an ideal point. Note that $t(k_y, \omega)$ at the DP is close to one even if the metamaterial has a small absorption.

To demonstrate the $1/L$ scaling law near the DP, a characteristic quantity $\xi = S_r L$ is defined to describe the light propagation inside the medium, where $S_r \equiv S(x, y = 0)/S_0$ is a relative energy flow along the x axis, and $S_0 \equiv S(x = 0, y = 0)$ depends on the coupling strength. Instead of the semi-infinite structure, we would like to discuss the propagation of light through a homogenous slab system which is the realistic case. Figure 13 (a) shows the change of the characteristic quantity ξ as a function of the distance inside different slabs with different thicknesses d. It is clearly seen that as d increases, the change of ξ inside the slab approaches to the limit of $d \to \infty$ (i. e., the semi-infinite structure). For the finite thickness d, the value ξ always initially increases and then gradually decays in order to match the second boundary

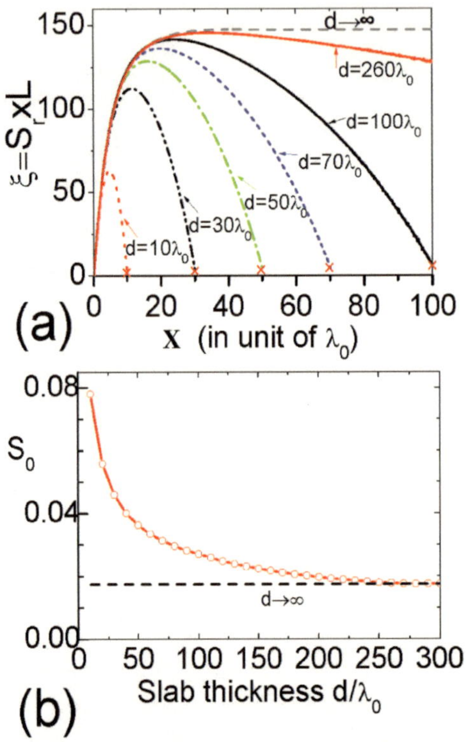

Fig. 13. (Color online) Change of ξ as a function of the propagating position x inside the finite slab systems with different thicknesses, with the light frequency $\omega = \omega_D = 10$ GHz. The crossed points denote the exit positions $x = d$ for different slabs. (b) Dependence of the energy flows S_0 on the slab thickness d. The incident beam is with $10\lambda_D$ and other parameters are the same as in Fig. 12 except for $\gamma = 10^{-4}$ GHz.

condition at $x = d$ [the crossed point in Fig. 13(a)]. From Fig. 13 (a), it is expected that for a sufficient thick slab, the light energy transport obeys the $1/L$ scaling law. Meanwhile, from Fig. 13 (b), it is found that for the small d, the coupling strength of the light field inside the finite slab is strong (with large S_0). With the increasing of the slab thickness d, the value S_0 gradually decreases to the limit of $d \to 0$ (the semi-infinite structure). It indicates that the effect of the second interface on S_0 becomes weaker and weaker, therefore the light energy transport obeys the $1/L$ scaling law.

We emphasize here that although the light field doesn't obey very well the $1/L$ scaling law inside the finite homogenous slab with small d, the static property of the light field is still clear demonstrated, as shown in Fig. 14. From Fig. 14, it turns out that the evolutions of the total electric field and the Poynting vector for a narrow Gaussian beam $E(k_y, 0) = W/\sqrt{2}\exp(-W^2k_y^2/4)$ with a half-width W inside the finite slab system possess the diffusive property. The light fields near the DP have no phase delay and diffuse inside the medium.

In addition, we have also demonstrated the Zitterbewegung effect for optical pulses during the propagation inside the a homogenous NZPIM slab. As we know, Zitterbewgung effect refers to the interference between the positive- and negative-states in the relativistic electron's wave packet. The initial finite pulse is considered to be Gaussian shape both in the transverse spatial domain and in the longitudinal temporal domain. The dynamics of the finite optical

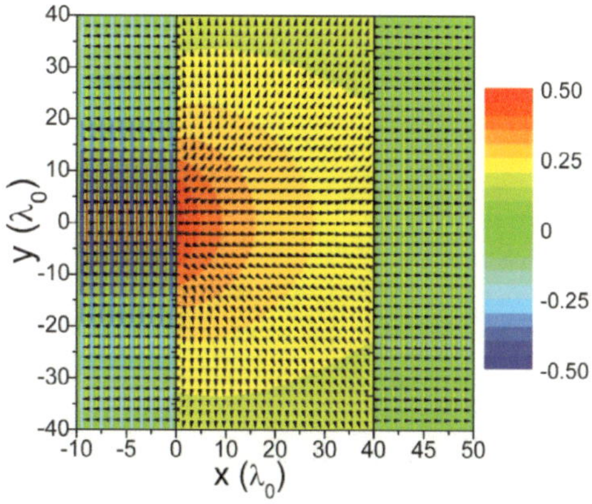

Fig. 14. (Color online) Evolutions of the total electric fields for a narrowed Gaussian beam passing through the finite slab system at $\omega_D = 10$ GHz with $W = 10\lambda_D$, where $\gamma = 0.05$ GHz other parameters are the same as in Fig. 12. The arrows denote the direction and relative magnitude of the energy flow $\vec{S}$.

pulse through the slab shows that before reaching the slab, the pulse propagates in the free space, and after entering the slab, it gradually diffuses and oscillates as the propagating distance increases. Thus, the output pulse at the exit of the slab has the oscillation with the characteristic frequency independent of the slab thickness. The physical origin of such oscillation comes from the interference between the upper and lower high-transmission bands near the DP of the NZPIM slab. Actually, the finite pulse can be divided into two parts: one ($\omega > \omega_D$) belongs to the upper band and the other one ($\omega < \omega_D$) belongs to the lower band (see Fig. 12), and each part undergoes the opposite phase change due to the opposite properties of the two bands. For instance, in the upper band the wave number k is positive so the phase shift is positive, while in the lower band the phase shift is negative due to negative k. Therefore the two parts interfere with each other, which leads to the oscillatory property. At the DP, the light field obeys the diffuse equation, thus the pulse behavior also has the diffusion property. Moreover, for the pulse with a fixed transverse spatial width W, the oscillation frequency is proportional to the pulse spectral width near the DP, and for the pulse with a fixed time duration, the oscillation frequency decreases with the increasing of W. From the experimental viewpoint, it is suggested that smaller transverse spatial width and/or shorter pulse duration of the finite pulse make the optical Zitterbewegung effect easier to be observed experimentally.

4.3 Optical Goos-Hänchen shift in Bragg-like reflection

In this subsection, we will take an example of the optical simulations in NZPIM with DP. Similar to the Sec. 2.2, we start to investigate the GH shifts of the light beam reflected from the slab of NZPI medium. Firstly, we assume that the incident plane wave is $E_z^{in}(x,y) = \exp\left[i(k_x x + k_y y)\right]$, where $k_x = k_0 \cos\theta_0$, $k_y = k_0 \sin\theta_0$, $k_0 = (\varepsilon_0\mu_0)^{1/2}\omega/c$ is the wave number in the air, ε_0 and μ_0 are the relative permittivity and permeability of the air, the reflected and transmitted plane waves can be expressed by $E_z^{ref}(x,y) = r\exp\left[i(-k_x x + k_y y)\right]$ and

$E_z^{tr}(x,y) = t\exp\{i[-k_x(x-d)+k_y y]\}$, where the reflection coefficient r

$$r = \frac{\exp(i\pi/2)}{4g^2}\left(\frac{\mu_0}{\mu_1}\frac{k_{1x}}{k_x} - \frac{\mu_1}{\mu_0}\frac{k_x}{k_{1x}}\right)\left[\sin 2k_{1x}d + i\left(\frac{\mu_1}{\mu_0}\frac{k_x}{k_{1x}} + \frac{\mu_0}{\mu_1}\frac{k_{1x}}{k_x}\right)\sin^2 k_{1x}d\right], \quad (37)$$

and the transmission coefficient is $t = e^{i\phi}/g$ with $ge^{i\phi} = \cos k_{1x}d + \frac{i}{2}\left(\frac{\mu_1}{\mu_0}\frac{k_x}{k_{1x}} + \frac{\mu_0}{\mu_1}\frac{k_{1x}}{k_x}\right)\sin k_{1x}d$, $k_{1x} = \sqrt{k_1^2 - k_y^2}$ and $k_1 = (\omega - \omega_D)/v_D$ near the DP. As mentioned above, the upper and lower bands have different properties. Thus, we will discuss the unique properties of reflection and transmission in two cases of $\omega > \omega_D$ and $\omega < \omega_D$. The details can be seen in the literature (Chen et al., 2009 , b).

Similar to the transmission in graphene barrier, the transmission as the function of frequency ω has a gap. The only difference is that the width of transmission gap has the following form: $\Delta\omega = 2k_y v_D$. To avoid the repetition, we concentrate on the reflection case here. The reflection behaves Bragg-like reflection. Figure 15 indicates the dependence of corresponding reflection probability R on the wavelength $\lambda = 2\pi\omega/c$, where $\theta_0 = 20°$ and $\omega_D = 10 \times 2\pi$ GHz. Solid and dashed line correspond to $a = 100$ mm and $a = 10$ mm. It is interesting that the light beam can be perfectly reflected by such single NZPIM slab at certain range of the wavelength. As indicated in Fig. 15, the wavelength window for perfect reflection will become narrower with the increase of the width of slab. These frequency or wavelength passing-bands in reflection discussed here are similar to but different from the Bragg reflection in the 1D PCs. This so-called Bragg-like reflection discussed here is exactly due to the linear Dirac dispersion, which results in the evanescent waves in two cases of $\omega > \omega_D$ and $\omega < \omega_D$, corresponding to the two eigenfunctions of electric fields with the same $k(\omega)$.

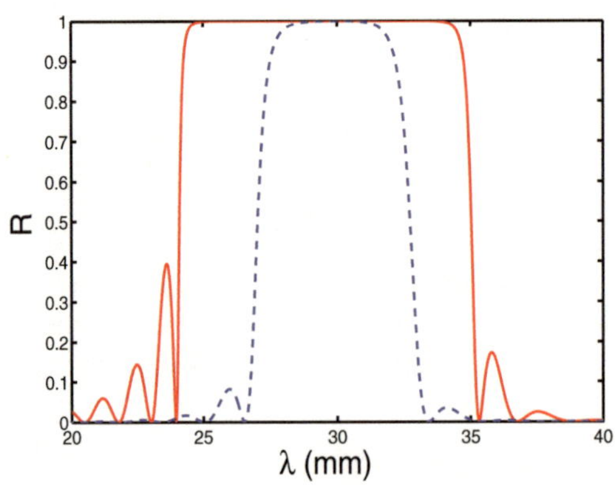

Fig. 15. (Color online) The reflection probability R as the function of the wavelength λ, where $\theta_0 = 20°$, $\omega_D = 10 \times 2\pi$ GHz, $d = 100$ mm (red solid line), and $d = 10$ mm (blue dashed line).

Next, we have a look at the optical GH shifts inside the optical NZPIM slab. It is noted that the GH shift in transmission is the equal to that in reflection inside such symmetric slab configuration, because the values of the derivation of the phase shifts with respect to k_y are the same. Here we just only consider the GH shift in reflection.

Figure 16 demonstrates that the GH shifts can be positive and negative. Due to the properties of DP, it is reasonable that the GH shifts are negative in the case of $\omega < \omega_D$, while they are positive in the case of $\omega > \omega_D$ is positive. More interestingly, the GH shifts near the DP can change from positive to negative with the increase (decrease) of the wavelength (frequency). In addition, it is also shown that the GH shifts near the DP have only the order of wavelength due to the evanescent waves. The smallness of the GH shifts are similar to those in total reflection or frustrate-total-internal-reflection structure (Chen et al., 2009 , a).

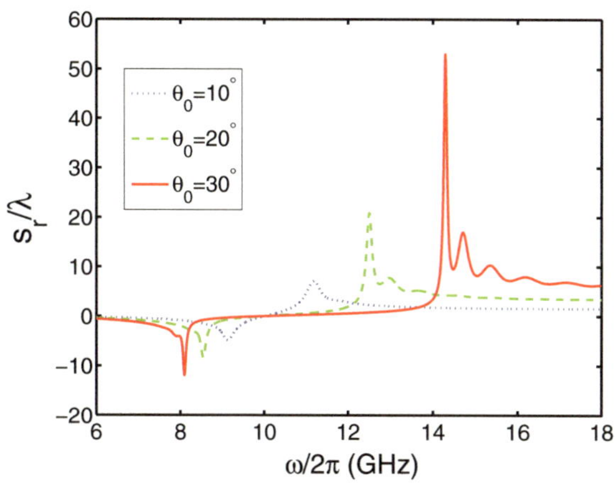

Fig. 16. (Color online) The GH shifts as the function of the wavelength, ω, where $d = 100$ mm, and other parameters are the same as in Fig. 15. Solid, dashed and dotted curves correspond to $\theta_0 = 30°$, $\theta_0 = 20°$, and $\theta_0 = 10°$.

However, these GH shifts don't change sign with increasing the angle of incidence. This is similar to the quantum GH effect in graphene on the total reflection (Zhao et al., 2010), but it is quite different from the result given in the literature (Beenakker et al., 2009), where the sublattice (or "pseudospin") degree of freedom is considered. Anyway, we have managed to simulate the quantum GH in graphene barrier by the optical metamaterial with the DP.

4.4 Further work on optical Dirac point in metamaterials

Finally, we would like to point out other interesting work on the optical DP in metamaterials. Motivated by the realization of the optical Dirac dispersion in the homogenous NZPIM, we make a theoretical investigation on the properties of thermal emission in layered structures containing the NZPI medium (Wang et al., 2010). When the thermal emission frequency is close to the DP, the spectral hemisperical power of thermal emission in such a structure is strongly suppressed and the emission can become a high directional source with large spatial coherence.

In addition, the guided modes and nonlinear surface waves near the DP have been also studied, respectively (Shen et al., 2010 , a;b). In the simple model of optical waveguide, our theoretical results show that due to the linear Dirac dispersion, the fundamental mode is absent when the angular frequency is smaller than the DP, while the behaviors of NZPIM waveguide are similar to the conventional dielectric waveguide when the angular frequency is larger than the DP. The unique properties of the guided modes are analogous to the propagation of electron waves in graphene waveguide (Zhang et al., 2009), corresponding to the classical motion and the Klein tunneling. It is amazing that electron guiding as the analogue of an optical fiber has been experimentally demonstrated in graphene by tuning the carrier type and density using local electrostatic fields (Williams et al., 2010). This timely work will definitely simulate us to design various novel optic-like devices in graphene according to the Dirac-like properties of graphene.

5. Conclusion

In summary, there are a lot of optic-like phenomena of electron wave in graphene, which leads to novel Dirac electron wave devices. The field covered here is vast, and we pay special attention to work done by the authors, while making effort to to offer a global perspective. In this chapter, we have presented the propagation of electron waves in monolayer graphene and optical simulations with NZPIM. The specific electronic analogies of Bragg-like reflection (transmission gap), zero-averaged wavenumber gap, and GH effect have been respectively discussed in single and multiple monlayer graphene barriers. The key point is that the transmission gap has great effect on the electronic transports including electric conductance and Fano factor. More importantly, all these results suggest that the electron wave propagation and their optical counterparts in NZPIM not only give the deeper understanding of several exotic phenomena in graphene, but also predict richer phenomena in different physical systems.

6. Acknowledgement

This work could not have been completed without the contributions of S.-Y. Zhu, C.-F. Li, Y. Ban, Z. G. Wang, J.-X. Zhang, M. Shen, Y. He and J.-W. Tao in the reviewed manuscripts. We acknowledge H. Q. Lin, C. W. J. Beenakker, and F. M. Peeters for discussions and suggestions. This work was partly supported by the National Natural Science Foundation of China (Grant Nos. 60806041 and 61078021), the Shanghai Rising-Star Program (Grant No. 08QA14030) the Shanghai Educational Development Foundation (Grant No.2007CG52) and the Shanghai Leading Academic Discipline Program (Grant No. S30105). We also thank funding by Juan de la Cierva Programme, the Basque Government (Grant IT472-10), the Ministerio de Ciencia e Innovación (FIS2009-12773-C02-01), and HongKong RGC 403609.

7. References

Arovas, D. P.; Brey, L.; Fertig, H. A.; Kim, E.-A. & Ziegler K. (2010), Dirac spectrum in piecewise constant one-dimensional potentials, arXiv:1002.3655

Bai C.-X. & Zhang, X.-D. (2007). Klein paradox and resonant tunneling in a graphene superlattice. *Phys. Rev. B*, Vol. 76, No. 7, Aug -2007, 075430 (1-4), ISSN 1098-0121

Barbier, M.; Peeters, F. M.; Vasilopoulos, P. & Milton Pereira, Jr., J. (2008). Dirac and Klein-Gordon particles in one-dimensional periodic potentials. *Phys. Rev. B*, Vol. 77, No. 11, Mar -2008, 115446 (1-4), ISSN 1098-0121

Barbier, M.; Vasilopoulos, P. & Peeters, F. M. (2009). Dirac electrons in a Kronig-Penney potential: Dispersion relation and transmission periodic in the strength of the barriers, *Phys. Rev. B*, Vol. 80, No. 11, Nov -2009, 205415 (1-5), ISSN 1098-0121

Barbier, M.; Vasilopoulos, P. & Peeters, F. M. (2010). Extra Dirac points in the energy spectrum for superlattices on single-layer graphene. *Phys. Rev. B*, Vol. 81, No. 7, Feb -2010, 075438 (1-7), ISSN 1098-0121

Beenakker, C. W. (2008). Colloquium: Andreev reflection and Klein tunneling in graphene.*Rev. Mod. Phys.*, Vol. 80, No. 4, Oct -2008, 1337 – 1354, ISSN 0034-6861

Beenakker, C. W. J.; Sepkhanov, R. A.; Akhmerov, A. R. & Tworzydło, J. (2009). Quantum Goos-Hänchen effect in graphene. *Phys. Rev. Lett.*, Vol. 102, No. 14, Jan -2009, 146804 (1-4), ISSN 0031-9007

Bliokh, Yury P.; Freilikher, V.; Savel'ev, S. & Nori, F. (2009). Transport and localization in periodic and disordered graphene superlattices. *Phys. Rev. B*, Vol. 79, No. 7, Feb -2009, 075123 (1-4), ISSN 1098-0121

Brey, L. & Fertig, H. A. (2009), Emerging zero modes for graphene in a periodic potential, *Phys. Rev. Lett.*, Vol.103, No.4, Jul.-2009, 046809, (1-4), ISSN 0031-9007

Castro Neto, A. H.; Guinea, F.; Peres, N. M. R.; Novoselov, K. S. & Geim, A. K. (2009). The electronic properties of graphene. *Rev. Mod. Phys.*, Vol. 81, No. 1, Jan -2009, 109 – 162, ISSN 0034-6861

Cheianov, V. V.; Fal'ko, V. & Altshuler, B. L. (2007). The focusing of electron flow and a veselago lens in graphene p-n junctions. *Science*, Vol. 315, No. 5816, Mar -2007, 1252 - 1255, ISSN 0036-8075

Chen, X.; Li, C.-F.; Wei, R.-R.; & Zhang, Y. (2009). Goos-Hänchen shifts in frustrated total internal reflection studied with wave-packet propagation. *Phys. Rev. A*, Vol. 80, No. 1, Jul.-2009, 015803 (1-4), ISSN 1050-2947

Chen, X.; Wang, L.-G. & Li, C.-F. (2009). Transmission gap, Bragg-like reflection, and Goos-Hänchen shifts near the Dirac point inside a negative-zero-positive index metamaterial slab. *Phys. Rev. A*, Vol. 80, No. 4, Oct -2009, 043839 (1-5), ISSN 1050-2947

Chen, X. & Tao, J.-W. (2009). Design of electron wave filters in monolayer graphene by tunable transmission gap. *Appl. Phys. Lett.*, Vol. 94, No. 26, Jun -2009, 262102 (1-3), ISSN 0003-6951

Chen, X.; Tao, J.-W. & Ban, Y. (2010). Goos-Hänchen-like shifts for Dirac fermions in monolayer graphene barrier. arXiv:1004.0350v1 (to appear in Eur. Phys. J. B).

Darancet, P.; Olevano, V. & Mayou, D. (2009). Coherent electronic transport through graphene constrictions: Subwavelength regime and optical analogy. *Phys. Rev. Lett.*, Vol. 102, No. 13, Mar -2009, 136803 (1-4), ISSN 0031-9007

Datta, S. (1996). *Electronic Transport in Mesoscopic Systems*, Cambridge University Press, ISBN 0-521-41604-3, New York.

Dragoman, D. & Dragoman, M. (2004). *Quantum-Classical Anologies*, Springer, ISBN 978-3-540-20147-2, Berlin.

Dell'Anna L. & De Martino, A. (2009). Multiple magnetic barriers in graphene. *Phys. Rev. B*, Vol. 79, No. 4, Jan -2009, 045420 (1-4), ISSN 1098-0121

Gaylord, T. K. & Brennan, K. F. (1989). Electron wave optics in semiconductors. *J. Appl. Phys.*, Vol. 65, No. 2, Jan -1989, 814 – 820, ISSN 0021-8979

Gaylord, T. K.; Glytsis, E. N.; Henderson, G. N.; Martin, K. P.; Walker, D. B.; Wilson, D. W. & Brennan, K. F. (1991). Quantum interference effects in semiconductors: a bibliography. *Proc. IEEE*, Vol. 79, No. 8, Aug -1991, 1159 – 1180, ISSN 0018-9219

Gerritsma, R.; Kirchmair, G.; Zähringer, F.; Solano, E.; Blatt, R. & Roos, C. F. (2010). Quantum simulation of the Dirac equation. *Nature*, Vol. 463, Jan -2010, 68 – 71, ISSN 0028-0836

Ghosh, S. & Sharma, M. (2009). Electron optics with magnetic vector potential barriers in graphene. *J. Phys.: Condens. Matter*, Vol. 21, No. 29, Jul -2009, 292204 (1-8), ISSN 0953-8984

Goos, F. & Hänchen, H. (1947), Ein neuer und fundamentaler Versuch zur Totalreflexion. *Ann. Phys. (Leipzig)* Vol. 436, No. 7 – 8, Oct-1943, 333-346, ISSN 0003-3804

Goos, F. & Hänchen, H. (1949), Neumessung des Strahlversetzungseffektes bei Totalreflexion. *Ann. Phys. (Leipzig)* Vol. 440, No. 3 – 5, Jul-1949, 251-252, ISSN 0003-3804

Güney D. Ö. & Meyer, D. A. (2009). Negative refraction gives rise to the Klein paradox. *Phys. Rev. A*, Vol. 79, No. 6, Jun -2009, 063834 (1-4), ISSN 1050-2947

Ho, J. H.; Chiu, Y. H.; Tsai, S. J. & Lin, M. F. (2009), Semimetallic graphene in a modulated electric potential, *Phys. Rev. B*, Vol.79, No.11, Mar.-2009, 115427, (1-7), ISSN 1098-0121

Ji, Y.; Chung, Y.; Sprinzak, D.; Heiblum, M.; Mahalu, D. & Shtrikman, H. (2003). An electronic Mach-Zehnder interferometer. *Nature*, Vol. 422, Mar -2003, 415 – 418, ISSN 0028-0836

Juzeliūnas, G.; Ruseckas, J.; Lingdberg, M.; Santos, L. & Öhberg, P. (2008). Quasirelativistic behavior of cold atoms in light fields. *Phys. Rev. A*, Vol. 77, No. 1, Jan -2008, 011802 (1-4), ISSN 1050-2947

Katsnelson, M. I.; Novoselov, K. S. & Geim, A. K. (2006). Chiral tunnelling and the Klein paradox in graphene. *Nature Phys.*, Vol. 2, Feb -2006, 620 – 625, ISSN 1745-2473

Katsnelson, M. I. (2006). Zitterbewegung, chirality, and minimal conductivity in graphene. *Eur. Phys. J. B* Vol.51, No. 2, May-2006, 157-160, ISSN 1434-6028

Lamata, L.; León, J.; Schätz, T. & Solano, E. (2007). Dirac equation and quantum relativistic effects in a single trapped ion. *Phys. Rev. Lett.*, Vol. 98, No. 25, Jan -2007, 253005 (1-4), ISSN 0031-9007

Li, J.; Zhou, L.; Chan, C. T. & Sheng, P. (2003), Photonic band gap from a stack of positive and negative index material, *Phys. Rev. Lett.*, Vol.90, No. 8, Feb-2003, 083901 (1-4), ISSN 0031-9007

Longhi, S. (2010). Klein tunneling of light in fiber Bragg gratings. *Physics Research International*, Vol. 2010, No. 645106, July -2010, 645106 (1-5), ISSN 20902220

De Martino, A.; Dell'Anna, L. & Egger, R. (2007). Magnetic confinement of massless Dirac fermions in graphene. *Phys. Rev. Lett.*, Vol. 98, No. 6, Feb -2007, 066802 (1-4), ISSN 0031-9007

Masir, M. R.; Vasilopoulos, P. & Peeters, F. M. (2008). Wavevector filtering through single-layer and bilayer graphene with magnetic barrier structures. *Appl. Phys. Lett.*, Vol. 93, No. 24, Dec -2008, 242103 (1-3), ISSN 0003-6951

Masir, M. R.; Vasilopoulos, P.; Matulis, A. & Peeters, F. M. (2008). Direction-dependent tunneling through nanostructured magnetic barriers in graphene. *Phys. Rev. B*, Vol. 77, No. 23, Jun -2008, 235443 (1-4), ISSN 1098-0121

Molenkamp, L. W.; Staring, A. A. M.; Beenakker, C. W. J.; Eppenga, R.; Timmering, C. E.; Williamson, J. G.; Harmans, C. J. P. M. & Foxon, C. T. (1990). Electron-beam collimation with a quantum point contact. *Phys. Rev. B*, Vol. 41, No. 2, Jan -1990, 1274 (1-4), ISSN 1098-0121

Novoselov, K. S.; Geim, A. K.; Morozov, S. V.; Jiang, D.; Zhang, Y.; Dubonos, S. V.; Grigorieva, I. V. & Firsov, A. A.(2004). Electric field effect in atomically thin carbon films. *Science*, Vol. 306, No. 5696, Oct -2004, 666 – 669, ISSN 0036-8075

Novoselov, K. S.; Geim, A. K.; Morozov, S. V.; Grigorieva, I. V.; Dubonos, S. V. & Firsov, A. A. (2005). Two-dimensional gas of massless Dirac fermions in graphene. *Nature (London)*, Vol. 438, Nov -2005, 197 – 200, ISSN 0028-0836

Palevski, A.; Heiblum, M.; Umbach, C. P.; Knoedler, C. M.; Broers, A. N. & Koch, R. H. (1989). Lateral tunneling, ballistic transport, and spectroscopy in a two-dimensional electron gas. *Phys. Rev. Lett.*, Vol. 62, No. 15, Apr -1989, 1776 (1-4), ISSN 0031-9007

Park, C.-H.; Yang, L.; Son, Y.-W.; Cohen, M. L. & Louie, S. G. (2008). Electron beam supercollimation in graphene superlattices. *Nano Lett.*, Vol. 8, No. 9, Aug -2008, 2920 – 2924, ISSN 1530-6984

Peraira, J. M.; Vasilopoulos, Jr., P. & Peeters, F. M. (2007). Graphene-based resonant-tunneling structures. *Appl. Phys. Lett.*, Vol. 90, No. 13, Mar -2007, 132122 (1-3), ISSN 0003-6951

Garcia-Pomar, J. L.; Cortijo, A. & Nieto-Vesperinas, M. (2008). Fully valley-polarized electron beams in graphene. *Phys. Rev. Lett.*, Vol. 100, No. 23, Jun -2008, 236801 (1-4), ISSN 0031-9007

Peleg, O.; Bartal, G.; Freedman,B.; Manela, O.; Segev, M. & Christodoulides, D. N. (2007). Conical diffraction and gap solitons in honeycomb photonic lattices. *Phys. Rev. Lett.*, Vol. 98, No. 10, Mar -2007, 103901 (1-4), ISSN 0031-9007

Shen, M.; Ruan, L.-X. & Chen, X. (2010). Guided modes near the Dirac point in negative-zero-positive index metamaterial waveguide. *Opt. Express*, Vol. 18, No. 12, May -2010, 12779 – 12787, ISSN 1094-4087

Shen, M.; Ruan, L.-X.; Chen, X.; Shi, J.-L.; Ding, H.-X.; Xi, N. & Wang Q. (2010). Nonlinear surface waves near the Dirac point in negative-zero-positive index metamaterial. *J. of Opt.* Vol. 12, N, 8, Aug- 2010, 085201 (1-5), ISSN 2040-8978

Sivan, U.; Heiblum, M.; Umbach, C. P. & Shtrikman, H. (1990). Electrostatic electron lens in the ballistic regime. *Phys. Rev. B*, Vol. 41, No. 11, Apr -1990, 7937 (1-4), ISSN 1098-0121

Spector, J.; Stormer, H. L.; Baldwin, K. W.; Pfeiffer, L. N. & West, K. W. (1990). Electron focusing in two-dimensional systems by means of an electrostatic lens. *Appl. Phys. Lett.*, Vol. 56, No. 13, Mar -1990, 1290 (1-3), ISSN 0003-6951

Sepkhanov, R. A.; Bazaliy, Y. B. & Beenakker, C. W. J. (2007). Extremal transmission at the Dirac point of a photonic band structure. *Phys. Rev. A*, Vol. 75, No. 6, Jun -2007, 063813 (1-5), ISSN 1050-2947

Steślicka, B.; Kucharczyk, R.; Akjoouj, A.; Rouhani, B. D.; Dobrzynski, L.; & Davison, S. G. (2002), Localised electronic states in semiconductor superlattices. *Surf. Sci. Rep.* Vol. 47, No.4 – 5, Aug-2002, 93-196, ISSN 0167-5729

Tworzydlo, J.; Trauzettel B., Titov, M.; Rycerz, A.; & Beenakker, C. W. J. (2006). Sub-Poissonian shot noise in graphene. *Phys. Rev. Lett* Vol. 96, No. 24, 246802 (1-4), Jun-2006, ISSN 0031-9007

Tsu, R. (2005), *Superlattice to Nanoelectronics*, Elsevier, ISBN: 978-0-08-044377-5, Oxford.

Van Houten, H.; van Wees, B. J.; Mooij, J. E.; Beenakker, C. W. J.; Williamson, J. G. & Foxon, C. T. (1988). Coherent electron focussing in a two-dimensional electron gas. *Europhys. Lett.*, Vol. 5, No. 8, Apr -1988, 721 – 725, ISSN 0295-5075

Wang, L.-G.; Wang, Z.-G.; Zhang, J.-X. & Zhu, S.-Y. (2009). Realization of Dirac point with double cones in optics. *Opt. Lett.*, Vol. 34, No. 10, May -2009, 1510-2, ISSN 0146-9592

Wang, L.-G.; Wang, Z.-G. & Zhu, S.-Y. (2009). Zitterbewegung of optical pulses near the Dirac point inside a negative-zero-positive index metamaterial. *EPL*, Vol. 86, No. 4, Jun -2009, 47008 – -5, ISSN 0295-5075

Wang, L.-G. & Zhu, S.-Y. (2010), The reversibility of the Goos-Hänchen shift near the band-crossing structure of one-dimensional photonic crystals containing left-handed metamaterials, *Appl. Phys. B: Lasers and Optics*, Vol.98, No.2 – 3, Feb.-2010, 459-463, ISSN 0946-2171

Wang, L.-G. & Zhu, S.-Y. (2010). Electronic band gaps and transport properties in graphene superlattices with one-dimensional periodic potentials of square barriers. *Phys. Rev. B*, Vol. 81, No. 20, May -2010, 205444 (1-9), ISSN 1098-0121

Wang, L.-G.; Li, G.-X. & Zhu, S.-Y. (2010). Thermal emission from layered structures containing a negative-zero-positive index metamaterial. *Phys. Rev. B*, Vol. 81, No. 7, Jan -2010, 073105 (1-4) , ISSN 1098-0121

Wang, L.-G. & Chen, X. (2010). Robust zero-averaged wave-number gap inside gapped graphene superlattices. arXiv:1008.0504. (to appear in J. Appl. Phys.)

Wang Z.-F. & Liu, F. (2010). Manipulation of electron beam propagation by hetero-dimensional graphene junctions. *ACS Nano*, Vol. 4, No. 4, Jan -2010, 2459 – 2465, ISSN 1936-0851

White, C. T. & Todornov, T. N. (2001). Quantum electronics: Nanotubes go ballistic. *Nature*, Vol. 411, Jun -2001, 649 – 651, ISSN 0028-0836

Williams, J. R.; Low, Tony; Lundstrom, M. S. & Marcus C. M. (2010). Fiber-optic and p-n guiding of electrons in graphene. arXiv:1008.3704

Yacoby, A.; Heiblum, M.; Umansky, V.; Shtrikman, H. & Mahalu, D. (1994). Unexpected periodicity in an electronic double slit interference experiment. *Phys. Rev. Lett.*, Vol. 73, No. 23, Dec -1994, 3149 (1-4), ISSN 0031-9007

Zhao, L. & Yelin, S. F. (2010). Proposal for graphene-based coherent buffers and memories. *Phys. Rev. B*, Vol. 81, No. 11, Mar -2010, 115441 (1-4), ISSN 1098-0121

Zhang, Y.; Tan, Y.-W.; Stormer, H. L. & Kim, P. (2005). Experimental observation of the quantum Hall effect and Berry's phase in graphene. *Nature (London)*, Vol. 438, Nov -2005, 201 – 204, ISSN 0028-0836

Zhang, X. (2008). Observing Zitterbewegung for photons near the Dirac point of a two-dimensional photonic crystal. *Phys. Rev. Lett.*, Vol. 100, No. 11, Mar -2008, 113903 (1-4), ISSN 0031-9007

Zhang, F.-M.; He, Y. & Chen, X. (2009). Guided modes in graphene waveguides. *Appl. Phys. Lett.* , Vol. 94, No. 21, May -2009, 212105 (1-3), ISSN 0003-6951

Zhu, S.-L.; Wang, B.-G. & Duan, L.-M. (2007). Simulation and detection of Dirac fermions with cold atoms in an optical lattice. *Phys. Rev. Lett.*, Vol. 98, No. 26, Jun -2007, 260402 (1-4), ISSN 0031-9007

22

Wave Packet Dynamics in a Monolayer Graphene

G. M. Maksimova, V. Ya. Demikhovskii, and E. V. Frolova
Nizhny Novgorod State University
Russian Federation

1. Introduction

In the last years the dynamics of wave packets in 2D electron gas and other systems in solids including the phenomenon of *Zitterbewegung* (ZB) or trembling motion has been the subject of numerous studies. (Shmueli et al.,1995; Schliemann et al.,2005; Schliemann et al.,2006; Schliemann,2008; Zawadzki,2005; Ferrari & Russo,1990; Zawadzki,2006; Katsnelson,2006; Cserti & David,2006; Winkler et al.,2007; Trauzettel et al.,2007; Rusin & Zawadzki,2007; Rusin & Zawadzki,2007; Rusin & Zawadzki,2008) Firstly the oscillatory motion analogous to the relativistic *Zitterbewegung* in two-dimensional systems with the structural and bulk inversion asymmetry was investigated by Schliemann *et al.*(Schliemann et al.,2005; Schliemann et al.,2006) In the recent work by authors the detailed studying of the electron wave packet dynamics in the semiconductor quantum well under the influence of the Rashba spin-orbit coupling was performed. (Demikhovskii et al.,2008) It was shown (analytically and numerically) that the initial wave packet splits into two parts with different spin polarizations propagating with unequal group velocities. It was demonstrated also that the splitting and overlapping of wave packets leads to the damping of *Zitterbewegung*.

As well known, the electron *Zitterbewegung* in relativistic physics at first time was predicted by Schrödinger (Schrödinger,1930) (see also (Barut & Bracken,1981)). This phenomenon is caused by the interference between positive and negative energy states in the wave packet. The frequency of ZB motion is determined by the gap between these two states and the amplitude of oscillations in a particle position is of the order of the Compton wave length. This phenomenon was discussed also in Refs.(Lock,1979; Thaller,2004; Braun et al.,1999).

The results of the first experimental observation of ZB phenomena were published recently in the paper by Gerritsma *et. al.*(Gerritsma et al.,2010) For the ZB simulation the experimentalists used a linear Paul trap where ion motion can be described by one-dimensional Dirac equation.(Lamata et al.,2007) The authors of Ref. (Gerritsma et al.,2010) study the motion of Ca^+ ion and determined its position as a function of time for different initial conditions. As was shown in Ref. (Lamata et al.,2007) the solution of the $3+1$ Dirac equation can also be simulated using a single trapped ion with four ionic internal states. In this case the ion position and momentum are associated with respective characteristics of 3D Dirac particle.

In the papers by Rusin and Zawadzki (Rusin & Zawadzki,2007; Rusin & Zawadzki,2008) the evolution of the wave packet in a monolayer and bilayer graphene as well as in carbon nanotubes was analyzed. The exact analytical expressions for two components of wave function and average value of position operator were found for bilayer graphene, which allowed to obtain analytical results for the ZB of Gaussian wave packet. It was shown that

the transient character ZB in bilayer graphene is due to the fact that wave subpackets related to positive and negative electron energies move in opposite directions, so their overlap diminishes with time. At the same time the dynamics of the wave packets in a monolayer graphene in Ref. (Rusin & Zawadzki,2007) was not investigated fully.
In this chapter we present the detailed description of wave packet evolution including the phenomenon of ZB of the packet center in a monolayer graphene (see also Ref. (Maksimova et al.,2008)). The analytical expressions for the components of wave function at $t > 0$ are found in the form of two-dimensional integrals. Using these equations we obtain the full electron probability density and such dynamical characteristics of the packet center as the average components of coordinate and velocity. We investigate the influence of the initial pseudospin polarization on the space-time evolution of the wave packet, in particular, on the trajectory of its center. As a result, the direction of the packet motion is determined not only by the orientation of the average momentum, but mainly by the phase difference between the up- and low- components of the wave functions. Our analytically results are illustrated by a graphic presentation. The obtained results can be useful for analysis of the functioning of graphene's electronic structures and devices.

2. Basic equations

Graphene is a single layer of carbon atom densely packed in a honeycomb lattice. The two-dimensional Hamiltonian describing its band structure has the form (Wallace,1947; Slonczewski & Weiss,1958; Novoselov et al.,2005; Novoselov et al.,2006; Zhang et al.,2005)

$$\hat{H} = u\vec{\sigma}\hat{\vec{p}}, \tag{1}$$

where $u \approx 10^8\ cm/s$, $\hat{\vec{p}} = (\hat{p}_x, \hat{p}_y)$ is the momentum operator defined with respect to the centre of the valley centered at the corner of the Brillouin zone with wave vector $\vec{K}$. Pauli matrices σ_i operate in the space of the electron amplitude on two sites (A and B) in the unit cell of a hexagonal crystal. This internal degree of freedom plays a role of a pseudospin. The Dirac-like Hamilton $\hat{H}$ determines the linear dispersion relation

$$E_{p,s} = sup. \tag{2}$$

Here $p = \sqrt{p_x^2 + p_y^2}$, $s = 1$ for the electron in the conduction band and $s = -1$ for the valence band ("hole" branch of quasiparticles). The corresponding eigenfunctions are given by

$$\varphi_{\vec{p},s}(\vec{r},t) = \frac{1}{2\sqrt{2}\pi\hbar}\exp(i\frac{\vec{p}\vec{r}}{\hbar} - i\frac{E_{p,s}t}{\hbar})\begin{pmatrix}1\\ se^{i\varphi}\end{pmatrix}, \tag{3}$$

with $e^{i\varphi} = \frac{p_x + ip_y}{p}$.
The time-evolution of an arbitrary initial state $\psi(\vec{r},0)$ in Shrödinger representation can be found with the help of Green's function $G_{\mu\nu}(\vec{r},\vec{r'})$

$$\psi_\mu(\vec{r},t) = \int G_{\mu\nu}(\vec{r},\vec{r'},t)\psi_\nu(\vec{r'},0)d\vec{r'}, \tag{4}$$

where $\mu, \nu = 1,2$ are matrix indices, corresponding to the upper and lower components of $\psi(\vec{r},t)$. These components are related to the probability of finding electron at the sites of the sublattices A and B correspondingly. The standard expression for Green's function is

$$G_{\mu\nu}(\vec{r},\vec{r'},t) = \sum_{s=\pm 1} \int d\vec{p}\; \varphi_{\vec{p},s;\mu}(\vec{r},t)\varphi^{*}_{\vec{p},s;\nu}(\vec{r'},0). \tag{5}$$

Using Eq.(3) for $\varphi_{\vec{p},s;\mu}(\vec{r},t)$ we find

$$G_{11}(\vec{r},\vec{r'},t) = G_{22}(\vec{r},\vec{r'},t) = \frac{1}{(2\pi\hbar)^2}\int \exp(i\frac{\vec{p}(\vec{r}-\vec{r'})}{\hbar})\cos(\frac{upt}{\hbar})d\vec{p}, \tag{6}$$

$$G_{21}(\vec{r},\vec{r'},t) = G^{*}_{12}(\vec{r},\vec{r'},t) = \frac{-i}{(2\pi\hbar)^2}\int \frac{p_x+ip_y}{p}\exp(i\frac{\vec{p}(\vec{r}-\vec{r'})}{\hbar})\sin(\frac{upt}{\hbar})d\vec{p}, \tag{7}$$

Let us represent the initial wave function by Gaussian wave packet having the width d and nonvanishing average momentum $p_{0y} = \hbar k_0$

$$\psi(\vec{r},0) = \frac{f(\vec{r})}{\sqrt{|c_1|^2+|c_2|^2}}\begin{pmatrix} c_1 \\ c_2 \end{pmatrix}, \tag{8a}$$

$$f(\vec{r}) = \frac{1}{d\sqrt{\pi}}\exp(-\frac{r^2}{2d^2}+ik_0y), \tag{8b}$$

where coefficients c_1 and c_2 determine the initial pseudospin polarization. We suppose that the packet width d is much greater than the lattice period and consequently $\psi(\vec{r},0)$ is smooth enveloping function. We suppose also that the most of the states in valence band are unfilled, that corresponds to negative Fermi level located far from Dirac point (see also Ref. (Rusin & Zawadzki,2008)). Substituting Eqs.(8a, 8b) in Eq.(4) and using the expressions (6) and (7) we obtain

$$\psi_1(\vec{r},t) = \frac{1}{\sqrt{|c_1|^2+|c_2|^2}}(c_1\phi_1(\vec{r},t) - c_2\phi_2(-x,y,t)), \tag{9}$$

$$\psi_2(\vec{r},t) = \frac{1}{\sqrt{|c_1|^2+|c_2|^2}}(c_2\phi_1(\vec{r},t) + c_1\phi_2(\vec{r},t)), \tag{10}$$

where, for notational convenience, $\phi_{1,2}(\vec{r},t)$ denote the functions

$$\phi_1(\vec{r},t) = \int G_{11}(\vec{r},\vec{r'},t)f(\vec{r'},0)d\vec{r'} = \frac{de^{-(k_0d)^2/2}}{2\hbar^2\sqrt{\pi^3}}\times$$
$$\times\int \exp(i\frac{\vec{p}\vec{r}}{\hbar} - \frac{p^2d^2}{2\hbar^2} + \frac{p_yk_0d^2}{\hbar})\cos(\frac{upt}{\hbar})d\vec{p}, \tag{11}$$

$$\phi_2(\vec{r},t) = \int G_{21}(\vec{r},\vec{r'},t)f(\vec{r'},0)d\vec{r'} = \frac{-ide^{-(k_0d)^2/2}}{2\hbar^2\sqrt{\pi^3}}\times$$
$$\times\int \frac{p_x+ip_y}{p}\exp(i\frac{\vec{p}\vec{r}}{\hbar} - \frac{p^2d^2}{2\hbar^2} + \frac{p_yk_0d^2}{\hbar})\sin(\frac{upt}{\hbar})d\vec{p}. \tag{12}$$

Using the cylindrical coordinates in Eqs.(11), (12) and integrating over the angular variable, we have

$$\phi_1(\vec{r},t) = \frac{e^{-\frac{a^2}{2}}}{d\sqrt{\pi}} \int_0^\infty e^{-\frac{q^2}{2}} \cos(qt) J_0(q\sqrt{r^2 - a^2 - 2iay}) q dq, \tag{13}$$

$$\phi_2(\vec{r},\tau) = \frac{e^{-\frac{a^2}{2}}}{d\sqrt{\pi}} \frac{x + a + iy}{\sqrt{r^2 - a^2 - 2iay}} \int_0^\infty e^{-\frac{q^2}{2}} \sin(qt) J_1(q\sqrt{r^2 - a^2 - 2iay}) q dq, \tag{14}$$

where $J_0(z)$, $J_1(z)$ are Bessel functions. For the sake of convenience we introduce in Eqs.(13), (14) and everywhere below the dimensionless variables, measuring the distance in the units of initial width of wave packet d and time in d/u units. Besides, instead of the wave vector k_0 we consider the parameter $a = k_0 d$.

3. *Zitterbewegung* of Gaussian wave packet with different pseudospin polarization

Now we describe the time dynamics of Gaussian wave packets, in particular, the ZB phenomenon and the influence of the initial pseudospin polarization on the characteristics of trembling motion.

i). Following Ref. (Rusin & Zawadzki,2007) let us firstly consider the model problem when the lower component of initial wave function is equal to zero, i.e. the parameters $c_1 = 1, c_2 = 0$ in Eq.(8a). That means that at the initial moment of time the electron probability is located at the sites of the sublattice A. It is not difficult to show that this packet is formed by the states with positive and negative energies. The relative weight of these states is equal to one. The wave function for $t > 0$ can be found using Eqs.(9), (10):

$$\psi(\vec{r},t) = \begin{pmatrix} \phi_1(\vec{r},t) \\ \phi_2(\vec{r},t) \end{pmatrix}, \tag{15}$$

where the functions $\phi_1(\vec{r},t), \phi_2(\vec{r},t)$ are defined by Eqs.(13),(14).

In Fig.1 we represent the full electron density at the moment $t = 7$ for initial wave packet, Eq.(8b) with width $d = 2\ nm$ and $k_0 = 0.6\ nm^{-1}$. As one can see, at $t > 0$ this packet splits in two parts moving along y axis with opposite velocities so that the electron probability density is symmetrical with respect to y: $\rho(x,y,t) = \rho(x,-y,t)$. Note that at the case $k_0 = 0$ the electron probability density has a cylindrical symmetry at all time. Indeed it is easy to see that the initial wave function is the eigenstate of total "angular momentum" operator $\hat{I}_z = \hat{l}_z + \hbar/2\ \sigma_z$ ($\hat{l}_z = -i\hbar\partial/\partial\alpha$) $\hat{I}_z\Psi(\vec{r},0) = \hbar/2\ \Psi(\vec{r},0)$. Since I_z is a conserved quantity the wave function $\Psi(\vec{r},t) = (\Psi_1(\vec{r},t), \Psi_2(\vec{r},t))^T$ obeys this equation too. It follows that only lower component depends on α: $\Psi_2(\vec{r},t) \sim e^{i\alpha} f_2(\varrho,t)$ (see Eqs.(13)-15) for $k_0 = 0$). Thus the probability density remains an axially symmetric at $t > 0$. For enough large time the width of both parts of the packet with $k_0 \neq 0$ increases with time due to effect of dispersion. One can check that in this situation the contributions of two components of wave functions $\psi_1(\vec{r},t)$ and $\psi_2(\vec{r},t)$ in full electron density are equal. In other words the electron probability distributes with the time on the sides of sublattice A and B. Note at the same time $\rho(x,y,t) \neq \rho(-x,y,t)$ and the packet center oscillates along x direction (*Zitterbewegung*).

To analyze this motion we find the average value of position operator. To do it, we use the momentum representation. The upper $(C_1(\vec{p},t))$ and lower $(C_2(\vec{p},t))$ components of wave function (15) in this representation can be easily obtained from Eqs.(11), (12). After that the usual definition

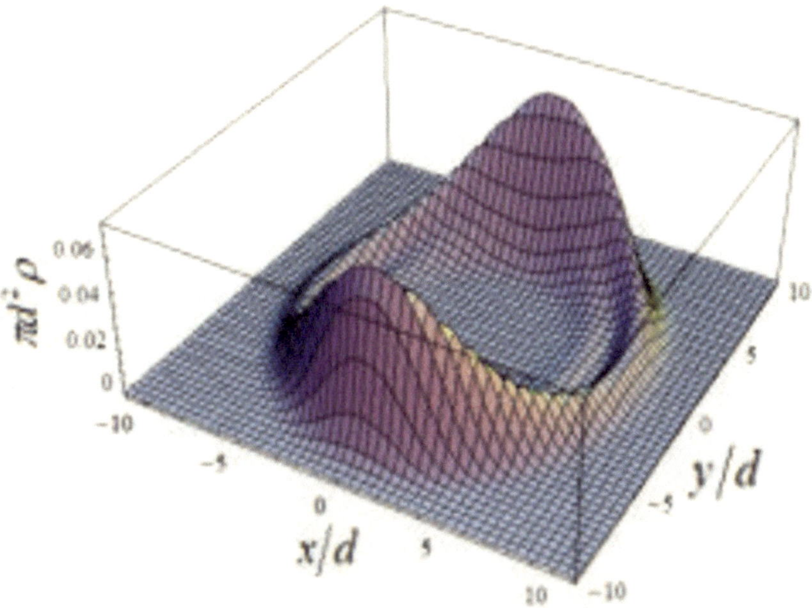

Fig. 1. (Color on line). The electron probability density $\rho(\vec{r},t) = |\psi_1|^2 + |\psi_2|^2$ for initial wave packet determined by Eqs.(8a), (8b) with $c_1 = 1$ and $c_2 = 0$ for $a = k_0 d = 1.2$ at the time $t = 7$ (in the units of d/u).

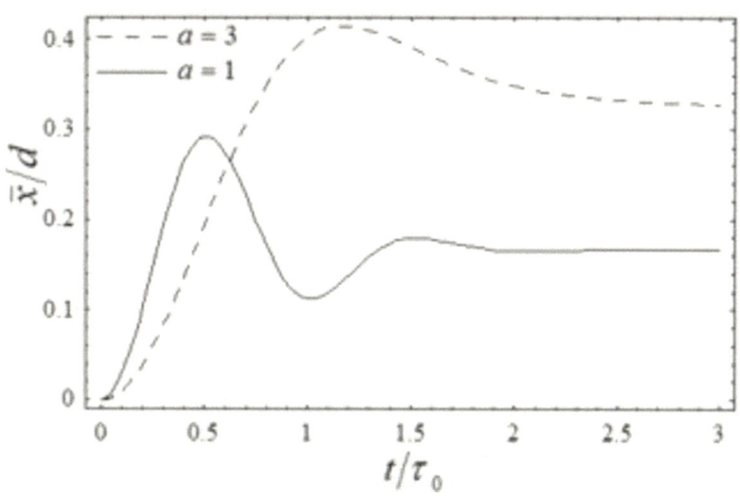

Fig. 2. The average coordinate $\bar{x}(t)$ versus time ($\tau_0 = d/u$) for the wave packet with initial pseudospin polarization along z axis for two values of a.

$$\bar{\vec{r}}(t) = \sum_{j=1}^{2} \int d\vec{p}\; C_j^*(\vec{p},t) i\hbar \frac{dC_j(\vec{p},t)}{d\vec{p}}, \tag{16}$$

readily yields

$$\bar{y}(t) = 0, \tag{17a}$$

$$\bar{x}(t) = \frac{1 - e^{-a^2}}{2a} - e^{-a^2} \int_0^\infty e^{-q^2} \cos(2qt) I_1(2aq) dq, \tag{17b}$$

where $I_1(z)$ is a modified Bessel function of the first order. In Eq.(17b) the integral term represents the *Zitterbewegung*. Note that average value $\bar{x}(t)$ depends on only one parameter a (in the dimensionless variables). The obtained functions $\bar{x}(t)$ which describes the typical transient *Zitterbewegung* are plotted in Fig.2. After the oscillation disappears the center of the packet is displaced by amount which equals to the first term Eq.(15). In the case when the wave packet width is large enough and the inequality $a = dk_0 \gg 1$ takes place, Eq.(17b) reduces to

$$\bar{x}(t) = \frac{1 - e^{-t^2} \cos(2at)}{2a}. \tag{18}$$

The details of the calculation of asymptotic formula similar to Eq. (18) are represented in Ref. (Demikhovskii et al.,2008).

As it follows from Eqs.(17), (18) for given initial polarization of wave packet the ZB occurs in the direction perpendicular to the initial momentum $p_{0y} = \hbar k_0$, just as for bilayer graphene (Rusin & Zawadzki,2007) and for the semiconductor quantum well in the presence of the Rashba spin-orbit coupling (Schliemann et al.,2005; Schliemann et al.,2006; Demikhovskii et al.,2008). One can see from Eq.(18) that the trembling motion has a transient character as it was described in Refs. (Rusin & Zawadzki,2007; Demikhovskii et al.,2008) and at $t \gg 1$ $x(t) \to 1/2a$. We should notice that Eqs.(17b), (18) coincide with corresponding formulas of Ref. (Demikhovskii et al.,2008). This is because the Hamiltonian for the system with Rashba-coupling

$$H_R = \frac{\hat{\vec{p}}^2}{2m} + \alpha(\hat{p}_y \hat{\sigma}_x - \hat{p}_x \hat{\sigma}_y), \tag{19}$$

where α is a Rashba coupling constant, transforms into Hamiltonian for monolayer graphene, Eq.(1), if we make the replacement in Eq.(19)

$$x \to -y', \quad y \to x', \quad \alpha \to u, \quad m \to \infty. \tag{20}$$

ii). Let us consider now the case when $c_1 = c_2 = 1$, that is pseudospin is directed along x axis at $t = 0$. Then from Eqs.(9), (10)

$$\psi(\vec{r}, t) = \frac{1}{\sqrt{2}} \begin{pmatrix} \phi_1(\vec{r}, t) - \phi_2(-x, y, t) \\ \phi_1(\vec{r}, t) + \phi_2(\vec{r}, t) \end{pmatrix}. \tag{21}$$

Fig.3 illustrates the corespondent electron probability density at the time moment $t = 7$ for initial wave packet, Eq.(8b), for the same parameters as in Fig.1. One can see that the initial wave packet at $t > 0$, as in previous case, splits into two parts propagating along y in opposite directions so that the symmetry concerning this axis, i.e. $\rho(x, y, t) = \rho(x, -y, t)$, retain during the time (as the case i)). The distribution of the probability density along x axis clearly demonstrates that its maximum is displaced in the positive direction that corresponds to the motion of the packet centre along x axis. The velocity of such motion $\bar{v}_x = \frac{d\bar{x}}{dt}$ consists

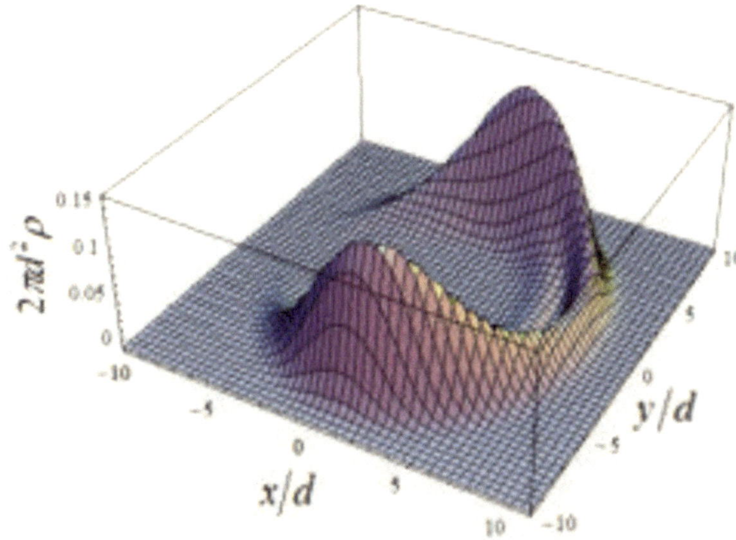

Fig. 3. (Color on line). The electron probability density $\rho(\vec{r},t) = |\psi_1|^2 + |\psi_2|^2$ for initial wave packet, Eqs.(8a), (8b) with $c_1 = 1$ and $c_2 = 1$ for $a = k_0 d = 1.2$ at the time $t = 7$ (in the units of d/u).

of both constant as well as oscillatory parts. Really, a straightforward calculation yields the average value of position operator x

$$\bar{x}(t) = \frac{1 - e^{-a^2}}{2a^2} t + \frac{e^{-a^2}}{2a} \int_0^\infty e^{-q^2} \sin(2qt) \frac{d}{dq} I_1(2aq) dq, \tag{22}$$

and $\bar{y}(t) = 0$ like for the case i). In Fig.4 we demonstrate the dependence $\bar{x}(t)$ for various values of parameter a. When the parameter a increases, the amplitude of ZB and the period of oscillations decrease. At $a \gg 1$ we have from Eq.(22)

$$\bar{x}(t) = \frac{t}{2a^2} + \frac{1}{2a} e^{-t^2} \sin(2at). \tag{23}$$

We see that the character of motion of wave packet is changed. Now the center of wave packet moves along x direction with constant velocity, which is determined by the first term in Eqs.(22), (23) and performs the damping oscillations. It is not difficult to find the constant component $\bar{\vec{V}}_0$ (in the units of u) of wave packet velocity for arbitrary initial wave function (8a) (see Appendix):

$$\bar{V}_{0x} = \frac{2}{|c_1|^2 + |c_2|^2} \int d\vec{p} |F(\vec{p})|^2 \left(\frac{p_x^2}{p^2} \mathrm{Re}(c_1^* c_2) + \frac{p_x p_y}{p^2} \mathrm{Im}(c_1^* c_2) \right), \tag{24}$$

$$\bar{V}_{0y} = \frac{2}{|c_1|^2 + |c_2|^2} \int d\vec{p} |F(\vec{p})|^2 \left(\frac{p_x p_y}{p^2} \mathrm{Re}(c_1^* c_2) + \frac{p_y^2}{p^2} \mathrm{Im}(c_1^* c_2) \right). \tag{25}$$

In particular for $F(\vec{p}) = F(-\vec{p})$ and $c_1 = c_2 = 1$ Eqs.(24),(25) give $\bar{V}_{x0} = 1/2$, $\bar{V}_{y0} = 0$. This result also follows from Eq.(22) when $k_0 = 0$ or for the narrow-width Gaussian wave packet ($a = k_0 d \to 0$). When the width of packet is increased the velocity of motion of its centre is

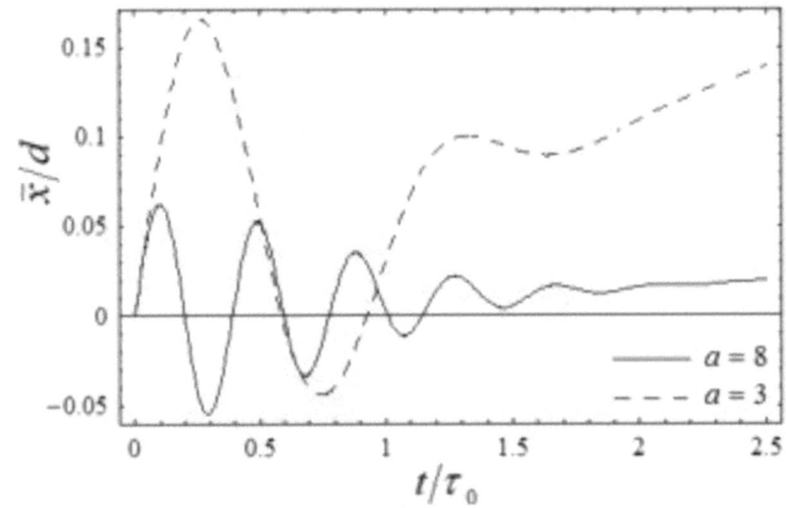

Fig. 4. The average coordinate $\bar{x}(t)$ versus time ($\tau_0 = d/u$) for the wave packet with initial pseudospin polarization along x axis for different values of a.

decreased. The frequency and amplitude of the *Zitterbewegung* for $a \gg 1$ are the same as in the case i). However, the first term in Eq.(22) corresponding to the motion of wave packet with constant velocity reduces the effect of ZB at least for $a \lesssim 1$ (Fig.4).

It is not difficult to show that as in the other two-band systems the phenomenon of ZB in graphene is a result of an interference of states corresponding to positive and negative eigenenergies of Hamiltonian, Eq.(1). For wide enough packet $a = k_0 d \gg 1$ and at time $t > 1$ when the ZB disappears two split parts of initial wave packet (see Fig. 3) move along y axis with opposite velocities $u/2$ and $-u/2$. In this situation the subpackets moving in the positive and negative directions consist of the states with positive and negative energies correspondingly.

iii). When the initial pseudospin is along y axis the wave function at $t > 0$ has the form

$$\psi(\vec{r},t) = \frac{1}{\sqrt{2}} \begin{pmatrix} \phi_1(\vec{r},t) - i\phi_2(-x,y,t) \\ i\phi_1(\vec{r},t) + \phi_2(\vec{r},t) \end{pmatrix}. \tag{26}$$

In Fig.5 the full electron density for the same moment of time and for the same parameters as in previous cases is shown. As one can see, the initial wave packet does not split into two parts at $t > 0$ unlike in the cases i) and ii). This result is confirmed by the straightforward calculations. Indeed, one can show that the eigenenergy states corresponding to propagation in the positive direction along y axis give the dominant contribution in total wave function, Eq.(26). For wide packets ($a \gg 1$) almost all of these states belong to the positive branch of energy.

The results of calculations of average values of x and y for this polarization lead to

$$\bar{x}(t) = 0, \tag{27}$$

$$\bar{y}(t) = (1 - \frac{1}{2a^2} + \frac{e^{-a^2}}{2a^2})\, t + \frac{e^{-a^2}}{2a} \int_0^\infty e^{-q^2} \sin(2qt) I_1(2aq) \frac{dq}{q}. \tag{28}$$

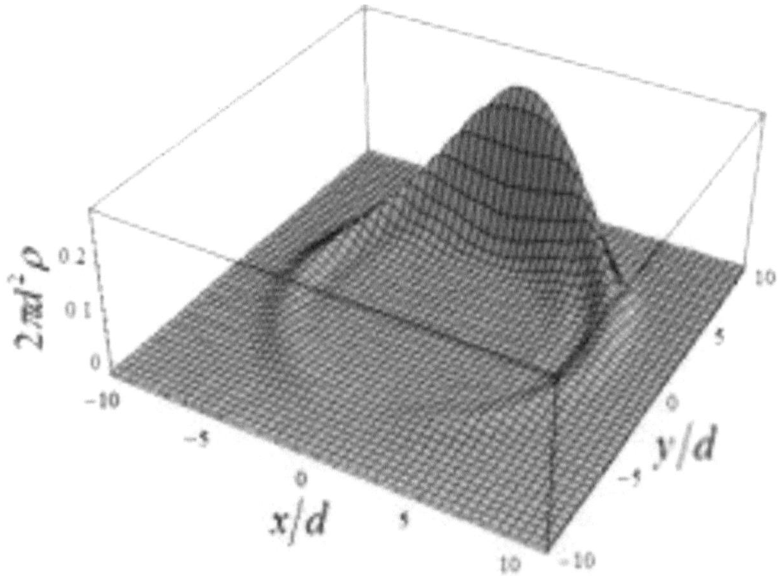

Fig. 5. (Color online). The electron probability density $\rho(\vec{r},t) = |\psi_1|^2 + |\psi_2|^2$ for initial wave packet, Eqs.(8a), (8b) with $c_1 = 1$ and $c_2 = i$ for $a = k_0 d = 1.2$ at time $t = 7$ (in the units of d/u).

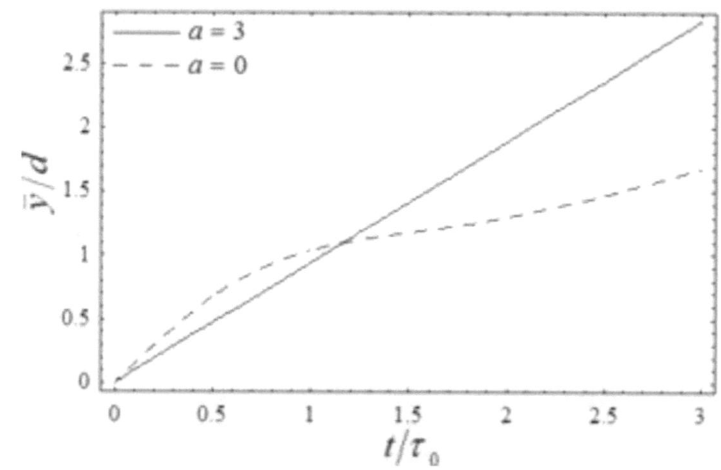

Fig. 6. The average coordinate $\bar{y}(t)$ versus time ($\tau_0 = d/u$) for the wave packet with initial pseudospin polarization along y axis for different values of a.

Thus in the considered case the wave packet propagates along y axis and the *Zitterbewegung* has the "longitudinal" character. It is interesting to note that $1D$ wave packet

$$\psi(y,0) = F(y)\begin{pmatrix} 1 \\ i \end{pmatrix},$$

where $F(y)$ is arbitrary function, propagates along y direction without changing of its form

$$\psi(y,t) = F(y-ut)\begin{pmatrix}1\\i\end{pmatrix}.$$

Note that in a numerical work (Thaller,2004) the author has observed similar oscillatory behavior of the expectation value of the position operator for one - dimensional relativistic electron in vacuum. In Ref.(Thaller,2004) it was also shown that apart from the rapid oscillation, the wave packet drifts slowly even when its average momentum is zero.

In Fig.6 we represent the dependence $\bar{y}(t)$ for different values of parameter a. As one can see, even at zero value of a the oscillations are absent. In this case, as it follows from Eqs.(25),(28) the drift velocity is equal to $1/2$ (in the units of u). As above, Eq.(28), takes more simple form at $a \gg 1$

$$\bar{y}(t) = t + \frac{1}{4a^3}e^{-t^2}\sin(2at). \tag{29}$$

Comparing Eqs.(18), (23), (29), we see that the amplitude for the "longitudinal" *Zitterbewegung* is much smaller than the amplitude of "transverse" *Zitterbewegung*. This fact can bee seen as a consequence of special form of the initial wave function, which in the given case consists of (at $a \gg 1$) the states with positive energy mostly. That makes the interference between the positive and negative components difficult, i.e. decreases the ZB. Moreover, at any values of the parameter a the integral term in Eq.(28), corresponding to the oscillating motion, is negligible in comparison with the first term, and one may neglect the effect of the "longitudinal" ZB.

As was demonstrated above, the direction of the average velocity depends not only on module of the components $\psi_1(\vec{r},0)$ and $\psi_2(\vec{r},0)$, but also on their phases. Specifically for the initial Gaussian packet

$$\psi(\vec{r},0) = \frac{f(\vec{r})}{\sqrt{2}}\begin{pmatrix}1\\e^{i\varphi}\end{pmatrix}, \tag{30}$$

the probability density becomes asymmetric and the average position operator has two components

$$\bar{\mathbf{r}}(t) = \bar{x}(t)\cos\varphi\,\vec{e}_x + \bar{y}(t)\sin\varphi\,\vec{e}_y, \tag{31}$$

where φ is an arbitrary phase difference between the up and low components of wave function and $\bar{x}(t)$, $\bar{y}(t)$ are determined by Eqs.(22), (28). For illustration we show in Fig.7 the electron probability density obtained for the initial packet, Eq.(30), with $\varphi = \pi/4$.

It is clear that the phase φ determines the direction of the average velocity of the packet center. Using the expression for velocity operator $\hat{\vec{v}} = u\vec{\sigma}$ and Eq.(30) we obtain (in the dimensionless variables) at $t = 0$:

$$\overline{v}_x(0) = \cos\varphi, \quad \overline{v}_y(0) = \sin\varphi. \tag{32}$$

The components of the velocity for a large enough time, when the trembling motion stops, can be found from Eqs.(22),(28) and (31) for arbitrary parameter a

$$\overline{v}_x = \frac{1-\exp(-a^2)}{2a^2}\cos\varphi,\ \overline{v}_y = \left(1 - \frac{1}{2a^2} + \frac{\exp(-a^2)}{2a^2}\right)\sin\varphi. \tag{33}$$

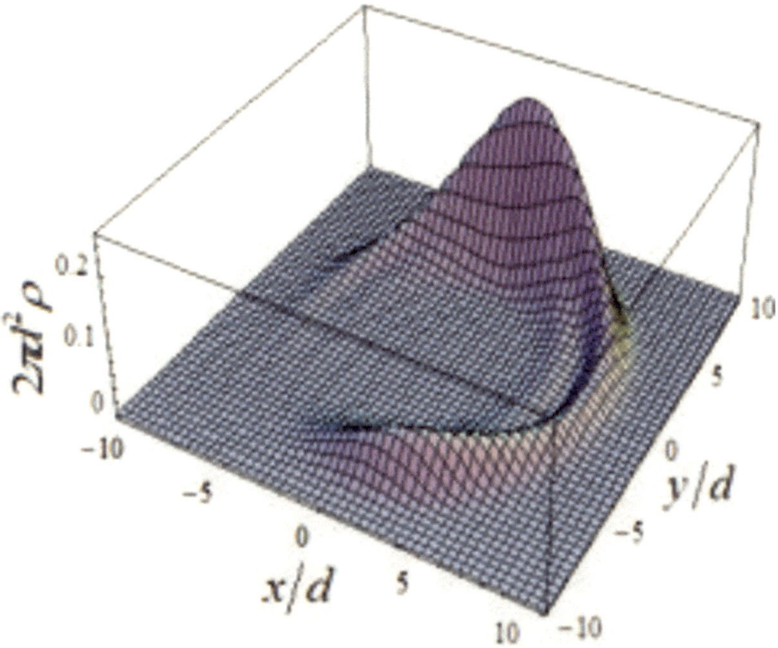

Fig. 7. (Color online). The electron probability density $\rho(\vec{r},t) = |\psi_1|^2 + |\psi_2|^2$ for initial wave packet, Eq.(8a), (8b) with $c_1 = 1$ and $c_2 = e^{i\pi/4}$ for $a = k_0 d = 1.2$ at the time $t = 7$ (in the units of d/u).

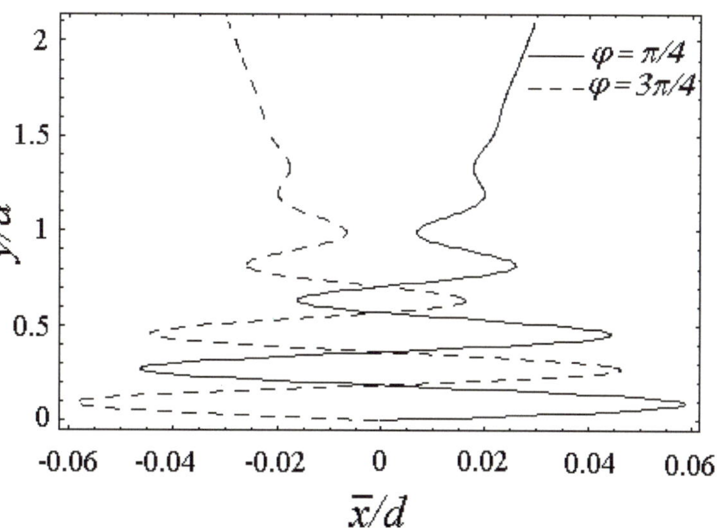

Fig. 8. The trajectories of the center of electron wave packet described by Eq.(30) for two initial phases $\varphi = \pi/4$ and $3\pi/4$. The parameter $a = 6$.

In particular, as it follows from Eq.(33) for $a \ll 1$, the direction of the motion of wave packet center at large time coincides with the initial one, Eq(32). In other limiting case $a \gg 1$ (and for not too small φ) asymptotic direction of the average velocity is along $0Y$ axis, i.e. along the

average momentum of wave packet $p_y = \hbar k_0$. Thus, by changing the initial phase φ, one can govern the packet motion and consequently the direction of dc current. To illustrate this, we plot in Fig.8 the packet center trajectories for two initial phases: $\varphi = \pi/4$ and $3\pi/4$. Note that the packet motion with the constant velocity predicted above (see Eqs.(22), (28)) should lead to the existence of the direct current in the corresponding direction.
To check our formalism let us consider the plane wave as the starting point. In this case it is easy to obtain the expression for the average value of electron velocity $\bar{\vec{v}}(t)$. Really, in the Heisenberg picture the kinetic velocity is (in the dimensional variables)

$$\hat{\vec{v}}(t) = \frac{1}{i\hbar}[\vec{r}, \hat{H}] = u\vec{\sigma}(t), \tag{34}$$

where

$$\frac{d\vec{\sigma}}{dt} = \frac{1}{i\hbar}[\vec{\sigma}, \hat{H}] = \frac{2u}{\hbar}[\hat{\vec{p}} \times \vec{\sigma}]. \tag{35}$$

In these equations $\vec{p}(t) = \vec{p}(0)$. Let the initial momentum $p_{oy} = \hbar k_0$. Then, using the solutions of Eqs.(34), (35) we find

$$\bar{v}_x(t) = u\bar{\sigma}_z(0)\sin\omega t + u\bar{\sigma}_x(0)\cos\omega t, \tag{36a}$$

$$\bar{v}_y(t) = u\bar{\sigma}_y(0), \tag{36b}$$

where $\omega = 2uk_0$ and $\sigma_i(0) = \sigma_i$ - Pauli matrixes ($i = 1, 2, 3$). So, if in the initial state pseudospin is along z direction, i.e. $\bar{\sigma}_z(0) = 1$ (case i)) we obtain from Eq.(34a) that $\bar{v}_x(t) = u\sin\omega t$ which leads to

$$\bar{x}(t) = const - \frac{u}{\omega}\cos\omega t. \tag{37}$$

Returning to the original variables in Eq.(18) and setting $d = \infty$ we see that this expression coincides with Eq.(37). We get similar results also for other initial polarizations.

4. Concluding remarks

We have studied the quantum dynamics of charge particles represented by Gaussian wave packets in two-dimensional single layer of carbon atoms (graphene). We investigated numerically also the spatial evolution of the initial wave packet and demonstrated the effect of the packet splitting for the pseudospin polarization perpendicular to the average momentum. The analytical expressions for the average values of position operators were obtained. These expressions clearly demonstrate that the evolution of wave function is accompanied by the *Zitterbewegung* and strongly depends on the initial pseudospin polarization. In particular, if the initial pseudospin polarization coincides with initial average momentum, the packet center moves and exhibits the ZB along the same direction. In this case the second term in Eq.(28) describing the longitudinal oscillations (the "longitudinal" ZB) is essentially smaller than the first one connected with the motion with constant velocity. As for other systems with two-band structure (Demikhovskii et al.,2008; Rusin & Zawadzki,2007; Thaller,2004), the ZB in monolayer graphene has a transient character.
It was also predicted that apart from the packet center exhibits the trembling motion it can move with constant velocity (for example, for the pseudospin polarization along x and y axis). The direction of this velocity depends on not only the orientation of average momentum $\vec{p}_0$,

but also on the module of the components $\psi_1(\vec{r},0)$, $\psi_2(\vec{r},0)$ and the differences of their phases (see Eqs.(30),(33)).
All above calculations have been done for the $\vec{K}$ point of the Brillouin zone in graphene. Similar results can be found for initial wave packet with wave vector $\vec{k}$ in the valley centered in inequivalent point $\vec{K}'$. The Dirac Hamiltonian around $\vec{K}'$ point can be written as

$$H_{K'} = u \begin{pmatrix} 0 & -\hat{p}_x - i\hat{p}_y \\ -\hat{p}_x + i\hat{p}_y & 0 \end{pmatrix}, \tag{38}$$

This expression can be obtained from Hamiltronian around $\vec{K}$ point given by Eq.(1) by replacement $\hat{p}_x \rightarrow -\hat{p}_x$. Other representation of Eq.(38) are common in the literature as well. In this connection see (Bena & Montambaux,2009; Gusynin et al.,2007). Thus values $\bar{x}(t)$ for the wave packet of different polarizations (and corresponding components of velocity) change sign while $\bar{y}(t)$ remain unchanged (see also(Rusin & Zawadzki,2008)).
In conclusion we would like to stress that the packet motion with the constant velocity (see Eqs.(22), (28)) leads to the appearance of the dc current. For the experimental detection of this current one needs sensitive current meters. Experimental observation of trembling motion is currently a more difficult task since it is necessary to use femtosecond techniques. (Zawadzki,2006; Rusin & Zawadzki,2007) If new methods of formation of wave packets with different pseudospin polarizations will be elaborated then their trajectories and spatial separations can be observed probably with the help of devices with quantum point contacts and gates (see for example (Castro et al.,2009)). The intresting experimental method allows an observation of ZB of electron in graphene exited by femtosecond laser pulse in the presence of magnetic field was proposed in Ref. (Rusin & Zawadzki,2009).

5. Appendix

Besides the rapid oscillations (ZB) the electron wave packet in graphene (as for other two-band systems) can drift with constant velocity $\vec{V}_0$ although its average momentum is zero. The existence of constant component in the wave packet center velocity depends on the form and symmetry of the initial wave function.
Let now find the drift velocity for the arbitrary initial state, Eq.(8a). At $t > 0$ the total wave function in the momentum space can be decomposed into positive- ($\Psi_+(\vec{p},t)$) and negative-energy ($\Psi_-(\vec{p},t)$) components.

$$\Psi(\vec{p},t) = \Psi_+(\vec{p},t) + \Psi_-(\vec{p},t) = a_+ U_+(\vec{p}) e^{-iE_+ t/\hbar} + l + a_- U_-(\vec{p}) e^{-iE_- t/\hbar}, \tag{A.1}$$

Using the expression (3) for the eigenspinors $U_\pm(\vec{p})$ we find the coefficients $a_\pm$

$$a_\pm = \frac{F(\vec{p})}{\sqrt{2(|c_1|^2 + |c_2|^2)}} (c_1 \pm \frac{(p_x - ip_y)}{p} c_2), \tag{A.2}$$

where $F(\vec{p})$ is to be determined from the Fourier expansion of coordinate wave function $f(\vec{r})$. Obviously the time-independent part $V_{\mu 0}$ of the density of velocity in the momentum space is defined as

$$V_{\mu 0}(\vec{p}) = \Psi_+^+(\vec{p},t) \hat{V}_\mu \Psi_+(\vec{p},t) + \Psi_-^+(\vec{p},t) \hat{V}_\mu \Psi_-(\vec{p},t), \tag{A.3}$$

where $\hat{V}_\mu = u\sigma_\mu$, $\mu = 1,2$ is velocity operator. The preceding expressions (A.1), (A.2), (A.3) immediately lead to the constant velocity of the wave packet center given in Eqs.(24), (25).

6. Acknowledgments

This work was supported by the Program of the Russian Ministry of Education and Science "Development of scientific potential of High education" (Project No. 2.1.1.2686) and Grant of Russian Foundation for Basic Research (No. 09-02-01241-a), and by the President of RF Grant for Young Researchers MK-1652.2009.2.

7. References

[Ferrari & Russo,1990] Ferrari, L. & Russo, G. (1990). Nonrelativistic *Zitterbewegung* in two-band systems.*Phys Rev. B*, Vol. 42, October 1990, 7454-7461.

[Shmueli et al.,1995] Shmueli, N.; Yacoby, A. & Imry, Y. (1995). Quantum Oscillatory Phenomena in a *p-n* Junction of Quantum Wells. *Europhys. Lett.*, Vol. 29, No. 9, March 1995, 711-716.

[Schliemann et al.,2005] Schliemann, J.; Loss, D. & Westervelt, R., M. (2005). *Zitterbewegung* of Electronic Wave Packets in III-V Zinc-Blende Semiconductor Quantum Wells. *Phys. Rev. Lett.*, Vol. 94, May 2005, 206801.

[Schliemann et al.,2006] Schliemann, J.; Loss, D. & Westervelt, R., M. (2006). *Zitterbewegung* of electrons and holes in III-V semiconductor quantum wells. *Phys. Rev. B*, Vol. 73, February 2006, 085323.

[Schliemann,2008] Schliemann, J. (2005). Cyclotron motion and magnetic focusing in semiconductor quantum wells with spin-orbit coupling. *Phys. Rev. B*, Vol. 77, March 2008, 125303.

[Zawadzki,2005] Zawadzki, W. (2005). *Zitterbewegung* and its effects on electrons in semiconductors. *Phys. Rev. B*, Vol. 72, August 2005, 085217.

[Zawadzki,2006] Zawadzki, W. (2006). One-dimensional semirelativity for electrons in carbon nanotubes. *Phys. Rev. B*, Vol. 74, November 2006, 205439.

[Katsnelson,2006] Katsnelson, M., I. (2006). *Zitterbewegung*, chirality, and minimal conductivity in graphene . *Europ. Phys. J. B*, Vol. 51, No. 2, May 2006, 157-160.

[Cserti & David,2006] Cserti, J. & David, G. (2006). Unified description of *Zitterbewegung* for spintronic, graphene, and superconducting systems. *Phys. Rev. B*, Vol. 74, November 2006, 172305.

[Winkler et al.,2007] Winkler, R.; Zulicke, U. & Bolter, J. (2007). Oscillatory multiband dynamics of free particles: The ubiquity of *Zitterbewegung* effects. *Phys. Rev. B*, Vol. 75, May 2007, 205314.

[Trauzettel et al.,2007] Trauzettel, B.; Blanter, Y., M. & Morpurgo, A., F. (2007). Photon-assisted electron transport in graphene: Scattering theory analysis. *Phys. Rev. B*, Vol. 75, January 2007, 035305.

[Rusin & Zawadzki,2007] Rusin, T., M. & Zawadzki, W. (2007). *Zitterbewegung* of nearly-free and tightly-bound electrons in semiconductors. *J. Phys. Cond. Matter*, Vol. 19, No.13, March 2007, 136219.

[Rusin & Zawadzki,2007] Rusin, T., M. & Zawadzki, W. (2007). Transient *Zitterbewegung* of charge carriers in mono- and bilayer graphene, and carbon nanotubes. *Phys. Rev. B*, Vol. 76, November 2007, 195439.

[Rusin & Zawadzki,2008] Rusin, T., M. & Zawadzki, W. (2008). *Zitterbewegung* of electrons in graphene in a magnetic field. *Phys. Rev. B*, Vol. 78, September 2008, 125419.

[Demikhovskii et al.,2008] Demikhovskii, V.,Ya.; Maksimova, G., M. & Frolova, E., V. (2008). Wave packet dynamics in a two-dimensional electron gas with spin orbit coupling: Splitting and *Zitterbewegung*. *Phys. Rev. B*, Vol. 78, September 2008, 115401.

[Schrödinger,1930] Schrödinger, E. (1930). Über die kräftefreie Bewegung in der relativistischen Quantenmechanik. *Sitzungsber. Peuss. Akad. Wiss. Phys. Math. Kl.*, Vol. 24, 1930, 418-428.

[Barut & Bracken,1981] Barut, A., O. & Bracken, A., J. (1981). *Zitterbewegung* and the internal geometry of the electron.*Phys. Rev. D*, Vol. 23, May 1981, 2454-2463.

[Lock,1979] Lock, J., A. (1979). The *Zitterbewegung* of a free localized Dirac particle. *Am. J. Phys.*, Vol. 47, September 1979, 797-802.

[Thaller,2004] Thaller, B. (2004). Visualizing the kinematics of relativistic wave packets. arXiv: quant-ph/0409079v1 , September 2004, 1-7.

[Braun et al.,1999] Braun, J., W.; Su, Q. & Grobe, R. (1999). Numerical approach to solve the time-dependent Dirac equation. *Phys. Rev. A*, Vol. 59, January 1999, 604-612.

[Demikhovskii et al.,2010] Demikhovskii,V.,Ya.; Maksimova,G., M.; Perov, A., A. & Frolova,E., V. (2010). Space-time evolution of Dirac wave packets. Phys.Rev A, Vol.82, November 2010, 052115.

[Gerritsma et al.,2010] Gerritsma, R.; Kirchmair, G.; Zähringer, F.; Solano, E.; Blatt, R. & Roos, C., F. (2010). Quantum simulation of the Dirac equation. *Nature*, Vol. 463, January 2010, 68-71.

[Lamata et al.,2007] Lamata, L.; Leon, J.; Schätz, T. & Solana, E. (2007). Dirac Equation and Quantum Relativistic Effects in a Single Trapped Ion. *Phys. Rev. Lett.*, Vol. 98, June 2007, 253005.

[Maksimova et al.,2008] Maksimova, G., M.; Demikhovskii, V., Ya. & Frolova, E., V. (2008). Wave packet dynamics in a monolayer graphene. *Phys. Rev. B*, Vol. 78, December 2008, 235321.

[Wallace,1947] Wallace, P., R. (1947). The Band Theory of Graphite. *Phys. Rev.*, Vol. 71, May 1947, 622-634.

[Slonczewski & Weiss,1958] Slonczewski, J., C. & Weiss, P., R. (1958). Band Structure of Graphite. *Phys. Rev.*, Vol. 109, January 1958, 272-279.

[Novoselov et al.,2005] Novoselov, K., S.; Geim, A., K.; Morozov, S., V.; Jiang, D., Katsnelson, M., I.; Grigorieva, I., V.; Dubonos, S., V. & Firsov, A., A. (2005). Two-dimensional gas of massless Dirac fermions in graphene. *Nature*, Vol. 438, November 2005, 197-200.

[Novoselov et al.,2006] Novoselov, K., S.; Cann, E., Mc.; Morozov, S., V.; Falko, V., I.; Katsnelson, M., I.; Zeitler, U., Yiang, D.; Schedin, F. & Geim, A., K. (2006). Unconventional quantum Hall effect and Berry's phase of 2n in bilayer graphene. Nature Phys., Vol. 2, February 2006, 177-180.

[Zhang et al.,2005] Zhang, Y.; Tan, Y., W.; Stormer, H., L. & Kim, P. (2005). Experimental observation of the quantum Hall effect and Berry's phase in graphene. *Nature*, Vol. 438, November 2005, 201-204.

[Rusin & Zawadzki,2008] Rusin, T., M. & Zawadzki, W. (2008). *Zitterbewegung* of electrons in graphene in a magnetic field. Phys. Rev. B, Vol. 78, September 2008, 125419.

[Bena & Montambaux,2009] Bena, C. & Montambaux, G. (2009). Remarks on the tight-binding model of graphene. New J. Phys., Vol. 11, September 2009, 095003.

[Gusynin et al.,2007] Gusynin, V., P.; Sharapov, S., G. & Carbotte, J., P. (2007). AC conductivity of graphene: from tight-binding model to 2+1-dimensional quantum electrodynamics. *Int.J.Mod.Phys.B*, Vol. 21, No. 27, October 2007, 4611-4658.

[Castro et al.,2009] Castro Neto, A., H.; Guinea, F., Peres, N., M.; Novoselov, K., S. & Geim, A., K. (2009). The electronic properties of graphene.*Rev. Mod. Phys.*, Vol. 81, January 2009, 109-162.

[Rusin & Zawadzki,2009] Rusin, T., M. & Zawadzki, W. (2009). Theory of electron *Zitterbewegung* in graphene probed by femtosecond laser pulses. *Phys. Rev. B*, Vol. 80, July 2009, 045416.

23

Graphene Field Effect Transistors: Diffusion-Drift Theory

G.I. Zebrev
Department of Micro and Nanoelectronics,
National Research NuclearUniversity,
"MEPHI", Moscow,
Russia

1. Introduction

Recently discovered stable monoatomic carbon sheet (graphene) which is comprised of field-effect structures has remarkable physical properties promising nanoelectronic applications (Novoselov, 2004). Practical semiconductor device simulation is essentially based on diffusion-drift approximation (Sze & Ng, 2007). This approximation remains valid for graphene field-effect transistors (GFET) due to unavoidable presence of scattering centers in the gate or the substrate insulators and intrinsic phonon scattering (Ancona, 2010). Traditional approaches to field-effect transistors modeling suffer from neglect of the key and indispensible point of transport description - solution of the continuity equation for diffusion-drift current in the channels. This inevitably leads to multiple difficulties connected with the diffusion current component and, consequently, with continuous description of the I-V characteristics on borders of operation modes (linear and saturation, subthreshold and above threshold regions). Many subtle and/or fundamental details (difference of behaviour of electrostatic and chemical potentials, specific form of the Einstein relation in charge-confined channels, compressibility of 2D electron system, etc.) are also often omitted in device simulations. Graphene introduces new peculiar physical details (specific electrostatics, crucial role of quantum capacitance etc.) demanding new insights for correct modeling and simulation (Zebrev, 2007). The goal of this chapter is to develop a consequent diffusion-drift description for the carrier transport in the graphene FETs based on explicit solution of current continuity equation in the channels (Zebrev, 1990) which contains specific and new aspects of the problem. Role of unavoidable charged defects near or at the interface between graphene and insulated layers will be also discussed.

Distinguishing features of approach to GFET operation modeling will be:

- diffusion-drift approach;
- explicit solution of current continuity equation in GFET channels;
- key role of quantum capacitance in the diffusion to drift current ratio and transport in GFETs;
- role of rechargeable near-interfacial defects and its influence on small-signal characteristics of GFETs.

2. General background

2.1 Carrier statistics in ideal graphene for nonzero temperature

The density of states is the number of discrete eigenenergy levels within a unit energy width per unit area (states/eV cm²). Taking into account valley and spin as well as angular degeneracy we have for two-dimensional density of states $g_{2D}(\varepsilon)$ in graphene

$$g_{2D}(\varepsilon)d\varepsilon = 4\frac{dp_x dp_y}{(2\pi\hbar)^2} = \frac{4}{(2\pi\hbar)^2}\frac{\varepsilon}{v_0^2}2\pi\, d\varepsilon\ , \tag{1}$$

and specifically for gapless graphene dispersion law $\varepsilon = v_0\sqrt{p_x^2 + p_y^2}$

$$g_{2D}(\varepsilon) = \frac{2\varepsilon}{\pi\hbar^2 v_0^2}\operatorname{sgn}\varepsilon = \frac{2|\varepsilon|}{\pi\hbar^2 v_0^2}\ , \tag{2}$$

where $\triangleq$ is the Plank constant, v_0 ($\cong 10^8$ cm/s) is the characteristic (Fermi) velocity in graphene. Using the equilibrium Fermi-Dirac function $f_{FD}(\varepsilon - \mu)$ the electron density per unit area n_e at a given chemical potential μ for nonzero temperature T reads

$$\begin{aligned} n_e(\mu) &= \int_0^{+\infty} d\varepsilon\, g_{2D}(\varepsilon) f_{FD}(\varepsilon - \mu) = \\ &= \frac{2(k_B T)^2}{\pi\hbar^2 v_0^2}\int_0^{+\infty} du \frac{u}{1 + \exp\left(u - \frac{\mu}{k_B T}\right)} = -\frac{2}{\pi}\left(\frac{k_B T}{\hbar v_0}\right)^2 Li_2\left(-e^{\frac{\mu}{k_B T}}\right), \end{aligned} \tag{3}$$

where T is absolute temperature, k_B is the Boltzmann constant, $Li_n(x)$ is the poly-logarithm function of n-th order (Wolfram, 2003)

$$Li_n(z) = \sum\nolimits_{k=1}^{\infty} z^k / k^n \tag{4}$$

Using electron-hole symmetry $g(\varepsilon) = g(-\varepsilon)$ we have similar relationship for the hole density n_h

$$n_h(\mu) = \int_{-\infty}^{0} d\varepsilon\, g_{2D}(\varepsilon)\left(1 - f_{FD}(\varepsilon - \mu)\right) = -\frac{2}{\pi}\left(\frac{k_B T}{\hbar v_0}\right)^2 Li_2\left(-e^{-\frac{\mu}{k_B T}}\right) = n_e(-\mu)\,. \tag{5}$$

Full charge density per unit area or the charge imbalance reads as

$$n_S \equiv n_e - n_h = \int_0^{+\infty} d\varepsilon\, g_{2D}(\varepsilon)\left(f(\varepsilon - \mu) - f(\varepsilon + \mu)\right) = \frac{2}{\pi}\left(\frac{k_B T}{\hbar v_0}\right)^2\left(Li_2\left(-e^{-\frac{\mu}{k_B T}}\right) - Li_2\left(-e^{\frac{\mu}{k_B T}}\right)\right). \tag{6}$$

Conductivity of graphene charged sheet is determined by the total carrier density

$$N_S = n_e + n_h = -\frac{2}{\pi}\left(\frac{k_B T}{\hbar v_0}\right)^2 \left(Li_2\left(-e^{-\frac{\mu}{k_B T}}\right) + Li_2\left(-e^{\frac{\mu}{k_B T}}\right)\right). \quad (7)$$

For ideal electrically neutral graphene without any doping (so called the charge neutrality point (NP) with the zero chemical potential $\mu = 0$) we have intrinsic density with equal densities of electrons and holes

$$N_S(\mu=0) \equiv 2n_i = -\frac{2}{\pi}\left(\frac{k_B T}{\hbar v_0}\right)^2 2Li_2(-1) = -\frac{2}{\pi}\left(\frac{k_B T}{\hbar v_0}\right)^2 2\sum_{k=1}^{\infty}\frac{(-1)^k}{2^k} = \frac{\pi}{3}\left(\frac{k_B T}{\hbar v_0}\right)^2. \quad (8)$$

Intrinsic carrier density at room temperature T = 300K is estimated to be of order $n_i \cong 8\times10^{10}$ cm^{-2} (slightly larger than in silicon). The Tailor series expansion in the vicinity of the $\mu = 0$

$$-Li_2\left(-e^{\frac{\mu}{k_B T}}\right) \cong \frac{\pi^2}{12} + \ln 2\frac{\mu}{k_B T} + \frac{\mu^2}{4(k_B T)^2} \quad (9)$$

yields a good approximation for only $\mu < 5k_B T$. It is convenient to use a following asymptotics

$$-Li_2\left(-e^z\right) \cong \frac{\pi^2}{12} + \frac{z^2}{2}, \; z >> 1; \quad (10)$$

$$-Li_2\left(-e^{\frac{\mu}{k_B T}}\right) \cong \frac{\pi^2}{12} + \frac{\mu^2}{2(k_B T)^2}; \quad (11)$$

$$n_e(\mu) \cong \frac{2}{\pi}\left(\frac{k_B T}{\hbar v_0}\right)^2\left(\frac{\pi^2}{12} + \frac{\mu^2}{2(k_B T)^2}\right) = \frac{\pi}{6}\left(\frac{k_B T}{\hbar v_0}\right)^2 + \frac{\mu^2}{\pi\hbar^2 v_0^2} = n_i + \frac{\mu^2}{\pi\hbar^2 v_0^2}. \quad (12)$$

This approximation yields both exact expression for electron charge concentration at the charge neutrality point and the correct asymptotics for $\mu >> k_B T$ aw well as good coincidence in the intermediate region $\mu \sim k_B T$. In spite of this fact this approximation is inappropriate for capacitance calculation at zero chemical potential point due to lack of linear terms in μ. In reality the region near the $\mu \sim 0$ should not be considered to be ideal because of inevitable disorder presence (Martin & Akerman, 2008).
The channel electron density per unit area for degenerate system ($\mu >> k_B T$) reads

$$n_S \cong \int_0^{\mu} d\varepsilon \, g_{2D}(\varepsilon) \simeq \frac{\mu^2}{\pi\hbar^2 v_0^2} \quad (13)$$

2.2 Quantum capacitance in graphene
Performing explicit differentiation of Eqs.(3,5) one reads

$$\frac{dn_e}{d\mu} = \frac{2}{\pi}\frac{k_BT}{\hbar^2 v_0^2}\ln\left(1+\exp\left(\frac{\mu}{k_BT}\right)\right), \quad \frac{dn_h}{d\mu} = -\frac{2}{\pi}\frac{k_BT}{\hbar^2 v_0^2}\ln\left(1+\exp\left(-\frac{\mu}{k_BT}\right)\right). \tag{14}$$

Exact expression for quantum capacitance (Luryi, 1988) of the graphene charge sheet may be defined as

$$C_Q \equiv e^2 \int_{-\infty}^{+\infty} g(\varepsilon)\left(-\frac{\partial f_0}{\partial \varepsilon}\right)d\varepsilon = \frac{ed(n_e - n_h)}{d\mu} = \frac{2}{\pi}\left(\frac{e^2}{\hbar v_0}\right)\frac{k_BT}{\hbar v_0}\ln\left(2+2\cosh\left(\frac{\mu}{k_BT}\right)\right). \tag{15}$$

Quantum capacitance for unbiased case (μ = 0) becomes formally exact ideal form

$$C_{Q\min} = \frac{2\ln 4}{\pi}\left(\frac{e^2}{\hbar v_0}\right)\frac{k_BT}{\hbar v_0}. \tag{16}$$

For a relatively high doping case ($|\mu| >> k_BT$) we have *approximate* relation for quantum capacitance

$$C_Q \cong e^2 \frac{dn_S}{d\mu} = \frac{2}{\pi}\left(\frac{e^2}{\hbar v_0}\right)\frac{|\mu|}{\hbar v_0} \tag{17}$$

For total density of free carriers we have relationship, which is valid for any μ

$$\frac{d(n_e + n_h)}{d\mu} = \frac{2}{\pi}\frac{|\mu|}{\hbar^2 v_0^2}. \tag{18}$$

In contrast to Eq. 17 the latter Eq.18 can be considered as an *exact* for ideal graphene for any chemical potential result connected to an exact form of the Einstein relation.

2.3 Einstein relation in graphene

Similar to the silicon MOSFETs, the transport properties of graphene are determined by scattering from the charged defects in the gate insulating oxide and from elastic (at least in low-field region) phonons (Das Sarma et al., 2010). The diffusion constant in 2D graphene sheet can be determined through the Fermi velocity v_0 and transport relaxation time τ_{tr} or mean free path $\ell = v_0\tau_{tr}$

$$D = \frac{1}{2}v_0^2\tau_{tr} = \frac{1}{2}v_0\ell. \tag{19}$$

Electron and hole mobility $\mu_{e/h}$ can be inferred from the Einstein relation in a following manner (e = $|e|$)

$$\mu_{e/h} = \frac{eD_{e/h}}{n_{e/h}}\frac{dn_{e/h}}{d\mu} \equiv \frac{eD_{e/h}}{\varepsilon_D}, \tag{20}$$

where a diffusion energy introduced (Ando et al. 1982)

$$\varepsilon_D \equiv n_{e/h}\Big/\left(dn_{e/h}/d\mu\right). \tag{21}$$

It is easy to show from Eq. 13 that rather far from the graphene charge neutrality point we have $\varepsilon_D = \varepsilon_F/2$. Bipolar conductivity is expressed formally with Eq.(20) through the sum of electron and hole components

$$\sigma_0 = e\mu_e n_e + e\mu_p n_p = e^2\left(D_e \frac{dn_e}{d\mu} + D_h \frac{dn_h}{d\mu}\right). \tag{22}$$

Using the exact Eq. 18 and the assumption of electron-hole symmetry ($D_e = D_h = D_0$), the total bipolar conductivity reads

$$\sigma_0 = e^2 D_0 \frac{d(n_e + n_p)}{d\mu} = \frac{2e^2}{h}\frac{\varepsilon_F \tau_{tr}}{\hbar} = \frac{2e^2}{h} k_F \ell, \tag{23}$$

where the Fermi wavevector is defined through the dispersion law in gapless graphene $\hbar v_0 k_F = \mu \cong \varepsilon_F$. The Einstein relation can be rewritten in an equivalent form via conductivity and quantum capacitance

$$D_0 C_Q = e\mu_0 N_S = \sigma_0 \tag{24}$$

The Einstein relation allows to easily obtain a relation for mobility of graphene carriers in highly doped ($|\mu| >> k_B T$) graphene

$$\mu_0 = \frac{e v_0 \tau_{tr}}{p_F} = \frac{e\ell}{p_F}. \tag{25}$$

Notice that in fact $\ell \propto p_F$ and μ_0 weakly depends on Fermi energy in graphene.

3. GFET electrostatics

3.1 Near-interfacial rechargeable oxide traps

It is widely known (particularly, from silicon-based CMOS practice) that the charged oxide defects inevitably occur nearby the interface between the insulated layers and the device channel. Near-interfacial traps (defects) are located exactly at the interface or in the oxide typically within 1-3 nm from the interface. These defects can have generally different charge states and capable to be recharged by exchanging carriers (electrons and holes) with device channel. Due to tunneling exchange possibility the near-interfacial traps sense the Fermi level position in graphene. These rechargeable traps tend to empty if their level ε_t are above the Fermi level and capture electrons if their level are lower the Fermi level.

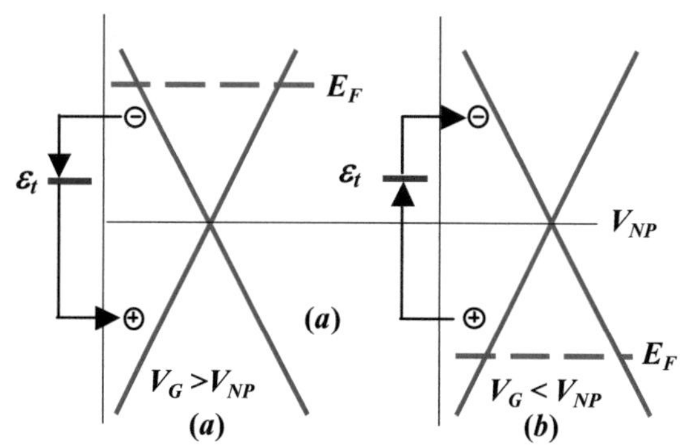

Fig. 1. Illustration of carrier exchange between graphene and oxide defects (a) filling; (b) emptying

There are two types of traps – donors and acceptors. Acceptor-like traps are negatively charged in a filled state and neutral while empty (- /0). Donor-like traps are positively charged in empty state and neutral in filled state (0/+). In any case, the Fermi level goes down with an increase V_G and the traps begin filled up, i.e. traps become more negatively charged (see Fig. 1). Each gate voltage corresponds to the respective position of the Fermi level at the interface with own "equilibrium" filling and with the respective density of equilibrium trapped charge $Q_t(\mu) = eN_t(\mu)$ which is assumed to be positive for definiteness. For traps with small recharging time the equilibrium with the substrate would establish fast. These traps rapidly exchanged with the substrate are often referred as to the interface traps (N_{it}) (Emelianov et al. 1996); (Fleetwood et al., 2008). Defects which do not have time to exchange charge with the substrate during the measurement time are referred to as oxide-trapped traps (N_{ot}). Difference between the interface and oxide traps is relative and depends, particularly, on the gate voltage sweep rate and the measurement's temperature. Interface trap capacitance per unit area C_{it} may be defined in a following way

$$C_{it} \equiv \frac{d}{d\mu}\left(-eN_t(\mu)\right) > 0 . \tag{26}$$

Note that the Fermi level dependent $eN_t(\mu)$ contains the charge on all traps, but for a finite voltage sweep time t_s only the "interface traps" with low recharging time constants $\tau_r < t_s$ contribute to the recharging process. Interface trap capacitance (F/cm^2) with accuracy up to the dimensional factor represents the energy density of the defect levels D_{it} (cm^{-2}eV^{-1}). It is easy to see that these values are related as

$$C_{it} = e^2 D_{it}(\mu) . \tag{27}$$

It is useful to note that 1 fF/μm^2 ≅ 6.25 × 10^{11} cm^{-2} eV^{-1}. The typical interface trap capacitance in modern silicon MOSFETs lies within the range D_{it} ~10^{11} -10^{12} cm^{-2} eV^{-1} and is rather sensitive (especially for thick (> 10 nm) insulated layers) to ionizing radiation impact (Fleetwood et al., 2008).

3.2 Electrostatics of graphene gated structures

Let us consider the simplest form of the gate-insulator-graphene (GIG) structure representing the two-plate capacitor capable to accumulate charges of the opposite signs. Without loss of generality we will reference the chemical potential in graphene from the level of charge neutrality E_{NP}. Electron affinity (or work function for Dirac point) of graphene with the reference of the vacuum energy level E_{vac} can be defined as

$$\chi_g = E_{vac} - E_{NP} . \tag{28}$$

Note that the graphene work function is of order of χ_g ~ 4.5 eV (Giovannetti et al., 2008). It is well known that voltage bias between any device's nodes is equivalent to applying of electrochemical potential bias. There are generally at least two contributions to the electrochemical potential

$$\mu = \zeta + U = \zeta - e\varphi \tag{29}$$

where ζ is proper electric charge independent chemical potential, U and φ are the electrostatic energy and potential $U = -e\varphi$. Neglecting voltage drop in the gate made routinely of good 3D conductors due to its extremely large quantum capacitance per unit area we get

$$\mu_{gate} = -e\varphi_{gate} - W_{gate}, \tag{30}$$

$$\mu_{graphene} = -\chi_g + \zeta - e\varphi_{graphene} = E_{NP} + \zeta, \tag{31}$$

where $\varphi_{graphene}$ is electrostatic potential of graphene sheet, W_{gate} is the work function of the gate material, and $E_{NP} = -\chi_g - e\varphi_{graphene}$ is the energy position of the charge neutrality (or, Dirac) point. Applying the gate voltage (to say, positive) with reference of grounded graphene plate we increase the chemical potential and electrostatic potential of the graphene sheet so as they exactly compensate each other keeping the electrochemical potential of the graphene sample unchanged (see Fig. 2).

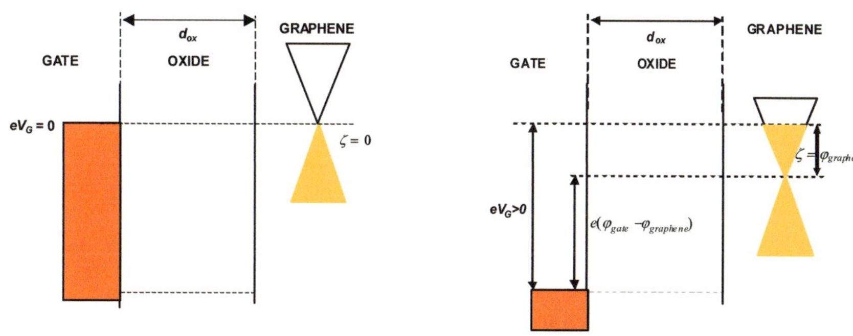

Fig. 2. Band diagram of gate-oxide- graphene structure at $V_G = 0$ (left) and $V_G > 0$ (right). Here, φ_{gg} =0, for simplicity.

Particularly, the electrical bias between the metallic (or almost metallic) gate and the graphene sample is equal to a difference between the electrochemical potentials in graphene ($\mu_{graphene}$)and the gate (μ_{gate})

$$eV_G = \mu_{graphene} - \mu_{gate} = e\varphi_{gg} + \zeta + e\left(\varphi_{gate} - \varphi_{graphene}\right). \tag{32}$$

where $e\varphi_{gg} \equiv W_{gate} - \chi_g$ is the work function difference between the gate and graphene. For zero oxide charge (or, for charged oxide defects located nearly the insulator-graphene interface) the electric field E_{ox} is uniform across the gate thickness (d_{ox}) and one reads

$$\varphi_{gate} - \varphi_{graphene} = E_{ox}\, d_{ox} = \frac{eN_{gate}}{\varepsilon_{ox}\varepsilon_0} d_{ox} \equiv \frac{eN_{gate}}{C_{ox}}, \tag{33}$$

where $N_{gate}(V_G)$ is the number of charge carriers on the metallic gate per unit area and the oxide (insulator) capacitance per unit area C_{ox} expressed through the dielectric constants of the insulator (ε_{ox}) is defined as

$$C_{ox} = \frac{\varepsilon_{ox}\varepsilon_0}{d_{ox}}. \tag{34}$$

3.3 Characteristic scales of gated graphene

The planar electric charge neutrality condition for the total gated structure can be written down as follows

$$N_G + N_t = n_S \,, \tag{35}$$

where N_G is the number of positive charges per unit area on the gate; n_S is the charge imbalance density per unit area (n_S may be positive or negative -), N_t is the defect density per unit area which is assumed to be positively charged (see Fig.3). Then total voltage drop (Eq.32) across the structure becomes modified as

$$eV_G = e\varphi_{gg} + e\varphi + \frac{e^2}{C_{ox}}\left(n_S(\zeta) - N_t(\zeta)\right). \tag{36}$$

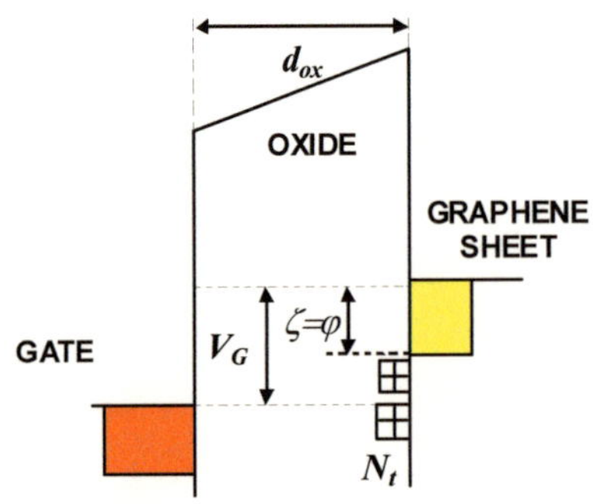

Fig. 3. Band diagram of graphene FET.

The voltage corresponding the electric charge neutrality point gate V_{NP} is defined in a natural way

$$V_{NP} \equiv V_G(\zeta = 0) = \varphi_{gg} - \frac{eN_t(\zeta = 0)}{C_{ox}}. \tag{37}$$

Chemical potential is positive (negative) at $V_G > V_{NP}$ ($V_G < V_{NP}$). Then we have

$$e(V_G - V_{NP}) = \zeta + \frac{e^2 n_S}{C_{ox}} + \frac{e^2\left(N_t(\zeta = 0) - N_t(\zeta)\right)}{C_{ox}}. \tag{38}$$

Taking for brevity without loss of generality V_{NP} =0 and assuming zero interface trap charge at the NP point as well as constant density of trap states we have

$$e^2\left(N_t(\zeta = 0) - N_t(\zeta)\right) \cong C_{it}\zeta. \tag{39}$$

Taking into account Eq.13 the basic equation of graphene planar electrostatics can be written down a in a form

$$eV_G = \varepsilon_F + \frac{e^2 n_S}{C_{ox}} + \frac{C_{it}}{C_{ox}}\varepsilon_F \equiv m\varepsilon_F + \frac{\varepsilon_F^2}{2\varepsilon_a}, \tag{40}$$

where we have introduced for convenience a dimensionless "ideality factor"

$$m \equiv 1 + \frac{C_{it}}{C_{ox}}, \tag{41}$$

and notation ε_F used instead of ζ. The specificity of the graphene-insulator-gate structure electrostatics is reflected in Eq.40 in appearance of the characteristic energy scale

$$\varepsilon_a = \frac{\pi \hbar^2 v_0^2 C_{ox}}{2e^2} = \frac{\varepsilon_{ox}}{8\alpha_G} \frac{\hbar v_0}{d_{ox}}, \tag{42}$$

where the graphene "fine structure constant" is defined as (in SI units)

$$\alpha_G = \frac{e^2}{4\pi\varepsilon_0 \hbar v_0}. \tag{43}$$

Fig.4 shows dependencies of characteristic electrostatic energy of gated graphene ε_a vs gate oxide thickness for typical dielectric constants 4 (SiO_2) and 16 (HfO_2).

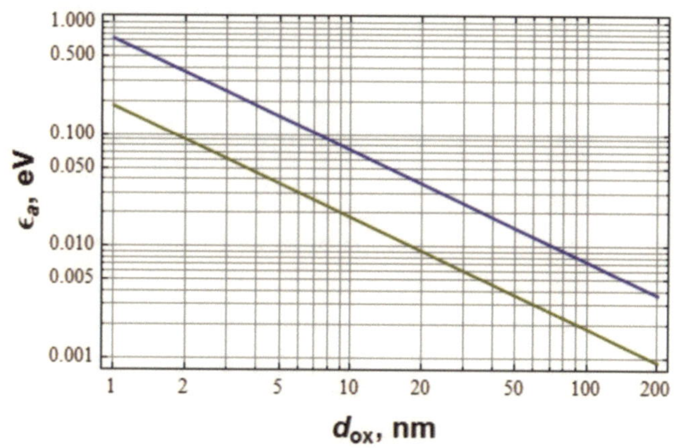

Fig. 4. The dependencies of the ε_a as functions of the insulator thickness d_{ox} for different dielectric permittivity equal to 4 (lower curve) and 16 (upper curve).

Energy scale ε_a bring in a natural spatial scale specific to the graphene gated structures

$$a_Q \equiv \frac{\hbar v_0}{\varepsilon_a} = \frac{2e^2}{\pi \hbar v_0 C_{ox}} = \frac{8\alpha_G}{\varepsilon_{ox}} d_{ox}, \tag{44}$$

and corresponding characteristic density

$$n_Q \equiv \frac{1}{\pi a_Q^2} = \frac{\varepsilon_a^2}{\pi \hbar^2 v_0^2} = n_S(\varepsilon_F = \varepsilon_a). \tag{45}$$

Due to the fact that graphene "fine structure constant" $\alpha_G \cong 2.0 - 2.2$ the characteristic length a_Q is occasionally of order of the oxide thickness for the insulators with ε_{ox}~16 (i.e. for HfO_2). Interestingly that the energy scale ε_a can be as well represented as functions of the Fermi energy and wavevector k_F, quantum capacitance and charge density

$$\frac{\varepsilon_a}{\varepsilon_F} = \frac{C_{ox}}{C_Q} = \frac{1}{k_F a_Q} = \frac{1}{\sqrt{\pi n_S a_Q^2}}. \tag{46}$$

3.4 Self-consistent solution of basic electrostatic equation

Solving algebraic Eq. (40) one obtains an explicit dependence (to be specific for $V_G > 0$) of the electron Fermi energy as function of the gate voltage

$$\varepsilon_F = \left(m^2\varepsilon_a^2 + 2\varepsilon_a e V_G\right)^{1/2} - m\varepsilon_a \tag{47}$$

This allows to immediately write the explicit relation for graphene charge density dependence on gate voltage

$$\frac{e^2 n_S}{C_{ox}} = eV_G - m\varepsilon_F = eV_G + m^2\,\varepsilon_a - m\left(m^2\varepsilon_a^2 + 2\varepsilon_a e V_G\right)^{1/2} \tag{48}$$

Restoring omitted terms the latter equation can be rewritten as (Zebrev, 2007); (Fang et al. 2007)

$$en_S\left(V_G\right) = C_{ox}\left(\left|V_G - V_{NP}\right| + V_0\left(1 - \left(1 + 2\frac{\left|V_G - V_{NP}\right|}{V_0}\right)^{1/2}\right)\right) \tag{49}$$

where the characteristic voltage $V_0 \equiv m^2\varepsilon_a / e$ is defined where interface trap capacitance is taken into account. Figs. 5-6 exhibit numerically the interrelation of V_0 with C_{it} and d_{ox}.

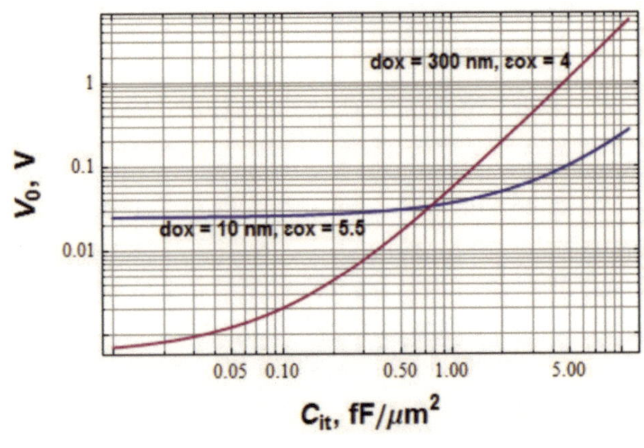

Fig. 5. Simulated dependencies of the characteristic voltage V_0 as functions of the interface trap capacitance C_{it} for different oxide parameters.

View of charge density dependence versus gate voltage is determined by relations of characteristic values (see Fig.5,6). At relatively high gate voltage $\left|V_G - V_{NP}\right| >> V_0$ (or, the same, for "thick" oxide) we have close to linear dependence

$$en_S \cong C_{ox}\left(\left|V_G - V_{NP}\right| - \left(2V_0\left|V_G - V_{NP}\right|\right)^{1/2}\right). \tag{50}$$

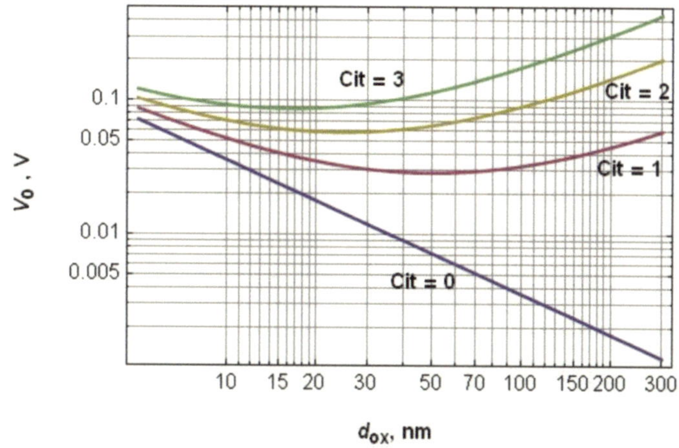

Fig. 6. Simulated dependencies of the characteristic voltage V_0 as functions of oxide thickness for different interface trap capacitance (in fF/μm²).

Most part of external gate voltage drops in this case on the oxide thickness. Such is the case of "standard" oxide thickness d_{ox} = 300 nm. Actually for not too small gate bias the charge density dependence on gate voltage is very close to linear (Novoselov et al., 2004). For future graphene FET the gate oxide thickness is assumed to be of order of few or ten of nanometers. For such case of much thinner oxides or under relatively small gate biases $C_{ox}|V_G - V_{NP}| < en_Q$ we have quadratic law for density dependence (see Fig. 2b)

$$en_S \cong C_{ox}\left(V_G - V_{NP}\right)\left(\frac{V_G - V_{NP}}{V_0}\right), \ V_G - V_{NP} < V_0 . \tag{51}$$

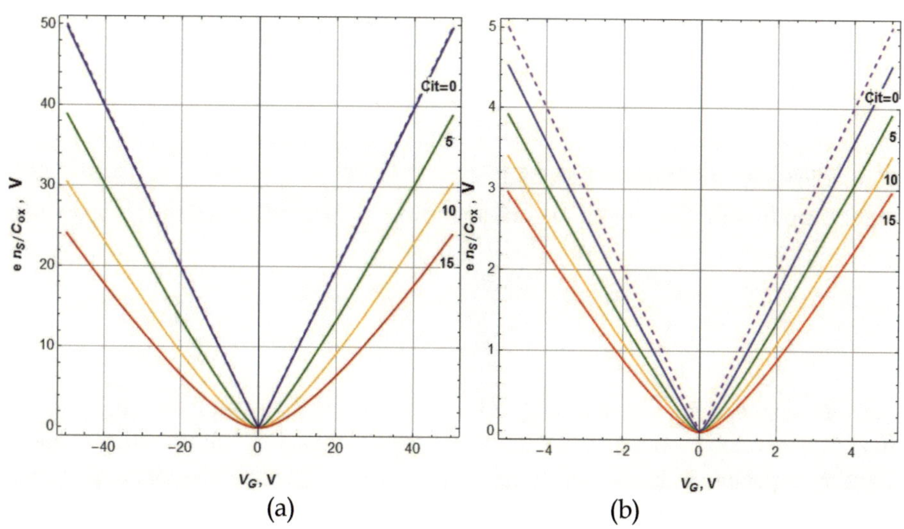

Fig. 7. Simulated charge density dependencies in reduced form $e\,n_S / C_{ox}$ as functions of gate voltage for ε_{ox} = 4 and different interface trap capacitance C_{it} = 0, 5, 10, 15 fF/μm²; (a) d_{ox} = 300 nm; (b) d_{ox} = 10 nm. Dashed curves correspond to $en_S/C_{ox} = V_G$.

Fig. 7 show that $n_S(V_G)$ curves are strongly affected by interface trap recharging even for relatively thin oxides.

3.5 Gate and channel capacitance

Capacitance-voltage measurements are very important in providing information about gated field-effect structures. Taking derivative of Eq. 36 with respect to chemical potential, we have

$$\frac{dV_G}{d\mu} = 1 + \frac{C_Q + C_{it}}{C_{ox}}. \tag{52}$$

Low-frequency gate capacitance can be defined as

$$C_G = e\left(\frac{\partial N_G}{\partial V_G}\right) = e\frac{dN_G/d\mu}{dV_G/d\mu} = \frac{C_Q + C_{it}}{1 + \frac{C_Q + C_{it}}{C_{ox}}} = \left(\frac{1}{C_{ox}} + \frac{1}{C_Q + C_{it}}\right)^{-1} \tag{53}$$

This relation corresponds to the equivalent electric circuit which is shown in Fig.8.

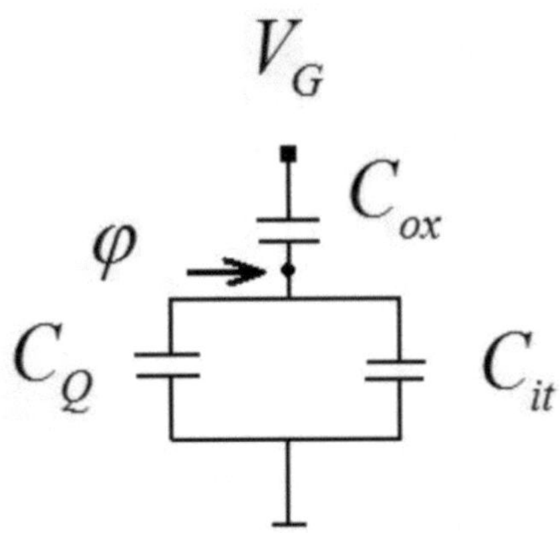

Fig. 8. Equivalent circuit of gated graphene.

One might introduce another relation corresponding to the intrinsic channel capacitance

$$C_{CH} = e\left(\frac{\partial N_S}{\partial V_G}\right) = e\frac{dN_S/d\mu}{dV_G/d\mu} = \frac{C_Q}{1 + \frac{C_Q + C_{it}}{C_{ox}}} = \frac{C_{ox}}{1 + \frac{C_{ox} + C_{it}}{C_Q}}. \tag{54}$$

where all capacitances are non-zero and assumed to be positive values for any gate voltage. Note that C_{CH} is often referred to as "total gate capacitance C_{tot}" in literature wherein the interface trap capacitance is frequently ignored. The gate and the channel capacitances are connected in graphene gated structures through exact relation

$$\frac{C_G}{C_{CH}} = 1 + \frac{C_{it}}{C_Q} \tag{55}$$

and can be considered to be coincided only for ideal devices without interface traps when C_{it} =0. All relationships for the differential capacitances remain valid for any form of interface trap energy spectrum. In an ideal case capacity-voltage characteristics $C_{CH}(V_G)$ should be symmetric with refer to the neutrality point implying approximately flat energy density spectrum of interface traps. For the latter case the channel capacity can be derived by direct differentiation of explicit dependence $n_S(V_G)$ in Eq.49

$$C_{CH} = e\frac{dn_S}{dV_G} = C_{ox}\left[1 - \frac{1}{\left[1+2|V_G - V_{NP}|/V_0\right]^{1/2}}\right]. \tag{56}$$

As can be seen in Fig.9 the capacitance-voltage characteristics $C_G(V_G)$ is strongly affected by the interface trap capacitance.

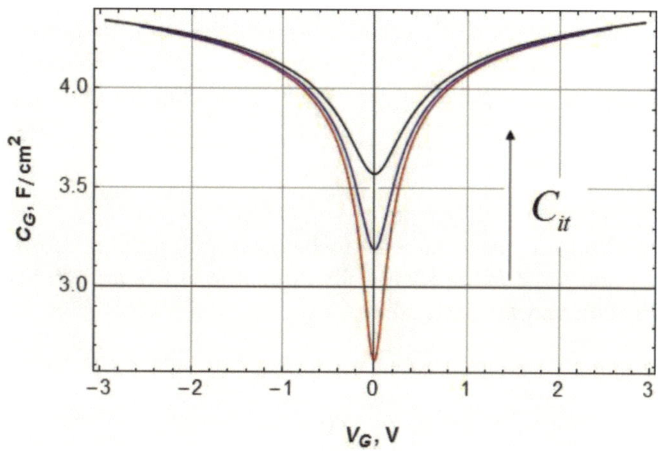

Fig. 9. Simulated dependencies of the gate capacitance $C_G(V_G)$ for different C_{it} = 1, 5, 10 fF/μm²; d_{ox} = 10 nm, ε_{ox} = 5.5 (Al_2O_3).

For the case C_{it} = 0 (i.e. m = 1) capacitance-voltage dependencies can be considered as to be universal curves depending on only thickness and permittivity of the gate oxide through the parameter ε_a. In practice one should discriminate the quantum and the interface trap capacitances and this is a difficult task since they are in a parallel connection in equivalent circuit. Comparison of "ideal" capacitance -voltage characteristics with real measured ones represents a standard method of interface trap spectra parameter extraction (Sze & Ng, 2007, Chap. 4,); (Nicollian & Brews, 1982).

4. Diffusion-drift current in graphene channels

4.1 Diffusion to drift current ratio

It is well-known that the channel electron current per unit width J_S can be expressed as a sum of drift and diffusion components

$$J_S = J_{DR} + J_{DIFF} = e\mu_0 n_S\frac{d\varphi}{dy} + eD_0\frac{dn_S}{dy}, \tag{57}$$

where μ_0 and D_0 are the electron mobility and diffusivity, y is a coordinate along the channel. This one can be rewritten in an equivalent form

$$J_D = \sigma_0 E\left(1-\left(\frac{D_0}{\mu_0}\right)\left(\frac{dn_S}{n_S d\zeta}\right)\left(\frac{d\zeta}{d\varphi}\right)\right), \tag{58}$$

where $E=-d\varphi/dy$ is electric field along the channel, $\sigma_0 = e\mu_0 n_S$ is the graphene sheet conductivity, $\zeta(y)$ and $\varphi(y)$ are the local chemical and electrostatic potential in the graphene channel , respectively. Using the Einstein relation for 2D system of non-interacting carrier as in Eq. 20 the diffusion-drift current reads (Zebrev & Useinov, 1990)

$$J_S \equiv e\mu_0 n_S E\left(1-\frac{d\zeta}{ed\varphi}\right) = e\mu_0 n_S E\,(1+\kappa)\,. \tag{59}$$

The ratio of the diffusion to the drift current is introduced in Eq.59 as the ratio of gradients of chemical (ζ) and electrostatic (φ) potentials along the channel, which are the components of electrochemical potential (or local Fermi energy for high doping case)

$$\kappa \equiv -\frac{d\zeta}{ed\varphi} = \frac{J_{DIF}}{J_{DR}} \tag{60}$$

Note that for equilibrium case the electrochemical potential is position independent ($\mu = \zeta - e\varphi$ =const) and $d\zeta/d\varphi$ is identically equals to unity and diffusion-drift current components exactly compensate each other

$$\left(\frac{\partial\zeta}{e\partial\varphi}\right)_\mu = -\frac{(\partial\mu/\partial\varphi)_\zeta}{e(\partial\mu/\partial\zeta)_\varphi} = 1\,. \tag{61}$$

On the contrary for non-equilibrium case both diffusion-drift components have the same direction ($d\zeta/d\varphi < 0$) and the parameter $\kappa > 0$. Unlike to the equilibrium case the electrostatic and chemical potential should considered as independent variables in non-equilibrium systems; e.g., the chemical potential controls particle (electron) density and is generally irrelevant to properly electric charge density and electrostatic potential. Two-dimensional electron density in the channel $n_S(\zeta)$ is a function exactly of the local chemical potential ζ rather than electrostatic (φ) or total electrochemical potential (μ). It is very important that the electrochemical potential distribution along the channel does not coincide in general with electrostatic potential distribution.

To properly derive explicit expression for control parameter κ we have to use the electric neutrality condition along the channel length in gradual channel approximation which is assumed to be valid even under non-equilibrium condition $V_{DS} > 0$. Differentiating Eq.36 with respect to chemical potential ζ (note that $V_G = const(y)$) and taking into consideration that $\varphi(y)$ and $\zeta(y)$ in the channel are generally non-equal and independent variables and n_S depends on only chemical potential ζ one can get

$$\kappa = -\left(\frac{\partial\zeta}{e\partial\varphi}\right)_{V_G} = \frac{(\partial V_G/\partial\varphi)_\zeta}{e(\partial V_G/\partial\zeta)_\varphi} = \frac{C_{ox}}{C_Q + C_{it}} \tag{62}$$

This dimensionless parameter κ is assumed to be constant along the channel for a given electric biases and expressed via the ratio of characteristic capacitances. For ideal graphene channel with low interface trap density the κ-parameter is a function of only ε_a and the Fermi energy

$$\kappa(C_{it}=0)=\frac{C_{ox}}{C_Q}=\frac{\varepsilon_a}{\varepsilon_F}=\frac{1}{k_F a_Q}=\frac{1}{\sqrt{\pi n_S a_Q^2}}. \tag{63}$$

For a high-doped regime (large C_Q) and/or thick gate oxide (low C_{ox}) when $C_Q >> C_{ox}$ we have $\kappa << 1$ by this is meant that the drift current component dominate the diffusion one and vice versa.

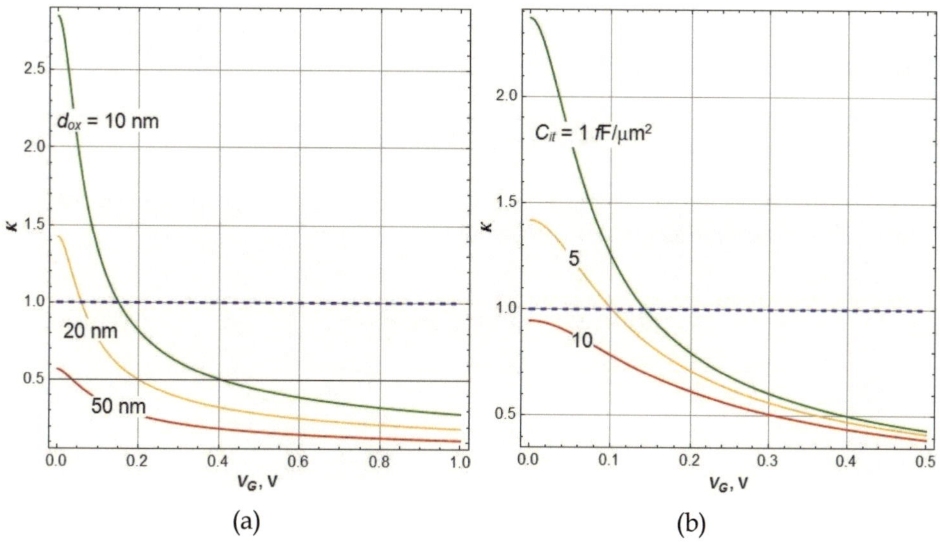

Fig. 10. Simulated κ curves as functions of gate voltage (a) for different oxide thicknesses, C_{it} = 0, ε_{ox} = 16; (b) for different interface trap capacitances C_{it} = 1, 5, 10 $fF/\mu m^2$, ε_{ox} = 16, d_{ox} = 10 nm.

Fig. 10 shows simulated dependencies of the parameter κ on gate voltage at variety of parameters.

4.2 Current continuity equation

The key point of this approach is an explicit analytical solution of continuity equation for channel current density. Total drain current $J_S = J_{DR} + J_{DIFF}$ should be conserved along the channel

$$\frac{dJ_S}{dy}=0 \Leftrightarrow \frac{d}{dy}(n_S E)=0 \tag{64}$$

that yields an equation for electric field distribution along the channel (Zebrev & Useinov, 1990)

$$\frac{dE}{dy}=\left(\frac{edn_S}{n_S d\zeta}\right)\left(-\frac{d\zeta}{ed\varphi}\right)\left(\frac{d\varphi}{dy}\right)=\frac{\kappa e}{\varepsilon_D}E^2 . \tag{65}$$

where κ and ε_D are assumed to be functions of only the gate voltage rather than the drain-source bias and position along the channel. Direct solution of ordinary differential Eq. 65 yields

$$E(y)=\frac{E(0)}{1-\frac{\kappa e E(0)}{\varepsilon_D}y}, \tag{66}$$

where $E(0)$ is electric field near the source, which should be determined from the condition imposed by a fixed electrochemical potential difference between drain and source V_D, playing a role of boundary condition

$$V_D=(1+\kappa)\int_0^L E(y)dy , \tag{67}$$

where L is the channel length. Using Eqs. (66) and (67) one obtains an expressions for $E(0)$ and electric field distribution along the channel

$$E(0)=\frac{\varepsilon_D/e}{\kappa L}\left(1-\exp\left(-\frac{\kappa}{1+\kappa}\frac{eV_D}{\varepsilon_D}\right)\right); \tag{68}$$

$$E(y)=\frac{\frac{\varepsilon_D/e}{\kappa L}\left(1-\exp\left(-\frac{\kappa}{1+\kappa}\frac{eV_D}{\varepsilon_D}\right)\right)}{1-\frac{y}{L}\left(1-\exp\left(-\frac{\kappa}{1+\kappa}\frac{eV_D}{\varepsilon_D}\right)\right)}. \tag{69}$$

4.3 Distributions of chemical and electrostatic potential along the channels

Integrating Eq. (69) we have obtained the explicit relationships for distributions of the chemical and electrostatic potentials along the channel length separately and electrochemical potential as a whole

$$\varphi(y)-\varphi(0)=-\frac{\varepsilon_D}{\kappa e}ln\left[1-\frac{y}{L}\left[1-\exp\left(-\frac{\kappa}{1+\kappa}\frac{eV_D}{\varepsilon_D}\right)\right]\right], \tag{70}$$

$$\zeta(y)-\zeta(0)=\varepsilon_D\, ln\left[1-\frac{y}{L}\left[1-\exp\left(-\frac{\kappa}{1+\kappa}\frac{eV_D}{\varepsilon_D}\right)\right]\right], \tag{71}$$

$$\mu(y)=\mu(0)+\varepsilon_D\frac{1+\kappa}{\kappa}ln\left[1-\frac{y}{L}\left[1-\exp\left(-\frac{\kappa}{1+\kappa}\frac{eV_D}{\varepsilon_D}\right)\right]\right], \tag{72}$$

where $\zeta(0)$, $\mu(0)$ and $\varphi(0)$ are the potentials nearby the source controlled by the gate-source bias V_{GS}. For any gate voltage V_{GS} (and corresponding $\kappa(V_G)$) the full drop of electrochemical potential μ on the channel length is fixed by the source-drain bias V_D

$$e\left(\varphi(L)-\varphi(0)\right)+\zeta(0)-\zeta(L)=\frac{eV_{DS}}{1+\kappa}+\frac{e\kappa V_{DS}}{1+\kappa}=eV_{DS} \tag{73}$$

Expanding Eqs. 70 at low drain bias and high carrier density case ($\kappa < 1$) we have familiar linear dependence of electrostatic potential on coordinate (as in any good conductor)

$$\varphi(y)-\varphi(0)\cong\frac{y}{L}V_D, \tag{74}$$

and negligible spatial change in chemical potential along the channel length $\Delta\zeta=\kappa\Delta\varphi<<\varphi$. Thus the full drop of chemical potential is negligible under high-doped channel compared to electrostatic potential but it becomes very important in saturation mode.

5. Channel current modeling

5.1 Current-voltage characteristics

The total drain current at constant temperature can be written as gradient of the electrochemical potential taken in the vicinity of the source

$$\begin{aligned} I_D &= -W\mu_0 n_S(0)\left(\frac{d\mu}{dy}\right)_{y=0} = eW\mu_0 n_S(0)(1+\kappa)E(0)= \\ &= e\frac{W}{L}D_0 n_S(0)\frac{1+\kappa}{\kappa}\left(1-\exp\left(-\frac{\kappa}{1+\kappa}\frac{eV_D}{\varepsilon_D}\right)\right), \end{aligned} \tag{75}$$

where W is the channel width, and the Einstein relation $D_0=\mu_0\varepsilon_D/e$ is employed. Notice that the total two-dimensional charge density $eN_S\cong en_S$ practically equals to charge imbalance density excepting the vicinity of the charge neutrality point where diffusion-drift approximation is failed.
Let us define the characteristic saturation source-drain voltage V_{DSAT} in a following manner

$$V_{DSAT}=2\frac{1+\kappa}{\kappa}\frac{\varepsilon_D}{e}=\frac{1+\kappa}{\kappa}\frac{\varepsilon_F}{e}, \tag{76}$$

where ε_F is the Fermi energy (the same chemical potential) nearby the source (recall that $\varepsilon_D\cong\varepsilon_F/2$ for $\zeta=\varepsilon_F>>k_BT$). Notice that employing this notation and Eq.71 one might write the chemical potential nearby the drain as

$$\zeta(L)=\left(1-V_D/V_{DSAT}\right)\varepsilon_F. \tag{77}$$

This implies that the condition $V_D=V_{DSAT}$ corresponds to zero of the chemical potential and current due to electrostatic blocking which is known as pinch-off for silicon MOSFETs (Sze & Ng, 2007). Actually, one might rewrite a general expression for the channel current as

$$I_D = \frac{W}{L}\sigma_0(0)\frac{V_{DSAT}}{2}\left(1 - \exp\left(-2\frac{V_D}{V_{DSAT}}\right)\right) \tag{78}$$

where σ_0 is the low-field conductivity nearby the source. It is evident from Eq.78 that V_{DSAT} corresponds to onset of drain current saturation. This expression describe I-V characteristics of graphene current in a continuous way in all operation modes (see Fig.11)

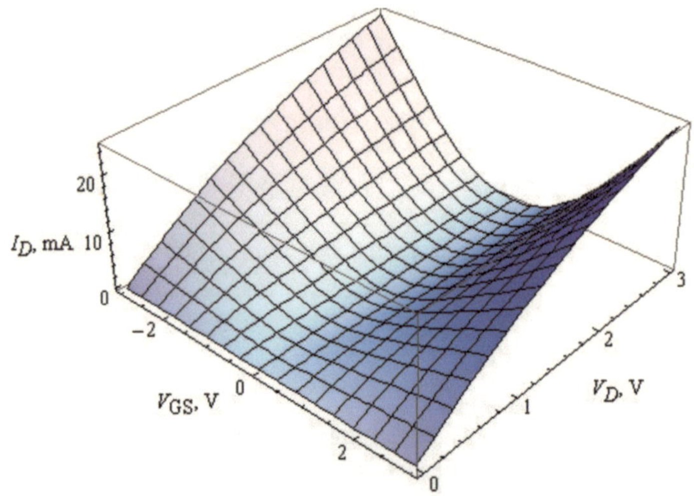

Fig. 11. Current voltage characteristics of graphene FET as function of gate and drain voltage.

5.2 Pinch-off (saturation) regime

Taking into account Eqs. 76, 62 and 63 one obtains

$$eV_{DSAT} = \varepsilon_F \frac{1+\kappa}{\kappa} = m\varepsilon_F + \frac{\varepsilon_F^2}{\varepsilon_a}. \tag{79}$$

Recall that $|V_G - V_{NP}| = m\varepsilon_F + \varepsilon_F^2 / 2\varepsilon_a$ one may derive an expression

$$V_{DSAT} = |V_G - V_{NP}| + \frac{\varepsilon_F^2}{2\varepsilon_a} = |V_G - V_{NP}| + \frac{en_S}{C_{ox}}, \tag{80}$$

which is specific for graphene field-effect transistors.

Notice that for thick oxide GFET we have very large $V_{DSAT} \cong 2|V_G - V_{NP}| >> 1$ V and pinch-off saturation is never observed. As can be seen in Fig. 12 the saturation voltage V_{DSAT} depends parametrically on the ε_a and on interface trap capacitance C_{it}. Under condition of high source-drain bias $V_D > V_{DSAT}$ the Eq.78 yields formal relationship for saturation current regime caused by electrostatic pinch-off.

$$I_{DSAT} \cong \frac{W}{L} D_0 n_S(0)\frac{1+\kappa}{\kappa} = \frac{W}{L}\sigma_0(0)\frac{V_{DSAT}}{2} \tag{81}$$

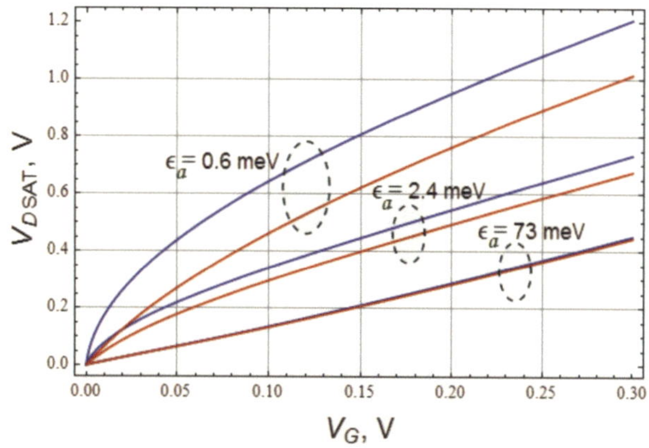

Fig. 12. Simulated V_{DSAT} curves as functions of gate voltage for $\varepsilon_a = 0.6$ meV (d_{ox} = 300 nm, $\varepsilon_{ox} = 4$); $\varepsilon_a = 2.4$ meV (d_{ox} = 300 nm, $\varepsilon_{ox} = 16$); $\varepsilon_a = 73$ meV (d_{ox} = 10 nm, $\varepsilon_{ox} = 16$); $C_{it} = 0$ $fF/\mu m^2$ (upper curve in the pairs) and $C_{it} = 1$ $fF/\mu m^2$ (lower curve).

5.3 Low-field linear regime

Linear (triode) operation mode corresponds to condition

$$V_D << V_{DSAT} = \varepsilon_F \frac{1+\kappa}{\kappa}. \tag{82}$$

For high doping regime when $\kappa << 1$ one has predominance of drift component of the channel current as in any metal. In contrast for $\kappa >> 1$ the diffusion current prevails. Equality of the current components occurs in ideal structure ($C_{it} = 0$) at $\varepsilon_F = \varepsilon_a$ or, equivalently, at the characteristic channel density $n_S = n_Q$, defined in Eq.45.

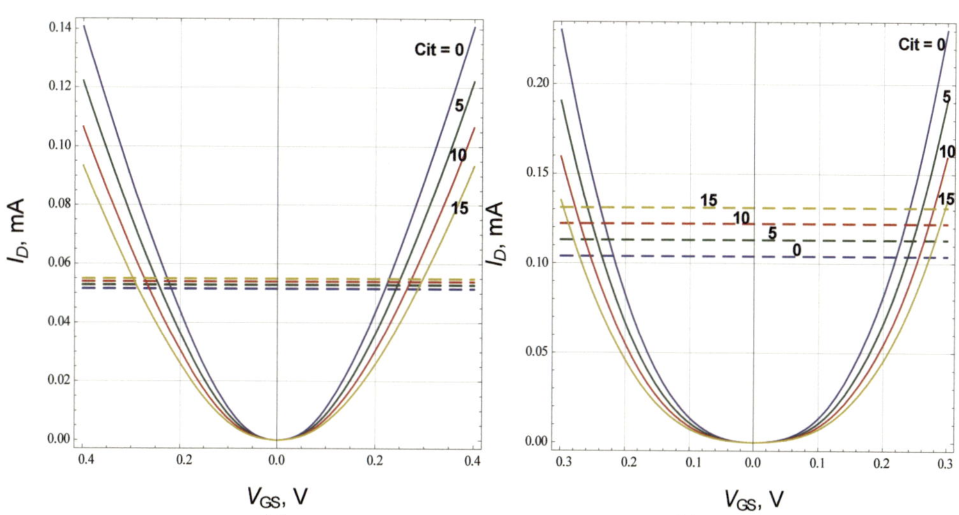

Fig. 13. Simulated drain channel currents as functions of gate voltage for different interface trap capacitances C_{it} = 0, 5, 10, 15 $fF/\mu m2$; d_{ox} = 5 nm, $\varepsilon_{ox} = 16$; W = 1 μm; L = 0.25 μm, μ_0 = 800 cm²/(V s); (a) V_D = 0.1 V; (b) V_D = 1 V. Dashed curves correspond to condition $\kappa = 1$.

Fig. 13 shows simulated transfer (I_D vs V_G) characteristics of graphene FET for different drain biases and interface trap capacitances. Portions of curves below the dashed curves correspond to predominance of diffusion current with pronounced current saturation, and the above dashed curves correspond mainly to drift current with linear dependence on the drain bias. Notice that the diffusion current region is negligible for dirty structures with thick oxides. For rather small drain bias one can get a usual linear expression expanding Eq.78 in series on V_D

$$I_D \cong e\frac{W}{L}\mu_0 n_S V_D . \tag{83}$$

Setting mobility μ_0 gate voltage independent the small-signal transconductance in the linear mode reads

$$g_m \equiv \left(\frac{\partial I_D}{\partial V_G}\right)_{V_D} = \frac{W}{L}\mu_0 C_{CH} V_D , \tag{84}$$

where the channel capacitance C_{CH} is defined in Eq.54. Field-effect mobility μ_{FE} can be defined from Eq.84 as

$$g_m = \frac{W}{L}\mu_0 V_D C_{CH} \equiv \frac{W}{L}\mu_{FE} V_D C_{ox} . \tag{85}$$

Eq.91 connects field-effect mobility μ_{FE} depending on charge exchange with extrinsic traps (defects in the gate oxides, chemical dopants etc.) and mobility μ_0 depending on only "microscopic" scattering mechanisms

$$\mu_{FE} = \frac{\mu_0}{1+\dfrac{C_{ox}+C_{it}}{C_Q}} = \frac{\mu_0\, \varepsilon_F}{m\varepsilon_a + \varepsilon_F} . \tag{86}$$

Note that the field-effect mobility, determined often immediately as a slope of the experimental conductivity vs gate voltage curves, is always less than truly microscopic mobility and significantly decreases nearby the charge neutrality point. In fact, μ_{FE} is close to μ_0 only if $C_Q >> mC_{ox}$ (or, equivalently, $\varepsilon_F >> m\varepsilon_a$), i.e. for a high doping regime.

Transconductance in field-effect transistors commonly degrades affected by electric stress, wear-out or ionizing radiation due to interface trap buildup. The field-effect mobility renormalization after externally induced interface trap capacitance alteration $C_{it} \to C_{it} + \Delta C_{it}$ can be expressed using Eq.86 via initial value $\mu_{FE}(C_{it})$

$$\mu_{FE}(C_{it}+\Delta C_{it}) = \frac{\mu_{FE}(C_{it})}{1+\dfrac{\Delta C_{it}}{C_{ox}+C_Q+C_{it}}} . \tag{87}$$

Logarithmic swing which characterizes the I_{ON}/I_{OFF} ratio and equals numerically to the gate voltage alteration needed for current change by an order can be computed using Eq.83 and Eq.54 as

$$S \equiv \left(\frac{d(\log_{10} I_D)}{dV_G} \right)^{-1} = \ln 10 \left(\frac{dn_S}{n_S dV_G} \right)^{-1} = \ln 10 \frac{en_S}{C_{CH}(V_G)}. \quad (88)$$

This formula can be written down in a form more familiar from silicon MOSFET theory

$$S = \ln 10 \left(\frac{en_S}{C_Q} \right) \left(1 + \frac{C_{it} + C_Q}{C_{ox}} \right) = \ln 10 \left(\frac{en_S}{C_{ox}} \right) \left(1 + \frac{C_{it} + C_{ox}}{C_Q} \right) = \ln 10 \, en_S \left(\frac{m}{C_Q} + \frac{1}{C_{ox}} \right). \quad (89)$$

Recall that the diffusion energy $\varepsilon_D = e^2 n_S / C_Q \cong \varepsilon_F / 2$ plays here role of the thermal potential $e\varphi_D = k_B T$ for the subthreshold (non-degenerate) operation mode of the silicon FETs wherein C_Q is negligible. Unlike the silicon FET case the subthreshold swing is a function of gate voltage. Excluding a small region nearby the Dirac point the latter expression yields an assessment of the logarithmic swing $S \geq \ln 10 \, en_S / C_{ox} >> 1\text{V/decade}$ for "thick" oxides and "clean" interface ($C_Q >> mC_{ox}$) and $S \cong \ln 10 m \, \varepsilon_F / 2e$ for "thin" oxide ($C_Q << mC_{ox}$).

5.4 Transit time through the channel length

Using electric field distribution (Eq. 69) the transit time through the whole channel length can be computed in a following way

$$\tau_{TT} = \int_0^L \frac{dy}{\mu_0 (1+\kappa) E(y)} \quad (90)$$

Performing direct integration one can explicitly get

$$\tau_{TT} = \frac{L^2}{2D_0} \frac{\kappa}{1+\kappa} \coth\left(\frac{\kappa}{1+\kappa} \frac{eV_D}{2\varepsilon_D} \right) = \frac{L^2}{\mu_0 V_{DSAT}} \coth\left(\frac{V_D}{V_{DSAT}} \right) \quad (91)$$

This expression yields the drift flight time for the linear regime (when $V_D << V_{DSAT}$)

$$\tau_{TT} = \frac{L^2}{\mu_0 V_D}, \quad (92)$$

and the diffusion time for $V_D > V_{DSAT}$ and low carrier density ($\kappa >> 1$)

$$\tau_{TT} = \frac{L^2}{2D_0} \frac{\kappa}{1+\kappa} \cong \frac{L^2}{2D_0}. \quad (93)$$

6. Conclusion

6.1 Applicability of diffusion-drift approximation

The theory presented in this chapter relies significantly on macroscopic diffusion-drift approximation which is still the ground of practical device simulation. Diffusion-drift approximation is semi-classical by its nature and valid for only small wave lengths and high carrier density. Diffusion-drift and Boltzmann equation approach validity in graphene depends on interrelation between basic spatial scales, namely, mean free path ℓ, the channel length L, carrier's wavelength at the Fermi energy $\lambda_F = hv_0 / \varepsilon_F$. The condition $L < \ell$ corresponds to ballistic transport. Inequalities $\lambda_F < \ell < L$ represent semi-classical case with weak scattering and well-defined dispersion law conditions. Using independence of mobility on carrier density n_S in graphene and recalling Eq. 25 one might rewrite a wavelength smallness requirement as a condition for n_S

$$\lambda_F < \ell \leftrightarrow n_S > \frac{2e}{\hbar \mu_0} \cong 3 \times 10^{12} \left(\frac{10^3}{\mu_0} \right) \text{cm}^{-2},$$

where carrier's mobility μ_0 is expressed in cm^2 /(V s). Thusly at low electric field the diffusion-drift approximation is valid for not too small carrier densities. In fact semi-classical description is rather suitable even for regions nearby the neutrality point due to presence of unavoidable disorder at the Dirac point with smooth potential relief. High transverse electric field near the drain leads to breaking of semi-classical approximation due to local lowering of charge density. Strong electric field near the drain can separate e-h pairs shifting equilibrium between generation and recombination and increasing electric field-induced non-equilibrium generation drain current. Quantum effects of inter-band interaction (so called "trembling" or "zitterbewegung") (Katsnelson, 2006) become significant for low carrier densities. These effects are similar to generation and recombination of virtual electron-hole pairs.

6.1 High-field effects

As carriers are accelerated in an electric field their drift velocity tends to saturate at high enough electric fields. Current saturation due to velocity saturation has been discussed in recent electronic transport experiments on graphene transistors (Meric et al., 2008). The validity of the diffusion-drift equations can be empirically extended by introduction of a field-dependent mobility obtained from empirical models or detailed calculation to capture effects such as velocity saturation at high electric fields due to hot carrier effects

$$\mu_0(E) = \frac{\mu_0}{1 + E / E_C}, \tag{94}$$

where μ_0 is the low field mobility, $v_{SAT} < v_0$ is saturation velocity, maintained mainly due to optical phonon emission , $E_{SAT} \sim v_{SAT} / \mu_0$... (1 – 5)×10^4 V/cm. Interrelation between electrostatic pinch-off discussed in the chapter and velocity saturation can be characterized with the dimensionless ratio (Zebrev, 1992)

$$a = \frac{V_{DSAT}}{2E_C L} = \frac{|V_G - V_{NP}| + en_S / C_{ox}}{2E_C L} \tag{95}$$

There are thusly two distinctly different current saturation mechanisms. Electrostatically induced current pinch-off dominates in the devices with long channels and large C_{ox} ($a \ll 1$) while in short-channel devices with thick gate oxides ($a \gg 1$) the channel current saturation $I_D = Wen_S v_{SAT}$ occurs due to drift velocity limitation.

Within the frame of diffusion-drift approximation validity the main qualitative difference between transport in graphene and in conventional silicon MOSFET is the specific form of dispersion law in graphene which lead to peculiarities in statistics and electrostatics of graphene field-effect transistor. All quantum and high electric field effects have remained beyond the scope of this chapter and should be subject of future works.

7. References

Ando T., Fowler A., Stern F.," Electronic properties of two-dimensional systems "*Rev. Mod. Phys*. Vol. 54, No.2, 1982, pp.437-462.

Ancona M.G., "Electron Transport in Graphene From a Diffusion-Drift Perspective," IEEE Transactions on Electron Devices, Vol. 57, No. 3, March 2010, pp. 681-689.

Das Sarma S., Shaffique Adam, Hwang E. H., and Rossi E. "Electronic transport in two dimensional graphene", 2010, *arXiv*: 1003.4731v1

Emelianov V.V., Zebrev, G.I., Ulimov, V.N., Useinov, R.G.; Belyakov V.V.; Pershenkov V.S., "Reversible positive charge annealing in MOS transistor during variety of electrical and thermal stresses, " *IEEE Trans. on. Nucl. Sci*., 1996, No.3, Vol. 43, pp. 805-809.

Fang, T., Konar A., Xing H., and Jena D., 2007, "Carrier statistics and quantum capacitance of graphene sheets and ribbons," *Appl. Phys. Lett.* Vol. 91, p. 092109.

Fleetwood D.M., Pantelides S.T., Schrimpf R.D. (Eds.) 2008, *Defects in Microelectronic Materials and Devices*, CRC Press Taylor & Francis Group, London - New York.

Giovannetti G., Khomyakov P. A., Brocks G., Karpan V. M., van den Brink J., and Kelly P. J. "Doping graphene with metal contacts," 2008, *arXiv*: 0802.2267.

Katsnelson M. I., "Zitterbewegung, chirality, and minimal conductivity in graphene," *Eur. Phys. J.* Vol. B 51, 2006, pp. 157-160.

Luryi S., "Quantum Capacitance Devices," *Applied Physics Letters*, Vol. 52, 1988, pp. 501-503.

Martin, J., Akerman N., Ulbricht G., Lohmann T., Smet J. H., Klitzing von K., and Yacobi A., "Observation of electron-hole puddles in graphene using a scanning single electron transistor," *Nature Physics*, 2008, No.4, 144

Meric I.; Han M. Y.; Young A. F.; Ozyilmaz B.; Kim P.; Shepard K. L. "Current saturation in zero-bandgap, top-gated graphene field-effect transistors," *Nat. Nanotechnol.* 2008, No. 3, pp. 654–659.

Nicollian E.H. & Brews J.R., 1982, *MOS (Metal Oxide Semiconductor) Physics and Technology*, Bell Laboratories, Murray Hill, USA.

Novoselov K. S., Geim A.K., et al. "Electric field effect in atomically thin carbon films," *Science*, Vol. 306, 2004, pp. 666-669.

Sze S. M. & Ng. K. K. , 2007, *Physics of Semiconductor Devices*, John Wiley & Sons, ISBN 978-0-471-14323-9, New Jersey, USA.

Wolfram S., (2003), *Mathematica Book*, Wolfram Media, ISBN 1–57955–022–3, USA.

Zebrev G. I., "Electrostatics and diffusion-drift transport in graphene field effect transistors," *Proceedings of 26th International Conference on Microelectronics* (MIEL), Nis, Serbia, 2008, pp. 159-162.

Zebrev G. I., "Graphene nanoelectronics: electrostatics and kinetics", *Proceedings SPIE*, 2008, Vol. 7025. - P. 70250M - 70250M-9, based on report to ICMNE-2007, October, 2007, Russia.

Zebrev G.I., Useinov R.G., "Simple model of current-voltage characteristics of a metal-insulator–semiconductor transistor", *Fiz. Tekhn. Polupr.* (*Sov. Phys. Semiconductors*), Vol. 24, No.5, 1990, pp. 777-781.

Zebrev G.I., "Current-voltage characteristics of a metal-oxide-semiconductor transistor calculated allowing for the dependence of mobility on longitudinal electric field, "*Fiz. Tekhn. Polupr.* (Sov. Phys. Semiconductors), No.1, Vol. 26, 1992, pp. 47-49.

Part 5

Optical Properties

24

Photo-Induced Carrier Density, Optical Conductance and Transmittance in Graphene in the Presence of Optic-Phonon Scattering

W. Xu and H.M. Dong

Department of Physics, Yunnan University & Key Laboratory of Materials Physics, Institute of Solid State Physics, Chinese Academy of Sciences P. R. China

1. Introduction

Since the experimental discovery of graphene in 2004 (Novoselov et al., 2004), the investigation of graphene-based electronics and optoelectronics has quickly become one of the most important research topics in condensed matter physics, nano-material science and nano-electronics (Zhang et al., 2005; Berger et al., 2006). Due to its excellent electronic transport, optical, and optoelectronic properties, such as high carrier density (up to 10^{13} cm^{-2}) and high carrier mobility at room temperature (up to 20 m^2/Vs) along with the high optical transmittance in the air-graphene-wafer systems, graphene has been proposed as an advanced material for new generation of electronic and optoelectronic devices. Graphene-based electronic devices exhibit high carrier mobility and quasi ballistic transport over sub-micron scales even at room temperature (Novoselov et al., 2005). It has already been used to realize high-speed and high-frequency electronic devices such as field-effect transistors (Castro et al., 2007), p-n junctions (González & Perfetto, 2008), high-frequency devices (Lin et al., 2009), to mention but a few. Very recently, graphene has also been proposed as an advanced transparent conducting material by utilizing its combined excellent transport and optical properties (Eda et al., 2008). It has been shown that graphene can be used to replace conventional indium tin oxide (ITO) transparent electrodes for making better and cheaper optical displays such the LCDs and LEDs (Hogan, 2008). Presently, graphene-based transparent electronics is a hot field of research for both fundamental studies and device applications (Eda et al., 2008).

For the usage of graphene as optoelectronic and transparent electronic devices, the investigation of its optical and optoelectronic properties is critical and essential. Recent experimental and theoretical work has demonstrated and predicted some particular and interesting optoelectronic properties in the infrared-to-visible spectral range for air-graphene-wafer systems. In particular, the results obtained from optical transmission (Kuzmenko et al., 2008) and infrared absorption (Li et al., 2008) measurements show the following features. (i) The optical conductance per graphene layer is a universal value $\sigma_0 = e^2/(4\hbar)$ in the visible frequency range (Kuzmenko et al., 2008; Li et al., 2008), which can be viewed as an intrinsic property of two-dimensional massless fermions. (ii) The corresponding light transmittance of monolayer and bilayer graphene on SiO_2 or Si wafers are, respectively, about 0.98 and 0.96 in the visible bandwidth (Li et al., 2008; Nair et al., 2008). (iii) There is an optical absorption window (Kuzmenko et al., 2008; Li et al., 2008; Choi et al., 2009) for radiation photon energy smaller than 0.2 eV. The width and depth of this window depend strongly on the temperature (Kuzmenko et al., 2008) and carrier density (or gate voltage) in graphene samples (Li et al., 2008). This interesting finding implies that graphene can also be used for infrared detection in ambient conditions. Further experimental investigation shows that graphene can have strong intra- and inter-band transitions which can be substantially modified through electrical gating, similar to the resistance tuning in graphene field-effect transistors (Li et al., 2008; Wang et al., 2008). The optical and optoelectronic properties of graphene-based electronic systems have also been studied theoretically (Vasko & Ryzhii, 2008; Falkovsky & Pershoguba, 2007; Satuber et al., 2008). Most of the theoretical results are in line with the main experimental findings but are valid for low temperatures and only electron scattering by acoustic phonons was considered (Vasko & Ryzhii, 2008). However, most of the experimental studies of the optical properties of graphene systems have been undertaken at relatively high temperatures and up to room temperature (Kuzmenko et al., 2008; Li et al., 2008). In particular, the published theoretical work has not yet given a fully satisfactory explanation of the optical absorption window, present in graphene in the infrared bandwidth, that is observed experimentally by different kinds of measurements (Kuzmenko et al., 2008; Li et al., 2008; Choi et al., 2009). Therefore, in this study we examine how carriers in graphene respond to the applied radiation field in order to have a better understanding of the optoelectronic features of graphene, especially in the infrared wavelength regime. Here we would like to point out that graphene is a gapless electronic system in which optical phonon energy is about 196 meV (Ando, 2007). Hence, in contrast to a conventional semiconductor materials which normally have a band-gap much lager than the phonon energy, phonon scattering is expected to play an important role in affecting the electronic transitions accompanied by the emission and absorption of photons in graphene, especially in the infrared bandwidth. For the gapless graphene, the electrons in the valence band can gain the energy from the radiation field via optical absorption and be excited into the conduction band, while the electrons in the conduction band can lose energy via emission of phonons and be relaxed into the valence band. Together with the fact that the electrons interact more strongly with phonons than with photons, carrier-phonon interaction is an important factor in affecting the inter-band electronic transitions and, therefore, in determining the optoelectronic properties of graphene at relatively high-temperatures. In this work we develop a systematic approach to calculate the electronic and optical coefficients of graphene in the presence of a radiation field. In such an approach, the electron-photon and electron-phonon couplings are included in the calculation and their consequences are examined in a consistently theoretical manner.

2. Theoretical approach

2.1 Electronic transition rate

We consider a configuration where the graphene sheet is placed on the (x, y) plane on top of a dielectric wafer such as SiO_2/Si substrate. A light field is applied perpendicular to the graphene layer and is polarized linearly along the x direction of the system. In the effective-mass approximation, a carrier (electron or hole) in a monolayer graphene can be described by Weyl's equation for a massless neutrino (McCann & Falko, 2006). The single-particle Hamiltonian that describes a Dirac quasi-particle in the π bands near the K point can be obtained from, e.g., the usual $\mathbf{k} \cdot \mathbf{p}$ approach. We assume that the system under study can be separated into the carriers of interest and the rest of the graphene crystal. Then the Hamiltonian that describes a carrier-photon-phonon system in graphene is:

$$H(t) = H_0 + H_{co}(t) + H_{ph} + H_{cp}(t). \tag{1}$$

Here,

$$H_0 = \gamma \begin{bmatrix} 0 & \vec{k}_x - i\vec{k}_y \\ \vec{k}_x + i\vec{k}_y & 0 \end{bmatrix} \tag{2}$$

is the Hamiltonian for a Dirac quasi-particle and $\gamma = \hbar v_F$ is the band parameter with $v_F = 10^8$ cm/s being the Fermi velocity for a carrier in graphene, and $\vec{k}_x = -i\partial/\partial x$ is the momentum operator along the x direction. The Schrödinger equation regarding H_0 can be solved analytically. The corresponding eigenvalue and eigenfunction are respectively

$$E_\lambda(\mathbf{k}) = \lambda\gamma|\mathbf{k}| = \lambda\gamma k \tag{3}$$

and

$$\psi_{\lambda\mathbf{k}}(\mathbf{r}) = |\mathbf{k}, \lambda >= 2^{-1/2}[1, \lambda e^{i\phi}]e^{i\mathbf{k}\cdot\mathbf{r}} \tag{4}$$

in the form of a row matrix. Here, $\mathbf{k} = (k_x, k_y)$ is the wavevector for a carrier, $k = \sqrt{k_x^2 + k_y^2}$, $\mathbf{r} = (x, y)$, $\lambda = +1$ for an electron and $\lambda = -1$ for a hole, and ϕ is the angle between $\mathbf{k}$ and the x direction. Using the usual coulomb gauge, the radiation field can be included by taking $k_x \to k_x - eA(t)/\hbar$ in Eq. (2) with $A(t)$ being the vector potential of the radiation field which is polarized along the x-direction. Thus, we can obtain the carrier-photon interaction Hamiltonian, which reads

$$H_{co}(t) = \frac{\gamma e A(t)}{\hbar} \begin{bmatrix} 0 & 1 \\ 1 & 0 \end{bmatrix}, \tag{5}$$

where $A(t) = (F_0/\omega)\sin(\omega t)$ with F_0 and ω being, respectively, the electric field strength and frequency of the radiation field. Furthermore, H_{ph} is the phonon Hamiltonian and

$$H_{cp}(t) = V_{\mathbf{q}} a_{\mathbf{q}} e^{i(\mathbf{q}\cdot\mathbf{r} - \omega_q t)} + V_{\mathbf{q}}^* a_{\mathbf{q}}^\dagger e^{-i(\mathbf{q}\cdot\mathbf{r} + \omega_q t)} \tag{6}$$

describes carrier interaction with 2D-like phonons, where $\mathbf{q} = (q_x, q_y)$ is the phonon wave vector in the (x, y) plane, $(a_{\mathbf{q}}^\dagger, a_{\mathbf{q}})$ are the canonical conjugate coordinates of the phonon system, $V_{\mathbf{q}}$ is the carrier-phonon interaction coefficient, and ω_q is the phonon frequency in graphene. For the case of a relatively weak radiation field and weak carrier-phonon interaction, $H_{co}(t)$ and $H_{cp}(t)$ can be treated as perturbations. Using Fermi's golden rule (Stroscio & Mitra, 2005), the first-order steady-state electronic transition rate is given by

$$W_{\lambda\lambda'}(\mathbf{k}, \mathbf{k}') = \sum_\nu W_{\lambda\lambda'}^\nu(\mathbf{k}, \mathbf{k}'); \tag{7}$$

it is the probability for scattering of a carrier from a state $|\mathbf{k}, \lambda >$ to a state $|\mathbf{k}', \lambda' >$ due to interaction with photons and phonons. Moreover, ν refers to different scattering mechanisms. In Eq. (7) the rate

$$W_{\lambda\lambda'}^{co,\pm}(\mathbf{k}, \mathbf{k}') = \frac{2\pi}{\hbar}\left(\frac{eF_0\gamma}{2\hbar\omega}\right)^2 \frac{1 + \lambda\lambda'\cos(2\phi)}{2} \delta_{\mathbf{k}',\mathbf{k}}\delta[E_\lambda(\mathbf{k}) - E_{\lambda'}(\mathbf{k}') \pm \hbar\omega]$$

is induced by direct carrier-photon interaction via absorption (+ sign) and emission (− sign) of a photon with an energy $\hbar\omega$, and ϕ is the angle between $\mathbf{k}$ and the x axis. The rate

$$W^{cp,\pm}_{\lambda\lambda'}(\mathbf{k},\mathbf{k}') = \frac{2\pi}{\hbar}\begin{bmatrix} N_q \\ N_q+1 \end{bmatrix} |U^{cp}_{\lambda\lambda'}(\mathbf{q},\theta)|^2 \delta_{\mathbf{k}',\mathbf{k}+\mathbf{q}} \delta[E_\lambda(\mathbf{k}) - E_{\lambda'}(\mathbf{k}') \pm \hbar\omega_q]$$

is induced by carrier-phonon interaction, where $N_q = (e^{\hbar\omega_q/k_BT} - 1)^{-1}$ is the phonon occupation number, $|U^{cp}_{\lambda\lambda'}(\mathbf{q},\theta)|^2 = |< \mathbf{k}',\lambda'|V_\mathbf{q}|\mathbf{k},\lambda >|^2$, θ is the angle between $\mathbf{k}'$ and $\mathbf{k}$, and the terms N_q and $N_q + 1$ correspond to the absorption and emission of a phonon, respectively. The rate

$$W^{cop}_{\lambda\lambda'}(\mathbf{k},\mathbf{k}') = \frac{4\pi}{\hbar}\begin{bmatrix} \sqrt{N_q} \\ \sqrt{N_q+1} \end{bmatrix} \frac{eF_0\gamma}{2\hbar\omega} \sum_{\mathbf{k}_1,\lambda_1} |U^{cp}_{\lambda_1\lambda'}(\mathbf{q},\theta)|^2 \sqrt{[1+\lambda\lambda_1\cos(2\phi)]/2}$$

$$\times\delta_{\mathbf{k},\mathbf{k}_1}\delta_{\mathbf{k}',\mathbf{k}_1+\mathbf{q}}\delta[E_\lambda(\mathbf{k}) - E_{\lambda_1}(\mathbf{k}_1) \pm \hbar\omega + E_{\lambda_1}(\mathbf{k}_1) - E_{\lambda'}(\mathbf{k}') \pm \hbar\omega_q]$$

describes the coupled carrier-photon-phonon scattering via absorption and emission of both photons and phonons. This term also represents the indirect optical and electronic transition channels.

Here we consider only carrier interaction with optical phonons. It is known that at relatively high-temperatures carriers in graphene interact more strongly with optical phonons (Xu et al., 2009). This coupling can result in a relatively large energy relaxation due to the inelastic nature of the scattering. Moreover, the published experimental results (Kuzmenko et al., 2008; Li et al., 2008; Choi et al., 2009) show that the optical absorption window in graphene occurs near the photon energy regime $\hbar\omega \sim 0.1$ eV. This energy is much larger than the acoustic phonon energy in graphene. As a result, at relatively high-temperatures we can neglect the effect of the quasi-elastic scattering by acoustic phonons. On the basis of the valence-force-field model, the coupling coefficient for carrier interaction with long-wavelength optical phonons in graphene is (Ando, 2007; Tse & Sarma, 2007)

$$V^\mu_\mathbf{q} = -g\gamma M^\mu_\mathbf{q}. \tag{8}$$

Here $g = (\hbar\mathcal{B}/b^2)/\sqrt{2\rho\hbar\omega_0}$, $\rho \simeq 6.5 \times 10^{-8}$ g/cm^2 is the areal density of the graphene sheet, $\omega_0 = 196$ meV the optical phonon frequency at the Γ-point, $\mathcal{B} = -d(\ln\gamma_0)/d(\ln b) \sim 2$ is a dimensionless parameter that relates to the change of the resonance integral γ_0 between nearest neighbor carbon atoms (Ando, 2007), and $b = a/\sqrt{3}$ is the equilibrium bond length. Furthermore,

$$M^l_\mathbf{q} = \begin{bmatrix} 0 & -e^{-i\phi_q} \\ e^{i\phi_q} & 0 \end{bmatrix} \text{ and } M^t_\mathbf{q} = \begin{bmatrix} 0 & ie^{-i\phi_q} \\ ie^{i\phi_q} & 0 \end{bmatrix}, \tag{9}$$

describe the coupling with, respectively, longitudinal (l) and transverse (t) phonon modes, and ϕ_q is the angle between $\mathbf{q}$ and the x axis. The carrier-phonon scattering matrix elements are

$$|U^l_{\lambda\lambda'}(\mathbf{q},\mathbf{k})| = (g\gamma/\sqrt{2})[1 - \lambda'\lambda\cos(\phi + \phi' - 2\phi_q)]^{1/2}$$

and

$$|U^t_{\lambda\lambda'}(\mathbf{q},\mathbf{k})| = (g\gamma/\sqrt{2})[1 + \lambda'\lambda\cos(\phi + \phi' - 2\phi_q)]^{1/2}.$$

2.2 Balance equations

In this research work we employ the semi-classical Boltzmann equation (BE) to study the response of the carriers in graphene to an applied radiation field. It is known that the Boltzmann equation is a powerful tool to study theoretically the linear and nonlinear responses of electrons in an electron gas system under the action of external driving ac and dc fields. In contrast to the Kubo formula based quantum theory which handles essentially the linear response, the BE can be used to study nonlinear transport and optical effects (Xu et al., 1991). In particular, the balance equation approach based on the BE can be used to study non-equilibrium electronic transport and to calculate corresponding coefficients self-consistently such as the electron density and electron drift velocity in different states (Xu, 2005). Hence, we would like to use a consistent and tractable theory to calculate both photo-excited carrier density and photo-induced electron energy loss in graphene. The BE based balance equation approach is therefore a good theoretical option. It should be noted that although the momentum- and energy-balance equations proposed by Lei and Ting (Lei, 1998) based on quantum approach can handle the nonlinear transport of electrons under strong dc and/or intense ac fields, they can only be used to calculate the averaged electron drift velocity and electron energy loss rate. This approach cannot be applied to evaluate the electron density in different states. Furthermore, by employing the balance equation approach on the basis of the BE to study graphene (Xu et al., 2009; Dong et al., 2008), we have already achieved a good agreement between theoretical results and experimental findings both qualitatively and quantitatively. The Boltzmann transport theory has also been applied to investigate graphenen system by other authors (Vasko & Ryzhii, 2008). It was found (Vasko & Ryzhii, 2008) that for a homogeneous graphene system with relatively high carrier density and long mean free path (which is indeed the case for graphene), such a theory can lead to the same results as those obtained from quantum transport theory (Falkovsky & Pershoguba, 2007; Satuber et al., 2008). For non-degenerate statistics, the BE can be written as

$$\frac{\partial f_\lambda(\mathbf{k})}{\partial t} = g_s g_v \sum_{\lambda',\mathbf{k}',\nu} [F^{\nu}_{\lambda'\lambda}(\mathbf{k}',\mathbf{k}) - F^{\nu}_{\lambda\lambda'}(\mathbf{k},\mathbf{k}')], \tag{10}$$

where $g_s = 2$ and $g_v = 2$ account, respectively, for spin and valley degeneracy, $f_\lambda(\mathbf{k})$ is the momentum distribution function for a carrier in a state $|\mathbf{k},\lambda>$, and $F^{\nu}_{\lambda'\lambda}(\mathbf{k}',\mathbf{k}) = f_{\lambda'}(\mathbf{k}')[1 - f_\lambda(\mathbf{k})]W^{\nu}_{\lambda'\lambda}(\mathbf{k}',\mathbf{k})$. Because the radiation field has been included in the electronic transition rate, the force term induced by this field does not appear in the drift term on the left-hand side of the BE to avoid double-counting. There is no simple and analytical solution to Eq. (10) with $W_{\lambda'\lambda}(\mathbf{k}',\mathbf{k})$ given by Eq. (7). In the present study we employ the usual balance-equation approach to solve the problem. For the first moment, the mass-balance equation (or rate equation) can be derived after operating with $g_s g_v \sum_{\mathbf{k}}$ on both sides of the BE. The obtained result is

$$\frac{\partial n_e}{\partial t} = \frac{\partial n_h}{\partial t} = \frac{n_h}{\tau^{\nu}_{-+}} - \frac{n_e}{\tau^{\nu}_{+-}}, \tag{11}$$

where $1/\tau^{\nu}_{\lambda'\lambda} = (16/n_{\lambda'})\sum_\nu F^{\nu}_{\lambda'\lambda}$ is the rate for scattering of a carrier from band λ' to band λ due to the νth scattering center, $F^{\nu}_{\lambda'\lambda} = \sum_{\mathbf{k}',\mathbf{k}} F^{\nu}_{\lambda'\lambda}(\mathbf{k}',\mathbf{k})$, and n_e and n_h are, respectively, the electron and hole densities in different bands. This equation implies that only inter-band scattering (i.e., $\lambda' \neq \lambda$) can alter the number of carries in a band of the graphene system. It also reflects the fact that the change of the electron number in the conduction band equals that of the hole number in the valence band, namely this equation expresses the charge number conservation in the system.

For the second moment, the energy-balance equation can be derived by operating with $\sum_{\mathbf{k},\lambda} E_\lambda(\mathbf{k})$ on both sides of the BE. From the energy-balance equation we obtain the energy transfer rate for a carrier, $P_\lambda = \sum_{\mathbf{k}} E_\lambda(\mathbf{k})\partial f_\lambda(\mathbf{k})/\partial t$, and the total energy transfer rate of the system is $P = P_+ + P_- = P_{ph} - P_{op}$, where

$$P_{op} = 4\hbar\omega \sum_{\lambda',\lambda} (F^{co,+}_{\lambda\lambda'} - F^{co,-}_{\lambda\lambda'} + F^{cop}_{\lambda\lambda'}) \tag{12}$$

is the energy transfer rate induced by optical absorption and emission via direct and indirect transition channels, and $F^{co,\pm}_{\lambda\lambda'}$ describes the absorption (+) and emission (-) of photons. Further,

$$P_{ph} = \pm 4 \sum_{\lambda',\lambda} \hbar\omega_q (F^{cp,\pm}_{\lambda\lambda'} + F^{cop}_{\lambda\lambda'}) \tag{13}$$

is the energy transfer rate induced by emission or absorption of optical phonons. In the steady state, $P = 0$ and $P_{op} = P_{ph}$ gives an energy conservation law, namely the carriers in the system gain energy from the radiation field via absorption of photons and phonons and lose energy via emission of optical phonons and photons.

For coupled carrier-photon-phonon scattering via absorption and emission of photons and optical phonons, we have

$$W^{cop}_{\lambda\lambda'}(\mathbf{k},\mathbf{k}') = \frac{4\pi}{\hbar}\left[\frac{\sqrt{N_0}}{\sqrt{N_0+1}}\right]\frac{eF_0 g\gamma^2}{2\hbar\omega}(\cos\Phi + \sin\Phi)(\cos\phi + \sin\phi)$$
$$\times \delta_{\mathbf{k}',\mathbf{k}+\mathbf{q}}\delta[E_\lambda(\mathbf{k}) - E_{\lambda'}(\mathbf{k}') + \hbar\omega \pm \hbar\omega_q],$$

with $\Phi = (\phi + \phi' - 2\phi_q)/2$. As a result, we explicitly obtain

$$F^{cop}_{+-} = F^{cop}_{-+} = 0. \tag{14}$$

This result implies that in graphene the coupled carrier-photon-phonon interaction via coupling with long wavelength optical phonons does not contribute to electronic transitions. In the steady state, i.e., for $dn_e/dt = dn_h/dt = 0$, the mass-balance equation becomes

$$F^{co,+}_{-+} + F^{cp,-}_{-+} = F^{cp,+}_{+-} + F^{co,-}_{+-}, \tag{15}$$

which reflects the fact that electrons pumped from the valence band into the conduction band are balanced by those relaxed from the conduction band into the valence band. Furthermore, the energy transfer rate induced by optical absorption in the steady state is

$$P_{op} = 4\hbar\omega \sum_{\lambda',\lambda} [F^{co,+}_{\lambda\lambda'} - F^{co,-}_{\lambda\lambda'}]. \tag{16}$$

When a graphene sheet is subjected to a radiation field, electrons in the occupied states, e.g., in the lower energy $\lambda = -$ band, are excited into the empty states, e.g., of the higher energy $\lambda = +$ band, so that an optical absorption occurs.

One of the advantages of the balance-equation approach is that we can avoid the difficulties to solve the BE directly and instead use a certain form of the carrier distribution function to calculate the physical quantities. Here we use the Fermi-Dirac type of statistical energy distribution as approximately the momentum distribution for a carrier, i.e., $f_\lambda(\mathbf{k}) \simeq f_\lambda(\lambda\gamma k)$, with $f_\lambda(x) = [1 + e^{(x-\mu^*_\lambda)/k_B T}]^{-1}$ where μ^*_λ is the quasi Fermi energy (or quasi chemical potential) for electrons or holes in the presence of the radiation field. For carrier-photon interaction via inter- and intra-band transition channels we have $F_{+-}^{co,+} = F_{-+}^{co,-} = F_{++}^{co,-} = F_{--}^{co,-} = 0$,

$$F_{-+}^{co,+} = \frac{e^2 F_0^2}{32\hbar^2\omega} f_-(-\frac{\hbar\omega}{2})[1 - f_+(\frac{\hbar\omega}{2})] \tag{17}$$

is for optical absorption from the valence band to the conduction band,

$$F_{+-}^{co,-} = \frac{e^2 F_0^2}{32\hbar^2\omega} f_+(\frac{\hbar\omega}{2})[1 - f_-(-\frac{\hbar\omega}{2})] \tag{18}$$

is for optical emission from the conduction band to the valence band, and

$$F_{\lambda\lambda}^{co,+} = \frac{e^2 F_0^2}{8\pi\hbar^4\omega^3} \frac{\omega\tau_\lambda}{1 + (\omega\tau_\lambda)^2} \int_0^\infty dE E f_\lambda(\lambda E)[1 - f_\lambda(\lambda E)] \tag{19}$$

is induced by intra-band optical absorption in the conduction band ($\lambda = 1$) and valence band ($\lambda = -1$) with τ_λ being the energy relaxation time for an electron or a hole in different bands. In fact, this term is caused by the usual free-carrier absorption channels. The energy relaxation time is used to describe the broadening of the scattering states, which can result in a spectrum structure for the intra-band optical absorption. It should be noted that for intra-band free-carrier absorption, the momentum conservation law still holds (i.e., for optical transitions the electron momentum at initial and final states must be the same during a scattering event, as given by Eq. (4)). However, in the presence of the external driving fields such as the radiation fields and of the scattering centers such as impurities and phonons, the scattering states are damped and broadened. As a result, the δ-function in Eq. (7) for intra-band optical transition can be replaced through Poisson Kernel: $\delta(E) \to (E_\lambda/\pi)(E^2 + E_\lambda^2)^{-1}$ with $E_\lambda = \hbar/\tau_\lambda$ being the energy broadening of the states.

For carrier-phonon interaction via different transition channels, we have $F_{+-}^{cp,-} = F_{-+}^{cp,+} = 0$,

$$F_{+-}^{cp,+} = \frac{g^2(N_0 + 1)}{2\pi\hbar\gamma^2} \int_0^{\hbar\omega_0} dx x(\hbar\omega_0 - x) f_+(x)[1 - f_-(x - \hbar\omega_0)] \tag{20}$$

is for phonon emission and corresponding electronic transition from the conduction band to the valence band, with $N_0 = [e^{\hbar\omega_0/k_BT} - 1]^{-1}$ and $\omega_q \to \omega_0$ a constant at the Γ-point for long-wavelength optical phonons, and

$$F_{-+}^{cp,-} = \frac{g^2 N_0}{2\pi\hbar\gamma^2} \int_0^{\hbar\omega_0} dx x(\hbar\omega_0 - x) f_-(-x)[1 - f_+(\hbar\omega_0 - x)], \tag{21}$$

is induced by the absorption of phonons and corresponding electronic transition from the valence band to the conduction band.

2.3 Photo-induced carriers

We now consider a graphene layer in which the conducting carriers are electrons (or a positive gate voltage is applied) in the absence of the radiation field (or in the dark). When a light field is applied to the system, the electrons in the valence band are excited into the conduction band so that photo-excited carriers can be induced. If n_0 is the electron density in the absence of the radiation field (or dark density) at $F_0 = 0$, the electron density at $F_0 \neq 0$ is $n_e = n_0 + \Delta n_e$. On account of charge number conservation we have $\Delta n_e = n_h$, the hole density in the presence of radiation field. Thus, we get

$$n_e = n_0 + n_h, \tag{22}$$

with

$$n_e = g_s g_v \sum_{\mathbf{k}} f_+(\mathbf{k}) = \frac{2}{\pi\gamma^2} \int_0^\infty \frac{dx\, x}{e^{(x-\mu_e^*)/k_B T} + 1} \tag{23}$$

and

$$n_h = n_e - n_0 = g_s g_v \sum_{\mathbf{k}} [1 - f_-(\mathbf{k})] = \frac{2}{\pi\gamma^2} \int_0^\infty \frac{dx\, x}{e^{(x+\mu_h^*)/k_B T} + 1}. \tag{24}$$

With the mass-balance equation given by Eq. (15) and the requirement of the charge number conservation shown as Eq. (22) we can determine the quasi chemical potentials μ_λ^* for electrons and holes. Then the electron density n_e and hole density n_h can be obtained in the presence of the radiation field F_0. We notice that this approach can also be applied to p-type graphene samples, when a negative gate voltage is applied so that the conducting carriers are holes in the valence band in the dark.

2.4 Optical conductance and transmission

With the obtained carrier chemical potential μ_λ^* we can calculate $F_{\lambda\lambda'}^{co,\pm}$. From the carrier energy transfer rate induced by optical absorption, described by Eq. (16), we can calculate the optical conductance $\sigma(\omega)$ for graphene using the expression (Wei et al., 2007)

$$\sigma(\omega) = 2P_{op}/F_0^2 = 8\hbar\omega \sum_{\lambda,\lambda'} (F_{\lambda,\lambda'}^{co,+} - F_{\lambda,\lambda'}^{co,-})/F_0^2. \tag{25}$$

Moreover, the transmission coefficient for a device with a graphene layer on top of a substrate, namely for an air-graphene-wafer system, is given by (Satuber et al., 2008)

$$T(\omega) = \sqrt{\frac{\epsilon_2}{\epsilon_1}} \frac{4(\epsilon_1\epsilon_0)^2}{|(\sqrt{\epsilon_1\epsilon_2} + \epsilon_1)\epsilon_0 + \sqrt{\epsilon_1}\sigma(\omega)/c|^2}, \tag{26}$$

where $\epsilon_1 \simeq 1$ for air, ϵ_2 is the effective high-frequency dielectric constant of the substrate, and ϵ_0 and c are, respectively, the dielectric constant and the speed of light in vacuum. It indicates that the light transmittance of the graphene layer in an air-graphene-substrate system depends on the dielectric constant of the substrate material. Moreover, a substrate with a larger dielectric constant can result in a smaller light transmittance for the graphene layer.

3. Numerical results and discussions

In the numerical calculations we consider a typical air-graphene-SiO_2/wafer system. Thus, $\epsilon_1 \simeq \epsilon_0 = 1$ and $\epsilon_2 \simeq 2.0$. The effect of the dielectric mismatch between the graphene layer and the SiO_2 substrate has been taken into account using the image charge method (Dong et al., 2008). Furthermore, it has been obtained experimentally (Sun et al., 2008) that in graphene the energy relaxation time is about $\tau_\lambda \sim 1$ ps for high-density samples. Thus, we take $\tau_\lambda \sim 1$ ps in the calculation for free-carrier absorption. A typical electric field strength of the radiation field $F_0 = 500$ V/cm is used in most of the calculations.

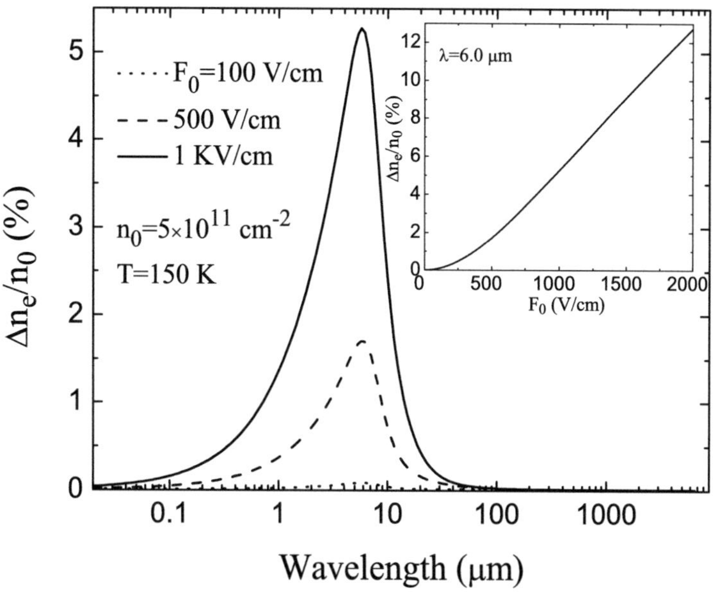

Fig. 1. Photo-excited electron density Δn_e as a function of the radiation wavelength, at temperature $T = 150$ K and a fixed dark electron density $n_0 = 5 \times 10^{11}$ cm^{-2}, for different strengths of the radiation field F_0. The inset shows Δn_e as a function of F_0 at a fixed radiation wavelength $\mathcal{L} = 6.0$ μm. Note that Δn_e is equal to n_h, the photo-induced hole density.

3.1 Photo-induced carrier densities

In Fig. 1 we show the dependence of the photon-excited electron density Δn_e on the radiation wavelength ($\mathcal{L}$) and radiation intensity for a fixed dark electron density n_0 and a temperature $T = 150$ K. Δn_e vs the strength of the radiation field F_0 is also shown in the inset for fixed radiation wavelength $\mathcal{L} = 6.0$ μm. It should be noted that $\Delta n_e = n_h$ is also the photo-induced hole density for a n-type graphene in the dark. As expected, the photo-induced carrier densities increase with radiation intensity $I \sim F_0^2$. For a typical radiation intensity with F_0 about 500 V/cm, several percents of photo-induced electron density can be achieved in graphene, similar to photo-excited carriers in conventional semiconductors. We notice that when the radiation wavelength is about $\mathcal{L} \sim 6$ μm Δn_e is maximum. Because the optical phonon wavelength in graphene is about 6 μm, the peak of Δn_e appears at about this wavelength. For $\mathcal{L} > 6$ μm, Δn_e decreases sharply with increasing $\mathcal{L}$. These results indicate that in graphene the photo-induced carrier densities can be observed clearly in the infrared bandwidth.

In Fig. 2 we show the photo-excited electron density Δn_e as a function of the radiation wavelength, at fixed dark electron density and radiation intensity, for different temperatures. We see that the photo-excited carrier densities are very sensitive to temperature. Δn_e decreases quickly with increasing temperature. Such a feature is typical when scattering by optical phonons is present. With increasing temperature, the phonon occupation number $N_0 = [e^{\hbar\omega_0/k_B T} - 1]^{-1}$ increases sharply so that a stronger phonon scattering occurs. The strong phonon scattering can bring electrons from the conduction band to the valence band via phonon emission. As a consequence, at relatively high temperatures less photo-excited electrons remain in the conduction band. Accordingly, Δn_e decreases with increasing

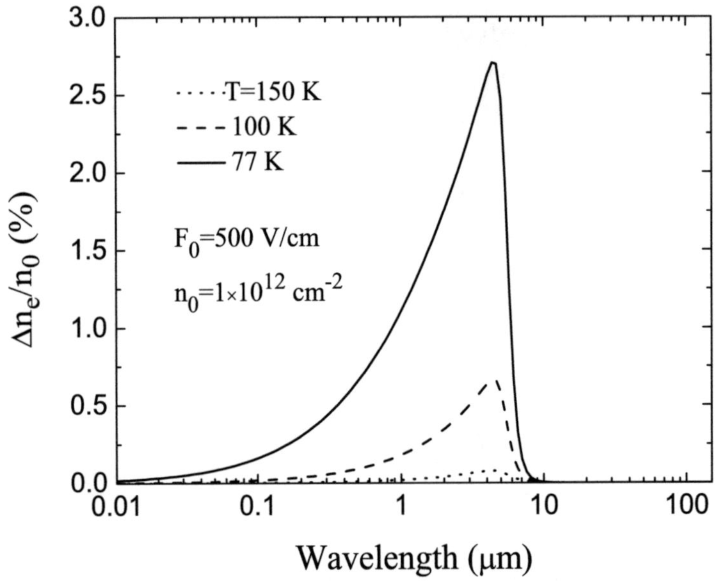

Fig. 2. Photo-excited electron density Δn_e as a function of the radiation wavelength, for fixed dark electron density $n_0 = 1 \times 10^{12}$ cm^{-2} and radiation intensity $F_0 = 500$ V/cm, at different temperatures.

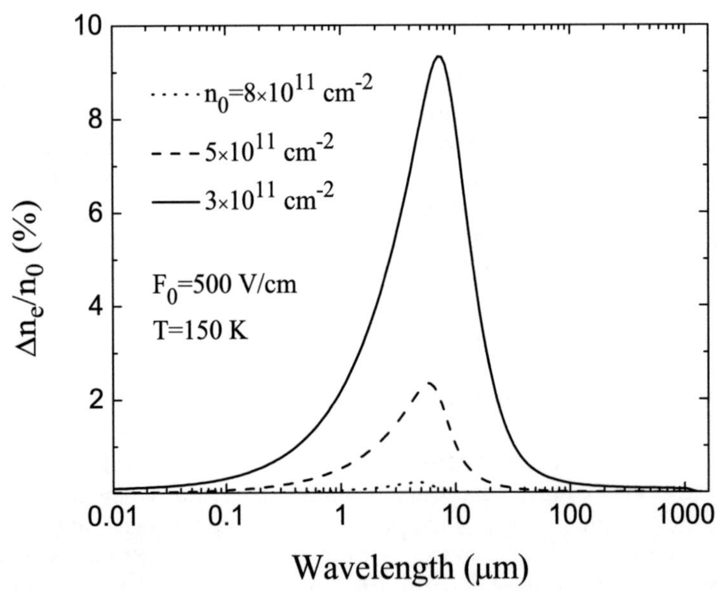

Fig. 3. Δn_e as a function of the radiation wavelength at a temperature $T = 150$ K and fixed strength of the radiation field $F_0 = 500$ V/cm, for different dark electron densities.

temperature. These results suggest that phonon scattering is an important mechanism in affecting the photo-excited carrier densities in graphene.

In Fig. 3 Δn_e is plotted as a function of the radiation wavelength, for $T = 150$ K and $F_0 = 500$ V/cm, for different dark electron densities. The main feature in Fig. 3 is that the photo-induced carrier density ratio $\Delta n_e/n_0$ decreases with increasing dark electron density n_0. At relatively weak levels of light excitation Δn_e changes only by several percents (see also Figs. 1 and 2). This implies that the change of the quasi Fermi levels in the system due to a weak excitation is not very significant. A larger n_0 means a higher Fermi level in the conduction band in the absence of the radiation field. As we know, the electrons make transitions mainly from the occupied states to the empty states in an electron gas system. For larger n_0 only the higher energy states above the Fermi level are available for photo-induced electrons to be excited to. Thus, larger photon energy (or shorter wavelength radiation) is required to excite electrons from the valence band to the conduction band. On the other hand, because of the linear energy spectrum for carriers in graphene, larger energy states correspond to larger momentum states in the conduction band. Since the electron-phonon scattering alters both the energy and momentum, as required by the corresponding energy and momentum conservation laws, larger momentum states can result in stronger electron-phonon coupling. Again, the strong phonon scattering can reduce the photo-induced electron density in the conduction band. Therefore, the densities shown in Fig. 3 are, at least partly, the result of electron-phonon interaction. Moreover, it should be noted that the dark carrier density in graphene can be modulated effectively by applying a gate voltage (Li et al., 2008; Wang et al., 2008). Hence, the densities of photo-induced carriers can be modulated electrically as well.

The main results from Figs. 1 - 3 are as follows. (i) The photo-induced electron density increases with radiation wavelength for $\mathcal{L} < 6.0$ μm. (ii) The peak of the photo-excited electron density can be observed around $\mathcal{L} \sim 6.0$ μm, which corresponds to the optical phonon energy $\hbar\omega_0 = 196$ meV in graphene. (iii) The photo-induced electron density decreases rapidly with increasing radiation wavelength for $\mathcal{L} > 6.0$ μm. When a graphene sample is subjected to a radiation field and the electron-phonon interaction is present, electrons in the valence band can gain energy from the radiation field through optical absorption and be excited into the conduction band, while electrons in the conduction band can lose energy via emission of phonons and be relaxed into the valence band. The balance of these two competing processes results in the photo-induced carriers in the system in the steady state. As we know, the optical phonon emission occurs when the electron energy is larger than the optical phonon energy (Xu et al., 1993). For $\mathcal{L} < 6.0$ μm electrons located around the top of the valence band can gain a photon energy $\hbar\omega > \hbar\omega_0$ via optical absorption and be excited into the conduction band. These electrons can lose energy $\hbar\omega_0$ through emission of optical phonons and be relaxed into the empty states of the valence band. This is why photo-excited carrier densities increase with the radiation wavelength for $\mathcal{L} < 6.0$ μm. At relatively long-wavelength radiations for $\mathcal{L} > 6.0$ μm, the electrons around the top of the valence band gain energy $\hbar\omega < \hbar\omega_0$ and are excited into the conduction band. However, when the electron energy is less than the optical phonon energy electronic transitions via phonon emission are much less likely (Xu et al., 1993). Thus, for $\hbar\omega < \hbar\omega_0$ the electrons in the conduction band are less likely to be relaxed into the valence band though electron-phonon coupling. This is the main reason why the photo-excited carrier densities decrease sharply with increasing radiation wavelength for $\hbar\omega < \hbar\omega_0$. When the photon energy is close to the optical phonon energy $\hbar\omega \sim \hbar\omega_0$, inter-band electronic transitions can occur through a process in which the electrons gain a photon energy $\hbar\omega$ from the radiation field and lose a phonon energy $\hbar\omega_0$ via phonon emission.

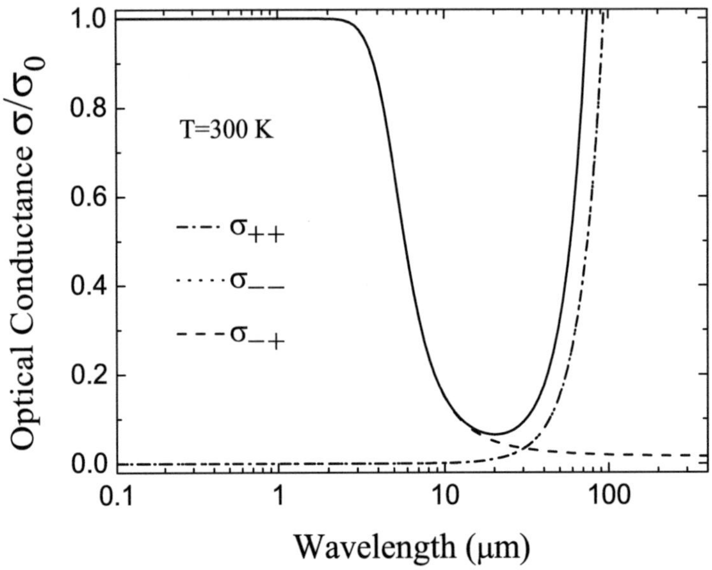

Fig. 4. Contribution to the total optical conductance (solid line) from different transition channels at room temperature for a fixed dark electron density $n_0 = 1 \times 10^{12}$ cm^{-2}. Here $\sigma_0 = e^2/(4\hbar)$ and σ_{-+} (dashed curve) is for transition from valence band ($-$) to conduction band ($+$). The curves for intra-band transitions, σ_{++} and σ_{--}, coincide roughly.

Thus, a resonant transition occurs in graphene. Such a mechanism is electrically equivalent to the electro-phonon resonance effect (Xu et al., 1993) proposed previously by us. As a result, the strongest inter-band electronic transitions can occur for $\hbar\omega \sim \hbar\omega_0$. That is, a peak of the photo-induced carrier density can be observed in graphene when the radiation frequency is close to the optical phonon frequency. The theoretical results discussed here indicate that in gapless graphene the scattering by optical phonons plays an important role in affecting photo-excited carrier densities.

3.2 Optical conductance and transmission

In Fig. 4, we show the contributions from different transition channels to the optical conductance $\sigma(\omega)$ or optical absorption for a fixed dark electron density n_0 at room temperature. We notice the following features. (i) Inter-band transitions contribute to the optical absorption in the short-wavelength regime ($\mathcal{L} < 3$ μm), whereas intra-band transitions give rise to the long-wavelength optical absorption. (ii) The optical absorption varies very little upon varying the radiation frequency in the short-wavelength regime ($\mathcal{L} < 3$ μm), whereas the optical conductance or absorption coefficient depends strongly on the radiation wavelength in the long-wavelength regime ($\mathcal{L} > 3$ μm). (iii) The optical conductance in the short-wavelength regime is a universal value $\sigma_0 = e^2/(4\hbar)$ in monolayer graphene as discovered experimentally (Kuzmenko et al., 2008; Li et al., 2008; Nair et al., 2008). (iv) More interestingly, there is an infrared absorption window in the $4 \sim 100$ μm wavelength range. As expected, inter-band transitions require larger photon energy. Intra-band transitions, which are caused by the usual free-carrier absorption, occur under low photon energy radiation. It is a common feature for free-carrier absorption that the strength of the optical absorption increases rapidly with radiation wavelength (Li, 2006). We find that the optical absorption window observed experimentally (Kuzmenko et al., 2008; Li et al., 2008; Choi et al., 2009) is

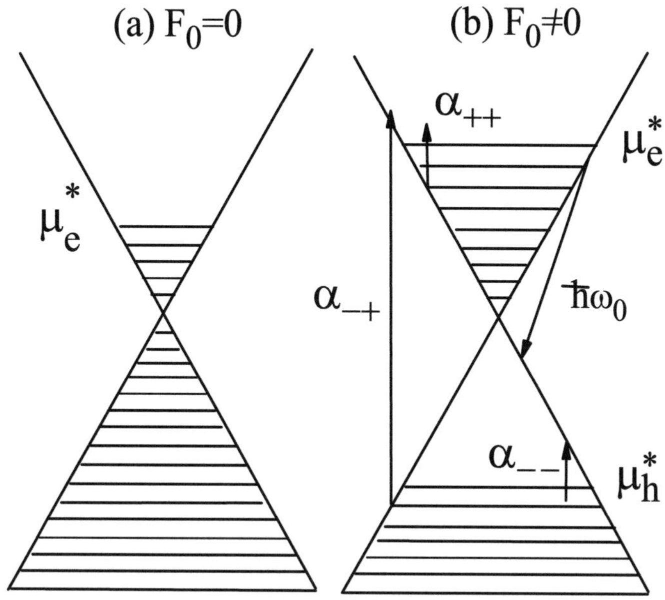

Fig. 5. (a) A graphene system in the absence of the radiation field ($F_0 = 0$). The conducting carriers are electrons with a Fermi energy μ_e^* in the conduction band. The hatched area shows the occupied states. (b) Optical absorption channels in the presence of a radiation field ($F_0 \neq 0$). Here μ_e^* and μ_h^* are the quasi Fermi energies for, respectively, electrons and holes and there are three optical absorption channels: α_{-+}, α_{++}, and α_{--}. $\hbar\omega_0$ is the energy of an emitted optical phonon.

induced by the competing absorption channels due to inter- and intra-band scattering events in graphene. This can explain and reproduce recent experimental findings (Kuzmenko et al., 2008; Li et al., 2008; Choi et al., 2009).

The interesting features of optical absorption in graphene can be understood with the help of Fig. 5. When the radiation field is absent, there is a single Fermi level (or chemical potential) in the conduction band in a n-type graphene sample (or in the presence of a positive gate voltage). In this case all states below μ_e^* are occupied by electrons as shown in Fig. 5 (a). When a radiation field is applied to the system (see Fig. 5 (b)), the electrons in the valence band can gain energy from the radiation field and be excited into the conduction band via absorption of photons. Thus, the electron density in the conduction band increases and so does the quasi Fermi level μ_e^* for electrons. Meanwhile, the holes are left in the valence band and a quasi Fermi level μ_h^* is established in this band for them. As shown in Fig. 5 (b), in the presence of a radiation field the intra-band electronic transition accompanied by the absorption of photons can occur not only in the conduction band via the α_{++} channel but also in the valence band via the α_{--} channel. The intra-band transitions are a direct consequence of the broadening of the scattering states in the conduction and valence bands. At the same time, the electrons in the conduction band can lose energy via emission of optical phonons and relax into the valence-band. The electrons in the valence band can also gain energy by absorption of optical phonons and be excited into the conduction band due to the inelastic character of the electron-phonon scattering, although such a process is much weaker than that for phonon emission. Because graphene is a gapless semiconductor, the electrons in the valence band can be more easily excited into the conduction band via optical absorption and

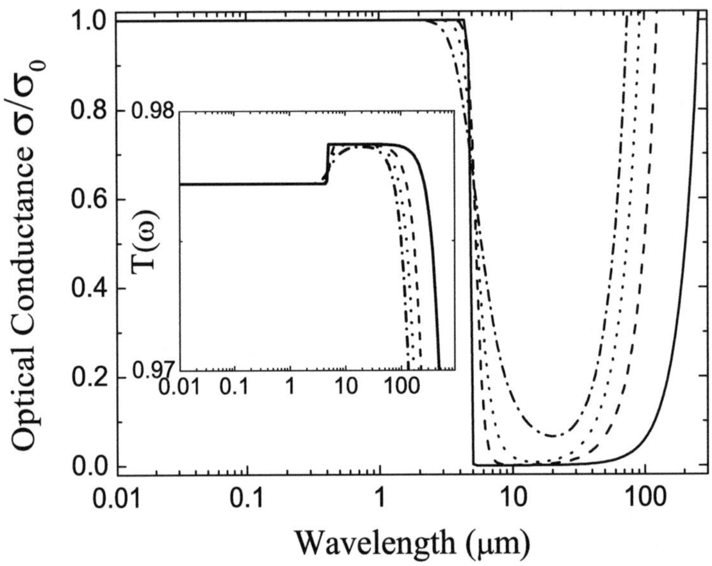

Fig. 6. Optical conductance and transmission coefficient (inset) as a function of the radiation wavelength, at a fixed dark electron density $n_0 = 1 \times 10^{12}$ cm^{-2}, for different temperatures $T = 10$ K (solid curve), 77 K (dashed curve), 150 K (dotted curve) and 300 K (dotted-dashed curve).

those in the conduction band can be easily relaxed into the valence band via phonon emission, in contrast to a conventional semiconductor. Thus, there is a strong inter-band optical and electronic transition channel (i.e., α_{-+} in Fig. 5 (b)) in graphene. Since optical absorption and phonon emission events describe transitions from occupied states to empty states, the intra-band transitions require less photon energy whereas a relatively larger photon energy is needed for inter-band transitions. Consequently, an optical absorption window can be induced through different energy requirements for inter- and intra-transition channels.

In Fig. 6 we show the optical conductance σ and transmission coefficient $T(\omega)$ as a function of the radiation wavelength, at fixed dark electron density n_0, for different temperatures. As can be seen, in the short-wavelength regime $\mathcal{L} < 3$ μm, both σ and $T(\omega)$ depend very little on the radiation wavelength. This confirms that σ does not depend on temperature under short-wavelength radiation in graphene. The corresponding transmission coefficient $T(\omega)$ is about 0.97 $\sim$ 0.98 in the short-wavelength regime and agrees quantitatively with the experimental data (Nair et al., 2008). In the long-wavelength regime, in which the optical absorption window can be observed, both the optical conductance and light transmittance depend sensitively on the temperature, which is in line with the experimental findings (Kuzmenko et al., 2008). It should be noted that for fixed electron and hole densities, the quasi chemical potential for electrons/holes decreases/increases with increasing temperature. Thus, due to the Pauli blockade effect (Krenner et al., 2006), a blue shift of the optical absorption window with decreasing temperature can be observed as shown in Fig. 6. We note that the strength of the optical absorption is proportional to the optical conductance. Therefore, the height of the optical absorption window decreases with increasing temperature. We find that a wider and deeper optical absorption window and a sharper cut-off of the optical absorption at the window edges can be observed at lower temperatures. These theoretical results can be used to explain the strong dependence of the optical conductance on temperature in the

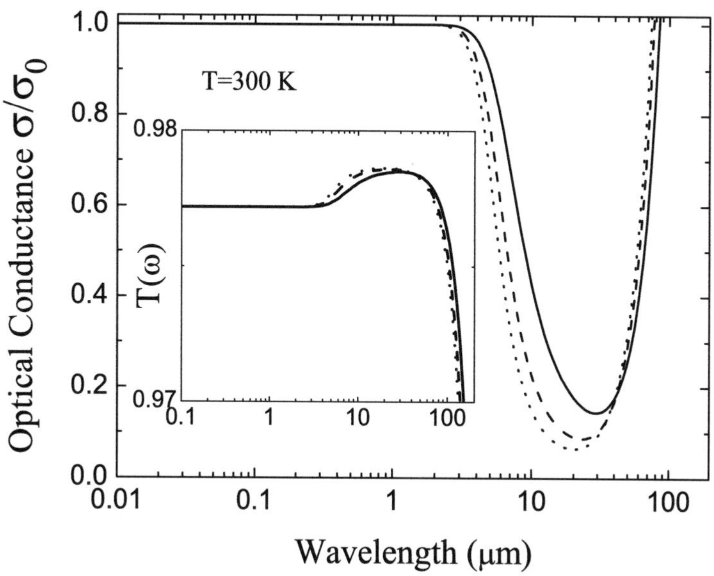

Fig. 7. Optical conductance and transmission coefficient (inset) as a function of the radiation wavelength, at room temperature, for different dark electron densities $n_0 = 5 \times 10^{11}$ cm^{-2} (solid curve), 8×10^{11} cm^{-2} (dashed curve), and 1×10^{12} cm^{-2} (dotted curve).

infrared bandwidth that was observed experimentally (Kuzmenko et al., 2008; Nair et al., 2008).

The optical conductance σ and transmission coefficient $T(\omega)$ are shown in Fig. 7 as a function of the radiation wavelength at room-temperature for different dark electron densities n_0. We note that the Fermi level for electrons becomes higher with increasing dark electron density. A higher Fermi level for electrons implies that the empty states in the conduction band have higher energies because of the linear shape of the energy spectrum for graphene. Since the optical transitions occur mainly via exciting electrons from occupied states to the empty states, a higher Fermi level corresponds to a higher transition energy. Thus, a blue shift of the optical absorption window can be observed in graphene samples with larger dark electron densities as shown in Fig. 7. This blue shift, with increasing gate voltage, has been observed experimentally (Li et al., 2008). Because in graphene the dark electron density increases almost linearly with increasing gate voltage (Novoselov et al., 2005), the theoretical results shown in Fig. 7 are in agreement with these experimental findings obtained from optical absorption measurements (Li et al., 2008). Furthermore, we find that the height of the absorption window increases with dark electron density and a sharper cut-off of the optical absorption at the window edges can be observed for larger electron densities. These theoretical results suggest that the width and height of the infrared absorption window in graphene can be controlled by applying a gate voltage. This feature can be utilized for making graphene-based and frequency-tunable infrared optoelectronic devices.

4. Conclusions

In this study we have examined theoretically the effect of optical phonon scattering on the optoelectronic properties of graphene. On the basis of the Boltzmann equation approach, we have derived the mass-balance and energy-balance equations for graphene in the presence

of a linearly polarized radiation field and of the electron-photon-phonon coupling. By solving these equations self-consistently, we have been able to determine the photo-excited carrier densities and optoelectronic coefficients, such as the optical conductance and light transmittance, for an air-graphene-wafer system. In particular, we have investigated the dependence of the photo-induced carrier densities, optical conductance, and transmission coefficient on intensity and wavelength of the radiation field, along with those on temperature and dark electron density. The main conclusions we have obtained are summarized as follows.
Because graphene is a gapless electronic system, electron-phonon interaction is an important mechanism in affecting the electronic transitions via both intra- and inter-band transition channels. In graphene the electrons in the valence band can gain the energy from the radiation field via optical absorption and be excited into the conduction band. At the same time, the electrons in the conduction band can lose energy via emission of phonons and be relaxed into the valence band. Thus, electron-phonon scattering can affect strongly the inter-band transition in contrast to conventional semiconductors. As a result, the electron-photon-phonon interaction is a major scattering mechanism to determine photo-induced carrier densities and optoelectronic properties of graphene.
In the presence of a radiation field, the photo-excited carrier densities in graphene first increase and then decrease with increasing radiation wavelength. The largest carrier densities caused by light radiation can be observed when the radiation photon energy equals to the optical phonon energy of graphene. Such resonant transitions are electrically equivalent to the electro-phonon resonance effect observed in conventional two-dimensional electron gas systems. The photo-excited carrier densities depend strongly on the radiation intensity and frequency, temperature, and dark carrier density.
In the short-wavelength regime ($\mathcal{L} < 3\ \mu$m), the universal optical conductance $\sigma_0 = e^2/(4\hbar)$ and light transmittance $T_0 \sim 0.98$ can be achieved for an air-graphene-SiO_2/wafer system. The optical conductance and transmission coefficient depend very little on temperature and dark electron density. These results agree with other theoretical works and with experimental findings.
We have found that there is an optical absorption window in the radiation wavelength range $4 \sim 100\ \mu$m. This infrared absorption window is induced by different transition energies required for inter- and intra-band optical absorption in the presence of the Pauli blockade effect. The depth and width of such an absorption window depend sensitively on the temperature and dark electron density in the sample due to the presence of a free-carrier absorption in this radiation wavelength regime. A prominent cut-off of the optical absorption can be observed at the edges of the window at lower temperatures and/or larger dark electron densities. These results can explain why experimentally the optical absorption window can be measured under long-wavelength radiation and why experimentally a blue shift of such a window can be observed when increasing the gate voltage.
The results obtained from this study indicate that in addition to the excellent optical properties of graphene in the visible regime, i.e., universal optical conductance, high light transmittance, etc., graphene can exhibit interesting and important features in the mid-infrared bandwidth, such as the optical absorption window. The width and the depth of this window can be tuned by varying the dark carrier density via, e.g., applying a gate voltage. This implies that graphene can be used as a frequency-tunable optoelectronic device operating in the mid-infrared bandwidth at room temperature for various applications. Together with the relevant phenomena discovered by very recent experimental work (Kuzmenko et al., 2008; Li et al., 2008; Choi et al., 2009), we hope that the present work sheds some light on the

application of graphene not only as a visible optoelectronic device but also as an infrared device in ambient conditions.

5. References

Novoselov, K.S.; Geim, A.K.; Morozov, S.V.; Jiang, D.; Zhang, Y.; Dubonos, S.V.; Grigoreva, I.V. & Firsov, A.A. (2004). Electric Field Effect in Atomically Thin Carbon Films, *Science* 306, pp. 666-669.

Zhang, Y.B.; Tan, Y.W.; Stormer, H.L.; & Kim, P. (2005). Experimental observation of the quantum Hall effect and Berry's phase in graphene, *Nature (London)* 438, pp. 201-204.

Berger C.; Song, Z.M.; Li, X.b.; Wu, X.S.; Brown, N.; Naud, C.; Mayou, D.; Li, T.B.; Hass, J.; Marchenkov, A.N.; Conrad, E. H.; First, P. N. & Heer, Walt A. de (2006). Electronic Confinement and Coherence in Patterned Epitaxial Graphene, *Science* 312, PP. 1191-1196.

Novoselov, K.S.; Geim, A.K.; Morozov, S.V.; Jiang, D.; Katsnelson, M.I.; Grigorieva, I.V.; Dubonos, S.V. & Firsov, A.A. (2005). Two-dimensional gas of massless Dirac fermions in graphene, *Nature* 438, PP. 197-200.

Castro, E. V.; Novoselov, K.S.; Morozov, S.V.; Peres, N.M.R.; Lopes dos Santos, J.M.B.; Nilsson, J.; Guinea, F.; Geim, A.K. & Castro Neto, A.H. (2007). Biased Bilayer Graphene: Semiconductor with a Gap Tunable by the Electric Field Effect, *Phys. Rev. Lett.* 99, PP. 216802-216805.

González, J. & Perfetto, E. (2008). Critical currents in graphene Josephson junctions, *J. Phys.: Condens. Matter* 20, pp. 145218-145226.

Lin, Y.M.; Jenkins, K. A.; Alberto V.G.; Small, J. P.; Farmer, D.B. & Avouris, P. (2009). Operation of Graphene Transistors at Gigahertz Frequencies, *Nano Letters* 9, pp. 422-426.

Eda, G.; Fanchini, G.; & Chhowalla, M. (2008). Large-area ultrathin films of reduced graphene oxide as a transparent and flexible electronic material, *Nature Nanotechnol.* 3, pp. 270-274.

Hogan, Hank (2008). *Photonics Spectra* 42, pp. 19.

Kuzmenko, A.B.; Heumen, E. van; Carbone, F.; & Marel, D. van der (2008). Universal Optical Conductance of Graphite, *Phys. Rev. Lett.* 100, pp. 117401-117404.

Li, Z.Q.; Henriksen, E.A.; Jiang, Z.; Hao, Z.; Martin, M.C.; Kim, P.; Stormer, H.L.; & Basov, D.N. (2008). Dirac charge dynamics in graphene by infrared spectroscopy, *Nat. Phys.* 4, pp. 532-535.

Nair, R.R.; Blake, P.; Grigorenko, A.N.; Novoselov, K.S.; Booth, T.J.; Stauber, T.; Peres, N.M.R. & Geim, A.K. (2008). Fine Structure Constant Defines Visual Transparency of Graphene, *Science* 320, pp. 1308.

Choi, H.; Borondics, F.; Siegel, D.A.; Zhou, S.Y.; Martin, M.C.; Lanzara, A. & Kaindl, R.A. (2009). Broadband electromagnetic response and ultrafast dynamics of few-layer epitaxial graphene, *Appl. Phys. Lett.* 94, pp. 172102-172104.

Wang, F.; Zhang, Yuanbo; Tian, Chuanshan; Giri, Caglar; Zettl, Alex; Crommie, Michael & Shen, Y. Ron (2008). Gate-Variable Optical Transitions in Graphene, *Science* 320, pp. 206-209.

Vasko, F.T. & Ryzhii, V. (2008). Photoconductivity of intrinsic graphene, *Phys. Rev. B* 77, pp. 195433-195440.

Falkovsky, L.A. & Pershoguba, S.S. (2007). Optical far-infrared properties of a graphene monolayer and multilayer, *Phys. Rev. B* 76, pp. 153410-153413.

Stauber, T.; Peres, N.M.R. & Geim, A.K. (2008). Optical conductivity of graphene in the visible region of the spectrum, *Phys. Rev. B* 78, pp. 085432-085439.

Ando, T. (2007). Magnetic Oscillation of Optical Phonon in Graphene, *J. Phys. Soc. Jpn.* 76, pp. 024712-024729.

McCann, E. & Falko, V.I. (2006). Landau-Level Degeneracy and Quantum Hall Effect in a Graphite Bilayer, *Phys. Rev. Lett.* 96, pp. 086805-086808.

Stroscio. Michael A. & Mitra Dutta (2005). *Phonons in Nanostructures*, Cambridge University Press, Cambridge.

Xu, W.; Peeters, F.M. & Lu, T.C. (2009). Dependence of resistivity on electron density and temperature in graphene, *Phys. Rev. B* 79, pp. 073403-073406.

Tse, Wang-Kong & Sarma S. Das (2007). Phonon-Induced Many- Body Renormalization of the Electronic Properties of Graphene, *Phys. Rev. Lett.* 99, pp. 236802-236805.

See, e.g., Xu, W.; Peeters, F.M.; & Devreese, J.T. (1991). Diffusion-to-streaming transition in a two-dimensional electron system in a polar semiconductor, *Phys. Rev. B* 43, pp. 14134-14141.

See, e.g., Xu, W. (2005). Screening length and quantum and transport mobilities of a heterojunction in the presence of the Rashba effect, *Phys. Rev. B* 71, pp. 245304-245312.

See, e.g., Lei, X.L. (1998). Balance-equation approach to hot- electron transport in semiconductors irradiated by an intense terahertz field, *J. Appl. Phys.* 84, pp. 1396-1404.

Dong, H.M.; Xu, W.; Zeng, Z.; Lu, T.C. & Peeters, F.M. (2008). Quantum and transport conductivities in monolayer graphene, *Phys. Rev. B 77*, pp. 235402-235410.

Wei, X.F.; Xu, W & Zeng, Z. (2007). Two-colour mid- infrared absorption in an InAs/GaSb-based type II and broken-gap quantum well, *J. Phys.: Condens. Matter* 19, pp. 506209-506215.

Sun, Dong; Wu, Zong-Kwei; Divin, Charles; Li, Xuebin; Berger, Claire; de Heer, Walt A.; First, Phillip N. & Norris, Theodore B. (2008). Ultrafast Relaxation of Excited Dirac Fermions in Epitaxial Graphene Using Optical Differential Transmission Spectroscopy, *Phys. Rev. Lett.* 101, pp. 157402-157405.

Xu, W.; Peeters, F.M. & Devreese, J.T. (1993). Electrophonon resonances in a quasi-two-dimensional electron system, *Phys. Rev. B* 48, pp. 1562-1570.

See, e.g., Li, Sheng S. (2006). *Semiconductor Physical Electronics* (2nd Edition) (Springer-Verlag, Berlin).

Krenner, H.J.; Clark, E.C.; Nakaoka, T.; Bichler, M.; Scheurer, C.; Abstreiter, G. & Finles, J.J.(2006). Optically Probing Spin and Charge Interactions in a Tunable Artificial Molecule, *Phys. Rev. Lett.* 97, pp. 076403-076406.

25

Frequency Mixing Effects in Graphene

Sergey Mikhailov
University of Augsburg
Germany

1. Introduction

Graphene is a new nanomaterial which has been discovered a few years ago, Novoselov et al. (2004); Novoselov et al. (2005); Zhang et al. (2005) and has demonstrated unique mechanical, electrical, thermal and optical properties, see review articles Katsnelson (2007); Castro Neto et al. (2009); Geim (2009). This is a one-atom-thick layer of carbon atoms arranged in a highly symmetric two-dimensional honey-comb lattice, Figure 1. Graphene exhibits many interesting fundamental physical properties such as the minimal electrical conductivity Novoselov et al. (2005); Zhang et al. (2005); Katsnelson (2006); Nomura & MacDonald (2007); Tan et al. (2007), unconventional quantum Hall effect Novoselov et al. (2005); Zhang et al. (2005) observable up to room temperatures Novoselov et al. (2007), Klein tunneling Stander et al. (2009); Young & Kim (2009), optical conductivity determined only by the fine structure constant Ando et al. (2002); Kuzmenko et al. (2008); Nair et al. (2008) and many other. Graphene promises many electronic applications like terahertz transistors, photodetectors, transparent electrodes for displays, gas and strain sensors and so on, Geim (2009).

Microscopically, the most distinguished feature of graphene is that, in contrast to other (semiconductor) materials with two-dimensional electron gases, electrons and holes in graphene have not a parabolic, but a linear energy spectrum near the Fermi level Wallace (1947); McClure (1956); Slonczewski & Weiss (1958). The Brillouin zone of graphene electrons has a hexagonal shape, Figure 2, and near the corners $\mathbf{K}_j$, $j = 1, \dots 6$, the electron and hole energy bands $E_{l\mathbf{p}}$ touch each other; here $\mathbf{p}$ is the quasi-momentum of an electron and l is the band index ($l = 1$ for holes and $l = 2$ for electrons). The spectrum $E_{l\mathbf{p}}$ near these, so called Dirac points is linear,

$$E_{l\mathbf{p}} = (-1)^l V|\mathbf{p} - \hbar\mathbf{K}_j| = (-1)^l V|\tilde{\mathbf{p}}| = (-1)^l \hbar V|\mathbf{k} - \mathbf{K}_j|, \quad (1)$$

where $\mathbf{k}$ is the quasi-wavevector, $\tilde{\mathbf{p}} = \mathbf{p} - \hbar\mathbf{K}_j$ and V is the Fermi velocity. In graphene $V \approx 10^8$ cm/s, so that electrons and holes behave like massless "relativistic" particles with the effective "velocity of light" $V \approx c/300$, where c is the real velocity of light. In the intrinsic graphene the chemical potential μ (or the Fermi energy E_F) goes through Dirac points, $\mu = 0$. If graphene is doped or if a dc (gate) voltage is applied between the graphene layer and a semiconductor substrate (in a typical experiment the graphene sheet lies on a substrate, e.g. on Si-SiO_2) the chemical potential can be shifted to the upper or lower energy band, so that the electron or hole density n_s can be varied from zero up to $\approx 10^{13}$ cm^{-2}. It is the unusual "relativistic" energy dispersion of graphene electrons (1) that leads to the unique physical properties of graphene.

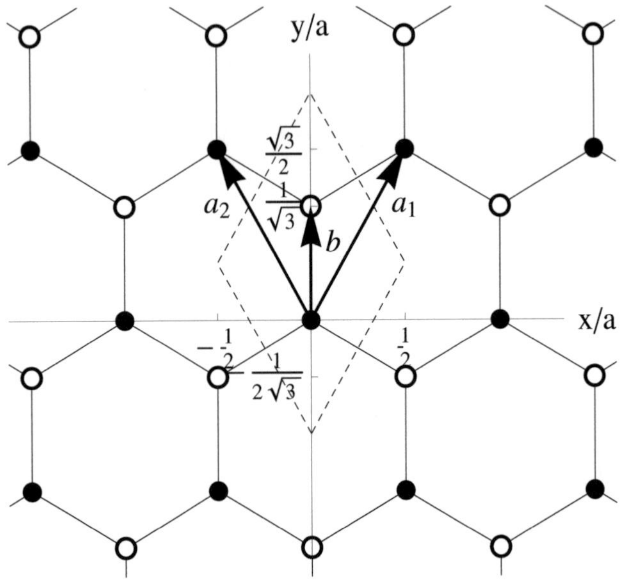

Fig. 1. The honey-comb lattice of graphene. All points of the sublattice A (black circles) are given by $n_1\mathbf{a}_1 + n_2\mathbf{a}_2$, of the sublattice B (open circles) by $n_1\mathbf{a}_1 + n_2\mathbf{a}_2 + \mathbf{b}$. Dashed lines show the boundaries of the elementary cell. a is the lattice constant.

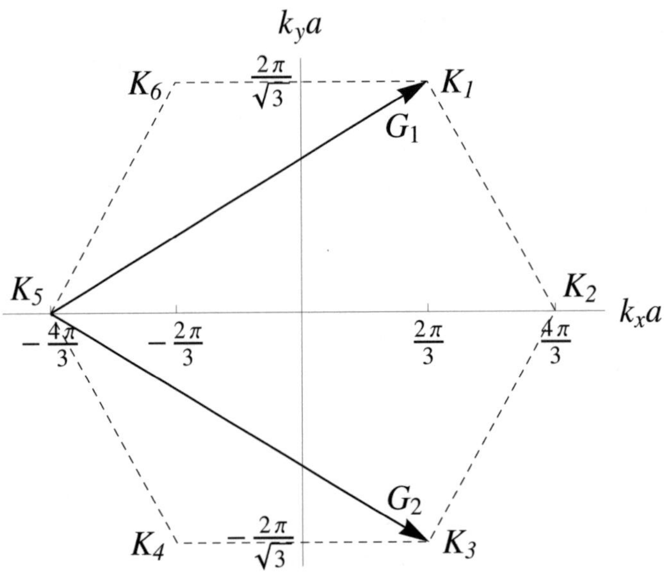

Fig. 2. The Brillouin zone of graphene. The basis vectors of the reciprocal lattice are $\mathbf{G}_1$ and $\mathbf{G}_2$. The vectors $\mathbf{K}_j$, $j = 1, \ldots, 6$, correspond to the corners of the Brillouin zone (the Dirac points). Here $\mathbf{K}_1 = -\mathbf{K}_4 = 2\pi a^{-1}(1/3, 1/\sqrt{3})$, $\mathbf{K}_2 = -\mathbf{K}_5 = 2\pi a^{-1}(2/3, 0)$, $\mathbf{K}_3 = -\mathbf{K}_6 = 2\pi a^{-1}(1/3, -1/\sqrt{3})$.

In 2007 it was predicted Mikhailov (2007) that the *linear* spectrum of graphene electrons (1) should lead to a *strongly nonlinear* electromagnetic response of this material. The physical origin of this effect is very simple. As seen from (1) the graphene electrons cannot stop and can move only with the velocity V. If such a particle is placed in the uniform oscillating electric field $\mathbf{E}(t) = \mathbf{E}_0 \cos \omega t$, its momentum will oscillate as $\mathbf{p}(t) \propto \sin \omega t$. The velocity of such particles $\mathbf{v} = \partial E_{l\mathbf{p}}/\partial \mathbf{p} = V\tilde{\mathbf{p}}/|\tilde{\mathbf{p}}|$ will then take the value of $+V$ or $-V$ dependent on the sign of the momentum $\tilde{\mathbf{p}}$, i.e. $\mathbf{v}(t) \propto \mathrm{sgn}(\sin \omega t)$. The electric current $\mathbf{j}(t)$ induced by the external field is determined by the velocity, therefore one will have [1]

$$\mathbf{j}(t) \propto \mathrm{sgn}(\sin \omega t) = \frac{4}{\pi}\left(\sin \omega t + \frac{1}{3}\sin 3\omega t + \frac{1}{5}\sin 5\omega t + \dots\right). \tag{2}$$

The induced current thus contains higher frequency harmonics $n\omega, n = 3, 5, \dots$ and hence a single graphene sheet should radiate electromagnetic waves not only at the frequency ω but also at $n\omega$ with $n = 3, 5, 7, \dots$ Mikhailov (2007; 2008); Mikhailov & Ziegler (2008). Graphene could thus serve as a simple and inexpensive frequency multiplier Mikhailov (2007; 2009). The nonlinear electromagnetic response of graphene has been also discussed by López-Rodríguez & Naumis (2008).

Apart from the frequency multiplication effect all other known nonlinear electromagnetic phenomena should be also observable in this material. For example, irradiation of the graphene layer by two electromagnetic waves $\mathbf{E}_1(t)$ and $\mathbf{E}_2(t)$ with the frequencies ω_1 and ω_2 should lead to the emission of radiation at the mixed frequencies $n_1\omega_1 + n_2\omega_2$ with integer numbers n_1 and n_2. Since the graphene lattice (Figure 1) has a central symmetry, the even order effects are forbidden in the infinite and uniform graphene layer, so that $n_1 + n_2$ must be an odd integer. In the third order in the external field amplitudes $\mathbf{E}_1$ and $\mathbf{E}_2$, apart from the frequencies $\omega_1, \omega_2, 3\omega_1$ and $3\omega_2$, the radiation at the mixed frequencies $\omega_1 \pm 2\omega_2$ and $2\omega_1 \pm \omega_2$ should be observed. In a recent experiment Hendry et al. (2010) the coherent emission from graphene at the frequency $2\omega_1 - \omega_2$ has indeed been discovered in the near-infrared and visible frequency range.

In this Chapter we develop a theory of the frequency mixing effect in graphene. We begin with a discussion of the electronic spectrum and the wave functions of graphene obtained in the tight-binding approximation (Section 2) and continue by a brief overview of the linear response theory of graphene in Section 3. Then we study the frequency mixing effects in graphene within the framework of the quasi-classical approach which works on low (microwave, terahertz) frequencies (Section 4). In Section 5 we introduce a quantum theory of the nonlinear electromagnetic response of graphene to high (infrared, optical) frequencies. A summary of results and the prospects for future research are discussed in Section 6.

2. Energy spectrum and wave functions of graphene

We calculate the spectrum and the wave functions of graphene electrons within the framework of the tight-binding approximation Wallace (1947); Reich et al. (2002) assuming that both the overlap and the transfer integrals are nonzero only for the nearest-neighbor atoms. The lattice of graphene, Figure 1, consists of two triangular sublattices A and B. For the basic vectors of the A lattice we choose the vectors $\mathbf{a}_1 = a(1/2, \sqrt{3}/2)$ and $\mathbf{a}_2 = a(-1/2, \sqrt{3}/2)$, where a is the lattice constant ($a \approx 2.46$ Å in graphene). The vector $\mathbf{b}$ connecting the sublattices is

[1] In conventional electron systems with the parabolic spectrum of charge carriers $E_\mathbf{p} = p^2/2m^\star$ the velocity $\mathbf{v} = \partial E_\mathbf{p}/\partial \mathbf{p} = \mathbf{p}/m^\star$ is proportional to the momentum, therefore $\mathbf{j} \propto \mathbf{v} \propto \sin \omega t$ and the higher frequency harmonics are not generated.

$\mathbf{b} = a(0, 1/\sqrt{3})$. The two-dimensional single-particle Hamiltonian $\hat{H}_0$ of graphene can then be written as

$$\hat{H}_0 = \frac{\hat{\mathbf{p}}^2}{2m} + \sum_{\mathbf{a}}[U_a(\mathbf{r}-\mathbf{a}) + U_a(\mathbf{r}-\mathbf{a}-\mathbf{b})], \tag{3}$$

where $\hat{\mathbf{p}} = -i\hbar(\partial_x, \partial_y)$ is the two-dimensional momentum operator, m is the free electron mass and U_a is the atomic potential. Following the standard procedure of the tight binding approximation we get the energy $E_{l\mathbf{k}}$ and the wave functions $|l\mathbf{k}\rangle$ of graphene electrons as

$$E_{l\mathbf{k}} = (-1)^l t|\mathcal{S}_{\mathbf{k}}|, \tag{4}$$

$$|l\mathbf{k}\rangle \equiv \Psi_{l\mathbf{k}}(\mathbf{r}) = \frac{1}{\sqrt{S}} e^{i\mathbf{k}\cdot\mathbf{r}} u_{l\mathbf{k}}(\mathbf{r}), \tag{5}$$

$$u_{l\mathbf{k}}(\mathbf{r}) = \sqrt{\frac{A}{2}} \sum_{\mathbf{a}} e^{-i\mathbf{k}\cdot(\mathbf{r}-\mathbf{a})} \left[(-1)^l \zeta_{\mathbf{k}} \psi_a(\mathbf{r}-\mathbf{a}) + \psi_a(\mathbf{r}-\mathbf{a}-\mathbf{b})\right], \tag{6}$$

where $l = 1, 2$, $\mathbf{k} = (k_x, k_y)$ is the quasi-wavevector, t is the transfer integral (in graphene $t \approx 3$ eV), S and A are the areas of the sample and of the elementary cell, respectively, and ψ_a is the atomic wave function. The functions $\mathcal{S}_{\mathbf{k}}$ and $\zeta_{\mathbf{k}}$ in (4) and (6) are defined as

$$\mathcal{S}_{\mathbf{k}} = 1 + e^{i\mathbf{k}\cdot\mathbf{a}_1} + e^{i\mathbf{k}\cdot\mathbf{a}_2} = 1 + 2\cos(k_x a/2) e^{i\sqrt{3}k_y a/2}, \tag{7}$$

$$\zeta_{\mathbf{k}} = \mathcal{S}_{\mathbf{k}}/|\mathcal{S}_{\mathbf{k}}|. \tag{8}$$

They are periodic in the $\mathbf{k}$-space and satisfy the equalities

$$\mathcal{S}_{-\mathbf{k}} = \mathcal{S}^\star_{\mathbf{k}}; \;\; \mathcal{S}_{\mathbf{k}+\mathbf{G}} = \mathcal{S}_{\mathbf{k}}; \;\; \zeta_{-\mathbf{k}} = \zeta^\star_{\mathbf{k}}; \;\; \zeta_{\mathbf{k}+\mathbf{G}} = \zeta_{\mathbf{k}}; \tag{9}$$

where $\mathbf{G}$ are the 2D reciprocal lattice vectors. Similar relations are valid for the energies $E_{l\mathbf{k}}$ and the wave functions $\Psi_{l\mathbf{k}}(\mathbf{r})$,

$$E_{l,-\mathbf{k}} = E_{l\mathbf{k}}; \;\; \Psi_{l,-\mathbf{k}}(\mathbf{r}) = \Psi^\star_{l\mathbf{k}}(\mathbf{r}); \;\; E_{l,\mathbf{k}+\mathbf{G}} = E_{l\mathbf{k}}; \;\; \Psi_{l,\mathbf{k}+\mathbf{G}}(\mathbf{r}) = \Psi_{l\mathbf{k}}(\mathbf{r}). \tag{10}$$

The basic reciprocal lattice vectors $\mathbf{G}_1$ and $\mathbf{G}_2$ can be chosen as $\mathbf{G}_1 = (2\pi/a)(1, 1/\sqrt{3})$, $\mathbf{G}_2 = (2\pi/a)(1, -1/\sqrt{3})$, see Figure 2.

The energy dispersion (4) is shown in Figure 3. At the corners of the Brillouin zone, in the Dirac points $\mathbf{k} = \mathbf{K}_j$, the function $\mathcal{S}_{\mathbf{k}}$ vanishes and at $|\mathbf{k} - \mathbf{K}_j|a = |\delta\mathbf{k}_j|a \ll 1$ one has

$$\mathcal{S}_{\mathbf{k}} \approx -\frac{\sqrt{3}a}{2}\left[(-1)^j \delta k_x^j + i\delta k_y^j\right], \tag{11}$$

$$\zeta_{\mathbf{k}} = -\frac{(-1)^j \delta k_x^j + i\delta k_y^j}{\sqrt{(\delta k_x^j)^2 + (\delta k_y^j)^2}}. \tag{12}$$

The energy (4) then assumes the form (1) with the velocity $V = \sqrt{3}ta/2\hbar \approx 10^8$ cm/s.

Using the wave functions (5)–(6) one can calculate the matrix elements of different physical quantities. For example, for the function $e^{i\mathbf{q}\cdot\mathbf{r}}$ we get

$$\langle l'\mathbf{k}'|e^{i\mathbf{q}\cdot\mathbf{r}}|l\mathbf{k}\rangle = \frac{1}{2}\delta_{\mathbf{k}',\mathbf{k}+\mathbf{q}}\left[1 + (-1)^{l'+l}\zeta^\star_{\mathbf{k}+\mathbf{q}}\zeta_{\mathbf{k}}\right]. \tag{13}$$

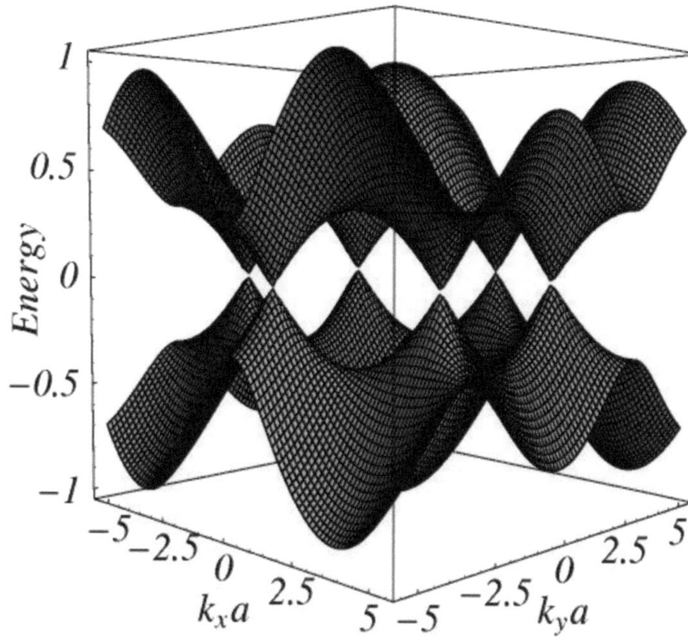

Fig. 3. The bandstructure of graphene electrons $E_{l\mathbf{k}}$ calculated in the tight-binding approximation, Eq. (4).

In the limit $\mathbf{q} \to \mathbf{0}$ we obtain from here the matrix elements of the coordinate

$$\langle l'\mathbf{k}|\hat{x}_\alpha|l\mathbf{k}\rangle = \frac{i}{2}(-1)^{l'+l}\zeta^\star_{\mathbf{k}}\frac{\partial \zeta_{\mathbf{k}}}{\partial k_\alpha} \tag{14}$$

and of the velocity operator $\hat{\mathbf{v}} = \hat{\mathbf{p}}/m$,

$$\langle l\mathbf{k}|\hat{v}_\alpha|l\mathbf{k}\rangle = \frac{1}{\hbar}\frac{\partial E_{l\mathbf{k}}}{\partial k_\alpha}, \quad \langle l\mathbf{k}|\hat{v}_\alpha|\bar{l}\mathbf{k}\rangle = \frac{E_{l\mathbf{k}} - E_{\bar{l}\mathbf{k}}}{2\hbar}\zeta^\star_{\mathbf{k}}\frac{\partial \zeta_{\mathbf{k}}}{\partial k_\alpha}. \tag{15}$$

Here $\bar{l}$ means *not* l, i.e. $\bar{l} = 2$ if $l = 1$ and $\bar{l} = 1$ if $l = 2$.

3. Linear response

Before presenting our new results on the nonlinear frequency mixing effects we will briefly overview the linear response theory of graphene. We will calculate the linear response dynamic conductivity of graphene $\sigma^{(1)}(\omega) \equiv \sigma(\omega)$ which has been studied theoretically in Refs. Gusynin & Sharapov (2006); Gusynin et al. (2006); Nilsson et al. (2006); Abergel & Fal'ko (2007); Falkovsky & Pershoguba (2007); Falkovsky & Varlamov (2007); Mikhailov & Ziegler (2007); Peres et al. (2008); Stauber, Peres & Castro Neto (2008) and experimentally in Refs. Dawlaty et al. (2008); Li et al. (2008); Mak et al. (2008); Stauber, Peres & Geim (2008).

3.1 Quantum kinetic equation

The system response is described by the quantum kinetic (Liouville) equation

$$i\hbar\frac{\partial \hat{\rho}}{\partial t} = [\hat{H}, \hat{\rho}] \tag{16}$$

for the density matrix $\hat{\rho}$, where

$$\hat{H} = \hat{H}_0 + \hat{H}_1 = \hat{H}_0 - e\phi(\mathbf{r}, t), \tag{17}$$

is the Hamiltonian of graphene in the presence of the external electric field $\mathbf{E}(\mathbf{r}, t) = -\nabla\phi(\mathbf{r}, t)$ and $e > 0$ is the electron charge. We will be interested in the response to the uniform electric field but at this step will describe the electric field by the potential

$$\phi(\mathbf{r}, t) = \phi_{\mathbf{q}\omega} e^{i\mathbf{q}\cdot\mathbf{r} - i\omega t + \gamma t}, \quad \gamma \to +0, \tag{18}$$

and will take the limit $\mathbf{q} \to \mathbf{0}$ later on. The unperturbed Hamiltonian $\hat{H}_0$ is given by Eq. (3). It has the eigenenergies and eigenfunctions,

$$\hat{H}_0 |l\mathbf{k}\sigma\rangle = E_{l\mathbf{k}} |l\mathbf{k}\sigma\rangle, \tag{19}$$

given by equations (4) and (5), respectively; we have also introduced the spin index σ here. Expanding the density matrix up to the first order in the electric field,

$$\hat{\rho} = \hat{\rho}_0 + \hat{\rho}_1, \tag{20}$$

where $\hat{\rho}_0$ satisfies the equation

$$\hat{\rho}_0 |l\mathbf{k}\sigma\rangle = f_0(E_{l\mathbf{k}}) |l\mathbf{k}\sigma\rangle, \tag{21}$$

and f_0 is the Fermi function, we get

$$\langle l'\mathbf{k}'\sigma' | \hat{\rho}_1 | l\mathbf{k}\sigma\rangle = \frac{f_0(E_{l'\mathbf{k}'}) - f_0(E_{l\mathbf{k}})}{E_{l'\mathbf{k}'} - E_{l\mathbf{k}} - \hbar(\omega + i0)} \langle l'\mathbf{k}'\sigma' | \hat{H}_1 | l\mathbf{k}\sigma\rangle, \tag{22}$$

$$\langle l'\mathbf{k}'\sigma' | \hat{H}_1 | l\mathbf{k}\sigma\rangle = -e\phi_{\mathbf{q}\omega} \delta_{\sigma\sigma'} \langle l'\mathbf{k}' | e^{i\mathbf{q}\cdot\mathbf{r}} | l\mathbf{k}\rangle. \tag{23}$$

Calculating the first order current $\mathbf{j}(\mathbf{r}, z, t) = \mathbf{j}_{\mathbf{q}\omega} e^{i\mathbf{q}\cdot\mathbf{r} - i\omega t + \gamma t} \delta(z)$ we obtain

$$\begin{aligned} \mathbf{j}_{\mathbf{q}\omega} &= -\frac{e}{2S} \mathrm{Sp}\left(\hat{\rho}_1 [\hat{\mathbf{v}}, e^{-i\mathbf{q}\cdot\mathbf{r}}]_+\right) \\ &= \frac{e^2 g_s}{2S} \phi_{\mathbf{q}\omega} \sum_{\mathbf{k}\mathbf{k}'ll'} \langle l\mathbf{k} | [\hat{\mathbf{v}}, e^{-i\mathbf{q}\cdot\mathbf{r}}]_+ | l'\mathbf{k}'\rangle \frac{f_0(E_{l'\mathbf{k}'}) - f_0(E_{l\mathbf{k}})}{E_{l'\mathbf{k}'} - E_{l\mathbf{k}} - \hbar(\omega + i\gamma)} \langle l'\mathbf{k}' | e^{i\mathbf{q}\cdot\mathbf{r}} | l\mathbf{k}\rangle, \end{aligned} \tag{24}$$

where $\mathbf{j} = (j_x, j_y)$ and $[\ldots]_+$ denotes the anti-commutator. Taking the limit $\mathbf{q} \to \mathbf{0}$ gives the frequency dependent conductivity $\sigma_{\alpha\beta}(\omega)$ which describes the linear response of graphene to a uniform external electric field. The conductivity $\sigma_{\alpha\beta}(\omega)$ consists of two contributions, the intra-band ($l = l'$) and the inter-band ($l \neq l'$) conductivities.

3.2 Intra-band conductivity

The intra-band conductivity reads

$$\sigma_{\alpha\beta}^{intra}(\omega) = \frac{-ie^2 g_s}{\hbar^2(\omega + i\gamma)S} \sum_{l\mathbf{k}} \frac{\partial E_{l\mathbf{k}}}{\partial k_\alpha} \frac{\partial f_0(E_{l\mathbf{k}})}{\partial E} \frac{\partial E_{l\mathbf{k}}}{\partial k_\beta}, \tag{25}$$

where $g_s = 2$ is the spin degeneracy and the summation over $\mathbf{k}$ is performed over the whole Brillouin zone. If the chemical potential μ lies within ~ 1 eV from the Dirac points (which is typically the case in the experiments) and if the photon energy $\hbar\omega$ does not exceed $1 - 2$ eV the main contribution to the integrals is given by the vicinity of Dirac points and one can use the linear (Dirac) approximation (11), (1). Then one gets $\sigma^{intra}_{\alpha\beta}(\omega) = \sigma^{intra}(\omega)\delta_{\alpha\beta}$ and

$$\sigma^{intra}(\omega) = \frac{ie^2 g_s g_v T}{2\pi\hbar^2(\omega + i\gamma)} \ln\left[2\cosh\left(\frac{\mu}{2T}\right)\right], \tag{26}$$

where T is the temperature and $g_v = 2$ is the valley degeneracy factor. At low temperatures $T \ll |\mu|$ the formula (26) gives

$$\sigma^{intra}(\omega) = \frac{e^2 g_s g_v |\mu|}{4\pi\hbar^2}\frac{i}{\omega + i\gamma} = \frac{n_s e^2 V^2}{|\mu|}\frac{i}{\omega + i\gamma} = \frac{n_s e^2}{m^\star}\frac{i}{\omega + i\gamma}, \tag{27}$$

where the last equalities are written in the Drude form with the phenomenological scattering rate γ, the "effective mass" of graphene quasiparticles at the Fermi level $m^\star = |\mu|/V^2$ and the charge carrier density

$$n_s = \frac{g_s g_v}{4}\frac{\mu^2}{\pi\hbar^2 V^2} = \frac{g_s g_v}{4}\frac{k_F^2}{\pi}. \tag{28}$$

The value k_F in (28) is the Fermi wavevector.

The intra-band conductivity has a standard Drude form. In the collisionless approximation $\omega \gg \gamma$ it is an imaginary function which falls down with the growing frequency as $1/\omega$. In the currently available graphene samples with the mobility $\mu_e \simeq 200000$ cm^2/Vs Orlita et al. (2008); Geim (2009) and the electron density $n_s \simeq 10^{12}$ cm^{-2} the condition $\omega \gtrsim \gamma$ is satisfied at $\omega/2\pi \gtrsim 0.1$ THz.

3.3 Inter-band conductivity

For the inter-band conductivity $\sigma^{inter}_{\alpha\beta}(\omega)$ which is dominant at high (infrared, optical) frequencies we get in the limit $\mathbf{q} \to \mathbf{0}$

$$\sigma^{inter}_{\alpha\beta}(\omega) = \frac{ie^2\hbar g_s}{S}\sum_{\mathbf{k},l\neq l'}\frac{f(E_{l'\mathbf{k}}) - f(E_{l\mathbf{k}})}{E_{l'\mathbf{k}} - E_{l\mathbf{k}} - \hbar(\omega + i\gamma)}\frac{\langle l\mathbf{k}|\hat{v}_\alpha|l'\mathbf{k}\rangle\langle l'\mathbf{k}|\hat{v}_\beta|l\mathbf{k}\rangle}{E_{l'\mathbf{k}} - E_{l\mathbf{k}}}. \tag{29}$$

Using the matrix elements (15), assuming that $\gamma \to 0$ and considering again only the vicinity of Dirac points one gets $\sigma^{inter}_{\alpha\beta}(\omega) = \sigma^{inter}(\omega)\delta_{\alpha\beta}$, where

$$\sigma^{inter}(\omega) = \frac{ie^2 g_s g_v}{16\pi\hbar}\int_0^\infty dk \frac{\sinh(\hbar V k/T)}{\cosh(\mu/T) + \cosh(\hbar V k/T)}\left(\frac{1}{k + \omega/2V + i0} - \frac{1}{k - \omega/2V - i0}\right). \tag{30}$$

The real part of the inter-band conductivity (30) is calculated analytically at all values of ω, μ and T,

$$\text{Re}\,\sigma^{inter}(\omega) = \frac{e^2 g_s g_v}{16\hbar}\frac{\sinh(\hbar|\omega|/2T)}{\cosh(\mu/T) + \cosh(\hbar|\omega|/2T)}. \tag{31}$$

At high frequencies $\hbar\omega \gg |\mu|, T$ this gives the remarkable result of the universal optical conductivity

$$\sigma^{opt}(\omega) = \frac{e^2 g_s g_v}{16\hbar} = \frac{e^2}{4\hbar}, \quad \hbar\omega \gg |\mu|, T, \tag{32}$$

which depends only on the fundamental constants e and $\hbar$, Ando et al. (2002); Kuzmenko et al. (2008); Nair et al. (2008). The imaginary part is determined by the principal value integral

$$\text{Im}\,\sigma^{inter}(\omega) = \frac{e^2 g_s g_v}{16\pi\hbar} \int_0^\infty dx \frac{\sinh x}{\cosh(|\mu|/T) + \cosh x} \left(\frac{1}{x + \hbar\omega/2T} - \frac{1}{x - \hbar\omega/2T} \right), \tag{33}$$

which can be analytically calculated at low temperatures $T \ll |\mu|$. Together with the real part this gives

$$\sigma^{inter}(\omega) = \frac{e^2 g_s g_v}{16\hbar} \left(\theta(\hbar\omega - 2\mu) + \frac{i}{\pi} \ln \left| \frac{1 - \hbar\omega/2|\mu|}{1 + \hbar\omega/2|\mu|} \right| \right), \tag{34}$$

see Figure 4. At high frequencies the inter-band conductivity tends to a (universal) constant and exceeds the intra-band contribution if $\hbar\omega \gtrsim |\mu|$. Finally, the total conductivity $\sigma(\omega) = \sigma^{intra}(\omega) + \sigma^{inter}(\omega)$ in the collisionless limit $\omega \gg \gamma$ has the form

$$\sigma(\omega) = \frac{e^2 g_s g_v}{16\hbar} \left(\frac{4i|\mu|}{\pi\hbar\omega} + \theta(\hbar\omega - 2|\mu|) + \frac{i}{\pi} \ln \left| \frac{1 - \hbar\omega/2|\mu|}{1 + \hbar\omega/2|\mu|} \right| \right) \tag{35}$$

and is shown in Figure 5.

Summarizing the linear response results on the dynamic conductivity of graphene one sees that at low frequencies $\hbar\omega \ll |\mu|$ the conductivity $\sigma(\omega)$ is imaginary and is described by the classical Drude formula. It corresponds to the intra-band response of the system. At high frequencies $\hbar\omega \gg |\mu|$ the real part of the quantum inter-band conductivity dominates. At the typical charge carrier densities of $n_s \simeq 10^{11} - 10^{13}$ cm^{-2} the transition between the two regimes $\hbar\omega \simeq |\mu|$ lies in graphene at the frequencies $\omega/2\pi \simeq 10 - 100$ THz. The low-frequency limit $\hbar\omega \ll |\mu|$ thus corresponds to the radio, microwave and terahertz frequencies, while the high-frequency limit – to the infrared and optical frequencies. The collisions can be ignored, in the high quality samples, at $\omega/2\pi \gtrsim 0.1$ THz.

4. Frequency mixing: Quasi-classical theory

In this Section we consider the frequency mixing effect using the quasi-classical approach based on the solution of the kinetic Boltzmann equation. The quasi-classical solution is simpler and, within the collisionless approximation, can be obtained non-perturbatively, at arbitrary values of the external electric field amplitudes Mikhailov (2007). The quasi-classical theory is valid at $\hbar\omega \ll |\mu|$, which corresponds to the very broad and technologically important range of radio, microwave and terahertz frequencies.

4.1 Boltzmann kinetic equation

Consider the classical motion of massless particles (1) under the action of the external electric field $\mathbf{E}(t)$. The evolution of the electron distribution function $f_\mathbf{p}(t)$ is determined by the Boltzmann kinetic equation which has the form

$$\frac{\partial f_\mathbf{p}}{\partial t} - e\mathbf{E}(t)\frac{\partial f_\mathbf{p}}{\partial \mathbf{p}} = 0 \tag{36}$$

in the collisionless approximation. Its exact solution is

$$f(\mathbf{p}, t) = \left[1 + \exp\left(\frac{V|\mathbf{p} - \mathbf{p}_0(t)| - \mu}{T} \right) \right]^{-1}, \tag{37}$$

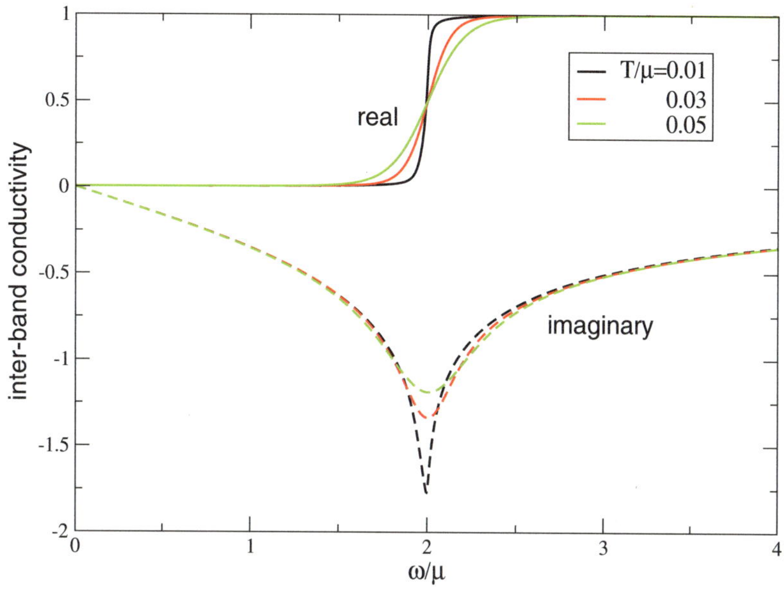

Fig. 4. The inter-band conductivity of graphene at $\gamma/\mu = 0.01$ and three different temperatures as shown in the Figure. The conductivity is measured in units $e^2 g_s g_s / 16\hbar$.

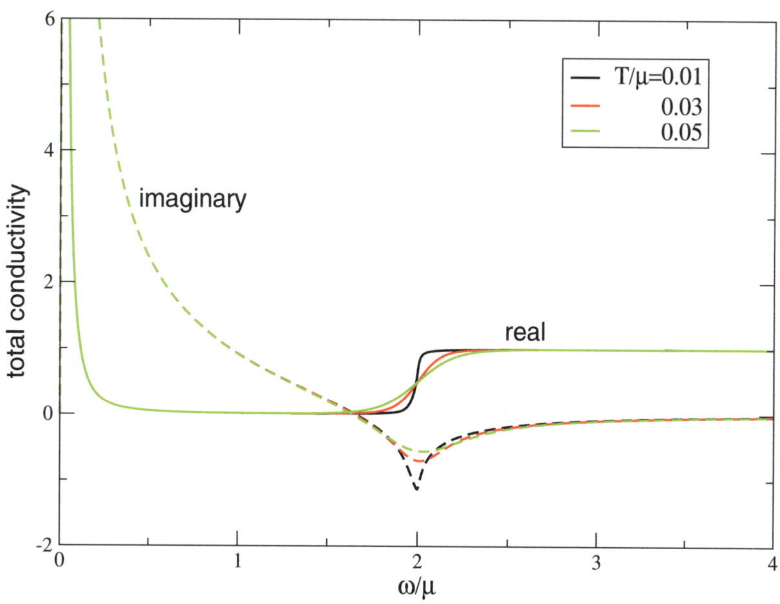

Fig. 5. The total conductivity of graphene at $\gamma/\mu = 0.01$ and three different temperatures as shown in the Figure. The conductivity is measured in units $e^2 g_s g_s / 16\hbar$. At low frequencies $\hbar\omega \lesssim |\mu|$ the conductivity is imaginary and corresponds to the intra-band classical contribution. At high frequencies $\hbar\omega \gtrsim |\mu|$ the quantum inter-band contribution dominates.

where $\mathbf{p}_0(t)$ satisfies the classical equation of motion $d\mathbf{p}_0(t)/dt = -e\mathbf{E}(t)$. The current can then be calculated as

$$\mathbf{j}(t) = -e\frac{g_s g_v V}{(2\pi\hbar)^2} \int dp_x dp_y \frac{\mathbf{p}}{\sqrt{p_x^2 + p_y^2}} \frac{1}{1 + \exp\left(\frac{V|\mathbf{p}-\mathbf{p}_0(t)|-\mu}{T}\right)}. \tag{38}$$

Assuming now that the temperature is low, $T \ll \mu$, and expanding the right-hand side of Eq. (38) we get

$$-\frac{\mathbf{j}(t)}{en_sV} = \frac{\mathbf{P}(t)}{\sqrt{P^2(t)+1}}\left[1+\frac{3}{8}\frac{P^2(t)}{[P^2(t)+1]^2}\right], \tag{39}$$

where

$$\mathbf{P}(t) = \frac{\mathbf{p}_0(t)}{p_F}. \tag{40}$$

Notice that the expansion parameter in Eq. (39) is $|P(t)/(P^2(t)+1)| \ll 1$, i.e. the result (39) is valid both at small and large $|P(t)|$. In the regime of low electric fields $|P(t)| \ll 1$ the current is

$$-\frac{\mathbf{j}(t)}{en_sV} \approx \mathbf{P}(t)\left(1-\frac{1}{8}P^2(t)\right). \tag{41}$$

4.2 Frequency mixing response

If the graphene layer is irradiated by two waves with the frequencies ω_1, ω_2 and the both waves are linearly polarized in the same direction,

$$\mathbf{E}(t) = \mathbf{E}_1\cos\omega_1 t + \mathbf{E}_2\cos\omega_2 t, \;\; \mathbf{E}_1 \parallel \mathbf{E}_2, \tag{42}$$

then

$$\mathbf{p}_0(t) = -e\left(\frac{\mathbf{E}_1}{\omega_1}\sin\omega_1 t + \frac{\mathbf{E}_2}{\omega_2}\sin\omega_2 t\right), \tag{43}$$

the current is parallel to the electric fields and equals

$$\begin{aligned}\frac{j(t)}{en_sV} &= \alpha_1\left(1-\frac{3}{32}\alpha_1^2-\frac{3}{16}\alpha_2^2\right)\sin\omega_1 t \\ &+ \frac{\alpha_1^3}{32}\sin 3\omega_1 t + \frac{3\alpha_1^2\alpha_2}{32}\Big[\sin(2\omega_1+\omega_2)t - \sin(2\omega_1-\omega_2)t\Big] + \dots,\end{aligned} \tag{44}$$

where

$$\alpha_j = \frac{eE_j}{p_F\omega_j} = \frac{eE_jV}{|\mu|\omega_j}, \;\; j = 1,2. \tag{45}$$

The omitted terms marked by the dots are obtained from the present ones by replacing $\omega_1 \leftrightarrow \omega_2$ and $\alpha_1 \leftrightarrow \alpha_2$. We will call α_j the *field parameters*. The field parameter $\alpha = eEV/\mu\omega$ is the work done by the electric field during one oscillation period (eEV/ω) divided by the average energy of electrons μ, Mikhailov (2007).

If the field parameters are small, $\alpha_j \ll 1$, Eq. (44) describes the low-frequency linear response, since $en_sV\alpha_j = \sigma^{intra}(\omega_j)E_j$, see Eq. (27). If α_j are not negligible, the first line in (44) represents the second order corrections to the linear conductivity, the first term in the second line gives the third harmonics generation effect, Mikhailov (2007), and the last terms in the second line – the frequency mixing. The amplitudes of the third-order mixed frequency current $j^{(3)}_{(2\omega_1\pm\omega_2)}(t) = j^{(3)}_{(2\omega_1\pm\omega_2)}\sin(\omega_2 \pm 2\omega_1)t$ can be rewritten as

$$j^{(3)}_{(2\omega_1\pm\omega_2)} = \frac{3}{32}en_sV\left(\frac{eE_1}{p_F\omega_1}\right)^2\frac{eE_2}{p_F\omega_2} = \frac{3}{32}\sigma^{intra}(\omega_2)E_2\alpha_1^2, \tag{46}$$

i.e., up to a numerical factor, the third order current is the product of the low-frequency Drude conductivity (27), the electric field E_2 and the squared field parameter α_1^2. One sees that the amplitude of the third order mixed-frequency current will be comparable with the linear response current if the field parameter α is of order unity (or larger). This means that the required electric field is determined by the inequality

$$E \gtrsim \frac{p_F\omega}{e} = \frac{\hbar\omega\sqrt{\pi n_s}}{e}. \tag{47}$$

The lower the charge carrier density n_s and the radiation frequency ω the smaller is the electric field needed for the observation of the nonlinear effects. If, for example, the density is $\simeq 10^{11}$ cm^{-2} and the frequency is $\simeq 0.5$ THz, Eq. (47) gives $E \gtrsim 1$ kV/cm which corresponds to the incident wave power $\simeq 2$ kW/cm^2.
If the linear polarizations of the two incident waves ω_1 and ω_2 are perpendicular to each other, $\mathbf{E}_1 \cdot \mathbf{E}_2 = 0$, the current at the mixed frequencies $2\omega_1 \pm \omega_2$ is three times smaller than for the parallel polarization. This is a general result which is also valid in the quantum regime.

5. Frequency mixing: Quantum theory

The full quantum theory of the nonlinear frequency mixing effects in graphene is substantially more complicated and is yet to be developed. In this paper we only consider the frequency mixing response at the frequency $\omega_e \equiv 2\omega_1 - \omega_2$ and calculate it under the conditions

$$\hbar\omega_1,\ \hbar\omega_2,\ \hbar\omega_e \gg |\mu|, \tag{48}$$

relevant for the experiment of Hendry et al. (2010).

5.1 Quantum kinetic (Liouville) equation

In order to investigate the nonlinear response problem in the quantum regime (48) we have to solve the quantum kinetic equation (16) in, at least, the third order in the external field amplitudes E_j, $j = 1,2$. We do this using the perturbation theory. Expanding the density matrix $\hat\rho$ in powers of the electric fields, $\hat\rho = \hat\rho_0 + \hat\rho_1 + \hat\rho_2 + \hat\rho_3 + \dots$, we get from (16) a set of recurrent equations

$$i\hbar\frac{\partial\hat\rho_n}{\partial t} = [\hat H_0, \hat\rho_n] + [\hat H_1, \hat\rho_{n-1}],\ \ n = 2,3,\dots \tag{49}$$

for $\hat\rho_n$. At high frequencies (48) we can write the Hamiltonian $\hat H_1$ in the form

$$\hat H_1 = (eE_1x\cos\omega_1 t + eE_2 x\cos\omega_2 t)\,e^{\gamma t} = h_{\omega_1}e^{i(\omega_1-i0)t} + h_{\omega_2}e^{i(\omega_2-i0)t} + \{\omega_j \to -\omega_j\}, \tag{50}$$

where $h_{\omega_1} = h_{-\omega_1} = eE_1x/2$, $h_{\omega_2} = h_{-\omega_2} = eE_2x/2$ and it is assumed that $\gamma \to 0$. In the first order in E_j we get

$$\langle\lambda|\hat\rho_1|\lambda'\rangle = \sum_{\omega=\{\pm\omega_1,\pm\omega_2\}} \frac{f_{\lambda'} - f_\lambda}{E_{\lambda'} - E_\lambda - \hbar\omega + i0}\langle\lambda|h_\omega|\lambda'\rangle e^{i\omega t} \equiv \sum_{\omega\{\pm\omega_1,\pm\omega_2\}} \rho_{1\omega}e^{i\omega t}, \tag{51}$$

where we have used a short notation $|\lambda\rangle$ for the set of three quantum numbers $|l\mathbf{k}\sigma\rangle$. The right hand side of Eq. (51) contains the oscillating exponents with the frequencies $\pm\omega_1$ and $\pm\omega_2$. For the matrix elements of $\hat\rho_2$ we obtain, similarly,

$$\langle\lambda|\hat\rho_2|\lambda'\rangle = \sum_{\omega_a,\omega_b} \frac{\langle\lambda|[h_{\omega_a}, \rho_{1\omega_b}]|\lambda'\rangle}{E_{\lambda'} - E_\lambda - \hbar\omega_a - \hbar\omega_b + i0} e^{i(\omega_a+\omega_b)t}. \tag{52}$$

The summation here is performed over the sets of frequencies $\omega_a = \{\pm\omega_1, \pm\omega_2\}$ and $\omega_b = \{\pm\omega_1, \pm\omega_2\}$, i.e. the right hand side in (52) contains the oscillating terms with the frequencies $\pm 2\omega_1$, $\pm 2\omega_2$, $\pm(\omega_1 + \omega_2)$, $\pm(\omega_1 - \omega_2)$ and a time independent term with $\omega = 0$. We will denote this set of frequencies as $\omega_c = \{\pm 2\omega_1, \pm 2\omega_2, \pm(\omega_1 + \omega_2), \pm(\omega_1 - \omega_2), 0\}$. The commutator $[h_{\omega_a}, \rho_{1\omega_b}]$ is calculated as

$$\langle\lambda|[h_{\omega_a}, \rho_{1\omega_b}]|\lambda'\rangle = \sum_{\lambda''} \Big(\langle\lambda|h_{\omega_a}|\lambda''\rangle\langle\lambda''|\rho_{1\omega_b}|\lambda'\rangle - \langle\lambda|\rho_{1\omega_b}|\lambda''\rangle\langle\lambda''|h_{\omega_a}|\lambda'\rangle \Big),$$

where the matrix elements $\langle\lambda|h_\omega|\lambda'\rangle$ and $\langle\lambda|\rho_{1\omega}|\lambda'\rangle$ are known from (14) and (51). Then, for the matrix elements of $\hat{\rho}_3$ we get

$$\langle\lambda|\hat{\rho}_3|\lambda'\rangle = \sum_{\omega_a, \omega_c} \frac{\langle\lambda|[h_{\omega_a}, \rho_{2\omega_c}]|\lambda'\rangle}{E_{\lambda'} - E_\lambda - \hbar\omega_a - \hbar\omega_c + i0} e^{i(\omega_a + \omega_c)t}. \tag{53}$$

Now the right hand side of Eq. (53) contains the terms with the frequencies $\pm\omega_1$, $\pm\omega_2$, $\pm 3\omega_1$, $\pm 3\omega_2$, $\pm(2\omega_1 \pm \omega_2)$ and $\pm(2\omega_2 \pm \omega_1)$.

The general formulas (51), (52) and (53) allows one, in principle, to calculate the time dependence of the matrix elements of the density matrix in the third order in E_j. Then, using the formula

$$j_\alpha(t) = -\frac{e}{S} \sum_{\lambda\lambda'} \langle\lambda'|\hat{v}_\alpha|\lambda\rangle\langle\lambda|\hat{\rho}_1 + \hat{\rho}_2 + \hat{\rho}_3 + \ldots|\lambda'\rangle \tag{54}$$

one can find the time dependence of the current in the same order of the perturbation theory. This general expression for the current will contain the terms with the frequencies $\omega_1, \omega_2, 3\omega_1$, $3\omega_2$, as well as $(2\omega_1 \pm \omega_2)$ and $(2\omega_2 \pm \omega_1)$.

5.2 Optical frequency mixing at $\omega_e = 2\omega_1 - \omega_2$

Being interested in this work only in the response at the frequency of the emitted light $\omega_e \equiv 2\omega_1 - \omega_2$ (see Hendry et al. (2010)) we get, after quite lengthy calculations,

$$\begin{aligned} j^{(3)}_{(\omega_e)}(t) = {} & \frac{e^4 g_s}{8\hbar^2} E_1^2 E_2 \frac{2\omega_1 - \omega_2}{\omega_1(\omega_1 - \omega_2)^2} e^{i\omega_e t} \frac{1}{S} \sum_{\mathbf{k}} (f_{2\mathbf{k}} - f_{1\mathbf{k}}) \langle 1\mathbf{k}|v_x|2\mathbf{k}\rangle\langle 1\mathbf{k}|x|2\mathbf{k}\rangle^3 \\ & \times \left(-\frac{2}{E_{2\mathbf{k}} - E_{1\mathbf{k}} + \hbar\omega_1 - i0} + \frac{2}{E_{2\mathbf{k}} - E_{1\mathbf{k}} - \hbar\omega_1 + i0} \right. \\ & + \frac{1}{E_{2\mathbf{k}} - E_{1\mathbf{k}} + \hbar\omega_2 - i0} - \frac{1}{E_{2\mathbf{k}} - E_{1\mathbf{k}} - \hbar\omega_2 + i0} \\ & \left. - \frac{1}{E_{2\mathbf{k}} - E_{1\mathbf{k}} - \hbar\omega_e + i0} + \frac{1}{E_{2\mathbf{k}} - E_{1\mathbf{k}} + \hbar\omega_e - i0} \right) + (\omega_j \to -\omega_j). \end{aligned} \tag{55}$$

The current $j^{(3)}_{(\omega_e)}(t)$ contains the resonant terms corresponding to the vertical optical transitions at $E_{2\mathbf{k}} - E_{1\mathbf{k}} = \hbar\omega_1, \hbar\omega_2$ and $\hbar\omega_e$. As it follows from the linear response theory (Section 3), the largest contribution to the current at the optical frequencies $\hbar\omega \gg |\mu|$ is given by the absorption terms proportional to $\delta(E_{2\mathbf{k}} - E_{1\mathbf{k}} - \hbar\omega)$. The same is valid in the nonlinear regime too. Taking into account in (55) only the terms $\propto \delta(E_{2\mathbf{k}} - E_{1\mathbf{k}} - \hbar\omega_{1,2,e})$ we finally get

$$j^{(3)}_{(\omega_e)}(t) \approx \frac{3e^4 g_s g_v V^2}{256\hbar^3} E_1^2 E_2 \frac{2\omega_1 - \omega_2}{\omega_1(\omega_1-\omega_2)^2}\left[\frac{2}{\omega_1^2} - \frac{1}{\omega_2^2} - \frac{1}{(2\omega_1-\omega_2)^2}\right]\cos(2\omega_1 - \omega_2)t$$

$$= -\frac{9}{8}\sigma^{opt}(\omega_2)E_2\beta_1^2 F\left(\frac{\omega_2}{\omega_1}\right)\cos(2\omega_1-\omega_2)t, \tag{56}$$

where

$$\beta_1 = \frac{eE_1V}{\hbar\omega_1^2} \tag{57}$$

is the *optical* field parameter and

$$F(x) = \frac{2+2x-x^2}{3x^2(2-x)}. \tag{58}$$

Eq. (56) is the main result of this work. It represents the ac electric current induced in the graphene layer at the frequency $\omega_e = 2\omega_1 - \omega_2$ by the two incident waves (42) polarized in the same direction (the current direction coincides with that of the fields). If the two incident waves are perpendicularly polarized, the numerical prefactor in (56) is reduced ($9/8 \to 3/8$). As seen from (56), at high frequencies the current depends neither on the chemical potential μ nor on the temperature T as it should be under the conditions when the vertical inter-band transitions play the main role. The second line of (56) is written in the form similar to (46): at the optical frequencies the current is the product of the high-frequency, universal optical conductivity (32), the electric field E_2 and the squared field parameter β_1^2. The *optical* field parameter $\beta = eEV/\hbar\omega^2$ is the work done by the electric field during one oscillation period (eEV/ω) divided by the photon energy $\hbar\omega$ (instead of the Fermi energy μ at low frequencies, Eq. (45)). In addition to the mentioned parameters, the current (56) weakly depends on the ratio of the two optical frequencies ω_2/ω_1 which is described by the function (58) shown in Figure 6. The function $F(x)$ is of order unity if the difference $\omega_2 - \omega_1$ is not very large. If ω_1, ω_2 or ω_e tend to zero the current (56) has a strong tendency to grow. These limits are very interesting for future studies but have been excluded from the current consideration by the conditions (48).

The formula (56) is in good quantitative agreement with the experimental results of Hendry et al. (2010). The comparison with other materials made in that paper showed that graphene has much stronger nonlinear properties than typical nonlinear insulators and some metals (Au). Comparing the experimental results of Hendry et al. (2010) with those of Erokhin et al. (1987) shows that graphene is also a stronger nonlinear material than a typical nonlinear semiconductor InSb. Further theoretical and experimental studies of the nonlinear electrodynamic and optical properties of graphene are therefore highly desirable.

6. Summary and conclusions

Due to the massless energy spectrum of the charge carriers graphene demonstrates strongly nonlinear electromagnetic properties. In this work we have developed a theory of the nonlinear frequency mixing effect in graphene. The two physically different regimes have been considered. At low frequencies, corresponding to the radio, microwave and terahertz range, the problem is solved within the quasi-classical approach which takes into account the intra-band response of the material. At high frequencies, corresponding to the infrared and

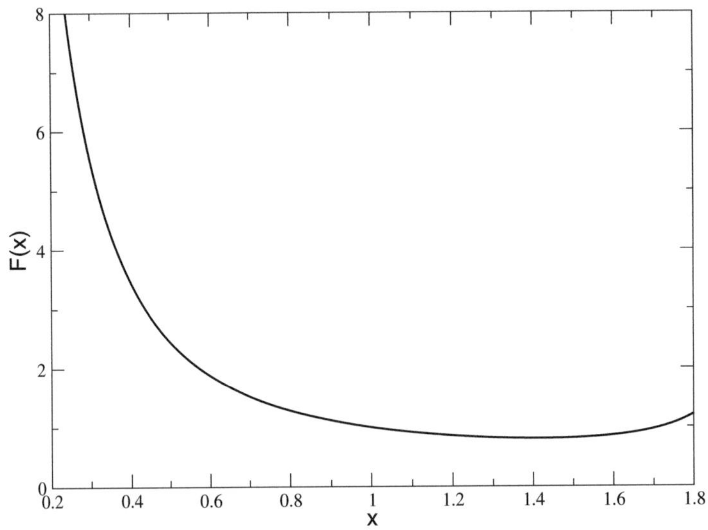

Fig. 6. The function $F(x)$ from Eq. (58).

visible light, a quantum theory is developed which takes into account the inter-band optical transitions.

At the optical frequencies our results quantitatively agree with the recent experimental findings of Hendry et al. (2010) who have observed the nonlinear electromagnetic response of graphene for the first time. The results of Hendry et al. (2010) show that in the visible and near-infrared frequency range the nonlinear parameters of graphene are much stronger than in many other nonlinear materials. Even more important conclusion is that, according to our theoretical predictions, the nonlinear response of graphene substantially grows at lower frequencies. One should expect therefore even stronger nonlinear properties of graphene at the mid-infrared, terahertz and microwave frequencies which would be of extreme importance for the future progress of the nonlinear terahertz- and optoelectronics.

The author thanks the Deutsche Forschungsgemeinschaft for support of this work.

7. References

Abergel, D. S. L. & Fal'ko, V. I. (2007). Optical and magneto-optical far-infrared properties of bilayer graphene, *Phys. Rev. B* 75: 155430.

Ando, T., Zheng, Y. & Suzuura, H. (2002). Dynamical conductivity and zero-mode anomaly in honeycomb lattices, *J. Phys. Soc. Japan* 71: 1318–1324.

Castro Neto, A. H., Guinea, F., Peres, N. M. R., Novoselov, K. S. & Geim, A. K. (2009). The electronic properties of graphene, *Rev. Mod. Phys.* 81: 109–162.

Dawlaty, J. M., Shivaraman, S., Strait, J., George, P., Chandrashekhar, M., Rana, F., Spencer, M. G., Veksler, D. & Chen, Y. (2008). Measurement of the optical absorption spectra of epitaxial graphene from terahertz to visible, *Appl. Phys. Lett.* 93: 131905.

Erokhin, A. I., Kovalev, V. I. & Shmelev, A. K. (1987). Nonlinear susceptibility of InSb at the wavelength of 10.6 μm, *Sov. J. Quantum Electron.* 17: 742–745.

Falkovsky, L. A. & Pershoguba, S. S. (2007). Optical far-infrared properties of a graphene monolayer and multilayer, *Phys. Rev. B* 76: 153410.

Falkovsky, L. A. & Varlamov, A. A. (2007). Space-time dispersion of graphene conductivity, *Europ. Phys. J. B* 56: 281–284.

Geim, A. K. (2009). Graphene: Status and prospects, *Science* 324: 1530–1534.

Gusynin, V. P. & Sharapov, S. G. (2006). Transport of Dirac quasiparticles in graphene: Hall and optical conductivities, *Phys. Rev. B* 73: 245411.

Gusynin, V. P., Sharapov, S. G. & Carbotte, J. P. (2006). Unusual microwave response of Dirac quasiparticles in graphene, *Phys. Rev. Lett.* 96: 256802.

Hendry, E., Hale, P. J., Moger, J., Savchenko, A. K. & Mikhailov, S. A. (2010). Coherent nonlinear optical response of graphene, *Phys. Rev. Lett.* 105: 097401.

Katsnelson, M. I. (2006). Minimal conductivity in bilayer graphene, *Europ. Phys. J. B* 52: 151–153.

Katsnelson, M. I. (2007). Graphene: carbon in two dimensions, *Materials Today* 10: 20–27.

Kuzmenko, A. B., van Heumen, E., Carbone, F. & van der Marel, D. (2008). Universal optical conductance of graphite, *Phys. Rev. Lett.* 100: 117401.

Li, Z. Q., Henriksen, E. A., Jiang, Z., Hao, Z., Martin, M. C., Kim, P., Stormer, H. L. & Basov, D. N. (2008). Dirac charge dynamics in graphene by infrared spectroscopy, *Nature Physics* 4: 532–535.

López-Rodríguez, F. J. & Naumis, G. G. (2008). Analytic solution for electrons and holes in graphene under electromagnetic waves: Gap appearance and nonlinear effects, *Phys. Rev. B* 78: 201406(R).

Mak, K. F., Sfeir, M. Y., Wu, Y., Lui, C. H., Misewich, J. A. & Heinz, T. F. (2008). Measurement of the optical conductivity of graphene, *Phys. Rev. Lett.* 101: 196405.

McClure, J. W. (1956). Diamagnetism of graphite, *Phys. Rev.* 104: 666–671.

Mikhailov, S. A. (2007). Non-linear electromagnetic response of graphene, *Europhys. Lett.* 79: 27002.

Mikhailov, S. A. (2008). Electromagnetic response of electrons in graphene: Non-linear effects, *Physica E* 40: 2626–2629.

Mikhailov, S. A. (2009). Non-linear graphene optics for terahertz applications, *Microelectron. J.* 40: 712–715.

Mikhailov, S. A. & Ziegler, K. (2007). New electromagnetic mode in graphene, *Phys. Rev. Lett.* 99: 016803.

Mikhailov, S. A. & Ziegler, K. (2008). Non-linear electromagnetic response of graphene: Frequency multiplication and the self-consistent field effects, *J. Phys. Condens. Matter* 20: 384204.

Nair, R. R., Blake, P., Grigorenko, A. N., Novoselov, K. S., Booth, T. J., Stauber, T., Peres, N. M. R. & Geim, A. K. (2008). Fine structure constant defines visual transparency of graphene, *Science* 320: 1308–1308.

Nilsson, J., Castro Neto, A. H., Guinea, F. & Peres, N. M. R. (2006). Electronic properties of graphene multilayers, *Phys. Rev. Lett.* 97: 266801.

Nomura, K. & MacDonald, A. H. (2007). Quantum transport of massless Dirac fermions, *Phys. Rev. Lett.* 98: 076602.

Novoselov, K. S., Geim, A. K., Morozov, S. V., Jiang, D., Katsnelson, M. I., Grigorieva, I. V., Dubonos, S. V. & Firsov, A. A. (2005). Two-dimensional gas of massless Dirac fermions in graphene, *Nature* 438: 197–200.

Novoselov, K. S., Geim, A. K., Morozov, S. V., Jiang, D., Zhang, Y., Dubonos, S. V., Grigorieva, I. V. & Firsov, A. A. (2004). Electric field effect in atomically thin carbon films, *Science* 306: 666–669.

Novoselov, K. S., Jiang, Z., Zhang, Y., Morozov, S. V., Stormer, H. L., Zeitler, U., Maan, J. C., Boebinger, G. S., Kim, P. & Geim, A. K. (2007). Room-temperature quantum Hall effect in graphene, *Science* 315: 1379.

Orlita, M., Faugeras, C., Plochocka, P., Neugebauer, P., Martinez, G., Maude, D. K., Barra, A.-L., Sprinkle, M., Berger, C., de Heer, W. A. & Potemski, M. (2008). Approaching the Dirac point in high-mobility multilayer epitaxial graphene, *Phys. Rev. Lett.* 101: 267601.

Peres, N. M. R., Stauber, T. & Castro Neto, A. H. (2008). The infrared conductivity of graphene, *Europhys. Lett.* 84: 38002.

Reich, S., Maultzsch, J., Thomsen, C. & Ordejon, P. (2002). Tight-binding description of graphene, *Phys. Rev. B* 66: 035412.

Slonczewski, J. C. & Weiss, P. R. (1958). Band structure of graphite, *Phys. Rev.* 109(2): 272–279.

Stander, N., Huard, B. & Goldhaber-Gordon, D. (2009). Evidence for Klein tunneling in graphene *p-n* junctions, *Phys. Rev. Lett.* 102: 026807.

Stauber, T., Peres, N. M. R. & Castro Neto, A. H. (2008). Conductivity of suspended and non-suspended graphene at finite gate voltage, *Phys. Rev. B* 78: 085418.

Stauber, T., Peres, N. M. R. & Geim, A. K. (2008). Optical conductivity of graphene in the visible region of the spectrum, *Phys. Rev. B* 78: 085432.

Tan, Y.-W., Zhang, Y., Bolotin, K., Zhao, Y., Adam, S., Hwang, E. H., Das Sarma, S., Stormer, H. L. & Kim, P. (2007). Measurement of scattering rate and minimum conductivity in graphene, *Phys. Rev. Lett.* 99: 246803.

Wallace, P. R. (1947). The band theory of graphite, *Phys. Rev.* 71(9): 622–634.

Young, A. F. & Kim, P. (2009). Quantum interference and Klein tunnelling in graphene heterojunctions, *Nature Physics* 5: 222–226.

Zhang, Y., Tan, Y.-W., Stormer, H. L. & Kim, P. (2005). Experimental observation of the quantum Hall effect and Berry's phase in graphene, *Nature* 438: 201–204.